邓小刚院士研究
论文选集

刘 刚 陈坚强 张来平 毛枚良 刘 伟 编

科学出版社
北 京

内 容 简 介

本书选编了作者及研究团队多年来在算法方面具有代表性的论文，主要涉及高精度有限差分格式构造方法及理论、高精度格式推广及应用技术等内容。前言简要介绍了作者本人以及所带领的研究团队开展研究工作的情况，第一部分主要是作者在高精度有限差分算法研究方面的综述性文章，第二部分为高精度格式及其构造理论，包括高精度有限差分格式的设计方法、频谱分析方法以及几何守恒律理论；第三部分为高精度格式推广和应用技术研究，主要包括如何将标量方程下设计的高精度格式推广到多维贴体坐标下的流动方程求解，如何在高性能计算机上进行高效并行计算等；第四部分为高精度格式的具体应用，主要介绍采用自主设计的 HDCS 格式和 WCNS 格式对典型复杂流动的精细模拟。附录罗列了邓小刚院士及合作者发表的部分论文。

本书可供从事计算流体力学的研究人员、高校流体力学专业的学生和教员阅读，也可作为航空航天相关领域工程技术人员的工具书和参考书。

图书在版编目(CIP)数据

邓小刚院士研究论文选集/刘刚等编. —北京：科学出版社，2018.10
ISBN 978-7-03-058923-1

Ⅰ. ①邓… Ⅱ. ①刘… Ⅲ. ①流体力学–数值模拟–文集 Ⅳ. ①O35-53

中国版本图书馆 CIP 数据核字 (2018) 第 219406 号

责任编辑：赵敬伟　/责任校对：胡庆家
责任印制：肖　兴　/封面设计：耕者工作室

科学出版社 出版
北京东黄城根北街 16 号
邮政编码：100717
http://www.sciencep.com

中国科学院印刷厂 印刷
科学出版社发行　各地新华书店经销

*

2018 年 10 月第 一 版　开本：720×1000　1/16
2018 年 10 月第一次印刷　印张：34 1/2　插页：24
字数：696 000

定价：280.00 元
(如有印装质量问题，我社负责调换)

前　言

本文集选编了邓小刚院士研究工作中具有代表性的论文，主要涉及高精度格式构造理论及方法、高精度格式推广应用技术等主题，重点展示了他在高精度数值模拟方法方面取得的丰硕成果。

邓小刚院士长期从事空气动力学数值模拟研究和应用工作，是我国高精度数值方法研究的杰出代表，在计算流体力学（CFD）理论方法研究、大型软件研制和解决国家重大项目中关键空气动力学问题等方面，做出了系统的创新成果。针对湍流、转捩、噪声和激波等跨尺度复杂流动数值模拟的需要，创立了高精度非线性紧致格式设计方法，解决了高精度紧致格式捕捉激波产生非物理波动的难题；在实现高精度数值方法工程化应用方面，发现并突破了几何守恒律、跨边界保持高精度和网格中奇点高精度处理等诸多难题；主持研制了高精度数值风洞（TH-HiNWT）及我国第一个具有自主知识产权的高超声速 CFD 软件平台（CHANT），其应用解决了我国航空航天飞行器大量关键气动问题。

邓小刚院士研究兴趣广泛，除高精度算法方面的研究外，其研究工作还涉及高超声速复杂流动机理、网格生成技术、湍流数值模拟、等离子体流动控制等诸多领域。在多个领域均取得了不少创新性成果，不完全统计，到目前为止，在国内外重要学术期刊发表论文 120 余篇。限于篇幅，本文集仅选编约 30 篇代表作，其他论文列于最后的发表论文目录之中。

目　录

前言

综述与进展

高精度加权紧致非线性格式的研究进展 ………(邓小刚, 刘昕, 毛枚良, 张涵信)　3

基于 HDCS-E8T7 格式的气动噪声数值模拟方法研究进展

…………………………………(姜屹, 邓小刚, 毛枚良, 刘化勇)　14

计算流体力学中的验证与确认……(邓小刚, 宗文刚, 张来平, 高树椿, 李超)　30

高精度格式及其构造理论

High-order accurate dissipative weighted compact nonlinear schemes……………

…………………………………………………………… (DENG Xiaogang)　43

Developing High-Order Weighted Compact Nonlinear Schemes……………………

…………………………………………… (Xiaogang Deng, Hanxin Zhang)　58

Geometric conservation law and applications to high-order finite difference

schemes with stationary grids ……………………………………………………

(Xiaogang Deng, Meiliang Mao, Guohua Tu, Huayong Liu, Hanxin Zhang)　81

A family of hybrid cell-edge and cell-node dissipative compact schemes satisfying

Geometric conservation law ……………………………………………………

(Xiaogang Deng, Yi Jiang, Meiliang Mao, Huayong Liu, Song Li, Guohua Tu)　97

Further studies on Geometric Conservation Law and applications to high-order

finite difference schemeswith stationary grids …………… (Xiaogang Deng,

Yaobing Min, Meiliang Mao, Huayong Liu, Guohua Tu, Hanxin Zhang) 114

On the freestream preservation of finite volume method in curvilinear coordinates

…… (Dan Xu, Xiaogang Deng, Yaming Chen, Yidao Dong, Guangxue Wang) 136

Reevaluation of high-order finite difference and finite volume algorithms with

freestream preservation satisfied…………………………………………………

…………………… (Yidao Dong, Xiaogang Deng, Dan Xu, Guangxue Wang) 149

高精度格式推广/应用技术研究

Extending Weighted Compact Nonlinear Schemes to Complex Grids with

Characteristic-Based Interface Conditions ·· (Xiaogang Deng,Meiliang Mao,Guohua Tu,Yifeng) 161

High-Order Behaviors of Weighted Compact Fifth-Order Nonlinear Schemes ··· (Liu Xin, Deng Xiaogang, Mao Meiliang) 173

Large eddy simulation on curvilinear meshes using seventh-order dissipative compact scheme ······ (Yi Jiang, Meiliang Mao, Xiaogang Deng, Huayong Liu) 178

Developing a hybrid flux function suitable for hypersonic flow simulation with high-order methods ·· (Dongfang Wang, Xiaogang Deng, Guangxue Wang, Yidao Dong) 189

Efficiency benchmarking of seventh-order tri-diagonal weighted compact nonlinear scheme on curvilinear mesh ··· (Shengye Wang, Xiaogang Deng, Guangxue Wang, Dan Xu, Dongfang Wang) 208

Developing a new mesh deformation technique based on support vector machine ··· (Xiang Gao, Yidao Dong, Chuanfu Xu, Min Xiong, Zhenghua Wang and Xiaogang Deng) 228

Large eddy simulation on curvilinear meshes using seventh-order dissipative compact scheme ······ (Yi Jiang, Meiliang Mao, Xiaogang Deng, Huayong Liu) 240

Effect of Nonuniform Grids on High-Order Finite Difference Method ······ (Xu Dan, Deng Xiaogang, Chen Yaming, Wang Guangxue, Dong Yidao) 251

Extending Seventh-Order Dissipative Compact Scheme Satisfying Geometric Conservation Law to Large Eddy Simulation on Curvilinear Grids ·············· (Yi Jiang, Meiliang Mao, Xiaogang Deng and Huayong Liu) 274

Effect of Geometric Conservation Law on Improving Spatial Accuracy for Finite Difference Schemes on Two-Dimensional Nonsmooth Grids ············· (Meiliang Mao, Huajun Zhu, Xiaogang Deng, Yaobing Min and Huayong Liu) 297

Further improvement of weighted compact nonlinear scheme using compact nonlinear interpolation ·· (Zhen-Guo Yan, Huayong Liu, Yankai Ma, Meiliang Mao, Xiaogang Deng) 331

New nonlinear weights for improving accuracy and resolution of weighted compact nonlinear scheme ·· (Zhenguo Yan, Huayong Liu, Meiliang Mao, Huajun Zhua, Xiaogang Deng) 342

Parallelizing a High-Order CFD Software for 3D, Multi-block, Structural Grids on the TianHe-1A Supercomputer ······ (Chuanfu Xu, Xiaogang Deng, Lilun Zhang, Yi Jiang, Wei Cao, Jianbin Fang, Yonggang Che, Yongxian Wang, Wei Liu) 357

Collaborating CPU and GPU for large-scale high-order CFD simulations with complex grids on the TianHe-1A supercomputer ·· (ChuanfuXu, XiaogangDeng, LilunZhang, etc) 371

Parallelizing and optimizing large-scale 3D multi-phase flow simulations on the

Tianhe-2 supercomputer ············ (Dali Li, Chuanfu Xu, Yongxian Wang, Zhifang Song, Min Xiong, Xiang Gao and Xiaogang Deng) 394

Performance modeling and optimization of parallel LU-SGS on many-core processors for 3D high-order CFD simulations ············
(Dali Li, Chuanfu Xu, Bin Cheng, Min Xiong, Xiang Gao and Xiaogang Deng) 409

高精度格式的具体应用

High-Order and High Accurate CFD Methods and Their Applications for Complex Grid Problems ············
· (Xiaogang Deng, Meiliang Mao, Guohua Tu, Hanxin Zhang, Yifeng Zhang) 431

Numerical investigation on body-wake flow interaction over rod-airfoil configuration ············ (Jiang Yi, Mao MeiLiang, Deng Xiaogang, Liu HuaYong) 453

Validation of a RANS transition model using a high-order weighted compact nonlinear scheme ············ (TU GuoHua, Deng Xiaogang, MAO MeiLiang) 488

Osher Flux with Entropy Fix for Two-Dimensional Euler Equations
(Huajun Zhu, Xiaogang Deng, Meiliang Mao, Huayong Liu and Guohua Tu) 495

基于雷诺应力模型的高精度分离涡模拟方法
············ (王圣业，王光学，董义道，邓小刚) 518

邓小刚院士及合作者发表的论文 ············ 535

综述与进展

高精度加权紧致非线性格式的研究进展

邓小刚　刘昕†　毛枚良　张涵信

中国空气动力研究与发展中心，绵阳　621000

摘　要　综述了高精度加权紧致非线性格式 WCNS 在理论分析以及复杂流动应用方面的研究进展. 首先回顾了国内外高精度格式研究的概况, 然后介绍了 WCNS 的研究与发展历程. 在对 WCNS 进行了 Fourier 分析和渐近稳定性分析后, 给出了 WCNS 求解多维复杂流动的算例.

关键词　高精度格式, 加权紧致非线性格式, Fourier 分析, 渐近稳定性理论

1 引言

近 30 年来, 计算流体力学 (computational fluid dynamics, CFD) 已进入到求解可压的雷诺平均 Navier-Stokes (N-S) 方程和三维定常及非定常完全黏性流动数值模拟阶段[1], 其发展的一个必然而且重要的途径在于对数值计算格式和方法的研究.

实际的流体运动是很复杂的, 各类波系在其中充当着不同的角色, 为真实地刻画流场内流动细节, 人们纷纷要求采用高 (阶) 精度格式. Gaitonde[2] 指出, 普通低阶格式导致的数值耗散与色散可通过高精度格式潜在地降低, 即使在受网格非均匀性、边界条件的低阶近似以及非线性现象的捕捉等相关因素的影响全场精度不得不降阶时, 建立在可靠的高阶公式基础上的高精度格式也比优化过的低阶格式计算结果更加精确. 即便高精度格式的运算量在每个迭代步有所增加, 但它整个过程的计算仍有着更高的效率. 因此开展高精度格式的研究近年来一直是 CFD 中的前沿研究课题.

本文就国内外高精度格式的发展做了详细的回顾, 重点介绍作者建立的高精度加权紧致非线性格式最新研究进展, 通过具体的大尺度流动算例研究高精度加权紧致非线性格式的分辨率与计算效率, 并对格式应用于湍流流动的前景作了展望.

2 国内外高精度格式研究概况

有关高精度、高分辨率捕捉激波流场的计算格式在过去几十年取得很大的进展. 好的格式应能得到原问题锐利的、逼真的物理图像, 它习惯地被称为具有高分辨率[3] (high resolution). Harten[4]1983 年首先提出高分辨率方法、总变差减小 (total variation diminishing, TVD) 或总变差不增 (total variation nonincreasing, TVNI) 的概念与限制器 (limiter) 技术, 并具体构造了具有 2 阶精度的高分辨率 TVD 格式. TVD 格式是非线性的保单调格式, 它为数值方法特别是差分方法的理论与构造开拓了一个崭新的方向. 同样具有 TVD 性质的格式 (monotonic upstream-centered scheme for conservation laws, MUSCL) 由 Van Leer 提出, 选择合适的限制器, 可以得到良好的色散与耗散特性, 激波等间断捕捉的分辨率高[5,6]. 1984 年, 张涵信通过小扰动分析的方法研究了色散误差对改善激波附近数值解行为的作用, 构造了无波动, 无自由参数的耗散差分 (non-oscillatory, containing no free parameters, and dissipative, NND) 格式[7,8], 现已广泛地应用于科学和工程领域并取得了比较好的效果.

在对激波捕捉的方法中, TVD 格式已被证实为可靠和有效, 但为满足其总变差减小的数学属性, 在极值点附近精度将退化为一阶. 为提高格式精度且极值点处不降阶, Harten[9,10] 提出实质无波动 (essentially non-oscillatory, ENO) 的概念, 它放宽了 TVD 中强调消除波动的限制, 允许格式有附加的高阶虚假波动, 这样格式的建立就具有较大的灵活性. 人们纷纷构造, 建立了各种具体形式的高阶 ENO 格式[11~16].

邓小刚, 刘昕, 毛枚良, 张涵信. 高精度加权紧致非线性格式的研究进展. 力学进展, 2007, 37(3): 417-427.

但是ENO格式在求解一维、多维守恒律方程组时，含间断的计算流场内存在着虚假波动，精度越高，这种波动越严重。1994年Liu[17]提出加权技术，建立了加权实质无波动(weighted essentially non-oscillatory, WENO)格式。WENO格式是基于ENO格式构造的高阶混合格式，它将ENO格式传统的通过逻辑判断选择最光滑模板(stencil)的方法，改进为采用所有模板的加权平均，而权值则可以度量模板的光滑程度。WENO格式在保持了ENO格式优点的同时，还使得计算流场中的虚假波动明显减少。Levy认为采用线性权值格式精度会随网格数变化而稍微变化，而采用非线性权值，在稀疏网格下获得的误差比线性权值方法的低，精度还能保持稳定，于是他采用含有非线性权值的4阶中心型WENO格式计算了二维Euler流动[18,19]。Lin[20]在耗散与色散关系基础上对WENO格式中的参数进行了研究，得到了不同形式的WENO格式，将这些格式推广到二维与轴对称情况，作了典型算例的数值实验研究。

WENO格式不总是保单调的，于是Suresh[21]提出了一类新的高阶保单调格式——MP5 (monotonicity preserving with fifth order)格式。2000年Balsara[22]将Suresh的保单调插值策略和WENO格式构造思路结合起来，构造了具有健壮激波捕捉能力的保单调加权实质无波动格式(monotonicity preserving weighted essentially non-oscillatory schemes, MPWENO)，应用于数值实验中取得不错的效果。

国内，张涵信进一步发展了建立光滑捕捉激波格式的原则，即耗散控制原则、色散控制原则、激波控制原则和频谱控制原则[23]。按照这些原则贺国宏[24]建立了3阶精度的实质无波动无自由参数格式(essentially non-oscillatory, containing no free parameters, ENN)，李沁[25,26]建立了激波处为2阶NND格式，其它地方为4阶加权非线性格式的混合格式，宗文刚[27]提出了双重加权实质无波动格式(double weighted essentially non-oscillatory, DWENO)。另外，FCT (flux-corrected transport)是一类通量修正方法，它以低阶格式为基础，对高阶格式所得的数值通量进行修正，使其满足局部单调性原则。沈孟育利用解析离散法，在导数计算公式中引入FCT以消除非物理解振荡，构造了高阶格式WENO-FCT[28]，并将其应用于一维和二维无黏流动当中。

另一方面，紧致格式也已逐渐成为高精度计算方法研究的主要方向。采用相同的网格模板构造出的紧致差分格式可以达到比传统差分格式更高阶的精度，同时它还具有更高的尺度分辨率以及更小的波相位误差。

1991年Clarksean[29]在二维、三维平板喷流的直接数值模拟中，在流向采用紧致差分格式，在其它方向则采用了谱方法，这样谱方法和紧致差分格式的优点就被有效地结合起来。次年Lele[30]发展了一系列高精度的对称型线性紧致格式，同时还推导出1阶以上导数的紧致逼近公式。Yu[31]采用6阶紧致格式求解了非定常Euler方程，数值模拟剪切层内波系运动。William[32]在论文中初步分析了中心型紧致格式的特征值、收敛性与并行计算的可行性等，利用构造的6阶格式计算了线性与非线性Poisson问题。Wilson[33]推出了高阶紧致格式并将它们应用于不可压N-S方程的计算当中，Smirnov[34]则建立任意网格下的紧致格式Pade'，计算了二维平板流与亚声速涡传播等问题。国内，王强[35]采用高精度紧致差分算法，对可压自由剪切层转捩区中的几种典型展向大尺度涡作用形态进行了直接数值模拟。沈孟育[36]提出1种3点3阶精度的紧致格式，成功地计算了无黏伯格斯方程与一维激波管流动。

采用对称型的这些紧致格式，它们的截断误差是色散型的，因而无法抑制高波数误差，格式的精度越高，数值振荡越大。于是Shang[37]通过加入空间过滤器(filter)发展了高阶紧致格式，有效地抑制住振荡。Koutsavdis[38]采用6阶紧致格式加8阶过滤器计算二维线性Euler方程，Visbal[39]则利用Pade'格式加上6阶与10阶精度的过滤器，求解动网格条件下的N-S方程。马延文和傅德薰[40,41]通过对模型方程的研究，提出了耗散比拟思想，依此构造了迎风紧致格式和带色散协调因子的迎风紧致格式并付诸应用。

Halt[42]认为紧致格式的进一步发展在于对无波动激波捕捉技术的研究。大部分高阶紧致格式采用的是线性形式，对于非线性紧致格式，相应的研究则进行得不多。采用合适的非线性形式的紧致格式能有效地解决由于激波间断处非线性特性的存在导致的点对点振荡。Cockburn[43]通过TVD与总变差限制TVB的概念建立了3阶与4阶精度的非线性紧致格式，但4阶格式在激波附近解产生明显的数值振荡。Ravichandran[44]将Minmod通量限制器应用于高阶紧致格式，构造了对于一维标量方程具有TVD性质的高阶紧致格式，数值模拟了多种含激波的流动。Adams[45]构造了高阶紧致格式和ENO格式的混合格式数值模拟了含激波的流动。Wang[46]在Pade'格式中加入ENO算法，发展了新的高阶ENO-Pade'格式，二维定常与非定常流动应用研究发现，此格式能消除Pade'格式跨间断的非物理振荡，并提高ENO在光滑区内的高精度特性。

针对构造的迎风紧致格式计算激波时仍有振荡这一问题,马延文[47]采用群速度直接控制方法重构紧致格式,计算了Sod问题以及二维激波反射问题. 袁湘江[48]构造了一种3阶迎风紧致格式,计算球锥体的超声速绕流,结果显示数值解在激波附近的非物理振荡得到改善. 宗文刚[49]以NND和ENN格式为基础构造了加权紧致格式WCNND和WCENN,并将之应用于多种复杂黏性流动当中.

湍流研究不断深入,人们逐渐意识到应用高精度格式的重要性. 湍流流动中存在着变化范围很大的空间尺度,而这些大小尺度结构相差悬殊,这要求高精度数值方法具有捕捉大小尺度流动结构的能力,具有低耗散、低色散的高分辨能力以及计算稳定的能力. 徐岚[50]采用4阶紧致格式,完成了充分发展槽道湍流的大涡模拟计算. 李新亮[51]运用高精度迎风差分方法及8阶精度群速度控制型差分格式对可压衰减湍流的流场及被动标量场进行了直接数值模拟. Rai[52]成功地用显式迎风偏置5阶格式(explicit upwind biased fifth-order scheme, EUW-5)模拟了湍流边界层的空间发展,但它是线性的,不能很好地计算激波流场. Kourta[53]采用4阶精度格式,直接模拟了对流马赫数为1.6的超声速剪切层. Gatski[54]采用高阶有限差分格式,直接模拟了来流马赫数2.25的平板湍流边界层的发展过程. 小激波串是可压缩湍流的常见结构,因此,直接数值模拟可压缩湍流,如激波与湍流的干扰,需要能光滑捕捉激波的高精度格式.

要保持高精度流场,黏性流动计算中N-S方程的黏性项也有必要离散为高阶精度. 而且高精度格式的应用不应只考虑内点格式,还须考虑采用何种边界及靠近边界的格式,这些格式对流场最终结果具有重要的影响作用. 目前大多数高阶格式对黏性项与边界格式处理为2阶精度以保证数值解的稳定性,全流场高精度格式还很少. Zingg[55~57]分析加上人工耗散的4阶线性格式,采用高阶差分近似的边界格式和黏性项,计算了二维翼型黏性绕流. Dong[58]采用对流项为5阶迎风显式格式,黏性项为6阶中心显式格式,比较并行算法中半隐式激波装配代码的效率,计算了三维超声速Couette流动稳定性. Chu对非周期边界使用了5阶边界格式,分别发展了3点6阶混合紧致格式(combined compact difference scheme, CCD[59]与staggered combined compact difference scheme, SCCD[60]),相对于CCD,SCCD在误差与时间精度方面有很大改善. Shen[61]提出了广义紧致方法(generalized compact difference scheme, GC),对边界格式处理为5阶精度,采用3点6阶精度内点格式很好地计算了定常激波反射、非定常激波散射及激波/涡干扰等问题.

3 高精度加权紧致非线性格式的研究发展历程

邓小刚等研究者一直致力于高阶精度内点紧致格式以及相关的高阶边界和靠近边界格式研究. 1996年建立了一类单参数的线性耗散紧致格式(dissipative compact schemes, DCS)[62],它只需在相应的中心型紧致格式中加入耗散项,选择合适的参数,内点精度可分别达到3阶、5阶、7阶与9阶,边界处格式也为3阶精度. 将5阶精度DCS应用于超声速平面Couette流动的特征值问题及其稳定性边值问题求解中[63],计算结果显示该格式具有优良的色散特性. 接着他们提出适应性插值的概念,以4阶精度单元中心型(cell-centered)紧致格式为基础,在单元边界进行高阶精度紧致自适应的非线性插值,构造了紧致非线性格式(compact nonlinear schemes, CNS)[64,65]. 对一维Euler方程进行计算,结果显示3阶与4阶精度的CNS能够较稳健地捕捉间断. Yameshite[66]使用CNS很好地计算了含强激波的高超声速流动,分析了马赫数为10条件下扰动波频率对椭球壁的影响. 然而紧致自适应插值同样在流场光滑区发挥作用,结果所选择的3个模板只有一个的信息被采用,这使得CNS的效率相对低,而且接触间断在某种程度上被抹平. 于是邓小刚等采用加权技术构造了一系列非线性紧致格式(weighted compact nonlinear schemes, WCNS)[67,68]. 采用4阶WCNS以及与之相容的4阶紧致边界和靠近边界格式,模拟了一维激波管问题和马赫数为3的无黏绕前台阶流动,激波和接触间断分辨得很清楚. 对于黏性流动,数值求解了马赫数为5.73、雷诺数为2050的绕圆柱流动,计算结果可以看出激波捕捉得很好,物面压力与热流分布与有激波装配的谱方法的结果吻合很好. 2001年邓小刚基于构造耗散紧致线性格式的方法,发展了新的高阶精度耗散加权紧致非线性格式(dissipative weighted compact nonlinear schemes, DWCNS)[69],通过一维和二维含激波的超声速流场的计算,数值结果显示了这些高精度格式捕捉间断的良好特性,对边界层的高分辨率与具有的良好收敛性[70],它们能够得到比低阶格式更准确的结果. WCNS与DWCNS具有同样的理论基础,高精度特性也表现得类似,但DWCNS的计算量要比WCNS的多,因而基于计算效率方面的考虑,本文重点对WCNS作了理论分析与应用研究.

4 高精度加权紧致非线性格式 WCNS 的理论分析

4.1 非线性格式的定义

考虑标量双曲型线性方程

$$\frac{\partial u}{\partial t} + c\frac{\partial u}{\partial x} = 0 \tag{1}$$

此处, $c > 0$ 为常数. 简单起见, 考虑均匀网格 $x_j = jh$ 上求解, h 为网格间距, $0 \leq j \leq N+1$, 并给定节点上的函数值 $u_j = u(x_j)$. 式 (1) 的半离散化方程为

$$\frac{\partial u}{\partial t} + cu'_j = 0 \tag{2}$$

式 (2) 空间导数 u'_j 的格式都可写成

$$\sum_{k=j-n_1}^{j+n_2} a_k u'_k = \sum_{k=j-m_1}^{j+m_2} b_k u_k$$

式中 a_k 和 b_k 为格式系数, $(x_{j-n_1}, \cdots, x_{j+n_2})$ 和 $(x_{j-m_1}, \cdots, x_{j+m_2})$ 为网格模板.

定义 对任何求解方程 (2) 的空间离散格式, 若格式的系数 a_k, b_k 或网格模板依赖于解 u_k 本身, 这样的格式就叫非线性格式, 否则为线性格式.

显然, TVD 格式使用限制器, ENO 格式的网格模板依赖于解本身, 所以它们都是非线性格式.

4.2 加权紧致非线性格式

将式 (1) 写成守恒形式, 即 $\left(\frac{\partial u}{\partial t}\right)_j = -f'_j$. 采用单元中心型的紧致格式计算 f'_j 得

$$kf'_{j+1} + f'_j + kf'_{j-1} = \frac{a}{h}\left(\tilde{f}_{j+1/2} - \tilde{f}_{j-1/2}\right) + \frac{b}{h}\left(\tilde{f}_{j+3/2} - \tilde{f}_{j-3/2}\right) \tag{3}$$

其中 $a = \frac{3}{8}(3-2k)$, $b = \frac{1}{24}(22k-1)$. $\tilde{f}_{j\pm 1/2} = f(\tilde{u}_{j\pm 1/2})$ 为单元边界的数值通量. 对上式进行 Taylor 级数展开, 可以得到

$$f'_j = \left(\frac{\partial f}{\partial x}\right)_j + \frac{9-62k}{1920}h^4 f_j^{(5)} + O(h^r) + O(h^6)$$

截断项 $O(h^r)$ 来自 $\tilde{u}$ 的插值. 若 $\tilde{u}_{j\pm\frac{1}{2}}$ 相对于 $u_{j\pm\frac{1}{2}}$ 具有 5 阶精度 ($r = 5$), 可推导出:

(1) 若 $k = \frac{9}{62}$, 则得到 5 阶隐式格式

$$\frac{9}{62}f'_{j+1} + f'_j + \frac{9}{62}f'_{j-1} = \frac{63}{62h}\cdot\left(\tilde{f}_{j+1/2} - \tilde{f}_{j-1/2}\right) + \frac{17}{186h}\left(\tilde{f}_{j+3/2} - \tilde{f}_{j-3/2}\right) \tag{4}$$

(2) 若 $k = 0$, 则可拓展为 5 阶显式格式

$$f'_j = \frac{75}{64h}\left(\tilde{f}_{j+1/2} - \tilde{f}_{j-1/2}\right) - \frac{25}{384h}\cdot\left(\tilde{f}_{j+3/2} - \tilde{f}_{j-3/2}\right) + \frac{3}{640h}\left(\tilde{f}_{j+5/2} - \tilde{f}_{j-5/2}\right) \tag{5}$$

应当指出的是, 如果 $\tilde{u}$ 代表单元边界的 u 只有 r 阶精度 ($r < 5$), 即 $\tilde{f}(u) = f(u) + O(h^r)$, 那么单元中心型紧致差分格式的色散与耗散特性由 $\tilde{u}$ 占主导作用, 因此必须让 $\tilde{u}$ 在单元边界获得高精度插值.

在流场光滑区单元边界 $x_{j\pm 1/2}$ 处 5 阶精度的插值公式为

$$\tilde{u}^{op}_{j+1/2} = u_j + \frac{1}{128}\cdot(3u_{j-2} - 20u_{j-1} - 38u_j + 60u_{j+1} - 5u_{j+2}) \tag{6}$$

$$\tilde{u}^{op}_{j-1/2} = u_j - \frac{1}{128}\cdot(5u_{j-2} - 60u_{j-1} + 38u_j + 20u_{j+1} - 3u_{j+2}) \tag{7}$$

但是激波附近这种处理将导致数值波动, 因此需采用加权技术, 它的主要思想是这样的: 3 个候选网格模板分别赋予权值, 该权值决定候选模板自身在最终单元边界插值的贡献, 也就是说, 权值在流场光滑区处于优化权值状态, 这样格式就达到 5 阶精度; 在间断附近的区域, 含有间断的模板被赋予的权值接近于零, 这样它对数值通量几乎没有贡献, 从而避免跨间断进行插值, 保证了这些区域内的计算精度也达到 3 阶

$$\tilde{u}^{\omega}_{Lj+1/2} = u_j + \frac{h}{2}g^*_{Lj} + \frac{1}{8}h^2 s^*_{Lj}$$
$$\tilde{u}^{\omega}_{Rj-1/2} = u_j - \frac{h}{2}g^*_{Rj} + \frac{1}{8}h^2 s^*_{Rj} \tag{8}$$

其中

$$g^*_{Lj} = \sum_{k=1}^{3}\omega_{Lk}g^k_j, \quad g^*_{Rj} = \sum_{k=1}^{3}\omega_{Rk}g^k_j$$

$$s^*_{Lj} = \sum_{k=1}^{3}\omega_{Lk}s^k_j, \quad s^*_{Rj} = \sum_{k=1}^{3}\omega_{Rk}s^k_j$$

$$g^1_j = \frac{1}{2h}(u_{j-2} - 4u_{j-1} + 3u_j)$$

$$g^2_j = \frac{1}{2h}(u_{j+1} - u_{j-1})$$

$$g^3_j = \frac{1}{2h}(-3u_j + 4u_{j+1} - u_{j+2})$$

$$s^1_j = \frac{1}{h^2}(u_{j-2} - 2u_{j-1} + u_j)$$

$$s^2_j = \frac{1}{h^2}(u_{j-1} - 2u_j + u_{j+1})$$

$$s^3_j = \frac{1}{h^2}(u_j - 2u_{j+1} + u_{j+2})$$

权值 $\omega_{L(R)k}$ 定义

$$\omega_{L(R)k} = \frac{\beta_{L(R)k}}{\sum_{m=1}^{3} \beta_{L(R)m}}, \quad \beta_{L(R)k} = \frac{C_{L(R)k}}{(\delta + IS_k)^2}$$

式中 $\delta = 10^{-6}$ 为一小量,是为了避免分母为零. IS_k 为模板光滑性的度量[71], $C_{L(R)k}$ 为优化权值,由式 (6) 与式 (7) 易得 $C_{L1} = C_{R3} = \frac{1}{16}$, $C_{L2} = C_{R2} = \frac{10}{16}$, $C_{L3} = C_{R1} = \frac{5}{16}$.

由于本文所用的是加权非线性插值,其插值模板依赖于解 u_k,所以将式 (3) 与式 (8) 相结合,这一完整的格式称为加权紧致非线性格式 WCNS. 因而式 (4) 简称为 WCNS-5; 式 (5) 简称为 WCNS-E-5.

4.3 WCNS 的 Fourier 分析

对于 1 阶导数格式的差分误差,可以通过比较从导数差分格式得到的 Fourier 系数 $\left(\hat{f}'_k\right)_{fd}$ 与精确的 1 阶导数 Fourier 系数 $\hat{f}'_k$ 来估算. 引进修正波数 $\omega^* = \omega_r^* + i\omega_i^*$, 任一差分格式均相应有

$$\left(\hat{f}'_k\right)_{fd} = i\omega^* \hat{f}_k \quad (9)$$

可见修正波数的虚部控制着解的振幅变化,而实部则影响解的相位变化. 格式的稳定性要求 $\omega_i^* \leq 0$.

把式 (4) 与式 (5) 和优化插值公式代入式 (9),得到修正波数

$$\omega^* h = \omega^* (\omega h) h \quad (10)$$

图 1 显示了 WCNS-5 和 WCNS-E-5 修正波数的实部和虚部. 可以看出,除了高阶 Pade'[30] 修正波数的虚部等于零外,其余都小于零,它们对应于格式的耗散特性. WCNS-5 和 WCNS-E-5 具有近乎相同的耗散与色散特性,它们的精度稍优于 EUW-5[52].

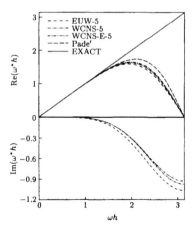

图 1 WCNS 的修正波数及其它高阶格式的比较

在多维计算中,差分格式的相位误差以各向异性的形式表现出来. 先研究二维情况,此时不同波数的相速度 ω^{**} 可表示为[30]

$$\omega^{**} = \frac{\omega^*(\omega h, \theta) h}{\omega h} = \left[\cos\theta\omega^*(\omega h \cos\theta) h + \sin\theta\omega^*(\omega h \sin\theta) h\right] / \omega h \quad (11)$$

图 2 为 $\frac{\omega h}{\pi} = \frac{1}{50}, \frac{5}{50}, \cdots, \frac{45}{50}, \frac{50}{50}$ 时几种高阶格式的各向异性相位误差分布. 最外面的曲线代表小波数 ωh 的波, 此时它的传播各向同性且只有一个相速度. 最里面的 "玫瑰形" 曲线表示在网格上能分辨到的最短波, 曲线形状越大意味着格式的各向异性相位误差限制于越窄的短波范围. 图中 WCNS、EUW-5 与 Pade' 的曲线形状相差不大, 这说明它们相位误差的各向异性特性表现得一致.

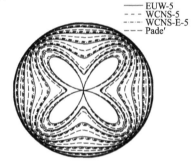

图 2 高阶格式在二维方向的相位速度各向异性图

同理,推出三维条件下不同波数的相速度 ω^{***}

$$\omega^{***} = \frac{\omega^*(\omega h, \theta, \varphi) h}{\omega h} = \left[\cos\theta\cos\varphi\omega^*(\omega h \cos\theta\cos\varphi) h + \cos\theta\sin\varphi\omega^*(\omega h \cos\theta\sin\varphi) h + \sin\theta\omega^*(\omega h \sin\theta) h\right] / \omega h \quad (12)$$

图 3 给出了 $\omega h = \pi$ 时这些有限差分格式的相位速度的各向异性图. 与二维图像类似, 格式分辨的最短波在三维方向传播表现为各向异性, 沿与 3 个坐标轴都成 $45°$ 位置波的相位速度最快.

现在规定一个误差容限 ε, 若有 $\omega^*(\omega h) h$ 使得 $\left|\frac{\omega^*(\omega h) h - \omega h}{\omega h}\right| \leq \varepsilon$, 则称具有这些波数的波是良好分辨率的, 否则就是不良分辨波. 设 ω_f 为具有良好分辨的最高波数, 则分数 $e_1 \equiv \frac{\omega_f}{\pi}$ 定义为差分格式的分辨效率. 表 1 给出了 ε 分别取 0.1、0.01 和 0.001 时 4 种高阶格式的分辨效率. 几种格式都表现出良好的分辨特性. 容限 ε 越大, 格式在更大波数范围逼近精确解, WCNS-5 与 WCNS-E-5 两者的分

辨效率非常一致,特别是在 ε 等于 0.01 和 0.001 时 WCNS-5 与 WCNS-E-5 的分辨效率好过其它格式.

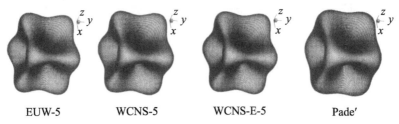

图 3 高阶格式在三维方向的相位速度各向异性图

表 1 高精度格式的分辨率

格式	$\varepsilon=0.1$	$\varepsilon=0.01$	$\varepsilon=0.001$
EUW-5	0.545	0.355	0.235
WCNS-5	0.570	0.370	0.250
WCNS-E-5	0.560	0.365	0.245
Pade′	0.595	0.360	0.205

4.4 WCNS 的渐近稳定性分析

对式 (2) 进行 WCNS 的差分离散可得出

$$A\frac{\mathrm{d}\boldsymbol{f}}{\mathrm{d}t} = \frac{c}{h}BC\boldsymbol{f} + \boldsymbol{g}(t)$$

这里 $\boldsymbol{f} = (u_1, u_2, ..., u_N)^\mathrm{T}$, $\boldsymbol{A}$ 为 1 阶导数的系数矩阵, $\boldsymbol{B}$ 为单元边界物理量的系数矩阵, $\boldsymbol{C}$ 为单元边界插值公式的系数矩阵. $\boldsymbol{g}(t)$ 对应于初始和边界条件, 不失一般性, 令 $\boldsymbol{g}(t) = 0$.

对于半离散化的方程, 渐近稳定性条件要求矩阵 $\boldsymbol{R} = \boldsymbol{A}^{-1}BC$ 的特征值实部非正. 图 4 给出了 WCNS 不同网格数下在均匀网格上的特征值. 由图 4 可见 WCNS-5 与 WCNS-E-5 是渐近稳定的, 随着 N 的增大 WCNS 越稳定.

在 WCNS 的实际计算当中, 经常需要采用非均匀网格. 为了分析非均匀网格对 WCNS 渐近稳定性的影响, 常使用指数函数

$$x(\xi) = 1 + \frac{\tilde{\beta}(1-\tilde{a})}{1+\tilde{a}}, \quad \tilde{a} = \left(\frac{\tilde{\beta}+1}{\tilde{\beta}-1}\right)^{1-\frac{\xi-1}{\xi N-1}}$$

式中 $\tilde{\beta}$ 为压缩因子, 可控制网格的疏密, 这里取 $\tilde{\beta} = 1.005$.

经过这一变换以后, 矩阵变成为 $\tilde{\boldsymbol{R}} = \boldsymbol{D}^{-1}\boldsymbol{R}$, 其中 $\boldsymbol{D} = \mathrm{diag}\{x_{\xi 1}, x_{\xi 2}, \cdots, x_{\xi N}\}$. 图 5 给出了 WCNS-5 和 WCNS-E-5 的特征值谱图. 图 5 中可以看到非均匀网格在网格数 N 为 30 与 40 时 WCNS 是渐近不稳定的, 但随着网格数增加, 拉伸程度减弱, 格式变得渐近稳定. 在 $N = 50$ 时, WCNS 是渐近稳定的, 而且 N 越大格式越稳定. 因此在实际计算中要注意 WCNS 的这一特性. 对于黏性流场的计算, 物理黏性具有稳定作用, 所以 WCNS 在稀疏网格上也可以是渐近稳定的. 特征值谱分析表明 WCNS-E-5 与 WCNS-5 的特性类似.

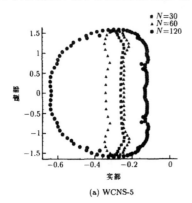

(a) WCNS-5

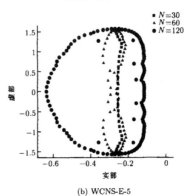

(b) WCNS-E-5

图 4 均匀网格上的特征值谱

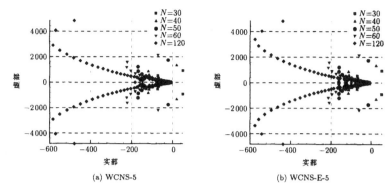

图 5 非均匀网格上的特征值谱

5 WCNS 应用于多维复杂流动研究

以上理论分析显示 WCNS-E-5 与 WCNS-5 具有相同的精度,相对 WCNS-5, WCNS-E-5 在计算中减少了 3 对角矩阵的求逆运算,提高了计算效率. 这里先就边界格式以及 2 阶导数的高阶离散作了说明,然后将它们与 WCNS-E-5 相结合,对复杂流动进行数值模拟,考察了格式的高精度特性.

5.1 WCNS-E-5 的边界格式与 2 阶导数的离散

高精度格式在实际应用中遇到的一个很重要问题是边界格式,边界格式对整个流场的计算结果有着重要影响. 倘若内点为 p 阶精度,为了不影响格式的全场精度则边界格式至少为 $(p-1)$ 阶. 对于 5 阶内点格式,导出如下 4 阶边界格式[72]

$$f'_1 = \frac{1}{24h}\left(-22\tilde{f}_{1/2} + 17\tilde{f}_{3/2} + 9\tilde{f}_{5/2} - 5\tilde{f}_{7/2} + \tilde{f}_{9/2}\right)$$

$$f'_2 = \frac{1}{24h}\left(\tilde{f}_{1/2} - 27\tilde{f}_{3/2} + 27\tilde{f}_{5/2} - \tilde{f}_{7/2}\right)$$

$$f'_{N-1} = -\frac{1}{24h}\left(\tilde{f}_{N+1/2} - 27\tilde{f}_{N-1/2} + 27\tilde{f}_{N-3/2} - \tilde{f}_{N-5/2}\right)$$

$$f'_N = \frac{1}{24h}\left(22\tilde{f}_{N+1/2} - 17\tilde{f}_{N-1/2} - 9\tilde{f}_{N-3/2} + 5\tilde{f}_{N-5/2} - \tilde{f}_{N-7/2}\right)$$

4 阶精度边界插值公式为

$$\tilde{u}_{1/2} = \frac{1}{16}(5u_0 + 15u_1 - 5u_2 + u_3)$$

$$\tilde{u}_{3/2} = \frac{1}{16}(-u_0 + 9u_1 + 9u_2 - u_3)$$

$$\tilde{u}_{N-1/2} = \frac{1}{16}(-u_{N+1} + 9u_N + 9u_{N-1} - u_{N-2})$$

$$\tilde{u}_{N+1/2} = \frac{1}{16}(5u_{N+1} + 15u_N - 5u_{N-1} + u_{N-2})$$

对于求解 N-S 方程的黏性项,有

$$f'_{vj} = \frac{75}{64h}\left(\tilde{f}_{vj+1/2} - \tilde{f}_{vj-1/2}\right) - \frac{25}{384h}\cdot\left(\tilde{f}_{vj+3/2} - \tilde{f}_{vj-3/2}\right) + \frac{3}{640h}\left(\tilde{f}_{vj+5/2} - \tilde{f}_{vj-5/2}\right)$$

单元边界上的黏性通量 $\tilde{f}_{vj+1/2}$ 含有原始变量本身以及它们的一阶偏导数. 4 阶精度的变量插值公式、1 阶偏导数的差分近似公式以及相应的 4 阶边界格式推导为:

(1) 单元边界的插值公式

$$u_{j+1/2} = \frac{9}{16}(u_{j+1} + u_j) - \frac{1}{16}(u_{j+2} + u_{j-1})$$

$$u_{1/2} = \frac{1}{16}(35u_1 - 35u_2 + 21u_3 - 5u_4)$$

$$u_{3/2} = \frac{1}{16}(5u_1 + 15u_2 - 5u_3 + u_4)$$

$$u_{N-1/2} = \frac{1}{16}(5u_N + 15u_{N-1} - 5u_{N-2} + u_{N-3})$$

$$u_{N+1/2} = \frac{1}{16}(35u_N - 35u_{N-1} + 21u_{N-2} - 5u_{N-3})$$

(2) 单元边界一阶偏导数的差分格式

$$u'_{j+1/2} = \frac{27}{24h}(u_{j+1} - u_j) - \frac{1}{24h}(u_{j+2} - u_{j-1})$$

$$u'_{1/2} = \frac{1}{24h}(-22u_0 + 17u_1 + 9u_2 - 5u_3 + u_4)$$

$$u'_{N+1/2} = -\frac{1}{24h}\cdot(-22u_{N+1} + 17u_N + 9u_{N-1} - 5u_{N-2} + u_{N-3})$$

5.2 WCNS-E-5 的高精度特性验证

5.2.1 格式捕捉激波和接触间断的能力

Levy[19] 采用 4 阶 WENO 研究了二维黎曼问

题, 这里 WCNS-E-5 也模拟相同的流动. 图 6 给出了两种高精度格式计算的结果, 图 6 中显示各类间断 (如激波、膨胀波和接触间断) 都很好地被捕捉到, 而这里的网格数仅为 Levy 的 1/4.

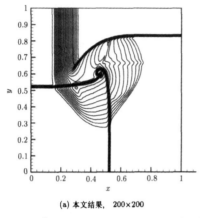 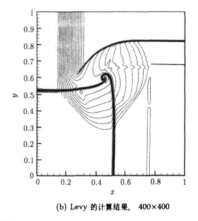

(a) 本文结果, 200×200　　　　　(b) Levy 的计算结果, 400×400

图 6　密度等值图

5.2.2 计算精度和效率

考虑双椭球外形, 计算状态选定为层流, 来流条件具体给定为: $M_\infty = 8.15$, $T_\infty = 56K$, $T_w = 288K$, $Re_\infty = 1.67 \times 10^5$. 这里提供了 3 套网格, 它们的分布按流向 × 周向 × 法向形式分别为: $126 \times 37 \times 61$ (网格 I); $126 \times 73 \times 61$ (网格 II); $126 \times 124 \times 61$ (网格 III). 它们的差别在于周向网格数目不等: 网格 I 是网格 II 的一半且与网格 II 一样均匀分布; 网格 III 由网格 II 的下部椭球网格数多一倍得到.

分别采用 MUSCL 与 WCNS-E-5 对这 3 种网格进行计算, 迭代相同步数后达到收敛. 图 7 得出了它们的物面流线侧视图, WCNS-E-5 准确地捕捉到下部椭球处的二次分离[73], MUSCL 的结果很不明显. 纵观比较, MUSCL 只有网格 III 的结果依稀反映了二次分离形成的趋势, 最多能与 WCNS-E-5 的网格 I 结果趋近. 从表 2 可以看出, 网格 III 的 MUSCL 计算时间为网格 I 的 WCNS-E-5 的 3.44 倍, 内存占用量为 2.66 倍.

图 8 为对称面沿流向计算的压力与热流分布, 并与实验值[74] 比较, 其中 WCNS-E-5 采用网格 I, MUSCL 采用的是网格 III. 分布曲线各以自己的驻点值归一化, 压力分布方面, WCNS-E-5 与 MUSCL 的结果在流向皆与实验数据接近, 在分离区略有差异, 热流分布方面, WCNS-E-5 得到的热流峰值大小、位置比 MUSCL 更趋近于实验值.

此算例充分说明稠密网格的低阶结果能通过稀疏网格的高阶格式计算更好求得, 这大大节省了计算内存和机时, 很大程度上提高了计算效率.

(a) MUSCL 网格 I　　　　(b) MUSCL 网格 II　　　　(c) MUSCL 网格 III

(d) WCNS-E-5 网格 I　　　(e) WCNS-E-5 网格 II　　　(f) WCNS-E-5 网格 III

图 7　双椭球体表面流线侧视图

表 2　两种格式不同网格计算量的比较

	WCNS-E-5			MUSCL		
	网格 I	网格 II	网格 III	网格 I	网格 II	网格 III
内存占用量 (M)	145	285	414	127	250	385
内存占用量比值	1.00	1.97	2.86	0.876	1.72	2.66
每迭代步的 CPU 花费时间 (s)	6.39	13.7	25.1	5.38	11.9	22.0
CPU 花费时间比值	1.00	2.14	3.93	0.842	1.86	3.44

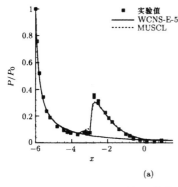
(a)

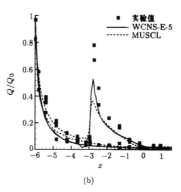
(b)

图 8　对称面沿流向压力与热流分布比较

6 结束语

本文总结了高精度加权紧致非线性格式 WCNS 的研究进展,已将该格式很好地应用于含激波的大尺度流动当中,接下来的目标是将 WCNS 计算小尺度涡流动,甚至直接数值模拟湍流. 我们认为,要高清晰分辨小尺度涡的结构,还需进一步在 4 个方面做研究: (1) 高质量的网格,开拓新思路,通过加密和合理布置生成高精度复杂网格; (2) 无反射边界的处理,流体介质的可压缩性对边界条件有重要影响,计算区域内的声波会影响到边界上的解,如果不考虑这种影响,声波将从边界反射到流场内部; (3) 时间的高精度,真实湍流流动属于非定常范畴,应保证得到时间高精度发展流场; (4) 涡流的精细处理,令人满意地显示涡产生、脱落、破裂及耗散等动态过程.

参 考 文 献

1 张涵信, 沈孟育. 计算流体力学 —— 差分方法的原理和应用. 北京: 国防工业出版社, 2003
2 Gaitonde D V, Visbal M R. Further development of a Navier-Stokes solution procedure based on high-order formulas. AIAA 99-0557, 1999
3 刘儒勋, 舒其望. 计算流体力学的若干新方法. 北京: 科学出版社, 2003
4 Harten A. High resolution schemes for hypersonic conservation laws. J Comp Phys, 1983, 49: 357~393
5 Billet G, Louedin O. Adaptive limiters for improving the accuracy of the MUSCL approach for unsteady flows. J Comp Phys, 2001, 170: 161~183
6 Yee H C, Klopfer G H, Montagne J L. High-resolution shock-capturing schemes for inviscid and viscous hypersonic flows. J Comp Phys, 1990, 88: 31~61
7 张涵信. 差分计算中激波上、下游出现波动的探讨. 空气动力学学报, 1984, 1: 12~19
8 张涵信. 无波动、无自由参数的耗散差分格式. 空气动力学学报, 1988, 6: 143~165
9 Harten A, Osher S. Uniformly high order accurate essentially non-oscillatory schemes I. SIAM J Num Anal, 1987, 24: 279~309
10 Harten A, Engquist B, Osher S, Chakravarthy S R. Uniformly high-order accurate essentially non-oscillatory shock-capturing schemes III. J Comp Phys, 1987, 71: 231~303
11 Shu C W, Osher S. Efficient implementation of essentially non-oscillatory shock capturing schemes I. J Comp Phys, 1988, 77: 439~471
12 Casper J. Finite-volume application of high-order ENO schemes to two-dimensional boundary-value problems. AIAA 91-0631, 1991
13 Godfrey A G, Mitchell C R, Walters R W. Practical aspects of spatially high accurate methods. AIAA 92-0054, 1992
14 Abgrall R. On essentially non-oscillatory schemes on unstructured meshes: Analysis and Implementation. J Comp

Phys, 1994, 114: 45~58

15 Garnier E, Sagaut P, Deville M. A class of explicit ENO filters with application to unsteady flows. *J Comp Phys*, 2001, 170: 184~204

16 Stiriba Y. A nonlinear flux split method for hyperbolic conservation laws. *J Comp Phys*, 2002, 176: 20~39

17 Liu X D, Osher S, Chan T. Weighted essentially nonoscillatory schemes. *J Comp Phys*, 1994, 115: 200~212

18 Levy D, Puppo G, Russo G. Central WENO schemes for hyperbolic systems of conservation laws. LMENS-98-17, April 1998

19 Levy D, Puppo G, Russo G. A fourth order central WENO scheme for multi-dimensional hyperbolic systems of conservation laws. *SIAM J Comput*, 2002, 24: 480~506

20 Lin S Y, Hu J J. Parametric study of weighted essentially nonoscillatory schemes for computational aeroacoustics. *AIAA Journal*, 2001, 39: 371~379

21 Suresh A, Huynk H T. Accurate monotonicity preserving scheme with Runge-Kutta time-stepping. *J Comp Phys*, 1997, 136: 83~99

22 Balsara D S, Shu C. Monotonicity preserving weighted essentially non-oscillatory schemes with increasingly high order of accuracy. *J Comp Phys*, 2000, 160: 405~452

23 袁先旭. 非定常流动数值模拟及飞行器动态特性分析研究. [博士论文]. 绵阳: 中国空气动力研究与发展中心, 2002

24 贺国宏. 三阶 ENN 格式及其在高超声速黏性复杂流场求解中的应用. [博士论文]. 绵阳: 中国空气动力研究与发展中心, 1994

25 张涵信, 李沁, 庄逢甘. 关于建立高阶差分格式的几个问题. 空气动力学学报, 1998, 16: 14~22

26 李沁, 张涵信, 高树椿. 关于超声速剪切流动的数值模拟. 空气动力学学报, 2000, 18: 67~76

27 宗文刚, 邓小刚, 张涵信. 双重加权实质无波动激波捕捉格式. 空气动力学学报, 2003, 21: 218~226

28 沈孟育, 李东海, 刘秋生. 用解析高散法构造 WENO-FCT 格式. 空气动力学学报, 1998, 16: 56~62

29 Clarksean R. Direct numerical simulation of a plane jet using the spectral-compact finite difference technique. AIAA-91-019, 1991

30 Lele S. Compact finite difference schemes with spectral-like resolution. *J Comp Phys*, 1992, 103: 16~42

31 Yu S T, Hultgren L S, Liu N S. Direct calculations of waves in fluid flows using high-order compact difference scheme. AIAA-93-0148, 1993

32 Spotz William Frederick. High-order compact finite difference schemes for computational mechanics. Dissertation for Ph.D. University of Texas at Austin, December 1995

33 Wilson R V, Demuren A Q, Carpenter M H. High-order compact schemes for numerical simulations of incompressible flows. ICASE Report 98-13, 1998

34 Smirnov S, Chrislacor, Baelmans M. A finite volume formulation for compact scheme with applications to LES. AIAA 2001-2546, 2001

35 王强, 傅德薰, 马延文. 平面可压基频涡列非线性演化行为数值研究. 力学学报, 2001, 33: 1~10

36 沈孟育, 蒋莉. 满足熵增原则的高精度高分辨率格式. 清华大学学报(自然科学版), 1998, 16: 56~62

37 Shang J S. High-order compact-difference schemes for time-dependent Maxwell equations. *J Comp Phys*, 1999, 153: 312~333

38 Koutsavdis E K, Blaisdell G A, Lyrintzis A S. On the use of compact schemes with spatial filtering in computational aeroacoustics. AIAA 99-0360, 1999

39 Visbal M R, Gordnier R E. A high-order flow solver for deforming and moving meshes. AIAA 2000-2619, 2000

40 马延文, 傅德薰. 计算空气动力学中一个新的激波捕捉法——耗散比拟法. 中国科学 (A 辑), 1992, 22: 263~271

41 傅德薰, 马延文. 高精度差分格式及多尺度流场特性的数值模拟. 空气动力学学报, 1998, 16: 24~34

42 Halt D W, Agarwal R K. Compact higher order characteristic-based Euler solver for unstructured grids. *AIAA Journal*, 1992, 30: 1993~1999

43 Cockburn B, Shu C W. Nonlinearly stable compact schemes for shock calculations. *SIAM J Numer Anal*, 1994, 31: 607~630

44 Ravichandran K S. Higher order KFVS algorithms using compact upwind difference operators. *J Comp Phys*, 1997, 130: 161~173

45 Adams N A, Shariff K. A high-resolution hybrid compact-ENO schemes for shock-turbulence interaction problems. *J Comp Phys*, 1996, 127: 27~51

46 Wang Z P, Huang G P. An essentially nonoscillatory high-order Pade'-type(ENO-Pade') scheme. *J Comp Phys*, 2002, 177: 37~58

47 马延文, 傅德薰. 群速度直接控制四阶迎风紧致格式. 中国科学 (A 辑), 2001, 31: 554~561

48 袁湘江, 周恒. 计算激波的高精度数值方法. 应用数学与力学, 2000, 21: 441~450

49 宗文刚. 高阶紧致格式及其在复杂流场求解中的应用. [博士论文]. 绵阳: 中国空气动力研究与发展中心, 2000

50 徐岚, 许春晓, 崔桂香, 陈乃祥. 四阶紧致格式有限体积法湍流大涡模拟. 清华大学学报(自然科学版), 2005, 45: 1122~1125

51 李新亮, 傅德薰, 马延文. 可压衰减湍流中被动标量场的直接数值模拟及谱分析. 中国科学 (G 辑), 2003, 33: 357~367

52 Rai M M, Moin P. Direct simulations of turbulent flow using finite-difference schemes. *J Comp Phys*, 1991, 96: 15~53

53 Sauvage K A. Computation of supersonic mixing layers. *Physics of Fluids*, 2002, 14: 3790~3797

54 Gatski T B, Erlebacher G. Numerical simulation of a spatially evolving supersonic turbulent boundary layer. NASDA-TM -2002-211934, 2002

55 Rango S D, Zingg D W. Aerodynamic computations using a high-order algorithm. AIAA 99-0167, 1999

56 Zingg D W, Rango S D, Nemec M, Pulliam T H. Comparison of several spatial discretizations for the Navier-Stokes equations. *J Comp Phys*, 2000, 160: 683~704

57 Rango S D, Zingg D W. High-order Aerodynamic computations on multi-block grids. AIAA 2001-2631, 2001

58 Dong H B, Zhong X L. A parallel high-order implicit algorithm for compressible Navier-Stokes equations. AIAA 2000-0275, 2000

59 Chu P C, Fan C W. A three-point combined compact difference scheme. *J Comp Phys*, 1998, 140: 370~399

60 Chu P C, Fan C W. A three-point sixth-order staggered combined compact difference scheme. *Mathematical and Computer Modeling*, 2000, 32: 323~340

61 Shen M Y, Zhang Z B, Niu X L. A new way for constructing high accuracy shock-capturing generalized compact difference schemes. *Comput Methods Appl Mech Engrg*, 2003, 192: 2703~2725
62 Deng X G, Maekawa H, Shen Q. A class of high order dissipative compact schemes. AIAA Paper 96-1972, 1996
63 邓小刚, 毛枚良. 超声速 Couette 流动稳定性直接数值模拟方法研究. 见: 湍流研究最新进展——中国科学技术协会青年科学家论坛第 41 次活动论文集. 北京: 科学出版社, 2001. 145~153
64 Deng X G, Maekawa H. Compact high-order accurate nonlinear schemes. *J Comp Phys*, 1997, 130: 77~91
65 Deng X G, Maekawa H. An uniform fourth-order compact scheme for discontinuities capturing. AIAA 96-1974, 1994
66 Yameshite K S, Maekawa H S. Direct numerical simulations of hypersonic boundary layer receptivity using compact convex combined schemes. AIAA 2001-1798, 2001
67 Deng X G, Mao M L. Weighted compact high-order nonlinear schemes for the Euler equations. AIAA paper 97-1941, 1997
68 Deng X G, Zhang H X. Developing high-order accurate nonlinear schemes. *J Comp Phys*, 2000, 165: 22~44
69 邓小刚. 高阶精度耗散加权紧致非线性格式. 中国科学 (A 辑), 2001, 31: 1104~1117
70 Deng X G, Mao M L, Liu J C. High-order dissipative weighted compact nonlinear schemes for Euler and Navier-Stokes equations. AIAA 2001-2626, 2001
71 Deng X G, Liu X, Mao M L, Zhang H X. Investigation on weighted compact fifth-order nonlinear scheme and applications to complex flow. AIAA-2005-5246, 2005
72 Liu X, Deng X G, Mao M L. Studying of weighted compact high-order nonlinear scheme WCNS-E-5 for complex flows. *CFD Journal*, 2004, 13: 173~180
73 Riedelbauch S, Brenner G, Muller B, Kordulla W. Numerical simulation of laminar hypersonic flow past a double-ellipsoid. AIAA 89-1840, 1989
74 Li Z W, Tao X C, Zhang H X. Numerical simulation of aerodynamic heating over a generic missile configuration. In: Proceedings of fourth Asia workshop on computational fluid dynamics, March 3-6, 2004, University of Tokyo. 2004. 141~146

ADVANCES IN HIGH-ORDER ACCURATE WEIGHTED COMPACT NONLINEAR SCHEMES*

DENG Xiaogang   LIU Xin†   MAO Meiliang   ZHANG Hanxin

China Aerodynamics Research and Development Center, Mianyang Sichuan 621000, China

Abstract The progress in the theoretical analysis of high-order accurate weighted compact nonlinear schemes (WCNS) and applications to the complicated flows is summarized in this paper. The development of high-order schemes is reviewed, focusing on WCNS. WCNS are analyzed with Fourier method and asymptotic stability theory. Some explicit WCNS are used to solve multi-dimensional complicated flows.

Keywords high-order accurate scheme, weighted compact nonlinear schemes, Fourier analysis, asymptotic stability theory

基于 HDCS-E8T7 格式的气动噪声数值模拟方法研究进展

姜屹[1,2], 邓小刚[3], 毛枚良[1,2], 刘化勇[1]

(1. 中国空气动力研究与发展中心 空气动力学国家重点实验室,四川 绵阳 621000;
2. 中国空气动力研究与发展中心 计算空气动力学研究所,四川 绵阳 621000;
3. 国防科学技术大学,湖南 长沙 410073)

摘　要:发展了气动噪声高精度数值模拟方法。空间导数离散采用了HDCS-E8T7格式及精度与之匹配的边界格式。时间离散方法采用了高精度多步龙格库塔方法和高精度隐式双时间步方法,发展了高精度对接边界算法,将方法推广至多块对接网格以满足解决复杂几何构型问题的需要,采用隐式大涡的概念处理可能出现的湍流问题。在此基础上,研究了几何守恒律对计算结果的影响,展示了复杂网格中高精度计算满足几何守恒律的重要性,完成了等熵涡、双圆柱散射、串列柱构型和喷嘴射流等典型噪声问题的求解。所得计算结果展示了所发展的模拟方法具有良好的预测精度和解决复杂构型气动噪声问题的潜力。

关键词:混合型线性耗散紧致格式;计算气动噪声;几何守恒律;隐式大涡模拟;复杂外形

中图分类号:V211.3　　　文献标识码:A　　　doi: 10.7638/kqdlxxb-2014.0103

0 引 言

气动噪声是人类生活和工业应用中常见的问题,会对人类健康和军事飞行器及潜艇的隐秘行动等多方面产生影响[1],因此,抑制气动噪声是一个非常重要课题[2]。然而,由于实际问题中流动现象的复杂性,目前有效的噪声预测手段是把复杂噪声系统分解为不同的子系统,分别研究其中的噪声机制,如射流噪声和流动/物体干扰噪声。

近十多年以来,计算机技术迅猛发展,计算机资源得到了极大的丰富,为数值研究噪声问题提供了强大的硬件支撑。在射流噪声数值模拟研究方面,Bogey等[3-4]数值模拟了马赫数和雷诺数对射流噪声的影响,同时围绕准确预测射流噪声在数值技术和数值处理方面开展了大量的研究工作,比如边界条件和大涡模拟(LES)的亚格子模型。Morris 研究了入口边界条件对噪声预测的影响[5]。Bodony 和 Lele 则研究了亚格子模型对射流噪声计算的影响[6]。Bogey 和 Bailly 在一系列文章中系统地研究了入口条件[7-8]和亚格子模型[9]对射流噪声计算的影响。最近,对射流噪声的研究已经把喷嘴考虑在计算域内,预测不同喷嘴的射流噪声,研究不同喷口形状对射流噪声的影响,其中包括双头喷嘴射流[10-12],具有锯齿边缘的双头喷嘴射流[13]以及含有V型边缘的亚声速射流[14-19]。Engblom[20]等针对V型锯齿单股冷喷和热喷流时,采用雷诺平均 Navier-Stokes(RANS)方法数值研究了锯齿内弯角度与射流流动的关系,尽管预测了远场噪声频谱的变化趋势,但是,他们的数值结果存在明显的不足[21]。事实上,对于射流噪声预测,RANS 方法还不能胜任。Tide 等在文献[21]中采用 SST 模型预测射流噪声,只能够得到不同位置噪声大小的变化趋势,而噪声绝对值与实验值差别则很大。采用 LES 方法可以很好地克服这个问题,比如Andersson[22-24]等采用 LES 方法预测射流噪声时,得到的数值解与实验值差异为±3dB。在流动与物体干扰噪声数值模拟研究方面,串列圆柱-翼型结构是研究这类噪声的基本模型,不同的计算方法被用于研究该干扰构型产生的气动噪声,包括 RANS 方法[25-26]、LES 方法[25-27]和分离涡模拟(DES)方法[28-30],而采用 RANS 方法预测串列圆柱-翼型干扰结构噪声没有得到可信赖的结果。Min 等[31]通过实验和 RANS 方法研究了串列圆柱-翼型结构中,圆柱与翼型的相对位置对气动噪声的影响,他们的计算值与实验值差别非常大。Boudet 等[27]通过对比发现采用 LES 计算串列圆柱-翼型干扰构型流动问题,不但可以得到比 RANS 更好的噪声频谱,而且可以得到更好的平均流场。出现这种现象的原因是 RANS 方法难于模拟这种湍流尾迹与壁面相互干扰的非定常现象,而这种现象是产生气动噪声的主要原因。

上面的研究实践表明,采用大涡模拟方法处理湍流效应非常有助于得到可靠的气动噪声预测结果。事实上,气动噪声问题与通常的动力学问题明显不同。首先,气动声学问题通常是非定常的波动问题,而且其脉动的频率范围非常宽,除了需要求解声源区域的流动特性外,还要探讨远场声波的指向性和声能量分布,这要求数值方法在全场都有一致的精度[32];其次,声学脉动量与流动量相比非常小,通常比流动量低 5~6 个量级,这可能使声学小量被数值误差掩盖[33]。为了正确和有效地计算气动噪声问题,气动噪声计算格式需要非常低的色散和耗散误差。由于高阶紧致有限差分格式的高分辨率和灵活性,它对于计算气动噪声而言是一种最具吸引力的方法[34]。

尽管高阶精度有限差分格式的研究与应用取得了很多进步,但是,其在复杂外形中的应用仍然存在许多问题,比如,网格敏感性和鲁棒性等[35-36]。这些问题可以通过满足几何守恒律得到明显的改善[37-41]。几何守恒律包括面积守恒律和体积守恒律,体积守恒律在动网格中已经被广泛地研究,而对于有限差分格式,面积守恒律则很少被研究。最近研究表明,几何守恒律对高精度有限差分格式在曲线网格中的应用非常重要。如果面积守恒律得不到满足,高精度有限差分格式有可能在计算中出现数值不稳定甚至是计算崩溃的现象。为了使高精度有限差分格式满足面积守恒律,邓小刚等在文献[40]中提出了守恒网格导数算法(CMM),该方法是通过使用同样的差分格式计算网格导数和通量导数而实现的,并证明了如果数值通量的一阶导数算子 δ_1 不分裂(加权紧致非线性格式(WCNS)[42]、紧致非线性格式(CNS)[43]和新的混合型加权紧致非线性格式(HWCNS)[44]等),CMM 方法在有限差分格式中很容易实现,而如果 δ_1 分裂为 δ_1^+ 和 δ_1^- (如加权本质无波动格式(WENO)[45]、耗散紧致格式(DCS)[46]等),CMM 方法在有限差分格式中则很难实现。为了使得 WENO 格式满足几何守恒律,一些学者把 WENO 格式分为两项:中心差分项和数值耗散项,其中中心差分项可以很方便地满足几何守恒律,而对于耗散项,他们通过冻结网格导数造迎风通量以达到满足几何守恒律的目的[47]。最近,邓小刚等在 CMM 算法的基础上,又进一步提出了可以明显改善不规则网格中计算精度的对称型守恒网格导数方法(SCMM)[48]。由于认识到了满足几何守恒律对高阶有限差分格式成功应用于复杂外形的重要性,基于高阶有限差分格式的计算气动声学方法已经逐渐地应用于预测复杂外形中的气动噪声,比如喷嘴射流噪声[49]。最近,Daude 等[50]采用 Lele 提出的六阶中心紧致格式[51],成功地预测了串列柱翼构型产生的气动噪声。Lele 提出的六阶中心紧致格式还被 Rizzetta 等[52]用于研究方腔气动噪声抑制。

耗散紧致格式(DCS)[46]是邓小刚等通过组合一阶导数格式和二阶导数格式引入数值耗散,得到的一类高分辨率线性紧致有限差分格式。DCS 格式在计算一些简单外形相关的流动问题和声学标准算例中取得了成功[53-56]。但是,由于它的 δ_1 分裂为 δ_1^+ 和 δ_1^-,SCMM 方法难于在其中实现[48],导致它在复杂外形中的应用会出现数值不稳定甚至是计算崩溃的现象。根据满足 SCMM 方法的原则,邓小刚等改进了 DCS 格式,并提出了一类可以适应复杂外形的新的混合型耗散紧致格式 HDCS,最高精度达十一阶。HDCS-E8T7 是其中的一种七阶混合型耗散紧致格式[57]。该类格式具有以下优点:(a)它能满足几何守恒律,以保持其在曲线网格中的精度和增强它应对复杂外形的能力;(b)它具备可以物理地耗散不能分辨波数的固有耗散;(c)可以在相对窄的模板里得到高精度,因此其边界和邻近边界格式构造相对简单,可以方便地应用于复杂外形。邓小刚等在文献[58]中系统地分析了 HDCS-E8T7 格式的性质。目前,HDCS-E8T7 格式已经成功地预测了一些典型气动噪声问题[59-60]。

为了把 HDCS-E8T7 格式应用于复杂外形,除了满足几何守恒律,还需要多块结构网格技术,因为一般情况下很难为复杂外形生成单块网格。对于多块结构网格而言,首要问题是实现网格块之间信息高精度地传播。Thomas[61] 在拼接网格上,发展了多维高精度插值方法,实现相邻网格块信息的传递,该方法在网格生成、计算效率等方面比对接网格具有明显的优势,但高精度多维插值方法比较复杂,涉及的插值模板点很多,如对于二维 3 阶插值,该文使用了 12 个点,特别是对于复杂情况,所希望得到的虚拟点可能并不在计算域中,实际计算精度并不能保证。Kim 等[62] 基于对接网格提出广义特征对接边界处理技术,能够消除几何奇性对计算精度的影响,实现网格块之间流动信息的高精度传递。邓小刚等基于流场信息沿特征线传播的原则,建立了对流信息沿特征方向传播的高精度对接边界算法[63]。在这种高精度对接边界算法的基础之上,HDCS-E8T7 格式被成功地推广应用于复杂外形中数值模拟[61]。

本文主要介绍七阶 HDCS-E8T7 格式预测气动噪声的进展,重点将集中在格式应用于气动噪声预测的数值模拟方法、几何守恒律对数值计算的影响以及其对典型气动噪声问题的模拟情况。文中下一节将详细介绍数值方法,第二节展示了几何守恒律对数值计算的影响,典型的数值算例将在第三节中给出。

1 数值方法

1.1 控制方程

考虑曲线坐标系下的 Navier-Stokes 方程:

$$\frac{\partial \widetilde{U}}{\partial t} + \frac{\partial \widetilde{E}}{\partial \xi} + \frac{\partial \widetilde{F}}{\partial \eta} + \frac{\partial \widetilde{G}}{\partial \zeta} =$$

$$\frac{\theta}{Re}\left(\frac{\partial \widetilde{E}_v}{\partial \xi} + \frac{\partial \widetilde{F}_v}{\partial \eta} + \frac{\partial \widetilde{G}_v}{\partial \zeta}\right) + \frac{S}{J} \quad (1)$$

当 $\theta=1$ 是 Navier-Stokes 方程, $\theta=0$ 是 Euler 方程, S 为源项,它只在双圆柱散射算例测试中用到。

$$\widetilde{U} = U/J$$
$$\widetilde{E} = (\xi_t U + \xi_x E + \xi_y F + \xi_z G)/J$$
$$\widetilde{F} = (\eta_t U + \eta_x E + \eta_y F + \eta_z G)/J$$
$$\widetilde{G} = (\zeta_t U + \zeta_x E + \zeta_y F + \zeta_z G)/J$$
$$\widetilde{E}_v = (\xi_x E_v + \xi_y F_v + \xi_z G_v)/J$$
$$\widetilde{F}_v = (\eta_x E_v + \eta_y F_v + \eta_z G_v)/J$$
$$\widetilde{G}_v = (\zeta_x E_v + \zeta_y F_v + \zeta_z G_v)/J$$

且

$$U = [\rho, \rho u, \rho v, \rho w, \rho e]^T$$
$$E = [\rho u, \rho u^2 + p, \rho v u, \rho w u, (\rho e + p)u]^T$$
$$F = [\rho v, \rho u v, \rho v^2 + p, \rho w v, (\rho e + p)v]^T$$
$$G = [\rho w, \rho u w, \rho v w, \rho w^2 + p, (\rho e + p)w]^T$$
$$E_v = [0, \tau_{xx}, \tau_{xy}, \tau_{xz}, u\tau_{xx} + v\tau_{xy} + w\tau_{xz} + \dot{q}_x]^T$$
$$F_v = [0, \tau_{yx}, \tau_{yy}, \tau_{yz}, u\tau_{yx} + v\tau_{yy} + w\tau_{yz} + \dot{q}_y]^T$$
$$G_v = [0, \tau_{zx}, \tau_{zy}, \tau_{zz}, u\tau_{zx} + v\tau_{zy} + w\tau_{zz} + \dot{q}_z]^T$$

粘性应力项可以写为:

$$\tau_{ij} = \mu\left(\frac{\partial u_i}{\partial x_j} + \frac{\partial u_j}{\partial x_i} - \frac{2}{3}\delta_{ij}\frac{\partial u_l}{\partial x_l}\right)$$

热传导项为:

$$\dot{q}_x = \frac{1}{(\gamma-1)p_\infty M_\infty^2}\mu\frac{\partial T}{\partial x}$$
$$\dot{q}_y = \frac{1}{(\gamma-1)p_\infty M_\infty^2}\mu\frac{\partial T}{\partial y}$$
$$\dot{q}_z = \frac{1}{(\gamma-1)p_\infty M_\infty^2}\mu\frac{\partial T}{\partial z}$$

其中 γ 为比热比, μ 为粘性系数(可以通过 Sutherland 公式求解)。状态方程和能量方程为:

$$p = \frac{1}{\gamma M_\infty^2}\rho T, \quad e = \frac{p}{(\gamma-1)\rho} + \frac{u^2+v^2+w^2}{2}$$

以上 u、v 和 w 分别为 x、y 和 z 方向的速度分量,p 为压力,ρ 为密度,T 为温度。无量纲量的定义分别为 $\rho = \rho^*/\rho_\infty^*$, $(u,v,w) = (u,v,w)^*/V_\infty^*$, $T = T^*/T_\infty^*$, $p = p^*/\rho_\infty^* V_\infty^{*2}$, $\mu = \mu^*/\mu_\infty^*$, 而且 $M_\infty = u_\infty^*/\sqrt{\gamma R T_\infty^*}$, $Re = \rho_\infty^* u_\infty^* r^*/\mu_\infty^*$, $Pr = \mu_\infty^* C_p/\kappa_\infty^*$ 分别为马赫数、雷诺数和普朗特数, r^* 为参考长度。J 是网格变换雅克比, $\xi_x, \xi_y, \xi_z, \eta_x, \eta_y, \eta_z, \zeta_x, \zeta_y, \zeta_z$ 为网格导数。其中的空间导数具有以下的守恒形式

$$\begin{cases}
\widetilde{\xi}_x = \xi_x/J = (y_\eta z)_\zeta - (y_\zeta z)_\eta \\
\widetilde{\xi}_y = \xi_y/J = (z_\eta x)_\zeta - (z_\zeta x)_\eta \\
\widetilde{\xi}_z = \xi_z/J = (x_\eta y)_\zeta - (x_\zeta y)_\eta \\
\widetilde{\eta}_x = \eta_x/J = (y_\zeta z)_\xi - (y_\xi z)_\zeta \\
\widetilde{\eta}_y = \eta_y/J = (z_\zeta x)_\xi - (z_\xi x)_\zeta \\
\widetilde{\eta}_z = \eta_z/J = (x_\zeta y)_\xi - (x_\xi y)_\zeta \\
\widetilde{\zeta}_x = \zeta_x/J = (y_\xi z)_\eta - (y_\eta z)_\xi \\
\widetilde{\zeta}_y = \zeta_y/J = (z_\xi x)_\eta - (z_\eta x)_\xi \\
\widetilde{\zeta}_z = \zeta_z/J = (x_\xi y)_\eta - (x_\eta y)_\xi
\end{cases} \quad (2)$$

1.2 空间离散方法

七阶 HDCS-E8T7 格式被用于离散控制方程(1)。考虑无粘项的离散:

综述与进展

$$\frac{\partial \widetilde{U}}{\partial t} + \frac{\partial \widetilde{E}}{\partial \xi} + \frac{\partial \widetilde{F}}{\partial \eta} + \frac{\partial \widetilde{G}}{\partial \zeta} = 0 \quad (3)$$

和它的半离散形式:

$$\frac{\partial \widetilde{U}}{\partial t} = -\delta_1^\xi \widetilde{E} - \delta_1^\eta \widetilde{F} - \delta_1^\zeta \widetilde{G} \quad (4)$$

其中 $\delta_1^\xi, \delta_1^\eta$ 和 δ_1^ζ 分别为 ξ, η 和 ζ 方向的有限差分算子,它们的形式是一致的,因此我们只讨论 ξ 方向。七阶 HDCS-E8T7 格式的 δ_1^ξ 算子为:

$$\delta_1^\xi \widetilde{E}_j = \frac{256}{175h}(\hat{E}_{j+1/2} - \hat{E}_{j-1/2}) - \frac{1}{4h}(\widetilde{E}_{j+1} - \widetilde{E}_{j-1}) + \frac{1}{100h}(\widetilde{E}_{j+2} - \widetilde{E}_{j-2}) - \frac{1}{2100h}(\widetilde{E}_{j+3} - \widetilde{E}_{j-3}), \quad (5)$$

其中, $\hat{E}_{j\pm 1/2} = \widetilde{E}(\hat{U}_{j\pm 1/2}, \hat{\xi}_{x,j\pm 1/2}, \hat{\xi}_{y,j\pm 1/2}, \hat{\xi}_{z,j\pm 1/2})$ 和 $\hat{E}_{j+m} = \widetilde{E}(U_{j+m}, \hat{\xi}_{x,j\pm m}, \hat{\xi}_{y,j\pm m}, \hat{\xi}_{z,j\pm m})$ 分别为半节点和节点上的数值通量。数值通量 $\hat{E}_{j\pm 1/2}$ 采用半节点上变量计算得到:

$$\hat{E}_{j\pm 1/2} = \widetilde{E}(\hat{U}^L_{j\pm 1/2}, \hat{U}^R_{j\pm 1/2}, \hat{\xi}_{x,j\pm 1/2}, \hat{\xi}_{y,j\pm 1/2}, \hat{\xi}_{z,j\pm 1/2}) \quad (6)$$

其中,$\hat{U}^L_{j\pm 1/2}, \hat{U}^R_{j\pm 1/2}$ 为半节点变量。

$$\frac{5}{14}(1-\alpha)\hat{U}^L_{j-1/2} + \hat{U}^L_{j+1/2} + \frac{5}{14}(1+\alpha)\hat{U}^L_{j+3/2} = \frac{25}{32}(U_{j+1} + U_j) + \frac{5}{64}(U_{j+2} + U_{j-1}) - \frac{1}{448}(U_{j+3} + U_{j-2}) + \alpha\left[\frac{25}{64}(U_{j+1} - U_j) + \frac{15}{128}(U_{j+2} - U_{j-1}) - \frac{5}{896}(U_{j+3} - U_{j-2})\right] \quad (7)$$

$\alpha<0$ 为控制 HDCS-E8T7 格式耗散的耗散参数。相应的 $\hat{U}^R_{j+1/2}$ 可以很容易地通过令 $\alpha>0$ 得到。耗散参数 α 对 HDCS-E8T7 格式精度没有影响,但是它对格式的分辨率有影响。与 Lele 提出的中心紧致格式[51]相比,图1画出了采用不同耗散参数的 HDCS-E8T7 格式的修正波数。图中可以看出 HDCS-E8T7 格式具有接近谱方法的分辨率。根据色散关系保持(DRP)[65]原则,HDCS-E8T7 格式可以通过调整耗散参数 α 得到优化。优化的耗散参数为 0.3[58],本文的 HDCS-E8T7 格式在应用过程中将采用该优化参数。

HDCS-E8T7 内点格式为七点模板,因此在左侧 $j=1,2,3$ 和右侧 $j=N,N-1,N-2$ 需要三点边界和临近边界格式。临近边界点 $j=3$ 和 $j=N-2$ 采用六阶格式:

$$\widetilde{E}'_j = \frac{64}{45h}(\hat{E}_{j+1/2} - \hat{E}_{j-1/2}) - \frac{2}{9h}(\widetilde{E}_{j+1} - \widetilde{E}_{j-1}) + \frac{1}{180h}(\widetilde{E}_{j+2} - \widetilde{E}_{j-2}) \quad (8)$$

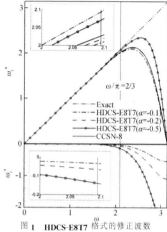

图 1 HDCS-E8T7 格式的修正波数
Fig. 1 Modified wavenumber of the HDCS-E8T7

临近边界点 $j=2$ 和 $j=N-1$ 采用四阶格式:

$$\widetilde{E}'_j = \frac{4}{3h}(\hat{E}_{j+1/2} - \hat{E}_{j-1/2}) - \frac{1}{6h}(\widetilde{E}_{j+1} - \widetilde{E}_{j-1}) \quad (9)$$

边界点 $j=1$ 和 $j=N$ 的四阶格式为:

$$\widetilde{E}'_1 = \frac{1}{6h}(48\hat{E}_{3/2} + 16\hat{E}_{5/2} - 25\widetilde{E}_1 - 36\widetilde{E}_2 - 3\widetilde{E}_3),$$

$$\widetilde{E}'_N = \frac{-1}{6h}(48\hat{E}_{N-1/2} + 16\hat{E}_{N-3/2} - 25\widetilde{E}_N - 36\widetilde{E}_{N-1} - 3\widetilde{E}_{N-2}) \quad (10)$$

边界格式中,$j=3/2$ 和 $j=N-1/2$ 的五阶插值:

$$\hat{U}^L_{3/2} = \frac{1}{128}(35U_1 + 140U_2 - 70U_3 + 28U_4 - 5U_5),$$

$$\hat{U}^L_{N-1/2} = \frac{1}{128}(35U_N + 140U_{N-1} - 70U_{N-2} + 28U_{N-3} - 5U_{N-4}) \quad (11)$$

同时 $j=5/2$ 和 $j=N-3/2$ 采用的五阶耗散插值为:

$$\frac{3}{10}(1-\alpha)\widetilde{U}^L_{j-1/2} + \hat{U}^L_{j+1/2} + \frac{3}{10}(1+\alpha)\widetilde{U}^L_{j+3/2} = \frac{3}{4}(U_{j+1} + U_j) + \frac{1}{20}(U_{j+2} + U_{j-1}) + \alpha\left[\frac{3}{8}(U_{j+1} - U_j) + \frac{3}{40}(U_{j+2} - U_{j-1})\right] \quad (12)$$

对于粘性项的计算,首先求解原始变量 u,v,w,T 导数以得到应力张量和热传导通量。然后再一次使用七阶 HDCS-E8T7 格式得到粘性通量导数。

1.3 时间积分方法

采用了两种时间积分格式:显式格式和隐式格式。显式时间积分采用 Hu 提出的 Runge-Kutta 方法[66]。如果 R 表示右端项,控制方程为:

$$\frac{\partial \widetilde{U}}{\partial t} = R = \frac{\theta}{Re}\left(\frac{\partial \widetilde{E}_v}{\partial \xi} + \frac{\partial \widetilde{F}_v}{\partial \eta} + \frac{\partial \widetilde{G}_v}{\partial \zeta}\right) +$$

$$\frac{S}{J} - \frac{\partial \widetilde{E}}{\partial \xi} - \frac{\partial \widetilde{F}}{\partial \eta} - \frac{\partial \widetilde{G}}{\partial \zeta} \quad (13)$$

则 Hu 的五步 Runge-Kutta 方法通过式(14)从时间 t_0(第 n 步)积分至 $t_0+\Delta t$(第 $n+1$ 步)。

$$R^i = \Delta t R(\widetilde{U}^n + \Delta t \alpha_i R^{i-1})$$
$$\widetilde{U}^{n+1} = \widetilde{U}^n + R^5 \quad (i=1,2,3,4,5) \quad (14)$$

其中,$\alpha_1=0, \alpha_2=\frac{1}{3}, \alpha_3=\frac{1}{4}, \alpha_4=\frac{1}{3}, \alpha_5=\frac{1}{2}$。

如果计算网格拉伸严重,显式时间推进方法稳定性不足,隐式时间推进方法非常必要。本文的噪声预测体系中的隐式时间推进格式为基于牛顿迭代[67]的隐式双时间步方法[68]。二阶双时间步方法可以表示为:

$$\left[J^{-1}\left(\frac{1}{\Delta \tau}+\frac{3}{2\Delta t}\right)I+M(U^p)\right]\Delta U^{p+1} = -\left[J^{-1}\frac{3U^p - 4U^n + U^{n-1}}{2\Delta t} + R(U^p)\right] \quad (15)$$

其中 $M(U) = \partial R(U)/\partial U$, τ 为虚拟时间。第一个子迭代时,$p=1, U^p=U^n$,而当 $p \to \infty$ 时,$U^p \to U^{n+1}$。隐式时间推进算子中,空间离散采用二阶格式,而对右端 R 的离散采用 HDCS-E8T7 格式。

1.4 网格导数计算方法

根据邓小刚等在文献[40]中的研究,面积守恒律在均匀网格中自动得到满足。但是,在曲线网格中则有可能不能满足。如果面积守恒律没有得到满足,则在复杂曲线网格计算中可以会出现数值不稳定,甚至是计算崩溃现象。本小节,我们将根据满足面积守恒律的原则,讨论网格导数的计算方法。守恒网格导数(2)具有以下的离散形式:

$$\begin{cases}
\widetilde{\xi}_x = \delta_{\text{II}}^\eta((\delta_{\text{III}}^\zeta y)z) - \delta_{\text{II}}^\zeta((\delta_{\text{III}}^\eta y)z) \\
\widetilde{\xi}_y = \delta_{\text{II}}^\eta((\delta_{\text{III}}^\zeta z)x) - \delta_{\text{II}}^\zeta((\delta_{\text{III}}^\eta z)x) \\
\widetilde{\xi}_z = \delta_{\text{II}}^\eta((\delta_{\text{III}}^\zeta x)y) - \delta_{\text{II}}^\zeta((\delta_{\text{III}}^\eta x)y) \\
\widetilde{\eta}_x = \delta_{\text{II}}^\zeta((\delta_{\text{III}}^\xi y)z) - \delta_{\text{II}}^\xi((\delta_{\text{III}}^\zeta y)z) \\
\widetilde{\eta}_y = \delta_{\text{II}}^\zeta((\delta_{\text{III}}^\xi z)x) - \delta_{\text{II}}^\xi((\delta_{\text{III}}^\zeta z)x) \\
\widetilde{\eta}_z = \delta_{\text{II}}^\zeta((\delta_{\text{III}}^\xi x)y) - \delta_{\text{II}}^\xi((\delta_{\text{III}}^\zeta x)y) \\
\widetilde{\zeta}_x = \delta_{\text{II}}^\xi((\delta_{\text{III}}^\eta y)z) - \delta_{\text{II}}^\eta((\delta_{\text{III}}^\xi y)z) \\
\widetilde{\zeta}_y = \delta_{\text{II}}^\xi((\delta_{\text{III}}^\eta z)x) - \delta_{\text{II}}^\eta((\delta_{\text{III}}^\xi z)x) \\
\widetilde{\zeta}_z = \delta_{\text{II}}^\xi((\delta_{\text{III}}^\eta x)y) - \delta_{\text{II}}^\eta((\delta_{\text{III}}^\xi x)y)
\end{cases} \quad (16)$$

其中 $\delta_{\text{I}}^\xi, \delta_{\text{I}}^\eta, \delta_{\text{I}}^\zeta$ 和 $\delta_{\text{III}}^\xi, \delta_{\text{III}}^\eta, \delta_{\text{III}}^\zeta$ 分别为 ξ, η 和 ζ 方向网格导数求解算子。在曲线网格中应用时,有限差分格式需要满足面积守恒律[40], $I_x = I_y = I_z = 0$:

$$\begin{cases}
I_x = \delta_{\text{I}}^\xi(\widetilde{\xi}_x) + \delta_{\text{I}}^\eta(\widetilde{\eta}_x) + \delta_{\text{I}}^\zeta(\widetilde{\zeta}_x) \\
I_y = \delta_{\text{I}}^\xi(\widetilde{\xi}_y) + \delta_{\text{I}}^\eta(\widetilde{\eta}_y) + \delta_{\text{I}}^\zeta(\widetilde{\zeta}_y) \\
I_z = \delta_{\text{I}}^\xi(\widetilde{\xi}_z) + \delta_{\text{I}}^\eta(\widetilde{\eta}_z) + \delta_{\text{I}}^\zeta(\widetilde{\zeta}_z)
\end{cases} \quad (17)$$

为了使高精度有限差分格式满足面积守恒律,邓小刚等在文献[40]中提出了守恒网格导数算法 CMM。CMM 方法是通过使用与求解通量导数相同的格式计算守恒型网格导数(16)实现的,即 $\delta_{\text{I}} = \delta_{\text{II}}$。文献[40]中证明了 CMM 方法可以很容易应用于通量导数算子 δ_{I} 不分裂的高阶有限差分格式,但是对于将 δ_{I} 分裂为迎风算子 δ_{I}^+ 和 δ_{I}^- 的格式则很难。虽然守恒型网格导数(16)中的内层算子 δ_{III} 对几何守恒律没有影响,邓小刚等在文献[40]中推荐关系式 $\delta_{\text{III}} = \delta_{\text{II}}$。不久之后,他们从几何的角度解释了关系式 $\delta_{\text{III}} = \delta_{\text{II}}$ 的意义[48],并且提出了可以显著提高不规则网格中计算准度的对称型守恒网格导数算法 SCMM[48]。为了满足面积守恒律,本文网格导数计算采用了 SCMM 方法[48]。半节点网格坐标等相关变量由高阶精度插值得到,其中内点 δ_{II} 和 δ_{III} 的八阶显式插值为:

$$\widehat{U}_{j+1/2} = \frac{1}{2048}[1225(U_{j+1}+U_j) - 245(U_{j+2}+U_{j-1}) + 49(U_{j+3}+U_{j-2}) - 5(U_{j+4}+U_{j-3})] \quad (18)$$

在边界点,δ_{II} 和 δ_{III} 的插值为:

$$\widehat{U}_{3/2} = \frac{1}{128}(35U_1 + 140U_2 - 70U_3 + 28U_4 - 5U_5)$$

$$\widehat{U}_{7/2} = \frac{1}{256}(3U_1 - 25U_2 + 150U_3 + 150U_4 - 25U_5 + 3U_6)$$

$$\widehat{U}_{5/2} = \frac{1}{128}(-5U_1 + 60U_2 + 90U_3 - 20U_4 + 3U_5)$$

$$\widehat{U}_{N-1/2} = \frac{1}{128}(35U_N + 140U_{N-1} - 70U_{N-2} + 28U_{N-3} - 5U_{N-4})$$

$$\widehat{U}_{N-3/2} = \frac{1}{128}(-5U_N + 60U_{N-1} + 90U_{N-2} - 20U_{N-3} + 3U_{N-4})$$

$$\widehat{U}_{N-5/2} = \frac{1}{256}(3U_N - 25U_{N-1} + 150U_{N-2} + 150U_{N-3} - 25U_{N-4} + 3U_{N-5}) \quad (19)$$

1.5 高精度对接边界处理方法

众所周知,针对复杂外形生成单块网格非常困难,一般需要多块网格技术,而且,网格线由于外形的原因通常会拐折。图 2 给出了这种网格的一个简单例子。从图中可以看出,网格线在奇点(图中以黑点表示)处突然发生拐折,显而易见,在这些点上左侧的网格导数不等于右侧的网格导数,这样就产生了奇性问题[62]。为了避免网格奇性,我们根据 Kim 等在文献[62]中的思想,把计算域沿着奇性线分解为两块。之后,在对接点和临近对接点为 HDCS-E8T7 格式构造特殊的对接边界格式。如图 2 所示,奇性点被分解为两个虚拟点。对于左侧和右侧网格分别计算导数,我们期望对接边界格式为不跨过对接面的单边格式,而且对接边界格式还需要满足几何守恒律以消除守恒

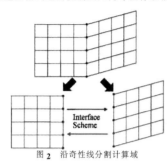

图 2 沿奇性线分割计算域
Fig. 2 Decomposing a computational domain along the singular line

误差即,$\delta_I = \delta_{II} = \delta_{III}$。为了满足这些要求,对接边界格式的构造分两步。

首先,构造 δ_I、δ_{II} 和 δ_{III} 的插值方法。为了避免网格奇性,需要构造单侧插值。如图 3(a) 所示,奇性点 j 被分解为空间位置相同,但网格导数值不同的两个虚拟点 j_l 和 j_r。在奇性点 j 的左右两侧,分别计算 j_l 和 j_r 的网格导数,j_l 和 j_r 的单侧插值为:

$$\hat{U}_{j_l-5/2} = \frac{1}{256}(3U_{j_l} - 25U_{j_l-1} + 150U_{j_l-2} + 150U_{j_l-3} - 25U_{j_l-4} + 3U_{j_l-5})$$

$$\hat{U}_{j_l-3/2} = \frac{1}{128}(-5U_{j_l} + 60U_{j_l-1} + 90U_{j_l-2} - 20U_{j_l-3} + 3U_{j_l-4})$$

$$\hat{U}_{j_l-1/2} = \frac{1}{128}(35U_{j_l} + 140U_{j_l-1} - 70U_{j_l-2} + 28U_{j_l-3} - 5U_{j_l-4})$$

$$\hat{U}_{j_l+1/2} = \frac{1}{128}(315U_{j_l} - 420U_{j_l-1} + 378U_{j_l-2} - 180U_{j_l-3} + 35U_{j_l-4})$$

$$\hat{U}_{j_l+3/2} = \frac{1}{128}(1155U_{j_l} - 2772U_{j_l-1} + 2970U_{j_l-2} - 1540U_{j_l-3} + 315U_{j_l-4})$$

$$\hat{U}_{j_l+5/2} = \frac{1}{128}(3003U_{j_l} - 8580U_{j_l-1} + 10010U_{j_l-2} - 5460U_{j_l-3} + 1155U_{j_l-4})$$

和

$$\hat{U}_{j_r-5/2} = \frac{1}{128}(3003U_{j_r} - 8580U_{j_r+1} + 10010U_{j_r+2} - 5460U_{j_r+3} + 1155U_{j_r+4})$$

$$\hat{U}_{j_r-3/2} = \frac{1}{128}(1155U_{j_r} - 2772U_{j_r+1} + 2970U_{j_r+2} - 1540U_{j_r+3} + 315U_{j_r+4})$$

$$\hat{U}_{j_r-1/2} = \frac{1}{128}(315U_{j_r} - 420U_{j_r+1} + 378U_{j_r+2} - 180U_{j_r+3} + 35U_{j_r+4})$$

$$\hat{U}_{j_r+1/2} = \frac{1}{128}(35U_{j_r} + 140U_{j_r+1} - 70U_{j_r+2} + 28U_{j_r+3} - 5U_{j_r+4})$$

$$\hat{U}_{j_r+3/2} = \frac{1}{128}(35U_{j_r} + 140U_{j_r+1} - 70U_{j_r+2} + 28U_{j_r+3} - 5U_{j_r+4})$$

$$\hat{U}_{j_r+5/2} = \frac{1}{256}(3U_{j_r} - 25U_{j_r+1} + 150U_{j_r+2} + 150U_{j_r+3} - 25U_{j_r+4} + 3U_{j_r+5})$$

第二步,采用半节点型 δ_I 构造单边格式计算网格导数。如图 3(b) 所示,我们把由奇性点分解得到的两个虚拟点重新合并在一起,从而对接点 j 的六阶半节点型 δ_I(等于 δ_{II} 和 δ_{III})可以写为:

$$\tilde{E}'_j = \frac{2250}{1920h}(\hat{E}_{j+1/2} - \hat{E}_{j-1/2}) - \frac{125}{1920h}(\hat{E}_{j+3/2} - \hat{E}_{j-3/2}) + \frac{9}{1920h}(\hat{E}_{j+5/2} - \hat{E}_{j-5/2})$$

(20)

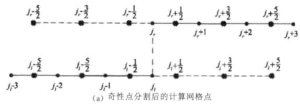

(a) 奇性点分割后的计算网格点

(b) 奇性点合并后的计算网格点
图 3 奇性点附近计算点示意图
Fig. 3 Sketch of computational grids near the singular point

对于网格导数计算,方程(20)将分别应用于奇性点的左侧和右侧。由于 HDCS-E8T7 格式采用七点模板,该对接边界格式将应用于 $[j-2,j+2]$。对于奇性点左右两侧的信息交换,将采用特征对接边界方法[63]。时间推进之后,对接点上的变量将通过平均左右两侧的值进行修正:

$$U_L^* = U_R^* = (U_L + U_R)/2 \qquad (21)$$

其中星号表示修正后的值。这种高精度对接边界处理方法可以保持 HDCS-E8T7 在复杂网格中的计算精度和稳定性[69]。

2 几何守恒律对数值计算的影响

为了展示面积守恒律的影响,我们设计了一种新的网格导数的算法,它通过加权守恒网格导数和传统网格导数得到计算网格导数值,如其中一个网格导数为:

$$\hat{\xi}_x^A = \sigma \hat{\xi}_x^T + (1-\sigma) \widetilde{\xi}_x \qquad (22)$$

其中 $\sigma \in [0,1]$ 为权值,且 $\hat{\xi}_x^T$ 与 $\widetilde{\xi}_x$ 分别为传统网格导数和守恒网格导数,均采用 HDCS-E8T7 格式计算得到,它们具有相同的离散精度。采用此网格导数的算法记为 HDCS-E8T7A,如果 $\sigma=0$,HDCS-E8T7A 将为满足面积守恒律的 HDCS-E8T7 算法,但是如果 $\sigma \neq 0$,面积守恒律将得不到满足。传统的网格导数具有以下表达形式:

$$\begin{cases}
\hat{\xi}_x^T = \delta_1^\eta(y)\delta_1^\zeta(z) - \delta_1^\eta(z)\delta_1^\zeta(y) \\
\hat{\xi}_y^T = \delta_1^\eta(z)\delta_1^\zeta(x) - \delta_1^\eta(x)\delta_1^\zeta(z) \\
\hat{\xi}_z^T = \delta_1^\eta(x)\delta_1^\zeta(y) - \delta_1^\eta(y)\delta_1^\zeta(x) \\
\hat{\eta}_x^T = \delta_1^\zeta(y)\delta_1^\xi(z) - \delta_1^\zeta(z)\delta_1^\xi(y) \\
\hat{\eta}_y^T = \delta_1^\zeta(z)\delta_1^\xi(x) - \delta_1^\zeta(x)\delta_1^\xi(z) \\
\hat{\eta}_z^T = \delta_1^\zeta(x)\delta_1^\xi(y) - \delta_1^\zeta(y)\delta_1^\xi(x) \\
\hat{\zeta}_x^T = \delta_1^\xi(y)\delta_1^\eta(z) - \delta_1^\xi(z)\delta_1^\eta(y) \\
\hat{\zeta}_y^T = \delta_1^\xi(z)\delta_1^\eta(x) - \delta_1^\xi(x)\delta_1^\eta(z) \\
\hat{\zeta}_z^T = \delta_1^\xi(x)\delta_1^\eta(y) - \delta_1^\xi(y)\delta_1^\eta(x)
\end{cases} \qquad (23)$$

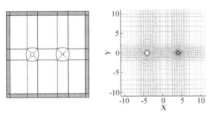

图 4 双圆柱散射计算网格示意图
Fig. 4 Computed domain and mesh of two-cylinder scattering case

2.1 双圆柱散射

该算例是第四届计算气动声学专题研讨会的标准算例,选用此算例的目的是展示几何守恒律在多块网格中对计算的影响。Euler 方程(1)右端的声源项只在能量方程中出现,它的形式为:

$$S = \exp[-2\ln2(x^2+y^2)]\sin(8\pi t) \qquad (24)$$

为了逐渐地把源项引入到计算域中,在计算开始阶段,一个立方律的修形函数加入到源项(24)中。壁面边界上,法向速度分量 $V = \eta_x u + \eta_y v + \eta_z w$ 为 0,切向速度分量 $U = \xi_x u + \xi_y v + \xi_z w$,压力和密度通过五阶外插得到。在远场边界处采用了远场无反射边界条件[70]和海绵层技术[71]。时间推进采用 1.3 小节中的五步 Runge-Kutta 方法,无量纲时间推进步长为 0.002(一个周期为 125 步),总共推进 20 000 步,其中最后 2 000 步用于统计。图 4 为计算网格示意图。计算采用了三套网格,它们的网格量分别为 469×409(粗网格),721×685(中等网格)和 841×775(细网格)。

在中等网格中,表 1 列出了 HDCS-E8T7 和 HDCS-E8T7A 格式的最大面积守恒误差。如图 5 所示,HDCS-E8T7A($\sigma=1$)格式的最大面积守恒误差出现在奇点附近。

表 1 HDCS-E8T7 和 HDCS-E8T7A 格式的最大 I_x 和 I_y 误差
Table 1 Maximal errors of I_x and I_y for the HDCS-E8T7 and HDCS-E8T7A

HDCS-E8T7	HDCS-E8T7A 权值 σ					
	10^{-5}	10^{-4}	10^{-3}	10^{-2}	10^{-1}	1
max(I_x) 1.37E-14	1.37E-8	1.37E-7	1.37E-6	1.37E-5	1.37E-4	1.37E-3
max(I_y) 1.47E-14	1.37E-8	1.37E-7	1.37E-6	1.37E-5	1.37E-4	1.37E-3

(a) 整个计算域   (b) 左边圆柱附近

图 5 $\sigma=1$ 时 HDCS-E8T7A 格式的 $|I_x+I_y|$ 误差(后附彩图)
Fig. 5 SCL error ($|I_x+I_y|$) of HDCS-E8T7A with $\sigma=1$

图 6 画出了 HDCS-E8T7 和 HDCS-E8T7A 计算得到的脉动压力均方根值等值线。图中显示当面积守恒误差足够小时($\sigma=10^{-5}$),声场没有被明显污染。但是当 σ 从 10^{-5} 变大到 10^{-3} 时,声场在面积守恒误差最大的奇点附近变得越来越差。如图 7 所示,当 σ 大于 10^{-2} 时,整个声场都被污染,而且最差的地方仍

综述与进展

然是在奇点附近。这清楚地说明了满足面积守恒律对 HDCS-E8T7 格式的应用非常重要。图 8 给出了 HDCS-E8T7 格式使用三套网格计算得到的圆柱壁面脉动压力。当网格加密时,数值解逐渐接近理论解。采用细网格得到的数值解与理论解非常吻合,这展示了满足几何守恒律的 HDCS-E8T7 格式在复杂网格中的高分辨率。

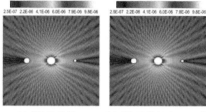
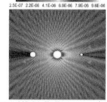
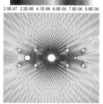

(a) HDCS-E8T7　(b) HDCS-E8T7A($\sigma=10^{-5}$)　(c) HDCS-E8T7A($\sigma=10^{-4}$)　(d) HDCS-E8T7A($\sigma=10^{-3}$)

图 6　HDCS-E8T7 和 HDCS-E8T7A 计算得到的脉动压力均方根值云图(后附彩图)

Fig. 6　Mean-squared fluctuating pressure contours of HDCS-E8T7 and HDCS-E8T7A

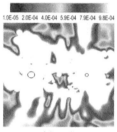

(a) $\sigma=10^{-2}$　(b) $\sigma=10^{-1}$　(c) $\sigma=1$

图 7　HDCS-E8T7A 格式计算得到的脉动压力均方根云图(后附彩图)

Fig. 7　Mean-squared fluctuating pressure contours of HDCS-E8T7A

积守恒误差。

无量纲计算域为 $(90\times 60\times \pi)$,网格大小为 $180\times 180\times 45$。图 9 展示了二维网格,网格在圆柱壁面附近进行了加密,法向最小的网格间距为 0.001。在外边界,采用了基于一维无粘近似的远场边界条件[70]和海绵层技术[71]。固壁面采用无滑移边界条件,等温壁,五阶精度外插实现压力法向梯度为 0。展向采用周期边界条件。

图 8　圆柱壁面脉动压力均方根分布

Fig. 8　RMS fluctuating pressure on the surface of cylinders

2.2　圆柱绕流

采用圆柱绕流算例,测试几何守恒律对 HDCS-E8T7 格式模拟湍流问题的影响。计算条件是来流马赫数为 0.2,基于圆柱直径 D 的雷诺数为 $Re_D = 3900$。本文的大涡模拟方法是基于七阶 HDCS-E8T7 格式的一种隐式大涡模拟方法(HILES)[72],虽然该方法与单调积分大涡模拟方法(MILES)[73]的思想是一致的,但是它具有以下两点特殊的性质[72,71]:(1) 空间离散采用的是具备固有耗散的七阶 HDCS-E8T7 格式;(2) HILES 可以消除可能污染流场的面

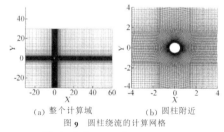

(a) 整个计算域　(b) 圆柱附近

图 9　圆柱绕流的计算网格

Fig. 9　Grid system of the cylinder test

表 2 列出了 HDCS-E8T7 和 HDCS-E8T7A 的最大面积守恒误差,其中可以看出当 $\sigma=1$ 时,面积守

表 2 HDCS-E8T7 和 HDCS-E8T7A 的最大 I_x 和 I_y 误差

Table. 2 The maximal errors of I_x and I_y for the HDCS-E8T7 and HDCS-E8T7A

格式	HDCS-E8T7	HDCS-E8T7A						
Weight σ	0	0.01	0.1	0.2	0.4	0.6	0.8	1.0
max(I_x)	4.83E-14	3.35E-6	3.35E-5	6.70E-5	1.34E-4	2.01E-4	2.68E-4	3.35E-4
max(I_y)	7.08E-14	3.36E-6	3.36E-5	6.72E-5	1.34E-4	2.02E-4	2.69E-4	3.36E-4

恒误差将达到最大。如图 10 所示，HDCS-E8T7A($\sigma = 1$)的最大面积守恒误差出现在奇点附近。

均匀流场作为初场开始计算，时间推进采用 1.3 小节中的二阶隐式双时间步方法，流动在无量纲时间 $t = 100$ 时达到全湍流状态。统计结果将由 $t = 100$ 到 $t = 200$ 的计算结果得到，计算中无量纲的时间推进步长为 $\Delta t = 0.01$。图 11 画出了采用 HDCS-E8T7 和 HDCS-E8T7A 计算得到的时均流向速度云图。从图中可以看出，面积守恒误差会污染计算流场。为了定量显示面积守恒误差的影响，我们在图 12 中画出了 $x/D = 1.06$ 处时均流向速度的垂直分布。图中清楚地显示，随着面积守恒误差的增大，非物理振荡越来越明显。这种现象同样在图 12 中 $S_1(1.06, 1.6)$ 和 $S_2(1.06, 1.7)$ 两点的速度分布中得到了体现。图 13 给出了瞬时涡量云图。图中显示了基于 HDCS-E8T7 格式的 HILES 模拟方法计算得到了非常细致的结果，特别是在圆柱附近的尾迹区，但涡量可能被面积守恒误差严重污染，因此，满足几何守恒律对高精度湍流模拟同样非常重要。

3 测试算例

3.1 等熵涡传播

流动设定为二维，且 $(u, v, p, T) = (1, 0, 1, 1)$。在初始均匀场中加入一个涡核位于 (x_c, y_c) 的等熵涡，它满足以下条件：

$$\begin{cases} (\delta u, \delta v) = \varepsilon r e^{\beta(1-r^2)}(\sin\theta, -\cos\theta), \\ \delta T = -(\gamma - 1)\varepsilon^2 e^{2\beta(1-r^2)}/(4\alpha\gamma), \\ \delta S = 0 \end{cases}$$

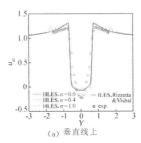

(a) 垂直线上

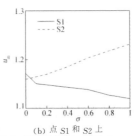

(b) 点 S_1 和 S_2 上

图 12 展向平均的时均流向速度分布($x/D = 1.06$)

Fig. 12 Spanwise averaged mean streamwise velocity distributions($x/D = 1.06$)

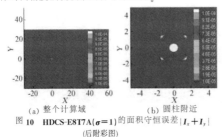

(a) 整个计算域 (b) 圆柱附近

图 10 HDCS-E8T7A($\sigma = 1$)的面积守恒误差 $|I_x + I_y|$（后附彩图）

Fig. 10 SCL error ($|I_x + I_y|$) of HDCS-E8T7A with $\sigma = 1$

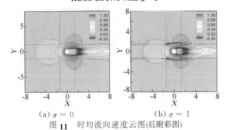

(a) $\sigma = 0$ (b) $\sigma = 1$

图 11 时均流向速度云图（后附彩图）

Fig. 11 Mean streamwise velocity contours

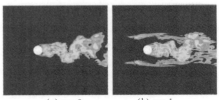

(a) $\sigma = 0$ (b) $\sigma = 1$

图 13 瞬时涡量云图（后附彩图）

Fig. 13 Instantaneous vorticity contours

其中 $\alpha=\beta=0.8$ 为涡衰减参数，$\varepsilon=0.3$ 表示涡强度，$\tau=r/r_c$，$r=[(x-x_c)^2+(y-y_c)^2]^{1/2}$，$r_c=1.0$ 为涡核半径。这里 $T=p/\rho$ 是温度，$S=p/\rho^\gamma$ 是熵。

计算采用均匀网格，且计算域为 $-10\leqslant x\leqslant10$，$-10\leqslant y\leqslant 10$。为了考察格式对波传播的分辨能力，本文设计不同的网格，使得涡核点数（PPVC），即涡核内部 x 和 y 方向上的网格点数为 $6\sim12$。根据这个原则，得到了 7 套不同的网格，网格量分别为 61×61，71×71，81×81，91×91，101×101，111×111 和 121×121，对应的 PPVC 分别为 6，7，8，9，10，11 和 12。

计算开始于 $t=0$ 和初始位置 $(x_c=0,y_c=0)$，一直持续至涡核运动至 $(4,0)$ 处，时间推进采用 1.3 小节中的五步 Runge-Kutta 方法。此时等熵涡还未到达外边界，远场无反射边界[70]对计算结果的影响可以忽略。为了定量地考察 PPVC 与计算结果的关系，我们比较了不同 PPVC 值的压力数值误差的 L_1 模、L_2 模和 L_∞ 模。图 14 展示了压力误差随 PPVC 的变化情况，从中可以看出 HDCS-E8T7 的计算值展现了非常好的网格收敛性。根据 L_1 模、L_2 模和 L_∞ 模，可以得到相应的数值精度。表 3 列出了计算得到的数值精度，从中可以看出 HDCS-E8T7 格式的设计精度得到了数值验证。

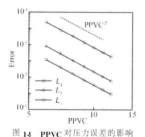

图 14　PPVC 对压力误差的影响

Fig. 14　Effect of PPVC on the errors of pressure

表 3　HDCS-E8T7 格式的数值精度

Table 3　Accuracy of the HDCS-E8T7

PPVC	L_1	精度	L_2	精度	L_∞	精度
6	1.189E-6	——	8.141E-6	——	2.425E-4	——
7	4.004E-7	7.06	2.697E-6	7.16	8.199E-5	7.03
8	1.558E-7	7.07	1.039E-7	7.14	3.201E-5	7.04
9	6.826E-8	7.01	4.497E-7	7.11	1.390E-5	7.08
10	3.267E-8	6.99	2.132E-7	7.08	6.591E-6	7.08
11	1.676E-8	7.00	1.089E-7	7.05	3.359E-6	7.07
12	9.106E-9	6.99	5.930E-8	6.99	1.818E-6	7.05

图 15 给出了 PPVC=6 时，得到的计算结果。从图中可以看出，由于 HDCS-E8T7 格式的高分辨率，等熵涡已经被很完美地捕捉，结合表 3 列出的压力误差值，我们可以看出七阶 HDCS-E8T7 格式对线性波传播问题具有很强的捕捉能力，这对预测气动噪声是非常有利的条件。

3.2　串列柱翼构型噪声

采用串列柱翼构型展示本文的方法预测复杂外形中气动噪声的能力。计算结果将与 Jacob 在文献[25]中给出的实验值进行比较，采用的计算条件与实验的相同，即：NACA0012 翼型（弦长为 $c=0.1$m）位于圆柱下游一倍弦长处，来流速度为 $U_\infty=72$m/s，基于圆柱直径 $d=0.1c$ 的雷诺数为 $Re_d=4.8\times10^4$。在这个构型中，圆柱后缘湍流尾迹与翼型前缘碰撞，期间伴随着涡的挤压、破裂等现象，这也是气动噪声产生的主要原因之一。HILES 将被用于模拟串列柱翼构型中的湍流流动现象，而对于远场噪声的预测，采用的是 FW-H 方法[75]。

图 16 画出了网格示意图，总网格量约为 1600

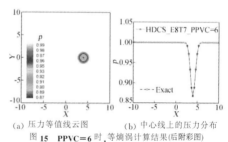

(a) 压力等值线云图　　(b) 中心线上的压力分布

图 15　PPVC=6 时，等熵涡计算结果（后附彩图）

Fig. 15　Numerical solutions of vortex convection test with PPVC=6

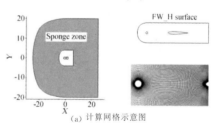

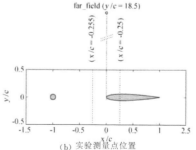

(a) 计算网格示意图

(b) 实验测量点位置

图 16　计算网格及实验测量点位置示意图

Fig. 16　Sectional view of the computational domain and sketch of the locations for the measurements

万,同时为了与实验数据对比,图 16 还展示了 Jacob 等在文献[25]中给出的实验测量点的位置示意图。为了进行大涡模拟计算,翼型和圆柱展向都延拓了 0.03m,且展向均匀分布的网格点数为 45。网格在圆柱和翼型壁面附近进行了加密处理,壁面距离为 1.0×10^{-5}(通过翼型弦长 c 无量纲得到的值)。计算中采用的边界条件为:远场采用无反射边界条件[70]和海绵层技术[71](图 16 中橙色部分);第 1.5 小节中的对接边界策略用于应对复杂多块结构网格,壁面采用无滑移边界条件,展向采用周期边界条件。时间推进采用 1.3 小节中的二阶隐式双时间步方法,无量纲的时间推进步长为 0.001,对应的真实时间步长约为 1.5×10^{-6}s,计算初始流场为均匀来流,共推进 30 000 时间步,其中最后 20 000 步用于统计平均,由于展向采用了周期边界条件,因此,统计时均变量值及脉动量均方根值时,对展向进行了平均。

为了展示 HILES 计算得到的流场,图 17 画出了中间截面上瞬时展向涡量和涡量大小等值线云图。从中我们可以看出圆柱后缘的尾迹碰撞翼型前缘后,破裂为更小的涡向翼型后缘传播。考察流场中典型位置的速度统计量,图 18 对比了计算得到的 $x/c = -0.255$ 和 $x/c = 0.25$ 处流向时均速度与相应的实验值。在 $x/c = -0.255$ 处,计算得到的中心线附近速度时均值比实验测量值低,这导致了 $x/c = 0.25$ 处,壁面附近的速度时均值高于实验测量值,这种现象在 Boudet 等[27]的计算结果中同样可以观察到。图 19 给出了 $x/c = -0.255$ 和 $x/c = 0.25$ 处流向速度均方根值分布与相应的实验值,从中可以看出本文的方法很好地预测了湍流脉动。圆柱后缘点的可分辨能谱同样证明了计算结果的可靠性,如图 20 所示。图中 $St = fd/U_\infty$ 为斯特鲁哈尔数,f 为频率,U_∞ 为来流速度。从中可以看出,数值计算达到了分辨惯性子区中湍动能的尺度要求,在此区间湍动能与 $St^{-5/3}$ 具有线性关系[76]。图 20 所展示的湍动能衰减斜率表明 HILES 捕捉到的湍动能能谱是可信赖的。图 21 画出了使用密度着色的第二不变量的等值面,第二不变量 Q 为:

$$Q = (\Omega_{ij}\Omega_{ij} + S_{ij}S_{ij})/2 \qquad (26)$$

其中 $\Omega_{ij} = (u_{i,j} - u_{j,i})/2$ 和 $S_{ij} = (u_{i,j} + u_{j,i})/2$ 分别为速度旋度的反对称分量和对称分量。由于 HDCS-E8T7 格式的高分辨率,我们可以看出许多涡结构细节。图 22 给出了 $(x=0.68m, y=1.74m)$ 处噪声功率谱密度。计算得到的声压峰值及其频率与实验值吻合得很好。但是,对于高频部分,本文的计算值与实验值相比偏高,这与 Boudet 等[27]的计算结果类

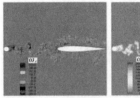

(a) 瞬时展向涡量等值线云图　(b) 瞬时涡量大小等值线云图

图 17　流场中涡量等值线云图(后附彩图)

Fig. 17　Contours of instantaneous vorticity

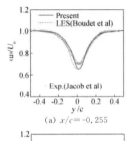

(a) $x/c = -0.255$

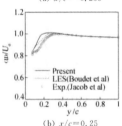

(b) $x/c = 0.25$

图 18　展向平均的流向时均速度分布

Fig. 18　Spanwise averaged mean streamwise velocity distributions

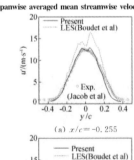

(a) $x/c = -0.255$

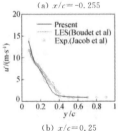

(b) $x/c = 0.25$

图 19　展向平均的流向脉动速度均方根值分布

Fig. 19　Spanwise averaged RMS value of fluctuating velocity

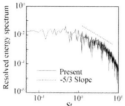

图 20 圆柱后缘点的可分辨能谱
Fig. 20 Resolved energy spectrum at a location behind the cylinder

$L/d=10$, $U_\infty=72m/s$, $Re_d=4.8\times10^4$

图 21 速度张量的第二不变量等值面(后附彩图)
Fig. 21 Iso-surfaces of the second invariant of velocity gradient tensor for instantaneous vortex structure

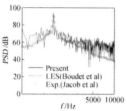

图 22 远场噪声功率谱密度图
Fig. 22 Power spectral density of the pressure perturbations measured in the farfield

似。Boudet 等在文献[27]中对这种现象给出的一个解释是:FW-H 积分面外的体积分没有用于计算气动噪声。不过,整体上还是很好地预测了串列柱翼构型辐射的宽频噪声。

3.3 喷嘴射流噪声

本文考虑的喷嘴与 Andersson 在文献[22]中的相同。喷嘴直径 D_j 为 $50mm$,基于喷嘴直径和喷嘴出口速度的雷诺数 Re_D 为 5.0×10^4,喷嘴出口马赫数为 M_a $=0.75$。计算的射流出口静温 T_j 等于背景温度 T_∞,射流入口条件与 Tide 在文献[21]中的条件相同。

图 23 给出了三维模拟的计算域平面示意图。为了减小计算域出口和远场反射对流动模拟的影响,出口和远场都加了海绵吸波层[71]。计算域在射流出口向前方向上延伸了 20 倍喷嘴直径,射流入口处计算域直径为 10 倍喷嘴直径,出口为 20 倍喷嘴直径,海绵层的长度为 40 倍喷嘴直径。计算域采用大约具有 1.8×10^8 个网格点的多块结构网格离散。图 24 展示了计算网格系统。

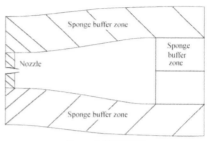

图 23 计算域示意图
Fig. 23 Sectional view of the computational domain

在射流入口边界处,给定了静压和总压。在射流出口边界处,即在海绵层的末端,给定了静压。对于外边界,采用了远场无反射边界条件[70]和海绵层技术[71]。固壁面采用无滑移边界条件,法向压力梯度为 0 和绝热壁。

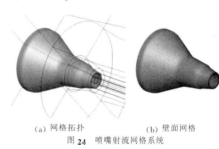

(a) 网格拓扑   (b) 壁面网格

图 24 喷嘴射流网格系统
Fig. 24 Mesh system for the nozzle

计算开始时,先采用定常算法直到计算域里的整体质量流量率变化小于射流出口质量流量率的 1%。在此基础上开始非定常计算,时间推进采用 1.3 小节中的二阶隐式双时间步方法。时间步长约为 $\Delta t = 4\mu s$,共推进 30 000 步,最后 20 000 步用于流动统计平均。然后采用 FW-H 方法[75]计算得到远场观测点的脉动声压。图 25 给出了这些观测点的位置。气动噪声通过 FW-H 积分面上的压力计算得到,这个积分面取得足够大以确保它包括了所有的声源,即这个

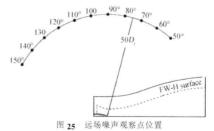

图 25 远场噪声观察点位置
Fig. 25 Locations of the far-field receivers

算例中的剪切层。

图 26 画出了瞬时轴向速度云图。图 27 比较了本文计算得到的中心线上的轴向平均速度与 Andersson 等在文献[22]中报告的计算值和实验值。可以看出本文计算得到的速度核心区长度和核心区后的速度衰减更好。图 28 画出了采用射流喷嘴出口速度无量纲化的流向 (u_{rms}) 和法向 (v_{rms}) 脉动速度均方根值。与实验值相比,本文的计算很好地捕捉了脉动速度 u_{rms} 和 v_{rms} 的峰值位置。但是在速度核心区 $x/D_j < 6$ 的位置,计算得到了强度非常低的湍流。Andersson 等在文献[22]中的计算结果同样观察到了这种现象。图 29 比较了计算得到的和实验测量得到的观察点上的声压级,从中可以看出本文的计算值与实验值吻合地很好。

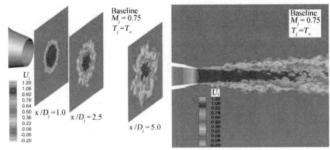

图 26　瞬时流向速度云图(后附彩图)
Fig. 26　Contours of instantaneous axial velocity

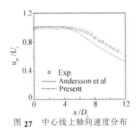

图 27　中心线上轴向速度分布
Fig. 27　Distributions of axial velocity along the centerline

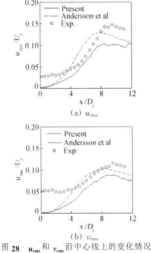

图 28　u_{rms} 和 v_{rms} 沿中心线上的变化情况
Fig. 28　Variation of u_{rms} and v_{rms} along the centerline

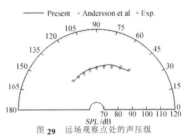

图 29　远场观察点处的声压级
Fig. 29　Sound pressure levels at observer locations

4　结束语

本文介绍了基于 HDCS-E8T7 格式的高精度数值模拟方法及其在气动噪声模拟方面的研究进展,得到以下结论:

(1) 所发展的算法具有预期的高阶精度;

(2) 在复杂网格中,几何守恒律误差对计算结果具有显著影响,消除它对模拟气动噪声问题尤其重要;

(3) 采用隐式大涡概念处理湍流,是高精度模拟湍流噪声的有效途径之一。

从模拟的对象和对问题的研究深度审视,本文对气动噪声模拟研究才刚刚起步。随着工作向求解实际工程问题迈进,将面临着诸多挑战,其中最基本的挑战有:

(1) 网格生成的困难。对接网格可能难于应对工程实际问题中的复杂几何特性,需要对复杂外形适应能力更强的网格生成方法,比如拼接网格和重叠网格等,以及相应的分区边界算法。

(2) 湍流模拟问题。采用隐式大涡概念可以有效地处理湍流,但求解效率还不足,目前还难以在工程实际问题中应用,需要发展计算效率更高的湍流求解方法。

(3) 流固耦合的气动噪声问题。目前求解都是流动非定常的气动噪声问题,对于飞行器几何构型变化(包括位置变化和变形)引起的动态非定常噪声问题尚未涉及。

在下一步工作中,我们将围绕这些挑战,继续完善算法,在解决工程气动噪声问题中不断进步。

参 考 文 献:

[1] WANG M, FREUND J B, LELE S K. Computational prediction of flow-generated sound[J]. Annu. Rev. Fluid Mech., 2006, 38: 483-512.

[2] JORDAN P, COLONIUS T. Wave packets and turbulent jet noise[J]. Annu. Rev. Fluid Mech., 2013, 45: 175-195.

[3] BOGEY C, BAILLY C. Investigation of subsonic jet noise using LES: Mach and Reynolds number effects[R]. AIAA 2004-3023.

[4] BOGEY C, BAILLY C. Direct computation of the sound radiated by a high-Reynolds number, subsonic round jet[R]. The CEAS Workshop From CFD to CAA, Confederation of European Aerospace Societies Paper 2, 2002.

[5] MORRIS P J, LONG L N, SCHEIDEGGER T E, et al. Simulations of supersonic jet noise [J]. International Journal of Aeroacoustics, 2002, 1(1): 17-41.

[6] BODONY D J, LELE S K. Large eddy simulation of turbulent jets and progress towards a subgrid scale turbulence model[C]. Proceedings of International Workshop on LES for Acoustics, DLR Göttingen, Göttingen, Germany, 2002: 1-12.

[7] BOGEY C, BAILLY C. LES of a high Reynolds, high subsonic jet: effects of the inflow conditions on flow and noise [R]. AIAA 2003-3170.

[8] BOGEY C, BAILLY C. Effects of inflow conditions and forcing on subsonic jet flows and noise[J]. AIAA J., 2005, 43(5): 1000-1007.

[9] BOGEY C, BAILLY C. LES of a high Reynolds, high subsonic jet: effects of the subgrid modellings on flow and noise [R]. AIAA 2003-3557.

[10] VUILLOT F, LUPOGLAZOFF N, RAHIER G. Double-stream nozzles flow and noise computations and comparisons to experiments[R]. AIAA 2008-0009.

[11] FAYARD B, RAHIER G, VUILLOT F, et al. Flow field analysis for double stream nozzle: application to jet noise[R]. AIAA 2008-2983.

[12] FAYARD B, RAHIER G, VUILLOT F. Modal analysis of jet flow from a coaxial nozzle with central plug[R]. AIAA 2009-3355.

[13] VISWANATHAN K, SHUR M, SPALART P, et al. Flow and noise predictions for single and dual-stream beveled nozzles [J]. AIAA J., 2008, 16(3): 601-626.

[14] SHUR M L, SPALART P R, STRELETS M K. Noise prediction for increasingly complex jets. Part II: Applications[J]. International Journal of Aeroacoustics, 2005, 4(3-4): 247-266.

[15] SHUR M L, SPALART P R, STRELETS M K, et al. Further steps in LES-based noise prediction for complex jets[R]. AIAA 2006-485.

[16] XIA H, PAUL G. Tucker and simon eastwood, towards jet flow LES of conceptual nozzles for acoustic predictions[C]. 46th AIAA Aerospace Sciences Meeting and Exhibit Reno, Nevada, 2008.

[17] XIA H, TUCKER P G, EASTWOOD S. Large-eddy simulations of chevron jet flows with noise prediction[J]. International Journal of Heat and Fluid Flow, 2009, 30: 1067-1079.

[18] UZUN A, HUSSAINI M Y. Simulation of noise generation in near-nozzle region of a chevron nozzle jet[J]. AIAA J., 2009, 47(8): 1793-1810.

[19] UZUN A, HUSSAINI M Y. High-fidelity numerical simulation of a chevron nozzle jet flow[R]. AIAA 2009-3191.

[20] ENGBLOM W A, KHAVARAN A, BRIDGES J. Numerical prediction of chevron nozzle noise reduction using WIND-MGBK methodology[R]. AIAA 2004-2979.

[21] TIDE P S, BABU V. A numerical predictions of noise due to subsonic jets from nozzles with and without chevrons[J]. Applied Acoustics, 2009, 70: 321-332.

[22] ANDERSSON N, ERIKSSON L E, DAVIDSON L. A study of Mach 0.75 jets and their radiated sound using large eddy simulation[R]. AIAA 2004-3024.

[23] ANDERSSON N, ERIKSSON L E, DAVIDSON L. Investigation of an isothermal Mach 0.75 jet and its radiated sound using large eddy simulation and Kirchhoff surface integration[J]. Int. J. Heat Fluid Flow, 2005, 26: 393-410.

[24] ANDERSSON N, ERIKSSON L E, DAVIDSON L. Large eddy simulation of a Mach 0.75 jet[R]. AIAA 2003-3312.

[25] JACOB M C, BOUDET J, CASALINO D, et al. A rod-airfoil experiment as benchmark for broadband noise modeling[J]. J. Theoret. Comput. Fluid. Dyn., 2005, 19(3): 171-96.

[26] CASALINO D, JACOB M C, ROGER M. Prediction of rod airfoil interaction noise using the FWH analogy[R]. AIAA 2002-2543.

[27] BOUDET J, GROSJEAN N, JACOB M C. Wake-airfoil interaction as broadband noise source: a large-eddy simulation study [J]. Int. J. Aeroacoustic, 2005, 4(1): 93-116.

[28] GEROLYMOS G A, VALLET I. Influence of temporal integration and spatial discretization on hybrid RSM-VLES computations[R]. AIAA 2007-4094.

[29] CARANI M, DAI Y, CARANI D. Acoustic investigation of rod airfoil configuration with DES and FWH[R]. AIAA 2007-4016.

[30] CRESCHNER B, THIELE F, CASALINO D, et al. Influence of turbulence modeling on the broadband noise simulation for complex flows[R]. AIAA 2004-2926.

[31] JIANG M, LI X D, ZHOU J J. Experimental and numerical investigation on sound generation from airfoil-flow interaction[J]. Appl. Math. Mech. -Engl. Ed., 2011, 32(6): 765-776.

[32] TAM C K W. Computational aeroacoustics: issues and methods [J]. AIAA J., 1995, 33(10): 1788-1796.

[33] TAM C K W. Recent advances in computational aeroacoustics[J]. Fluid Dynamics Research, 2006, 38: 591-615.

[34] COLONIUS T, LELE S K. Computational aeroacoustics: progress on nonlinear problems of sound generation[J]. Progress in Aerospace Sciences, 2004, 40: 345-416.

[35] WANG Z J. High-order methods for the Euler and Navier-Stokes equations on unstructured grids[J]. Progress in Aero-

[36] EKATERINARIS J A. High-order accurate, low numerical diffusion methods for aerodynamics[J]. Progress in Aerospace Sciences, 2005, 41(3-4): 192-300.

[37] NONOMURA T, IIZUKA N, FUJII K. Freestream and vortex preservation properties of high-order WENO and WCNS on curvilinear grids[J]. Computers and Fluids, 2010, 39: 197-214.

[38] THOMAS P D, LOMBARD C K. Geometric conservation law and its application to flow computations on moving grids[J]. AIAA J., 1979, 17(10): 1030-1037.

[39] PULLIAM T H, STEGER J L. On implicit finite-difference simulations of three-dimensional flow[R]. AIAA 78-10, 1978.

[40] DENG X G, MAO M L, TU G H, et al. Geometric conservation law and applications to high-order finite difference schemes with stationary grids[J]. J. Comput. Phys, 2011, 230: 1100-1115.

[41] VISBAL R M, GAITONDE D V. On the use of higher-order finite-difference schemes on curvilinear and deforming meshes[J]. J. Comput. Phys, 2002, 181: 155-185.

[42] DENG X, ZHANG H. Developing high-order weighted compact nonlinear schemes[J]. J. Comput. Phys, 2000, 165: 22-44.

[43] DENG X G, MAEKAWA H. Compact high-order accurate nonlinear schemes[J]. J. Comput. Phys, 1997, 130: 77.

[44] DENG X G, MAO M L, JIANG Y, et al. New high-order hybrid cell-edge and cell-node weighted compact nonlinear schemes[R]. AIAA 2011-3857.

[45] JIANG G, SHU C. Efficient implementation of weighted ENO[J]. J. Comput. Phys., 1996, 181: 202-228.

[46] DENG X, MAEKAWA H, SHEN Q. A class of high order dissipative compact schemes[R]. AIAA 96-1972.

[47] TAKU NONOMURA, DAIKI TERAKADO, YOSHIAKI A B E, et al. A new technique for finite difference WENO with geometric conservation law[R]. AIAA 2013-2569.

[48] DENG X G, MIN YAOBING, MAO MEILIANG, et al. Further studies on geometric conservation law and applications to high-order finite difference schemes with stationary grids[J]. J. Comput. Phys., 2013, 239: 90-111.

[49] PALIATH U, SHEN H, AVANCHA R, et al. Large eddy simulation for jets from chevron and dual flow nozzles[R]. AIAA 2011-2881.

[50] DAUDE F, BERLAND J, EMMERT T, et al. A high-order finite-difference algorithm for direct computation of aerodynamic sound[J]. Computers & Fluids, 2012, 61: 46-63.

[51] LELE S K. Compact finite difference schemes with spectral-like resolution[J]. J. Comput. Phys., 1992, 103: 16-42.

[52] RIZZETTA D P, VISBAL M R, MORGAN P E. A high-order compact finite-difference scheme for large-eddy simulation of active flow control[J]. Progress in Aerospace Sciences, 2008, 44: 397-426.

[53] MAO M L, DENG X G. Boundary schemes and asymptotic stability for high-order dissipative compact schemes[J]. ACTA Aerodynamica Sinica, 2000, 18(2): 165-171. (in Chinese)
毛枚良, 邓小刚. 高阶精度线性耗散紧致格式的渐近稳定性[J]. 空气动力学学报, 2000, 18(2): 165-171.

[54] MAO M L, DENG X G, LI S. Spectrum characteristic of dissipative compact schemes and application to couette flow[J]. Chinese Journal of Computational Physics, 2009, 26(3): 371-377. (in Chinese)
毛枚良, 邓小刚, 李松. 耗散紧致格式的频谱特性研究与应用[J]. 计算物理, 2009, 26(3): 371-377.

[55] MAO M L, JIANG Y, DENG X G. Study of LDDRK schemes for DCS5 scheme[J]. Chinese Journal of Computational Physics, 2010, 27(2): 159-167. (in Chinese)
毛枚良, 姜屹, 邓小刚. 基于DCS5格式的LDDRK算法[J]. 计算物理, 2010, 27(2): 159-167.

[56] JIANG Y, MAO M L, DENG X G. Application of DCS5 scheme in CAA[J]. ACTA Aerodynamica Sinica, 2012, 30: 431-436. (in Chinese)
姜屹, 毛枚良, 邓小刚. DCS5在计算气动声学中的应用研究[J]. 空气动力学学报, 2012, 30(4): 431-436.

[57] DENG X G, JIANG Y, MAO M L, et al. Developing hybrid cell-edge and cell-node dissipative compact scheme for complex geometry flows[J]. Sci. China. Tech. Sci., 2013, 56: 2361-2369.

[58] DENG X G, JIANG Y, MAO M L, et al. A family of hybrid cell-edge and cell-node dissipative compact schemes satisfying geometric conservation law[J]. submitted to Computers & Fluids.

[59] JIANG Y, MAO M L, DENG X G, et al. Numerical prediction of jet noise from nozzle using seventh-order dissipative compact scheme satisfying geometric conservation law[J]. Applied Mechanics and Materials, 2014, 574: 259-270.

[60] MAO M L, JIANG Y, DENG X G, et al. Noise prediction in subsonic flow using seventh-order dissipative compact scheme on curvilinear mesh[J]. Submitted to Advances in Applied Mathematics and Mechanics.

[61] THOMAS L G, et al. Multi-size-mesh, multi-time-step algorithm for noise computation around an airfoil in curvilinear meshes[R]. AIAA 2007-3504.

[62] KIM J W, LEE D J. Characteristic interface conditions for multiblock high-order computation on singular structured grid[J]. AIAA Journal, 2003, 41(2): 2341-2348.

[63] DENG X G, MAO M L, TU G H, et al. Extending weighted compact nonlinear schemes to complex grids with characteristic-based interface conditions[J]. AIAA J., 2010, 48(12): 2840-2851.

[64] JIANG Y, MAO M L, DENG X G, et al. Extending seventh-order dissipative compact scheme satisfying geometric conservation law to large eddy simulation on curvilinear grids[J]. Submitted to Advances in Applied Mathematics and Mechanics.

[65] TAM C K W, WEBB J C. Dispersion-relation-preserving finite difference schemes for computational acoustics[J]. Journal of Computational Physics, 1993: 107: 262-281.

[66] HU F Q, HUSSAINI M Y, MANTHEY J L. Low-dissipation and low-dispersion Runge-Kutta schemes for computational acoustics[J]. J. Comput. Phys., 1996, 124: 177-191.

[67] JOHN M HSU, ANTONY JAMESON. An implicit-explicit hybrid scheme for calculating complex unsteady flows[R]. AIAA 2002-0714.

[68] GORDNIER R E, VISBAL M R. Numerical simulation of delta-wing roll[R]. AIAA 93-0554.

[69] JIANG Y, MAO M L, DENG X G, et al. Extending seventh-order hybrid cell-edge and cell-node dissipative compact scheme to complex grids[C]. The 4th Asian Symposium on Computational Heat Transfer and Fluid Flow, Hong Kong, 2013.

[70] POINSOT T, LELE S K. Boundary conditions for direct simulations of compressible viscous flows[J]. J. Comput. Phys., 1992, 101: 104-129.

[71] DANIEL J Bodony. Analysis of sponge zones for computational fluid mechanics[J]. J. Comput. Phys., 2006, 212: 681-702.

[72] JIANG Y, MAO M L, DENG X G, et al. Large eddy simulation on curvilinear meshes using seventh-order dissipative com-

pact scheme[J]. *Computers and Fluids* (in press). doi:10.1016/j.compfluid.2014.08.003.

[73] BORIS J P, GRINSTEIN F F, ORAN E S, et al. New insights into large eddy simulation[J]. *Fluid Dyn. Res.*, 1992, 10: 199.

[74] JIANG Y, MAO M L, DENG X G, et al. Effect of surface conservation law on large eddy simulation based on seventh-order dissipative compact scheme[J]. *Applied Mechanics and Materials*, 2013, 419: 30-37.

[75] LYRINTZIS A S. Surface integral methods in computational aeroacoustics-from the (CFD) near-field to the (acoustic) farfield[J]. *Int. J. Aeroacoust*, 2003, 2(2): 95-128.

[76] XU C Y, CHEN L W, LU X Y. Large-eddy simulation of the compressible flow past a wavy cylinder[J]. *J. Fluid Mech.*, 2010, 665: 238-273.

Progress of aeroacoustic simulation method based on HDCS-E8T7 scheme

JIANG Yi[1,2], DENG Xiaogang[3], MAO Meiliang[1,2], LIU Huayong[1]

(1. State Key Laboratory of Aerodynamics, China Aerodynamics Research and Development Center, Mianyang 621000, China;
2. Institute of Computational Aerodynamics, China Aerodynamics Research and Development Center, Mianyang 621000, China;
3. National University of Defense Technology, Changsha 410073, China)

Abstract: Noise prediction using high-order numerical method is one of the most interested research topics in CFD. HDCS-E8T7 scheme is a seventh-order hybrid linear compact dissipative scheme. This scheme overcomes the disadvantage that many high-order finite difference schemes may have when geometric conservation law (GCL) is not satisfied, the numerical instability phenomenon may be avoided when the HDCS-E8T7 is applied to solve aeroacoustic problem in complex geometry. In the present high-order numerical noise prediction method, the spatial discretization adopts HDCS-E8T7 scheme and its boundary schemes have suitable accuracy comparing with the interior scheme. High-order multi-step Runge-Kutta and dual stepping scheme are employed for time integration, high-order interface boundary scheme is developed to extend the present method to multi-block point matched grid which meets the need of solving problem in complex geometry and the turbulence is handled by the concept of implicit large eddy simulation. Based on this method, the effect of GCL on the numerical solutions is studied. The importance of satisfying the GCL on complex grid for high-order simulation is illuminated. Some typical noise cases, such as vortex convection, scattering of acoustic waves by multiple cylinder, sound radiated by a rod-airfoil configuration and jet noise from nozzle, are investigated. The solutions of these tests show that the presented method has high prediction accuracy and potential application for handling aeroacoustic problem on complex geometry.

Key words: hybrid linear compact dissipative scheme; computational aeroacoustic; geometric conservation law; implicit large eddy simulation; complex geometry.

计算流体力学中的验证与确认

邓小刚[†]　宗文刚　张来平　高树椿　李超

中国空气动力研究与发展中心, 绵阳 621000

摘　要　计算流体力学 (CFD) 在航空航天等诸多领域的应用越来越广泛. 特别是近年来, CFD 在实际飞行器的设计中扮演着越来越重要的角色, 许多设计参数直接来源于 CFD 的计算结果. 由此, 飞行器设计师对 CFD 提供结果的可信度提出了更高的要求. 验证 (verification) 与确认 (validation) 是评价数值解精度和可信度的主要手段. 本文综述了国内外开展 CFD 验证与确认研究的进展. 在引言中论述了开展 CFD 验证与确认的重要性和必要性, 简述了国内外 CFD 验证与确认研究的历史和发展现状. 第 2 节中讨论了 CFD 验证与确认的一些基本概念, 以及这些概念定义的形成过程, 并指出了进行 CFD 验证与确认的基本步骤. 第 3 节和第 4 节分别讨论了 CFD 验证与确认的方法, 如 CFD 验证中的精确解比较方法, 制造解比较方法, 网格收敛性研究; CFD 确认中的层次结构, 流动分类法, 确认实验指南. 在第 5 节中我们列举了几个 CFD 验证与确认的应用实例. 最后, 对我国开展 CFD 验证与确认研究工作提出了若干建议, 包括: (1) 开展流动分类法研究, (2) 推行软件质量工程方法, (3) 开展规范精细的实验, 建立国内的网络数据库.

关键词　计算流体力学, 实验流体力学, 验证, 确认, 可信度, 校准, 认证, 软件质量工程

1　引　言

以往飞行器的设计和研发主要依赖于风洞实验和飞行试验. 随着计算机技术的迅猛发展, 计算流体力学 (computational fluid dynamics, CFD) 在飞行器设计和研发过程中扮演着越来越重要的角色. 特别是现在, 随着先进的网格技术和计算方法的发展, CFD 被广泛应用于模拟各种真实外形的复杂流动. 但是, 长期以来, CFD 工作者对 CFD 软件的验证与确认工作一直没有给予足够的重视. 因此, 对于计算结果的可信度, CFD 研究人员并不能给出明确的回答. 这使得 CFD 软件的使用者对 CFD 也持一种矛盾的心态——既想利用 CFD 这种快捷经济的设计工具, 又对 CFD 的计算结果心存疑虑. 为了促进 CFD 本身的发展, 更为了给飞行器设计部门提供高效可靠的 CFD 工具, 我们必须开展 CFD 的可信度研究. 当前, 在空气动力学预研基金的资助下, 国内正在开展 CFD 的软件化工作, CFD 软件的可信度研究就显得尤其重要. 而可信度研究的基本内容和方法就是 CFD 的验证 (verification) 和确认 (validation)(简称为 V&V).

CFD 的验证确认和可信度评价在国外一直受到高度重视. 例如文献 [1] 中指出: 1992 年 NASA 对 CFD 的投资范围包括 CFD 算法、应用、网格生成、可视化、转捩、湍流模型、验证和确认, 总投资 1 399.8 万美元, 而 CFD 验证确认一项就投资 713.5 万美元, 占对 CFD 总投资的 50.97%, 可见 CFD 验证和确认在 CFD 技术中的重要地位. 实际上, 从 1987 年开始, 美国、欧洲就开展了大规模、有组织、有计划的 CFD 验证和确认工作, 例如 J. of Fluid Engineering[2]、AIAA Journal[3]、Journal of Heat Transfer[4] 等专业杂志发布了涉及数值模拟可信度的编辑方针; CFD 的研究和应用机构、航空航天飞器发展计划等组织进行了大量 CFD 验证确认工作, 如 Boeing 公司的全机 CFD 验证确认、美国 HSCT 计划中的 CFD 阻力专题验证等; 1998 年, AIAA 发布了 "Guide for the Verification and Validation of Computational Fluid Dynamics Simulations"[5], 这是目前世界上关

邓小刚, 宗文刚, 张来平, 高树椿, 李超. 计算流体力学中的验证与确认. 力学进展, 2007, 37(2): 279-288.

于 CFD 验证和确认、可信度评价的第一个系统、深入的指南. 在国内, 关于 CFD 可信度的研究也正逐步受到重视, 相关单位安排组织了若干气动外形的数值计算和试验对比研究, 空气动力学预研基金也设立专题开展 CFD 可信度研究.

实际上, CFD 的可信度研究是 CFD 学科的方法论, 是科学哲学思想的体现. 它着眼于 CFD 数值模拟实施过程的各个环节, 分析每一个环节的不确定性和误差, 并研究这些不确定性和误差的评估手段. 采用这些评估手段, 我们完全可以定性甚至定量地考察数值模拟的计算精度, 评价模型问题和物理真实问题的逼近程度, 以此为基础, 我们就可以对数值计算结果的可信度做出判断. 这种判断是立足于科学分析基础之上的, 它不再仅仅是个人的看法和意见, 因此它完全可以被整个 CFD 界所接受, 更进一步地, 使用这些数值计算结果的工业界也会对数据的可靠性有明确的认识.

2 CFD 验证与确认过程中的基本概念

2.1 基本概念的发展历程

从 20 世纪 30 年代开始, 在运筹学的相关领域, 人们就开始了模型的可信度研究. 其主要内容就是模型的验证和确认. 1979 年, 美国计算机模拟协会首次定义了验证和确认, 以及计算模型 (computerized model or computational model) 和概念模型 (conceptual model) 的概念. 20 世纪 80 年代末期, 全美电气工程师协会定义了协会自身的验证和确认, 美国核工业协会和国际标准组织也采用了该定义. 1996 年, 美国国防部下属的国防建模与仿真办公室吸取以往这些定义的精华, 给出了更为简洁、明确的定义. 1998 年, AIAA 的 CFD 标准委员会首次全面定义了 CFD 的验证和确认等概念, 这标志着 CFD 的可信度研究进入了一个新的发展阶段.

尽管如此, 概念的准确定义仍需深入的研究, 其完全统一还有待时日, 本文主要介绍了 AIAA 指南中这些概念的定义. 应该指出的是, 这些概念的汉语翻译目前比较混乱, 在国内学术界还没有形成统一的认识 (不仅概念的定义不统一, 而且概念的汉语名称本身也不统一). 在本文中, 我们根据自己的理解给出了一组概念的汉语名称和定义的翻译, 希望能对我国的 CFD 可信度研究起到抛砖引玉的作用. 我们也希望有关的学会和组织重视这一现象, 及早开展与 CFD 可信度相关的术语学研究.

2.2 基本概念

2.2.1 计算流体力学中的模型

计算流体力学数值模拟过程中, 使用了两种模型, 概念模型和计算模型. 概念模型是由描述物理现象或者物理过程的数学方程和物理参数组成, 它是在分析和观察物理系统的基础上提出来的. 在计算流体力学中, 概念模型就是指表征质量、动量以及能量守恒的偏微分方程组, 还包括湍流模型, 材料的本构方程等, 当然也包括这些方程组的初始条件和边界条件. 计算模型是实现概念模型的计算方法, 比如有限差分法和有限体积法, 时间推进方法, TVD 格式, 高阶格式等. Metha[6] 提出了模拟模型 (simulation model) 的概念, 它是指通过计算模型实现的概念模型.

2.2.2 不确定性 (uncertainty) 和误差 (error)

AIAA 指南定义不确定性为由于知识的缺乏而导致的、模型化过程中所出现的潜在缺陷. 这个定义中的关键词是潜在, 这表明因为知识的缺乏所造成的缺陷可能存在也可能不存在, 实际上也就是 "不确定" 这 3 个字的含义. 误差则被定义为模型化和数值模拟过程中公认的缺陷, 它不是由于知识的缺乏导致的. 误差的显著特征是它可以通过审查来确认. Roache[7] 对不确定性和误差的定义与 AIAA 指南有较大的不同, 他把误差定义为计算值或试验值与真实值的差别. 当真实值不确定或不可知的时候, 计算值或试验值的误差就不能确定, 这时不确定性就定义为误差的估计.

2.2.3 验证和确认

美国计算机模拟协会给出的最初定义分别为: 验证是在给定的精度范围内, 计算模型表述概念模型的证实过程; 确认是在一定的应用范围内, 与应用目的相容的, 计算模型所具有的精度满意域的确定过程. 图 1 为美国计算机模拟协会提出的概念模型、计算模型、真实世界、验证、确认等概念的关系图. 从图 1 中可以看出, 从概念模型到计算模型是通过编程完成的, 它们之间需要经过模型的验证; 从计算模型到真实世界通过计算机模拟来实现, 它们之间需要经过模型的确认; 从真实世界返回到概念模型的改进, 需要经过分析, 结果是对概念模型的认证.

在验证中, 精度一般是由简单模型问题的精确解或高精度的数值解来确定的, 它是离散数学的计算机代码正确求解概念模型的证实过程, 要求正确地求解问题, 强调求解过程是否正确, 重点考察计算模型的误差, 因此验证的重点不在于建立概念模型与真实世界之间的关系. 确认过程中, 精度一般是由实验数据来确定的, 它是确定计算模型模拟真实世界的精确度, 要求求解正确的问题, 强调求解问题是否正确, 考察的是模拟模型的误差.

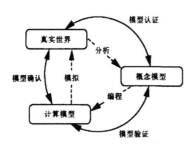

图 1 模型和模拟的 3 个方面以及验证和确认的作用 [8]

2.2.4 校准 (calibration) 和认证 (certification)

校准是指计算模型中调整数值模型参数或者物理模型参数的过程，它的目标是改进计算结果和实验结果的符合程度，因此这一过程不涉及误差和不确定性的评估。认证是一个软件评估过程，它从软件系统本身的客观规律来考察计算流体力学软件，它要包括验证和确认，也包括软件的文档化、质量保证以及版本控制等。

2.3 建立 CFD 可信度的方法论

为了建立计算流体力学数值模拟的可信度，我们在数值模拟过程中必须坚持一定的准则，只有这样，才有可能建立数值模拟的可信度，并更进一步地树立整个计算流体力学学科的可信度。以下的 3 个步骤是实施 CFD 可信度研究的方法论基础：

(1) 采用验证方法，来确定计算模型的误差，并对数值模拟结果做出精度估计。如果计算模型误差可以忽略，那么数值模拟结果就可以进入确认过程；

(2) 数值模拟结果和实验测量结果进行对比，这要求实验结果具有已知的测量不确定性，并且两者必须在相同的流动条件下。如果数值模拟结果和测量结果具有可接受的一致程度，那么模拟结果就可以确认；

(3) 数值模拟的验证和确认过程完成以后，必须把数值计算过程的所有信息，以及计算结果和实验结果的对比完全公开，允许感兴趣的专家对结果进行审查和评论。

经过以上的步骤，CFD 模拟的精度可以得到有效评估，结果的可信度评价也可以建立，并且非常重要的是，模拟过程的完全公开化，使得 CFD 模拟结果的可信度能够得到整个业界的赞同与否定。

3 CFD 数值模拟的验证方法

3.1 验证的方法学原理

计算流体力学数值模拟过程中，要产生很多的误差，这些误差中有些是不可避免的，如物理模型误差、计算模型误差以及机器舍入误差，而有些误差则是可以避免的，如编码误差以及使用误差。计算模型误差主要是物理模型的离散误差，如空间离散误差和时间离散误差，数值计算的迭代过程的不完全收敛也会造成误差。

验证环节是对计算模型做精度估计，它的目标就是识别和量化计算模型误差，数值解与精确解的比较研究以及数值解的网格收敛性研究是其主要方法。图 2 示意了验证过程的原理，即验证过程是比较和测试计算解与解析解或基准微分方程解的误差。通过纯粹的数值试验，验证过程可以建立数值解的精度等级，以及数值解对计算模型参数的敏感程度等级。在计算流体力学中，网格收敛性研究也是数值解验证的主要方法，而且可以确定为了得到特定精度等级的数值解所需的网格，这足以显示出物理现象本身数值化的难度。

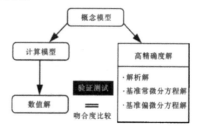

图 2 验证过程的原理图 [5]

3.2 验证的方法

3.2.1 精确解比较方法

很早以前就采用非常简单的精确解来验证计算模型。选择的精确解一定要包括足够多的结构，使得方程中的所有项，以及离散化中产生的所有误差项都能体现出来，因此，分析和鉴别好的、敏感的精度估计算例是非常重要的。在捕捉激波格式的研究中，计算流体力学就经常选用 Burges 方程问题，Buckley-Leverett 问题以及 Riemann 问题等来验证和确认计算方法的精度 [9,10]。这些问题的主要特征都是在一定的条件下，精确解内包含间断，就如同气体动力学中的激波和接触间断，因此对这些模型问题中间断的计算精度就代表了计算方法模拟真实激波问题的计算精度。

3.2.2 制造解比较方法

制造解比较方法 [11,12] 是精确解方法更加一般化、简单化的替代技术。它的核心是针对代码求解的偏微分方程假设一个解，然后将这个解带入原方程，并确定为了使方程能够成立所必须添加的源项以及边界条件。这个过程一般是通过符号运算软件来实现的,并且添加的源项以及边界条件都可以采用符号运算软件直接生成 Fortran 源代码,以避免人为的编码错误.在这样的源项以及边界条件下,数值求解原来的偏微分方程,就可以得到数值解,这个数值解是假设精确解的近似,两个解之间的比较,就可以确定代码的误差.制造解比较方法的应用范围显然要比精确解比较方法宽广的多,而且采用制造解方法可以发现代码编制中的任何错误,可以说通过制造解方法验证的代码将具有非常高的可信度.

3.2.3 网格收敛性研究

网格收敛性研究是数值计算验证的有效手段,众多杂志的编辑方针都认为严格定义的网格细化或者粗化研究是计算结果精度评估的一种有效措施,并且要求作者尽可能的结合 Richardson 外插方法来判断数值解的收敛性以及实际的计算精度.很多人认为进行网格收敛性研究的计算机花费过大,以至于不能进行这样的研究. Roache[13]指出,如果说网格细化的研究确实困难,那么可以进行网格粗化的研究,即沿每一方向网格减半,这样以来在三维情况下计算工作量将降低到原来网格下的 1/8. 显然这是一个非常小的工作量,但是已经可以给出一个明确的数值解精度估计.

4 CFD 数值模拟的确认方法

4.1 确认的方法学原理

确认是模型精确表示的物理状态与模型预期用途逼近程度的测度过程,强调求解正确的问题.确认的基本内容是指出和量化在概念模型和计算模型中的误差和不确定度,评估实验的不确定度,并进行计算结果和实验数据的比较.确认过程中并不假设实验数据比计算结果具有更高的精度,只是认为实验测量忠实地反映了实际状态.图 3 给出了确认的基本过程,即将数值解同实验数据进行比较并测试符合程度.

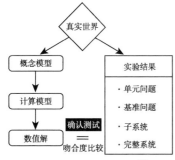

图 3 确认过程的原理图[5]

对于绝大多数的复杂系统进行整体的确认是不现实的,这些复杂系统往往包含了复杂的流动现象和物理化学过程.因此,我们必须采用分解的办法将复杂的工业系统分解为若干子系统,而将子系统又分解为基准问题,从基准问题中又可提取出若干单元问题.图 4 给出了 CFD 确认的这种层次结构.只有将单元问题和基准模型问题完全确认了,才有可能谈得上对子系统、乃至整个复杂系统的确认.

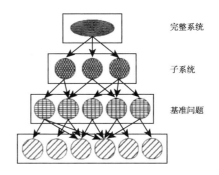

图 4 确认的层次结构[5]

4.2 流动分类法

CFD 应用领域相当广泛,每个应用领域的流动现象都各具特色,互不统属,因此为了达到完整系统的验证与确认,必须对这些庞杂的流动现象进行分类归纳,然后针对每一类流动现象进行验证确认.事实上,建立一个航空领域的完整流动分类方法就是一项艰巨的工作.按照层次结构原理,Rizzi[14]提出了如下的航空领域流动分类方法:

(1) 流动状态.在飞行包线的不同点上,飞行器的流动状态可能分类为低速流、定常流、非定常流、跨声速流、颤振、高升力和失速、超声速、高超声速流动等.

(2) 飞行器全机和部件.层次结构中系统和子系统的定义比较直接.系统是指带完整子系统的飞行器类型,如民机、军机、导弹等;子系统是指主要的气动部件,如机翼、进气道等.

(3) 流动特征.在层次结构的底层,最感兴趣的是流动的科学层面,即按照流动现象、流动物理以及流动化学,将流动的主要特征鉴别出来.

以一个巡航飞行的商业运输机为例,流动状态是定常跨声速流,外形(系统)是一个宽体飞机,而主要部件则是大展弦比的机翼.机翼绕流包含如下特征:三维、黏性、高 Reynolds 数可压缩湍流外流、有中等强度的激波、包含小翼的复杂外形、分离流动.

为了进一步说明确认的层次结构和流动分类

法，Oberkampf 等[15]介绍了一个吸气式高超声速巡航导弹的层次结构，这个实例对我们开展确认的层次结构和流动分类法的研究具有一定的指导意义. 图 5 绘制了该巡航导弹的层次结构图. 从图 5 中可以看到，在整个系统中，包含了推进系统、气动布局、导航系统和武器系统等大的系统. 在子系统层中，包括气动 / 热防护子系统、结构子系统、电动力学子系统等. 在基准问题层，与气动 / 热防护子系统相关的物理化学过程包括：带烧蚀的高超声速层流和湍流、边界层转捩、热防护涂层的烧蚀、金属子结构的热传导等. 在单元问题层，又可以进一步分解出：绕简单外形的层流和湍流、带壁面吹气的层流和湍流、激波 / 湍流边界层干扰、简单外形的边界层转捩、低温升华、各向异性热传导等.

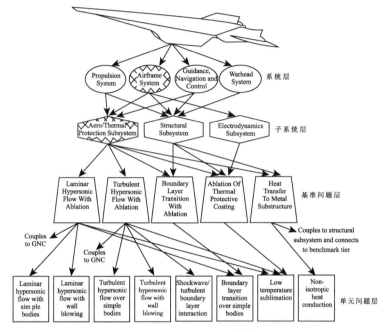

图 5 确认的层次结构举例[15]

4.3 确认实验指南

为了满足 CFD 确认的要求，精细的确认实验应该遵循以下原则[16]：

(1) 确认实验必须由实验人员、模型提出人员、CFD 代码研发人员、以及 CFD 代码的使用人员共同设计，从实验的设计阶段到最终的文件归档，所有人员都必须密切合作. 除了要发挥实验和计算二者的优势外，更重要的是要能坦诚地暴露各自的不足.

(2) 确认实验的设计要能捕捉感兴趣的本质物理特性.

(3) 确认实验应该致力于强调实验与计算方法内在的协同. 实验和计算的结合将有利于双方的发展，一方面计算可以指导实验，另一方面实验又可以确认计算的结果.

(4) 尽管确认实验的设计必须由实验和计算两方面的研究人员共同设计，但是实验数据和计算结果的获取必须"背靠背"地完成. 在实验方案确定以后，实验和计算人员应该互不影响地开展工作. 在双方都得到最终结果后，再进行数据的对比，并共同研究误差产生的原因.

(5) 确认实验的测量也需要建立一个由简到难的层次结构，遵循从全局量到局部量的原则.

(6) 确认实验必须分析并评估随机误差和系统误差，给出明确的误差带.

5 验证与确认的应用

为了具体说明 CFD 验证和确认的关系与作用，以下我们介绍 3 个应用实例：

5.1 算例 1——Lax 问题[17]

Lax 问题是一个一维 Riemann 问题. Riemann 问题是在初始时刻，给定一个物理量的间断面，进而研究这个初始间断面随时间发展，逐步分解产生的解, 在考虑到熵增的条件下, 这个解是唯一确定的, 并且可以求得精确的理论解. 许多计算格式研究人员都用这个标准算例来考核自己方法的精度. 这里, 我们比较了 WENO 格式和我们自己发展的双重加权 ENO 格式 (DWENO 格式) 求解 Lax 问题的精度. 理论上, 这两个格式都可以在光滑区达到 5 阶精度,

在间断附近则降为 3 阶精度. 图 6 给出了 $t=0.2$ 时刻 WENO 格式和 DWENO 格式的数值计算结果, 实线是精确解. 显然, 通过将数值解和精确解的比较, 我们可以看出 DWENO 格式对激波、接触间断的分辨率都高于 WENO 格式. 这便是验证方法中的精确解比较方法.

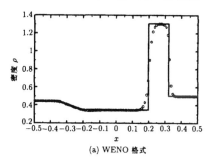
(a) WENO 格式

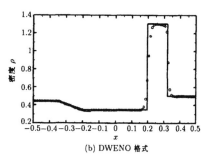
(b) DWENO 格式

图 6 Lax 问题, $t = 0.2$

5.2 算例 2——空心圆柱裙问题和尖双锥问题

近年, 针对空心圆柱裙问题和尖双锥问题做了大量的实验和数值计算研究[17,18], 这些研究的首要目的就是要使这两个问题成为数值模拟和计算软件在高超声速条件下的验证、确认范例. 实际上, 这两个问题代表的流动特征主要是激波 - 激波干扰, 激波 - 边界层干扰, 研究这一类问题, 对准确评估空间飞行器控制面在大偏折角下的效率和热环境有重要意义.

Gnoffo[18] 采用 LAURA(langley aerothermodynamic upwind relaxation algorithm) 软件数值模拟了这两个问题, 图 7 示意了问题的计算区域以及 Gnoffo 计算的压力等值线. 在这篇文章中, Gnoffo 详细阐述了数值计算的无黏项、黏性项的处理方法; 描述了采用的气体模型, 特别是针对流动介质是氮气的特点, 给出了计算黏性系数的 Sutherlands 公式; 说明了网格生成方法, 指出为了更有效的使用计算网格, 采用了一个针对激波的网格对准算法, 并强调在尖双锥问题上, 这一算法引发了大尺度的不稳定性, 激波位置是振荡的; 边界条件的论述较为简单, 这符合算例计算区域简单的实际情况.

图 8 引用了 Gnoffo 计算的空心圆柱裙问题物面压力系数和 Stanton 数分布. 在这个问题上, 随着网格加密, 物理量的变化范围不大, 峰值出现的位置几乎保持了不变, 分离区的大小也只是有细微差别. 图 9 引用了 Gnoffo 计算的尖双锥问题物面压力系数

和 Stanton 数分布. 从这两个图看来, 随着网格加密, 物理量的变化范围很大, 峰值位置移动非常明显, 分离区的大小也变化较大, 只有在非常密的网格上才显示出了峰值位置和分离区大小的网格无关性.

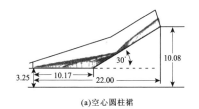
(a) 空心圆柱裙

(b) 尖双锥

图 7 空心圆柱裙问题和尖双锥问题计算区域和压力等值线[18]

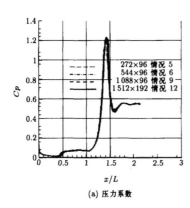
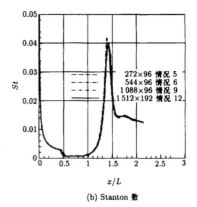

图 8　空心圆柱裙问题物面物理量分布 [18]

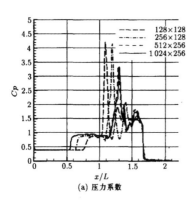
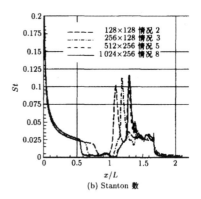

图 9　尖双锥问题物面物理量分布 [18]

这两个问题网格收敛性研究的对比表明, 对于不同问题, 网格收敛性研究的结果会有较大差别. 同时尖双锥问题鲜明地说明了一点, 如果不进行数值计算的网格收敛性研究, 数值计算结果的变异性将会非常显著, 在这个问题上, 如果采用最稀疏网格的结果作为问题的最终计算结果, 难免会得到非常荒谬的结论. 推而广之, 我们可以知道没有进行网格收敛性研究的数值计算结果, 其可信度是难以保证的. 当然, 在 Gnoffo 的文章中, 还没有将数值计算结果和实验结果进行对比, 也就是说还没有完成数值模拟结果的确认工作, 因此计算结果的可信度还不能完全建立.

5.3　算例 3——尖拱柱体侧向喷流干扰数值模拟

作为一个实际复杂问题的应用实例, 我们利用前面介绍的验证和确认方法, 对我们的侧向喷流计算软件进行了验证和确认. 针对侧向喷流问题, Brandeis[19] 做了大量的实验, 系统研究了弹体头部钝度、喷口形状、喷流流量、喷射角度、来流马赫数、来流攻角等对喷流干扰效应的影响. 如此充分、翔实的实验对验证和确认侧喷干扰流动数值模拟结果是非常有益的 [20]. 在整个计算过程中, 我们采用了粗、中、细 3 套网格, 分别数值计算了单圆喷口垂直喷射下层流流态和湍流流态的侧向喷流干扰流动. 这 3 套网格依次为 (流向 × 周向 × 法向): 52×21×81, 103×41×81, 205×81×81, 其中较粗的网格都是在较密的网格上隔一取一抽取出来的. 这里, 我们重点研究的是喷口模拟精度与数值模拟精度的关系, 因此主要考察了流向网格和周向网格变化对计算结果的影响, 而将法向网格固定不变.

图 10 给出了 3 套网格上, 数值计算的单圆喷口背风面中心线上的压力分布. 显然, 层流数值模拟的分离点位置随网格的变化而变化, 没有显示出丝毫的网格收敛性, 同时物面压力分布与实验结果有很大差别, 这说明层流数值模拟的结果不能得到验证和确认, 因此计算结果不能建立可信度. 细网格下湍

流数值模拟的物面压力分布与实验结果的符合程度很好,而且具有可接受的网格收敛性,当然我们必须承认喷流前方分离区内的压力分布还有很大差别,这表明在目前的网格尺度下,分离区内的压力分布还不能得到有效验证,但是很明显的一点就是,不同网格下湍流数值计算的分离点位置是非常吻合的.

表1给出了实验、文献以及我们数值计算的干扰气动力、干扰力矩以及喷流推力放大因子[20,21]. 单圆喷口垂直喷射层流计算的干扰力和干扰力矩都有很好的网格收敛性,和实验结果的对比却显示了较大的差异,尤其是喷流干扰力的符号都是和实验结果相反的,干扰力矩的差别也达到倍数关系,这一点充分表明,即使单圆喷口垂直喷射层流计算的干扰力和干扰力矩通过网格不断加密可以得到验证,但是和实验结果的对比却表明该计算结果是不能得到确认的,因此这个结果的可信度值得怀疑. 单圆喷口垂直喷射湍流计算的干扰力矩有很好的网格收敛性,但是干扰力还不能认为得到了有效验证,细网格和中等网格的计算结果还有明显差别,为了验证计算结果,必须进行更密网格上的数值计算. 和实验结果以及文献计算结果的对比表明,本文计算的干扰力矩与实验、文献结果吻合,完全可以得到确认. 干扰力与文献计算结果已经很接近,但是两者与实验结果还有较大差别,因此干扰力的计算结果还不能得到充分确认.

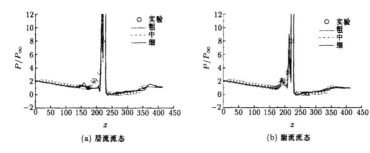

(a) 层流流态   (b) 湍流流态

图 10 尖拱柱侧向喷流干扰流动数值模拟的物面压力分布

表 1 喷流干扰气动力、干扰力矩以及喷流推力放大因子

	网格	干扰力	干扰力矩	推力放大因子
单圆喷口垂直喷射 (实验)		−0.070	−0.210	0.935
单圆喷口垂直喷射 (文献)		−0.029	−0.202	0.970
单圆喷口垂直喷射 (层流)	粗	−0.004	−0.116	0.996
	中	0.008	−0.092	1.007
	细	0.007	−0.095	1.006
单圆喷口垂直喷射 (湍流)	粗	0.010	−0.190	1.010
	中	−0.016	−0.202	0.985
	细	−0.026	−0.198	0.975

综上所述,单圆喷口垂直喷射层流数值计算结果和实验结果以及文献计算结果的对比表明物面压力分布和实验结果不能吻合,干扰力和干扰力矩的计算结果也有很大差距,这表明这些计算结果都是不能得到确认的,因此这个结果的可信度不高. 单圆喷口垂直喷射湍流计算结果既和实验结果在物面压力分布上有相当好的吻合程度,又在干扰力和干扰力矩的计算上和文献计算结果有相当大的符合程度,充分表明了这一计算结果的高可信度.

在这个算例研究中,采用的验证方法是网格收敛性研究,即通过网格的不断加密来考察计算模型的误差情况,以此来判断网格是否达到了求解问题需要的水平,显然,就目前的细网格来看也还有一定的差距. 确认方法采用的是和实验结果以及文献结果的对比来达到的,通过这种比较,我们可以评估物理模型的误差. 在这个算例上,我们可以得到结论,层流流动模型是不能确认的,而湍流模型可以得到一定程度的确认.

6 建 议

前面综述了国内外关于 CFD 验证和确认的研究进展,根据这些情况,结合我国的实际,我们对在我国 CFD 界开展进一步的验证和确认工作,提出如下建议：

(1) 开展流动分类法研究. 正如第 4.2 节所述,进行基本的流动分类是做好 CFD 验证和确认的基础. 特别是在基准问题层和单元层的建立过程中,要求我们对流动的物理和化学特征有较深的认识. 只有这样,才能在一个复杂系统中,分解出有代表性的典型流动,从而建立相应的简化模型. 这本身能促进流体力学自身的发展.

(2) 推行软件质量工程方法. CFD 软件作为应用软件的一类,在研发过程中必须遵循软件工程的一些基本原则. 有文献统计指出：在每 1 000 行可执行的 C 程序中有 8 个严重的静态错误,而对于一个大型的 CFD 软件系统,代码的行数可能长达 10 万行,甚至几百万行,因此其中的错误是在所难免的. 国内 CFD 软件研制正处在系统化、商业化的转型阶段,如果能够利用软件质量工程方法进行全程控制,那么就可以将代码中的编程错误降至最低.

(3) 开展规范精细的实验研究,建立国内的网络数据库. CFD 确认实验较一般的风洞试验有更高的要求,我们应该依照前文述及的原则,开展规范精细的 CFD 确认实验的设计与测量,并逐步建立国内的网络数据库. 另外,国内各型风洞均做过为数众多的标模实验,可以将这些标模实验数据进行整理,逐步纳入网络数据库系统.

最后,计算流体力学的验证与确认在国外属于热点问题,涉及计算流体力学的航空航天诸机构都参与其中,资金投入和支持力度都很大. 而国内在这一领域的研究工作还很少,虽然已经引起了各方面的高度重视,但是还需要开展更加广泛的研究工作,以推动计算流体力学可信度研究的发展,为我国计算流体力学的研究和应用打下坚实的基础.

参 考 文 献

1　MacCormack R W. A perspective on a quarter century of CFD research. AIAA-93-3291-CP, 1993
2　Roache P J, Chia K, White F. Editorial policy statement on the control of numerical accuracy. *Journal of Fluids Engineering*, 1986, 108(1): 2
3　AIAA. Editorial policy statement on numerical accuracy and experimental uncertainty. *AIAA Journal*, 1994, 32(1): 3
4　Editorial Board. Journal of heat transfer editorial policy statement on numerical accuracy. *Journal of Heat Transfer*, 1994, 116: 797~798
5　AIAA. Guide for the verification and validation of computational fluid dynamics simulations. AIAA G-077-1998, 1998
6　Metha U B. Guide to credible computational fluid dynamics simulations. AIAA-95-2225, 1995
7　Roache P J. Verification of codes and calculations. *AIAA Journal*, 1998, 36(5): 696~702
8　Schlesinger S. Terminology for model credibility. *Simulation*, 1979, 32(3): 103~104
9　Harten A, Engquist B, Osher S, et al. Uniformly high-order accurate essentially non-oscillatory shock-capturing schemes III. *J of Comp Phys*, 1987, 71: 231~323
10　Shu C, Osher S. Efficient implementation of essentially non-oscillatory shock-capturing schemes. *J of Comp Phys*, 1988, 77: 439~471
11　Roache P J, Knupp P, Steinberg S, et al. Experience with benchmark test cases for groundwater flow. In: Celik I, Freitas C J, eds. Benchmark Test Cases for Computational Fluid Dynamics, FED-vol. 93. New York: The American Society of Mechanical Engineers, 1990. 49~56
12　Roache P J. Code verification by the method of manufactured solutions. *ASME Journal of Fluids Engineering*, 2002, 114(1): 4~10
13　Roache P J. Need for control of numerical accuracy. *Journal of Spacecraft and Rockets*, 1990, 27(2): 98~102
14　Rizzi J V. Toward establishing credibility in computational fluid dynamics simulations. *AIAA Journal*, 1998, 36(5): 668~675
15　Oberkampf W L, Trucano T G. Validation methodology in computational fluid dynamics. AIAA-2000-2549, 2000
16　Oberkampf W L, Trucano T G. Verification and validation in computational fluid dynamics. SAND 2002-0529, 2002
17　Holden M, Harvey J. Comparisons between experimental measurements over cone/cone and cylinder/flare configurations and predictions employing DSMC and Navier-Stokes solvers. AIAA-2001-1031, 2001
18　Gnoffo P. CFD validation studies for hypersonic flow prediction. AIAA-2001-1025, 2001
19　Brandeis J, Gill J. Experimental investigation of side-jet steering for supersonic and hypersonic missiles. *Journal of Spacecraft and Rockets*, 1996, 33(3): 346~352
20　Graham M J, Weinacht P. Numerical investigation of supersonic jet interaction for axisymmetric bodies. *Journal of Spacecraft and Rockets*, 2000, 37(5): 675~683

VERIFICATION AND VALIDATION IN COMPUTATIONAL FLUID DYNAMICS*

DENG Xiaogang[†]   ZONG Wengang   ZHANG Laiping   GAO Shuchun   LI Chao

China Aerodynamics Research and Development Center, Mianyang 621000, China

Abstract   During the last two or three decades, computational fluid dynamics (CFD) has been widely used in the scientific research and the analysis and design of engineering systems. Especially in recent years, CFD has been playing a more and more important role in the design of aeronautical and astronautical vehicles. So many design conditions for realistic aircraft have been determined directly from the results of CFD. However, users and developers of CFD today face a critical question: How can the confidence in modeling and simulations be critically assessed? Verification and validation (V&V) of CFD are the primary methods for building and qualifying this confidence. In this paper, the state of the art of V&V of CFD is reviewed. In the introduction, we first discuss the background and importance of V&V in CFD, and then review briefly the history of V&V. Secondly, the basic terminology and methodology of V&V are discussed, including several fundamental concepts, the definitions and the basic processes. Section three discusses the primary processes and methods of verification, including analytic solution method, manufactured solution method and grid convergence analysis. Section four discusses the primary processes and methods of validation, including validation tier hierarchy, flow classification, and some guidelines for validation experiments. In the fifth section, several applications of verification and validation are presented, including the typical Lax problem solved by WENO scheme and our DWENO scheme, hypersonic flows over the hollow cylinder/truncated flare and sharp double cone configurations presented by P. Gnoffo, supersonic side-jet interaction for an axissymmetric body. Finally, some comments are made with respect to the development of V&V of CFD in China: (1) the importance of flow classification research, (2) on software quality assurance, (3) on the international academic activities and successful databases, (4) on standardizing our experiment data and documents, and building our own database for CFD validation, (5) on an organization to manage the issues of V&V in China, (6) on the editorial policy on numerical accuracy in related Chinese academic publications.

Keywords   computational fluid dynamics, experimental fluid dynamics, verification, validation, credibility, calibration, certification, software quality assurance

高精度格式及其构造理论(格式设计/频谱分析/几何守恒律)

High-order accurate dissipative weighted compact nonlinear schemes

DENG Xiaogang (邓小刚)

China Aerodynamics Research and Development Center, Mianyang 621000, China
(email: xgdeng@my-public.sc.cninfo.net)

Received September 5, 2000; revised June 7, 2001

Abstract Based on the method deriving dissipative compact linear schemes (DCS), novel high-order dissipative weighted compact nonlinear schemes (DWCNS) are developed. By Fourier analysis, the dissipative and dispersive features of DWCNS are discussed. In view of the modified wave number, the DWCNS are equivalent to the fifth-order upwind biased explicit schemes in smooth regions and the interpolations at cell-edges dominate the accuracy of DWCNS. Boundary and near boundary schemes are developed and the asymptotic stabilities of DWCNS on both uniform and stretching grids are analyzed. The multi-dimensional implementations for Euler and Navier-Stokes equations are discussed. Several numerical inviscid and viscous results are given which show the good performances of the DWCNS for discontinuities capturing, high accuracy for boundary layer resolutions, good convergent rates (the root-mean-square of residuals approaching machine zero for solutions with strong shocks) and especially the damping effect on the spurious oscillations which were found in the solutions obtained by TVD and ENO schemes.

Keywords: numerical calculation, compact schemes, nonlinear schemes, Euler equations, Navier-Stokes equations.

In recent years high-order compact schemes have attracted much attention in various fields such as computational aero acoustics (CAA), large eddy simulations (LES) and direct numerical simulations (DNS) of turbulences. A class of central linear compact schemes were systematically discussed[1]. It is well known that finite difference schemes cannot approximate derivatives in the high wave number range. Inevitably there are numerical dispersions in this range. Spurious grid-to-grid oscillations, referred to as parasite waves, were reported[2,3]. These numerical contaminants are most undesirable. They degrade the quality of the numerical solutions. The main features of central type linear finite difference schemes are their non-dissipative characteristics, hence they cannot suppress the oscillation mentioned above. Fu and Ma[4] derived the 3rd and 5th order upwind compact schemes. We also developed a class of high-order dissipative compact linear schemes (DCS)[5]. These schemes have the properties that they keep the high accuracy for the low and intermediate wave numbers, and dissipate the errors in the unresolved high wave numbers. Furthermore, a general method to construct high-order dissipative compact schemes was suggested.

For the discontinuities capturing, such as shock waves and contact discontinuity, the nonlinear schemes, e.g., TVD, ENO schemes, work well. Recently we developed compact high-order nonlinear schemes (CNS)[6,7]. These schemes achieve high-order accuracy by cell-centered fourth order compact schemes and compact high-order interpolations at cell edges were designed. The numerical solutions for the Euler equations showed that CNS can capture discontinuities robustly. One advantage of these schemes is that various variables, primitive variables, conservative variables and flux themselves can be used for the calculations of numerical flux at the cell-edges though they are finite difference schemes. Based on the weighted techniques[8,9], we further developed the fourth and fifth weighted compact nonlinear schemes (WCNS) as well as the boundary and near boundary schemes[10,11], applied them to Euler and Navier-Stokes equations calculations and compared four flux splitting methods for WCNS[12]. The numerical tests showed that WCNS can capture discontinuities well and have good convergent property. Very recently, however, Lee and Zhong[13] reported the spurious oscillations in the vorticity field for the inviscid Mach 4 flow around a circular cylinder calculated by nonlinear TVD and ENO schemes. The oscillations were caused by the skewness of shock wave with the grid lines. Our calculations of the same flow using fifth-order WCNS also found these oscillations. Furthermore, for high-order scheme applications, the convergent problems are severe. Most shock-capturing schemes can hardly obtain steady-state solutions because of the minor shock waves oscillations.

The objective of the present work is to develop novel high-order dissipative weighted compact nonlinear schemes (DWCNS) with the method suggested in ref. [5]. By Fourier analysis, the dispersive and dissipative features of DWCNS are discussed. Boundary and near boundary schemes are derived and the asymptotic stabilities are analyzed on both uniform and stretching grids. The extension of DWCNS to multi-dimensional Euler and Navier-Stokes equations is given. Several inviscid and viscous numerical results are obtained, which show the good performances of DWCNS, especially the good convergent rate (the root-mean-square of residuals approaching machine zero for solutions with strong shocks), high accuracy for the boundary layer resolutions and the removal of spurious oscillations in the inviscid flow.

1 Definitions of dissipative schemes and nonlinear schemes

We consider the scalar linear hyperbolic equation,

$$\frac{\partial u}{\partial t} + c \frac{\partial u}{\partial x} = 0, \tag{1}$$

where $c > 0$ is a constant speed. For simplicity, consider a uniform spaced mesh where $x_j = jh$ (h is the space interval) for $0 \leq j \leq N+1$ and the function values at the nodes $u_j = u(x_j)$ are given. The semi-discrete form of eq. (1) is

$$\frac{\partial u_j}{\partial t} + c u'_j = 0, \tag{2}$$

where u'_j is the spatial discretization of $\partial u / \partial x$ at grid point x_j. For a sinusoidal function,

$$u_j(t) = A(t) \exp(iwx_j), \tag{3}$$

with w the wave number, we define the modified wave number $w^* = w_r^* + i w_i^*$ as

$$u'_j = iw^* A(t) \exp(iwx_j). \tag{4}$$

The modified wave number $w^* = w^*(w)$ depends on the spatial schemes which approximate the derivative $\partial u / \partial x$. Inserting (3) and (4) into (2), we get the solution

$$u_j = \exp(cw_i^* t) \exp\left[iw\left(x_j - c \frac{w_r^*}{w} t\right)\right]. \tag{5}$$

When $w_i^*(w) < 0$ for all w, there is an amplitude decay of the corresponding sinusoidal solution as

$$\exp(cw_i^* t) < 1.$$

Definition 1. For any spatial discretization schemes, if
$$w_i^r(w) < 0,$$
i.e. the imaginary part of modified wave number is negative, the schemes are dissipative schemes. Vichnevetsky[14] gave a detailed discussion about this feature of semi-discretization schemes.

Any spatial schemes to solve the linear equation (2) can be written as
$$\sum_{k=j-n1}^{j+n2} a_k u'_k = \sum_{k=j-m1}^{j+m2} b_k u_k, \qquad (6)$$
where a_k and b_k are the coefficients and the intervals $(x_{j-n1}, \cdots, x_{j+n2})$ and $(x_{j-m1}, \cdots, x_{j+m2})$ are the stencils of the schemes.

Definition 2. If the coefficients a_k and b_k or the stencils of the spatial discretization (6) to solve a linear equation are dependent on the solutions u_k themselves, the schemes are nonlinear schemes, otherwise they are linear schemes.

As we know, TVD schemes use limiter functions, hence they are nonlinear schemes; the stencils of ENO schemes are dependent on the solutions, they are also nonlinear schemes.

2 Dissipative weighted compact nonlinear schemes

In order to construct compact nonlinear schemes, we consider cell-centered compact schemes to calculate the derivative u'_j. Based on the method deriving DCS[5], a novel class of cell-centered dissipative compact schemes can be easily derived,
$$\sum_{l=-1}^{1} a_l u'_{j+l} = \frac{1}{h} \sum_{m=0}^{2} b_m (u_{j+m+1/2} - u_{j-m-1/2}) + \frac{1}{h} \sum_{m=0}^{2} c_m (u_{j+m+1/2} - 2u_j + u_{j-m-1/2}), \qquad (7)$$
where $u_{j\pm1/2}$ are the cell edge values and can be obtained by the 5th order weighted nonlinear interpolation from node values. As the stencils of the interpolation depend on the solutions, these schemes are nonlinear ones. The coefficients in (7) can be easily obtained by Taylor expanding to corresponding accuracy orders. In the case of $c_m = 0$, the cell-centered central compact schemes are recovered, which were used in our previous work[6,7,10,11,15].

2.1 The cell-centered dissipative weighted compact nonlinear schemes for Euler equations

Let $U_j = U(x_j, t)$, $x_j = jh$, denote a numerical approximation to the solution of Euler equations,
$$\frac{\partial U}{\partial t} + \frac{\partial E}{\partial x} = 0. \qquad (8)$$
At every node x_j, we approximate eq.(8) by the semi-discrete finite difference scheme,
$$\left(\frac{\partial U}{\partial t}\right)_j = -E'_j, \qquad (9)$$
where E'_j is the approximation to the spatial derivative. In order to apply the dissipative schemes (7) to calculate E'_j, the flux E must be split as $E = E^+ + E^-$. The cell-centered dissipative weighted compact nonlinear schemes have the forms,
$$\kappa_1 E^{+'}_{j-1} + E^{+'}_j + \kappa_2 E^{+'}_{j+1} = \frac{a}{h}(\tilde{E}^+_{j+1/2} - \tilde{E}^+_{j-1/2}) + \frac{b}{h}(\tilde{E}^+_{j+3/2} - \tilde{E}^+_{j-3/2})$$
$$+ \frac{c}{h}(\tilde{E}^+_{j+5/2} - \tilde{E}^+_{j-5/2}) + \frac{a_r}{h}(\tilde{E}^+_{j+1/2} - 2\tilde{E}^+_j + \tilde{E}^+_{j-1/2})$$

$$+ \frac{b_r}{h}(\tilde{E}^+_{j+3/2} - 2E^+_j + \tilde{E}^+_{j-3/2}) + \frac{c_r}{h}(\tilde{E}^+_{j+5/2} - 2E^+_j + \tilde{E}^+_{j-5/2}),$$

$$\kappa_2 E^{-\prime}_{j-1} + E^{-\prime}_j + \kappa_1 E^{-\prime}_{j+1} = \frac{a}{h}(\tilde{E}^-_{j+1/2} - \tilde{E}^-_{j-1/2}) + \frac{b}{h}(\tilde{E}^-_{j+3/2} - \tilde{E}^-_{j-3/2})$$

$$+ \frac{c}{h}(\tilde{E}^-_{j+5/2} - \tilde{E}^-_{j-5/2}) - \frac{a_r}{h}(\tilde{E}^-_{j+1/2} - 2E^-_j + \tilde{E}^-_{j-1/2})$$

$$- \frac{b_r}{h}(\tilde{E}^-_{j+3/2} - 2E^-_j + \tilde{E}^-_{j-3/2}) - \frac{c_r}{h}(\tilde{E}^-_{j+5/2} - 2E^-_j + \tilde{E}^-_{j-5/2}), \tag{10}$$

and $E'_j = E^{+\prime}_j + E^{-\prime}_j$. In this paper, we discuss two cases:

1) the fifth-order dissipative weighted compact nonlinear schemes (DWCNS-5),

$$\frac{9}{62}(1 + \alpha) E^{+\prime}_{j-1} + E^{+\prime}_j + \frac{9}{62}(1 - \alpha) E^{+\prime}_{j+1} = \frac{63}{62h}(\tilde{E}^+_{j+1/2} - \tilde{E}^+_{j-1/2})$$

$$+ \frac{17}{186h}(\tilde{E}^+_{j+3/2} - \tilde{E}^+_{j-3/2}) - \frac{9\alpha}{62h}(\tilde{E}^+_{j+1/2} - 2E^+_j + \tilde{E}^+_{j-1/2})$$

$$- \frac{21\alpha}{186h}(\tilde{E}^+_{j+3/2} - 2E^+_j + \tilde{E}^+_{j-3/2}),$$

$$\frac{9}{62}(1 - \alpha) E^{-\prime}_{j-1} + E^{-\prime}_j + \frac{9}{62}(1 + \alpha) E^{-\prime}_{j+1} = \frac{63}{62h}(\tilde{E}^-_{j+1/2} - \tilde{E}^-_{j-1/2})$$

$$+ \frac{17}{186h}(\tilde{E}^-_{j+3/2} - \tilde{E}^-_{j-3/2}) + \frac{9\alpha}{62h}(\tilde{E}^-_{j+1/2} - 2E^-_j + \tilde{E}^-_{j-1/2})$$

$$+ \frac{21\alpha}{186h}(\tilde{E}^-_{j+3/2} - 2E^-_j + \tilde{E}^-_{j-3/2}), \tag{11}$$

where the dissipative parameter $\alpha \geq 0$. As $\alpha = 0$, the weighted compact nonlinear scheme (WCNS-5) is recovered, which was discussed in our previous paper[11];

2) the second case is the fifth-order dissipative weighted explicit scheme (DWCNS-E-5),

$$E'_j = \frac{75}{64h}(\tilde{E}_{j+1/2} - \tilde{E}_{j-1/2}) - \frac{25}{384h}(\tilde{E}_{j+3/2} - \tilde{E}_{j-3/2}) + \frac{3}{640h}(\tilde{E}_{j+5/2} - \tilde{E}_{j-5/2})$$

$$- \frac{75\alpha}{128h}(\tilde{E}^+_{j+1/2} - 2E^+_j + \tilde{E}^+_{j-1/2}) + \frac{25\alpha}{256h}(\tilde{E}^+_{j+3/2} - 2E^+_j + \tilde{E}^+_{j-3/2})$$

$$- \frac{3\alpha}{256h}(\tilde{E}^+_{j+5/2} - 2E^+_j + \tilde{E}^+_{j-5/2}) + \frac{75\alpha}{128h}(\tilde{E}^-_{j+1/2} - 2E^-_j + \tilde{E}^-_{j-1/2})$$

$$- \frac{25\alpha}{256h}(\tilde{E}^-_{j+3/2} - 2E^-_j + \tilde{E}^-_{j-3/2}) + \frac{3\alpha}{256h}(\tilde{E}^-_{j+5/2} - 2E^-_j + \tilde{E}^-_{j-5/2}). \tag{12}$$

It is also required that $\alpha \geq 0$. As $\alpha = 0$, we obtain a new explicit fifth-order weighted scheme (WCNS-E-5),

$$E'_j = \frac{75}{64h}(\tilde{E}_{j+1/2} - \tilde{E}_{j-1/2}) - \frac{25}{384h}(\tilde{E}_{j+3/2} - \tilde{E}_{j-3/2}) + \frac{3}{640h}(\tilde{E}_{j+5/2} - \tilde{E}_{j-5/2}). \tag{13}$$

It is obvious that DWCNS-E-5 and WCNS-E-5 are not compact schemes. We use "DWCNS-E-5" and "WCNS-E-5" here because they can be taken as the special case of the general compact schemes (10).

In (11)—(13), $\tilde{E}_{j\pm 1/2} = E(\tilde{U}_{j\pm 1/2})$ is the numerical flux at the cell-edges. There are various ways to get them:

(1) the Roe's flux-difference splitting scheme,

$$\tilde{E}_{j+1/2} = \frac{1}{2}[E(\tilde{U}_{Rj+1/2}) + E(\tilde{U}_{Lj+1/2}) - |A|(\tilde{U}_{Rj+1/2} - \tilde{U}_{Lj+1/2})]; \quad (14)$$

(2) the flux vector splitting schemes,

$$\tilde{E}_{j+1/2} = E^-(\tilde{U}_{Rj+1/2}) + E^+(\tilde{U}_{Lj+1/2}). \quad (15)$$

Here, $E^{\pm}$ may be obtained by Steger-Warming's[16], Van Leer's[17] splitting or other splitting methods.

The numerical flux in the dissipative terms is simply calculated by

$$\tilde{E}^+_{j+1/2} = E^+(\tilde{U}_{Lj+1/2}), \tilde{E}^-_{j+1/2} = E^-(\tilde{U}_{Rj+1/2}), E^{\pm}_j = E^{\pm}(U_j),$$

thus in the case of $\alpha \neq 0$, only flux vector splitting schemes can be used in DWCNS-5. On the other hand, however, both (14) and (15) can be used in DWCNS-E-5, WCNS-E-5 and WCNS-5. Comparing the computation cost, DWCNS-E-5 is much more efficient than DWCNS-5, so we concentrate on DWCNS-E-5 (WCNS-E-5) in the following sections. In ref. [12], we gave a detailed comparison of four flux splitting methods (Steger-Warming and Van Leer's flux vector splittings, Roe's flux difference splitting and AUSM+) for WCNS-E-5.

In the above formulas, $\tilde{U}_{Lj+1/2}$ and $\tilde{U}_{Rj+1/2}$ are left and right cell edge values which can be obtained by the 5th order weighted nonlinear interpolations. Thus the overall schemes are the weighted nonlinear ones.

2.2 Weighted interpolations at cell-edges

In our previous paper[11], the fifth-order weighted interpolations at cell edges have been derived, which can be directly used in DWCNS. To the completeness of this paper, we give the main results of the interpolation. Denote l^p (row vector) and r^p (column vector) as the pth left and right eigenvectors of matrix $A = \partial E/\partial U$, consider the interpolation in cell $[x_{j-1/2}, x_{j+1/2}]$ for the pth characteristic variables $Q_{j,p}$,

$$Q_{j,p} = l_n^p \cdot U_j, \quad (16)$$

where the subscript n is a function of the cell $[x_{j-1/2}, x_{j+1/2}]$. n is fixed as j for the interpolation of $\tilde{Q}_j(x)$ in the cell $[x_{j-1/2}, x_{j+1/2}]$. The fifth-order weighted interpolation of $\tilde{Q}_{Lj+1/2}$ and $\tilde{Q}_{Rj+1/2}$ is

$$\tilde{Q}^\omega_{Lj+1/2,p} = Q_j + \frac{h}{2}f^*_{Lj} + \frac{1}{8}h^2 s^*_{Lj}, \tilde{Q}^\omega_{Rj-1/2,p} = Q_j - \frac{h}{2}f^*_{Rj} + \frac{1}{8}h^2 s^*_{Rj}, \quad (17)$$

where

$$f^*_{Lj} = \sum_{k=1}^{3} \omega_{Lk} f^k_j, \quad s^*_{Lj} = \sum_{k=1}^{3} \omega_{Lk} s^k_j,$$
$$f^*_{Rj} = \sum_{k=1}^{3} \omega_{Rk} f^k_j, \quad s^*_{Rj} = \sum_{k=1}^{3} \omega_{Rk} s^k_j, \quad (18)$$

and

$$f^1_j = \frac{1}{2h}(Q_{j-2} - 4Q_{j-1} + 3Q_j),$$
$$f^2_j = \frac{1}{2h}(Q_{j+1} - Q_{j-1}), \quad (19)$$
$$f^3_j = \frac{1}{2h}(-3Q_j + 4Q_{j+1} - Q_{j+2}),$$

$$s_j^1 = \frac{1}{h^2}(Q_{j-2} - 2Q_{j-1} + Q_j),$$
$$s_j^2 = \frac{1}{h^2}(Q_{j-1} - 2Q_j + Q_{j+1}), \tag{20}$$
$$s_j^3 = \frac{1}{h^2}(Q_j - 2Q_{j+1} + Q_{j+2}).$$

The weights are defined as,
$$\omega_{Lk} = \frac{\beta_{Lk}}{\sum_{m=1}^{3} \beta_{Lm}}, \quad \omega_{Rk} = \frac{\beta_{Rk}}{\sum_{m=1}^{3} \beta_{Rm}},$$

where
$$\beta_{Lk} = \frac{C_{Lk}}{(\varepsilon + IS_k)^2}, \quad \beta_{Rk} = \frac{C_{Rk}}{(\varepsilon + IS_k)^2},$$

$\varepsilon = 10^{-6}$ is a small number to prevent the denominator from becoming zero and IS_k is the smooth measures which are given by
$$IS_k = (hf_j^k)^2 + (h^2 s_j^k)^2.$$

C_{Lk} and C_{Rk} are optimal weights,
$$C_{L1} = C_{R3} = \frac{1}{16}, \quad C_{L2} = C_{R2} = \frac{10}{16}, \quad C_{L3} = C_{R1} = \frac{5}{16}.$$

If $\omega_{Lk} = C_{Lk}$ and $\omega_{Rk} = C_{Rk}$ in (18), the interpolants (17) are the optimal ones.

Finally the dependent variables at cell-edges can be obtained by
$$\tilde{U}_{Lj+1/2} = \sum_{p=1}^{3} \tilde{Q}_{Lj+1/2,p}^{\omega} r_n^p, \quad \tilde{U}_{Rj-1/2} = \sum_{p=1}^{3} \tilde{Q}_{Rj-1/2,p}^{\omega} r_n^p. \tag{21}$$

In refs. [10,11], we give a detailed analysis of this interpolation.

2.3 Boundary and near boundary schemes

For high order finite difference schemes development, it is important to derive boundary and near boundary schemes. As we know, for a pth order interior scheme, the accuracy of boundary schemes can be $(p-1)$th order accurate without reducing the global accuracy of the interior scheme. For the fifth-order DWCNS interior schemes, we derive the following fourth order boundary schemes,

$$E'_1 = \frac{1}{24h}(-22\tilde{E}_{1/2} + 17\tilde{E}_{3/2} + 9\tilde{E}_{5/2} - 5\tilde{E}_{7/2} + \tilde{E}_{9/2}),$$
$$E'_2 = \frac{1}{24h}(\tilde{E}_{1/2} - 27\tilde{E}_{3/2} + 27\tilde{E}_{5/2} - \tilde{E}_{7/2}),$$
$$E'_{N-1} = -\frac{1}{24h}(\tilde{E}_{N+1/2} - 27\tilde{E}_{N-1/2} + 27\tilde{E}_{N-3/2} - \tilde{E}_{N-5/2}), \tag{22}$$
$$E'_N = -\frac{1}{24h}(-22\tilde{E}_{N+1/2} + 17\tilde{E}_{N-1/2} + 9\tilde{E}_{N-3/2} - 5\tilde{E}_{N-5/2} + \tilde{E}_{N-7/2}).$$

For boundary and near boundary interpolation, the explicit fourth order interpolants are derived,
$$\tilde{U}_{1/2} = \frac{1}{16}(5U_0 + 15U_1 - 5U_2 + U_3),$$
$$\tilde{U}_{3/2} = \frac{1}{16}(-U_0 + 9U_1 + 9U_2 - U_3),$$
$$\tilde{U}_{N-1/2} = \frac{1}{16}(-U_{N+1} + 9U_N + 9U_{N-1} - U_{N-2}),$$

$$\tilde{U}_{N+1/2} = \frac{1}{16}(5U_{N+1} + 15U_N - 5U_{N-1} + U_{N-2}). \tag{23}$$

These schemes can be directly used in DWCNS-E-5. Only the first and fourth schemes in (22) are needed for DWCNS-5.

2.4 The Fourier analysis and asymptotic stability of DWCNS

For the Fourier analysis of DWCNS, we insert (11), (12), (13) and the optimal interpolants into (4), and obtain the modified wave number functions

$$w^* h = w^*(wh)h. \tag{24}$$

Fig. 1 shows the real and imaginary parts of $w^* h$ of DWCNS-E-5. It may be noted that in view of the modified wave numbers, DWCNS-E-5 are a little superior to Explicit UpWind biased 5th (EUW5) schemes which were successfully applied to the direct numerical simulation of turbulences. In view of fig. 1, a question may arise: WCNS-E-5 is also a dissipative scheme, is it necessary to further develop DWCNS-E-5? It was found by numerical tests that WCNS-E-5 as well as TVD and ENO (WENO) cannot eliminate the vorticity oscillations in the inviscid flow around a circular cylinder. The DWCNS-E-5, on the other hand, can damp out these oscillations as the grid lines do not skew the shock wave severely (see example 3 in sec. 5).

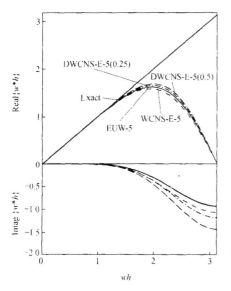

Fig 1. The modified wave numbers of DWCNS-E-5 compared with 5th explicit upwind biased scheme. ——, Exact; -----, EUW-5; ———, WCNS-E-5; -·-·-, DWCNS-E-5 (0.25); ----, DWCNS-E-5 (0.5).

Substituting (12), (22), (23) and the optimal interpolants at cell edges into (2) leads to

$$A\frac{du}{dt} = \frac{c}{h}BCu + g(t),$$

where $u = (u_1, u_2, \cdots, u_N)^T$, and $g(t)$ corresponds to initial and boundary conditions (without losing generality, we set $g(t) = 0$ in the following analysis). As discussed in ref. [1], the asymptotic stability condition for the semi-discrete equation is that all eigenvalues of the matrix $R = A^{-1}BC$ have no positive real parts.

Fig. 2 shows the eigenvalue spectra for the DWCNS-E-5 on uniform grids with $\alpha = 0$ (WCNS-E-5), 0.25 and 0.5. It can be seen that DWCNS-E-5 is asymptotic stable and the larger the parameter α, the more stable the schemes.

For practical calculations, especially for viscous flow simulations, the stretching grids are necessary. In order to investigate the effect of grid stretching, we transform eq. (1) from the stretching physical space x to the uniform computational domain ξ by

$$x(\xi) = \frac{A(\beta + 2\alpha) + 2\alpha - \beta}{(2\alpha + 1)(1 + A)}, \quad A = \left(\frac{\beta + 1}{\beta - 1}\right)^{(1-\xi-\alpha)/(\alpha-1)}, \tag{25}$$

where $\alpha = 0.5$ and $\beta = 1.01015$ are constants[18]. Based on this transformation, the matrix R is changed as $\tilde{R} = D^{-1}R$ and $D = \text{diag}\{x_{\xi 0}, x_{\xi 1}, \cdots, x_{\xi N+1}\}$. The eigenvalue spectra of DWCNS-

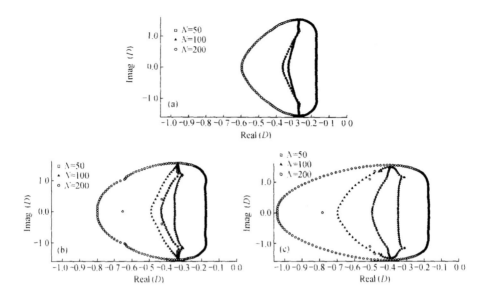

Fig. 2. The eigenvalues spectra of DWCNS-E-5 on a uniform grid. (a) $\alpha = 0$ (WCNS-E-5); (b) $\alpha = 0.25$; (c) $\alpha = 0.5$.

E-5 on this stretching grid is showed in fig. 3. We know from the figure that DWCNS-E-5 becomes unstable as the grid number is equal to 50, a very strong stretched grid. With increasing of the grid number, the relative stretching of the grid reduces and the schemes become stable. On the other hand, if the dissipative parameter α is larger, DWCNS-E-5 will also be asymptotic stable on the 50 grids. Therefore, in practical simulations we should pay attention to the strong stretching grids which may have destabilizing effect on DWCNS-E-5. For the viscous flow simula-

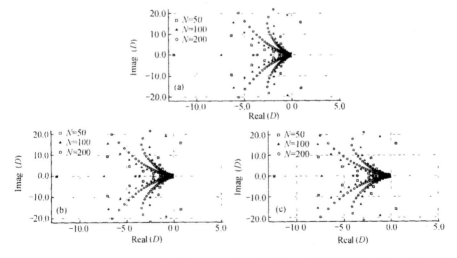

Fig. 3. The eigenvalues spectra of DWCNS-E-5 on a stretching grid (a) $\alpha = 0$ (WCNS-E-5); (b) $\alpha = 0.25$; (c) $\alpha = 0.5$.

tions, however, the physical viscosity has the stabilizing effect, such that the DWCNS-E-5 on a coarse grid may be stable.

2.5 Time discretization

We treat (9) as an ordinary differential equation,

$$\frac{\partial U}{\partial t} = R(U), \tag{26}$$

and the third-order TVD Runge-Kutta method [19] is employed for the time integration

$$\begin{aligned}
U^{(1)} &= U^n + \Delta t R(U^n), \\
U^{(2)} &= \frac{3}{4} U^n + \frac{1}{4} U^{(1)} + \frac{1}{4} \Delta t R(U^{(1)}), \\
U^{n+1} &= \frac{1}{3} U^n + \frac{2}{3} U^{(2)} + \frac{2}{3} \Delta t R(U^{(2)}).
\end{aligned} \tag{27}$$

For steady viscous flow calculations, the implicit schemes are more efficient than the explicit ones. In this paper we try LU-SGS[20] implicit scheme for steady flow calculations,

$$LD^{-1}U = \Delta t R(U^n), \tag{28}$$

where

$$\begin{aligned}
L &= D - \Delta t (A^+_{i-1,j,k} + B^+_{i,j-1,k} + C^+_{i,j,k-1}), \\
U &= D + \Delta t (A^-_{i+1,j,k} + B^-_{i,j+1,k} + C^-_{i,j,k+1}), \\
D &= I + \Delta t [\tilde{\rho}(A) + \tilde{\rho}(B) + \tilde{\rho}(C)], \\
A^{\pm} &= \frac{1}{2} (A \pm \tilde{\rho}(A)), B^{\pm} = \frac{1}{2} (B \pm \tilde{\rho}(B)), C^{\pm} = \frac{1}{2} (C \pm \tilde{\rho}(C)).
\end{aligned}$$

And $\tilde{\rho}(A)$, $\tilde{\rho}(B)$ and $\tilde{\rho}(C)$ are the spectral radius of the Jacobian matrices A, B and C, respectively.

3 Implementation of DWCNS in multi-dimensional Euler and Navier-Stokes equations

In this section we discuss the method of implementation of DWCNS in multi-dimensional Euler and Navier-Stokes equations. The nondimensional Navier-Stokes equations in general coordinates have the forms

$$\frac{\partial U}{\partial t} + \frac{\partial E_l}{\partial \xi_l} = \frac{\partial E_{vl}}{\partial \xi_l}, \tag{29}$$

where the vector U, E_l and E_{vl} are dependent variables, convective flux and viscous flux in the lth spatial coordinate respectively ($l = 1,2,3$), i.e.

$$U = J \begin{pmatrix} \rho \\ \rho u_1 \\ \rho u_2 \\ \rho u_3 \\ e \end{pmatrix}, \quad E_l = J \begin{pmatrix} \rho V_l \\ \rho V_l u_1 + \xi_{lx} p \\ \rho V_l u_2 + \xi_{ly} p \\ \rho V_l u_3 + \xi_{lz} p \\ (e+p) V_l \end{pmatrix}, \quad E_{vl} = J \begin{pmatrix} 0 \\ \xi_{lx} \tau_{11} + \xi_{ly} \tau_{12} + \xi_{lz} \tau_{13} \\ \xi_{lx} \tau_{21} + \xi_{ly} \tau_{22} + \xi_{lz} \tau_{23} \\ \xi_{lx} \tau_{31} + \xi_{ly} \tau_{32} + \xi_{lz} \tau_{33} \\ \xi_{lx} \beta_1 + \xi_{ly} \beta_2 + \xi_{lz} \beta_3 \end{pmatrix},$$

with

$$p = \frac{1}{\gamma M_\infty^2} \rho T, \quad e = \frac{p}{(\gamma - 1)} + \frac{\rho}{2} u_m u_m,$$

$$\tau_{ij} = \frac{\mu}{Re} \left(\frac{\partial u_i}{\partial x_j} + \frac{\partial u_j}{\partial x_i} - \frac{2}{3} \delta_{ij} \frac{\partial u_m}{\partial x_m} \right),$$

$$\beta_l = \tau_{im}u_m + \frac{\mu}{(\gamma-1)ReM_\infty^2 \Pr}\frac{\partial T}{\partial x_l},$$

where γ is the ratio of specific heats, and the viscous coefficient μ can be calculated by the Sutherland's law. The non-dimensional variables are defined as $\rho = \rho^*/\rho_\infty^*$, $u_l = u_l^*/u_\infty^*$, $T = T^*/T_\infty^*$, $p = p^*/\rho_\infty^* u_\infty^{*2}$, respectively, and $M_\infty = u_\infty^*/\sqrt{\gamma R T_\infty^*}$, $Re = \rho_\infty^* u_\infty^* r^*/\mu_\infty^*$ and $\Pr = \mu_\infty^* C_p/\kappa_\infty^*$ are the Mach number, Reynolds number and Prandtl number, r^* the characteristic length. J is the Jacobian of grid transformation, ξ_{lt}, ξ_{lx}, ξ_{ly}, ξ_{lz} are grid derivatives and $V_l = \xi_{lt} + u\xi_{lx} + v\xi_{ly} + w\xi_{lz}$.

Based on the method of line, DWCNS can be directly applied to each of the inviscid fluxes E_l. For the viscous fluxes calculation, the central 6th order scheme

$$E'_{vij} = \frac{75}{64h}(\tilde{E}_{vij+1/2} - \tilde{E}_{vij-1/2}) - \frac{25}{384h}(\tilde{E}_{vij+3/2} - \tilde{E}_{vij-3/2}) + \frac{3}{640h}(\tilde{E}_{vij+5/2} - \tilde{E}_{vij-5/2}) \quad (30)$$

can be used. The viscous fluxes $\tilde{E}_{vij+1/2}$ contain first order derivatives of the dependent variables as well as the variables themselves at cell edges. We use the high-order central type schemes to calculate these values. In Appendix, we give the detailed schemes required for the viscous flux calculations at cell edges.

4 Numerical tests

To test the behaviors of the DWCNS-E-5 and WCNS-E-5, we carried out some numerical solutions of Euler and Navier-Stokes equations. The tests contain two groups: one is the Euler equations solutions which are selected to show the ability of DWCNS-E-5 for the discontinuities resolution. Another is the Navier-Stokes solutions to show the accuracy of DWCNS-E-5 in smooth regions, especially for the boundary layer resolutions. The Roe's flux difference scheme with the entropy fixing technique in ref. [21] is used in one-dimensional cases and Steger-Warming flux splitting is used for two-dimensional solutions. WCNS-E-5 is used for one-dimensional test example.

Example 1. One-dimensional Riemann problem for the Euler equations is solved with the initial conditions

$$U^0(x) = \begin{cases} U_L, & 0 \leq x < 0.1 \\ U_M, & 0.1 \leq x < 0.9 \\ U_R, & 0.9 \leq x < 1 \end{cases} \quad (31)$$

where

$$\rho_L = \rho_M = \rho_R = 1, \quad u_L = u_M = u_R = 0, p_L = 10^3, \quad p_M = 10^{-2}, \quad p_R = 10^2.$$

The boundaries at $x = 0$ and $x = 1$ are solid walls with the reflecting boundary conditions. At the final time $t = 0.038$, the flow field has three contact discontinuities. The middle one, which is generated by the two shocks interacting with each other, is very difficult to resolve. Fig. 4 shows the results with contact discontinuity sharpening technique[10] with CFL = 0.6, 600 grid points. It can be seen that the contact discontinuities are resolved well. Compared with fifth-order WENO results[9] with the Yang's artificial compressing technique, the middle contact discontinuity of our results is captured more sharply.

Example 2. The second example is the two-dimensional wind tunnel flow with a step at Mach 3. Woodward and Colella[22] investigated this flow carefully. In our calculations, the grid

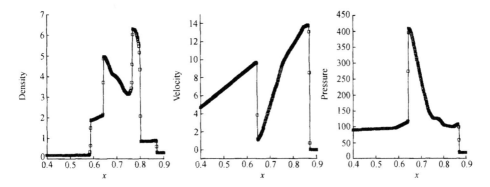

Fig. 4. The blast wave results at $T = 0.038$ with 600 grids calculated by WCNS-E-5 with contact sharpening.

contains 241×81 meshes and CFL = 0.6. Fig. 5 shows the density contours calculated by WCNS-E-5. The Steger-Warming's flux vector splitting method was used here. The results show that WCNS-E-5 performs well for this example. It has good resolutions for the shock and contact discontinuity.

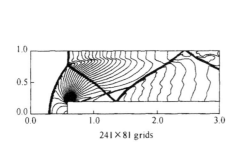

Fig. 5. Density contours of flow past a step calculated by WCNS-E-5 with Steger-Warming flux splitting.

Fig. 6. Mesh grid.

Example 3. This example is the steady Mach 4 inviscid flow around a circular cylinder. Kopriva[23] previously computed this flow using a spectral shock-fitting scheme. Very recently Lee and Zhong[13] studied the vorticity distributions of this flow by various TVD and ENO shock-capturing schemes. They found the spurious oscillations in the vorticity field behind the shock wave. We calculated this flow by WCNS-E-5 and DWCNS-E-5 with Steger-Warming flux splitting. Fig. 6 shows the 61×61 mesh grid. Fig. 7 gives the pressure, Mach number and vorticity contours obtained by WCNS-E-5. The spurious oscillations also appear as reported by Lee and Zhong in the vorticity contours behind the shock wave though the residuals approach machine zero (Fig. 8). We repeated the calculation on the same grid using the DWCNS-E-5 with the dissipative parameter $\alpha = 0.25$. The spurious oscillations have been eliminated (Fig. 9), which shows the advantage of DWCNS developed in this paper.

Example 4. The last example is the steady hypersonic viscous flow around a circular cylinder. We select this problem because there is a shock-fitting based spectral solution[24] that

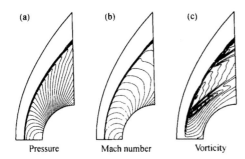

Fig. 7. Contours of Mach 4 inviscid flows calculated by WCNS-E-5.

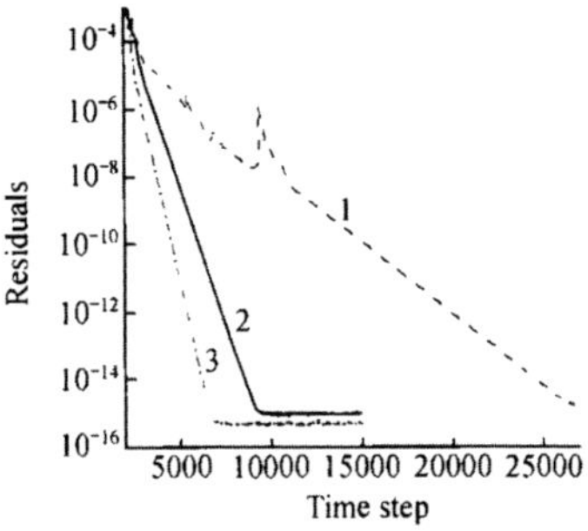

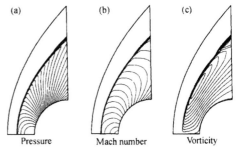

Fig. 8. The convergent history of Mach 4 inviscid flow. 1, WCNS-E-5 (RK3); 2, DWCNS-E-5 (RK3); 3, DWCNS-E-5 (LU-SGS).

can be used for the comparison, especially for the comparison of heat transfer on the body surface. The flow conditions are $M_\infty = 5.73$, $Re = 2050$, $T_\infty^* = 39.6698\ K$, $T_w^* = 210.2\ K$, $\gamma = 1.4$, $Pr = 0.77$ and the cylinder radius $r^* = 0.0061468$ m.

Fig. 10 gives the results by WCNS-E-5 with Steger-Warming flux splitting on 61×61 grids. It can be seen that the bow shock is captured well and the pressure coefficient and heat transfer on the wall (fig. 11) compare well with the spectral solution. The vorticity contours show that there are no spurious oscillations. Fig. 12 shows the convergent history. All the residuals also converge to machine zero.

Fig. 9. Contours of Mach 4 inviscid flows calculated by DWCNS-E-5.

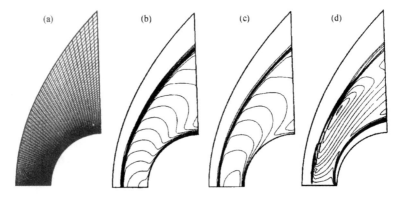

Fig. 10. Contours calculated by WCNS-E-5 with 61×61 grids. (a) Grid; (b) Mach number; (c) temperature; (d) vorticity.

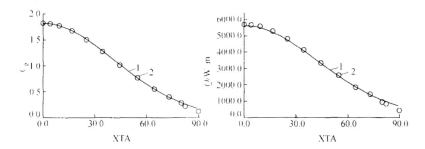

Fig 11. The surface pressure (C_p) and heat transfer (Q) distribution with 61 × 61 grids compared with the spectral method. 1, WCNS-E-5; 2, spectral

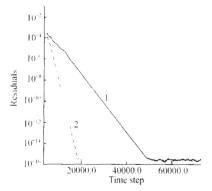

Fig. 12. Convergent history of WCNS-E-5 with 61 × 61 grids 1, WCNS-E-5 (RK3); 2, WCNS-E-5 (LUSGS)

5 Concluding remarks

The fifth order accurate dissipative weighted compact nonlinear schemes (DWCNS-E-5 and DWCNS-5) are developed in this paper. By Fourier analysis, the dispersive and dissipative features of DWCNS-E-5 in smooth regions were discussed, which showed that the interpolations of variables at cell-edges dominate the accuracy of the schemes and DWCNS-E-5 are almost the same accurate as the fifth order upwind biased explicit linear scheme in smooth regions which was successfully used in the DNS of turbulences. The asymptotic stability analysis shows that the boundary and near boundary schemes developed in this paper, as well as interior schemes are asymptotic stable on uniform grid, and on the stretching grids the schemes are also asymptotic stable as the stretching is not very strong or the dissipative parameter α is large enough.

Based on our numerical tests, it was found that WCNS-E-5 work well in one-dimensional flows. For the two-dimensional inviscid flow calculations, it is necessary to use DWCNS-E-5 to suppress the spurious vorticity oscillations. The physical viscosity of Navier-Stokes equations has the effect to suppress the oscillations, hence for viscous flow calculations we suggest WCNS-E-5. In view of the computational costs, the fifth-order explicit scheme DWCNS-E-5 (WCNS-E-5) are more efficient than the compact ones. Therefore in practical calculations, it is better to use DWCNS-E-5 and WCNS-E-5.

In this paper, we only show the performance of DWCNS-E-5 and WCNS-E-5 for one- and two-dimensional calculations. Further researches will apply them to three-dimensional Navier-Stokes equations for the DNS of hypersonic stability problems, Large Eddy Simulations (LES) of complex turbulent flows to show the accuracy for the resolutions of fine structures.

Appendix

For the viscous terms calculations, the following derivative schemes and interpolation schemes are needed.

(1) The first order derivatives at cell edges

$$u'_{1/2} = \frac{1}{24h}(-22u_0 + 17u_1 + 9u_2 - 5u_3 + u_4),$$

$$u'_{j+1/2} = \frac{27}{64h}(u_{j+1} - u_j) - \frac{1}{24h}(u_{j+2} - u_{j-1}), \quad (A1)$$

$$u'_{N+1/2} = -\frac{1}{24h}(-22u_{N+1} + 17u_N + 9u_{N-1} - 5u_{N-2} + u_{N-3}).$$

(2) The first order derivatives at grid nodes

$$u'_j = \frac{8}{12h}(u_{j+1} - u_{j-1}) - \frac{1}{12h}(u_{j+2} - u_{j-2}), \quad (A2)$$

with two sets of boundary schemes,

$$u'_1 = \frac{1}{12h}(-25u_1 + 48u_2 - 36u_3 + 16u_4 - 3u_5),$$

$$u'_2 = \frac{1}{12h}(-3u_1 - 10u_2 + 18u_3 - 6u_4 + u_5),$$

$$u'_{N-1} = -\frac{1}{12h}(-3u_N - 10u_{N-1} + 18u_{N-2} - 6u_{N-3} + u_{N-4}), \quad (A2.1)$$

$$u'_N = -\frac{1}{12h}(-25u_N + 48u_{N-1} - 36u_{N-2} + 16u_{N-3} - u_{N-4}),$$

and

$$u'_1 = \frac{1}{12h}(-3u_0 - 10u_1 + 18u_2 - 6u_3 + u_4),$$

$$u'_N = -\frac{1}{12h}(-3u_{N+1} - 10u_N + 18u_{N-1} - 6u_{N-2} + u_{N-3}). \quad (A2.2)$$

(3) The interpolation at cell edges

$$u_{j+1/2} = \frac{9}{16}(u_{j+1} + u_j) - \frac{1}{16}(u_{j+2} + u_{j-1}), \quad (A3)$$

with two sets of boundary schemes,

$$u_{1/2} = \frac{1}{16}(35u_1 - 35u_2 + 21u_3 - 5u_4),$$

$$u_{3/2} = \frac{1}{16}(5u_1 + 15u_2 - 5u_3 + u_4),$$

$$u_{N-1/2} = \frac{1}{16}(5u_N + 15u_{N-1} - 5u_{N-2} + u_{N-3}), \quad (A3.1)$$

$$u_{N+1/2} = \frac{1}{16}(35u_N - 35u_{N-1} + 21u_{N-2} - 5u_{N-3}),$$

and

$$u_{1/2} = \frac{1}{16}(5u_0 + 15u_1 - 5u_2 + u_3),$$

$$u_{N+1/2} = \frac{1}{16}(5u_{N+1} + 15u_N - 5u_{N-1} + u_{N-2}). \quad (A3.2)$$

(4) The interpolation at grid nodes

$$u_1 = \frac{1}{16}(5u_{1/2} + 15u_{3/2} - 5u_{5/2} + u_{7/2}),$$

$$u_j = \frac{9}{16}(u_{j+1/2} + u_{j-1/2}) - \frac{1}{16}(u_{j+3/2} + u_{j-3/2}),$$
$$u_N = \frac{1}{16}(5u_{N+1/2} + 15u_{N-1/2} - 5u_{N-3/2} + u_{N-5/2}). \tag{A4}$$

In ref. [9], we gave the compact forms of the above schemes.

Acknowledgements This work was supported by the project of Basic Research on Frontier Problems in Fluid and Aerodynamics in China and the National Natural Science Foundation of China (Grant No. 19772072).

References

1. Lele, S. K., Compact finite difference schemes with spectral-like resolution, J. Comp. Phys., 1992, 103: 16—42.
2. Trefethen, L. N., Group velocity in finite difference schemes, SIAM Rev., 1982, 24: 113—136.
3. Yu, S. T., Hsieh, K. C., Tsai, Y. P., Direct calculations of waves in fluid flows using high-order compact difference schemes, AIAA Journal, 1994, 32: 1766—1733.
4. Fu, D. X., Ma, Y. W., A high order accurate difference scheme for complex flow fields, J. Comp. Phys., 1997, 134: 1—45.
5. Deng, X. G., Maekawa, H., Shen, Q., A class of high order dissipative compact schemes, AIAA paper 96-1972, 27th AIAA Fluid Dynamics Meeting, New Orleans, LA, 1996.
6. Deng, X. G., Maekawa, H., Compact high-order accurate nonlinear schemes, J. Comp. Phys., 1997, 130: 77—91.
7. Deng, X. G., Maekawa, H., A uniform fourth-order nonlinear compact schemes for discontinuities capturing, AIAA paper 96-1974, 27th AIAA Fluid Dynamics Meeting, New Orleans, LA, 1996.
8. Liu, X. D., Osher, S., Chan, T., Weighted essentially non-oscillatory schemes, J. Comp. Phys., 1994, 115: 200—212.
9. Jiang, G. S., Shu, C. W., Efficient Implementation of Weighted ENO Schemes, ICASE Report No. 95-73, 1995, also NASA CR 198228.
10. Deng, X. G., Mao, M. L., Weighted compact high-order nonlinear schemes for the euler equations, AIAA paper 97-1941, 13th AIAA Computational Fluid Dynamic Conference, Snowmass, 1997.
11. Deng, X. G., Zhang, H. X., Developing high-order weighted compact nonlinear schemes, J. Comp. Phys., 2000, 165: 22.
12. Deng, X. G., Maekawa, H., Applications of fifth-order weighted compact nonlinear schemes to Euler and Navier-Stokes equations, Proceedings of 4th Asian Computational Fluid Dynamics Conference (ed. Zhang Hanxin), Mianyang: University of Electronic Science & Technology of China Press, 2000.
13. Lee, C. K., Zhong, X. L., Spurious numerical oscillations in simulation of supersonic flows using shock-capturing schemes, AIAA Journal, 1999, 37: 313.
14. Vichnevetsky, R., Bowles, J. B., Fourier Analysis of Numerical Approximations of Hyperbolic Equations, Philadelphia: SIAM, 1982.
15. Mao, M. L., Deng, X. G., Applications of high-order dissipative compact schemes to compressible stability studies, Aerodynamica Sinica (in Chinese), 1994, (4).
16. Steger, J. L., Warming, R. F., Flux vector splitting of inviscid gasdynamic equations with application to finite-difference methods, J. Comp. Phys., 1981, 40: 263.
17. Van Leer, B., Flux-vector splitting for the Euler equations, ICASE Report 82-30, 1982.
18. Roberts, G. O., Computational meshes for the boundary problems, in Proc. Second International Conf. Numerical Methods Fluid Dyn., Lecture Notes in Physics, New York: Springer-Verlag, 1971, vol. 8, 171.
19. Shu, C. W., Osher, S., Efficient implementation of essentially non-oscillatory shock-capturing schemes II, J. Comp. Phys., 1989, 83: 32—78.
20. Yoon, S., Jamson, A., Lower-upper symmetric Gauss Seidel method for the Euler and Navier-Stokes equations, AIAA Journal, 1987, 26(9): 1025—1026.
21. Gaitonde, D., Shang, J. S., Accuracy of flux-split algorithms in high-speed viscous flows, AIAA Journal, 1993, 31: 1215.
22. Woodward, P., Colella, P., The numerical simulation of two-dimensional fluid flow with strong shocks, J. Comp. Phys., 1984, 54: 115.
23. Kopriva, D. A., Zang, T. A., Hussaini, M. Y., Spectral methods for the Euler equations: The blunt body problem revisited, AIAA Journal, 1991, 29: 1458.
24. Kopriva, D. A., Spectral solution of the viscous blunt-body problem, AIAA Journal, 1993, 31: 1235.

Developing High-Order Weighted Compact Nonlinear Schemes

Xiaogang Deng and Hanxin Zhang

Computational Aerodynamic Institute of CARDC, P.O. Box 211, Mianyang, Sichuan 621000, People's Republic of China

Received November 1, 1999; revised May 17, 2000; published online November 3, 2000

The weighted technique is introduced in the compact high-order nonlinear schemes (CNS) and three fourth- and fifth-order weighted compact nonlinear schemes (WCNS) are developed in this paper. By Fourier analysis, the dissipative and dispersive features of WCNS are discussed. In view of the modified wave number, the WCNS are equivalent to fifth-order upwind biased explicit schemes in smooth regions and the interpolations at cell-edges dominate the properties of WCNS. Both flux difference splitting and flux vector splitting methods can be applied in WCNS, though they are finite difference schemes. Boundary and near boundary schemes are developed and the asymptotic stability of WCNS is analyzed. Several numerical results are given which show the good performances of WCNS for discontinuity capture high accuracy for boundary layer calculation, and good convergent rate. We also compare WCNS with MUSCL scheme and spectral solutions. WCNS are more accurate than MUSCL, as expected, especially for heat transfer calculations. © 2000 Academic Press

Key Words: compact schemes; nonlinear schemes; finite difference schemes; Euler equations; Navier–Stokes equations.

1. INTRODUCTION

In the 1990s, compact schemes for the direct numerical simulation of turbulence and aeroacoustic calculations have received much attention. Lele [1] analyzed a series of compact schemes and derived compact schemes with spectral resolutions. Wilson *et al.* [2] proposed high-order compact schemes and discussed then application to incompressible Navier–Stokes equation calculations. Leslie and Purser [3] derived cell-centered fourth-order compact schemes and solved a regional forecast model. Garanzha and Konshin [4] developed numerical algorithms for viscous incompressible fluid flows based on cell-centered compact schemes. Gaitonde and Shang [5] and Kobayashi [6] proposed and analyzed finite-volume compact schemes and so forth. Despite their differences these schemes are all linear

---

Deng Xiaogang, Zhang Hanxin. Developing high-order weighted compact nonlinear schemes. Journal of Computational Physics, 2000, 165(1): 22–44.

ones. For nonlinear compact schemes, the research work is less compared to linear schemes. Cockburn and Shu [7] proposed third- and fourth-order compact nonlinear schemes based on TVD and TVB concepts, but the solutions near shock waves oscillate obviously for the fourth-order schemes.

In this paper, we continue the development of compact high-order nonlinear schemes (CNS). Third- and fourth-order compact nonlinear schemes (CNS3, CNS4) were derived in our previous work [8, 9]. These schemes achieve high-order accuracy by cell-centered fourth-order compact schemes and compact high-order interpolations at cell-edges are designed. Two propositions, which guarantee the uniform high order of the interpolations, were proved. The numerical solutions for the Euler equations showed that CNS3 and CNS4 can capture discontinuities robustly. One advantage of these schemes is that primitive variables, conservative variables, and flux themselves can be used for the calculation of numerical flux at the cell-edges though they are finite difference formulations. The efficiency of CNS3 and CNS4, however, is relatively low due to the three tridiagonal inversions that are needed for the calculation of the derivative, and the contact discontinuity is somewhat smeared. Furthermore, in smooth regions only parts of the useful information are used because the compact adaptive interpolations also play a role in these regions. Recently a weighted technique was introduced in ENO schemes by Liu *et al.* [12] and improved by Jiang and Shu [13], such that weighted ENO schemes were developed. The analytic work shows that WENO schemes are more efficient and more accurate in a smooth region than ENO schemes [13]. Moreover, no logical statements, which perform poorly on vector supercomputers, are required in WENO schemes. However, in view of the one-dimensional blast wave calculations, the WENO schemes of Jiang and Shu did not resolve well the contact discontinuity resulting from wave interactions, though Yang's artificial compression method was used in their calculations. Furthermore, in [13] only the interior schemes are discussed. It is not known how the boundary and near boundary conditions have been applied for WENO, which is very important for high-order scheme applications.

The objective of the present paper is to develop high-order CNS based on the weighted technique, such that weighted compact nonlinear schemes (WCNS) are obtained, which need only one tridiagonal inversion for derivative calculations and are fourth- or fifth-order accurate in smooth regions. By Fourier analysis, the dispersive and dissipative features of WCNS are discussed. Boundary and near boundary schemes are derived and asymptotic stability is analyzed on both uniform and stretching grids. For contact discontinuity sharpening, the method of Huynh [15] is adopted. In Section 3, the WCNS are applied to Euler and Navier–Stokes equations and several numerical results are obtained which show the good performances of WCNS, especially the good convergence rate and high accuracy for the boundary layer simulations.

2. NUMERICAL METHODOLOGY

In this paper we consider numerical approximations to solutions of three-dimensional Navier–Stokes equations in general coordinates,

$$\frac{\partial U}{\partial t} + \frac{\partial E_l}{\partial \xi_l} = \frac{\partial E_{vl}}{\partial \xi_l}, \qquad (1)$$

where the vectors U, E_l, and E_{vl} are dependent variables, convective flux, and viscous flux in the lth spatial coordinate, respectively ($l = 1, 2, 3$), i.e.,

$$U = J \begin{pmatrix} \rho \\ \rho u_1 \\ \rho u_2 \\ \rho u_3 \\ e \end{pmatrix}, \quad E_l = J \begin{pmatrix} \rho V_l \\ \rho V_l u_1 + \xi_{lx} p \\ \rho V_l u_2 + \xi_{ly} p \\ \rho V_l u_3 + \xi_{lz} p \\ (e + p) V_l \end{pmatrix}, \quad E_{vl} = J \begin{pmatrix} 0 \\ \xi_{lx}\tau_{11} + \xi_{ly}\tau_{12} + \xi_{lz}\tau_{13} \\ \xi_{lx}\tau_{21} + \xi_{ly}\tau_{22} + \xi_{lz}\tau_{23} \\ \xi_{lx}\tau_{31} + \xi_{ly}\tau_{32} + \xi_{lz}\tau_{33} \\ \xi_{lx}\beta_1 + \xi_{ly}\beta_2 + \xi_{lz}\beta_3 \end{pmatrix},$$

with

$$p = \frac{1}{\gamma M_\infty^2} \rho T,$$

$$e = \frac{p}{(\gamma - 1)} + \frac{\rho}{2} u_m u_m,$$

$$\tau_{ij} = \frac{\mu}{\text{Re}} \left(\frac{\partial u_i}{\partial x_j} + \frac{\partial u_j}{\partial x_i} - \frac{2}{3} \delta_{ij} \frac{\partial u_m}{\partial x_m} \right),$$

$$\beta_i = \tau_{im} u_m + \frac{\mu}{(\gamma - 1) \, \text{Re} \, M_\infty^2 \, \text{Pr}} \frac{\partial T}{\partial x_i},$$

where γ is the ratio of specific heats and the viscous coefficient μ can be calculated by Sutherland's law. The nondimensional variables are defined as $\rho = \rho^*/\rho_\infty^*$, $u_l = u_l^*/u_\infty^*$, $T = T^*/T_\infty^*$, and $p = p^*/\rho_\infty^* u_\infty^{*2}$, respectively, and $M_\infty = u_\infty^*/\sqrt{\gamma R T_\infty^*}$, $\text{Re} = \rho_\infty^* u_\infty^* r^*/\mu_\infty^*$ and $\text{Pr} = \mu_\infty^* C_p / \kappa_\infty^*$ are the Mach number, Reynolds number, and Prandtl number, and r^* is the characteristic length. J is the Jacobian of grid transformation, ξ_{lt}, ξ_{lx}, ξ_{ly}, and ξ_{lz} are grid derivatives, and $V_l = \xi_{lt} + u \xi_{lx} + v \xi_{ly} + w \xi_{lz}$.

The Cell-centered Compact Schemes

The governing Eqs. (1) can be discretized line by line in the computational space (ξ_1, ξ_2, ξ_3). Therefore, in this section, we only consider one-dimensional convective flux discretization for simplicity. Let $U_j = U(x_j, t)$, let $x_j = jh$, and denote a numerical approximation to the solution of the equation

$$\frac{\partial U}{\partial t} + \frac{\partial E}{\partial x} = 0. \qquad (2)$$

At every node x_j, we discrete Eq. (2) by the semi-discrete finite difference scheme

$$\left(\frac{\partial U}{\partial t} \right)_j = -E'_j, \qquad (3)$$

where E'_j is the approximation to the spatial derivative. As discussed in Refs. [1, 8, 11], the cell-centered finite difference compact schemes (CCS) are used to calculate E'_j which have the form

$$\kappa E'_{j-1} + E'_j + \kappa E'_{j+1} = \frac{a}{h} (\tilde{E}_{j+1/2} - \tilde{E}_{j-1/2}) + \frac{b}{h} (\tilde{E}_{j+3/2} - \tilde{E}_{j-3/2}), \qquad (4)$$

with the parameters

$$a = \frac{3}{8}(3 - 2\kappa), \quad b = \frac{1}{24}(22\kappa - 1),$$

where $\tilde{E}_{j\pm 1/2} = E(\tilde{U}_{j\pm 1/2})$ is the numerical flux at the cell-edges. By the Taylor series expansion of (4), we may get

$$E'_j = \left(\frac{\partial E}{\partial x}\right)_j + \frac{9 - 62\kappa}{1920} h^4 E_j^{(5)} + O(h^r) + O(h^6). \tag{5}$$

The truncation term $O(h^r)$ comes from the interpolations of $\tilde{U}$. If $\tilde{U}_{j\pm 1/2}$ approximates $U_{j\pm 1/2}$ to the fifth-order ($r = 5$), which is the case in the present paper, the schemes are fourth- or fifth-order accurate depending on the parameter κ. In this paper, the following three cases are discussed:

(1) $\kappa = \frac{1}{22}$ (WCNS-4),

$$\kappa E'_{j-1} + E'_j + \kappa E'_{j+1} = \frac{a}{h}(\tilde{E}_{j+1/2} - \tilde{E}_{j-1/2}). \tag{6}$$

This is the most compact fourth-order scheme and was used in CNS [8, 9]. Leslie and Purse [3] also derived this scheme.

(2) $\kappa = \frac{9}{62}$ (WCNS-5),

$$\kappa E'_{j-1} + E'_j + \kappa E'_{j+1} = \frac{a}{h}(\tilde{E}_{j+1/2} - \tilde{E}_{j-1/2}) + \frac{b}{h}(\tilde{E}_{j+3/2} - \tilde{E}_{j-3/2}). \tag{7}$$

In this case the scheme is fifth-order accurate. If $\tilde{E} = E$, i.e., without interpolation errors of U at cell edges, it is sixth-order.

(3) $\kappa = 0$ (WCNS-E-4),

$$E'_j = \frac{a}{h}(\tilde{E}_{j+1/2} - \tilde{E}_{j-1/2}) + \frac{b}{h}(\tilde{E}_{j+3/2} - \tilde{E}_{j-3/2}). \tag{8}$$

This is an explicit scheme and more efficient than WCNS-4 and WCNS-5, because non-tridiagonal inversion is required.

Cell-centered compact schemes have the advantage over unstaggered schemes that their dispersive errors are considerably lower [1]. This feature can be seen clearly from Fig. 1, which plots the modified wave numbers for CCS-4, CCS-6, and CCS-E-4, which correspond to WCNS-4, WCNS-5, and WCNS-E-4 as $\tilde{E} = E$, respectively. To compare, the cell-centered with the unstaggered compact schemes, the Pàde schemes (fourth-order) and sixth-order compact scheme (CS-6) of Lele [1] are also shown in this figure.

It should be pointed out if $\tilde{U}$ are only rth order accurate approximations of U at cell-edges, i.e., $\tilde{E} = E + O(h^r)$, the dispersive and dissipative properties of CCS are dominated by $\tilde{U}$. Thus, it is important to get high-order interpolation of $\tilde{U}$ at cell-edges. We will further discuss this feature in the next section.

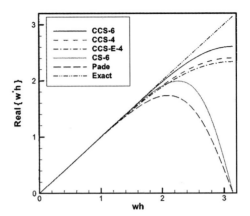

FIG. 1. The modified wave numbers for various compact schemes.

Weighted Interpolations at Cell Edges

In [8, 9], the third- and fourth-order compact adaptive interpolations at cell-edges were derived and compact nonlinear schemes (CNS-3, CNS-4) were obtained. The main idea in designing these interpolations is to prevent them from crossing the discontinuities, so that uniform high order is achieved even for discontinuous data. Though CNS worked well, two problems arose. First CNS contain three tridiagonal inversions that made CNS less efficient than explicit schemes. Second CNS was involved in three grid stencils for each grid point derivative calculation, but only one stencil information was finally selected, which resulted in a waste of information in smooth regions. We used the weighted technique of Jiang and Shu [13] to solve these problems. The idea is that each of the three stencils is assigned a weight which determines the contribution of the stencil to the final approximations of the cell-edge value. The weights are designed in such a way that in a smooth region they approach the optimal weights to achieve fifth-order accuracy and require nontridiagonal inversion. In regions near discontinuities, the stencils which contain the discontinuities are assigned nearly zero weights, so that third-order interpolations are achieved in these regions. Thus, the weighted interpolations are also prevented from crossing discontinuities. This technique completely removes the logical statements that appeared in CNS. It may be expected that the weighted compact nonlinear schemes are more efficient than CNS on vector machines.

For the convective flux calculations, the interpolations are usually approximated in the characteristic fields. Denotes as $\mathbf{l}^p$(row vector) and $\mathbf{r}^p$(column vector) the pth left and right eigenvectors of matrix $A = \partial E/\partial U$, considering the interpolation in cell $[x_{j-1/2}, x_{j+1/2}]$ for the pth characteristic variables $Q_{j,p}$,

$$Q_{j,p} = \mathbf{l}_n^p \cdot U_j, \tag{9}$$

where the subscript n is a function of the cell $[x_{j-1/2}, x_{j+1/2}]$. n is fixed as j for the interpolation of $\tilde{Q}_j(x)$ in the cell $[x_{j-1/2}, x_{j+1/2}]$. The general third-order interpolation of Q_j (the subscript p is omitted here) in $[x_{j-1/2}, x_{j+1/2}]$ can be written as

$$\tilde{Q}_j(x) = Q_j + (x - x_j)f_j + \frac{1}{2}(x - x_j)^2 s_j, \tag{10}$$

where f_j and s_j approximate first and second derivatives Q'_j and Q''_j, respectively. Assume

$$f_j = Q'_j + O(h^2), \quad s_j = Q''_j + O(h), \tag{11}$$

and we may get third-order cell-edge interpolated values,

$$\begin{aligned}\tilde{Q}_{Lj+1/2} &= \tilde{Q}_j\left(x_j + \frac{h}{2}\right) = Q_j + \frac{h}{2}f_j + \frac{1}{8}h^2 s_j, \\ \tilde{Q}_{Rj-1/2} &= \tilde{Q}_j\left(x_j - \frac{h}{2}\right) = Q_j - \frac{h}{2}f_j + \frac{1}{8}h^2 s_j.\end{aligned} \tag{12}$$

The subscripts L and R are introduced for upwind consideration.

In the region $[x_{j-2}, x_{j-1}, x_j, x_{j+1}, x_{j+2}]$, three kinds of f_j and s_j can be obtained at three different stencils, i.e.,

$$\begin{aligned} f_j^1 &= \frac{1}{2h}(Q_{j-2} - 4Q_{j-1} + 3Q_j), \\ f_j^2 &= \frac{1}{2h}(Q_{j+1} - Q_{j-1}), \\ f_j^3 &= \frac{1}{2h}(-3Q_j + 4Q_{j+1} - Q_{j+2}) \end{aligned} \tag{13}$$

and

$$\begin{aligned} s_j^1 &= \frac{1}{h^2}(Q_{j-2} - 2Q_{j-1} + Q_j), \\ s_j^2 &= \frac{1}{h^2}(Q_{j-1} - 2Q_j + Q_{j+1}), \\ s_j^3 &= \frac{1}{h^2}(Q_j - 2Q_{j+1} + Q_{j+2}). \end{aligned} \tag{14}$$

Note that Eqs. (13) and (14) satisfy conditions (11). When they are inserted into Eq. (12), the corresponding interpolations are $\tilde{Q}^1_{L(R)j\pm1/2}$, $\tilde{Q}^2_{L(R)j\pm1/2}$, and $\tilde{Q}^3_{L(R)j\pm1/2}$ respectively, which are third-order accurate. In order to get fifth-order accurate interpolations, we may combine these three values together with the weights ω_k,

$$\tilde{Q}^{\omega}_{Lj+1/2} = \sum_{k=1}^{3} \omega_{Lk} \tilde{Q}^k_{Lj+1/2}, \quad \tilde{Q}^{\omega}_{Rj-1/2} = \sum_{k=1}^{3} \omega_{Rk} \tilde{Q}^k_{Rj-1/2}, \tag{15}$$

and the requirements for $\omega_{L(R)k}$ are

$$\sum_{k=1}^{3} \omega_{Lk} = 1, \quad \sum_{k=1}^{3} \omega_{Rk} = 1. \tag{16}$$

On the other hand, in the region $[x_{j-2}, x_{j-1}, x_j, x_{j+1}, x_{j+2}]$, a fifth-order interpolation at cell-edge $x_{j\pm1/2}$ may be obtained as

$$\begin{aligned} \tilde{Q}^{op}_{j+1/2} &= Q_j + \frac{1}{128}(3Q_{j-2} - 20Q_{j-1} - 38Q_j + 60Q_{j+1} - 5Q_{j+2}) \\ \tilde{Q}^{op}_{j-1/2} &= Q_j - \frac{1}{128}(5Q_{j-2} - 60Q_{j-1} + 38Q_j + 20Q_{j+1} - 3Q_{j+2}) \end{aligned} \tag{17}$$

Thus, the optimal weights of $\omega_{L(R)k}$, i.e., $C_{L(R)k}$, can be obtained by the equivalents

$$\tilde{Q}^{op}_{j+1/2} = \sum_{k=1}^{3} C_{Lk}\tilde{Q}^{k}_{Lj+1/2}, \quad \tilde{Q}^{op}_{j-1/2} = \sum_{k=1}^{3} C_{Rk}\tilde{Q}^{k}_{Rj-1/2}, \tag{18}$$

and we get

$$C_{L1} = C_{R3} = \frac{1}{16}, \quad C_{L2} = C_{R2} = \frac{10}{16}, \quad C_{L3} = C_{R1} = \frac{5}{16}.$$

It is obvious that

$$\sum_{k=1}^{3} C_{Lk} = 1, \quad \sum_{k=1}^{3} C_{Rk} = 1. \tag{19}$$

By (15) and (18), the weighted interpolations (15) can be rewritten as

$$\tilde{Q}^{\omega}_{Lj+1/2} = \tilde{Q}^{op}_{Lj+1/2} + \sum_{k=1}^{3} (\omega_{Lk} - C_{Lk})\tilde{Q}^{k}_{Lj+1/2},$$
$$\tilde{Q}^{\omega}_{Rj-1/2} = \tilde{Q}^{op}_{Rj-1/2} + \sum_{k=1}^{3} (\omega_{Rk} - C_{Rk})\tilde{Q}^{k}_{Rj-1/2}, \tag{20}$$

and the last terms represent the high-order truncation errors. We may see this clearly by further writing these terms as follows by means of (16) and (19),

$$\sum_{k=1}^{3}(\omega_{Lk} - C_{Lk})\tilde{Q}^{k}_{Lj+1/2} = \sum_{k=1}^{3}(\omega_{Lk} - C_{Lk})(\tilde{Q}^{k}_{Lj+1/2} - Q(x_{j+1/2})),$$
$$\sum_{k=1}^{3}(\omega_{Rk} - C_{Rk})\tilde{Q}^{k}_{Rj-1/2} = \sum_{k=1}^{3}(\omega_{Rk} - C_{Rk})(\tilde{Q}^{k}_{Rj-1/2} - Q(x_{j-1/2})). \tag{21}$$

With (21), we note that if

$$\omega_{Lk} = C_{Lk} + O(h^2), \quad \omega_{Rk} = C_{Rk} + O(h^2), \tag{22}$$

the weighted interpolations (15) or (20) are fifth-order accurate, i.e.,

$$\tilde{Q}^{\omega}_{Lj+1/2} = Q(x_{j+1/2}) + O(h^5), \quad \tilde{Q}^{\omega}_{Rj-1/2} = Q(x_{j-1/2}) + O(h^5).$$

The weights are defined by

$$\omega_{Lk} = \frac{\beta_{Lk}}{\sum_{m=1}^{3}\beta_{Lm}}, \quad \omega_{Rk} = \frac{\beta_{Rk}}{\sum_{m=1}^{3}\beta_{Rm}},$$

where

$$\beta_{Lk} = \frac{C_{Lk}}{(\epsilon + IS_k)^2}, \quad \beta_{Rk} = \frac{C_{Rk}}{(\epsilon + IS_k)^2}$$

$\epsilon = 10^{-6}$ is a small number to avoid the denominator becoming zero, and IS_k is a smooth measure. In this paper, we simply define the smooth measures IS_k as

$$IS_k = \left(hf_j^k\right)^2 + \left(h^2 s_j^k\right)^2.$$

It is obvious that in smooth regions

$$IS_k = (Q'h)^2(1 + O(h^2))$$

and the conditions (22) are satisfied. Thus, the interpolation values

$$Q^{\omega}_{Lj+1/2} = Q_j + \frac{h}{2}f^*_{Lj} + \frac{1}{8}h^2 s^*_{Lj}, \quad Q^{\omega}_{Rj-1/2} = Q_j - \frac{h}{2}f^*_{Rj} + \frac{1}{8}h^2 s^*_{Rj} \quad (23)$$

are fifth-order accurate in smooth regions, where

$$f^*_{Lj} = \sum_{k=1}^{3} \omega_{Lk} f^k_j, \quad s^*_{Lj} = \sum_{k=1}^{3} \omega_{Lk} s^k_j,$$

$$f^*_{Rj} = \sum_{k=1}^{3} \omega_{Rk} f^k_j, \quad s^*_{Rj} = \sum_{k=1}^{3} \omega_{Rk} s^k_j.$$

Finally, the dependent variables at cell edges can be obtained as

$$\tilde{U}_{Lj+1/2} = \sum_{p=1}^{3} \tilde{Q}^{\omega}_{Lj+1/2,p} \mathbf{r}^p_n,$$

$$\tilde{U}_{Rj-1/2} = \sum_{p=1}^{3} \tilde{Q}^{\omega}_{Rj-1/2,p} \mathbf{r}^p_n. \quad (24)$$

In practical calculations, the characteristic interpolations given above have high costs. In [11], a simple method was used to save computing time. Setting

$$a_j = \left(\frac{4}{5}\delta_j^2 \rho - \delta_{j-1}^2 \rho\right)\left(\frac{5}{4}\delta_j^2 \rho - \delta_{j-1}^2 \rho\right),$$

$$b_j = \left(\frac{4}{5}\delta_j^2 \rho - \delta_{j+1}^2 \rho\right)\left(\frac{5}{4}\delta_j^2 \rho - \delta_{j+1}^2 \rho\right),$$

where ρ is the density and $\delta_j^2 \rho = \rho_{j-1} - 2\rho_j + \rho_{j+1}$, if

$$\max(a_j, b_j) > \epsilon \rho, \quad (25)$$

the characteristic interpolations (10)–(24) are adopted ($\epsilon = 10^{-5}$). Otherwise, the grid point x_j is located in a smooth region, and the optimal interpolations (17) are used for the dependent variables U directly. It was found in our calculations that this technique is efficient, but the residuals can only converge to 10^{-6}–10^{-7}, though the solutions seem acceptable. Therefore, this technique is not recommended for accurate calculations.

The Fourier Analysis of WCNS

In this section, we further discuss the dissipative and dispersive properties of WCNS by Fourier analysis. The scalar linear hyperbolic equation is considered here

$$\frac{\partial u}{\partial t} + c\frac{\partial u}{\partial x} = 0, \quad (26)$$

where $c > 0$ is a constant speed. The semi-discrete form of Eq. (26) can be written as

$$\frac{\partial u_j}{\partial t} + cu'_j = 0. \tag{27}$$

By differencing a periodic function,

$$u \equiv \exp(iwx), \tag{28}$$

we define the modified wave number $w^* = w^*_r + iw^*_i$ as

$$u' = iw^* \exp(iwx). \tag{29}$$

The accurate solutions of Eq. (27) with the initial condition (28) can be obtained by inserting $u(x, t) = A(t) \exp(iwx)$ into Eq. (27) to get

$$u_j = \exp(cw^*_i t) \exp\left[iw\left(x_j - c\frac{w^*_r}{w}t\right)\right]. \tag{30}$$

The cell-centered schemes (4) for the u'_j have the form

$$\kappa u'_{j-1} + u'_j + \kappa u'_{j+1} = \frac{a}{h}\left(\tilde{u}_{Lj+1/2} - \tilde{u}_{Lj-1/2}\right) + \frac{b}{h}\left(\tilde{u}_{Lj+3/2} - \tilde{u}_{Lj-3/2}\right). \tag{31}$$

Note that only $\tilde{u}_L$ need to be interpolated because $c > 0$. What we are most interested in is the solution on smooth regions. So the optimal interpolation

$$\tilde{u}_{j+1/2} = u_j + \frac{1}{128}(3u_{j-2} - 20u_{j-1} - 38u_j + 60u_{j+1} - 5u_{j+2}). \tag{32}$$

is used for analysis. Inserting Eqs. (28) and (29) into (32) and (31), we get the modified wave number function

$$w^*h = w^*(wh)h. \tag{33}$$

Figure 2 shows the real and imaginary parts of w^*h of WCNS compared with the explicit up wind biased fifth-order scheme (EUW5) and fourth-order Páde schemes. It may be noted that in terms of the modified wave numbers, the WCNS-5 scheme is a little superior to the EUW5 scheme which was successfully applied to the direct numerical simulation of turbulences, WCNS-4 and WCNS-E-4 have almost the same dispersive and dissipative features, and the WCNS schemes approach the fourth-order Páde scheme except that the dissipative errors are confined to intermediate and high wave numbers. Comparing Fig. 2 with Fig. 1, it is known that the interpolations of the variables at cell edges change the dispersive features of WCNS. It may be observed that $w^*_i h < 0$ for all cases in Fig. 2, which just corresponds to the dissipative property of the schemes. In Ref. [10], we gave a detailed discussion of this feature.

Boundary Schemes and Asymptotic Stability of WCNS

For high-order finite difference schemes development, it is important to derive boundary and near boundary schemes. As shown by Gustaffson [16], for a pth order interior scheme, the accuracy of boundary schemes can be $(p-1)$th order accurate without reducing the

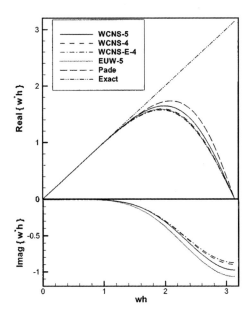

FIG. 2. The modified wave number of WCNS compared with fifth-order explicit upwind biased scheme and Pade scheme.

global accuracy of the interior scheme. For the fourth and fifth WCNS interior schemes, we derive the fourth-order boundary schemes

$$u'_1 + \alpha_1 u'_2 = \frac{1}{h}\left(a_1 \tilde{u}_{1/2} + b_1 \tilde{u}_{3/2} + c_1 \tilde{u}_{5/2} + d_1 \tilde{u}_{7/2} + e_1 \tilde{u}_{9/2}\right),$$
$$u'_N + \alpha_1 u'_{N-1} = -\frac{1}{h}\left(a_1 \tilde{u}_{N+1/2} + b_1 \tilde{u}_{N-1/2} + c_1 \tilde{u}_{N-3/2} + d_1 \tilde{u}_{N-5/2} + e_1 \tilde{u}_{N-7/2}\right), \quad (34)$$

where

$$a_1 = \frac{1}{24}(\alpha_1 - 22), \quad b_1 = \frac{1}{24}(17 - 27\alpha_1), \quad c_1 = \frac{1}{24}(9 + 27\alpha_1),$$
$$d_1 = -\frac{1}{24}(5 + \alpha_1), \quad e_1 = \frac{1}{24}.$$

As $\alpha = -71/31$, this is fifth-order accurate. For boundary and near boundary interpolation, the explicit fourth-order interpolants are derived:

$$\tilde{u}_{1/2} = \frac{1}{16}(5u_0 + 15u_1 - 5u_2 + u_3),$$
$$\tilde{u}_{3/2} = \frac{1}{16}(-u_0 + 9u_1 + 9u_2 - u_3),$$
$$\tilde{u}_{N-1/2} = \frac{1}{16}(5u_N + 15u_{N-1} - 5u_{N-2} + u_{N-3}),$$
$$\tilde{u}_{N+1/2} = \frac{1}{16}(35u_N - 35u_{N-1} + 21u_{N-2} - 5u_{N-3}).$$
(35)

Substituting (31), (32), (34), and (35) into (27) leads to

$$\mathbf{A}\frac{d\mathbf{u}}{dt} = \frac{c}{h}\mathbf{BCu} + \mathbf{g}(t),$$

where $\mathbf{u} = (u_1, u_2, \ldots, u_N)^T$, $\mathbf{g}(t)$ corresponds to initial and boundary conditions (without losing generality, we set $\mathbf{g}(t) = 0$ in the following analysis), and

$$\mathbf{A} = \begin{pmatrix} 1 & \alpha_1 & & & & & \\ \kappa & 1 & \kappa & & & & \\ & \ddots & \ddots & \ddots & & & \\ & & \kappa & 1 & \kappa & & \\ & & & \ddots & \ddots & \ddots & \\ & & & & \kappa & 1 & \kappa \\ & & & & & \alpha_1 & 1 \end{pmatrix}_{N \times N},$$

$$\mathbf{B} = -\begin{pmatrix} a_1' & b_1 & c_1 & d_1 & e_1 & & & & \\ -b & -a & a & b & & & & & \\ & \ddots & \ddots & \ddots & \ddots & & & & \\ & & -b & -a & a & b & & & \\ & & & \ddots & \ddots & \ddots & \ddots & & \\ & & & & -b & -a & a & b & \\ & & & & -e_1 & -d_1 & -c_1 & -b_1 & -a_1 \end{pmatrix}_{N \times (N+1)},$$

$$\mathbf{C} = \begin{pmatrix} \frac{15}{16} & -\frac{5}{16} & \frac{1}{16} & & & & & & \\ \frac{9}{16} & \frac{9}{16} & -\frac{1}{16} & & & & & & \\ -\frac{20}{128} & \frac{90}{128} & \frac{60}{128} & -\frac{5}{128} & & & & & \\ \frac{3}{128} & -\frac{20}{128} & \frac{90}{128} & \frac{60}{128} & -\frac{5}{128} & & & & \\ & \ddots & \ddots & \ddots & \ddots & \ddots & & & \\ & & \frac{3}{128} & -\frac{20}{128} & \frac{90}{128} & \frac{60}{128} & -\frac{5}{128} & & \\ & & & \ddots & \ddots & \ddots & \ddots & & \\ & & & & \frac{1}{16} & -\frac{5}{16} & \frac{15}{16} & \frac{5}{16} \\ & & & & & -\frac{5}{16} & \frac{21}{16} & -\frac{35}{16} & \frac{35}{16} \end{pmatrix}_{(N+1) \times N}$$

As discussed in [1], the asymptotic stability condition for the semi-discrete equation is that all eigenvalues of the matrix $\mathbf{R} = \mathbf{A}^{-1}\mathbf{BC}$ have no positive parts.

Figure 3 shows the eigenvalue spectra for the WCNS-E-4, WCNS-4, and WCNS-5 on the uniform grid. It can be seen that these schemes are all asymptotically stable. Further calculations show that the parameter α_1 in the boundary scheme (34) has little effect on the asymptotic stability.

For practical calculations, especially for viscous flow simulations, the stretching grids are necessary. In order to investigate the effect of grid stretching, we transform Eq. (26) from the stretching physical space x to the uniform computational domain ξ by

$$x(\xi) = \frac{A(\beta + 2\alpha) + 2\alpha - \beta}{(2\alpha + 1)(1 + A)}, \quad A = \left(\frac{\beta + 1}{\beta - 1}\right)^{(1-\xi-\alpha)/(\alpha-1)}, \tag{36}$$

where $\alpha = 0.5$ and $\beta = 1.01015$ are constants [17]. Based on this transformation, the matrix $\mathbf{R}$ is changed as $\tilde{\mathbf{R}} = \mathbf{D}^{-1}\mathbf{R}$ and $\mathbf{D} = \text{diag}(x_{\xi 1}, x_{\xi 2}, \cdots, x_{\xi N})$. The eigenvalue spectra

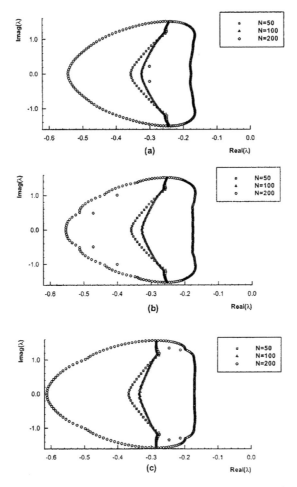

FIG. 3. The eigenvalue spectra on uniform grids: (a) WCNS-E-4, (b) WCNS-4, and (c) WCNS-5.

of WCNS on this stretching grid is showed in Fig. 4. We know from the figure that WCNS become unstable as the grid number equal to 50, a very strong stretched grid. With the increasing of the grid number, the relative stretching of the grid reduces and the schemes become stable. Therefore, in practical simulations we should pay attention to the strong stretching grids which may have destabilizing effect on WCNS. On the other hand, for the viscous flow simulations, the physical viscosity has the stabilizing effect, such that the overall schemes may be stable.

In practical calculations, the last two interpolants in Eq. (35) may be replaced by

$$\tilde{u}_{N-1/2} = \frac{1}{16}(-u_{N+1} + 9u_N + 9u_{N-1} - u_{N-2}),$$
$$\tilde{u}_{N+1/2} = \frac{1}{16}(5u_{N+1} + 15u_N - 5u_{N-1} + u_{N-2}). \tag{37}$$

The boundary and near boundary schemes (34), (35), and (37) can be used in the convective

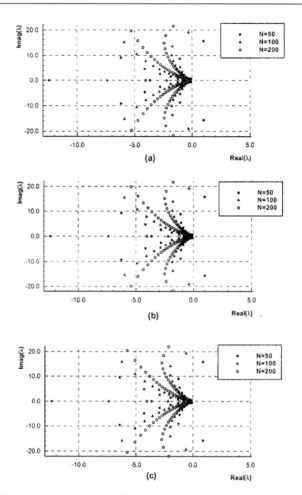

FIG. 4. The eigenvalues spectra on stretching grids: (a) WCNS-E-4, (b) WCNS-4, and (c) WCNS-5.

flux of Euler and Navier–Stokes equations directly, with u' replaced by E', $\tilde{u}$ by $\tilde{U}$, and u by U, respectively.

Upwind Flux Splitting Schemes

One advantage of WCNS is that the interpolation procedure is independent of the numerical flux calculations. As the cell-edge values have been interpolated, there are various ways to construct the numerical flux. We have used Roe's flux difference scheme,

$$\tilde{E}^{\text{Roe}}_{j+1/2} = \frac{1}{2}\left[E\left(\tilde{U}_{Rj+1/2}\right) + E\left(\tilde{U}_{Lj+1/2}\right) - |\tilde{A}|\left(\tilde{U}_{Rj+1/2} - \tilde{U}_{Lj+1/2}\right)\right], \quad (38)$$

in the CNS [8, 9]. In fact, the flux vector splitting schemes,

$$\tilde{E}^{\text{FV}}_{j+1/2} = E^{-}\left(\tilde{U}_{Rj+1/2}\right) + E^{+}\left(\tilde{U}_{Lj+1/2}\right), \quad (39)$$

can also be used. Here, $E^{\pm}$ may be obtained by Steger and Warming's or Van Leer's splitting or other flux vector splitting methods. Furthermore, the flux themselves can be used to interpolate the cell-edge values by (9)–(24) and get

$$\tilde{E}^F_{j+1/2} = \tilde{E}^+_{j+1/2} + \tilde{E}^-_{j+1/2};$$
$$\tilde{E}^+_{j+1/2} = E^+_j + \frac{h}{2}F^{+*}_j + \frac{1}{8}h^2 S^{+*}_j, \quad (40)$$
$$\tilde{E}^-_{j+1/2} = E^-_{j+1} - \frac{h}{2}F^{-*}_{j+1} + \frac{1}{8}h^2 S^{-*}_{j+1},$$

where $F^\pm = \partial E^\pm/\partial x$ and $S^\pm = \partial^2 E^\pm/\partial x^2$. In this paper, we compared Eqs. (38) and (39) in the Euler equation calculations.

Discretizations of Viscous Flux for Navier–Stokes Equations

In the above section, we only discussed the convective flux discretization. For the Navier–Stokes equations calculations, the cell-centered compact scheme (4) can also be applied to viscous flux, i.e.,

$$\kappa E'_{vj-1} + E'_{vj} + \kappa E'_{vj+1} = \frac{a}{h}(\tilde{E}_{vj+1/2} - \tilde{E}_{vj-1/2}) + \frac{b}{h}(\tilde{E}_{vj+3/2} - \tilde{E}_{vj-3/2}). \quad (41)$$

The viscous flux $\tilde{E}_{vj+1/2}$ at cell edges contains first-order derivatives of the dependent variables as well as the dependent variables themselves. We first use the high-order central type compact schemes to calculate these values, and then substitute them into Eq. (41) to obtain the derivatives of the viscous flux. In Appendix, we give the detailed schemes required for the viscous flux calculation at the cell edges.

Time Discretization

We treat Eq. (3) as an ordinary differential equation,

$$\frac{\partial U}{\partial t} = R(U), \quad (42)$$

and employ the third-order TVD Runge–Kutta method [14] for the time integration:

$$U^{(1)} = U^n + \Delta t R(U^n)$$
$$U^{(2)} = \frac{3}{4}U^n + \frac{1}{4}U^{(1)} + \frac{1}{4}\Delta t R(U^{(1)}) \quad (43)$$
$$U^{n+1} = \frac{1}{3}U^n + \frac{2}{3}U^{(2)} + \frac{2}{3}\Delta t R(U^{(2)}).$$

A fourth-order Runge–Kutta scheme is

$$U^{(1)} = U^n + \frac{1}{2}\Delta t R(U^n)$$
$$U^{(2)} = U^n + \frac{1}{2}\Delta t R(U^{(1)})$$
$$U^{(3)} = U^n + \Delta t R(U^{(2)}) \quad (44)$$
$$U^{n+1} = \frac{1}{3}(-U^n + U^{(1)} + 2U^{(2)} + U^{(3)}) + \frac{1}{6}\Delta t R(U^{(3)}).$$

It is not TVD type. In this paper, only Eq. (43) is used in the numerical tests.

used to interpolate the cell-edge values by (9)–(24) and get

$$\tilde{E}^F_{j+1/2} = \tilde{E}^+_{j+1/2} + \tilde{E}^-_{j+1/2};$$
$$\tilde{E}^+_{j+1/2} = E^+_j + \frac{h}{2} F^{+*}_j + \frac{1}{8} h^2 S^{+*}_j, \qquad (40)$$
$$\tilde{E}^-_{j+1/2} = E^-_{j+1} - \frac{h}{2} F^{-*}_{j+1} + \frac{1}{8} h^2 S^{-*}_{j+1},$$

where $F^{\pm} = \partial E^{\pm}/\partial x$ and $S^{\pm} = \partial^2 E^{\pm}/\partial x^2$. In this paper, we compared Eqs. (38) and (39) in the Euler equation calculations.

Discretizations of Viscous Flux for Navier–Stokes Equations

In the above section, we only discussed the convective flux discretization. For the Navier–Stokes equations calculations, the cell-centered compact scheme (4) can also be applied to viscous flux, i.e.,

$$\kappa E'_{vj-1} + E'_{vj} + \kappa E'_{vj+1} = \frac{a}{h} \left(\tilde{E}_{vj+1/2} - \tilde{E}_{vj-1/2} \right) + \frac{b}{h} \left(\tilde{E}_{vj+3/2} - \tilde{E}_{vj-3/2} \right). \qquad (41)$$

The viscous flux $\tilde{E}_{vj+1/2}$ at cell edges contains first-order derivatives of the dependent variables as well as the dependent variables themselves. We first use the high-order central type compact schemes to calculate these values, and then substitute them into Eq. (41) to obtain the derivatives of the viscous flux. In Appendix, we give the detailed schemes required for the viscous flux calculation at the cell edges.

Time Discretization

We treat Eq. (3) as an ordinary differential equation,

$$\frac{\partial U}{\partial t} = R(U), \qquad (42)$$

and employ the third-order TVD Runge–Kutta method [14] for the time integration:

$$\begin{aligned} U^{(1)} &= U^n + \Delta t R(U^n) \\ U^{(2)} &= \frac{3}{4} U^n + \frac{1}{4} U^{(1)} + \frac{1}{4} \Delta t R(U^{(1)}) \\ U^{n+1} &= \frac{1}{3} U^n + \frac{2}{3} U^{(2)} + \frac{2}{3} \Delta t R(U^{(2)}). \end{aligned} \qquad (43)$$

A fourth-order Runge–Kutta scheme is

$$\begin{aligned} U^{(1)} &= U^n + \frac{1}{2} \Delta t R(U^n) \\ U^{(2)} &= U^n + \frac{1}{2} \Delta t R(U^{(1)}) \\ U^{(3)} &= U^n + \Delta t R(U^{(2)}) \\ U^{n+1} &= \frac{1}{3} \left(-U^n + U^{(1)} + 2U^{(2)} + U^{(3)} \right) + \frac{1}{6} \Delta t R(U^{(3)}). \end{aligned} \qquad (44)$$

It is not TVD type. In this paper, only Eq. (43) is used in the numerical tests.

3. NUMERICAL TESTS

In this section, we test the behaviors of the WCNS for several examples. The tests contain two groups. One is the Euler equation solutions, which are selected to show the ability of WCNS to resolve discontinuities. The another is the Navier–Stokes solutions which show the accuracy of WCNS in smooth regions, especially for the boundary layer simulations. The Roe flux difference scheme is used in one-dimensional cases.

EXAMPLE 1. The one-dimensional Riemann problem for the Euler equations of gas dynamics is solved with the initial conditions

$$(\rho_L, u_L, p_L) = (1, 0, 1), \quad (\rho_R, u_R, p_R) = (0.125, 0, 0.1). \tag{45}$$

In the calculation, we used the characteristic interpolations with 100 grid points, CFL = 0.2, and 200 time steps. The calculated density results are given in Fig. 5. Three WCNS(-S) give almost the same results (the suffix "-S" means the sharpening technique being used). The sharpening technique with $\sigma = 12$ can steepen the contact discontinuity at 3–4 grid points; here σ is a sharpening parameter. In [11], the details of the contact discontinuity sharpening technique can be found. Compared with the results calculated by WENO schemes [13], our resolution of the corner of rarefaction waves (discontinuities in derivatives) are improved.

EXAMPLE 2. This is the same equations as in Example 1 with the initial conditions

$$U^0(x) = \begin{cases} U_L, & 0 \leq x < 0.1 \\ U_M, & 0.1 \leq x < 0.9 \\ U_R, & 0.9 \leq x < 1, \end{cases} \tag{46}$$

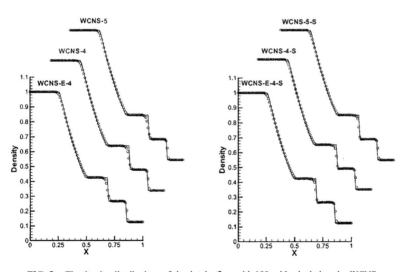

FIG. 5. The density distributions of shock tube flow with 100 grid calculations by WCNS.

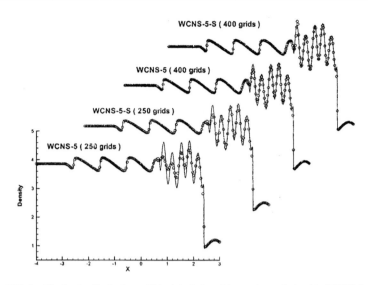

FIG. 8. The density distributions of "shock/turbulence" interactions calculated by WCNS-5.

middle contact discontinuity, are resolved well. Compared with fifth-order WENO results [13] with Yang's artificial compressing technique, the middle contact discontinuity of our results is captured more sharply.

EXAMPLE 3. The third example is a model problem for the shock/turbulence interaction. It originates from the following initial conditions:

$$(\rho, u, p) = \begin{cases} (3.857143, 2.629369, 10.33333) & x < 4, \\ (1 + 0.2 \sin 5x, 0, 1) & x \geq 4. \end{cases} \quad (47)$$

Figure 8 shows the density results calculated by WCNS-5-(S) with 250 and 400 grid points and $\sigma = 7$. The main fine structure was resolved on 250 grid points by the sharpening technique. The solid lines in this figure were calculated by CNS [8, 9] with 1600 grid points. It can be regarded as the exact solution.

EXAMPLE 4. The fourth example is two-dimensional wind tunnel flow with a step at Mach 3. Woodward and Colella [19] investigated this flow carefully. The wind tunnel is 1 length unit wide and 3 length units long. The step is 0.2 length units high and located 0.6 length unit from the left-hand end of the tunnel. The problem is initialized by a right-moving Mach 3 flow. Reflective boundary conditions are imposed at wind tunnel walls and step surface. In-flow and out-flow boundary conditions are applied at the left-hand and right-hand sides, respectively. For the treatment of the corner of the step, the method suggested in [19] is adopted.

In the calculations, two grids were used. The first one is the medium grid with 121×41 grid points and the fine grid contains 241×81 grid meshes. CFL $= 0.6$ is used for all the calculations. Figures 9 and 10 show the density contours calculated by WCNS-E-4, WCNS-4, and WCNS-5 on these two grids. The Steger–Warmings flux vector splitting method was used here. The results show that WCNS perform well for this example. They have good

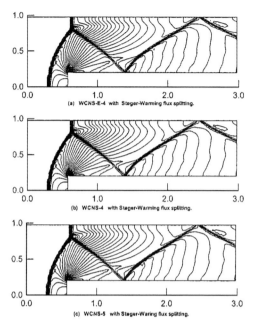

FIG. 9. Density contrours of flow past a step on a medium grid, 121×41, at $t = 4$: (a) WCNS-E-4, (b) WCNS-4, and (c) WCNS-5.

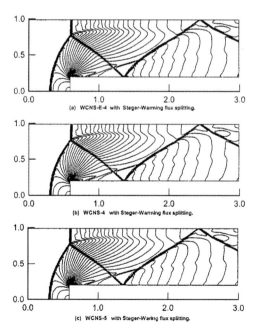

FIG. 10. Density contours of flow past a step on a fine grid, 241×81, at $t = 4$: (a) WCNS-E-4, (b) WCNS-4, and (c) WCNS-5.

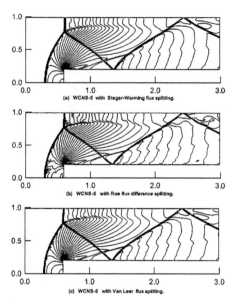

FIG. 11. Density contours of flow past a step on a fine grid, 241×81, at $t = 4$ with: (a) WCNS-5 Steger–Warming flux splitting, (b) WCNS-5 Roe flux difference splitting, and (c) WCNS-5 Van Leer flux splitting.

resolutions for the shock and contact discontinuity. The results on the fine grid are almost the same as that of the PPM scheme [19]. Furthermore, there are almost no differences between WCNS-4 and WCNS-5, which confirms that the weighted interpolation dominates the accuracy of WCNS.

Figure 11 shows the results calculated by WCNS-5 on a fine grid with different flux splitting methods. We know from this figure that Steger–Warming and Van Leer flux vector splitting gave more smooth solutions than Roe flux difference scheme for this problem.

EXAMPLE 5. The last example is steady hypersonic viscous flow around a circular cylinder. We select this problem because there is a spectral solution [20] that can be used for the comparison, especially for the heat transfer on the body surface. The flow conditions are $M_\infty = 5.73$, Re $= 2050$, $T_\infty^* = 39.6698$ K, $T_w^* = 210.2$ K, $\gamma = 1.4$, Pr $= 0.77$, and the cylinder radius $r^* = 0.0061468$ m.

Figure 12 gives the results by WCNS-5 with Steger–Warming flux splitting on a 61×61 grid. It can be seen that the bow shock is captured well and the pressure coefficient and heat transfer in Fig. 13 compare well with the spectral solution. In order to investigate the grid convergence, we repeated the calculations on two other grids 31×31 and 15×21. The pressure coefficient and heat transfer on the body surface shown in Fig. 14 compare well with the spectral solution on the three grids, and Fig. 15 shows the convergent history. All the residuals on these three grids converge to machine zero.

For the comparison of the WCNS with the high resolution TVD type or MUSCL schemes, we calculated this flow by MUSCL scheme on 15×21 grid. Figure 16 shows the surface pressure coefficient and heat transfer. Though the MUSCL scheme can give a good pressure coefficient, the heat transfer compares poorly with the WCNS and spectral solutions. This confirms the high-order accuracy of WCNS.

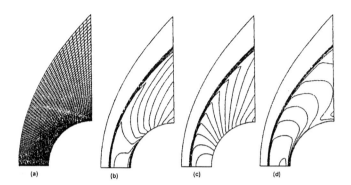

FIG. 12. Contours calculated by WCNS-5 with 61×61 grids: (a) grid, (b) density, (c) pressure, and (d) temperature.

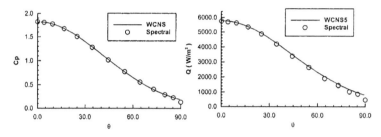

FIG. 13. The surface pressure (C_p) and heat transfer (Q) distribution on grid 61×61 compared with the spectral method.

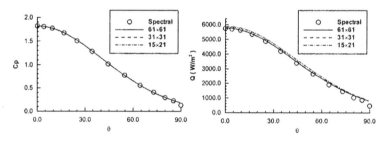

FIG. 14. The surface pressure (C_p) and heat transfer (Q) distribution on various grids compared with the spectral method.

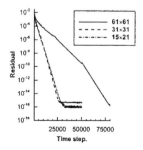

FIG. 15. The convergent history on various grids.

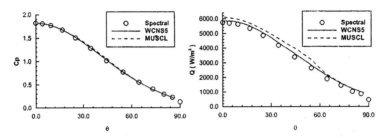

FIG. 16. The surface pressure (C_p) and heat transfer (Q) distribution on grid 15×21 compared with MUSCL scheme.

4. CONCLUDING REMARKS

Weighted compact nonlinear schemes (WCNS) are developed in this paper. By Fourier analysis, the dispersive and dissipative features of WCNS in smooth regions were discussed, which showed that the interpolations of variables at cell edges dominate the accuracy of the schemes and WCNS have almost same accuracy as the fifth-order upwind biased explicit linear scheme in smooth regions. Asymptotic stability analysis shows that the boundary and near boundary schemes developed in this paper, as well as the interior schemes, are asymptotically stable on the uniform grid. On the stretching grids, the schemes are also asymptotically stable, as the stretching is not very strong. One of the advantages of WCNS is that the dependent variable interpolation is independent of the flux splitting schemes, so that we have the freedom to select various flux splitting schemes according to various flows. One- and two-dimensional Euler equation calculations showed that WCNS can capture discontinuities robustly. Hypersonic steady viscous flow simulations showed that WCNS can give results comparable to the spectral method solution and the residuals can converge to machine zero. Furthermore, the viscous numerical results confirmed that the WCNS are more accurate than the MUSCL scheme, as expected.

In this paper, we only showed the performance of WCNS for one- and two-dimensional calculations. Further research is to apply them to three-dimensional Navier–Stokes equations for the DNS of hypersonic stability problems, large eddy simulations (LES) of complex turbulent flows to show the accuracy of WCNS for the resolution of the fine structures.

APPENDIX

For the viscous term calculations, the following derivative schemes and interpolation schemes are needed.

(1) The compact first-order derivatives at cell edges:

$$u'_{1/2} + \alpha_1 u'_{3/2} = \frac{1}{h}(a_1 u_0 + b_1 u_1 + c_1 u_2 + d_1 u_3 + e_1 u_4),$$

$$\kappa u'_{j-1/2} + u'_{j+1/2} + \kappa u'_{j+3/2} = \frac{a}{h}(u_{j+1} - u_j) + \frac{b}{h}(u_{j+2} - u_{j-1}), \qquad (A1)$$

$$u'_{N+1/2} + \alpha_1 u'_{N-1/2} = -\frac{1}{h}(a_1 u_N + b_1 u_{N-1} + c_1 u_{N-1} + d_1 u_{N-2} + e_1 u_{N-3}).$$

The parameters in these schemes are the same as in (31) and (34).

(2) The compact first-order derivatives at grid nodes:

$$u'_1 + 3u'_2 = \frac{1}{6h}(-17u_1 + 9u_2 + 9u_3 - u_4),$$

$$\frac{1}{4}u'_1 + u'_2 + \frac{1}{4}u'_3 = \frac{3}{4h}(u_3 - u_1),$$

$$\frac{1}{3}u'_{j-1} + u'_j + \frac{1}{3}u'_{j+1} = \frac{1}{36h}(u_{j+2} - u_{j-2}) + \frac{7}{9h}(u_{j+1} - u_{j-1}), \quad (A2)$$

$$\frac{1}{4}u'_{N-2} + u'_{N-1} + \frac{1}{4}u'_N = \frac{3}{4h}(u_N - u_{N-2}),$$

$$u'_N + 3u'_{N-1} = -\frac{1}{6h}(-17u_N + 9u_{N-1} + 9u_{N-2} - u_{N-3}).$$

(3) The compact interpolation at cell edges:

$$\alpha u_{j-1/2} + u_{j+1/2} + \alpha u_{j+3/2} = \frac{10\alpha + 9}{16}(u_{j+1} + u_j) + \frac{6\alpha - 1}{16}(u_{j+2} + u_{j-1}). \quad (A3)$$

As $\alpha = \frac{3}{10}$ the interpolant is sixth-order accurate. Two sets of boundary schemes can be derived:

$$u_{1/2} = \frac{1}{16}(35u_1 - 35u_2 + 21u_3 - 5u_4),$$

$$u_{3/2} = \frac{1}{16}(5u_1 + 15u_2 - 5u_3 + u_4),$$

$$u_{N-1/2} = \frac{1}{16}(5u_N + 15u_{N-1} - 5u_{N-2} + u_{N-3}), \quad (A3.1)$$

$$u_{N+1/2} = \frac{1}{16}(35u_N + 35u_{N-1} + 21u_{N-2} - 5u_{N-3}),$$

and

$$u_{1/2} = \frac{1}{16}(5u_0 + 15u_1 - 5u_2 + u_3),$$

$$u_{N+1/2} = \frac{1}{16}(5u_{N+1} + 15u_N - 5u_{N-1} + u_{N-2}), \quad (A3.2)$$

(4) The compact interpolation at grid nodes:

$$u_1 = \frac{1}{16}(5u_{1/2} + 15u_{3/2} - 5u_{5/2} + u_{7/2}),$$

$$\alpha u_{j-1} + u_j + \alpha u_{j+1} = \frac{10\alpha + 9}{16} + (u_{j+1/2} + u_{j-1/2}) + \frac{6\alpha - 1}{16}(u_{j+3/2} + u_{j-3/2}),$$

$$u_N = \frac{1}{16}(5u_{N+1/2} + 15u_{N-1/2} - 5u_{N-3/2} + u_{N-5/2}). \quad (A4)$$

Also, as $\alpha = \frac{3}{10}$ the interior interpolant is sixth-order accurate.

ACKNOWLEDGMENTS

This work was supported by the project of Basic Research on Frontier Problems in Fluid and Aerodynamics in China and the China National Natural Science Foundation. The first author acknowledges associate professor M. L. Mao discussing the problem with him.

REFERENCES

1. S. K. Lele, Compact finite difference schemes with spectral-like resolution, *J. Comput. Phys.* **103**, 16 (1992).
2. R. V. Wilson, A. Q. Demuren, and M. H. Carpenter, High-order compact schemes for numerical simulations of incompressible flows, ICASE Report 98-13 ICASE, Langley Research Center (1998).
3. L. M. Leslie and R. J. Purser, Three-dimensional mass-conserving semi-Lagrangian scheme employing forward tajectories, *Mon. Weather Rev.* **123**, 2551 (1995).
4. V. A. Garanzha and Konshin, Numerical algorithms for viscous incompressible fluid flows based on the high-order conservative compact schemes, *Comput. Math. Math. Phys.* **8**, 1321 (1999).
5. D. Gaitonde and J. S. Shang, Optimized compact-difference-based finite-volume schemes for linear wave phenomena, *J. Comput. Phys.* **138**, 617 (1997).
6. M. H. Kobayashi, On a class of Pade finite volume methods, *J. Comput. Phys.* **156**, 137 (1999).
7. B. Cockburn and C. W. Shu, Nonlinearly stable compact schemes for shock calculations, *SIAM J. Numer. Anal.* **31**, 607 (1994).
8. X. G. Deng and H. Maekawa, Compact high-order accurate nonlinear schemes, *J. Comput. Phys.* **130**, 77 (1997).
9. X. G. Deng and H. Maekawa, An uniform fourth-order nonlinear compact schemes for discontinuities capturing, in *AIAA paper 96-1974, 27th AIAA Fluid Dynamics Meeting* (New Orleans, LA, AIAA, 1996).
10. X. G. Deng, H. Maekawa, and Q. Shen, A class of high-order dissipative compact schemes, in *AIAA paper 96-1972, 27th AIAA Fluid Dynamics Meeting* (New Orleans, LA, AIAA, 1996).
11. X. G. Deng and M. L. Mao, Weighted compact high-order nonlinear schemes for the Euler equations, AIAA Paper 97-1941, presented of the 13th AIAA Computational Fluid Dynamic Conference (Snowmass).
12. X. D. Liu, S. Osher, and T. Chan, Weighted essentially non-oscillatory schemes, *J. Comput. Phys.* **115**, 200 (1994).
13. G. S. Jiang and C. W. Shu, Efficient implementation of weighted ENO schemes, ICASE Report, No. 95-73, 1995; also NASA CR 198228 (1995).
14. C. W. Shu and S. Osher, Efficient implementation of essentially non-oscillatory shock-capturing schemes II, *J. Comput. Phys.* **83**, 32 (1989).
15. H. T. Huynh, Accurate upwind methods for the Euler equations, *SIAM J. Numer. Anal.* **32**(5), 1565 (1995).
16. B. Gustaffson, The convergence rate for difference approximations to mixed initial boundary value problems, *Math. Comput.* **49**(130), 396 (1975).
17. G. O. Roberts, Computational meshes for the boundary problems, in *Proc. Second International Conf. Numerical Methods Fluid Dyn.*, Lecture Notes in Physics (Springer–Verlag, New York, 1971), p. 171.
18. D. Gaitonde and J. S. Shang, Accuracy of flux-split algorithms in high-speed viscous flows, *AIAA J.* **31**, 1215 (1993).
19. P. Woodward and P. Collela, The numerical simulation of two-dimensional fluid flow with strong shocks, *J. Comput. Phys.* **54**, 115 (1984).
20. D. A. Kopriva, Spectral solution of the viscous blunt-body problem, *AIAA J.* **31**, 1235 (1993).

Geometric conservation law and applications to high-order finite difference schemes with stationary grids

Xiaogang Deng *, Meiliang Mao, Guohua Tu, Huayong Liu, Hanxin Zhang

State Key Laboratory of Aerodynamics, P.O. Box 211, Mianyang 621000, PR China
China Aerodynamics Research & Development Center, P.O. Box 211, Mianyang 621000, PR China

ARTICLE INFO

Article history:
Received 11 October 2009
Received in revised form 16 October 2010
Accepted 20 October 2010
Available online 17 November 2010

Keywords:
Geometric conservation laws
Surface conservation law
High-order finite difference schemes
WCNS high-order schemes
Curvilinear coordinates

ABSTRACT

The geometric conservation law (GCL) includes the volume conservation law (VCL) and the surface conservation law (SCL). Though the VCL is widely discussed for time-depending grids, in the cases of stationary grids the SCL also works as a very important role for high-order accurate numerical simulations. The SCL is usually not satisfied on discretized grid meshes because of discretization errors, and the violation of the SCL can lead to numerical instabilities especially when high-order schemes are applied. In order to fulfill the SCL in high-order finite difference schemes, a conservative metric method (CMM) is presented. This method is achieved by computing grid metric derivatives through a conservative form with the same scheme applied for fluxes. The CMM is proven to be a sufficient condition for the SCL, and can ensure the SCL for interior schemes as well as boundary and near boundary schemes. Though the first-level difference operators δ_3 have no effects on the SCL, no extra errors can be introduced as $\delta_3 = \delta_2$. The generally used high-order finite difference schemes are categorized as central schemes (CS) and upwind schemes (UPW) based on the difference operator δ_1 which are used to solve the governing equations. The CMM can be applied to CS and is difficult to be satisfied by UPW. Thus, it is critical to select the difference operator δ_1 to reduce the SCL-related errors. Numerical tests based on WCNS-E-5 show that the SCL plays a very important role in ensuring free-stream conservation, suppressing numerical oscillations, and enhancing the robustness of the high-order scheme in complex grids.

© 2010 Elsevier Inc. All rights reserved.

1. Introduction

With the gradually wide applications of high-order finite difference schemes for the Large Eddy Simulations (LES), Detached Eddy Simulations (DES) and Direct Numerical Simulations (DNS) of turbulences and aeroacoustics, it is important to study and minimize any discretization errors which are not noticeable or could be neglected for lower-order schemes. Recent researches [9,19] showed that if the geometric conservation law (GCL) has not been satisfied, it will cause large errors and even results in numerical instabilities for high-order schemes. Furthermore, as it is well know, one of the most difficult problems which hinder high-order schemes from complex configurations is the grid qualities. However, it is extremely difficult to generate high quality single-block grids suitable for high-order schemes when the configuration is complex. This problem can also be largely mitigated by the fulfillment of the GCL in high-order multi-block grid computations.

The concept of the GCL was first introduced by Trulio and Trigger [1]. Pulliam and Steger [2] observed that the metric discretization will lead to the nonconservation of the governing equations. They pointed out that the discretization-based errors of metric derivatives shall be avoided by some special strategy. Thomas and Lombard [3] extended the conception of the GCL to a general application and showed that the grid motion-induced errors are responsible for some numerical oscillations. Zhang et al. [16] pointed out that the GCL includes the surface conservation law (SCL) and the volume conservation law (VCL). The numerical violation of the SCL will lead to a misrepresentation of the convective velocities, while the errors from the unsatisfied VCL will produce extra sources or sinks in the physically conservative media. Étienne, et al. [4] concluded that a numerical method satisfying the GCL will generally allow a much large computational time step. Viviand and Ghazzi [10] showed that numerical instabilities appeared as the conservative form of governing equations were solved. The VCL has been widely discussed by many authors, for example, Étienne et al. [4], Mavriplis and Yang [5] and Farhat et al. [6] discussed the GCL for Arbitrary Lagrangian Eulerian (ALE) methods. In order to satisfy the GCL, Pulliam and Steger [18] employed a simple averaging method while Vinokur [20] proposed a finite-volume like method. These two methods, which work well for their special second-order schemes, are not suitable for high-order schemes [9]. Cai and Ladeinde [21] modified the governing equations to preserve the free-stream conditions by high-order Weighted Essentially Non-Oscillatory (WENO) scheme. The errors, however, are large for the high-order accuracy computations though the conservative forms of the metric derivatives suggested by Thomas and Lombard [3] were adopted. Visbal and Gaitonde [9] carefully studied the metric evaluation errors for the high-order central type compact schemes and found the errors can be largely decreased by means of the conservative forms of metric derivatives and calculating the metric derivatives with the same scheme as calculating the flow-flux derivatives. Recently, Nonomura et al. [19] successfully applied Visbal and Gaitonde's technique to Weighted Compact Nonlinear Schemes (WCNS) derived by Deng and Zhang [11] and showed the excellent free-stream and vortex preservation properties of WCNS compared with those of WENO [15] for the first time. Visbal and Gaitonde [9] and Nonomura et al. [19], however, did not give theoretical analysis about their techniques.

The WCNS were first derived by Deng and Zhang [11] for Euler and Navier–Stokes equations. Recently Deng [12,13], Zhang et al.[25] and Nonomura et al. [26,27] have constructed various high-order WCNSs based on the thoughts of [11]. The WCNS-E-5 [11–13], which is a typical explicit one of WCNSs, has been successfully applied to a wide range of flow simulations so far to show its flexibility and robustness by Liu et al. [17], Nonomura et al.[19], Ishiko et al. [28], and Deng et al. [29,36]. In present work, we continue our efforts to apply the WCNS-E-5 schemes for complex flows over complex configurations. In Section 2, we give a theoretical analysis to satisfy the GCL for finite difference schemes at stationary grids. The method based on this analysis is referred as "conservative metric method (CMM)". In Section 3, applications of the CMM for various typical high-order finite difference linear and nonlinear schemes are discussed. The generally used finite difference schemes are categorized as central schemes (CS) and upwind schemes (UPW) based on the difference operators δ_1 which are used to solve the derivatives of flow fluxes. Some numerical tests including a spherical Riemann problem and turbulent flows over a blunt configuration with dual backward-facing steps were solved by the WCNS-E-5 with the CMM in Section 4. Furthermore, a method to ensure free-stream preservation for high-order upwind compact scheme is tested in Section 4.

2. The conservative metric method (CMM) for the surface conservation law (SCL)

2.1. The analysis of the geometric conservation law (GCL) for finite difference schemes

In this section, we give a detailed analysis about the GCL for finite difference schemes. In Cartesian coordinates (x,y,z), the three-dimensional nondimensional Navier–Stokes equations are

$$\frac{\partial Q}{\partial t} + \frac{\partial E}{\partial x} + \frac{\partial F}{\partial y} + \frac{\partial G}{\partial z} = \frac{\partial E_v}{\partial x} + \frac{\partial F_v}{\partial y} + \frac{\partial G_v}{\partial z}, \qquad (1)$$

where the vectors

$$Q = [\rho, \rho u, \rho v, \rho w, \rho e]^T,$$

$$\begin{cases} E = [\rho u, \rho u^2 + p, \rho uv, \rho uw, (\rho e + p)u]^T, \\ F = [\rho v, \rho uv, \rho v^2 + p, \rho vw, (\rho e + p)v]^T, \\ G = [\rho w, \rho uw, \rho vw, \rho w^2 + p, (\rho e + p)w]^T, \end{cases}$$

$$\begin{cases} E_v = \frac{1}{Re}[0, \tau_{xx}, \tau_{xy}, \tau_{xz}, \beta_x]^T, \\ F_v = \frac{1}{Re}[0, \tau_{xy}, \tau_{yy}, \tau_{yz}, \beta_y]^T, \\ G_v = \frac{1}{Re}[0, \tau_{xz}, \tau_{yz}, \tau_{zz}, \beta_z]^T \end{cases}$$

are dependent variables, convective fluxes and viscous fluxes respectively, and

$$p = \frac{1}{\gamma M_\infty^2} \rho T, \quad e = \frac{p}{(\gamma - 1)\rho} + \frac{u_m u_m}{2},$$

$$\tau_{ij} = \mu\left(\frac{\partial u_i}{\partial x_j} + \frac{\partial u_j}{\partial x_i} - \frac{2}{3}\delta_{ij}\frac{\partial u_m}{\partial x_m}\right), \quad \beta_i = \tau_{im}u_m + \frac{\mu}{(\gamma-1)M_\infty^2 \Pr}\frac{\partial T}{\partial x_i},$$

where, γ is the ratio of specific heats, and the viscous coefficient μ can be calculated by the Sutherland's law. The nondimensional variables are defined as $\rho = \rho^*/\rho_\infty^*$, $u_i = u_i^*/u_\infty^*$, $T = T^*/T_\infty^*$, $p_i = p_i^*/\rho_\infty^* u_\infty^{*2}$, and $M_\infty = u_\infty^*/\sqrt{\gamma R T_\infty^*}$, $\mathrm{Re} = \rho_\infty^* u_\infty^* r^*/\mu_\infty^*$, and $\Pr = \mu_\infty^* C_p/\kappa_\infty^*$ are the Mach number, Reynolds number and Prandtl number, respectively. r^* is the characteristic length.

The governing Eq. (1) can be transformed in curvilinear coordinates (ξ, η, ζ, τ)

$$\frac{\partial \tilde{Q}}{\partial \tau} + \frac{\partial(\tilde{E}-\tilde{E}_v)}{\partial \xi} + \frac{\partial(\tilde{F}-\tilde{F}_v)}{\partial \eta} + \frac{\partial(\tilde{G}-\tilde{G}_v)}{\partial \zeta} = 0, \tag{2}$$

where,

$$\tilde{Q} = Q/J,$$

$$\tilde{E} = \tilde{\xi}_t Q + \tilde{\xi}_x E + \tilde{\xi}_y F + \tilde{\xi}_z G, \quad \tilde{F} = \tilde{\eta}_t Q + \tilde{\eta}_x E + \tilde{\eta}_y F + \tilde{\eta}_z G, \quad \tilde{G} = \tilde{\zeta}_t Q + \tilde{\zeta}_x E + \tilde{\zeta}_y F + \tilde{\zeta}_z G,$$

$$\tilde{E}_v = \tilde{\xi}_x E_v + \tilde{\xi}_y F_v + \tilde{\xi}_z G_v, \quad \tilde{F}_v = \tilde{\eta}_x E_v + \tilde{\eta}_y F_v + \tilde{\eta}_z G_v, \quad \tilde{G}_v = \tilde{\zeta}_x E_v + \tilde{\zeta}_y F_v + \tilde{\zeta}_z G_v.$$

The Jacobian J and the traditional standard metrics are

$$J^{-1} = \left|\frac{\partial(x,y,z)}{\partial(\xi,\eta,\zeta)}\right| = \begin{vmatrix} x_\xi & x_\eta & x_\zeta \\ y_\xi & y_\eta & y_\zeta \\ z_\xi & z_\eta & z_\zeta \end{vmatrix},$$

$$\begin{cases} \tilde{\xi}_x = \xi_x/J = y_\eta z_\zeta - z_\eta y_\zeta, & \tilde{\xi}_y = \xi_y/J = z_\eta x_\zeta - x_\eta z_\zeta, & \tilde{\xi}_z = \xi_z/J = x_\eta y_\zeta - y_\eta x_\zeta, \\ \tilde{\eta}_x = \eta_x/J = y_\zeta z_\xi - z_\zeta y_\xi, & \tilde{\eta}_y = \eta_y/J = z_\zeta x_\xi - x_\zeta z_\xi, & \tilde{\eta}_z = \eta_z/J = x_\zeta y_\xi - y_\zeta x_\xi, \\ \tilde{\zeta}_x = \zeta_x/J = y_\xi z_\eta - z_\xi y_\eta, & \tilde{\zeta}_y = \zeta_y/J = z_\xi x_\eta - x_\xi z_\eta, & \tilde{\zeta}_z = \zeta_z/J = x_\xi y_\eta - y_\xi x_\eta. \end{cases} \tag{3}$$

In uniform flow regions, i.e., Q, E, F, and G are constants, and $E_v = F_v = G_v = 0$. Eq. (2) are simplified as

$$\frac{\partial \tilde{Q}}{\partial \tau} = -[Q(I_t - (1/J)_\tau) + I_x E + I_y F + I_z G], \tag{4}$$

where,

$$\begin{cases} I_t = (1/J)_\tau + \left(\tilde{\xi}_t\right)_\xi + (\tilde{\eta}_t)_\eta + \left(\tilde{\zeta}_t\right)_\zeta, \\ I_x = \left(\tilde{\xi}_x\right)_\xi + (\tilde{\eta}_x)_\eta + \left(\tilde{\zeta}_x\right)_\zeta, \\ I_y = \left(\tilde{\xi}_y\right)_\xi + (\tilde{\eta}_y)_\eta + \left(\tilde{\zeta}_y\right)_\zeta, \\ I_z = \left(\tilde{\xi}_z\right)_\xi + (\tilde{\eta}_z)_\eta + \left(\tilde{\zeta}_z\right)_\zeta. \end{cases} \tag{5}$$

As discussed by Zhang et al. [16], $I_t = 0$ means the volume conservation law (VCL), and $I_x = I_y = I_z = 0$ means the surface conservation law (SCL). In this paper, we only analyze the stationary grids, i.e., $\tilde{\xi}_t = \tilde{\eta}_t = \tilde{\zeta}_t = 0$, and $\tau = t$. By substituting (3) into (5), we can see

$$\begin{cases} I_x = (y_\eta z_\zeta - z_\eta y_\zeta)_\xi + (y_\zeta z_\xi - z_\zeta y_\xi)_\eta + (y_\xi z_\eta - z_\xi y_\eta)_\zeta = 0, \\ I_y = (z_\eta x_\zeta - x_\eta z_\zeta)_\xi + (z_\zeta x_\xi - x_\zeta z_\xi)_\eta + (z_\xi x_\eta - x_\xi z_\eta)_\zeta = 0, \\ I_z = (x_\eta y_\zeta - y_\eta x_\zeta)_\xi + (x_\zeta y_\xi - y_\zeta x_\xi)_\eta + (x_\xi y_\eta - y_\xi x_\eta)_\zeta = 0. \end{cases} \tag{6}$$

Then

$$\frac{\partial \tilde{Q}}{\partial \tau} = 0. \tag{7}$$

So the uniform flow conditions are hold. The condition (6) is called the surface conservation law (SCL) [16]. Although the SCL (6) is hold theoretically, i.e., all the derivatives in (2) and (3) are calculated analytically, they may not be true numerically. Following we give a detailed analysis about these features.

Suppose numerical derivative operators δ_1^ξ, δ_1^η, and δ_1^ζ, which may be different finite difference schemes in ξ, η and ζ coordinate direction respectively, are used for the discretization of the fluxes in Eq. (2), and δ_2^ξ, δ_2^η, and δ_2^ζ, are used for the metric calculations of (3). At grid node (i,j,k), Eqs. (2), (4) and (5) are

$$\left(\frac{\partial \tilde{Q}}{\partial \tau}\right)_{i,j,k} + \delta_1^\xi \left(\tilde{E} - \tilde{E}_v\right)_{i,j,k} + \delta_1^\eta \left(\tilde{F} - \tilde{F}_v\right)_{i,j,k} + \delta_1^\zeta \left(\tilde{G} - \tilde{G}_v\right)_{i,j,k} = 0, \tag{8}$$

$$\left(\frac{\partial \tilde{Q}}{\partial \tau}\right)_{i,j,k} = -\left(I_x^N E + I_y^N F + I_z^N G\right)_{i,j,k}, \tag{9}$$

$$\begin{cases} I_x^N = \delta_1^\xi \left(\tilde{\xi}_x^N\right) + \delta_1^\eta \left(\tilde{\eta}_x^N\right) + \delta_1^\zeta \left(\tilde{\zeta}_x^N\right), \\ I_y^N = \delta_1^\xi \left(\tilde{\xi}_y^N\right) + \delta_1^\eta \left(\tilde{\eta}_y^N\right) + \delta_1^\zeta \left(\tilde{\zeta}_y^N\right), \\ I_z^N = \delta_1^\xi \left(\tilde{\xi}_z^N\right) + \delta_1^\eta \left(\tilde{\eta}_z^N\right) + \delta_1^\zeta \left(\tilde{\zeta}_z^N\right). \end{cases} \tag{10}$$

The superscript "N" denotes the numerical values, For example

$$\begin{cases} \tilde{\xi}_x^N = (\delta_2^\eta y)(\delta_2^\zeta z) - (\delta_2^\zeta y)(\delta_2^\eta z), \\ \tilde{\eta}_x^N = (\delta_2^\zeta y)(\delta_2^\xi z) - (\delta_2^\xi y)(\delta_2^\zeta z), \\ \tilde{\zeta}_x^N = (\delta_2^\xi y)(\delta_2^\eta z) - (\delta_2^\eta y)(\delta_2^\xi z). \end{cases} \tag{11}$$

By substituting (11) into (10), it is obvious that it can not ensure $I_x^N = 0$ numerically for curvilinear grids and nonuniform grids. Thus the SCL is usually violated numerically. To settle this problem, Thomas and Lombard [3] had suggested the following conservative metrics,

$$\begin{cases} \tilde{\xi}_x = (y_\eta z)_\zeta - (y_\zeta z)_\eta, & \tilde{\xi}_y = (z_\eta x)_\zeta - (z_\zeta x)_\eta, & \tilde{\xi}_z = (x_\eta y)_\zeta - (x_\zeta y)_\eta, \\ \tilde{\eta}_x = (y_\zeta z)_\xi - (y_\xi z)_\zeta, & \tilde{\eta}_y = (z_\zeta x)_\xi - (z_\xi x)_\zeta, & \tilde{\eta}_z = (x_\zeta y)_\xi - (x_\xi y)_\zeta, \\ \tilde{\zeta}_x = (y_\xi z)_\eta - (y_\eta z)_\xi, & \tilde{\zeta}_y = (z_\xi x)_\eta - (z_\eta x)_\xi, & \tilde{\zeta}_z = (x_\xi y)_\eta - (x_\eta y)_\xi. \end{cases} \tag{12}$$

Thus Eq. (10) are changed as, for example,

$$I_x^N = \delta_1^\xi \delta_2^\eta \left((\delta_3^\zeta y)z\right) - \delta_1^\xi \delta_2^\zeta \left((\delta_3^\eta y)z\right) + \delta_1^\eta \delta_2^\zeta \left((\delta_3^\xi y)z\right) - \delta_1^\eta \delta_2^\xi \left((\delta_3^\zeta y)z\right) + \delta_1^\zeta \delta_2^\xi \left((\delta_3^\eta y)z\right) - \delta_1^\zeta \delta_2^\eta \left((\delta_3^\xi y)z\right). \tag{13}$$

Here, the derivative operators δ_3 are introduced to calculate the first-level metric derivatives. It can be seen clearly, if the cross-derivative operators, δ_1 and δ_2 in (13), are exchangeable, i.e.,

$$\delta_1^\xi \delta_2^\eta = \delta_1^\eta \delta_2^\xi, \quad \delta_1^\xi \delta_2^\zeta = \delta_1^\zeta \delta_2^\xi, \quad \text{and} \quad \delta_1^\eta \delta_2^\zeta = \delta_1^\zeta \delta_2^\eta, \tag{14}$$

then $I_x^N = 0$. Following we discuss on what numerical conditions Eq. (14) can be hold.

For any function $\phi_{i,j,k} = \phi(\xi_i, \eta_j, \zeta_k)$, the numerical derivate operators $\delta_1^\xi, \delta_1^\zeta, \delta_2^\xi$, and δ_2^ζ can be written as the following general forms,

$$\delta_1^\xi \phi_{i,j,k} = \sum_{p=M_1}^{N_1} a_p \left(\phi_{i+p+1,j,k} - \phi_{i+p,j,k}\right), \quad \delta_1^\zeta \phi_{i,j,k} = \sum_{q=M_2}^{N_2} b_q \left(\phi_{i,j,k+q+1} - \phi_{i,j,k+q}\right),$$

$$\delta_2^\xi \phi_{i,j,k} = \sum_{p=M_3}^{N_3} c_p \left(\phi_{i+p+1,j,k} - \phi_{i+p,j,k}\right), \quad \delta_2^\zeta \phi_{i,j,k} = \sum_{q=M_4}^{N_4} d_q \left(\phi_{i,j,k+q+1} - \phi_{i,j,k+q}\right).$$

Thus, the cross-derivative terms are

$$\begin{aligned} \delta_1^\xi \delta_2^\zeta \phi_{i,j,k} &= \sum_{p=M_1}^{N_1} a_p \left[\sum_{q=M_4}^{N_4} d_q \left(\phi_{i+p+1,j,k+q+1} - \phi_{i+p+1,j,k+q}\right) - \sum_{q=M_4}^{N_4} d_q \left(\phi_{i+p,j,k+q+1} - \phi_{i+p,j,k+q}\right)\right] \\ &= \sum_{p=M_1}^{N_1} a_p \left[\sum_{q=M_4}^{N_4} d_q \left(\phi_{i+p+1,j,k+q+1} - \phi_{i+p+1,j,k+q} - \phi_{i+p,j,k+q+1} + \phi_{i+p,j,k+q}\right)\right] \\ &= \sum_{p=M_1}^{N_1} \sum_{q=M_4}^{N_4} \left[a_p d_q \left(\phi_{i+p+1,j,k+q+1} - \phi_{i+p+1,j,k+q} - \phi_{i+p,j,k+q+1} + \phi_{i+p,j,k+q}\right)\right] \end{aligned} \tag{15}$$

$$\delta_1^\varsigma \delta_2^\varsigma \phi_{i,j,k} = \sum_{q=M_2}^{N_2} b_q \left[\sum_{p=M_3}^{N_3} c_p (\phi_{i+p+1,j,k+q+1} - \phi_{i+p,j,k+q+1}) - \sum_{p=M_3}^{N_3} c_q (\phi_{i+p+1,j,k+q} - \phi_{i+p,j,k+q}) \right]$$

$$= \sum_{q=M_2}^{N_2} b_q \left[\sum_{p=M_3}^{N_3} c_p (\phi_{i+p+1,j,k+q+1} - \phi_{i+p,j,k+q+1} - \phi_{i+p+1,j,k+q} + \phi_{i+p,j,k+q}) \right]$$

$$= \sum_{q=M_2}^{N_2} \sum_{p=M_3}^{N_3} [b_q c_p (\phi_{i+p+1,j,k+q+1} - \phi_{i+p,j,k+q+1} - \phi_{i+p+1,j,k+q} + \phi_{i+p,j,k+q})]$$

$$= \sum_{p=M_3}^{N_3} \sum_{q=M_2}^{N_2} [c_p b_q (\phi_{i+p+1,j,k+q+1} - \phi_{i+p+1,j,k+q} - \phi_{i+p,j,k+q+1} + \phi_{i+p,j,k+q})]. \tag{16}$$

From (15) and (16), we may see that $\delta_1^\varsigma \delta_2^\varsigma \phi_{i,j,k} = \delta_1^\varsigma \delta_2^\varsigma \phi_{i,j,k}$ if

$$a_p = c_p, \quad M_1 = M_3, \quad N_1 = N_3, \quad \text{and} \quad d_q = b_q, \quad M_4 = M_2, \quad N_4 = N_2, \tag{17}$$

respectively, which means that the numerical derivative operators satisfy

$$\delta_1^\xi = \delta_2^\xi \quad \text{and} \quad \delta_1^\varsigma = \delta_2^\varsigma, \tag{18}$$

respectively.

Other cross-derivative terms in (14) and (12) can be proved in the same manner, such that $I_x^N = I_y^N = I_z^N = 0$ numerically.

Based on above analysis, we conclude that in order to satisfy the GCL numerically at stationary grids, i.e., the SCL, it is needed that

(1) the conservative metric forms (12) are used, and
(2) the numerical derivative operators δ_2 in one coordinate direction to calculate the metric derivatives in (12) have to be the same with the numerical derivative operators δ_1 in the same coordinate direction to calculate the derivatives of the fluxes in Eq. (2), i.e., $\delta_2 = \delta_1$.

The numerical methods which satisfy the above conditions are called the conservative metric methods (CMM) in the present work.

For practical applications of the CMM, the following remarks are given,

Remark 1. It is better that the numerical derivative operators δ_1 are linear ones, i.e., they are independent from the flow variables Q, such that δ_2 and grid metrics can be easily calculated. Otherwise, it is somehow difficult or even impossible to apply the CMM. This feature will be discussed in the next section with the WENO's implementation.

Remark 2. From the conditions (18), we may see that the CMM does not require $\delta_1^\xi = \delta_1^\varsigma$, which means that different operators δ_1 can be used in different coordinate directions. This feature is very useful for boundary and near boundary grids, as boundary and near boundary schemes are usually different from the interior point schemes. Thus the CMM is exactly satisfied at boundary and near boundary grids, provided $\delta_1 = \delta_2$ in the same coordinate direction.

2.2. Discussions about the first-level metric difference operators δ_3

As discussed in the former subsection, the forms of the operators δ_3 to calculate the first-level metric derivatives in (13) have no effects on the SCL, so that any high-order schemes can be used for δ_3. In practical calculations, however, we found that if $\delta_3 \neq \delta_2$, the solutions may have large errors, and sometimes the calculating may break down. Following we give an analysis about this feature.

The difference between the traditional metric forms (11) and the conservative metric forms (12) can be indicated by

$$\varepsilon_{\delta_2}^{xy} = \delta_2(xy) - (x\delta_2 y + y\delta_2 x). \tag{19}$$

It is clear that $\varepsilon_{\delta_2}^{xy}$ is equal to zero if δ_2 is a differential factor. The metric coefficient $\tilde{\xi}_x$ in (11) and (12), for example, has the following forms, respectively

$$\tilde{\xi}_x^T = \delta_2^\eta y \delta_2^\varsigma z - \delta_2^\varsigma y \delta_2^\eta z, \tag{20}$$

$$\tilde{\xi}_x^{CMM} = \delta_2^\varsigma ((\delta_3^\eta y)z) - \delta_2^\eta ((\delta_3^\varsigma y)z) = z \delta_2^\varsigma \delta_3^\eta y + \delta_3^\eta y \delta_2^\varsigma z + \varepsilon_{\delta_2}^{(\delta_3^\eta y)z} - z \delta_2^\eta \delta_3^\varsigma y - \delta_3^\varsigma y \delta_2^\eta z - \varepsilon_{\delta_2}^{(\delta_3^\varsigma y)z}. \tag{21}$$

Let $\sigma(\tilde{\xi}_x) = \tilde{\xi}_x^{CMM} - \tilde{\xi}_x^T$, then

$$\sigma(\tilde{\xi}_x) = (\delta_3^\eta y \delta_2^\varsigma z - \delta_3^\varsigma y \delta_2^\eta z) - (\delta_2^\eta y \delta_2^\varsigma z - \delta_2^\varsigma y \delta_2^\eta z) + z(\delta_2^\varsigma \delta_3^\eta y - \delta_2^\eta \delta_3^\varsigma y) + \varepsilon_{\delta_2}^{(\delta_3^\eta y)z} - \varepsilon_{\delta_2}^{(\delta_3^\varsigma y)z}. \tag{22}$$

If $\delta_3 = \delta_2$, we know the first three terms in (22) are zero by the proof of (14), thus

$$\sigma(\tilde{\xi}_x) = \varepsilon_{\delta_2}^{(\delta_3^y y)z} - \varepsilon_{\delta_2}^{(\delta_3^z y)z}, \tag{23}$$

which just is the difference between the traditional metric method and the CMM. That means there are no extra errors for the metric calculations except the differences between the traditional metric forms (11) and the conservative metric forms (12) if $\delta_3 = \delta_2$. It should be noted that the above analysis is suited not only to interior schemes, but also to boundary and near boundary schemes. Thus we recommend that $\delta_3 = \delta_2$ in all the calculating regions in practical applications. Numerical tests for this feature are given in Section 4.2.

3. Applications of the CMM on various typical high-order finite difference schemes

In this section, we discuss the applications of the CMM on four typical schemes: the 6th-order linear central compact schemes used by Visbal and Gaitonde [9] and Rizzetta et al. [22] in their series papers, the linear dissipative compact schemes (DCS) [14], the 5th-order nonlinear WCNS [11,12], and the nonlinear WENO [15].

(1) *The 6th-order linear central compact scheme based on cell-node (CCSN6)*
This scheme was first derived by Lele [8], and was widely used in subsonic flow calculations. The governing equations, including the metric derivatives (12), are calculated by the same scheme, i.e., the linear operators δ_1, δ_2 and δ_3 are the same 6th-order central compact scheme based on nodes (CCSN6) [8],

$$\delta_{1,2,3}^\xi : \frac{1}{3}\tilde{E}'_{i-1,j,k} + \tilde{E}'_{i,j,k} + \frac{1}{3}\tilde{E}'_{i+1,j,k} = \frac{1}{36\Delta\xi}\left(\tilde{E}_{i+2,j,k} - \tilde{E}_{i-2,j,k}\right) + \frac{7}{9\Delta\xi}\left(\tilde{E}_{i+1,j,k} - \tilde{E}_{i-1,j,k}\right), \tag{24}$$

so that the CMM is hold. Visbal and Gaitonde's results [9] showed that the errors calculated by these schemes reach 10^{-12}, which are much lower than those by standard metrics. However, they only emphasized in their paper that the conservative metric forms (12) are very important to reduce errors.

(2) *The 5th-order linear dissipative compact schemes (DCS5)*
The DCS are a class of dissipative compact schemes derived by Deng et al. [14]. The flux splitting technique is used as $\tilde{E} = \tilde{E}^+ + \tilde{E}^-$ and the flux derivatives in (2) are also splitted as,

$$\delta_1^\xi \tilde{E} = \delta_1^{\xi+}\tilde{E}^+ + \delta_1^{\xi-}\tilde{E}^- = \tilde{E}'^+ + \tilde{E}'^-, \tag{25}$$

where, the operators $\delta_1^{\xi\pm}$ are the 5th-order DCS (DCS5) [14],

$$\delta_1^{\xi\pm} : \frac{(1\pm\alpha)}{3}\tilde{E}'^\pm_{i-1,j,k} + \tilde{E}'^\pm_{i,j,k} + \frac{(1\mp\alpha)}{3}\tilde{E}'^\pm_{i+1,j,k}$$
$$= \frac{1}{36\Delta\xi}\left(\tilde{E}^\pm_{i+2,j,k} - \tilde{E}^\pm_{i-2,j,k}\right) + \frac{7}{9\Delta\xi}\left(\tilde{E}^\pm_{i+1,j,k} - \tilde{E}^\pm_{i-1,j,k}\right) \mp \frac{\alpha}{18\Delta\xi}\left(\tilde{E}^\pm_{i+2,j,k} - 2\tilde{E}^\pm_{i,j,k} + \tilde{E}^\pm_{i-2,j,k}\right)$$
$$\mp \frac{4\alpha}{9\Delta\xi}\left(\tilde{E}^\pm_{i+1,j,k} - 2\tilde{E}^\pm_{i,j,k} + \tilde{E}^\pm_{i-1,j,k}\right). \tag{26}$$

In (26), $\alpha > 0$ is the dissipative coefficient. If $\alpha = 0$, the CCSN6 (24) is recovered, and the 5th-order upwind compact scheme (UCDA5) of Fu and Ma [35] can also be obtained as $\alpha = 1$. It is obvious $\delta_1^{\xi+} \neq \delta_1^{\xi-}$ as $\alpha \neq 0$. Then it is difficult to get (4) at uniform flow regions by (26) with $\alpha \neq 0$, thus the CMM cannot be satisfied. In Section 4.4, a delta form of the DCS5 is derived and tested, which can preserve the free-stream conditions.

(3) *The 5th-order weighted compact nonlinear scheme (WCNS-E-5)*
One aim of our studies is to apply the WCNS-E-5 [11,12] to flows with complex geometries, for example, two-dimensional multi-element airfoil flows, three-dimensional wing-body transonic configuration flows, subsonic delta-wing flows and hypersonic flows over complex geometries. Recently Nonomura et al. [19] carefully compared WCNS-E-5 with WENO5 for free-stream and vortex preservation properties. The operator δ_1 for the WCNS-E-5 is

$$\delta_1^\xi : \tilde{E}'_{i,j,k} = \frac{75}{64\Delta\xi}\left(\tilde{E}_{i+1/2,j,k} - \tilde{E}_{i-1/2,j,k}\right) - \frac{25}{384\Delta\xi}\left(\tilde{E}_{i+3/2,j,k} - \tilde{E}_{i-3/2,j,k}\right) + \frac{3}{640\Delta\xi}\left(\tilde{E}_{i+5/2,j,k} - \tilde{E}_{i-5/2,j,k}\right), \tag{27}$$

where the flux at cell-edges is

$$\tilde{E}_{i+1/2,j,k} = \tilde{E}^*\left(Q^*_{Li+1/2,j,k}, Q^*_{Ri+1/2,j,k}, \tilde{\xi}_{xi+1/2,j,k}, \tilde{\xi}_{yi+1/2,j,k}, \tilde{\xi}_{zi+1/2,j,k}\right) \tag{28}$$

and the cell edge variables $Q^*_{Li+1/2,j,k}$ and $Q^*_{Ri+1/2,j,k}$ are 5th-oder weighted interpolations and have been derived by Deng and Zhang [11], $\tilde{E}^*$ are the flux splitting schemes which can also be found in [11], for example, the Steger-Warming fluxes are used as

$$\begin{cases} \widetilde{E}_{i+1/2,j,k} = \widetilde{E}^+_{j+1/2,j,k} + \widetilde{E}^-_{j+1/2,j,k}, \\ \widetilde{E}^+_{j+1/2,j,k} = \widetilde{E}^+\left(Q^*_{Li+1/2,j,k}\widetilde{\xi}_{xi+1/2,j,k}, \widetilde{\xi}_{yi+1/2,j,k}, \widetilde{\xi}_{zi+1/2,j,k}\right), \\ \widetilde{E}^-_{j+1/2,j,k} = \widetilde{E}^-\left(Q^*_{Ri+1/2,j,k}\widetilde{\xi}_{xi+1/2,j,k}, \widetilde{\xi}_{yi+1/2,j,k}, \widetilde{\xi}_{zi+1/2,j,k}\right). \end{cases} \quad (29)$$

The cell edge metrics can be obtained by the 6th-order Lagrange interpolation [19],

$$\widetilde{\xi}_{xi+1/2,j,k} = \frac{75}{128}\left(\widetilde{\xi}_{xi,j,k} + \widetilde{\xi}_{xi+1,j,k}\right) - \frac{25}{256}\left(\widetilde{\xi}_{xi-1,j,k} + \widetilde{\xi}_{xi+2,j,k}\right) + \frac{3}{256}\left(\widetilde{\xi}_{xi-2,j,k} + \widetilde{\xi}_{xi+3,j,k}\right), \quad (30)$$

or by the following 4th-order interpolation

$$\widetilde{\xi}_{xi+1/2,j,k} = \frac{1}{16}\left(-\widetilde{\xi}_{xi-1,j,k} + 9\widetilde{\xi}_{xi,j,k} + 9\widetilde{\xi}_{xi+1,j,k} - \widetilde{\xi}_{xi+2,j,k}\right). \quad (31)$$

In order to satisfy the CMM, the operator δ_2 should be the same as δ_1. For example, from (12),

$$\widetilde{\xi}_x = a_\zeta - b_\eta, \quad \text{and} \quad a = y_\eta z, \quad b = y_\zeta z, \quad (32)$$

where $a_\zeta = \partial a/\partial \zeta$ and $b_\eta = \partial b/\partial \eta$. The operator δ_2 is, for example,

$$\widetilde{\delta_2}: a_{\zeta i,j,k} = \frac{75}{64\Delta\zeta}(a_{i,j,k+1/2} - a_{i,j,k-1/2}) - \frac{25}{384\Delta\zeta}(a_{i,j,k+3/2} - a_{i,j,k-3/2}) + \frac{3}{640\Delta\zeta}(a_{i,j,k+5/2} - a_{ijk-5/2}) \quad (33)$$

and the cell-edge values $a_{i,j,k+1/2}$ can be interpolated by (30) or (31), i.e.,

$$a_{i,j,k+1/2} = \frac{75}{128}(a_{i,j,k} + a_{i,j,k+1}) - \frac{25}{256}(a_{i,j,k-1} + a_{i,j,k+2}) + \frac{3}{256}(a_{i,j,k-2} + a_{i,j,k+3}), \quad (34)$$

or

$$a_{i,j,k+1/2} = \frac{1}{16}(-a_{i,j,k-1} + 9a_{i,j,k} + 9a_{i,j,k+1} - a_{i,j,k+2}). \quad (35)$$

The same schemes can be used for δ_3, i.e., $\delta_3 = \delta_1$. Although the WCNS-E-5 is a nonlinear scheme, the operators δ_1 and δ_2, however, are linear ones, thus the CMM can be easily applied. Furthermore, it should be noted the difference between the scheme (26) and (27)–(28). Though flux splitting schemes are also adopted in (28), the grid metrics, such as $\widetilde{\xi}_{xi+1/2,j,k}$, are at the same cell edges $i + 1/2$ in $\widetilde{E}^+$ and $\widetilde{E}^-$ in (29), and scheme (27) is a central one. Thus, we can get (4) in uniform regions.

(4) *The 5th-order WENO Scheme (WENO5)*

The original WENO5 use flux splitting as the above DCS5, and furthermore nonlinear weighted reconstructions to $\widetilde{E}^\pm$ are adopted, such that the operators δ_1 are nonlinear ones. As discussed above, it is difficult to fulfill the CMM in WENO schemes. Nonomura et al. [19] also discussed this difficulty and tried the non-conservative governing equations to satisfy the free-stream conditions for WENO5. Cai and Ladeinde [21] suggested solving the following equations to overcome this difficulty,

$$\frac{\partial \widetilde{Q}}{\partial \tau} + \frac{\partial\left(\widetilde{E} - \widetilde{E}_v\right)}{\partial \xi} + \frac{\partial\left(\widetilde{F} - \widetilde{F}_v\right)}{\partial \eta} + \frac{\partial\left(\widetilde{G} - \widetilde{G}_v\right)}{\partial \zeta} = -\left[I_x(E - E_v) + I_y(F - F_v) + I_z(G - G_v)\right]. \quad (36)$$

This technique seems to be a remedy to reduce the errors of the SCL violation and preserve the free-stream conditions approximately for WENO5. It is obvious by our analysis that the errors arising from the violation of the SCL, however, cannot be eliminated with this technique.

From the above applications, we may categorize the generally used finite difference schemes as two types based on δ_1: central schemes (CS) and upwind schemes (UPW). CS can be further divided into cell-node type (CS-Node) and cell-edge type (CS-Edge). UPW can be divided into linear upwind type (UPWL) and nonlinear upwind type (UPWN). The generally used finite difference schemes are listed in Table 1. Based on the derivation of the CMM and the above

Table 1
The categorization of usual high-order finite difference schemes based on δ_1.

δ_1			CMM
CS	CS-Node	4th-order or 6th-order explicit central schemes (ECS4, ECS6), Compact central schemes based on cell-node (CCSN) by Lele [8], Dispersion Relation Preserving (DRP) schemes [30], Optimized central compact schemes [31], …	Yes
	CS-Edge	Compact central schemes based on cell-edge (CCSE) by Lele [8], WCNS [11,12], …	Yes
UPW	UPWL	Explicit 3-point 2nd-order upwind scheme, upwind compact schemes (UCDA5) [35], dissipative compact schemes (DCS) [14], …	Difficult
	UPWN	ENO schemes [23], WENO schemes [15], compact-ENO [32], compact-WENO schemes [33], WCS by Xie and Liu [34],…	Difficult

applications, we may see that the central schemes (CS), including CS-Node schemes and CS-Edge schemes, can satisfy the CMM. It is difficult, however, to apply the CMM for the upwind schemes (UPW) because $\delta_1^+ \neq \delta_1^-$. Thus, it is critical to select the difference operator δ_1 to reduce the SCL-related errors. Further researches for UPW to satisfy the SCL are under going.

4. Numerical tests

The CMM is tested in this section for the calculations of Euler and Reynolds-averaged Navier–Stokes (RANS) equations to show its features with high-order finite difference schemes in solving complex shock wave problems and turbulence problems. We also give an example to show the property of δ_3. The WCNS-E-5 is applied with the explicit third-order TVD type Runge–Kutta method presented in [23] and the implicit LU-SGS for temporal discretization of Euler equations and RANS equations, respectively. At last, a delta form of the DCS5 which can preserve free-stream condition is tested.

4.1. A spherical Riemann problem

In this test, a spherical Riemann problem between two parallel walls at $z = 0$ and $z = 1$ is considered. The sphere is centered at $(x,y,z) = (0,0,0.4)$ with radius 0.2. Initially the gas is at rest with density $\rho = 1$ and pressure

$$\begin{cases} p = 1 & \text{if } r > 0.2, \\ p = 5 & \text{else}. \end{cases} \tag{37}$$

As reported by Langseth and LeVeque [7], the jump in pressure results in a strong outward moving shock wave and contact discontinuity and an inward moving rarefaction wave. This inward moving wave causes a local "implosion", and a second outward moving shock wave is created. Due to the symmetry, the computational domain is chosen to be $(x,y,z) \in [0,1.5] \times [0,1.5] \times [0,1]$. The grids are generated by the following equations

$$\begin{cases} x = x_0 + A\Delta x[\sin(8\pi x_0)\sin(64\pi(y_0 + z_0))], \\ y = y_0 + A\Delta y[\sin(8\pi y_0)\sin(64\pi(x_0 + z_0))], \\ z = z_0 + A\Delta z[\sin(16\pi z_0)\sin(64\pi(x_0 + y_0))], \end{cases} \tag{38}$$

where $x_0 = \Delta x(i-1)$, $y_0 = \Delta y(j-1)$, $z_0 = \Delta z(k-1)$, $i = 1,\ldots,I$; $j = 1,\ldots,J$; $k = 1,\ldots,K$; $\Delta x = 1.5/(I-1)$; $\Delta y = 1.5/(J-1)$; $\Delta z = 1/(K-1)$, $I = J = 76$, $K = 51$. Here $A = 0.2$ is the coefficient to control the nonuniformity of the grids. The 3D grids are sketchily shown in Fig. 1. Table 2 lists the maximum of I_x, I_y and I_z, which are calculated by five different schemes, the 2nd-order explicit central scheme (ECS2), the 4th-order explicit central scheme (ECS4), the 4th-order Padé-type central compact scheme (CCSN4), the 6th-order Padé-type central compact scheme (CCSN6), and the WCNS-E-5. It is easy to find that, if δ_2 are different from δ_1, the metric cancellation errors are much greater than those with the CMM.

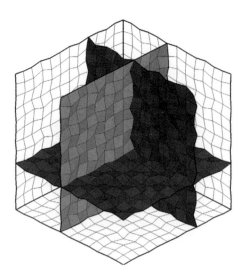

Fig. 1. Grid sketch for the Riemann problem.

Table 2
The maximal errors of I_x, I_y and I_z computed by different δ_1 and δ_2 schemes with the conservative form of Eq. (12).

		δ_1 Schemes				
		ECS2	ECS4	CCSN4	CCSN6	WCNS-E-5
δ_2 Schemes	ECS2	2.52E−17	6.55E−04	4.43E−04	7.11E−04	2.87E−04
	ECS4	2.67E−03	3.50E−16	1.15E−03	7.43E−04	1.78E−03
	CCSN4	1.11E−03	7.90E−04	2.65E−16	7.73E−04	8.13E−03
	CCSN6	2.86E−03	7.53E−03	1.23E−03	6.05E−16	2.11E−03
	WCNS-E-5	3.99E−04	6.77E−04	4.88E−04	7.99E−04	9.95E−17

Numerical simulations are carried out by the WCNS-E-5 with the CMM as well as the traditional standard metric method where the 6th-order explicit central scheme (ECS6) is applied to calculate the grid metrics. Part of the grids in the $j = 11$ plane are given in Fig. 2, and the pressure and density in the $z = 0$ plane at $t = 0.7$ are shown in Fig. 3. We may see that the CMM is much superior to the traditional standard metrics in the condition of low quality grids and projects new sight on applying high-order schemes for complex grids. Some successful engineering-oriented applications of the high-order WCNS-E-5 scheme for complex 2D and 3D problems can be found in [36].

4.2. Numerical tests about the difference operator δ_3

We repeated the calculations of the former example also by the WCNS-E-5 with the CMM, except that the 6th-order explicit central scheme (ECS6) is used for δ_3, i.e., $\delta_3 \neq \delta_2$. The computing, however, breaks down after five time steps as $\Delta t = 0.0001$. We know by Eq. (38), the smaller the parameter A, the more smooth the grids, and $A = 0$ corresponds to the uniform grid. Further numerical tests show that the calculation also breaks down even as $A = 0.05$. This numerical test gives an evidence that the forms of δ_3 are important for practical calculations as discussed in Section 2.2. In order to show the importance of δ_3 more clearly on interior grids and near boundary grids, other two grids are generated by Eq. (38) with $A = 0.2$ and $A = 0.1$, except that five grids near the boundaries are uniform. The computing is finished on the grids $A = 0.1$ and still breaks down as $A = 0.2$. Fig. 4 shows part of the grids in the $j = 11$ plane for the grids $A = 0.1$. Fig. 5 gives the pressure contours for both $\delta_3 = \delta_2$ and $\delta_3 \neq \delta_2$. We can see that the results of "$\delta_3 = \delta_2$" are more smooth than that of "$\delta_3 \neq \delta_2$".

4.3. Turbulent flows over a blunt configuration with dual backward-facing steps

A two-dimensional blunt configuration with dual backward-facing steps is shown in Fig. 6(a) and (b). The governing equations here are the RANS equations, and the turbulence model is the Spalart–Allmaras one-equation model [24] without the 'trip' function of f_{t1} and f_{t2}. The physical parameters for which the computations are performed are: $M_\infty = 0.3$, $Re = 2.7 \times 10^6$ and $T_\infty = 300K$. Adiabatic wall condition is used for the solid walls.

The convective terms of the governing Eq. (2) are solved by the WCNS-E-5, and the viscous terms are solved by a fourth-order central treatment [12]. Two methods are applied to calculate the metric derivatives: the first one is the CCSN6 with the standard metric relations; the second one adopts the CMM with the WCNS-E-5 and the 4th-order interpolation (35). As

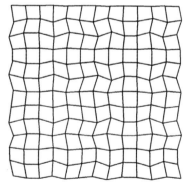

Fig. 2. Part of the grids in the $j = 11$ plane. $A = 0.2$.

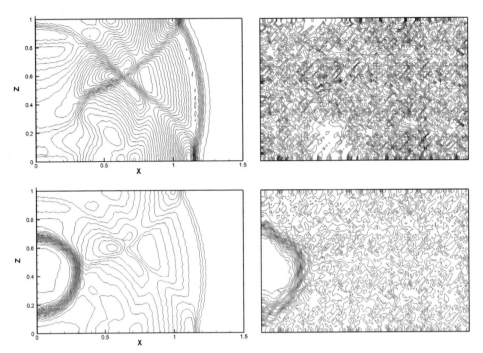

Fig. 3. Pressure (upper) and density (lower), 30 equally spaced contours. Left: the conservative metric method (CMM); right: the traditional standard metric method.

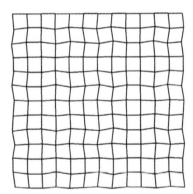

Fig. 4. Part of the grids in the $j = 11$ plane. $A = 0.1$.

shown in Fig. 7(a) and (b), the standard metrics with the CCSN6 produce very lager errors in I_x, I_y, and I_z, and these errors are dependent on the grid sizes, stretch rates, distortion and skewness of a mesh. It is obvious that the SCL-related errors play a vital role in this test case, and the numerical oscillations shown in Fig. 8 are evident if the CMM is not followed. Fig. 9 shows the results given by the CMM. It may be seen that the free-stream condition is preserved and the contour lines are still smooth even when the mesh is extremely skewed around the corners. This case indicates that the WCNS-E-5 may be applied to complicated flows over complex configurations with relatively low quality meshes as the CMM is adopted.

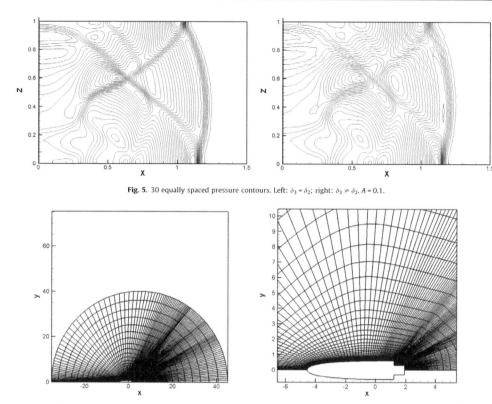

Fig. 5. 30 equally spaced pressure contours. Left: $\delta_3 = \delta_2$; right: $\delta_3 \neq \delta_2$. $A = 0.1$.

(a) The computed domain and the mesh. (b) The enlarged part near the solid walls.

Fig. 6. Configuration and mesh.

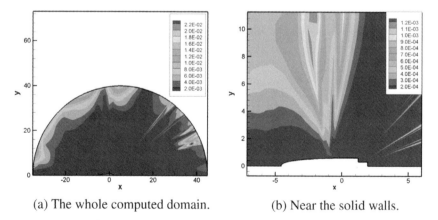

(a) The whole computed domain. (b) Near the solid walls.

Fig. 7. The SCL errors ($|I_x| + |I_y| + |I_z|$) of 6th-order traditional standard metrics.
(colored picture attached at the end of the book)

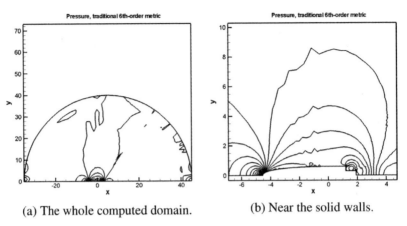

(a) The whole computed domain. (b) Near the solid walls.

Fig. 8. The results with the 6th-order traditional standard metric, 50 equally spaced pressure contours.

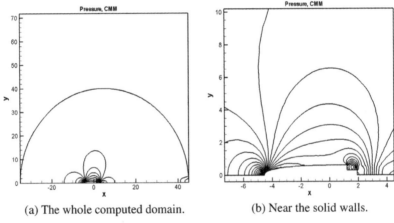

(a) The whole computed domain. (b) Near the solid walls.

Fig. 9. The results with the CMM, 50 equally spaced pressure contours.

4.4. A delta form of the DCS5 to preserve the free-stream conditions

As we know from Section 3, the DCS5, as well as other upwind schemes, are difficult to satisfy the CMM. A delta-form of the DCS5 (DCS5-Δ) is derived which can preserve the free-stream conditions.

In one dimensional Cartesian coordinates, the DCS5 (26) can be rewritten as,

$$\frac{(1\pm\alpha)}{3}E'^{\pm}_{i-1} + E'^{\pm}_{i} + \frac{(1\mp\alpha)}{3}E'^{\pm}_{i+1} = \frac{1}{36\Delta\xi}\left[(1\mp 2\alpha)\Delta E^{\pm}_{i+3/2} + (29\mp 18\alpha)\Delta E^{\pm}_{i+1/2} + (29\pm 18\alpha)\Delta E^{\pm}_{i-1/2} + (1\pm 2\alpha)\Delta E^{\pm}_{i-3/2}\right], \tag{39}$$

where, $\Delta E^{\pm}_{i+1/2} = E^{\pm}_{i+1} - E^{\pm}_{i}$. We get the flux Jacobian $A = \partial E/\partial Q$, thus,

$$\Delta E^{\pm}_{i+1/2} = \overline{A}^{\pm}(Q_{i+1}, Q_i)(Q_{i+1} - Q_i) = \overline{A}^{\pm}_{i+1/2}\Delta Q_{i+1/2}, \tag{40}$$

where, $\overline{A}^{\pm}_{i+1/2} = \overline{A}^{\pm}(Q_{i+1}, Q_i)$ is the flux Jacobian calculated by the Roe's averages of Q_i and Q_{i+1} [37]. Submitting Eq. (40) into Eq. (39), we can get the DCS5-Δ,

$$\frac{(1\pm\alpha)}{3}E'^{\pm}_{i-1} + E'^{\pm}_{i} + \frac{(1\mp\alpha)}{3}E'^{\pm}_{i+1} = \frac{1}{36\Delta\xi}\Big[(1\mp 2\alpha)\bar{A}^{\pm}_{i+3/2}\Delta Q_{i+3/2} + (29\mp 18\alpha)\bar{A}^{\pm}_{i+1/2}\Delta Q_{i+1/2} + (29$$
$$\pm 18\alpha)\bar{A}^{\pm}_{i-1/2}\Delta Q_{i-1/2} + (1\pm 2\alpha)\bar{A}^{\pm}_{i-3/2}\Delta Q_{i-3/2}\Big]. \tag{41}$$

For curvilinear coordinates, the Jacobian matrixes in (41) are $\bar{A}^{\pm}_{i+1/2} = \bar{A}^{\pm}(Q_{i+1}, Q_i, \tilde{\xi}_{xi+1/2}, \tilde{\xi}_{yi+1/2}, \tilde{\xi}_{zi+1/2})$, and $\tilde{\xi}_{x,i+1/2} = (\tilde{\xi}_{xi} + \tilde{\xi}_{xi+1})/2$, $\tilde{\xi}_{yi+1/2} = (\tilde{\xi}_{yi} + \tilde{\xi}_{yi+1})/2$, and $\tilde{\xi}_{zi+1/2} = (\tilde{\xi}_{zi} + \tilde{\xi}_{zi+1})/2$. In uniform flow regions, it is obvious $\Delta Q = 0$ in (41), thus the free-stream conditions are preserved.

DCS5-Δ is a remedy to preserve the free-stream conditions for DCS, so that $\delta_3 = \delta_2$ does not needed. Further studies for upwind schemes to satisfy CMM are under going.

A two-dimensional moving vortex problem is computed. An isentropic vortex centered at $(x_c, y_c) = (6,6)$ is superposed to a uniform flow. The vortex is described as a perturbation to the uniform flow as follows:

$$\begin{cases} (\delta u, \delta v) = \varepsilon\tau e^{\beta(1-\tau^2)}(\sin\theta, -\cos\theta), \\ \delta T = -(\gamma-1)\varepsilon^2 e^{2\beta(1-\tau^2)}/(4\alpha\gamma), \\ \delta S = 0, \end{cases} \tag{42}$$

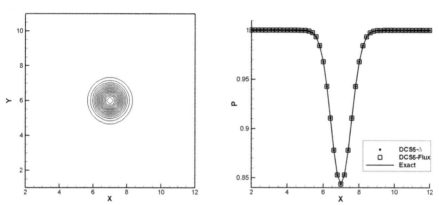

(a) 15 equally spaced pressure contours by DCS5-Δ. (b) Pressure along the central stream-wise grid line.

Fig. 10. Results on the uniform 61×61 grids.

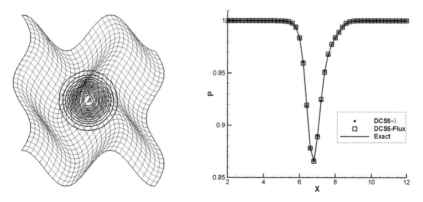

(a) 15 equally spaced pressure contours. (b) Pressure along the central stream-wise grid line.

Fig. 11. Results on the wavy 61×61 grids.

where $\beta = 0.204$, $\varepsilon = 0.3$, $\tau = r/r_c$, $r = [(x - x_c)^2 + (y - y_c)^2]^{1/2}$, and $r_c = 0.5$. Here $T = p/\rho$ is the temperature and $S = p/\rho^\gamma$ is the entropy.

The first grids are the uniform 61×61 Cartesian grids, and the computed domain is $(x,y) \in [0,12] \times [0,12]$. The dissipative coefficient $\alpha = 0.25$ is chosen in the calculations. Fig. 10(a) shows the computed pressure contours of the DCS5-Δ. It can be found from Fig. 10(b) that the results of the DCS5-Δ and the DCS5-Flux are almost identical in Cartesian coordinates. In this section, the DCS5-Flux is the original DCS5, i.e., Eq. (26).

The second grids are wavy ones generated by the same equations in [9]. The grid metrics are calculated by the CCSN6 (24). The results are shown in Fig. 11. Fig. 11(a) shows the pressure contours of the DCS5-Δ on the wavy mesh. Fig. 11(b) shows the pressure distributions along the central stream-wise grid line, which may indicate that the DCS5-Δ and DCS5-Flux are almost identical for this case. The L_1, L_2 and L_∞ errors of density as well as the accuracy order are listed in Table 3. It can be found that the results calculated by the DCS5-Δ have the 5th-order accuracy on the wavy grids.

A randomly disturbed grids similar to that of [9] are chosen to further study the property of the DCS5-Δ. Part of the grids is shown in Fig. 12. The computed pressure and cross velocity are shown in Figs. 13 and 14, respectively. It is very clear that

Table 3
The density errors of the DCS5-Δ on wavy grids.

Grid	L_1 errors	Accuracy order	L_2 errors	Accuracy order	L_∞ errors	Accuracy order
61×61	1.90E−05	–	1.70E−04	–	4.83E−03	–
121×121	2.07E−07	6.52	8.73E−07	7.61	1.57E−05	8.27
241×241	6.44E−09	5.00	2.59E−08	5.07	3.35E−07	5.55

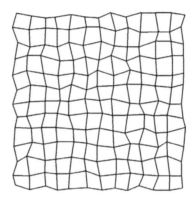

Fig. 12. Randomly disturbed grids.

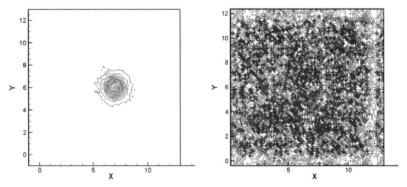

Fig. 13. Pressure contours, left: DCS5-Δ; right: DCS5-Flux.

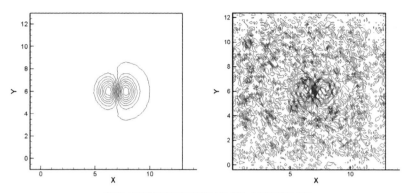

Fig. 14. Cross velocity (v) contours, left: DCS5-Δ; right: DCS5-Flux.

the result calculated by the DCS5-Δ are largely improved comparing with that of the DCS5-Flux because the free-stream conditions are preserved for the DCS5-Δ.

Further studies to ensure the SCL for upwind schemes are under going. It should be pointed out that this Delta form of upwind schemes cannot be used by WENO5, for WENO schemes have the nonlinear properties.

5. Conclusions

The geometric conservation law (GCL) is important for the applications of high-order finite difference schemes in curvilinear coordinates. The GCL includes the volume conservation law (VCL) and the surface conservation law (SCL). We made great efforts to study the SCL in this paper for stationary grids. The importance of the SCL for free-stream preservation is theoretical analyzed, and a conservative metric method (CMM) is given so as to ensure the SCL to be satisfied numerically. We also proved that the CMM is a sufficient condition for the SCL and can be applied on the interior, boundary and near boundary grids. The first-level difference operators δ_3 are also discussed. Though δ_3 have no effects on the SCL, it is important for practical calculations. As $\delta_3 = \delta_2$, no extra errors are introduced in the grid metric calculations. The numerical tests also show that $\delta_3 = \delta_2$ can give better results than other high-order forms of δ_3.

The generally used high-order difference schemes are divided into central schemes (CS) and upwind schemes (UPW) based on the difference operator δ_1. The CMM can be applied to CS, including cell-node type schemes (cs-node) and cell-edge type schemes (cs-edge), but can not be applied to upwind/dissipative ones (UPW) no matter they are linear or nonlinear schemes. Thus, it is critical to select the difference operator δ_1 to reduce the SCL-related errors. The implementations of the CMM on a linear compact central scheme (CCSN), a dissipative compact scheme (DCS), WCNS and WENO are discussed. Numerical tests based on the WCNS-E-5 for a complex shock wave problem as well as a turbulence problem demonstrate that the SCL plays a vital role for high-order finite difference schemes and is also useful to prevent numerical oscillations. The proposed CMM, which can effectively avoid the violation of the SCL numerically, may be used to overcome the barriers which hinder the applications of high-order finite difference schemes on complex grids. A remedy to preserve the free-stream conditions for high-order upwind schemes is tested.

Acknowledgments

This study was supported by the project of National Natural Science Foundation of China (Grant 10621062 and 11072259) and National Basic Research Program of China (Grant 2009CB723800). The authors would like to thank Prof. Laiping Zhang, Dr. Yifeng Zhang and Assistant researcher Yaobin Min of China Aerodynamics Research & Development Center for their contributions and useful discussions.

References

[1] J.G. Trulio, K.R. Trigger, Numerical solution of the one-dimensional hydrodynamic equations in an arbitrary time-dependent coordinate system, Technical Report UCLR-6522, University of California Lawrence Radiation laboratory, 1961.
[2] T.H. Pulliam, J.L. Steger, On implicit finite-difference simulations of three-dimensional flow, in: AIAA Paper 78-10, 1978.
[3] P.D. Thomas, C.K. Lombard, Geometric conservation law and its application to flow computations on moving grids, AIAA Journal 17 (10) (1979) 1030–1037.
[4] S. Étienne, A. Garon, D. Pelletier, Perspective on the geometric conservation law and finite element methods for ALE simulations of incompressible flow, Journal of Computational Physics 228 (2009) 2313–2333.

[5] D.J. Mavriplis, Z. Yang, Achieving higher-order time accuracy for dynamic unstructured mesh fluid flow simulations: role of the GCL, in: AIAA Paper 2005-5114, 2005.
[6] C. Farhat, P. Geuzaine, C. Grandmont, The discrete geometric conservation law and the nonlinear stability of ALE schemes for the solution of flow problems on moving grids, Journal of Computational Physics 174 (2001) 669-694.
[7] J.O. Langseth, R.J. LeVeque, A wave propagation method for three-dimensional hyperbolic conservation laws, Journal of Computational Physics 165 (2000) 126-166.
[8] S.K. Lele, Compact finite difference schemes with spectral-like resolution, Journal of Computational Physics 103 (1992) 16-42.
[9] R.M. Visbal, D.V. Gaitonde, On the use of higher-order finite-difference schemes on curvilinear and deforming meshes, Journal of Computational Physics 181 (2002) 155-185.
[10] H. Viviand, W. Ghazzi, Numerical solution of the compressible Navier–Stokes equations at high Reynolds numbers with applications to the blunt body problem, Lecture Notes in Physics, vol. 59, Springer-Verlag, (1976) 434-439.
[11] X. Deng, H. Zhang, Developing high-order weighted compact nonlinear schemes, Journal of Computational Physics 165 (2000) 22-44.
[12] X. Deng, High-order accurate dissipative weighted compact nonlinear schemes, Science in China (Series A) 45 (3) (2002) 356-370.
[13] X. Deng, X. Liu, M. Mao, Advances in high-order accurate weighted compact nonlinear schemes, Advances in Mechanics 37 (3) (2007) 417-427.
[14] X. Deng, H. Maekawa, C. Shen, A class of high-order dissipative compact schemes, in: AIAA Paper 96-1972, 1996.
[15] G. Jiang, C. Shu, Efficient implementation of weighted ENO, Journal of Computational Physics 181 (1996) 202-228.
[16] H. Zhang, M. Reggio, J.Y. Trépanier, R. Camarero, Discrete form of the GCL for moving meshes and its implementation in CFD schemes, Computers and Fluids 22 (1) (1993) 9-31.
[17] X. Liu, X. Deng, M. Mao, High-order behaviors of weighted compact fifth-order nonlinear schemes, AIAA Journal 45 (8) (2007) 2093-2097.
[18] T.H. Pulliam, J.L. Steger, Implicit finite-difference simulation of three-dimensional compressible flow, AIAA Journal 18 (2) (1980) 159-167.
[19] T. Nonomura, N. Iizuka, K. Fujii, Freestream and vortex preservation properties of high-order WENO and WCNS on curvilinear grids, Computers and Fluids 39 (2010) 197-214.
[20] M. Vinokur, An analysis of finite-difference and finite-volume formulations of conservation laws, Journal of Computational Physics 81 (1989) 1-52.
[21] X. Cai, F. Ladeinde, Performance of WENO scheme in generalized curvilinear coordinate systems, in: AIAA Paper 2008-36, 2008.
[22] D.P. Rizzetta, M.R. Visbal, P.E. Morgan, A high-order compact finite-difference scheme for large-eddy simulation of active flow control, in: AIAA Paper 2008-526, 2008.
[23] C.W. Shu, S. Osher, Efficient implementation of essentially non-oscillatory shock capturing schemes, Journal of Computational Physics 77 (1988) 439-471.
[24] P.R. Spalart, S.R. Allmaras, A one-equation turbulence model for aerodynamic flows, in: AIAA Paper 92-0439, 1992.
[25] S. Zhang, S. Jiang, C.-W. Shu, Development of nonlinear weighted compact schemes with increasingly higher order accuracy, Journal of Computational Physics 227 (2008) 7294-7321.
[26] T. Nonomura, K. Fujii, Effects of difference scheme type in high-order weighted compact nonlinear schemes, Journal of Computational Physics 228 (2009) 3533-3539.
[27] T. Nonomura, Y. Goto, K. Fujii, Improvements of efficiency in seventh-order weighted compact nonlinear scheme, in: 6th Asia Workshop on Computational Fluid Dynamics (6AWCFD) AW6-16, Tokyo, Japan, 2010.
[28] K. Ishiko, N. Ohnishi, K. Ueno, et al, Implicit large eddy simulation of two-dimensional homogeneous turbulence using weighted compact nonlinear scheme, Journal of Fluid Engineering 131 (6) (2009) 1-14.
[29] X. Deng, M. Mao, G. Tu, et al., High-order and high-accurate CFD methods and their applications, in: The 8th Asian Computational Fluid Dynamics Conference, Paper No. ACFD0126-T002-A-001, Hong Kong, 2010.
[30] C.K.W. Tam, J.C. Webb, Dispersion-relation-preserving finite difference schemes for computational acoustics, Journal of Computational Physics 107 (1993) 262-281.
[31] J.W. kim, D.J. Lee, Optimized compact finite difference schemes with maximum resolution, AIAA Journal 34 (5) (1996) 887-893.
[32] N.A. Adams, K. Shariff, A high-resolution hybrid compact-ENO scheme for shock-turbulence interaction problems, Journal of Computational Physics 127 (1996) 27-51.
[33] S. Pirozzoli, Conservative hybrid compact-WENO schemes for shock-turbulence interaction, Journal of Computational Physics 178 (2002) 81-117.
[34] P. Xie, C. Liu, Weighted compact and non-compact scheme for shock tube and shock entropy interaction, in: AIAA Paper 2007-509, 2007.
[35] D. Fu, Y. Ma, A high-order accurate difference schemes for complex flow fields, Journal of Computational Physics 134 (1) (1997) 1-15.
[36] X. Deng, M. Mao, G. Tu, Y. Zhang, et al, Extending weighted compact nonlinear schemes to complex grids with characteristic-based interface conditions, AIAA Journal 48 (2010), doi:10.2514/1.J050285.
[37] P.L. Roe, Approximate Riemann solvers, parameter vectors, and difference schemes, Journal of Computational Physics 43 (1981) 357-372.

A family of hybrid cell-edge and cell-node dissipative compact schemes satisfying geometric conservation law

Xiaogang Deng [a,b], Yi Jiang [b,c,*], Meiliang Mao [b,c], Huayong Liu [b], Song Li [b], Guohua Tu [b]

[a] *National University of Defense Technology, Changsha, Hunan 410073, PR China*
[b] *State Key Laboratory of Aerodynamics, China Aerodynamics Research and Development Center, P.O. Box 211, Mianyang 621000, PR China*
[c] *Computational Aerodynamics Institute, China Aerodynamics Research and Development Center, P.O. Box 211, Mianyang 621000, PR China*

ARTICLE INFO

Article history:
Received 10 June 2014
Received in revised form 8 April 2015
Accepted 10 April 2015
Available online 22 April 2015

Keywords:
High-order compact scheme
Hybrid cell-edge and cell-node dissipative compact scheme
Geometric conservation law
Dissipative scheme
High lift trapezoidal wing

ABSTRACT

Growing evidences show that the Symmetrical Conservative Metric Method (SCMM) is essential in preserving freestream conservation and orders of accuracy for high-order finite difference schemes to simulate flows with complex geometries. In this paper, a new family of Hybrid cell-edge and cell-node Dissipative Compact Schemes (HDCSs) has been developed for geometry-complex flows by fulfilling the SCMM as well as by introducing dissipation according to the concept adopted in the construction of the high-order Dissipative Compact Schemes (DCSs). The resolution and dissipation properties of HDCSs are investigated by the Fourier analysis, and the stability property of HDCSs is also investigated by asymptotic stability analysis and amplification factor analysis. HDCSs are validated by computing several benchmark test cases. The vortex convection test case demonstrates that the orders of accuracy of the HDCSs are preserved unless the GCL is satisfied. Although high resolution of HDCSs is observed in the test of acoustic wave scattering of multiple cylinders, the solutions can be contaminated if the GCL is not satisfied. Moreover, the numerical solutions of flow past a high lift trapezoidal wing demonstrate the promising ability of the newly developed HDCSs in solving complex flow problems.

© 2015 Elsevier Ltd. All rights reserved.

1. Introduction

Nowadays, compact high-order finite difference schemes have been widely used for a broad range of problems with multiple spatial and temporal scales such as turbulence and aeroacoustics [1]. The advantages of compact schemes using a compact stencil over traditional explicit finite difference schemes are mainly regarded as the relatively higher order of accuracy and higher resolution [2]. Although much advance has been achieved in constructing high-order compact schemes, applications of these schemes are still challenged by complex geometry. When the accurate numerical simulation of a broad spectrum phenomena is performed by high-order compact schemes on a complex mesh, it is argued that there are many obstacles such as robustness and grid-quality sensitivity [3,4]. This argument can be largely settled by considering the researches on the Geometric Conservation Law (GCL) [5–9]. The GCL consists of Surface Conservation Law (SCL) and Volume Conservation Law (VCL). The VCL has been widely studied for time-dependent grids, while the SCL is rarely discussed for high-order finite difference schemes. Recent research [7] shows that the SCL plays an important role in the applications of high-order finite difference schemes on complex grids. If the SCL is satisfied, the improvement of the finite difference scheme is obvious in handling complex geometry [10].

In order to satisfy the SCL for high-order finite difference schemes, a Conservative Metric Method (CMM) is derived by Deng et al. [7]. The CMM is achieved by computing grid metric derivatives through a conservative form with the same scheme applied for spatial fluxes. It has been proved in Ref. [7] that the CMM can be easily implemented into high-order schemes if the finite differencing operator δ_1 of the first derivative is not split, while it is difficult to apply the CMM to the schemes where the δ_1 is split into two upwind parts as δ_1^+ and δ_1^-. Recently, a Symmetrical Conservative Metric Method (SCMM) [11], which can evidently increase the numerical accuracy on irregular grids, is proposed based on further studies on the GCL. In the circumstances, the SCMM is suggested to be fulfilled to satisfy the GCL. Besides the SCMM, multi-block structured grid technique may be another requirement for the applications of high-order finite difference schemes on complex geometries, because it is difficult to generate a single-block structured grid for a complex configuration. The characteristic-based interface conditions (CBICs) in Ref. [12] and the nonlinear patched grid method in Ref. [13] are quite useful in handling muti-block structured grids, so as to enable the use of high-order schemes on complex geometries. In this paper, the CBICs will be employed to handle the mutiblock structured grids.

With the help of satisfying the GCL, Central Compact Schemes (CCSs) have been successfully applied to various flow simulations on complex grids [5]. A shortcoming of the CCSs proposed by Lele [14] is that they do not have inherent dissipation and then some extra techniques may be required to prevent the notorious Gibbs phenomenon, for example, high-order filters are generally added to these schemes to suppress numerical contaminants. However, a successful usage of filters strongly relies on specialized experiences. A satisfactory way of dissipating unresolvable wavenumbers is to use a scheme with inherent dissipation [15]. The Dissipative Compact Schemes (DCSs) derived by Deng et al. [16] are such kind of schemes. The DCSs have favorable spectrum characteristic [17], furthermore,some aeroacoustic benchmark problems have been simulated successfully by the fifth-order DCS [18,19]. However, applications of the DCSs on complex grids may have serious problem because the SCL is dissatisfied [7].

In this paper, a new family of Hybrid cell-edge and cell-node Dissipative Compact Schemes (HDCSs) is developed to overcome the problem of the DCSs in fulfilling the SCMM. In the next section, the design principle of HDCS is described. In Section 3, details of the HDCSs are given and the HDCSs up to eleventh-order formal accuracy have been developed. In Section 4, the HDCSs have been optimized to achieve the perfect resolution $2/3\omega_m$ corresponding with Orszag's Two-Thirds Rule [20], and two optimized HDCSs are suggested for high-order accurate simulations. Fourier analysis, asymptotic stability analysis and amplification factor analysis are also given in this section. In Section 5, a short introduce is given for the terminologies: GCL, VCL and SCL, and then detailed implementation of the CMM and SCMM is given for the purpose of explaining that the HDCSs can satisfy the SCL strictly. Several numerical tests for the HDCSs are presented in Section 6.

2. The methodology to derive high-order dissipative schemes fulfilling the SCMM

Let us consider a one-dimensional conservation system,

$$\frac{\partial U}{\partial t} + \frac{\partial E(U)}{\partial x} = 0, \quad (1)$$

and its semi-discrete approximation,

$$\left(\frac{\partial U}{\partial t}\right)_j = -\delta_1 E_j, \quad (2)$$

where the operator δ_1 denotes the finite difference scheme used to approximate the first spatial derivative. The Cell-Node type Central Compact Schemes (CCSNs) proposed by Lele [14] for δ_1 have the forms,

$$\delta_1 : E'_j + \sum_{n=1}^{r} \kappa_n^* \left(E'_{j+n} + E'_{j-n}\right) = \frac{1}{h}\sum_{m=1}^{p} a_m^*(E_{j+m} - E_{j-m}), \quad (3)$$

where h is the grid size. The κ_n^* and a_m^* are coefficients which can be acquired by matching the Taylor series coefficients of various orders, and more details are given in Ref. [14]. As well known, central type approximations of the first derivatives have no dissipation and the dispersive errors dominate the algorithms. These errors may destroy the numerical solutions and should be damped out. As demonstrated by Deng et al. [16], the method of adding even-order derivatives into Eq. (3) is proven to be an effective way of damping out numerical contaminants, and the DCSs [16] have been derived as

$$\delta_1 : E'_j + \sum_{n=1}^{r}\left(\kappa_n^{**}E'_{j+n} + \kappa_{-n}^{**}E'_{j-n}\right) = \frac{1}{h}\sum_{m=1}^{p*}a_m^{**}(E_{j+m} - E_{j-m})$$
$$+ \sum_{m=1}^{q*}b_m^{**}(E_{j+m} - 2E_j + E_{j-m}). \quad (4)$$

In order to ensure that the schemes are dissipative for both left-propagating and right-propagating waves, the differencing operator of DCSs needs to be split into two parts as $\delta_1 = \delta_1^+ + \delta_1^-$.

As discussed in Ref. [11], the positive and negative flux reconstructions have difficulties in fulfilling the SCMM, because grid metrics can not be split into two upwind parts. In order to avoid the splitting of the differencing operator, cell-edge terms are added into the central compact schemes (3), and a new class of Hybrid cell-node and cell-edge Compact Schemes (HCSs) is derived as

$$\delta_1 : E'_j + \sum_{n=1}^{r}\kappa_n\left(E'_{j+n} + E'_{j-n}\right) = \frac{1}{h}\sum_{m=1}^{p}a_m(E_{j+m} - E_{j-m})$$
$$+ \frac{1}{h}\sum_{l=1}^{q}\varphi_l(E_{j+(l-1/2)} - E_{j-(l-1/2)}) + T_{error}. \quad (5)$$

The coefficients in (5) can be derived by matching the Taylor series coefficients of various orders. For a family of HCSs with $r \leq 2, p \leq 3$, and $q = 1$, the consistent condition requires $1 + 2\kappa_1 + 2\kappa_2 = \varphi_1 + 2a_1 + 4a_2 + 6a_3$, and the truncation errors of (5) are

$$T_{error} = \frac{-1}{24}[\varphi_1 + 8(a_1 + 8a_2 + 27a_3 - 3\kappa_1 - 12\kappa_2)]h^2\frac{\partial^3 E_j}{\partial x^3}$$
$$- \left(\frac{\varphi_1}{1920} + \frac{a_1}{60} + \frac{8a_2}{15} + \frac{81a_3}{20} - \frac{\kappa_1}{12} - \frac{4\kappa_2}{3}\right)h^4\frac{\partial^5 E_j}{\partial x^5}$$
$$- \left(\frac{\varphi_1}{322560} + \frac{a_1}{2520} + \frac{16a_2}{315} + \frac{243a_3}{280} - \frac{\kappa_1}{360} - \frac{8\kappa_2}{45}\right)h^6\frac{\partial^7 E_j}{\partial x^7}$$
$$- \left(\frac{\varphi_1}{92897280} + \frac{a_1}{181440} + \frac{18a_2}{2835} + \frac{243a_3}{2240} - \frac{\kappa_1}{20160} - \frac{4\kappa_2}{315}\right)h^8\frac{\partial^9 E_j}{\partial x^9}$$
$$- \left(\frac{\varphi_1}{40874803200} + \frac{a_1}{19958400} + \frac{16a_2}{155925} + \frac{2187a_3}{246400}\right)$$
$$- \frac{\kappa_1}{1814400} - \frac{8\kappa_2}{14175}\right)h^{10}\frac{\partial^{11} E_j}{\partial x^{11}}$$
$$- \left(\frac{\varphi_1}{25505877196800} + \frac{a_1}{3113510400} + \frac{16a_2}{60810725}\right.$$
$$\left. + \frac{6561a_3}{12812800} - \frac{\kappa_1}{239500800} - \frac{8\kappa_2}{467775}\right)h^{12}\frac{\partial^{13}E_j}{\partial x^{13}} + \mathcal{O}(h^{14}). \quad (6)$$

It can be seen that the truncation errors contain only odd-order derivatives, then the HCSs are non-dissipative and only have dispersive errors. For the purpose of constructing the dissipative scheme, we use the variable reconstructions to add dissipation into the HCSs instead of the flux ones used in the construction of the DCSs, and the numerical cell-edge fluxes for Eq. (5) are evaluated as,

$$\tilde{E}_{j\pm 1/2} = E\left(\tilde{U}_{j\pm 1/2}^L, \tilde{U}_{j\pm 1/2}^R\right), \quad (7)$$

where the left $\tilde{U}_{j\pm 1/2}^L$ and right $\tilde{U}_{j\pm 1/2}^R$ cell-edge variables are reconstructed by the dissipative interpolations,

$$\begin{cases} \tilde{U}_{j+1/2}^L = \tilde{U}_{j+1/2}^C + \alpha h^{2k-1}\frac{\partial^{2k-1}U_{j+1/2}}{\partial x^{2k}} + \mathcal{O}(h^{2k}), \\ \tilde{U}_{j+1/2}^R = \tilde{U}_{j+1/2}^C - \alpha h^{2k-1}\frac{\partial^{2k-1}U_{j+1/2}}{\partial x^{2k}} + \mathcal{O}(h^{2k}). \end{cases} \quad (8)$$

In (8), $\tilde{U}_{j\pm1/2}^C$ are the central type interpolations with an order of accuracy $2k$ and α is the dissipative parameter. It should be noted that the accuracy order of the truncation errors of (8) is an odd one so that the cell-edge term $\left(\tilde{E}_{j+1/2} - \tilde{E}_{j-1/2}\right)/h$ can produce an even-order derivative in the truncation errors of (5), and the cell-edge term can be deemed as a dissipation source. Various flux splitting methods, such as Roe flux splitting and Steger–Warming flux splitting, can be used to calculate the fluxes $\tilde{E}_{j+1/2}$ in (7).

Since $\tilde{U}_{j\pm1/2}^L$ and $\tilde{U}_{j\pm1/2}^R$ are $(2k-1)$-th-order approximations of $U_{j\pm1/2}$, $\tilde{E}_{j\pm1/2}$ may be rewritten as

$$\tilde{E}_{j\pm1/2} = E_{j\pm1/2} + d(x_{j\pm1/2})h^{2k-1} + \mathcal{O}(h^{2k}), \tag{9}$$

where $d(x)$ is Lipschitz continuous. When the $E_{j\pm1/2}$ of (5) are replaced by $\tilde{E}_{j\pm1/2}$, the errors of (5) are

$$\tilde{T}_{error} = T_{error} + \frac{1}{h}\sum_{l=1}^{q}\varphi_l(d_{j+(l-1/2)} - d_{j-(l-1/2)})h^{2k-1} + \mathcal{O}(h^{2k}). \tag{10}$$

It is obvious that the accuracy order of the new schemes with the truncation errors $\tilde{T}_{error}$ is $(2k-1)$ if the accuracy order of T_{error} is not less than $2k-1$; otherwise, the accuracy order of these new schemes is determined by the truncation errors T_{error}. These new schemes defined by Eqs. (5), (8), (7) can be regarded as developing generations of the DCSs and are called HDCSs.

In order to show the dissipative property of the HDCSs clearly, we consider the $(2k-1)$th-order explicit type HDCSs for scalar linear wave equation,

$$\frac{\partial u}{\partial t} + c\frac{\partial u}{\partial x} = 0, \tag{11}$$

where $c > 0$ for right propagating wave. A class of $(2k)$th-order explicit type HCSs approximating Eq. (11) is obtained by setting $q = 1$ and $\kappa_n = 0$ in (5),

$$\left(\frac{\partial u}{\partial x}\right)_j = \frac{1}{h}\sum_{m=1}^{p}a_m(u_{j+m} - u_{j-m}) + \frac{1}{h}\varphi(\tilde{u}_{j+1/2} - \tilde{u}_{j-1/2})$$
$$+ T_{2k}^0 h^{2k}\frac{\partial^{2k+1}u_j}{\partial x^{2k+1}} + \mathcal{O}(h^{2k+2}), \tag{12}$$

where T_{2k}^0 is the coefficient of the truncation errors and can be obtained by Taylor series expansion. We can construct a $(2k-1)$th-order dissipative interpolation by combining the $(2k)$th-order central interpolation with the $(2k-1)$th derivative,

$$\tilde{u}_{j+1/2} = \tilde{u}_{j+1/2}^{C,(2k)th} + \alpha h^{2k-1}\tilde{u}_{j+1/2}^{(2k-1)}, \tag{13}$$

where $\tilde{u}_{j+1/2}^{C,(2k)th}$ is acquired by the $(2k)$th-order central interpolation, and $\tilde{u}_{j+1/2}^{(2k-1)}$ is the numerical approximation of $\frac{\partial^{2k-1}u_{j+1/2}}{\partial x^{2k-1}}$. Eq. (13) can be rewritten as

$$\tilde{u}_{j+1/2} = u_{j+1/2} + \alpha h^{2k-1}\frac{\partial^{2k-1}u_{j+1/2}}{\partial x^{2k-1}} + \mathcal{O}(h^{2k}). \tag{14}$$

Substituting (14) into (12), a $(2k-1)$th-order HDCS is obtained, and then substituting (12) into (11), we obtain

$$\left(\frac{du}{dt}\right)_j = -c\left[\left(\frac{\partial u}{\partial x}\right)_j + \varphi\alpha h^{2k-1}\frac{\partial^{2k}u_j}{\partial x^{2k}} + \mathcal{O}(h^{2k})\right]. \tag{15}$$

It is obvious that the $(2k-1)$th-order HDCS with $\varphi\alpha \cdot (-1)^{k-1} < 0$ is a dissipative one for the calculating of Eq. (11).

We comment on the characteristics of the HDCSs in the following three remarks.

Remark 1. It is worth to note that the construction of the HDCSs is based on flow variable interpolations other than flux reconstructions, therefore the SCL can be satisfied by the HDCSs with the implementation of the SCMM [11]. For the construction of finite difference scheme, the variable interpolations replacing the flux reconstructions are the key to fulfill the SCMM [11].

Remark 2. The method of adding dissipation in the HDCSs is somewhat similar to that in MUSCL [21] and WCNS [22]. For the WCNS, the δ_1 is chosen as (5) with $p = 0$ and $q > 1$, then the grid stencil is wide due to the interpolations involved in the calculations of the numerical fluxes. For instance, the stencil width of the fifth-order WCNS is nine. On the other hand, the δ_1 for the HDCS is chosen as (5) with $p > 1$ and $q = 1$, as a result, relatively less interpolations are involved and the grid stencil of the HDCS is more compact. For example, a fifth-order HDCS can be derived on a five-point stencil. It is more convenient to construct the boundary and near boundary schemes for the HDCS.

Remark 3. If the dissipative variable interpolations (8) are replaced by the high-order weighted nonlinear ones derived in Ref. [22], the HDCSs can be recognized as the high-order Hybrid Weighted Compact Nonlinear Schemes (HWCNSs) which have been successfully applied for discontinuity capturing [23].

The high-order explicit and compact dissipative interpolations will be derived in the next section. In Section 5, we will discuss the implementation of the SCMM for the HDCS.

3. Details of hybrid cell-edge and cell-node dissipative compact schemes

3.1. Dissipation and dissipative interpolations

According to the methodology discussed in Section 2, the construction of the HDCSs contains two steps. The first step is to compute the first derivatives by a central-type (or non-split) operator. Eq. (5) presents a variety of high-order schemes. The fourth-order explicit δ_1, named E4 is given by

$$E_j' = \frac{4}{3h}\left(\tilde{E}_{j+1/2} - \tilde{E}_{j-1/2}\right) - \frac{1}{6h}(E_{j+1} - E_{j-1}) + \frac{1}{480}h^4\frac{\partial^5 E_j}{\partial x^5} + \mathcal{O}(h^6). \tag{16}$$

A sixth-order explicit δ_1, named E6 is,

$$E_j' = \frac{64}{45h}\left(\tilde{E}_{j+1/2} - \tilde{E}_{j-1/2}\right) - \frac{2}{9h}(E_{j+1} - E_{j-1})$$
$$+ \frac{1}{180h}(E_{j+2} - E_{j-2}) - \frac{1}{5040}h^6\frac{\partial^7 E_j}{\partial x^7} + \mathcal{O}(h^8). \tag{17}$$

A eighth-order explicit δ_1, named E8 is written as,

$$E_j' = \frac{256}{175h}\left(\tilde{E}_{j+1/2} - \tilde{E}_{j-1/2}\right) - \frac{1}{4h}(E_{j+1} - E_{j-1})$$
$$+ \frac{1}{100h}(E_{j+2} - E_{j-2}) - \frac{1}{2100h}(E_{j+3} - E_{j-3}) + \frac{1}{40320}h^8$$
$$\times \frac{\partial^9 E_j}{\partial x^9} + \mathcal{O}(h^{10}). \tag{18}$$

The coefficients for the implicit schemes up to twelfth-order of accuracy are listed in Appendix A.

The second step is to compute $\tilde{U}_{j+1/2}^L$, $\tilde{U}_{j+1/2}^R$ by appropriate dissipative interpolations. A class of $(2k)$th-order explicit central interpolations can be written as

$$\tilde{U}_{j+1/2}^C = \sum_{m=1}^{k}a_{1m}(U_{j+m} + U_{j+1-m}). \tag{19}$$

The detailed expressions of various central interpolations are given in Appendix B.1.

Substituting the interpolation (19) into the differencing operator (5), the explicit type HDCSs is obtained, and it can be easily proved that these schemes are non-dissipative, and only have dispersion errors which may destroy the numerical solutions. In order to introduce dissipation, the $(2k-1)$th derivative is added to the central interpolation as

$$\widetilde{U}^L_{j\pm 1/2} = \widetilde{U}^C_{j\pm 1/2} + \alpha h^{2k-1} U^{(2k-1)}_{j\pm 1/2}. \tag{20}$$

In Appendix B.2, the detailed expressions are given for various high-order numerical derivatives.

To obtain higher resolution in more compact stencils, it is desirable to construct high-order compact interpolations with implicit dissipation. We consider a class of central compact interpolations of $U_{j+1/2}$,

$$\beta \widetilde{U}^C_{j-1/2} + \widetilde{U}^C_{j+1/2} + \beta \widetilde{U}^C_{j+3/2} = \sum_{m=1}^{n} a_{2m}(U_{j+m} + U_{j+1-m}). \tag{21}$$

The coefficients for the central compact interpolations up to twelfth-order of accuracy are listed in Appendix B.1. According to the concept of constructing the explicit dissipative interpolations, high-order compact dissipative interpolations have the form as,

$$A(1-\alpha)\widetilde{U}^L_{j-1/2} + \widetilde{U}^L_{j+1/2} + A(1+\alpha)\widetilde{U}^L_{j+3/2}$$
$$= \sum_{m=1}^{n} a_{3m}(U_{j+m} + U_{j+1-m}) + \alpha \sum_{m=1}^{n} b_{3m}(U_{j+m} - U_{j+1-m}), \tag{22}$$

where α is also the parameter controlling the dissipation intensity. By matching the Taylor series coefficients of various orders, we can get fifth-order dissipative compact interpolation,

$$\frac{3}{10}(1-\alpha)\widetilde{U}^L_{j-1/2} + \widetilde{U}^L_{j+1/2} + \frac{3}{10}(1+\alpha)\widetilde{U}^L_{j+3/2}$$
$$= \frac{3}{4}(U_{j+1} + U_j) + \frac{1}{20}(U_{j+2} + U_{j-1})$$
$$+ \alpha\left[\frac{3}{8}(U_{j+1} - U_j) + \frac{3}{40}(U_{j+2} - U_{j-1})\right] - \frac{3h^5 \alpha U^{(5)}_{j+1/2}}{640} + \mathcal{O}(h^6), \tag{23}$$

seventh-order dissipative compact interpolation,

$$\frac{5}{14}(1-\alpha)\widetilde{U}^L_{j-1/2} + \widetilde{U}^L_{j+1/2} + \frac{5}{14}(1+\alpha)\widetilde{U}^L_{j+3/2}$$
$$= \frac{25}{32}(U_{j+1} + U_j) + \frac{5}{64}(U_{j+2} + U_{j-1}) - \frac{1}{448}(U_{j+3} + U_{j-2})$$
$$+ \alpha\left[\frac{25}{64}(U_{j+1} - U_j) + \frac{15}{128}(U_{j+2} - U_{j-1}) - \frac{5}{896}(U_{j+3} - U_{j-2})\right]$$
$$+ \frac{5h^7 \alpha U^{(7)}_{j+1/2}}{7168} + \mathcal{O}(h^8), \tag{24}$$

ninth-order dissipative compact interpolation,

$$\frac{7}{18}(1-\alpha)\widetilde{U}^L_{j-1/2} + \widetilde{U}^L_{j+1/2} + \frac{7}{18}(1+\alpha)\widetilde{U}^L_{j+3/2}$$
$$= \frac{1225}{1536}(U_{j+1} + U_j) + \frac{49}{512}(U_{j+2} + U_{j-1})$$
$$- \frac{7}{1536}(U_{j+3} + U_{j-2}) + \frac{1}{4608}(U_{j+4} + U_{j-3})$$
$$+ \alpha\left[\frac{1225}{3072}(U_{j+1} - U_j) + \frac{147}{1024}(U_{j+2} - U_{j-1}) - \frac{35}{3072}(U_{j+3} - U_{j-2})\right.$$
$$\left. + \frac{7}{9216}(U_{j+4} - U_{j-3})\right] + \frac{35 h^9 \alpha U^{(9)}_{j+1/2}}{294912} + \mathcal{O}(h^{10}), \tag{25}$$

and eleventh-order dissipative compact interpolation,

$$\frac{9}{22}(1-\alpha)\widetilde{U}^L_{j-1/2} + \widetilde{U}^L_{j+1/2} + \frac{9}{22}(1+\alpha)\widetilde{U}^L_{j+3/2}$$
$$= \frac{6615}{8192}(U_{j+1} + U_j) + \frac{441}{4096}(U_{j+2} + U_{j-1}) - \frac{27}{4096}(U_{j+3} + U_{j-2})$$
$$+ \frac{9}{16384}(U_{j+4} + U_{j-3}) - \frac{5}{180224}(U_{j+5} + U_{j-4})$$
$$+ \alpha\left[\frac{6615}{16384}(U_{j+1} - U_j) + \frac{1323}{8192}(U_{j+2} - U_{j-1}) - \frac{135}{8192}(U_{j+3} - U_{j-2})\right]$$
$$+ \alpha\left[\frac{63}{32768}(U_{j+4} - U_{j-3}) - \frac{45}{360448}(U_{j+5} - U_{j-4})\right]$$
$$+ \frac{63 h^{11} \alpha U^{(11)}_{j+1/2}}{2883584} + \mathcal{O}(h^{12}). \tag{26}$$

By substituting these interpolations into (5), a series of HDCSs can be obtained. The $\widetilde{U}^R_{j+1/2}$ can be easily obtained by setting $\alpha = -\alpha$ in above interpolations (23)–(26). Other types of interpolations can also be derived for the HDCSs, for instance, some biased compact dissipative interpolations are given in Appendix B.3.

Although the above interpolations are dissipative, they are not suitable for the problems which contain strong discontinuities such as shock waves. Nonlinear interpolations are generally required to capture the discontinuities. If the fifth-order weighted nonlinear interpolation derived in [22] is combined with the sixth-order explicit scheme E6, a fifth-order HWCNS [23] being suitable for shock wave problems is obtained.

3.2. Notation of the HDCS

We would like to point out that the differencing operator δ_1 can be combined freely with different interpolations. However, in order to formulate a reasonable HDCS, the accuracy of the δ_1 is suggested to be one order higher than that of interpolations. For the fifth-order explicit interpolations, the sixth-order explicit δ_1, namely E6, is recommended to complete the HDCS. This HDCS is called HDCS-E6E5 for briefness, and the two numbers in the abbreviation sequentially denote the accuracy order of the differencing operator δ_1 and the accuracy order of the interpolation, respectively, meanwhile, the first letter behind the connector denotes the formula type of the δ_1, and the second letter denotes the formula type of the interpolation. The letter E represents explicit type. We also use the letter T to express tri-diagonal compact type and letter P to express penta-diagonal compact type. For example, HDCS-E8T7 denotes the HDCS scheme consists of the explicit 8th-order δ_1, namely E8 (18) and the tri-diagonal 7th-order interpolation (namely Eq. (24)).

3.3. Boundary closures

It is important to derive boundary and near boundary schemes for the HDCSs. The seven-point stencil is used for the operator δ_1 in this paper, thus three levels of boundary and near-boundary schemes are required for the left-side boundary nodes $j = 1, 2, 3$ as well as the right-side boundary nodes $j = N, N-1, N-2$. According to the construction procedure of interior schemes, the construction of boundary schemes also contains the following two steps. First of all, numerical schemes are formulated to approximate the first derivative; Secondly, boundary and near boundary interpolations are derived to compute cell-edge variables.

To compute the first derivative for the HDCSs in this paper, the E6 (17) is used for the near boundary nodes $j = 3$ and $j = N-2$, the E4 (16) is employed for the nodes $j = 2$ and $j = N-1$, and the fourth-order schemes at $j = 1$ and $j = N$ are,

$$E'_1 = \frac{1}{6h}(48\tilde{E}_{3/2} + 16\tilde{E}_{5/2} - 25E_1 - 36E_2 - 3E_3),$$

$$E'_N = \frac{-1}{6h}(48\tilde{E}_{N-1/2} + 16\tilde{E}_{N-3/2} - 25E_N - 36E_{N-1} - 3E_{N-2}).$$

Considering the calculations of the variables at cell-edge, the fifth-order interpolations for $j = 3/2$ and $j = N - 1/2$ are,

$$\tilde{U}^L_{3/2} = \tilde{U}^R_{3/2} = \frac{1}{128}(35U_1 + 140U_2 - 70U_3 + 28U_4 - 5U_5),$$

$$\tilde{U}^L_{N-1/2} = \tilde{U}^R_{N-1/2} = \frac{1}{128}(35U_N + 140U_{N-1} - 70U_{N-2} + 28U_{N-3} - 5U_{N-4}),$$

synchronously, the $\tilde{U}^{L/R}_{5/2}$ and $\tilde{U}^{L/R}_{N-3/2}$ can be calculated by the fifth-order dissipative compact interpolation (23).

4. Fourier analysis, optimization and stability analysis of the HDCS

4.1. Fourier analysis

Fourier analysis is widely used to analyze spectral property of finite difference schemes because it provides an effective way to quantify the resolution and dissipation characteristics. Furthermore, this quantification can be used to further guide an optimization of the scheme. For the purpose of Fourier analysis, the dependent variables are assumed to be periodic over the domain $[0, L]$ of the independent variables, i.e. $u_1 = u_{N+1}$, $h = L/N$, and $x_j = h(j - 1)$. The dependent variables may be decomposed into their Fourier coefficients as

$$U(x_j) = \sum_{k=-N/2}^{N/2} \hat{U}_k exp(i\omega j), \quad (27)$$

where $i = \sqrt{-1}$ and $\omega = 2\pi k h/L$. The differencing errors of the first derivative scheme may be assessed by comparing the Fourier coefficients of the numerical derivative $\hat{U}'^{,Num}_k$ obtained from the differencing schemes with the exact Fourier coefficients $\hat{U}'^{,Exa}_k$. Substituting Eq. (27) into (5) and supposing $\tilde{U}_{j\pm 1/2} = U_{j\pm 1/2}$, we can get $\hat{U}'^{,Num}_k = i\omega^* \hat{U}_k$. ω^* is named as modified wavenumber and is a function of ω, for example, the function for the 4th-order scheme E4 is

$$\omega^*(\omega) = \frac{8sin(\omega/2) - sin\omega}{3}. \quad (28)$$

According to the discussion in Ref. [14], the real part of ω^*, namely, ω^*_r denotes the resolution power of a scheme, and the imaginary part of ω^*, namely, ω^*_i denotes the dissipation intensity of a scheme. High resolution requires ω^*_r close to ω in a broad range of wavenumbers, and dissipative schemes usually require $\omega^*_i \leq 0$. In Eq. (28), $\omega^*_i = 0$ means that the scheme E4 is non-dissipative. The non-dissipative property is common to central schemes. Fig. 1 plots of the modified wavenumbers ω^* against wavenumber ω for several explicit HCSs. Despite that the distributions shown in Fig. 1 belong to the explicit type HCSs, it is indicated that spectral-like resolution is obtained by the hybrid cell-edge and cell-node method. The explicit schemes mean inversion of matrix is not required, and they are more convenient and efficient than tri- or penta-diagonal compact schemes. In this paper, the explicit type schemes are recommended as the δ_1 to construct the HDCSs.

The dissipative interpolations are employed to introduce the dissipation for the HDCSs. Considering the dissipative interpolations, the HDCSs have the following general expression,

$$A\frac{d\vec{U}}{dx} = \frac{1}{h}E\vec{U}_E + \frac{1}{h}B\vec{U}, \quad (29)$$

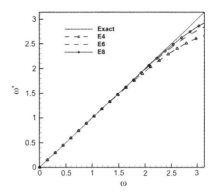

Fig. 1. The modified wavenumbers of explicit δ_1.

where, $\vec{U} = (U_1, U_2, \cdots, U_N)^T$ are values on the cell nodes, and $\vec{U}_E = \left(\tilde{U}_{3/2}, \tilde{U}_{5/2}, \cdots, \tilde{U}_{N-1/2}\right)^T$ are approximation values on the cell edges. The equations used to solve $\vec{U}_E$ can be written as,

$$A_E\vec{U}_E = B_E\vec{U}. \quad (30)$$

If the compact dissipative interpolations (22) are adopted for the HDCSs, the matrix A_E can be fixed by the coefficients on the left-hand side of Eq. (22) and B_E is the matrix consisting of the coefficients on the right-hand side of Eq. (22). It should be noted that the boundary closures in Section 3.3 are applied to make the matrices closed. Substituting (30) into (29), we can get

$$\frac{d\vec{U}}{dx} = \frac{1}{h}A^{-1}\left(B + EA_E^{-1}B_E\right)\vec{U}, \quad \text{or}$$

$$\frac{dU_j}{dx} = \frac{1}{h}\sum_{m=1}^{N}\left(A^{-1}\left(B + EA_E^{-1}B_E\right)\right)_{jm} U_m. \quad (31)$$

The Fourier transforms of the left and right sides of (31) yield

$$i\omega^*_j \hat{U}_j = \sum_{m=1}^{N}\left(A^{-1}\left(B + EA_E^{-1}B_E\right)\right)_{jm} \hat{U}_j FL_m, \quad (32)$$

where $FL_m = e^{i(j\cdots m)\omega}$, $\hat{U}_j$ is the Fourier transform of U_j. Comparing the two sides of (32), it is clear that

$$\omega^*_j = -i\sum_{m=1}^{N}\left(A^{-1}\left(B + EA_E^{-1}B_E\right)\right)_{jm} FL_m, \quad (33)$$

and ω^*_j is the modified wavenumber at node j, it is in general a complex quantity. In order to focus on the resolution and dissipation characteristics of the HDCSs at the interior point, j is set to be $N/2$.

In order to show the improvement of compact dissipative interpolations over explicit dissipative ones on the resolution property, the E6 is chosen as the δ_1 operator, while two different interpolations are adopted to calculate the variables at the cell-edge. The first one is the explicit fifth-order interpolation, and the second one is (23) which also has fifth-order of accuracy but a compact form. Then, it can be figured out that the two schemes shall be marked as HDCS-E6E5 and HDCS-E6T5, respectively. The modified wavenumbers of the two HDCSs are compared in Fig. 2. It is clear that the resolution power is largely improved by applying the compact interpolation. The resolution of fifth-order DCS (DCS-5) is also shown in Fig. 2.

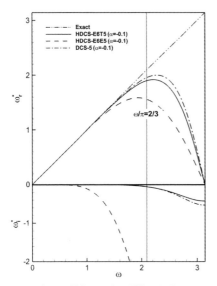

Fig. 2. Modified wavenumbers of different HDCS.

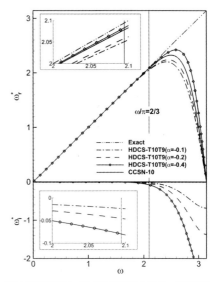

Fig. 4. Modified wavenumbers of HDCS-T10T9 comparing with CCSN-10.

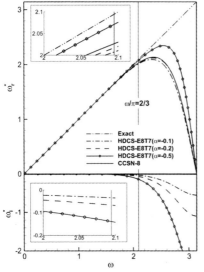

Fig. 3. Modified wavenumbers of HDCS-E8T7 comparing with CCSN-8.

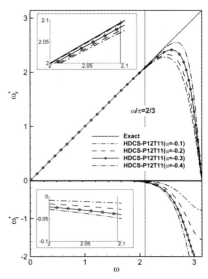

Fig. 5. Modified wavenumbers of HDCS-P12T11 with different α.

The Fourier analysis results of the schemes up to eleventh-order of accuracy are shown and compared to the corresponding CCSN in Figs. 3–5. As shown in Fig. 3, the resolution power of the HDCS-E8T7 may be superior to the CCSN-8. Comparing with the CCSN-10 which is the spectral-like scheme proposed by Lele et al. [14], the HDCS-T10T9 has comparable high resolution (see Fig. 4). The super high resolution of the HDCS-P12T11 can be observed in Fig. 5.

4.2. Optimization of the HDCS

It has been reported that the suitable resolution for finite difference scheme is $(2/3)\omega_m$ if aliasing error exists [20]. In Fig. 3, it can be seen that the HDCS-E8T7 has the potential to achieve this suitable resolution. Furthermore, the stencil of the interpolation of the HDCS-E8T7 is just the same as that of the differencing operator E8. Therefore the HDCS-E8T7 is a pretty choice for the optimization of the HDCS. If much better resolution is required, the HDCS-T8T7 is another choice. The HDCS-E8T7 and the HDCS-T8T7 can be optimized within the scaled wavenumber region of $[0, 2\pi/3]$ by following the concept of the Dispersion-Relation-Preserving (DRP) [24]. The optimization relation is given by

$$\int_0^{2\pi/3} (\omega_r^*)^2 d\omega = MIN. \tag{34}$$

In order to make the schemes to be dissipative ones, the following inequality should be satisfied,

$$\omega_i^* \leq 0 \quad \forall \omega \in [0, \pi]. \tag{35}$$

The HDCS can be further optimized by adopting the method of Ramboer et al. [25], then the following special restrained conditions are added for the optimization,

$$\omega_i^*(\omega = 2\pi/3) < \epsilon_1,$$
$$\omega_i^*(\omega = 2\pi/3 + 0.1) < \epsilon_2, \tag{36}$$
$$2\pi/3 - \epsilon_3 < \omega_r^*(\omega = 2\pi/3) < 2\pi/3.$$

Here ϵ_1 and ϵ_2 are negative numbers to ensure that the optimized schemes are dissipative in the high wavenumber range, ϵ_3 is an arbitrary positive number tuning the dispersion error at $w = 2\pi/3$. In this paper, the parameters are chosen as $\epsilon_1 = -0.05$, $\epsilon_2 = -0.1$ and $\epsilon_3 = 0.04$, respectively.

Since there is only one free parameter in the HDCS-E8T7, i.e., α, the optimized parameter for the HDCS-E8T7 is

$$\alpha = -0.300, \tag{37}$$

and the optimized scheme is named HDCS-E8T7-O which consists of the E8 and the seventh-order dissipative compact interpolation (24) with the optimized α.

There are two optimized parameters for the HDCS-T8T7,

$$\kappa_1 = 0.263, \alpha = -0.801, \tag{38}$$

and the optimized scheme is named HDCS-T8T7-O which consists of the T8 in Appendix A and the seventh-order dissipative compact interpolation (24) with the optimized κ_1 and α, respectively.

The resolution of these two optimized schemes together with that of the CCSN-8 is shown in Fig. 6. Although the HDCS-T8T7-O provides higher resolution, the HDCS-E8T7-O is recommended, because it provides a somewhat optimal balance between efficiency and accuracy. Of course, the δ_1 can be freely combined with any high-order interpolations, for example, HDCS-E8T9. The resolution of the HDCS-E8T9 is also plotted in Fig. 6.

4.3. Stability analysis of HDCS

Supposing the linear convection equation

$$\frac{\partial U}{\partial t} + c \frac{\partial U}{\partial x} = 0, \quad c > 0 \tag{39}$$

is discretized by Eq. (31), its semi-discrete equation can be written as

$$\frac{d\vec{U}}{dt} = \frac{c}{h} R \vec{U} + g(t), \tag{40}$$

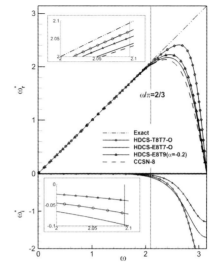

Fig. 6. Modified wavenumbers of the optimal schemes comparing with CCSN-8.

where $R = -A^{-1}(B + EA_E^{-1}B_E)$, and $g(t)$ corresponds to initial and boundary conditions. Without losing generality, $g(t) = 0$ in the following analysis. Eq. (40) is therefore solved for U_j, where j varies from 2 to N.

As discussed in [14], the asymptotic stability condition for the semi-discrete equation is that all eigenvalues of the matrix R have no positive parts. The eigenvalues $\lambda_j = \lambda_j^R + i\lambda_j^I$ are obtained by eigenvalue solver from the IMSL library. Fig. 7 shows the eigenvalue spectra of the HDCS-E6E5, the HDCS-E8T7-O and the HDCS-T8T7-O on uniform grids. It can be seen that these schemes are all asymptotic stable.

It worth to notice that the above asymptotic stable analysis is based on the spatial semi-discrete Eq. (40). Considering the global stability, the effect of particular time discretization should be analyzed. The matrix R can be decomposed as

$$R = L^{-1} \Lambda L, \tag{41}$$

where $\Lambda = diag(\lambda_2 \cdots \lambda_N)$ and L are the diagonal eigenvalue matrix and the left eigenmatrix of R, respectively. The corresponding characteristic variable $\vec{W} = [W_2, \cdots, W_N]^T$ can be written as

$$\vec{W} = L \vec{U}. \tag{42}$$

and Eq. (40) can be transformed to a characteristic form as

$$\frac{dW_j}{dt} = \frac{c}{h} \lambda_j W_j \quad (j = 2, \cdots, N). \tag{43}$$

When the first-order backward time integration is adopted, the following equation is obtained,

$$\frac{W_j^{n+1} - W_j^n}{\Delta t} = \frac{c}{h} \lambda_j W_j^{n+1}, \tag{44}$$

and the amplification factor $|g_j| = \left| W_j^{n+1}/W_j^n \right|$ is written as

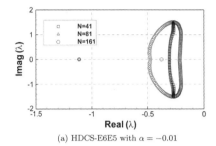

(a) HDCS-E6E5 with $\alpha = -0.01$

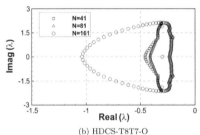

(b) HDCS-T8T7-O

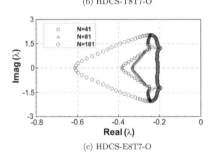

(c) HDCS-E8T7-O

Fig. 7. The eigenvalues spectra on uniform grids.

$$|g_j| = \frac{1}{\sqrt{\left(1 - CFL \cdot \lambda_j^R\right)^2 + \left(CFL \cdot \lambda_j^I\right)^2}}, \quad (45)$$

where $CFL = c\frac{\Delta t}{h}$ is the CFL number.

The asymptotic stability results in Fig. 7 show that the λ_j^Rs of the HDCS-E6E5, the HDCS-E8T7-O and the HDCS-T8T7-O are always negative, which means $|g_j| < 1$, namely the three schemes are numerically stable for all positive CFL numbers if they are employed together with the first-order backward time integration.

Since multi-stage Runge–Kutta time integrations are very popular in CFD community for their high-order accuracy and simplicity, it is worth to find out the amplification factor $|g_j|$ for some widely used Runge–Kutta schemes together with the HDCSs. If a classical pth-order Runge–Kutta scheme 26] is employed, the following equation is obtained,

$$\frac{U_j^{n+1} - U_j^n}{\Delta t} = \frac{c}{h} R^T U_j^n, \quad (46)$$

where $R^T = R + \frac{1}{2!}R^2 \cdot CFL + \cdots + \frac{1}{p!}R^p \cdot CFL^{p-1}$. Therefore the amplification factor $|g_j|$ can be written as

$$|g_j| = \sqrt{\left(1 + CFL \cdot \lambda_j^{TR}\right)^2 + \left(CFL \cdot \lambda_j^{TI}\right)^2}, \quad (47)$$

where λ_j^{TR} and λ_j^{TI} are real and imaginary part of the eigenvalue λ_j^T of R^T, and

$$\lambda_j^T = \sum_{k=1}^{p} \frac{1}{k! \cdot CFL}(CFL \cdot \lambda_j)^k. \quad (48)$$

Fig. 8 plots the maximum $|g_j|$ for the HDCSs together with the fifth-order Runge–Kutta time integration method [26]. It can be noticed that the value of N has slight influence on the maximum allowable CFL number, especially for the seventh-order schemes. Considering the third- and fifth-order Runge–Kutta schemes in Ref. [26], Fig. 9 illustrates the maximum $|g_j|(1 \leq j \leq 41)$ vs CFL number for HDCS-E6E5 ($\alpha = -0.01$), HDCS-E8T7-O and HDCS-T8T7-O. It can be seen that for every given Runge–Kutta method

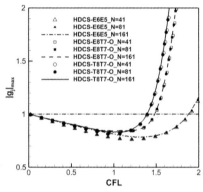

Fig. 8. The amplification factor for HDCS schemes combined with fifth-order Runge–Kutta time integration method.

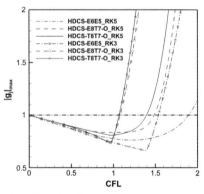

Fig. 9. The amplification factor for HDCS schemes combined with third or fifth-order Runge–Kutta time integration method.

the maximum allowable CFL number of the HDCS-E6E5 is the largest, and that of the HDCS-T8T7 is the smallest. If the same HDCS is employed, it is also shown that maximum allowable CFL number for the fifth-order Runge–Kutta scheme is larger than that for the third-order Runge–Kutta scheme.

5. Governing equations and Implementation of the SCMM

When the fluid flow is simulated by finite difference schemes, the governing equations in Cartesian coordinates (t,x,y,z) are usually transformed into ones in curvilinear (or computational) coordinates (τ,ξ,η,ζ). The three dimensional Euler/Navier–Stokes equations in curvilinear coordinates can be written as

$$\frac{\partial \tilde{U}}{\partial \tau} + \frac{\partial \tilde{E}}{\partial \xi} + \frac{\partial \tilde{F}}{\partial \eta} + \frac{\partial \tilde{G}}{\partial \zeta} = \frac{\theta}{R_e}\left(\frac{\partial \tilde{E}_v}{\partial \xi} + \frac{\partial \tilde{F}_v}{\partial \eta} + \frac{\partial \tilde{G}_v}{\partial \zeta}\right) + \frac{\vec{S}}{J}, \quad (49)$$

$\tilde{U} = U/J,$
$\tilde{E} = (\xi_t U + \xi_x E + \xi_y F + \xi_z G)/J, \quad \tilde{E}_v = (\xi_x E_v + \xi_y F_v + \xi_z G_v)/J,$
$\tilde{F} = (\eta_t U + \eta_x E + \eta_y F + \eta_z G)/J, \quad \tilde{F}_v = (\eta_x E_v + \eta_y F_v + \eta_z G_v)/J,$
$\tilde{G} = (\zeta_t U + \zeta_x E + \zeta_y F + \zeta_z G)/J, \quad \tilde{G}_v = (\zeta_x E_v + \zeta_y F_v + \zeta_z G_v)/J,$

where $\theta = 0$ and 1 for Euler equations and Navier–Stokes equations, respectively; $\vec{S}$ is a source term; J is the Jacobian of grid transformation; $\xi_t, \xi_x, \xi_y, \xi_z, \eta_t, \eta_x, \eta_y, \eta_z, \zeta_t, \zeta_x, \zeta_y, \zeta_z$ are grid derivatives. The details of the equations can be found in Ref. [7]. In this paper, $\vec{S}$ is non-zero only for the computation of acoustic wave scattering of multiple cylinders (Section 6.3).

The coordinate transformation brings about the conception of the GCL which consists of the VCL and the SCL.

The VCL means

$$I_t = (1/J)_\tau + \left(\tilde{\xi}_t\right)_\xi + (\tilde{\eta}_t)_\eta + \left(\tilde{\zeta}_t\right)_\zeta = 0, \quad (50)$$

and the SCL means

$$I_x = \left(\tilde{\xi}_x\right)_\xi + (\tilde{\eta}_x)_\eta + \left(\tilde{\zeta}_x\right)_\zeta = 0,$$
$$I_y = \left(\tilde{\xi}_y\right)_\xi + (\tilde{\eta}_y)_\eta + \left(\tilde{\zeta}_y\right)_\zeta = 0, \quad (51)$$
$$I_z = \left(\tilde{\xi}_z\right)_\xi + (\tilde{\eta}_z)_\eta + \left(\tilde{\zeta}_z\right)_\zeta = 0.$$

In this paper the time-constant grids are considered, thus the VCL is satisfied automatically. If the governing equations are discretized by differencing schemes δ_1, the SCL can be written as

$$\begin{cases} I_x = \delta_1^\xi\left(\tilde{\xi}_x\right) + \delta_1^\eta(\tilde{\eta}_x) + \delta_1^\zeta\left(\tilde{\zeta}_x\right) = 0, \\ I_y = \delta_1^\xi\left(\tilde{\xi}_y\right) + \delta_1^\eta(\tilde{\eta}_y) + \delta_1^\zeta\left(\tilde{\zeta}_y\right) = 0, \\ I_z = \delta_1^\xi\left(\tilde{\xi}_z\right) + \delta_1^\eta(\tilde{\eta}_z) + \delta_1^\zeta\left(\tilde{\zeta}_z\right) = 0. \end{cases} \quad (52)$$

where $\delta_1^\xi, \delta_1^\eta$ and δ_1^ζ are the finite differencing operators used to discretize the flow fluxes in the ξ, η and ζ coordinate directions, respectively.

The SCL can be satisfied by applying the SCMM proposed by Deng et al. [11]. In order to fulfill the SCMM, the metrics and Jacobian are required to be calculated through the following conservative form [11]

$$\begin{cases} \tilde{\xi}_x = \frac{1}{2}\left[\delta_2^\eta(\delta_3^\zeta(y)z) - \delta_2^\zeta(\delta_3^\eta(y)z) + \delta_2^\eta(y\delta_3^\zeta(z)) - \delta_2^\zeta(y\delta_3^\eta(z))\right], \\ \tilde{\eta}_x = \frac{1}{2}\left[\delta_2^\zeta(\delta_3^\xi(y)z) - \delta_2^\xi(\delta_3^\zeta(y)z) + \delta_2^\zeta(y\delta_3^\xi(z)) - \delta_2^\xi(y\delta_3^\zeta(z))\right], \\ \tilde{\zeta}_x = \frac{1}{2}\left[\delta_2^\xi(\delta_3^\eta(y)z) - \delta_2^\eta(\delta_3^\xi(y)z) + \delta_2^\xi(y\delta_3^\eta(z)) - \delta_2^\eta(y\delta_3^\xi(z))\right], \end{cases} \quad (53)$$

$$J^{-1} = \frac{1}{3}\left[\delta_1^\xi\left(x\tilde{\xi}_x + y\tilde{\xi}_y + z\tilde{\xi}_z\right) + \delta_1^\eta\left(x\tilde{\eta}_x + y\tilde{\eta}_y + z\tilde{\eta}_z\right) + \delta_1^\zeta\left(x\tilde{\zeta}_x + y\tilde{\zeta}_y + z\tilde{\zeta}_z\right)\right] \quad (54)$$

where, $\delta_2^\xi, \delta_2^\eta, \cdots, \delta_3^\zeta$ are finite differencing operators used to compute the grid derivatives.

In order to satisfy the SCL, the finite differencing operators δ_1^ξ, δ_1^η and δ_1^ζ shall not be split into two upwind parts [7]. It is true that the operator δ_1 for the HDCS is not split, but the cell-edge flow variables $U_{j+1/2}^L$ and $U_{j+1/2}^R$ are computed by dissipative interpolations, and $U_{j+1/2}^L$ is usually not equal to $U_{j+1/2}^R$. Since the grid metrics are not split, the dissipative interpolation shall be avoided in the process of computing metric derivatives. Instead, central-type interpolations are used to compute the grid metrics at the cell-edge of the interior points. For boundary and near boundary points, biased interpolations can be used without splitting. Then it is figured out that, in curvilinear coordinates, Eq. (7) may be rewritten as

$$\tilde{E}_{j+1/2} = E\left(\tilde{U}_{j+1/2}^L, \tilde{U}_{j+1/2}^R, \tilde{\chi}_{xj\pm1/2}, \tilde{\chi}_{yj\pm1/2}, \tilde{\chi}_{zj\pm1/2}\right), \chi = \xi, \eta, \text{or}\zeta. \quad (55)$$

The central interpolations defined in Eq. (21) (also given in Appendix B.1) and the biased interpolations in Appendix B.3 can be applied here to compute $\tilde{\chi}_{xj\pm1/2}$, $\tilde{\chi}_{yj\pm1/2}$, and $\tilde{\chi}_{zj\pm1/2}$ for interior points and boundary points, respectively.

Following the process deriving the CMM given by Deng et al. [7], we can proved that the SCL is satisfied if $\delta_1^\xi = \delta_2^\xi$, $\delta_1^\eta = \delta_2^\eta$, and $\delta_1^\zeta = \delta_2^\zeta$. Although the inner-level differencing operators δ_3 in Eq. (53) have no effect on the SCL, the constraint $\delta_3 = \delta_2$ is recommended by Deng et al. [7]. More recently, the constraint $\delta_3 = \delta_2$ has been explained from geometry viewpoint, and the SCMM, which can evidently increase the numerical accuracy on irregular grids, is proposed in Ref. [11]. To eliminate the SCL errors on curvilinear mesh, we calculate the grid metrics with the conservative form (53) by the same schemes used for flux derivative calculations, i.e. $\delta_I = \delta_{II} = \delta_{III}$, to implement the SCMM.

6. Numerical results

6.1. Vortex convection on curvilinear mesh

This case is chosen to test the accuracy of the HDCS. The governing equations are the Euler equations. The flow is set to be two-dimensional with $(u,v,p,T) = (1.0,1,1)$. The initial condition is imposed by prescribing a vortex, centered at the location (x_c, y_c), and satisfying the following relations:

$$\begin{cases} (\delta u, \delta v) = \varepsilon \tau e^{\beta(1-\tau^2)}(\sin\theta, -\cos\theta), \\ \delta T = -(\gamma-1)\varepsilon^2 e^{2\beta(1-\tau^2)}/(4\alpha\gamma), \\ \delta S = 0 \end{cases}$$

where $\alpha = \beta = 0.8$, $\varepsilon = 0.3$, $\tau = r/r_c$, $r = \left[(x-x_c)^2 + (y-y_c)^2\right]^{1/2}$, and $r_c = 0.2$. Here $T = p/\rho$ is the temperature and $S = p/\rho^\gamma$ is the entropy.

Six meshes spanning the domain $-4 \leqslant x \leqslant 4$, $-4 \leqslant y \leqslant 4$ are used to calculate the accuracy order of the HDCS, the grid sizes are $61 \times 61, 71 \times 71, 81 \times 81, 91 \times 91, 101 \times 101$ and 111×111, respectively. The computational grid (61×61) is shown in Fig. 10a, the mesh is curvilinear due to the protuberances on the top-bound and bottom-bound. The LODI (Local One-Dimensional Inviscid) approximations [27] are applied to all the boundaries. Starting from $t = 0$ with initial position $x_c = 0$ and $y_c = 0$, the

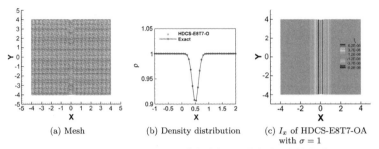

(a) Mesh   (b) Density distribution   (c) I_x of HDCS-E8T7-OA with $\sigma = 1$

Fig. 10. Performance on a curvilinear mesh. (colored picture attached at the end of the book)

computation is carried out until the vortex core arrives at the location $(0.5, 0)$. A fifth-order five-stage Runge–Kutta scheme [26] is used for the time integration, a series of the HDCSs is examined in this test. The accuracy order of all these HDCSs satisfying the SCL is preserved, and we only give the result of the HDCS-E8T7-O here. For the mesh (61×61), Fig. 10b plots the density distributions along the central line. To show the influence of the SCL, we design a new scheme HDCS-E8T7-OA, which is same as the HDCS-E8T7-O except that the grid metric derivatives are calculated by the operator with $\delta_3 = \delta_2 \neq \delta_1$. δ_3 and δ_2 are the combination of the E8 and E6. For example, the $\tilde{\xi}_x$ for the HDCS-E8T7-OA can be written as

$$\tilde{\xi}_x^A = \sigma \tilde{\xi}_x^{E6} + (1-\sigma)\tilde{\xi}_x^{E8}, \tag{56}$$

where $0 \leq \sigma \leq 1$ is the weight coefficient. $\tilde{\xi}_x^{E6}$ and $\tilde{\xi}_x^{E8}$ are the grid metric derivatives obtained by the E6 and the E8, respectively. If $\sigma = 0$, the HDCS-E8T7-O satisfying the SCL is recovered. However, the SCL is dissatisfied if $\sigma \neq 0$. The SCL errors in I_x with $\sigma = 1$ on the mesh (61×61) are plotted in Fig. 10c. The L_1 errors of ρ as well as the order of accuracy are listed in Table 1. It can be seen that the accuracy order of the HDCS-E8T7-O satisfying the SCL is preserved. However, if the SCL is dissatisfied, the accuracy order is diminished, or even disappeared.

6.2. Acoustic waves generated by oscillating cylinder

When the cylinder oscillates acoustically left and right in the static inviscid flow, it will generate a dipole sound field in the direction of the oscillation. Let the amplitude and the frequency of the oscillation be ε and ω, respectively, then the normal velocity of the cylinder surface can be written as $\tilde{U} = \varepsilon c_\infty cos\phi sin(\omega t)$, where ϕ is measured from the axis of oscillation. In this test, the axis of oscillation is equal to x axis. Therefore, the time derivative of normal velocity is given by,

$$\frac{d\tilde{U}}{dt}\bigg|_{wall} = \varepsilon \omega c_\infty cos\phi cos(\omega t), \tag{57}$$

Table 1
Accuracy order of the HDCS-E8T7-O and HDCS-E8T7-OA on curvilinear mesh.

Mesh	HDCS-E8T7-O		HDCS-E8T7-OA($\sigma = 1$)	
	L_1	Accuracy	L_1	Accuracy
61 × 61	1.606E−5	–	3.14E−5	–
71 × 71	6.682E−6	5.69	1.91E−5	3.22
81 × 81	2.738E−6	6.68	1.91E−5	0.00
91 × 91	1.253E−6	6.64	9.39E−6	6.03
101 × 101	6.187E−7	6.70	9.81E−6	−0.42
111 × 111	3.242E−7	6.78	6.19E−6	4.83

which is imposed into the characteristic inviscid wall conditions [28] for the wall boundary conditions of the oscillating cylinder. The CBICs [12] and far-field boundary conditions based on the LODI approximations [27] are also used in the present work. The analytic solution is available for this case in Ref. [29]. The analytic solution for the sound pressure level (SPL) expressed by root-mean-square (RMS) acoustic pressure $p'(x,t) = p(x,t) - p_\infty$ is given by,

$$p'_{rms}(x) = \overline{p'^2}^{1/2}(x) = \frac{1}{\sqrt{2}} \varepsilon \rho_\infty c_\infty^2 cos\phi \left|\frac{H'_0(\omega r/c_\infty)}{H''_0(\omega R/c_\infty)}\right|, \tag{58}$$

where $H_0(z)$ is a zeroth-order Hankel function, $H'_o(z) = dH_0/dz$, r is distance from the center of the cylinder, R is the cylinder radius, ρ_∞, p_∞ and c_∞ are the reference density, pressure and sound speed, respectively.

A fifth-order five-stage Runge–Kutta scheme [26] is used for the time integration. The dipole sound field is calculated with the initial velocity is set to be zero. The cylinder oscillates with $\varepsilon = 0.0001$ and $\omega = \pi c_\infty/R$. The grid system used for this problem is shown in Fig. 11. The mesh is clustered toward the cylinder, and the smallest size of the grid is $\Delta_{min} = 0.02$, the number of grid points is 200×139.

In order to show the improvement of the HDCS over the DCS, the solutions obtained by the HDCS-E6E5 and the DCS-5 are

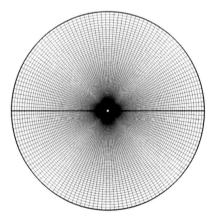

Fig. 11. Computed domain and mesh system of oscillating cylinder case.

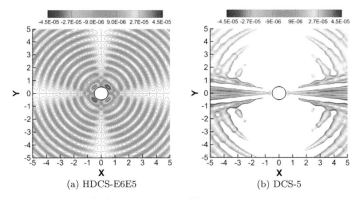

Fig. 12. Contours of v-velocity around an inviscid cylinder. (colored picture attached at the end of the book)

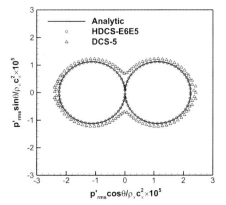

Fig. 13. Directivity patterns of SPLs around an inviscid cylinder.

compared with each other. The contours of instantaneous v-velocity in Fig. 12 clearly show that the sound field calculated by the HDCS-E6E5 is much better. To compare the calculated solutions with the analytic solution quantitatively, the directivity patterns of the SPL at the distance $r/R = 10$ are given in Fig. 13. It is obvious that the solution of the HDCS-E6E5 is almost identical to the analytic solution and is much better than that of the DCS-5. It should be noticed that both the two schemes have fifth-order accuracy and the DCS-5 has better resolution than the HDCS-E6E5 (shown in Fig. 2). The only reason for the worse solutions of DCS-5 is that the SCL is not satisfied.

6.3. Scattering of acoustic waves by multiple cylinder

The case examined here is specified in Category 2 of the Fourth Computational Aeroacoustics Workshop on Benchmark Problems. The source term on the right-hand side of Euler Eq. (49) is applied to the energy equation only, and it is specified as

$$S = exp[-2ln2(x^2 + y^2)]sin(8\pi t). \tag{59}$$

A cubic ramping function is added to the above Eq. (59) to gradually introduce the source into the computational domain over some finite amount of time at the beginning of the computation. At the

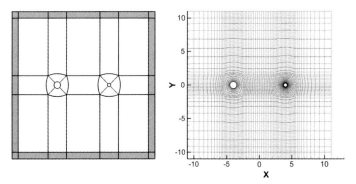

Fig. 14. Computed domain and mesh of two-cylinder scattering case.

Table 2
The maximal errors of l_x and l_y for the HDCS-E8T7-O and HDCS-E8T7-OA.

	HDCS-E8T7-O	HDCS-E8T7-OA Weight σ					
		10^{-5}	10^{-4}	10^{-3}	10^{-2}	10^{-1}	1
max(l_x)	1.37E − 14	1.37E − 8	1.37E − 7	1.37E − 6	1.37E − 5	1.37E − 4	1.37E − 3
max(l_y)	1.47E − 14	1.37E − 8	1.37E − 7	1.37E − 6	1.37E − 5	1.37E − 4	1.37E − 3

 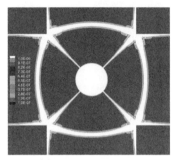

(a) The whole computed domain   (b) Near the left cylinder

Fig. 15. The SCL error ($|l_x| + |l_y|$) of HDCS-E8T7-OA with $\sigma = 1$. (colored picture attached at the end of the book)

solid surfaces, the physical velocity components are found from the contravariant velocity components by setting $V = \eta_x u + \eta_y v + \eta_z w$ at the surface equal to zero and extrapolating for $U = \xi_x u + \xi_y v + \xi_z w$ on the surface using a fifth-order formulation. The pressure and density at the wall are also calculated by using the fifth-order extrapolation. At the far-field boundaries, sponge technique [30] is employed. The computation is advanced with a time step of 0.002 (125 time steps per source period) for a total of 20,000 time steps, with mean values computed over the last 2000 time steps. Fig. 14 shows the computational domain. Three meshes have been used to solve this problem. The numbers of the grid points are 469 × 409 (coarse mesh), 721 × 685 (medium mesh), 841 × 775 (fine mesh), respectively.

The time integration is still the fifth-order five-stage Runge–Kutta scheme [26]. The HDCS-E8T7-O and the HDCS-E8T7-OA are employed to calculate numerical flux derivatives. It should be reminded that there is a parameter σ ($0 \leqslant \sigma \leqslant 1$) which controls the deviation from the SCL for the HDCS-E8T7-OA. The maximal SCL errors of the HDCS-E8T7-O and HDCS-E8T7-OA on the medium mesh are listed in Table 2. It can be seen that the SCL error is highest when $\sigma = 1$. As shown in Fig. 15, the largest SCL error of the HDCS-E8T7-OA with $\sigma = 1$ appears at the singular point.

Fig. 16 plots the computed RMS value of the fluctuating pressure for the HDCS-E8T7-O and HDCS-E8T7-OA. It is shown that acoustic field is not contaminated when the SCL error is low enough ($\sigma = 10^{-5}$). However, when σ is enlarged from 10^{-5} to 10^{-3}, the field is getting worse near the singular points. As shown in Fig. 17, the whole acoustic field is contaminated when σ is larger than 10^{-2}, and the worst area is still near the singular point. It is clear that the SCL errors shall be controlled to be low enough for accurate numerical simulations. In Fig. 18, we plot the RMS value of the fluctuating pressure on the surface of the cylinder calculated by the HDCS-E8T7-O for the three meshes. Numerical solution is approaching the analytic solution as the number of grid points increases. On the fine mesh, excellent agreement has been observed between the analytic and the numerical solutions, which manifests the high resolution of the HDCS-E8T7-O.

6.4. Flow over a high lift trapezoidal wing

The performance of this newly developed high-order scheme HDCS on complex geometry is examined in this subsection for the case of flow past a high lift trapezoidal wing [31]. Trapezoidal wing layout and key geometric have been presented in Ref [32]. In this paper, Configuration 1 (slat deflected at 30° and flap deflected at 25°) is chosen to show the potential ability of the HDCS-E8T7-O scheme for complex configurations. The flow conditions are Mach number $M = 0.2$, $Re = 4.3 \times 10^6$ (corresponding to mean aerodynamic chord of the wing), and the solutions have been computed at an angle-of-attack of 13°. The reference quantities are half-model reference area $S/2 = 2.0465$ m², mean aerodynamic chord $c_{ref} = 1.0067$ m. For this case, the Shear Stress Transport (SST) turbulence model [33] is adopted for the turbulence simulation, and the lower/upper symmetric Gauss–Seidel method is used for temporal integration.

The computation is performed on a grid with 7.0 million points which consists of 372 structured blocks. Fig. 19 shows the grid system and the close-up views near the area where the grid topology is complex. The CBICs [12] are used to handle this complex multi-block structured grid here. Solution convergence is evaluated by monitoring variations in the force coefficients. A solution is considered converged when fluctuations in the lift coefficient is reduced to less than 0.5% of its average value calculated over the previous 6000 iterations. The experimental lift coefficient and drag

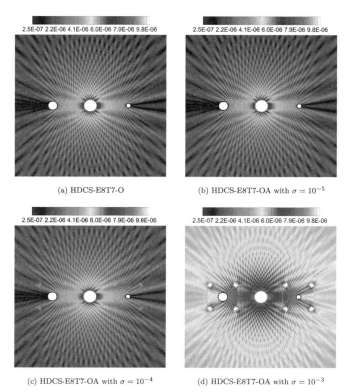

Fig. 16. Mean-squared fluctuating pressure contours of HDCS-E8T7-O and HDCS-E8T7-OA.
(colored picture attached at the end of the book)

(a) HDCS-E8T7-OA with $\sigma = 10^{-2}$ (b) HDCS-E8T7-OA with $\sigma = 10^{-1}$ (c) HDCS-E8T7-OA with $\sigma = 1$

Fig. 17. Mean-squared fluctuating pressure contours of HDCS-E8T7-OA.
(colored picture attached at the end of the book)

coefficient are 2.0468 and 0.3330, respectively. The computed average results are 2.0095 and 0.3213, correspondingly. The differences between the experimental results and the computational results are about 1.8% for lift coefficient and 3.5% for drag coefficient. Fig. 20 shows the computational surface pressures comparing with experimental data at spanwise locations (non-

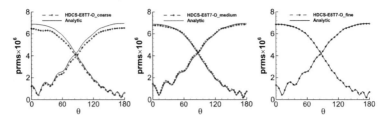

Fig. 18. RMS fluctuating pressure on the surface of cylinders.

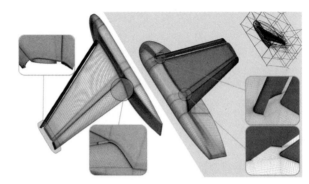

Fig. 19. Gird system of trapezoidal wing case.

(colored picture attached at the end of the book)

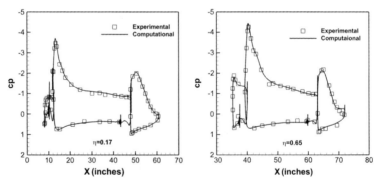

Fig. 20. Pressure comparison at two spanwise locations.

dimensionalized by the semi-span length) $\eta = 0.17$ and $\eta = 0.65$. Good agreement between the experimental and the numerical solutions has been observed. Fig. 21 shows the pressure contours and streamlines in the flow-field. The solutions of this case demonstrate the ability of the newly developed high-order scheme HDCS-E8T7-O in solving flow problems with complex configuration. To illustrate the importance of the SCL in the application of high-order finite difference schemes on complex geometry, we continue the calculation using the HDCS-E8T7-OA with $\sigma = 1$ from the solution of the HDCS-E8T7-O. The computing is collapse anticipatively after 11253 iterations. In Fig. 22, we plot the Mach number calculated by the HDCS-E8T7-OA on the symmetry surface at the 5000 iteration step. Comparing with the HDCS-E8T7-O, it is clear that the solution of the HDCS-E8T7-OA is unreasonable.

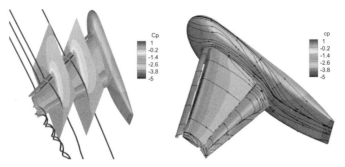

Fig. 21. Pressure contours and streamlines.(colored picture attached at the end of the book)

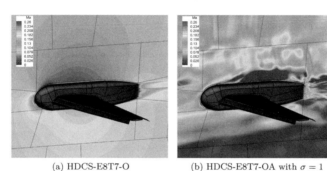

(a) HDCS-E8T7-O   (b) HDCS-E8T7-OA with $\sigma = 1$

Fig. 22. The Mach number on the symmetry surface. (colored picture attached at the end of the book)

Table A.1
Coefficients of the implicit HCSs.

$$\kappa_2 E'_{j-2} + \kappa_1 E'_{j-1} + E'_j + \kappa_1 E'_{j+1} + \kappa_2 E'_{j+2} = \frac{\varphi_1}{h}\left(\widetilde{E}_{j+1/2} - \widetilde{E}_{j-1/2}\right) + \frac{1}{h}\sum_m a_m(E_{j+m} - E_{j-m})$$

Marker symbol	Accuracy order	κ_2	κ_1	φ_1	a_1	a_2	a_3
T4	4th-order	0	κ_1	$\frac{4(1-4\kappa_1)}{3}$	$-\frac{1-22\kappa_1}{6}$	0	0
T6	6th-order	0	κ_1	$\frac{64(1-3\kappa_1)}{45}$	$-\frac{2-27\kappa_1}{9}$	$\frac{1+12\kappa_1}{180}$	0
T8	8th-order	0	κ_1	$\frac{256(3-8\kappa_1)}{525}$	$-\frac{1-11\kappa_1}{4}$	$\frac{3+32\kappa_1}{300}$	$-\frac{1+9\kappa_1}{2100}$
T10	10th-order	0	$-\frac{1}{8}$	$\frac{1024}{525}$	$-\frac{19}{32}$	$-\frac{5}{300}$	$\frac{1}{16800}$
P6	6th-order	κ_2	κ_1	$\frac{64(1-3\kappa_1+12\kappa_2-60a_3)}{45}$	$-\frac{2-27\kappa_1+120\kappa_2-525a_3}{9}$	$\frac{1+12\kappa_1+522\kappa_2-1680a_3}{180}$	a_3
P8	8th-order	κ_2	κ_1	$\frac{256(3-8\kappa_1+20\kappa_2)}{525}$	$-\frac{3-33\kappa_1+100\kappa_2}{12}$	$\frac{3+32\kappa_1+630\kappa_2}{300}$	$-\frac{1+9\kappa_1-180\kappa_2}{2100}$
P10	10th-order	$\frac{1+8\kappa_1}{100}$	κ_1	$\frac{4096(1-2\kappa_1)}{2625}$	$-\frac{4-25\kappa_1}{12}$	$\frac{93+824\kappa_1}{3000}$	$-\frac{4+27\kappa_1}{10500}$
P12	12th-order	$-\frac{1}{300}$	$-\frac{1}{6}$	$\frac{16384}{7875}$	$-\frac{49}{72}$	$-\frac{133}{9000}$	$-\frac{1}{21000}$

Here, "T" stands for tri-diagonal scheme, and "P" stands for penta-diagonal scheme.

Table B.1
High-order central interpolations.

$$\beta \widetilde{U}^c_{j-1/2} + \widetilde{U}^c_{j+1/2} + \beta \widetilde{U}^c_{j+3/2} = \sum_{m=1}^{5} a_{2m}(U_{j+m} + U_{j+1-m})$$

	β	a_{21}	a_{22}	a_{23}	a_{24}	a_{25}
4th-order, explicit	0	$\frac{9}{16}$	$-\frac{1}{16}$	0	0	0
4th-order, compact	$\frac{1}{6}$	$\frac{2}{3}$	0	0	0	0
6th-order, explicit	0	$\frac{75}{128}$	$-\frac{25}{256}$	$\frac{3}{256}$	0	0
6th-order, compact	$\frac{3}{10}$	$\frac{3}{4}$	$\frac{1}{20}$	0	0	0
8th-order, explicit	0	$\frac{1225}{2048}$	$-\frac{245}{2048}$	$\frac{49}{2048}$	$-\frac{5}{2048}$	0
8th-order, compact	$\frac{5}{14}$	$\frac{25}{32}$	$\frac{5}{64}$	$-\frac{1}{448}$	0	0
10th-order, compact	$\frac{7}{18}$	$\frac{1225}{1536}$	$\frac{49}{512}$	$-\frac{7}{1536}$	$\frac{1}{4608}$	0
12th-order, compact	$\frac{9}{22}$	$\frac{6615}{8192}$	$\frac{441}{4096}$	$-\frac{27}{4096}$	$\frac{9}{16384}$	$-\frac{5}{180224}$

Table B.2
High-order numerical derivatives.

High-order derivatives: $\widetilde{U}^{(n)}_{j+1/2} = \frac{1}{h^n}\sum_{m=1}^{4} b_{2m}(U_{j+m} - U_{j+1-m})$

	b_{21}	b_{22}	b_{23}	b_{24}
3rd derivative	-3	1	0	0
5th derivative	10	-5	1	0
7th derivative	-35	21	-7	1

Table B.3
Biased compact interpolations.

Left side: $\alpha_L \tilde{U}_{j-1/2} + \tilde{U}_{j+1/2} + \beta_L \tilde{U}_{j+3/2} = \sum_{m=-3}^{3} L_m U_{j+m}$

	α_L	β_L	L_{-3}	L_{-2}	L_{-1}	L_0	L_1	L_2	L_3
5th-order	$\frac{1}{2}$	$\frac{1}{10}$	0	0	$\frac{1}{10}$	1	1	0	0
7th-order	$\frac{1}{2}$	$\frac{3}{14}$	0	$-\frac{1}{224}$	$\frac{1}{8}$	$\frac{15}{16}$	$\frac{5}{8}$	$\frac{1}{12}$	0
9th-order	$\frac{1}{2}$	$\frac{5}{18}$	$\frac{1}{2304}$	$-\frac{1}{128}$	$\frac{35}{256}$	$\frac{175}{192}$	$\frac{125}{256}$	$\frac{7}{128}$	$-\frac{1}{768}$

Right side: $\alpha_R \tilde{U}_{j-1/2} + \tilde{U}_{j+1/2} + \beta_R \tilde{U}_{j+3/2} = \sum_{m=-3}^{3} R_m U_{j+m}$

	α_R	β_R	R_{-3}	R_{-2}	R_{-1}	R_0	R_1	R_2	R_3
5th-order	$\frac{1}{10}$	$\frac{1}{2}$	0	0	$\frac{1}{2}$	1	$\frac{1}{10}$	0	0
7th-order	$\frac{3}{14}$	$\frac{1}{2}$	0	$\frac{1}{32}$	$\frac{5}{8}$	$\frac{15}{16}$	$\frac{1}{8}$	$-\frac{1}{224}$	0
9th-order	$\frac{5}{18}$	$\frac{1}{2}$	$-\frac{1}{768}$	$\frac{7}{128}$	$\frac{175}{256}$	$\frac{175}{192}$	$\frac{35}{256}$	$-\frac{1}{128}$	$\frac{1}{2304}$

7. Conclusions

Although a variety of high-order finite difference schemes (such as compact schemes) have been constructed for accurate computations of partial differential equations, it is still difficult to find one that can possess inherent dissipation and fulfill the SCMM concurrently. Dissipation is useful in preventing high-wavenumber oscillations which are common in central scheme applications. Meanwhile, growing evidences show that fulfilling SCMM is a crucial precondition for successful applications of high-order finite difference schemes in curvilinear grids.

A new class of high-order accurate and high resolution finite difference schemes named HDCSs has been derived by following the principle of holding dissipation and fulfilling the SCMM simultaneously. Due to the dissipation, the HDCSs can be used without extra oscillation-preventing methods such as filters. Since the SCMM is satisfied, these schemes are suitable for complex geometries and their accuracy orders also can be preserved on curvilinear meshes. Another advantage of the HDCSs is that high-order accuracy can be obtained in more compact stencils, thus the HDCSs can be applied on complex multi-block structured grid conveniently.

The construction of the HDCSs contains two steps: finite differencing and dissipative interpolation. The finite differencing operators, δ_1, are non-split (generally central type), then the SCMM can be satisfied. The second step is also vital to the HDCSs because dissipation should be added into HDCSs. A new method is given to construct dissipative interpolations including explicit-type ones and compact-type ones. Different HDCSs can be obtained by combining differencing operators with dissipative interpolations freely, however, in order to formulate a reasonable HDCS, the differencing operator is suggested to be one order of accuracy higher than the interpolation. Two optimized schemes HDCS-E8T7-O and HDCS-T8T7-O are demonstrated capable of achieving the perfect resolution $(2/3)\omega_m$ by using the Fourier analysis method. The stability property of the HDCSs is investigated by asymptotic stability analysis and amplification factor analysis. The results show that the interior schemes as well as the boundary and near boundary schemes tend to be stable.

Several test cases are examined by the new HDCSs. According to the solutions of vortex convection, the SCMM is very important for the HDCSs on curvilinear meshes, because the accuracy order may be diminished, or even disappeared when the SCMM is not satisfied. It has been demonstrated by the solutions of Euler tests that the HDCS-E6E5 has manifest improvement in handling complex meshes over the DCS-5. Excellent agreement between the analytic and numerical solutions of acoustic wave of multiple cylinder indicates that the HDCS-E8T7-O has fine resolution, however, the solutions may be contaminated if the SCMM is not satisfied. Furthermore, the ability of the newly developed high-order scheme HDCS-E8T7-O in solving complex configuration problems has been demonstrated by simulating the flow past high lift trapezoidal wing.

It should be noted that the HDCSs can not handle strong discontinuities such as shock waves. For problems containing discontinuities, some kinds of nonlinear interpolations may be applied instead of the linear dissipative interpolations. An example can be found in Ref. [23] where the fifth-order nonlinear interpolation of Deng and Zhang [22] is applied to compute the cell-edge values.

Acknowledgments

This study was supported by the project of National Basic Research Program of China (Grant No. 2009CB723800), National Natural Science Foundation of China (Grant Nos. 11072259 and 11202226) and the foundation of State Key Laboratory of Aerodynamics (Grant Nos. JBKY11030902 and JBKY11010100). The authors are grateful to Assistant researcher Yaobing Min for his useful discussions.

Appendix A. Coefficient of Eq. (5)

The construction of the differencing operator δ_1 is a key issue for the construction of the HDCSs satisfying the GCL. Following the principle of fulfilling the SCMM, a series of HCSs is designed as the δ_1 for the HDCSs satisfying the GCL. Table A.1 lists the coefficients for the implicit type of the HCSs.

Appendix B. Interpolations of the HDCS

For the construction of the HDCSs, different interpolations may be derived. In this appendix, we give the formulations for the high-order numerical derivatives (which are usually added into central interpolations), explicit interpolation, central compact interpolation and biased compact interpolation.

B.1. High-order central interpolations

Table B.1 lists some explicit and compact interpolations. In order to obtain higher resolution, compact interpolations are recommended.

B.2. High-order numerical derivatives

Actually, different dissipative interpolations of $U_{j-1/2}$ can be obtained by combining a central interpolation with a high-order derivative. In Table B.2, several explicit forms of high-order numerical derivatives are given.

B.3. Biased compact interpolation

The biased compact interpolations are given in Table B.3. Although high resolution can be achieved by the HDCSs using these interpolations, it is undesirable that $\tilde{U}^L_{j+1/2}$ and $\tilde{U}^R_{j+1/2}$ of these interpolations adopt different grid stencils.

References

[1] Rizzetta DP, Visbal MR, Morgan PE. A high-order compact finite-difference scheme for large-eddy simulation of active flow control. Progress Aerospace Sci 2008;44:397–426.
[2] Shen YQ, Zha G Ch. Generalized finite compact difference scheme for shock/complex flowfield interaction. J Comput Phys 2011;230:4419–36.
[3] Wang ZJ. High-order methods for the Euler and Navier–Stokes equations on unstructured grids. Progress Aerospace Sci 2007;43:1–41.
[4] Ekaterinaris JA. High-order accurate, low numerical diffusion methods for aerodynamics. Progress Aerospace Sci 2005;41(3-4):192–300.

[5] Visbal MR, Gaitonde DV. On the use of higher-order finite-difference schemes on curvilinear and deforming meshes. J Comput Phys 2002;181:155–85.
[6] Nonomura T, Iizuka N, Fujii K. Freestream and vortex preservation properties of high-order WENO and WCNS on curvilinear grids. Comput Fluids 2010;39:197–214.
[7] Deng XG, Mao ML, Tu GH, Liu HY, Zhang HX. Geometric conservation law and applications to high-order finite difference schemes with stationary grids. J Comput Phys 2011;230:1100–15.
[8] Thomas PD, Lombard CK. Geometric conservation law and its application to flow computations on moving grids. AIAA J 1979;17(10):1030–7.
[9] Pulliam TH, Steger JL. On implicit finite-difference simulations of three-dimensional flow. AIAA Paper 78-10.
[10] Deng XG, Jiang Y, Mao ML, Liu HY, Tu GH. Developing hybrid cell-edge and cell-node dissipative compact scheme for complex geometry flows. Sci China Tech Sci 2013;56:2361–9.
[11] Deng XG, Min YB, Mao ML, Liu HY, Tu GH, Zhang HX. Further studies on geometric conservation law and applications to high-order finite difference schemes with stationary grids. J Comput Phys 2013;239:90–111.
[12] Deng XG, Mao ML, Tu GH, Zhang YF, Zhang HX. Extending weighted compact nonlinear schemes to complex grids with characteristic-based interface conditions. AIAA J 2010;48(12):2840–51.
[13] Tu GH, Deng XG, Mao ML. Implementing high-order weighted compact nonlinear scheme on patched grids with a nonlinear interpolation. Comput Fluids 2013;77:181–93.
[14] Lele SK. Compact finite difference schemes with spectral-like resolution. J Comput Phys 1992;103:16–42.
[15] Zhou Q, Yao Zh H, He F, Shen MY. A new family of high-order compact upwind difference schemes with good spectral resolution. J Comput Phys 2007;227:1306–39.
[16] Deng XG, Maekawa H, Shen Q, A class of high-order dissipative compact schemes. AIAA paper 96-1972.
[17] Mao ML, Deng XG, Li S. Spectrum characteristic of dissipative compact schemes and application to Couette flow. Chin J Computat Phys 2009;26(3):371–7.
[18] Mao ML, Jiang Y, Deng XG. LDDRK schemes based on DCS5 scheme. Chin J Computat Phys 2010;27(2):159–67.
[19] Jiang Y, Mao ML, Deng XG. Application of DCS5 scheme in CAA. Acta Aerodynamica Sinica 2012;30(4):431–6.
[20] Orszag SA. On the elimination of aliasing in finite-difference schemes by filtering high-wavenumber components. J Atmos Sci 1971;28:1074–82.
[21] Van Leer B. Towards the ultimate conservative difference scheme V. A second-order sequel to Godunov's method. J Comput Phys 1979;32:101–36.
[22] Deng XG, Zhang HX. Developing high-order weighted compact nonlinear schemes. J Comput Phys 2000;165:22–44.
[23] Deng XG, Mao ML, Jiang Y, Liu HY. New high-order hybrid cell-edge and cell-node weighted compact nonlinear schemes. AIAA paper 2011–3857.
[24] Tam CKW, Webb JC. Dispersion-relation-preserving finite difference schemes for computational acoustics. J Comput Phys 1993;107:262–81.
[25] Ramboer J, Tim B, Sergey S, Chris L. Optimization of time integration schemes coupled to spatial discretization for use in CAA applications. J Comput Phys 2006;220:270–89.
[26] Hu FQ, Hussaini MY, Manthey JL. Low-dissipation and low-dispersion Runge-Kutta schemes for computational acoustics. J Comput Phys 1996;124:177–91.
[27] Poinsot T, Lele SK. Boundary conditions for direct simulations of compressible viscous flows. J Comput Phys 1992;101:104–29.
[28] Kim JW, Lee DJ. Generalized characteristic boundary conditions for computational aeroacoustics. Part 2, AIAA J 2004;42:47–55.
[29] Dowling AP, Ffowcs Williams JE. Sound and sources of sound, 1st ed., Vol. 2. New York: Wiley; 1983. p. 60–1.
[30] Daniel JB. Analysis of sponge zones for computational fluid mechanics. J Comput Phys 2006;212:681–702.
[31] 1st AIAA CFD high lift prediction workshop, <http://hiliftpw.larc.nasa.gov/>, email: hiliftpw@gmail.com, June 2010.
[32] Slotnick JP, Hannon JA, Chaffin M. Overview of the first AIAA CFD high lift prediction workshop. AIAA Paper 2011–862.
[33] Menter FR. Two-equation eddy viscosity turbulence models for engineering applications. AIAA J 1994;32(8):1598–605.

Further studies on Geometric Conservation Law and applications to high-order finite difference schemes with stationary grids

Xiaogang Deng [a,b,*], Yaobing Min [a,c], Meiliang Mao [a,c], Huayong Liu [a], Guohua Tu [a], Hanxin Zhang [c]

[a] State Key Laboratory of Aerodynamics, China Aerodynamics Research and Development Center, P.O. Box 211, Mianyang, Sichuan 621000, People's Republic of China
[b] National University of Defense Technology, Changsha, Hunan 410073, People's Republic of China
[c] Computational Aerodynamics Institute, China Aerodynamics Research and Development Center, P.O. Box 211, Mianyang, Sichuan 621000, People's Republic of China

ARTICLE INFO

Article history:
Received 19 January 2012
Received in revised form 31 August 2012
Accepted 1 December 2012
Available online 19 December 2012

Keywords:
Geometric Conservation Law
Surface Conservation Law
Symmetrical Conservative Metric Method
Symmetrical conservative Jacobian
Weighted Compact Nonlinear Scheme
Curvilinear coordinates

ABSTRACT

The metrics and Jacobian in the fluid motion governing equations under curvilinear coordinate system have a variety of equivalent differential forms, which may have different discretization errors with the same difference scheme. The discretization errors of metrics and Jacobian may cause serious computational instability and inaccuracy in numerical results, especially for high-order finite difference schemes. It has been demonstrated by many researchers that the Geometric Conservation Law (GCL) is very important for high-order Finite Difference Methods (FDMs), and a proper form of metrics and Jacobian, which can satisfy the GCL, can considerably reduce discretization errors and computational instability. In order to satisfy the GCL for FDM, we have previously developed a Conservative Metric Method (CMM) to calculate the metrics [1] and the difference scheme δ^3 in the CMM is determined with the suggestion $\delta^3 = \delta^2$. In this paper, a Symmetrical Conservative Metric Method (SCMM) is newly proposed based on the discussions of the metrics and Jacobian in FDM from geometry viewpoint by following the concept of vectorized surface and cell volume in Finite Volume Methods (FVMs). Interestingly, the expressions of metrics and Jacobian obtained by using the SCMM with second-order central finite difference scheme are equivalent to the vectorized surfaces and cell volumes, respectively. The main advantage of SCMM is that it makes the calculations based on high-order WCNS schemes aroud complex geometry flows possible and somewhat easy. Numerical tests on linear and nonlinear problems indicate that the quality of numerical results may be largely enhanced by utilizing the SCMM, and the advantage of the SCMM over other forms of metrics and Jacobian may be more evident on highly nonuniform grids.

© 2012 Elsevier Inc. All rights reserved.

1. Introduction

In our previous paper, a Conservative Metric Method (CMM) has been proposed in order to ensure the Geometry Conservation Law (GCL) for high-order finite difference schemes [1]. The CMM is achieved by computing grid metric derivatives through a conservative form with the same scheme applied for fluxes. It has also been shown that the CMM is suitable not only to interior schemes but also to boundary and near boundary schemes on the condition that the difference operators δ^1 for the first-order derivatives of fluxes are linear ones. A special case of the CMM can be found in Visbal and Gaitonde's numerical technique [12], where the same central compact scheme is applied to both flow fluxes and metrics. Visbal and Gaitonde's numerical technique was also followed by Nonomura et al. [8] in implementing the Weighted Compact Nonlinear Scheme (WCNS) derived by Deng and Zhang [15] on curvilinear grids. Nonomura et al. [8] also showed that the free-stream and vortex preservation properties of WCNS is superior to that of the WENO scheme [9], and the essential reason analysed by Deng et al. [1] is that CMM can be introduced to satisfy the GCL with WCNS. Although the inner-level difference operators δ^3 in the conservative metrics have no effect on the GCL, the constraint $\delta^3 = \delta^2$ is recommended by Deng et al. [1] where δ^2 is the outer-level difference operators in the conservative metrics. Deng et al. [1] also showed that extra errors in metrics may appear if $\delta^3 \neq \delta^2$.

In this paper, we further study the GCL from the geometry viewpoint. By analyzing the surface calculation of Finite Volume Method (FVM), we obtained a general metrics calculation method for Finite Difference Method (FDM), i.e., the metrics calculation of FDM for second order accuracy is equivalent to the surface calculation of FVM. This method is extended to high-order metrics calculation which not only satisfies CMM, but also requires $\delta^3 = \delta^2$. Thus we know why the errors may be large as $\delta^3 \neq \delta^2$ in [1]. Beyond the metrics, the calculation of Jacobian introduced by the transformation from Cartesian coordinates to curvilinear coordinates is traditionally only considered for moving or deforming grids as the Volume Conservation Law (VCL) [5,6], very little attention is paid for time-constant grids. In this paper, it will be verified that different expressions of the Jacobian may have quite different influences on numerical results with irregular grids. With an eye to the Jacobian obtained by a second-order central different scheme, from geometry viewpoint, we also expect that the Jacobian might describe the cell volume enclosed by its vectorized surfaces and should be uniquely determined by its vectorized surfaces as that in FVM. Unfortunately, the traditional definition of the Jacobian cannot be enclosed by its corresponding vectorized surfaces, though many endeavors are done to develop appropriate metrics. In the present work, two new symmetrical forms of Jacobian are derived from the geometry viewpoint, one is non-conservative, the other is conservative provided that the Surface Conservative Law (SCL) has already been satisfied. Special attentions should be paid to the symmetrical conservative Jacobian, because it can produce a volume-like Jacobian which is completely enclosed by its vectorized surfaces when the second-order central different scheme is used. Furthermore, the errors introduced by this conservative form are smaller than that by other forms as we will see through the numerical tests with irregular grids. Based on the analysis, a new Symmetrical Conservative Metric Method (SCMM) is derived. It is proved that the SCMM is equivalent to the vectorized surfaces and volume in second order accuracy as that in Finite Volume Method (FVM) in second order accuracy, thus the calculations with SCMM by FDM may be as robust as that by FVM. Furthermore, the main advantage of SCMM is that it makes the high-order WCNS schemes [15,18] calculations aroud complex geometry flows possible and somewhat easy.

In Section 2, the discretizations of governing equations by FDM and FVM with second-order accuracy is presented, and the SCL in both FDM and FVM are briefly introduced. The surface and volume in FDM and FVM are discussed from geometry viewpoint in Section 3, and SCMM is proposed via the discussions. In Section 4 the implementation of the SCMM on high-order procedure is presented in details. Some numerical tests including the linear wave propagation problems and nonlinear isentropic vortex moving problems are simulated with different forms of metrics and Jacobian on randomized grids, and the numerical results are discussed in Section 5 with a test on a typical flow around ONERA M6 wing. Finally, concluding remarks are presented in Section 6.

2. Discretizations of governing equations in FDM and FVM

2.1. Discretizations of governing equations in FDM and SCL

The strong conservative three-dimensional Euler equation in Cartesian coordinates (t,x,y,z) are

$$\frac{\partial Q}{\partial t} + \frac{\partial E}{\partial x} + \frac{\partial F}{\partial y} + \frac{\partial G}{\partial z} = 0, \tag{1}$$

where Q is dependent variable and E, F and G are all convective fluxes

$$Q = \begin{pmatrix} \rho \\ \rho u \\ \rho v \\ \rho w \\ \rho e \end{pmatrix}, \quad E = \begin{pmatrix} \rho u \\ \rho u u + p \\ \rho v u \\ \rho w u \\ (\rho e + p)u \end{pmatrix}, \quad F = \begin{pmatrix} \rho v \\ \rho u v \\ \rho v v + p \\ \rho w v \\ (\rho e + p)v \end{pmatrix}, \quad G = \begin{pmatrix} \rho w \\ \rho u w \\ \rho v w \\ \rho w w + p \\ (\rho e + p)w \end{pmatrix},$$

where

$$p = \frac{1}{\gamma M_\infty^2}\rho T, \quad e = \frac{p}{(\gamma-1)\rho} + \frac{u^2+v^2+w^2}{2}$$

and γ is the specific heat ratio. The non-dimensional variables are defined as $\rho = \rho^*/\rho_\infty^*$, $(u,v,w) = (u^*,v^*,w^*)/U_\infty^*$, $T = T^*/T_\infty^*$ and $p = p^*/\left(\rho_\infty^* U_\infty^{*2}\right)$, and $M_\infty = U_\infty^*/\sqrt{\gamma R T_\infty^*}$ is the free-stream Mach number.

Transformation from Cartesian coordinates (t,x,y,z) to curvilinear coordinates (τ,ξ,η,ζ) is

$$\begin{cases} \tau = t, \\ \xi = \xi(t,x,y,z), \\ \eta = \eta(t,x,y,z), \\ \zeta = \zeta(t,x,y,z) \end{cases} \tag{2}$$

and Eq. (1) in three-dimensional curvilinear coordinates (τ,ξ,η,ζ) are

$$\frac{\partial \hat{Q}}{\partial \tau} + \frac{\partial \hat{E}}{\partial \xi} + \frac{\partial \hat{F}}{\partial \eta} + \frac{\partial \hat{G}}{\partial \zeta} = 0, \tag{3}$$

where

$$\hat{Q} = J^{-1}Q,$$
$$\hat{E} = \hat{\xi}_t Q + \hat{\xi}_x E + \hat{\xi}_y F + \hat{\xi}_z G,$$
$$\hat{F} = \hat{\eta}_t Q + \hat{\eta}_x E + \hat{\eta}_y F + \hat{\eta}_z G,$$
$$\hat{G} = \hat{\zeta}_t Q + \hat{\zeta}_x E + \hat{\zeta}_y F + \hat{\zeta}_z G.$$

Here, the traditional forms of Jacobian and metrics are all listed as follows.

$$J^{-1} = \left|\frac{\partial(x,y,z)}{\partial(\xi,\eta,\zeta)}\right| = x_\xi y_\eta z_\zeta - x_\xi y_\zeta z_\eta + x_\eta y_\zeta z_\xi - x_\eta y_\xi z_\zeta + x_\zeta y_\xi z_\eta - x_\zeta y_\eta z_\xi, \tag{4}$$

$$\hat{\xi}_x = J^{-1}\xi_x = y_\eta z_\zeta - y_\zeta z_\eta, \quad \hat{\xi}_y = J^{-1}\xi_y = z_\eta x_\zeta - z_\zeta x_\eta, \quad \hat{\xi}_z = J^{-1}\xi_z = x_\eta y_\zeta - x_\zeta y_\eta,$$
$$\hat{\eta}_x = J^{-1}\eta_x = y_\zeta z_\xi - y_\xi z_\zeta, \quad \hat{\eta}_y = J^{-1}\eta_y = z_\zeta x_\xi - z_\xi x_\zeta, \quad \hat{\eta}_z = J^{-1}\eta_z = x_\zeta y_\xi - x_\xi y_\zeta, \tag{5}$$
$$\hat{\zeta}_x = J^{-1}\zeta_x = y_\xi z_\eta - y_\eta z_\xi, \quad \hat{\zeta}_y = J^{-1}\zeta_y = z_\xi x_\eta - z_\eta x_\xi, \quad \hat{\zeta}_z = J^{-1}\zeta_z = x_\xi y_\eta - x_\eta y_\xi,$$

$$\hat{\xi}_t = J^{-1}\xi_t = -x_\tau \hat{\xi}_x - y_\tau \hat{\xi}_y - z_\tau \hat{\xi}_z,$$
$$\hat{\eta}_t = J^{-1}\eta_t = -x_\tau \hat{\eta}_x - y_\tau \hat{\eta}_y - z_\tau \hat{\eta}_z, \tag{6}$$
$$\hat{\zeta}_t = J^{-1}\zeta_t = -x_\tau \hat{\zeta}_x - y_\tau \hat{\zeta}_y - z_\tau \hat{\zeta}_z.$$

Although only stationary grids are considered in the present work, Eqs. (6) is still list here for completeness.

If the grids are time-constant, Eq. (3) can be discretized as

$$\delta_\tau Q = -\frac{1}{J^{-1}}\left(\delta_\xi^1 \hat{E} + \delta_\eta^1 \hat{F} + \delta_\zeta^1 \hat{G}\right), \tag{7}$$

where δ^1 is the spatial difference operator acting on the fluxes in Eq. (3) and the subscripts indicate the partial directions, δ_τ represents the time difference operator acting on the dependent variable Q. If δ^1 is the second-order central finite difference scheme and the grid points illuminated in Fig. 1 are used, Eq. (7) can be rewritten as

$$\delta_\tau Q_{i,j,k} = -\frac{1}{J_{i,j,k}^{-1}}\left(\frac{\hat{E}_{i+1,j,k} - \hat{E}_{i-1,j,k}}{2\Delta\xi} + \frac{\hat{F}_{i,j+1,k} - \hat{F}_{i,j-1,k}}{2\Delta\eta} + \frac{\hat{G}_{i,j,k+1} - \hat{G}_{i,j,k-1}}{2\Delta\zeta}\right). \tag{8}$$

In uniform flow field, Q, E, F and G are all constants, Eqs. (7) can be simplified as

$$\delta_\tau Q = -\frac{1}{J^{-1}}\left(I_x^N E + I_y^N F + I_z^N G\right), \tag{9}$$

where

$$I_x^N = \delta_\xi^1\left(\hat{\xi}_x\right) + \delta_\eta^1\left(\hat{\eta}_x\right) + \delta_\zeta^1\left(\hat{\zeta}_x\right),$$
$$I_y^N = \delta_\xi^1\left(\hat{\xi}_y\right) + \delta_\eta^1\left(\hat{\eta}_y\right) + \delta_\zeta^1\left(\hat{\zeta}_y\right), \tag{10}$$
$$I_z^N = \delta_\xi^1\left(\hat{\xi}_z\right) + \delta_\eta^1\left(\hat{\eta}_z\right) + \delta_\zeta^1\left(\hat{\zeta}_z\right),$$

and the superscript "N" denotes the numerical values. In uniform flow field, the free-stream preservation property presents as the cancellation of the right hand side of Eq. (9), namely

$$I_x^N = I_y^N = I_z^N = 0. \tag{11}$$

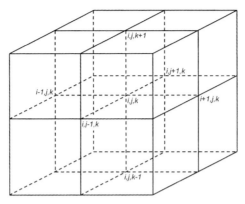

Fig. 1. The grid points in three-dimensional coordinates.

Actually the metrics cancellation of Eq. (11) is the different form of the Surface Conservation Law (SCL), which indicates that the vectorized surfaces should enclose a specific cell volume and must be guaranteed strictly from geometry viewpoint. If δ^1 is also the second-order central difference scheme, Eqs. (10) can be rewritten as follows.

$$I_{x;i,j,k}^N = \frac{\hat{\xi}_{x;i+1,j,k} - \hat{\xi}_{x;i-1,j,k}}{2\Delta\xi} + \frac{\hat{\eta}_{x;i,j+1,k} - \hat{\eta}_{x;i,j-1,k}}{2\Delta\eta} + \frac{\hat{\zeta}_{x;i,j,k+1} - \hat{\zeta}_{x;i,j,k-1}}{2\Delta\zeta},$$

$$I_{y;i,j,k}^N = \frac{\hat{\xi}_{y;i+1,j,k} - \hat{\xi}_{y;i-1,j,k}}{2\Delta\xi} + \frac{\hat{\eta}_{y;i,j+1,k} - \hat{\eta}_{y;i,j-1,k}}{2\Delta\eta} + \frac{\hat{\zeta}_{y;i,j,k+1} - \hat{\zeta}_{y;i,j,k-1}}{2\Delta\zeta}, \quad (12)$$

$$I_{z;i,j,k}^N = \frac{\hat{\xi}_{z;i+1,j,k} - \hat{\xi}_{z;i-1,j,k}}{2\Delta\xi} + \frac{\hat{\eta}_{z;i,j+1,k} - \hat{\eta}_{z;i,j-1,k}}{2\Delta\eta} + \frac{\hat{\zeta}_{z;i,j,k+1} - \hat{\zeta}_{z;i,j,k-1}}{2\Delta\zeta}.$$

By substituting the metrics defined by Eq. (5) into Eq. (12), the SCL terms are not identically zero. This phenomenon was first detected by Steger [2,3]. In order to strictly satisfy the SCL within finite difference frame, many endeavors have been tried, and some of those will be briefly introduced in Section 3.

2.2. Discretizations of governing equations in FVM and SCL

The integral formulation of Eq. (1) in three-dimensional Cartesian coordinates can be gained by using Gauss's divergence theorem.

$$\frac{\partial}{\partial t}\iiint_\Omega Q dV + \oiint_{\partial\Omega} \vec{F} \cdot d\vec{S} = 0, \quad (13)$$

where the flux vector $\vec{F} = E\vec{i} + F\vec{j} + G\vec{k}$, Ω and $\partial\Omega$ are the whole domain and surface of the super cell respectively illuminated in Fig. 1. On three-dimensional time-constant structured curvilinear grids, Eq. (13) may be discretized by a cell-vertex FVM [13,14]

$$\delta_\tau \overline{Q}_{i,j,k} = -\frac{1}{V_{i,j,k}} \begin{pmatrix} \vec{F}_{i+1,j,k} \cdot \vec{\xi}_{i+1,j,k}^V - \vec{F}_{i-1,j,k} \cdot \vec{\xi}_{i-1,j,k}^V + \\ \vec{F}_{i,j+1,k} \cdot \vec{\eta}_{i,j+1,k}^V - \vec{F}_{i,j-1,k} \cdot \vec{\eta}_{i,j-1,k}^V + \\ \vec{F}_{i,j,k+1} \cdot \vec{\zeta}_{i,j,k+1}^V - \vec{F}_{i,j,k-1} \cdot \vec{\zeta}_{i,j,k-1}^V \end{pmatrix} \quad (14)$$

where $\overline{Q}$ is the volumetric averaging variable $\overline{Q} = \frac{1}{V}\int\int_\Omega Q dV$, and $\vec{\xi}^V$, $\vec{\eta}^V$ and $\vec{\zeta}^V$ are the vectorized surfaces in ξ, η and ζ directions respectively in FVM. For example, the vectorized surface at point "O" in the ζ direction illustrated in Fig. 2 can be given as

$$\vec{\zeta}_O^V = \vec{S}_{OGDH} + \vec{S}_{OHAE} + \vec{S}_{OEBF} + \vec{S}_{OFCG} = \frac{1}{2}\left(\vec{AO} \times \vec{EH} + \vec{HG} \times \vec{OD} + \vec{OC} \times \vec{FG} + \vec{EF} \times \vec{BO}\right). \quad (15)$$

And $V_{i,j,k}$ is the volume of the super cell volume illuminated in Fig. 1, which is formed by the eight cells surrounding the cell vertex (i,j,k). In FVM the cell volume is enclosed by its vectorized surfaces, and so can be uniquely determined by them. In

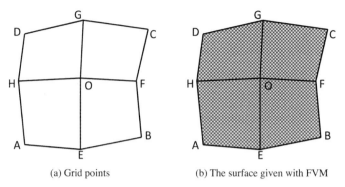

(a) Grid points (b) The surface given with FVM

Fig. 2. The vectorized surface at point "O" in ζ direction, the ξ and η directions are the same as the vector $\vec{AE}$ and $\vec{AH}$, respectively, and the ξ, η and ζ directions are components of a right-hand coordinate.

order to correctly carry the information about geometry, the volume of the super cell $V_{i,j,k}$ can be calculated by the following expression.

$$V_{i,j,k} = V_{i,j,k}\left(\vec{\xi}^V_{i+1,j,k}, \vec{\xi}^V_{i-1,j,k}, \vec{\eta}^V_{i,j+1,k}, \vec{\eta}^V_{i,j-1,k}, \vec{\zeta}^V_{i,j,k+1}, \vec{\zeta}^V_{i,j,k-1}\right).$$

In FVM, in order to calculate the volume of the super cell enclosed by its vectorized surfaces, the super cell is usually partitioned into several pyramids, and to sum all the volumes of pyramids one can get the volume $V_{i,j,k}$. As we will see in Fig. 6, an easy way to partition the super cell is to form pyramids with each of the vectorized surfaces calculated by Eq. (15) as bottom surface and with the same apex point "O" in Fig. 6. In such situation, the volume $V_{i,j,k}$ can be represented as

$$V_{i,j,k} = V^O_{A_L E_L B_L F_L C_L G_L D_L H_L} + V^O_{AA_L H_L D_L D D_R H_R A_R} + V^O_{AA_R E_R B_R B B_L E_L A_L} + V^O_{A_R H_R D_R G_R C_R F_R B_R E_R} + V^O_{B_R F_R C_R C C_L F_L B_L B} + V^O_{D D_L G_L C_L C C_R G_R D_R}$$
$$= V_{A_L A_R B_L B_R C_L C_R D_L D_R}, \quad (16)$$

where the subscripts of the uppercase "V" denote the vertex points of the bottom surface of a pyramid and the superscript denotes the apex point of a pyramid, the last volume in Eq. (16) represents the volume as its eight apexes.

The free-stream preservation property of Eq. (13) in uniform flow fields requires

$$\oiint_{\partial\Omega} d\vec{S} = 0.$$

which is the integral form of the SCL. The following discretized form of the SCL in FVM can be gotten with the grid points illuminated in Fig. 1

$$\vec{\xi}^V_{i+1,j,k} - \vec{\xi}^V_{i-1,j,k} + \vec{\eta}^V_{i,j+1,k} - \vec{\eta}^V_{i,j-1,k} + \vec{\zeta}^V_{i,j,k+1} - \vec{\zeta}^V_{i,j,k-1} = 0. \quad (17)$$

The SCL in Eq. (17) is always satisfied profiting from the vectorized surfaces calculation method in Eq. (15), and cell volume V in Eq. (16) is completely enclosed by its corresponding vectorized surfaces. In FVM the vectorized surfaces enclose the cell volume, and the cell volume is uniquely determined by its vectorized surfaces, thus the information about geometry is correctly presented.

3. Discussions on the metrics and Jacobian in FDM from the geometry viewpoint

It is well known that the vectorized surfaces and cell volume within finite volume frame are uniquely determined and correctly carry information about grid geometry, while the metrics and Jacobian within finite difference frame have a variety of equivalent partial-differential expressions, which may have different discretization errors despite that a same difference scheme is applied to discrete the partial derivatives. It can be seen that the right hand sides of Eq. (8) and Eq. (14) are similar to each other, it is naturally expected that some indications might be obtained from FVM to determine the method for calculating the metrics and Jacobian in FDM.

3.1. Discussions on metrics in FDM and vectorized surfaces in FVM

With an eye on the numerical metrics in Eq. (5) from geometry viewpoint, the numerical metrics such as $\hat{\zeta}_x$, $\hat{\zeta}_y$ and $\hat{\zeta}_z$ are supposed to describe the vectorized surface in the ζ direction. However, the numerical metrics calculated by Eq. (5) can not exactly convey this message of the vectorized surfaces of a cell volume, neither do the conservative form in Refs. [1,6,12]. These two forms of metrics will be further discussed in the following texts of the present subsection.

Considering the grid points illuminated in Fig. 2a, within finite difference frame the vectorized surface at point "O" in the ζ direction can be gained from different metric forms such as the non-conservative form in Eq. (5) and the conservative form in Eq. (21). Within finite volume frame, the vectorized surface at point "O" in the ζ direction is illustrated in Fig. 2b. This vectorized surface is expected to be obtained within finite difference frame and marked as $\vec{S}_{FVM}$.

$$\vec{S}_{FVM} = \vec{\zeta}_O^v = \frac{1}{2}\left(\vec{AO} \times \vec{EH} + \vec{HG} \times \vec{OD} + \vec{OC} \times \vec{FG} + \vec{EF} \times \vec{BO}\right). \tag{18}$$

For the grids in Fig. 2a the vectorized surface at point "O" in the ζ direction within finite difference frame can be given from the definition in Eqs. (5).

$$\begin{cases} \hat{\zeta}_x^n = \left(\delta_\zeta^2 y\right)\left(\delta_\eta^2 z\right) - \left(\delta_\eta^2 y\right)\left(\delta_\zeta^2 z\right), \\ \hat{\zeta}_y^n = \left(\delta_\zeta^2 z\right)\left(\delta_\eta^2 x\right) - \left(\delta_\eta^2 z\right)\left(\delta_\zeta^2 x\right), \\ \hat{\zeta}_z^n = \left(\delta_\zeta^2 x\right)\left(\delta_\eta^2 y\right) - \left(\delta_\eta^2 x\right)\left(\delta_\zeta^2 y\right), \end{cases} \tag{19}$$

where the superscript "n" denotes the non-conservative form of metrics and δ^2 is the numerical difference operator in Eqs. (5). If the second-order central difference scheme in all coordinate directions is used as δ^2, the numerical results of the non-conservative form of metrics in Eq. (19) are

$$\begin{cases} \hat{\zeta}_x^n = \frac{1}{4}[(y_F - y_H)(z_G - z_E) - (z_F - z_H)(y_G - y_E)] = \frac{1}{4}\left(\vec{HF} \times \vec{EG}\right)_x, \\ \hat{\zeta}_y^n = \frac{1}{4}[(z_F - z_H)(x_G - x_E) - (x_F - x_H)(z_G - z_E)] = \frac{1}{4}\left(\vec{HF} \times \vec{EG}\right)_y, \\ \hat{\zeta}_z^n = \frac{1}{4}[(x_F - x_H)(y_G - y_E) - (y_F - y_H)(x_G - x_E)] = \frac{1}{4}\left(\vec{HF} \times \vec{EG}\right)_z, \end{cases} \tag{20}$$

$$\vec{\zeta}^n = \hat{\zeta}_x^n \vec{i} + \hat{\zeta}_y^n \vec{j} + \hat{\zeta}_z^n \vec{k} = \frac{1}{4}\vec{HF} \times \vec{EG} = \frac{1}{2}\vec{S}_{EFGH}.$$

The vectorized surface $\vec{S}_{EFGH}$ in Eq. (20) is illuminated by the black shadow in Fig. 3a. If the vectorized surface defined by Eqs. (20) is used as metrics in Eqs. (12), the SCL cannot be guaranteed due to the abnormity of the vectorized surfaces calculated by the non-conservative metrics of Eqs. (5). At the same time the vectorized surfaces given in Eqs. (20) cannot enclose any cell volume at all, which is one of reasons for the violation of the SCL in FDM.

In order to ensure the SCL with second-order accuracy, Pulliam and Steger [4] presented a special weighted averaging, and the corresponding vectorized surface gained from the averaging procedure is illuminated in Fig. 3b. The same vectorized surface was derived by Vinokur in his review literature [11] within finite volume frame, and his metric method is called as finite-volume-like method in subsequent literatures [1,8]. The substantial reason for the satisfaction of the SCL is that the vectorized surfaces gained from Pulliam and Steger's averaging procedure can enclose a cell volume despite that the cell volume is more or less different from the control volume enclosed by the surfaces in Eq. (14).

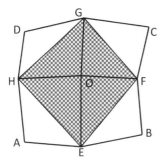

(a) The surface S_{ERGH} in Eq. (20)

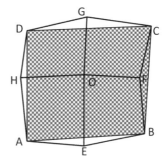

(b) The surface gained from Pulliam and Steger's averaging procedure

Fig. 3. The surfaces S_{ERGH} in Eq. (20) and Pulliam and Steger's averaging procedure.

The averaging procedure presented by Pulliam and Steger [4,10] is central type and only second-order accurate. Two momentous shortcomings must be faced, one is the difficulty to implement on boundary and near boundary points where central type numerical operators cannot be applied, and the other is that the averaging procedure is hard to be implemented with high-order difference schemes.

3.1.1. The conservative form of metrics S1

Considering the complexity and hard implementation with high-order difference schemes of the averaging procedure proposed by Pulliam and Steger [4], Thomas and Lombard [6] proposed a conservative form of metrics which has been commended as a milestone in the history of the GCL in FDM and is distinctively marked as S1 in the present work. The terminology "Geometric Conservation Law (GCL)" was also coined by Thomas and Lombard [5,6] who focus on the computation of Jacobian on dynamic grids. The conservative form of metrics S1 may be written as

$$\hat{\xi}_x^{S1} = (y_\eta z)_\zeta - (y_\zeta z)_\eta, \quad \hat{\xi}_y^{S1} = (z_\eta x)_\zeta - (z_\zeta x)_\eta, \quad \hat{\xi}_z^{S1} = (x_\eta y)_\zeta - (x_\zeta y)_\eta,$$
$$\hat{\eta}_x^{S1} = (y_\zeta z)_\xi - (y_\xi z)_\zeta, \quad \hat{\eta}_y^{S1} = (z_\zeta x)_\xi - (z_\xi x)_\zeta, \quad \hat{\eta}_z^{S1} = (x_\zeta y)_\xi - (x_\xi y)_\zeta, \quad (21)$$
$$\hat{\zeta}_x^{S1} = (y_\xi z)_\eta - (y_\eta z)_\xi, \quad \hat{\zeta}_y^{S1} = (z_\xi x)_\eta - (z_\eta x)_\xi, \quad \hat{\zeta}_z^{S1} = (x_\xi y)_\eta - (x_\eta y)_\xi.$$

These metrics can be discretized by finite difference operators as

$$\hat{\xi}_x^{S1} = \delta_\zeta^2\left(z\delta_\eta^3 y\right) - \delta_\eta^2(z\delta_\zeta^3 y), \quad \hat{\xi}_y^{S1} = \delta_\zeta^2\left(x\delta_\eta^3 z\right) - \delta_\eta^2(x\delta_\zeta^3 z), \quad \hat{\xi}_z^{S1} = \delta_\zeta^2\left(y\delta_\eta^3 x\right) - \delta_\eta^2(y\delta_\zeta^3 x) \quad (22)$$

and with similar expressions for the other metrics. Hence two-level spatial derivatives are introduced in Eqs. (22), where the outer-level numerical difference operator is marked as δ^2 and the inner-level numerical difference operator is marked as δ^3. Thomas and Lombard [6] pointed out that if the same second-order central difference operator is used to evaluate all the spatial derivatives in Eq. (21), the SCL terms in Eqs. (12) vanish, namely $I_x^N = I_y^N = I_z^N = 0$.

For high-order FDM, Deng et al. [1] have recently presented a theoretical analysis of the SCL and have derived a sufficient condition to satisfy the SCL based on Thomas and Lombard's conservative form of metrics S1. Deng et al. [1] named the methods satisfying the sufficient condition as the Conservative Metric Method (CMM), which can be exactly presented as: (1) the conservative form of metrics in Eq. (21) is used, and (2) the condition $\delta^1 = \delta^2$ is satisfied. Although δ^3 has no effect on the SCL, the constraint $\delta^3 = \delta^2$ is suggested in Ref. [1], where the extra errors that may be introduced by $\delta^3 \neq \delta^2$ is analyzed. In the present work, the constraint $\delta^3 = \delta^2$ is further discussed from geometry viewpoint by the similarity between metrics in FDM and vectorized surface in FVM.

It can be found out that Thomas and Lombard's [5,6] numerical metrics is a second-order case of the CMM. A high-order case of the CMM for central compact schemes can be found in Visbal and Gaitonde's numerical technique [12]. It is worth to be noted that the CMM can ensure the SCL not only for interior grid points but also for boundary and near boundary grid points. However, if the same difference scheme, such as second-order central difference scheme, is applied to calculate the spatial derivatives in Eqs. (21), it is very difficult to define its corresponding vectorized surface. In order to endow the metrics with geometric information namely the vectorized surfaces in FVM, two other forms of metrics are presented and discussed next.

3.1.2. The conservative form of metrics S2

It is obvious that the metrics defined by the traditional non-conservative form of Eqs. (5) are identical to the metrics defined by the conservative form of Eqs. (21) if

$$(y_\zeta)_\eta = (y_\eta)_\zeta, \quad (z_\zeta)_\eta = (z_\eta)_\zeta \quad (23)$$

and with similar conditions for the other cross derivatives. The conditions in the cross derivatives may lead to another conservative form of metrics which is marked as S2.

$$\hat{\xi}_x^{S2} = (yz_\zeta)_\eta - (yz_\eta)_\zeta, \quad \hat{\xi}_y^{S2} = (zx_\zeta)_\eta - (zx_\eta)_\zeta, \quad \hat{\xi}_z^{S2} = (xy_\zeta)_\eta - (xy_\eta)_\zeta,$$
$$\hat{\eta}_x^{S2} = (yz_\xi)_\zeta - (yz_\zeta)_\xi, \quad \hat{\eta}_y^{S2} = (zx_\xi)_\zeta - (zx_\zeta)_\xi, \quad \hat{\eta}_z^{S2} = (xy_\xi)_\zeta - (xy_\zeta)_\xi, \quad (24)$$
$$\hat{\zeta}_x^{S2} = (yz_\eta)_\xi - (yz_\xi)_\eta, \quad \hat{\zeta}_y^{S2} = (zx_\eta)_\xi - (zx_\xi)_\eta, \quad \hat{\zeta}_z^{S2} = (xy_\eta)_\xi - (xy_\xi)_\eta.$$

The sufficient condition proved by Deng et al. [1] for the metrics S1 in Eqs. (21) can also satisfy the SCL for the metrics S2 with high-order WCNS, by the reason of the same conservative property of both S1 and S2. However, similar to that of S1, it is also difficult to define the vectorized surfaces for the metrics S2.

3.1.3. The symmetrical conservative form of metrics S3

Before the introduction of the symmetrical conservative metrics, some important properties of linear difference operator δ are given without proof.

$$\delta(c\phi) = c\delta\phi, \tag{25a}$$

$$\delta(\phi + \varphi) = \delta\phi + \delta\varphi, \tag{25b}$$

$$\delta_\zeta(\delta_\eta \phi) = \delta_\eta(\delta_\zeta \phi). \tag{25c}$$

Here c is an arbitrary constant. The properties of Eq. (25a) and Eq. (25b) can be seen straightly, only the one in Eq. (25c) requires proof, which was given by Deng et al. [1].

Because of the properties of linear difference operator in Eqs. (25), it can be figured out that an arbitrary linear combination of S1 and S2 can satisfy the SCL on the sufficient condition $\delta^1 = \delta^2$ with high-order WCNS. Then it is hard to say which conservative form of the metrics is better than the others from the SCL viewpoint. The following symmetrical conservative form of metrics is derived from the vectorized surface viewpoint and marked as S3.

$$\begin{aligned}
\hat{\xi}_x^{S3} &= \frac{1}{2}\left(\hat{\xi}_x^{S1} + \hat{\xi}_x^{S2}\right), & \hat{\xi}_y^{S3} &= \frac{1}{2}\left(\hat{\xi}_y^{S1} + \hat{\xi}_y^{S2}\right), & \hat{\xi}_z^{S3} &= \frac{1}{2}\left(\hat{\xi}_z^{S1} + \hat{\xi}_z^{S2}\right), \\
\hat{\eta}_x^{S3} &= \frac{1}{2}\left(\hat{\eta}_x^{S1} + \hat{\eta}_x^{S2}\right), & \hat{\eta}_y^{S3} &= \frac{1}{2}\left(\hat{\eta}_y^{S1} + \hat{\eta}_y^{S2}\right), & \hat{\eta}_z^{S3} &= \frac{1}{2}\left(\hat{\eta}_z^{S1} + \hat{\eta}_z^{S2}\right), \\
\hat{\zeta}_x^{S3} &= \frac{1}{2}\left(\hat{\zeta}_x^{S1} + \hat{\zeta}_x^{S2}\right), & \hat{\zeta}_y^{S3} &= \frac{1}{2}\left(\hat{\zeta}_y^{S1} + \hat{\zeta}_y^{S2}\right), & \hat{\zeta}_z^{S3} &= \frac{1}{2}\left(\hat{\zeta}_z^{S1} + \hat{\zeta}_z^{S2}\right).
\end{aligned} \tag{26}$$

As we will see, if the same second-order central difference scheme in all the coordinate directions is applied to both δ^2 and δ^3, the numerical result of metrics in the ζ direction in Eqs. (26) can be represented by

$$\vec{\zeta}^{S3} = \frac{1}{8}\left(\vec{AO} \times \vec{EH} + \vec{HG} \times \vec{OD} + \vec{OC} \times \vec{FG} + \vec{EF} \times \vec{BO}\right) = \frac{1}{4}\vec{S}_{FVM}. \tag{27}$$

It is very interesting that the numerical result of S3 in Eqs. (26) represents exactly one fourth of the vectorized surface in Fig. 2b, which is derived within finite volume frame. The same conservative form of metrics S3 in Eq. (26) was also proposed by Vinokur and Yee [7] as the coordinate invariant form of metrics, but the geometry meanings were not illuminated.

Some remarks are valuable to be added here as complementarity for the new symmetrical conservative form of metrics S3 in Eqs. (26).

Remark 1. From geometry viewpoint, $\delta^3 = \delta^2$ is a necessary condition for an essential connection or a bridge between the metrics in FDM and the vectorized surfaces in FVM. Then the symmetrical conservative form of metrics S3 with the constraint $\delta^3 = \delta^2$ are suggested to be utilized in FDM.

Remark 2. If δ^2 and δ^3 are the same first-order backward difference scheme in both ξ and η directions, the numerical vectorized surface can be represented in Eq. (28a), and similarly Eq. (28b) represents the result in the situation of the first-order forward difference scheme for both δ^2 and δ^3 in both ξ and η coordinate directions.

$$\vec{\zeta}_b^{S3} = \frac{1}{2}\vec{AO} \times \vec{EH} = \vec{S}_{AEOH}, \tag{28a}$$

$$\vec{\zeta}_f^{S3} = \frac{1}{2}\vec{OC} \times \vec{FG} = \vec{S}_{OFCG}. \tag{28b}$$

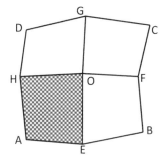
(a) The surface S_{AEOH} in Eq. (28a)

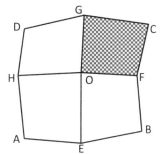
(b) The surface S_{OFCG} in Eq. (28b)

Fig. 4. The surfaces S_{AEOH} and S_{OFCG} in Eq. (28).

The two corresponding vectorized surfaces $\vec{S}_{AEOH}$ and $\vec{S}_{OFCG}$ are illuminated by the black shadow in Fig. 4a and Fig. 4b respectively. Thus if δ^2 and δ^3 are not central type difference scheme, the vectorized surfaces gained with $\delta^3 = \delta^2$ are not central type either.

Remark 3. Similar property as that in Eq. (27) can be obtained by the S3 for the other metrics in the ξ and η directions, and more exactly, the constraint $\delta^3 = \delta^2$ shall be written as

$$\delta^3_\xi = \delta^2_\xi, \quad \delta^3_\eta = \delta^2_\eta, \quad \delta^3_\zeta = \delta^2_\zeta. \tag{29}$$

This constraint can be conveniently realized at boundary and near boundary grid points, where the numerical difference scheme used in one direction is usually different from the ones used in the other directions. For example, if the first-order backward difference scheme is applied in the ξ direction for both δ^2_ξ and δ^3_ξ, while the second-order central difference scheme in the η direction for both δ^2_η and δ^3_η, the numerical result can be represented in Eq. (30a). Eq. (30b) represents another case that the second-order central difference scheme used in the ξ direction for both δ^2_ξ and δ^3_ξ while the first-order forward difference scheme used in the η direction for both δ^2_η and δ^3_η.

$$\vec{\zeta}^{\,3}_{bc} = \frac{1}{4}\left(\vec{AO} \times \vec{EH} + \vec{HG} \times \vec{OD}\right) = \frac{1}{2}\vec{S}_{AEOGDH}, \tag{30a}$$

$$\vec{\zeta}^{\,3}_{cf} = \frac{1}{4}\left(\vec{HG} \times \vec{OD} + \vec{OC} \times \vec{FG}\right) = \frac{1}{2}\vec{S}_{HOFCGD}. \tag{30b}$$

The two corresponding vectorized surfaces $\vec{S}_{AEOGDH}$ and $\vec{S}_{HOFCGD}$ of Eq. (30a) and Eq. (30b) are illuminated in Fig. 5a and Fig. 5b, respectively. Thus at boundary and near boundary grid points, different numerical schemes can be used in different coordinate directions, but when Eq. (29) is hold, the numerical metrics can represent as vectorized surfaces.

3.2. Discussions on Jacobian in FDM and cell volume in FVM

Similar to the metrics, the Jacobian also has a variety of equivalent partial-differential expressions. From geometry viewpoint, the proper form of Jacobian is expected to correctly represent the volume of a cell. However, as discussed in the followings, the numerical result of the Jacobian defined by Eq. (4) is hard to convey the message of cell volume.

3.2.1. The traditional form of Jacobian V1

Eq. (4) is a mathematic definition of Jacobian, which is marked as V1 in this paper.

$$J^{-1}_{V1} = x_\xi y_\eta z_\zeta - x_\xi y_\zeta z_\eta + x_\eta y_\zeta z_\xi - x_\eta y_\xi z_\zeta + x_\zeta y_\xi z_\eta - x_\zeta y_\eta z_\xi. \tag{31}$$

The Jacobian can be written in a vector form.

$$J^{-1}_{V1} = \left(\vec{r}_\xi \times \vec{r}_\eta\right) \cdot \vec{r}_\zeta, \tag{32}$$

where $\vec{r}$ represents the coordinate vector, and the subscripts indicate the partial derivatives. If the second-order central difference scheme in all coordinate directions is utilized to approximate the derivatives in Eq. (31), and the grids utilized to calculate the Jacobian is illuminated in Fig. 6a, the Jacobian of V1 can be written as

$$8J^{-1}_O = \left(\vec{O_L O_R} \times \vec{HF}\right) \cdot \vec{EG} = \left(\vec{O_L O_R} \times \vec{HF}\right) \cdot \vec{EO} + \left(\vec{O_L O_R} \times \vec{HF}\right) \cdot \vec{OG} = 6\left(V^G_{O_L HO_R F} + V^E_{O_L FO_R H}\right).$$

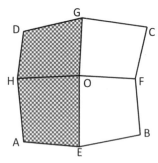
(a) The surface S_{AEOGDH} in Eq. (30a)

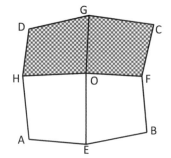
(b) The surface S_{HOFCGD} in Eq. (30b)

Fig. 5. The surfaces S_{AEOGDH} and S_{HOFCGD} in Eq. (30).

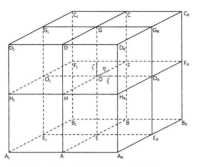
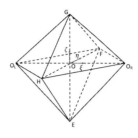

(a) The grid used in Jacobin calculation.   (b) The volume $V^G_{O_L H O_R F}$ and $V^E_{O_L F O_R H}$ in Eq. (33).

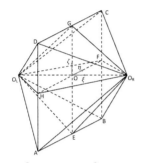
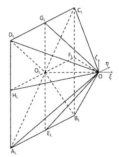

(c) The volume $V^{O_L}_{AHDGCFBE}$ and $V^{O_R}_{AEBFCGDH}$ in Eq. (38).   (d) The volume $V^O_{A_L E_L B_L F_L C_L G_L D_L H_L}$ in Eq. (45).

Fig. 6. The grids used in Jacobian calculation and the numerical volumes.

namely

$$J_O^{-1} = (\vec{r}_\xi \times \vec{r}_\eta) \cdot \vec{r}_\zeta = \frac{3}{4} \left(V^G_{O_L H O_R F} + V^E_{O_L F O_R H} \right). \tag{33}$$

Here the subscripts of the uppercase "V" denote the vertex points of the bottom surface of a pyramid and the superscript denotes the apex point.

The volume $V^G_{O_L H O_R F}$ and $V^E_{O_L F O_R H}$ in Eq. (33) is illuminated in Fig. 6b. Although the shape of this volume is symmetrical, it is not enclosed by the vectorized surfaces of the cell volume illuminated in Fig. 1.

3.2.2. The symmetrical form of Jacobian V2

The Jacobian (volume) can also be expressed as a function of the metrics (vectorized surfaces)

$$\begin{aligned}
J^{-1} &= J^{-1}_{V2}(x) = x_\xi \widehat{\xi}_x + x_\eta \widehat{\eta}_x + x_\zeta \widehat{\zeta}_x, \text{ or} \\
J^{-1} &= J^{-1}_{V2}(y) = y_\xi \widehat{\xi}_y + y_\eta \widehat{\eta}_y + y_\zeta \widehat{\zeta}_y, \text{ or} \\
J^{-1} &= J^{-1}_{V2}(z) = z_\xi \widehat{\xi}_z + z_\eta \widehat{\eta}_z + z_\zeta \widehat{\zeta}_z.
\end{aligned} \tag{34}$$

Unfortunately, no specific cell volume can be gained by any one of $J^{-1}_{V2}(x)$, $J^{-1}_{V2}(y)$ and $J^{-1}_{V2}(z)$ alone, no matter what kinds of difference schemes and vectorized surfaces are adopted. The three forms of Jacobian in Eqs. (34) can be summed to form a symmetrical Jacobian.

$$J^{-1}_{V2} = \frac{1}{3}\left(x_\xi \widehat{\xi}_x + x_\eta \widehat{\eta}_x + x_\zeta \widehat{\zeta}_x + y_\xi \widehat{\xi}_y + y_\eta \widehat{\eta}_y + y_\zeta \widehat{\zeta}_y + z_\xi \widehat{\xi}_z + z_\eta \widehat{\eta}_z + z_\zeta \widehat{\zeta}_z \right). \tag{35}$$

This form of Jacobian is marked as V2. Eq. (35) can be rearranged into a vector form as

$$J^{-1}_{V2} = \frac{1}{3}\left(\vec{r}_\xi \cdot \vec{\xi}^D + \vec{r}_\eta \cdot \vec{\eta}^D + \vec{r}_\zeta \cdot \vec{\zeta}^D \right). \tag{36}$$

where $\vec{\xi}^D, \vec{\eta}^D$ and $\vec{\zeta}^D$ are the vectors of metrics in ξ, η and ζ directions respectively in FDM. If the second-order central difference scheme is utilized to approximate the vectors of symmetrical conservative metrics S3 in Eq. (26), similar to Eq. (27) one can get

$$\vec{\xi}^D = \frac{1}{4}\vec{\xi}^V, \quad \vec{\eta}^D = \frac{1}{4}\vec{\eta}^V, \quad \vec{\zeta}^D = \frac{1}{4}\vec{\zeta}^V \tag{37}$$

and so

$$\frac{1}{3}\vec{r}_\xi \cdot \vec{\xi}^D = \frac{1}{6}\overrightarrow{O_L O_R} \cdot \vec{\xi}_0^D = \frac{1}{8}\left(V_{AHDGCFBE}^{O_L} + V_{AEBFCGDH}^{O_R}\right). \tag{38}$$

Its corresponding volumes $V_{AHDGCFBE}^{O_L}$ and $V_{AEBFCGDH}^{O_R}$ are illuminated in Fig. 6c. Similarly the other two volumes in Eq. (36) can be gained as

$$\frac{1}{3}\vec{r}_\eta \cdot \vec{\eta}^D = \frac{1}{6}\overrightarrow{HF} \cdot \vec{\eta}_0^D = \frac{1}{8}\left(V_{GG_LO_LE_LEE_RO_RG_R}^{H} + V_{GG_RO_RE_REE_LO_LG_L}^{F}\right),$$

$$\frac{1}{3}\vec{r}_\zeta \cdot \vec{\zeta}^D = \frac{1}{6}\overrightarrow{EG} \cdot \vec{\zeta}_0^D = \frac{1}{8}\left(V_{HH_LO_LF_LFF_RO_RH_R}^{E} + V_{HH_RO_RF_RFF_LO_LH_L}^{G}\right)$$

and the volume calculated by Eq. (35) can be written as

$$J_0^{-1} = \frac{1}{3}\left(\vec{r}_\xi \cdot \vec{\xi}^D + \vec{r}_\eta \cdot \vec{\eta}^D + \vec{r}_\zeta \cdot \vec{\zeta}^D\right) = \frac{1}{8}\begin{pmatrix} V_{AHDGCFBE}^{O_L} + V_{GG_LO_LE_LEE_RO_RG_R}^{H} + V_{HH_LO_LF_LFF_RO_RH_R}^{E} + \\ V_{AEBFCGDH}^{O_R} + V_{GG_RO_RE_REE_LO_LG_L}^{F} + V_{HH_RO_RF_RFF_LO_LH_L}^{G} \end{pmatrix}.$$

However, these volumes are still irregular and cannot yet fulfill the control volume enclosed by the vectorized surfaces (metrics) due to its non-conservative form in Eq. (35). In order to get a volume-like Jacobian, a new symmetrical conservative Jacobian will be derived in the next subsection.

3.2.3. The symmetrical conservative form of Jacobian V3

Let us consider a conservative form of Jacobian, which can be obtained based on Eqs. (34) and written as

$$J^{-1} = J_{V3}^{-1}(x) = \left(x\hat{\xi}_x\right)_\xi + (x\hat{\eta}_x)_\eta + \left(x\hat{\zeta}_x\right)_\zeta - xI_x, \text{ or}$$
$$J^{-1} = J_{V3}^{-1}(y) = \left(y\hat{\xi}_y\right)_\xi + (y\hat{\eta}_y)_\eta + \left(y\hat{\zeta}_y\right)_\zeta - yI_y, \text{ or} \tag{39}$$
$$J^{-1} = J_{V3}^{-1}(z) = \left(z\hat{\xi}_z\right)_\xi + (z\hat{\eta}_z)_\eta + \left(z\hat{\zeta}_z\right)_\zeta - zI_z,$$

where I_x, I_y and I_z are exactly the SCL terms. If the SCL is satisfied exactly, namely $I_x = I_y = I_z = 0$, Eq. (39) may be simplified to the following completely conservative forms

$$J_{V3}^{-1}(x) = \left(x\hat{\xi}_x\right)_\xi + (x\hat{\eta}_x)_\eta + \left(x\hat{\zeta}_x\right)_\zeta, \text{ or}$$
$$J_{V3}^{-1}(y) = \left(y\hat{\xi}_y\right)_\xi + (y\hat{\eta}_y)_\eta + \left(y\hat{\zeta}_y\right)_\zeta, \text{ or} \tag{40}$$
$$J_{V3}^{-1}(z) = \left(z\hat{\xi}_z\right)_\xi + (z\hat{\eta}_z)_\eta + \left(z\hat{\zeta}_z\right)_\zeta.$$

Similar to that of V2 no specific cell volume can be gained by each one of $J_{V3}^{-1}(x), J_{V3}^{-1}(y)$ and $J_{V3}^{-1}(z)$ alone. The three forms of Jacobian in Eqs. (40) can also be summed to form a symmetrical conservative form as Eq. (41), and the form is marked as V3.

$$J_{V3}^{-1} = \frac{1}{3}\left[\left(x\hat{\xi}_x + y\hat{\xi}_y + z\hat{\xi}_z\right)_\xi + (x\hat{\eta}_x + y\hat{\eta}_y + z\hat{\eta}_z)_\eta + \left(x\hat{\zeta}_x + y\hat{\zeta}_y + z\hat{\zeta}_z\right)_\zeta\right] = \frac{1}{3}\left[\left(\vec{r}\cdot\vec{\xi}^D\right)_\xi + \left(\vec{r}\cdot\vec{\eta}^D\right)_\eta + \left(\vec{r}\cdot\vec{\zeta}^D\right)_\zeta\right]. \tag{41}$$

With the second-order central difference scheme, Eq. (41) may be written as

$$6J_{i,j,k}^{-1} = \vec{r}_{i+1,j,k}\cdot\vec{\xi}_{i+1,j,k}^D - \vec{r}_{i-1,j,k}\cdot\vec{\xi}_{i-1,j,k}^D + \vec{r}_{i,j+1,k}\cdot\vec{\eta}_{i,j+1,k}^D - \vec{r}_{i,j-1,k}\cdot\vec{\eta}_{i,j-1,k}^D + \vec{r}_{i,j,k+1}\cdot\vec{\zeta}_{i,j,k+1}^D - \vec{r}_{i,j,k-1}\cdot\vec{\zeta}_{i,j,k-1}^D. \tag{42}$$

The metrics calculated by the S3 with proper finite difference schemes can exactly express the vectorized surfaces as well as ensure the SCL, i.e.

$$\vec{\xi}_{i+1,j,k}^D - \vec{\xi}_{i-1,j,k}^D + \vec{\eta}_{i,j+1,k}^D - \vec{\eta}_{i,j-1,k}^D + \vec{\zeta}_{i,j,k+1}^D - \vec{\zeta}_{i,j,k-1}^D = 0. \tag{43}$$

Then we can get

$$6J_{i,j,k}^{-1} = (\vec{r}_{i+1,j,k} - \vec{r}_{i,j,k})\cdot\vec{\xi}_{i+1,j,k}^D + (\vec{r}_{i,j,k} - \vec{r}_{i-1,j,k})\cdot\vec{\xi}_{i-1,j,k}^D + (\vec{r}_{i,j+1,k} - \vec{r}_{i,j,k})\cdot\vec{\eta}_{i,j+1,k}^D + (\vec{r}_{i,j,k} - \vec{r}_{i,j-1,k})\cdot\vec{\eta}_{i,j-1,k}^D$$
$$+ (\vec{r}_{i,j,k+1} - \vec{r}_{i,j,k})\cdot\vec{\zeta}_{i,j,k+1}^D + (\vec{r}_{i,j,k} - \vec{r}_{i,j,k-1})\cdot\vec{\zeta}_{i,j,k-1}^D, \tag{44}$$

where

$$(\vec{r}_{i,j,k}-\vec{r}_{i-1,j,k})\cdot\vec{\xi}^D_{i-1,j,k}=\frac{1}{8}\begin{bmatrix}\overrightarrow{O_LO}\cdot\left(\overrightarrow{A_LE_L}\times\overrightarrow{E_LO_L}\right)+\overrightarrow{O_LO}\cdot\left(\overrightarrow{A_LO_L}\times\overrightarrow{O_LH_L}\right)+\\ \overrightarrow{O_LO}\cdot\left(\overrightarrow{H_LO_L}\times\overrightarrow{O_LG_L}\right)+\overrightarrow{O_LO}\cdot\left(\overrightarrow{H_LG_L}\times\overrightarrow{G_LD_L}\right)+\\ \overrightarrow{O_LO}\cdot\left(\overrightarrow{O_LF_L}\times\overrightarrow{F_LC_L}\right)+\overrightarrow{O_LO}\cdot\left(\overrightarrow{O_LC_L}\times\overrightarrow{C_LG_L}\right)+\\ \overrightarrow{O_LO}\cdot\left(\overrightarrow{E_LB_L}\times\overrightarrow{B_LF_L}\right)+\overrightarrow{O_LO}\cdot\left(\overrightarrow{E_LF_L}\times\overrightarrow{F_LO_L}\right),\end{bmatrix}=\frac{3}{4}V^O_{A_LE_LB_LF_LC_LG_LD_LH_L}. \tag{45}$$

The corresponding volume $V^O_{A_LE_LB_LF_LC_LG_LD_LH_L}$ in Eq. (45) is illuminated in Fig. 6d. Analogously, the other volumes in Eq. (44) can be represented as follows.

$$(\vec{r}_{i+1,j,k}-\vec{r}_{i,j,k})\cdot\vec{\xi}^D_{i+1,j,k}=\frac{3}{4}V^O_{A_RH_RD_RG_RC_RF_RB_RE_R},$$

$$(\vec{r}_{i,j,k}-\vec{r}_{i,j-1,k})\cdot\vec{\eta}^D_{i,j-1,k}=\frac{3}{4}V^O_{AA_LH_LD_LDD_RH_RA_R},$$

$$(\vec{r}_{i,j+1,k}-\vec{r}_{i,j,k})\cdot\vec{\eta}^D_{i,j+1,k}=\frac{3}{4}V^O_{B_RF_RC_RCC_LF_LB_LB},$$

$$(\vec{r}_{i,j,k}-\vec{r}_{i,j,k-1})\cdot\vec{\zeta}^D_{i,j,k-1}=\frac{3}{4}V^O_{AA_RE_RB_RBB_LE_LA_L},$$

$$(\vec{r}_{i,j,k+1}-\vec{r}_{i,j,k})\cdot\vec{\zeta}^D_{i,j,k+1}=\frac{3}{4}V^O_{DD_LG_LC_LCC_RG_RD_R}.$$

So the volume calculated by the V3 can be got by gathering all the volume parts together.

$$J_O^1=\frac{1}{3}\left[\left(\vec{r}\cdot\vec{\xi}^D\right)_\xi+\left(\vec{r}\cdot\vec{\eta}^D\right)_\eta+\left(\vec{r}\cdot\vec{\zeta}^D\right)_\zeta\right]=\frac{1}{8}V_{A_LA_RB_LB_RC_LC_RD_LD_R}. \tag{46}$$

Thus the volume calculated by the symmetrical conservative form of Jacobian V3 is exactly one eighth of the cell volume of FVM in Eq. (16).

It is well known that, in FVM, a cell volume is always enclosed by its vectorized surfaces and simultaneously bounded by these surfaces, in other words, the volume should be uniquely determined by its vectorized surfaces. Interestingly, in FDM, the symmetrical conservative form of Jacobian V3 is just enclosed by its vectorized surfaces if the spatial derivatives in Eqs. (26) and (41) are all calculated by the same second-order central difference scheme. Then V3 in Eq. (41) is suggested to be a preferable form of Jacobian from the geometry viewpoint.

If all the spatial derivatives are discreted by the second-order central difference scheme, the Jacobian V1, V2 and V3 would involve different grid points and the corresponding numerical values generally may be different. But when all the quadrilaterals in the grid are parallelograms, the Jacobian V1, V2 and V3 will be equivalent. The differences between V1, V2 and V3 exactly represent the deviation of quadrilaterals from parallelograms, thus the relative error of V1, V2 and V3 can be somewhat devised to detect the quality of the grid, such as

$$V_M=\max\left(V^{12},V^{23},V^{13}\right),\quad V^{12}=\frac{|V^1-V^2|}{V^2},\quad V^{23}=\frac{|V^2-V^3|}{V^3},\quad V^{13}=\frac{|V^1-V^3|}{V^3}.$$

For the grid of high quality, V_M is small, otherwise V_M is relatively large.

3.3. The derivation of Symmetrical Conservative Metric Method (SCMM)

By substituting the vectorized surfaces in Eq. (27) and the Jacobian in Eq. (46) into the discretized governing equations, namely Eq. (8), we can see that the physical variable $\delta_\tau Q$ calculated by the FDM with second-order accuracy is actually similar to that of Eq. (14) which is discretized by the second-order FVM. In other words, if the metrics are calculated with the symmetrical conservative form of S3 and Jacobian is computed with the symmetrical conservative form of V3, the same numerical results would be achieved for both FDM and FVM with the same second-order accuracy.

Though this similarity only exists when second-order central scheme is used, it give us an inspiration that we may get a general method to calculate the geometry metrics and Jacobian for high-order finite difference schemes. In this paper, this method is named as Symmetrical Conservative Metric Method (SCMM) which can be exactly defined by:

(1) the symmetrical conservative form of metrics S3 in Eqs. (26) is adopted,
(2) the symmetrical conservative form of Jacobian V3 in Eq. (41) is adopted, and
(3) the numerical derivative operators δ^1 to calculate the flux derivatives in Eq. (3) and the spatial derivatives in Eq. (41), the numerical derivative operators δ^2 to calculate the outer-level derivatives of metrics in Eqs. (26) and the numerical derivative operators δ^3 to calculate the inner-level derivatives of metrics in Eqs. (26) have to be all the same difference scheme in the same coordinate direction, i.e.

$$\delta^1_\xi=\delta^2_\xi=\delta^3_\xi,\quad \delta^1_\eta=\delta^2_\eta=\delta^3_\eta,\quad \delta^1_\zeta=\delta^2_\zeta=\delta^3_\zeta. \tag{47}$$

4. Implementation to high-order finite difference procedure

The Symmetrical Conservative Metric Method (SCMM) proposed in previous section needs the symmetrical conservative metrics and Jacobian, which may be rewritten completely as follows,

Symmetrical conservative metrics S3

$$\begin{cases} \hat{\xi}_x^{S3} = \frac{1}{2}\left[(zy_\eta)_\zeta + (yz_\zeta)_\eta - (zy_\zeta)_\eta - (yz_\eta)_\zeta\right], \\ \hat{\xi}_y^{S3} = \frac{1}{2}\left[(xz_\eta)_\zeta + (zx_\zeta)_\eta - (xz_\zeta)_\eta - (zx_\eta)_\zeta\right], \\ \hat{\xi}_z^{S3} = \frac{1}{2}\left[(yx_\eta)_\zeta + (xy_\zeta)_\eta - (yx_\zeta)_\eta - (xy_\eta)_\zeta\right]. \end{cases}$$
$$\begin{cases} \hat{\eta}_x^{S3} = \frac{1}{2}\left[(zy_\zeta)_\xi + (yz_\xi)_\zeta - (zy_\xi)_\zeta - (yz_\zeta)_\xi\right], \\ \hat{\eta}_y^{S3} = \frac{1}{2}\left[(xz_\zeta)_\xi + (zx_\xi)_\zeta - (xz_\xi)_\zeta - (zx_\zeta)_\xi\right], \\ \hat{\eta}_z^{S3} = \frac{1}{2}\left[(yx_\zeta)_\xi + (xy_\xi)_\zeta - (yx_\xi)_\zeta - (xy_\zeta)_\xi\right]. \end{cases} \quad (48)$$
$$\begin{cases} \hat{\zeta}_x^{S3} = \frac{1}{2}\left[(zy_\xi)_\eta + (yz_\eta)_\xi - (zy_\eta)_\xi - (yz_\xi)_\eta\right], \\ \hat{\zeta}_y^{S3} = \frac{1}{2}\left[(xz_\xi)_\eta + (zx_\eta)_\xi - (xz_\eta)_\xi - (zx_\xi)_\eta\right], \\ \hat{\zeta}_z^{S3} = \frac{1}{2}\left[(yx_\xi)_\eta + (xy_\eta)_\xi - (yx_\eta)_\xi - (xy_\xi)_\eta\right]. \end{cases}$$

The delta forms of S3 in Eqs. (48) are

$$\begin{cases} \hat{\xi}_x^{S3} = \frac{1}{2}\left[\delta_\zeta^2(z\delta_\eta^3 y) + \delta_\eta^2(y\delta_\zeta^3 z) - \delta_\zeta^2(z\delta_\zeta^3 y) - \delta_\zeta^2(y\delta_\eta^3 z)\right], \\ \hat{\xi}_y^{S3} = \frac{1}{2}\left[\delta_\zeta^2(x\delta_\eta^3 z) + \delta_\eta^2(z\delta_\zeta^3 x) - \delta_\eta^2(x\delta_\zeta^3 z) - \delta_\zeta^2(z\delta_\eta^3 x)\right], \\ \hat{\xi}_z^{S3} = \frac{1}{2}\left[\delta_\zeta^2(y\delta_\eta^3 x) + \delta_\eta^2(x\delta_\zeta^3 y) - \delta_\eta^2(y\delta_\zeta^3 x) - \delta_\zeta^2(x\delta_\eta^3 y)\right]. \end{cases}$$
$$\begin{cases} \hat{\eta}_x^{S3} = \frac{1}{2}\left[\delta_\xi^2(z\delta_\zeta^3 y) + \delta_\zeta^2(y\delta_\xi^3 z) - \delta_\xi^2(z\delta_\xi^3 y) - \delta_\zeta^2(y\delta_\zeta^3 z)\right], \\ \hat{\eta}_y^{S3} = \frac{1}{2}\left[\delta_\xi^2(x\delta_\zeta^3 z) + \delta_\zeta^2(z\delta_\xi^3 x) - \delta_\xi^2(x\delta_\xi^3 z) - \delta_\zeta^2(z\delta_\zeta^3 x)\right], \\ \hat{\eta}_z^{S3} = \frac{1}{2}\left[\delta_\xi^2(y\delta_\zeta^3 x) + \delta_\zeta^2(x\delta_\xi^3 y) - \delta_\xi^2(y\delta_\xi^3 x) - \delta_\zeta^2(x\delta_\zeta^3 y)\right]. \end{cases} \quad (49)$$
$$\begin{cases} \hat{\zeta}_x^{S3} = \frac{1}{2}\left[\delta_\eta^2(z\delta_\xi^3 y) + \delta_\xi^2(y\delta_\eta^3 z) - \delta_\xi^2(z\delta_\eta^3 y) - \delta_\eta^2(y\delta_\xi^3 z)\right], \\ \hat{\zeta}_y^{S3} = \frac{1}{2}\left[\delta_\eta^2(x\delta_\xi^3 z) + \delta_\xi^2(z\delta_\eta^3 x) - \delta_\xi^2(x\delta_\eta^3 z) - \delta_\eta^2(z\delta_\xi^3 x)\right], \\ \hat{\zeta}_z^{S3} = \frac{1}{2}\left[\delta_\eta^2(y\delta_\xi^3 x) + \delta_\xi^2(x\delta_\eta^3 y) - \delta_\xi^2(y\delta_\eta^3 x) - \delta_\eta^2(x\delta_\xi^3 y)\right]. \end{cases}$$

Symmetrical conservative form of Jacobian V3

$$J_{V3}^{-1} = \frac{1}{3}\left[\left(x\hat{\xi}_x + y\hat{\xi}_y + z\hat{\xi}_z\right)_\xi + (x\hat{\eta}_x + y\hat{\eta}_y + z\hat{\eta}_z)_\eta + \left(x\hat{\zeta}_x + y\hat{\zeta}_y + z\hat{\zeta}_z\right)_\zeta\right] = \frac{1}{3}\left[\left(\vec{r}\cdot\hat{\vec{\xi}}\right)_\xi + (\vec{r}\cdot\hat{\vec{\eta}})_\eta + \left(\vec{r}\cdot\hat{\vec{\zeta}}\right)_\zeta\right] \quad (50)$$

and its delta form can be described as:

$$J_{V3}^{-1} = \frac{1}{3}\left[\delta_\xi^1\left(x\hat{\xi}_x + y\hat{\xi}_y + z\hat{\xi}_z\right) + \delta_\eta^1(x\hat{\eta}_x + y\hat{\eta}_y + z\hat{\eta}_z) + \delta_\zeta^1\left(x\hat{\zeta}_x + y\hat{\zeta}_y + z\hat{\zeta}_z\right)\right] = \frac{1}{3}\left[\delta_\xi^1\left(\vec{r}\cdot\hat{\vec{\xi}}\right) + \delta_\eta^1(\vec{r}\cdot\hat{\vec{\eta}}) + \delta_\zeta^1\left(\vec{r}\cdot\hat{\vec{\zeta}}\right)\right]. \quad (51)$$

The implementation of the SCMM on high-order linear schemes are trivial, where the only requirement is that the same difference scheme is applied for all the spatial derivatives in Eqs. (3), (48) and (50). However, some special cares shall be paid on other kind of implementation, and a sixth-order WCNS [15] is chosen as an example to show the procedures on implementing the SCMM on WCNS and MUSCL-type[1] schemes.

With the concept of WCNS proposed by Deng and Zhang [15], three steps would be taken to calculate the spatial derivatives in Eq. (3). The first step is interpolation, which means to interpolate the metrics and the flow-field variables to cell edges from the values at cell nodes, usually different schemes, which satisfy the accuracy demand, may be used in this step, generally the interpolating schemes used for metrics shall be central type, while the ones for the flow-field variables may be upwind biased or even nonlinear. The second step is calculating the fluxes at cell edges, in this paper only the standard Roe flux [16] is applied. The last step is calculating the cell-node spatial derivatives from the cell-edge fluxes, and the numerical schemes applied in this step are usually central type. By inserting the interpolated cell-edge values in the first step into the numerical spatial derivatives in the last step, one can obtain the virtual numerical spatial derivative operators acting on the metrics, and the interpolation method for flow-field variables has no effect on the SCL, neither dose the flux splitting method used in the second step.

Let us only consider the ξ direction without loss of generality, the sixth-order WCNS may be expressed as

$$\frac{\partial \hat{E}_i}{\partial \xi} = \frac{75}{64\Delta\xi}\left(\hat{E}_{i+1/2} - \hat{E}_{i-1/2}\right) - \frac{25}{384\Delta\xi}\left(\hat{E}_{i+3/2} - \hat{E}_{i-3/2}\right) + \frac{3}{640\Delta\xi}\left(\hat{E}_{i+5/2} - \hat{E}_{i-5/2}\right), \tag{52}$$

where

$$\hat{E}_{i+1/2} = \hat{E}\left(U^L_{i+1/2}, U^R_{i+1/2}, \hat{\xi}_{x,i+1/2}, \hat{\xi}_{y,i+1/2}, \hat{\xi}_{z,i+1/2}\right) \tag{53}$$

is the cell-edge flux and calculated by the Roe flux method [16] in this paper. The cell-edge metrics in Eq. (53) are calculated from cell-node metrics by the following sixth-order interpolation

$$\hat{\xi}_{x,i+1/2} = \frac{150}{256}\left(\hat{\xi}_{x,i+1} + \hat{\xi}_{x,i}\right) - \frac{25}{256}\left(\hat{\xi}_{x,i+2} + \hat{\xi}_{x,i-1}\right) + \frac{3}{256}\left(\hat{\xi}_{x,i+3} + \hat{\xi}_{x,i-2}\right) \tag{54}$$

and with similar relations for the other metrics.

In order to satisfy the SCL as well as to endow the metrics and the Jacobian with their corresponding geometry information as vectorized surfaces and cell volume, respectively, the symmetrical conservative form of metrics S3 in Eqs. (26), the symmetrical conservative form of Jacobian V3 in Eq. (50), and the constraints on the difference operators $\delta^{1,2,3}$ shall all be followed on calculating the metrics and Jacobian for the WCNS.

To summarize the procedure on implementing the SCMM to the WCNS, the difference operators for the metrics are

$$\delta^1 = \delta^2 = \delta^3 : \begin{cases} a_{i+1/2} = \frac{150}{256}(a_{i+1} + a_i) - \frac{25}{256}(a_{i+2} + a_{i-1}) + \frac{3}{256}(a_{i+3} + a_{i-2}), \\ \frac{\partial a_i}{\partial \xi} = \frac{75}{64\Delta\xi}(a_{i+1/2} - a_{i-1/2}) - \frac{25}{384\Delta\xi}(a_{i+3/2} - a_{i-3/2}) + \frac{3}{640\Delta\xi}(a_{i+5/2} - a_{i-5/2}) \end{cases} \tag{55}$$

and the spatial derivatives in Eqs. (21, 24, 26, 41) are all calculated by the difference operator defined by Eqs. (55) except boundary and near boundary points where different operators may be applied in different coordinate directions provided that $\delta^1_\xi = \delta^2_\xi = \delta^3_\xi$, $\delta^1_\eta = \delta^2_\eta = \delta^3_\eta$ and $\delta^1_\zeta = \delta^2_\zeta = \delta^3_\zeta$.

It should be pointed out that the methods of calculating the cell-edge flow-field variables $U_{i\pm 1/2}$ in Eqs. (7) have no effects on the SCL, then any upwind-based linear or nonlinear interpolation/reconstruction methods, such as that in the WCNS and the MUSCL-type schemes, may be used to calculate $U^L_{i+1/2}$ and $U^R_{i+1/2}$. In this paper, $U^L_{i+1/2}$ and $U^R_{i+1/2}$ are calculated by the optimal version of the fifth-order nonlinear weighted interpolation proposed by Deng and Zhang [15].

$$U^L_{i+\frac{1}{2}} = \frac{1}{128}(+3U_{i-2} - 20U_{i-1} + 90U_i + 60U_{i+1} - 5U_{i+2}),$$

$$U^R_{i+\frac{1}{2}} = \frac{1}{128}(-5U_{i-1} + 60U_i + 90U_{i+1} - 20U_{i+2} + 3U_{i+3}).$$

5. Numerical tests

In this section, the numerical tests including a linear wave propagation problem and an Euler isentropic vortex moving problem in randomized grids with different perturbation amplitudes are conducted in order to investigate the influence of the above forms of metrics and Jacobian on the computational results. Finally, the three dimensional flow around the typical ONERA M6 wing is calculated.

5.1. Linear wave propagation problems

The linear wave propagation problem is chosen to clearly investigate the property of different forms of metrics and Jacobian. The governing equation in Cartesian coordinates is

$$\frac{\partial u}{\partial t} + a\frac{\partial u}{\partial x} + b\frac{\partial u}{\partial y} + c\frac{\partial u}{\partial z} = 0. \tag{56}$$

where a, b and c are all constants.

In three-dimensional curvilinear coordinates, Eq. (56) may be written as

$$\frac{\partial \hat{Q}}{\partial \tau} + \frac{\partial \hat{E}}{\partial \xi} + \frac{\partial \hat{F}}{\partial \eta} + \frac{\partial \hat{G}}{\partial \zeta} = 0, \tag{57}$$

where

$$\hat{Q} = J^{-1}u,$$
$$\hat{E} = \hat{\xi}_t u + \hat{\xi}_x au + \hat{\xi}_y bu + \hat{\xi}_z cu,$$
$$\hat{F} = \hat{\eta}_t u + \hat{\eta}_x au + \hat{\eta}_y bu + \hat{\eta}_z cu,$$
$$\hat{G} = \hat{\zeta}_t u + \hat{\zeta}_x au + \hat{\zeta}_y bu + \hat{\zeta}_z cu.$$

In this paper the constants are set to $a = 4, b = 2, c = 1$.
A nominal $51 \times 51 \times 51$ uniform mesh is generated in the domain $-1 \leqslant x \leqslant 1, -1 \leqslant y \leqslant 1$ and $-1 \leqslant z \leqslant 1$. The interior grids are then perturbed by random disturbances with different amplitudes A of the grid size in each coordinate direction, while the boundary points on the edges of the computational field are left unperturbed for easy boundary treatment. The grids illuminated in Fig. 7 is the sub-zone with $12 \leqslant i \leqslant 40, 12 \leqslant j \leqslant 40$ and $k = 26$. All the boundaries are treated to be

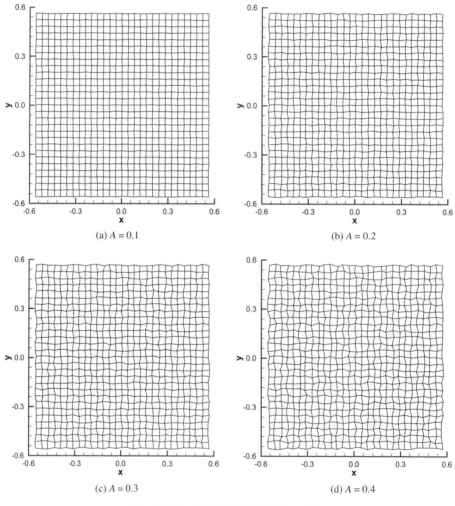

Fig. 7. Randomized grids with different perturbation coefficients A.

periodical for flow-field variables, metrics and Jacobian. The third-order total variation diminishing Runge–Kutta scheme presented by Shu and Osher [17] is adopted for time integration, and the non-dimensional time step $\Delta t = 0.001$ is used for all the numerical tests on this grid in the present work. The computations are carried out until time $t = 2$, at which moment the flow fields move back to its original position of the initial conditions.

The initial condition is

$$u_0 = \begin{cases} sin[\pi(x+0.5)]sin[\pi(y+0.5)]sin[\pi(z+0.5)] & \text{if } (x,y,z) \in [-0.5, 0.5], \\ 0 & \text{else}. \end{cases} \tag{58}$$

The numerical results of the linear wave propagation problem with sixth-order accuracy procedure are shown in Table 1–4. The maximum value of $\left|I_x^N\right|$, $\left|I_y^N\right|$ and $\left|I_z^N\right|$ which are calculated by three different metrics of S1, S2 and S3 are listed in these tables, and the L_1, L_2 and L_∞ error norms of u between the numerical results and the analytic solutions are included also. The numerical results of I_x^N, I_y^N and I_z^N are all machine zero, while it is interesting that the numerical errors of metrics

Table 1
Errors of u for linear wave propagation problem in randomized grid with $A = 0.1$.

$A = 0.1$		S1	S2	S3
$\max\left(\left|I_x^N\right|, \left|I_y^N\right|, \left|I_z^N\right|\right)$		0.84184E−15	0.74457E−15	0.43117E−15
L_1	V1	0.15167E−02	0.15047E−02	0.14152E−02
	V2	0.14075E−02	0.14015E−02	0.13636E−02
	V3	0.13937E−02	0.13968E−02	0.13588E−02
L_2	V1	0.41481E−02	0.41065E−02	0.38022E−02
	V2	0.37496E−02	0.37333E−02	0.35656E−02
	V3	0.36818E−02	0.36866E−02	0.35168E−02
L_∞	V1	0.48627E−01	0.46782E−01	0.41781E−01
	V2	0.39592E−01	0.39562E−01	0.36934E−01
	V3	0.37278E−01	0.36999E−01	0.34034E−01

Table 2
Errors of u for linear wave propagation problem in randomized grid with $A = 0.2$.

$A = 0.2$		S1	S2	S3
$\max\left(\left|I_x^N\right|, \left|I_y^N\right|, \left|I_z^N\right|\right)$		0.86827E−15	0.93648E−15	0.68993E−15
L_1	V1	0.25907E−02	0.25396E−02	0.20464E−02
	V2	0.18583E−02	0.18491E−02	0.15585E−02
	V3	0.17471E−02	0.17611E−02	0.14950E−02
L_2	V1	0.76280E−02	0.75196E−02	0.58867E−02
	V2	0.53492E−02	0.53230E−02	0.43218E−02
	V3	0.49927E−02	0.50294E−02	0.40481E−02
L_∞	V1	0.90407E−01	0.84460E−01	0.72626E−01
	V2	0.62491E−01	0.62843E−01	0.55163E−01
	V3	0.56980E−01	0.56785E−01	0.47270E−01

Table 3
Errors of u for linear wave propagation problem in randomized grid with $A = 0.3$.

$A = 0.3$		S1	S2	S3
$\max\left(\left|I_x^N\right|, \left|I_y^N\right|, \left|I_z^N\right|\right)$		0.80606E−15	0.73219E−15	0.52614E−15
L_1	V1	0.45555E−02	0.44788E−02	0.33241E−02
	V2	0.28239E−02	0.28165E−02	0.20111E−02
	V3	0.25671E−02	0.26006E−02	0.17955E−02
L_2	V1	0.13548E−01	0.13383E−01	0.98583E−02
	V2	0.84581E−02	0.84547E−02	0.58798E−02
	V3	0.76473E−02	0.77480E−02	0.51448E−02
L_∞	V1	0.13837E+00	0.13101E+00	0.10977E+00
	V2	0.90781E−01	0.92510E−01	0.78112E−01
	V3	0.82397E−01	0.81654E−01	0.64272E−01

Table 4
Errors of u for linear wave propagation problem in randomized grid with $A = 0.4$.

$A = 0.4$		S1	S2	S3
$\max\left(\left|I_x^N\right|,\left|I_y^N\right|,\left|I_z^N\right|\right)$		0.79046E−15	0.68669E−15	0.54120E−15
L_1	V1	0.71376E−02	0.70266E−02	0.50898E−02
	V2	0.42686E−02	0.42787E−02	0.27345E−02
	V3	0.38336E−02	0.39053E−02	0.23108E−02
L_2	V1	0.20931E−01	0.20746E−01	0.15145E−01
	V2	0.12783E−01	0.12821E−01	0.82311E−02
	V3	0.11487E−01	0.11670E−01	0.68770E−02
L_∞	V1	0.18432E+00	0.17713E+00	0.14644E+00
	V2	0.12137E+00	0.12470E+00	0.10402E+00
	V3	0.11161E+00	0.10886E+00	0.84649E−01

cancellation calculated with S3 are the smallest among the three forms of metrics on the randomized grids with a variety of perturbation amplitudes A. Although the forms of Jacobian have no effect on the SCL, i.e. the values of I_x^N, I_y^N and I_z^N only depend on the methods of calculating the metrics, different forms of Jacobian have quite different influences on the numerical errors of the flow fields.

The last column in Table 5 denotes the accuracy improvements from S1 and V1 to S3 and V3, especially the improvements are quite obvious for highly irregular grids. At the same time Table 5 also shows that the accuracy improvements become smaller as the grid getting smoother. This means that the SCMM for metrics and Jacobian may possess more advantages than the CMM on low-quality grids.

From Table 1 to Table 5, the results of the linear wave propagation problem show that S3 is better than both S1 and S2 in calculating the metrics for the randomized grids. At the same time V2 is better than V1, and V3 is the best on computing the Jacobian. The advantage of S3 and V3 can be also demonstrated from Fig. 8 which shows u contours with the same grid domain as illustrated in Fig. 7. Only the results computed in randomized grid with $A = 0.4$ are illustrated in Fig. 8 in order to condense the paper. The improvement of S3 and V3 over the other forms of metrics and Jacobian can be obviously observed from Fig. 8a to Fig. 8i, while the results calculated with S1 are almost equivalent to the results with S2 when the same form of Jacobian is used.

5.2. Isentropic moving vortex problems

The initial conditions for the isentropic moving vortex problems are

$$(u', v', w') = \varepsilon\tau e^{\alpha(1-\tau^2)}\left(\frac{y-y_c}{r}, -\frac{x-x_c}{r}, 0\right),$$

$$T' = -\frac{(\gamma-1)\varepsilon^2}{4\alpha\gamma}e^{2\alpha(1-\tau^2)}, \quad S' = 0,$$

$$\tau = \frac{r}{r_c}, \quad r = \sqrt{(x-x_c)^2 + (y-y_c)^2}.$$

The specific heat ratio of the fluid is set to $\gamma = 1.4$, and $r_c = 0.2$ denotes the vortex core width, $\varepsilon = 0.3$ represents vortex strength, $\alpha = 0.4$ is the parameter of the length scale of vortex decay. $T = p/\rho$ is the temperature implied that the freestream Mach number is set to $M_\infty = 1/\sqrt{\gamma}$, and $S = p/\rho^\gamma$ is the entropy. The center of the vortex is initially located at $(x_c, y_c) = (0, 0)$. The mesh and time marching method used in this test are the same as linear wave propagation ones.

Table 5
Errors of u for linear wave propagation problem in randomized grids.

Randomized grids		$error_{V1S1}$	$error_{V3S3}$	$\frac{error_{V1S1}}{error_{V3S3}}$
L_1	$A = 0.1$	0.15167E−02	0.13588E−02	1.12
	$A = 0.2$	0.25907E−02	0.14950E−02	1.73
	$A = 0.3$	0.45555E−02	0.17955E−02	2.54
	$A = 0.4$	0.71376E−02	0.23108E−02	3.09
L_2	$A = 0.1$	0.41481E−02	0.35168E−02	1.18
	$A = 0.2$	0.76280E−02	0.40481E−02	1.88
	$A = 0.3$	0.13548E−01	0.51448E−02	2.63
	$A = 0.4$	0.20931E−01	0.68770E−02	3.04
L_∞	$A = 0.1$	0.48627E−01	0.34034E−01	1.43
	$A = 0.2$	0.90407E−01	0.47270E−01	1.91
	$A = 0.3$	0.13837E+00	0.64272E−01	2.15
	$A = 0.4$	0.18432E+00	0.84649E−01	2.18

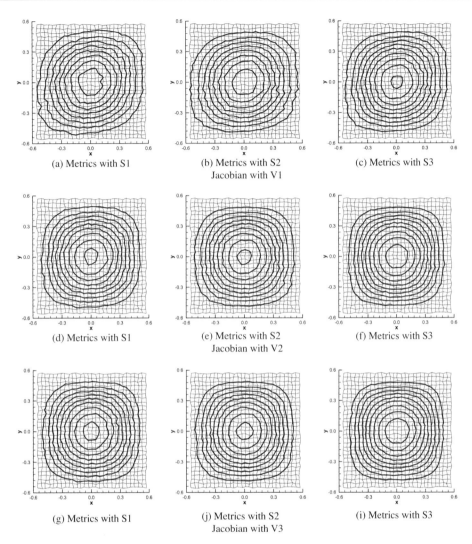

Fig. 8. Nine equally spaced contours from 0.1 (outer) to 0.9 (inner) of u calculated in randomized grid with $A = 0.4$.

The numerical results of the nonlinear Euler equation in randomized grids with sixth-order procedure are shown in Tables 6–9 with different perturbation amplitudes A. The equally spaced contours of ρ are showed in Fig. 9. The maximum of $|I_x^N|$, $|I_y^N|$ and $|I_z^N|$ in Tables 6–9 which are the same to those in Tables 1–4 are listed for completeness, and three error norms of ρ are included also. From Tables 6–9, the same conclusions can be gained for the nonlinear Euler equation as those for the linear wave propagation equation. The errors of ρ in Table 10 indicate that the combination of S3 and V3 is the best choice for metrics and Jacobian respectively. Similar conclusion can also be drawn from the previous test case as shown from Tables 1–5.

Fig. 9 shows the contours of density for different forms of metrics and Jacobian on the same randomized grid with $A = 0.4$. It is obvious that the combination of S3 and V3 can produce the best result. The error differences between V1S1 and V3S3 in Table 10 seem to be larger than those in Table 5 though the same randomized grids are used. In other words, on irregular

Table 6
Errors of ρ for moving vortex problems in randomized grid with $A = 0.1$.

$A = 0.1$		S1	S2	S3
$\max\left(\left|I_x^N\right|,\left|I_y^N\right|,\left|I_z^N\right|\right)$		0.84184E−15	0.74457E−15	0.43117E−15
L_1	V1	0.16065E−03	0.13706E−03	0.11577E−03
	V2	0.11337E−03	0.97785E−04	0.76867E−04
	V3	0.96876E−04	0.10232E−03	0.69005E−04
L_2	V1	0.32358E−03	0.28872E−03	0.24060E−03
	V2	0.23129E−03	0.20632E−03	0.16137E−03
	V3	0.20108E−03	0.20698E−03	0.14111E−03
L_∞	V1	0.43806E−02	0.37414E−02	0.32572E−02
	V2	0.33676E−02	0.26946E−02	0.23176E−02
	V3	0.31954E−02	0.27164E−02	0.18761E−02

Table 7
Errors of ρ for moving vortex problems in randomized grid with $A = 0.2$.

$A = 0.2$		S1	S2	S3
$\max\left(\left|I_x^N\right|,\left|I_y^N\right|,\left|I_z^N\right|\right)$		0.86827E−15	0.93648E−15	0.68993E−15
L_1	V1	0.34893E−03	0.31708E−03	0.25453E−03
	V2	0.24133E−03	0.21757E−03	0.16319E−03
	V3	0.20750E−03	0.21589E−03	0.14327E−03
L_2	V1	0.68171E−03	0.62903E−03	0.50559E−03
	V2	0.47771E−03	0.43526E−03	0.33007E−03
	V3	0.41792E−03	0.42640E−03	0.28700E−03
L_∞	V1	0.94541E−02	0.85188E−02	0.65796E−02
	V2	0.68400E−02	0.57185E−02	0.46375E−02
	V3	0.65495E−02	0.55125E−02	0.40477E−02

Table 8
Errors of ρ for moving vortex problems in randomized grid with $A = 0.3$.

$A = 0.3$		S1	S2	S3
$\max\left(\left|I_x^N\right|,\left|I_y^N\right|,\left|I_z^N\right|\right)$		0.80606E−15	0.73219E−15	0.52614E−15
L_1	V1	0.56699E−03	0.54005E−03	0.42227E−03
	V2	0.38704E−03	0.36178E−03	0.26312E−03
	V3	0.33639E−03	0.34433E−03	0.22611E−03
L_2	V1	0.11004E−02	0.10500E−02	0.81511E−03
	V2	0.75421E−03	0.70358E−03	0.51495E−03
	V3	0.66489E−03	0.67050E−03	0.44323E−03
L_∞	V1	0.15476E−01	0.14236E−01	0.11194E−01
	V2	0.10233E−01	0.92216E−02	0.69364E−02
	V3	0.10048E−01	0.84305E−02	0.65072E−02

Table 9
Errors of ρ for moving vortex problems in randomized grid with $A = 0.4$.

$A = 0.4$		S1	S2	S3
$\max\left(\left|I_x^N\right|,\left|I_y^N\right|,\left|I_z^N\right|\right)$		0.79046E−15	0.68669E−15	0.54120E−15
L_1	V1	0.81119E−03	0.79601E−03	0.61839E−03
	V2	0.54851E−03	0.52656E−03	0.37718E−03
	V3	0.48192E−03	0.48645E−03	0.31867E−03
L_2	V1	0.15903E−02	0.15568E−02	0.11793E−02
	V2	0.10689E−02	0.10186E−02	0.72158E−03
	V3	0.95062E−03	0.94747E−03	0.61404E−03
L_∞	V1	0.22677E−01	0.20641E−01	0.16874E−01
	V2	0.14402E−01	0.13207E−01	0.94600E−02
	V3	0.14379E−01	0.11777E−01	0.92103E−02

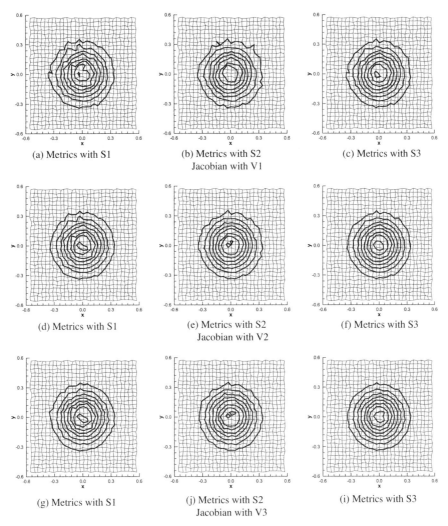

Fig. 9. Eight equally spaced contours of density from 0.92 (inner) to 0.99 (outer) calculated in randomized grid with $A = 0.4$.

grids, the accuracy improvements benefited from S3 and V3 seems to be more evident for nonlinear problems than linear problems.

5.3. ONERA M6 wing

The current simulations is based on the flow conditions of Test 2308 of Ref. [20]. The free stream flow is characterized as $Ma_\infty = 0.8395$, $\alpha = 3.06°$, $Re_c = 11.72 \times 10^6$. The Reynolds number Re_c is based on the mean aerodynamic chord $c = 0.64607m$. Fig. 10a shows the computational grid around wingtip of ONERA M6 wing as well as the magnified grid around rear edge of wingtip, and the total grid points are about 0.32 million. In this test case, the high-order WCNS-E-5 [15] scheme is applied to discretize the spatial derivatives, and the Spalart–Allmaras model [21] is adopted. Based on the

Table 10
Errors of ρ for moving vortex problems in randomized grids.

Randomized grids		$error_{V1S1}$	$error_{V3S3}$	$\frac{error_{V1S1}}{error_{V3S3}}$
L_1	$A = 0.1$	0.16065E−03	0.69005E−04	2.33
	$A = 0.2$	0.34893E−03	0.14327E−03	2.44
	$A = 0.3$	0.56699E−03	0.22611E−03	2.51
	$A = 0.4$	0.81119E−03	0.31867E−03	2.55
L_2	$A = 0.1$	0.32358E−03	0.14111E−03	2.29
	$A = 0.2$	0.68171E−03	0.28700E−03	2.38
	$A = 0.3$	0.11004E−02	0.44323E−03	2.48
	$A = 0.4$	0.15903E−02	0.61404E−03	2.59
L_∞	$A = 0.1$	0.43806E−02	0.18761E−02	2.33
	$A = 0.2$	0.94541E−02	0.40477E−02	2.34
	$A = 0.3$	0.15476E−01	0.65072E−02	2.38
	$A = 0.4$	0.22677E−01	0.92103E−02	2.46

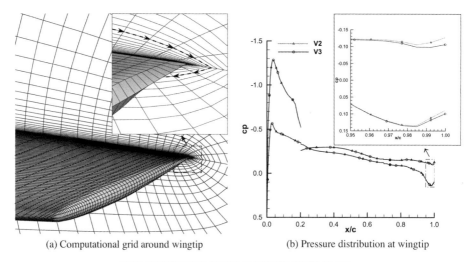

(a) Computational grid around wingtip (b) Pressure distribution at wingtip

Fig. 10. Computational grid and pressure distribution of ONERA M6 wing.

SCMM derived in this paper, the symmetrical conservative form of metric S3 in Eq. (26) is utilized. To calculate the Jacobian, all three volume-like forms, i.e. V1, V2 and V3 derived in this paper are tested. As the traditional form of Jacobian V1 adopted, however, the calculation blows up, though the V1 Jacobians at every grid point are positive. The pressure distributions around wingtip calculated by Jacobian V2 and V3 are showed in Fig. 10b. We may see the difference of pressure distribution near the rear edge of wingtip, where the difference of V2 and V3 Jacobians $V^{23} = \frac{|V^2 - V^3|}{V^3}$ is obvious.

6. Concluding remarks

In FDM, there is a variety of equivalent differential forms of metrics and Jacobian which may introduce different discretization errors. In order to settle the problem on developing a proper form, a Symmetrical Conservative Metric Method (SCMM) is newly derived to calculate metrics and Jacobian. The SCMM is derived from the geometry analogy between FDM and FVM and is based on the CMM [1] which can ensure the SCL with the condition $\delta^1 = \delta^2$. The constraint $\delta^3 = \delta^2$, which is recommended by Deng et al. [1], is proved from geometry viewpoint.

In FDM, the metrics can be computed with various non-conservative or conservative forms, but only the symmetrical conservative form of metrics S3 can simultaneously produces the vectorized-surface-like metrics and ensures the SCL. The Jacobian can also be computed with various forms, but only the symmetrical conservative form of Jacobian V3 with the proper difference scheme can be enclosed by its vectorized surfaces. Then the SCMM is expressed as the combination of S3 and V3, together with $\delta^1_\zeta = \delta^2_\zeta = \delta^3_\zeta$, $\delta^1_\eta = \delta^2_\eta = \delta^3_\eta$ and $\delta^1_\varsigma = \delta^2_\varsigma = \delta^3_\varsigma$. The similarity between second-order FDM and FVM is essential for that metrics and Jacobian calculated by the SCMM in FDM is identical to the vectorized surfaces and cell volume in FVM, respectively. The main advantage of the SCMM is that it makes the high-order FDM such as WCNS calculations around complex geometry flows possible and somewhat easy.

The numerical tests in the present work show that the SCMM can evidently increase the numerical accuracy and robustness on irregular grids with high-order WCNS schemes. Thus the SCMM is mightily suggested as a preferable method to calculate the metrics and Jacobian in high-order FDM simulations of flows around complex geometries.

Acknowledgments

This study was supported by the project of National Basic Research Program of China (Grant No. 2009CB723800), National Natural Science Foundation of China (Grant No. 11072259) and the foundation of State Key Laboratory of Aerodynamics (Grant Nos. JBKY11030902 and JBKY11010101). The authors are grateful to Assistant researcher Yi Jiang for his useful discussions. The vectorized surface and volume calculation method and the grid check method based on three different forms of Jacobian derived in this paper have been applied for patents in the People's Republic of China, and the patent application numbers are 2012102848054 and 2012103109115, respectively.

References

[1] X. Deng, M. Mao, G. Tu, H. Liu, H. Zhang, Geometric conservation law and applications to high-order finite difference schemes with stationary grids, Journal of Computational Physics 230 (2011) 1100–1115.
[2] J.L. Steger, Implicit finite difference simulation of flow about arbitrary geometries with application to airfoils, AIAA Paper 77-665, 1977.
[3] J.L. Steger, Implicit finite difference simulation of flow about arbitrary two-dimensional geometries, AIAA Journal 16 (1978) 679–686.
[4] T.H. Pulliam ,J.L. Steger, On implicit finite-difference simulations of three-dimensional flow, AIAA Paper 78-10, 1978.
[5] P.D. Thomas, C.K. Lombard, The Geometric conservation law-a link between finite difference and finite volume methods of flow computation on moving grids, AIAA paper 78-1208, 1978.
[6] P.D. Thomas, C.K. Lombard, Geometric conservation law and its application to flow computations on moving grids, AIAA Journal 17 (1979) 1030–1037.
[7] M. Vinokur, H.C. Yee, Extension of efficient low dissipation high-order schemes for 3-D Curvilinear moving grids, in: David A. Caughey, Mohamed M. Hafez (Eds.), Frontiers of Computational Fluid Dynamics, World Scientific, 2002 (Chapter 8).
[8] T. Nonomura, N. Iizuka, K. Fujii, Free-stream and vortex preservation properties of high-order WENO and WCNS on curvilinear grids, Computers and Fluids 39 (2010) 197–214.
[9] G.S. Jiang, C.W. Shu, Efficient implementation of weighted ENO schemes, Journal of Computational Physics 126 (1996) 202–228.
[10] M. Vinokur, An analysis of finite-difference and finite-volume formulations of conservation laws, Journal of Computational Physics 81 (1989) 1–52.
[11] R.M. Visbal, On the use of higher-order finite-difference schemes on curvilinear and deforming meshes, Journal of Computational Physics 181 (2002) 155–185.
[12] R. Radespiel, C. Rossow, An efficient cell-vertex multigrid scheme for the three-dimensional Navier–Stokes equations, AIAA paper-1989-1953, 1989.
[13] R. Radespiel, C. Rossow, Efficient cell-vertex multigrid scheme for the three-dimensional Navier–Stokes equations, AIAA Journal 28 (1990) 1464–1472.
[14] X. Deng, H. Zhang, Developing high-order weighted compact nonlinear schemes, Journal of Computational Physics 165 (2000) 22–44.
[15] P.L. Roe, Approximate Riemann solvers, parameter vectors, and difference schemes, Journal of Computational Physics 43 (1981) 357–372.
[16] C.W. Shu, S. Osher, Efficient implementation of essentially non-oscillatory shock capturing schemes, Journal of Computational Physics 77 (1988) 439–471.
[17] X. Deng, High-order accurate dissipative weighted compact nonlinear schemes, Science in China (Series A) 45 (2002) 356–370.
[18] B. Van Leer, Towards the ultimate conservative difference scheme V: a second-order sequel to Godunov's method, Journal of Computational Physics 32 (1979) 101–136.
[19] V. Schmitt, F. Charpin, Pressure distributions on the ONERA M6 wing at transonic Mach numbers, experimental data base for computer program assessment, Report of the Fluid Dynamics Panel Working Group 04, AGARD AR 138, May 1979.
[20] P.R. Spalart ,S.R. Allmaras, A one-equation turbulence model for aerodynamic flows, AIAA paper-1992-0439, 1992.

On the freestream preservation of finite volume method in curvilinear coordinates

Dan Xu[a,*], Xiaogang Deng[a], Yaming Chen[b], Yidao Dong[a], Guangxue Wang[a,c]

[a] College of aerospace science and engineering, National University of Defense Technology, Changsha, Hunan 410073, People's Republic of China
[b] College of Science, National University of Defense Technology, Changsha, Hunan 410073, People's Republic of China
[c] Computational Aerodynamics Institute, China Aerodynamics Research and Development Center, Mianyang, Sichuan 621000, People's Republic of China

ARTICLE INFO

Article history:
Received 10 June 2015
Revised 31 December 2015
Accepted 30 January 2016
Available online 6 February 2016

Keywords:
Finite volume method
Freestream preservation
Curvilinear coordinates

ABSTRACT

As the importance of freestream preservation is widely discussed in the finite difference (FD) method, it is found in this paper that this problem also exists in the finite volume (FV) method based on the curvilinear coordinate system. If the freestream is not well preserved, serious errors will be introduced in numerical results. In this paper, the conditions of the freestream preservation are studied in the FV method using the dimension-by-dimension reconstruction. In two space dimensions, it is demonstrated that the order of the polynomials in evaluating metrics should not be higher than the order of accuracy of the Gaussian quadrature. In addition, the Jacobian at each Gaussian point needs to be obtained by the same reconstruction algorithm. In three space dimensions, the metrics can be written in three different forms: the cross product, conservative and symmetric conservative forms. For guaranteeing the freestream preservation, it is proved that except for the conditions proposed in two space dimensions, the metrics should be evaluated by the conservative or symmetric conservative forms, but not the cross product form. Numerical tests are presented to illustrate the effect of the conditions proposed in this paper.

© 2016 Elsevier Ltd. All rights reserved.

1. Introduction

The development of novel numerical algorithms will be critical in CFD [1]. However, no matter what kind of algorithm is developed, some basic rules should be satisfied to guarantee the designed features. As discussed in previous studies [2–4], the freestream preservation is an important requirement for the discretization of conservation laws; if not satisfied, discretizations of the governing equations can catastrophically degrade the fidelity of numerical algorithms. Actually, this problem has attracted much attention and widely studied in the FD method. Visbal and Gaitonde [5] found the conservative metric technique could largely decrease the numerical errors. The conservative form was extended to the symmetric conservative one by Vinokur and Yee [6]. Nonomura et al. [7] demonstrated that the WENO scheme could not satisfy the freestream preservation, but the WCNS [8] could by using the conservative metrics. Recently, they also proposed a new technique [9] for the finite-difference WENO scheme to preserve the freestream on stationary curvilinear grids. The method of solving this problem on moving and deforming grids was developed by Abe et al. [10]. Based on the weighted compact nonlinear scheme (WCNS), Deng et al. [11] developed the symmetrical conservative metric method (SCMM) that not only fulfilled the requirements of the freestream preservation, but also presented the geometric meanings of the metrics and Jacobian, establishing a link between the FV and FD methods [12]. Abe et al. [13] also presented the geometric interpretations of the metrics for the high-order FD schemes. All these studies have well solved the problem and made the FD method applicable in complex configurations [14,15].

In contrast, the issue of freestream preservation is seldom discussed in the FV method. The main reason is that when the FV method is carried out in the physical space, whether on structured or unstructured meshes, the freestream preservation can be automatically satisfied. Considering this characteristic and the good conservation property, the FV method is popular in CFD simulations and widely adopted by commercial softwares, but most of the applications and softwares using this technique are limited to be 2nd-order [2,16]. With the development of computational aeroacoustics (CAA), large eddy simulations (LES) and direct numerical simulations (DNS), high-order schemes become essential tools. Although many high-order FV schemes using the ENO/WENO method have been proposed (see [17-19] for instance), the direct implementation in the physical space brings big challenges: the complexity of algorithms and codes increases dramatically because of the difficulty in choosing stencils in the multi-dimensional reconstruction. As a result, these schemes are much more expensive in CPU cost [20], which makes them hardly be used in practical simulations.

To overcome the difficulties, an efficient way is to apply the FV method in the curvilinear coordinate system, where structured meshes should be used. Although the flexibility may be reduced in comparison with the unstructured meshes, the good numerical characteristics are preserved. As the high-order FD method can also be used on the curvilinear meshes, there are still some advantages that make the FV method a choice. The first one is the exact global conservation property, which is only approximately satisfied in the high-order FD method [21]. The second one is the strict adherence to the integral form that makes it preferred in numerical simulations [22,23]. Furthermore, based on the structured meshes, the standard dimension-by-dimension reconstruction [24] can be used, and the requirements for CPU and memory are considerably less [25]. In fact, the FV method in curvilinear coordinates has been widely used in the simulations of turbulence [26], radiative transport problems [27], shallow water equations [28] and even astrophysics [29]. As the coordinate transformation is needed, the freestream preservation again becomes a problem to be considered. Moreover, it should be noted that this problem is as important as that in the FD method. If the freestream is not preserved, the numerical results will also be degraded by extra errors. When using the FV method in curvilinear coordinates, many researchers have paid attention to this problem. Casper et al. [30] pointed out that the analytical metrics could not satisfy the metric identities. Collela et al. [31] developed a method for computing the averages of the metric terms on faces such that the freestream preservation was automatically satisfied, but his FV method is not based on the dimension-by-dimension reconstruction. Gallerano et al. [28] studied the computation of metrics in solving the shallow water equations. However, they did not give detailed analysis about their work.

In this paper, we adopt the dimension-by-dimension reconstruction in the FV method and study how to satisfy the freestream preservation. Based on the curvilinear coordinates, we first describe the discretization procedure of the FV method and the evaluation of the metrics and Jacobian. In two space dimensions, the metric identities are presented and the conditions that satisfy the metric identities discretely are studied. It is proved that for preserving the freestream, the order of the polynomials in evaluating metrics should not be higher than the order of accuracy of the Gaussian quadrature. In addition, Jacobian should be obtained by the same reconstruction algorithm. When extending the conditions to three space dimensions, it is found that the metrics can be written in three different forms that are equivalent in the analytical sense, but behave differently in numerical simulations. Furthermore, it is shown that the conservative and symmetric conservative forms should be used, while the cross product form cannot satisfy the metric identities discretely.

The rest of the paper is organized as follows. The governing equations in the curvilinear coordinates and a brief description of the reconstruction procedure are presented in Section 2. In Section 3 the conditions that satisfy the freestream preservation in two space dimensions are studied. The conditions are extended to three space dimensions in Section 4 and an extra requirement in the metrics is proposed. In Section 5, numerical tests are presented to demonstrate the above conditions and the effect on numerical results. Conclusions are drawn in Section 6.

2. Finite volume method in curvilinear coordinates

Consider the three-dimensional hyperbolic systems in conservative form
$$\frac{\partial Q}{\partial t} + \frac{\partial E}{\partial x} + \frac{\partial F}{\partial y} + \frac{\partial G}{\partial z} = 0. \tag{1}$$

By introducing a coordinate transformation $(x, y, z) \to (\xi, \eta, \zeta)$, the governing equations in the curvilinear coordinates can be expressed as
$$\frac{\partial}{\partial t}\left(\frac{Q}{J}\right) + \left(\frac{\partial \hat{E}}{\partial \xi} + \frac{\partial \hat{F}}{\partial \eta} + \frac{\partial \hat{G}}{\partial \zeta}\right) = 0, \tag{2}$$
where
$$\hat{E} = \hat{\xi}_x E + \hat{\xi}_y F + \hat{\xi}_z G,$$
$$\hat{F} = \hat{\eta}_x E + \hat{\eta}_y F + \hat{\eta}_z G,$$
$$\hat{G} = \hat{\zeta}_x E + \hat{\zeta}_y F + \hat{\zeta}_z G. \tag{3}$$

The metrics and Jacobian originating from the transformation can be evaluated by
$$J^{-1} = \left|\frac{\partial(x,y,z)}{\partial(\xi,\eta,\zeta)}\right| = \begin{vmatrix} x_\xi & x_\eta & x_\zeta \\ y_\xi & y_\eta & y_\zeta \\ z_\xi & z_\eta & z_\zeta \end{vmatrix}, \tag{4}$$

$$\hat{\xi}_x = \xi_x/J = y_\eta z_\zeta - z_\eta y_\zeta,$$
$$\hat{\xi}_y = \xi_y/J = z_\eta x_\zeta - x_\eta z_\zeta,$$
$$\hat{\xi}_z = \xi_z/J = x_\eta y_\zeta - y_\eta x_\zeta,$$
$$\hat{\eta}_x = \eta_x/J = y_\zeta z_\xi - z_\zeta y_\xi,$$
$$\hat{\eta}_y = \eta_y/J = z_\zeta x_\xi - x_\zeta z_\xi,$$
$$\hat{\eta}_z = \eta_z/J = x_\zeta y_\xi - y_\zeta x_\xi,$$
$$\hat{\zeta}_x = \zeta_x/J = y_\xi z_\eta - z_\xi y_\eta,$$
$$\hat{\zeta}_y = \zeta_y/J = z_\xi x_\eta - x_\xi z_\eta,$$
$$\hat{\zeta}_z = \zeta_x/J = x_\xi y_\eta - y_\xi x_\eta. \tag{5}$$

As shown in Fig. 1, a control volume is defined as $I_{ijk} = [\xi_{i-1/2}, \xi_{i+1/2}] \times [\eta_{j-1/2}, \eta_{j+1/2}] \times [\zeta_{k-1/2}, \zeta_{k+1/2}]$, where the flow information is stored. The subscripts i, j and k represent the position of the control volumes and the coordinate values of mesh points are located at the cell vertices. For example, the grid coordinate value of the vertex $(\xi_{i+1/2}, \eta_{j+1/2}, \zeta_{k-1/2})$ as shown in Fig. 1 is $(x_{i+1/2,j+1/2,k-1/2}, y_{i+1/2,j+1/2,k-1/2}, z_{i+1/2,j+1/2,k-1/2})$. Integrating Eq. (2) over I_{ijk}, we obtain

$$\frac{d}{dt}\overline{\left(\frac{Q}{J}\right)}_{ijk} + \frac{1}{\Delta\xi}\left(\hat{E}_{i+1/2,j,k} - \hat{E}_{i-1/2,j,k}\right) + \frac{1}{\Delta\eta}\left(\hat{F}_{i,j+1/2,k} - \hat{F}_{i,j-1/2,k}\right)$$
$$+ \frac{1}{\Delta\zeta}\left(\hat{G}_{i,j,k+1/2} - \hat{G}_{i,j,k-1/2}\right) = 0. \tag{6}$$

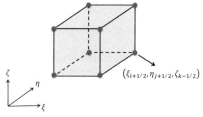

Fig. 1. Schematic of the control volume and the cell vertices.

where $\overline{\overline{\overline{(\frac{Q}{J})}}}_{ijk}$ is the spatial average in the control volume I_{ijk},

$$\overline{\overline{\overline{\left(\frac{Q}{J}\right)}}}_{ijk} = \frac{1}{\Delta\xi\Delta\eta\Delta\zeta}\int_{\xi_{i-1/2}}^{\xi_{i+1/2}}\int_{\eta_{j-1/2}}^{\eta_{j+1/2}}\int_{\zeta_{k-1/2}}^{\zeta_{k+1/2}}\left(\frac{Q}{J}\right)d\xi d\eta d\zeta, \quad (7)$$

and $\hat{E}_{i+1/2,j,k}$, $\hat{F}_{i,j+1/2,k}$, $\hat{G}_{i,j,k+1/2}$ are spatial averages of fluxes over the cell faces,

$$\hat{E}_{i+1/2,j,k} = \frac{1}{\Delta\eta\Delta\zeta}\int_{\eta_{j-1/2}}^{\eta_{j+1/2}}\int_{\zeta_{k-1/2}}^{\zeta_{k+1/2}}\hat{E}\left(\xi_{i+1/2},\eta,\zeta\right)d\eta d\zeta,$$

$$\hat{F}_{i,j+1/2,k} = \frac{1}{\Delta\xi\Delta\zeta}\int_{\xi_{i-1/2}}^{\xi_{i+1/2}}\int_{\zeta_{k-1/2}}^{\zeta_{k+1/2}}\hat{F}\left(\xi,\eta_{j+1/2},\zeta\right)d\xi d\zeta,$$

$$\hat{G}_{i,j,k+1/2} = \frac{1}{\Delta\xi\Delta\eta}\int_{\xi_{i-1/2}}^{\xi_{i+1/2}}\int_{\eta_{j-1/2}}^{\eta_{j+1/2}}\hat{G}\left(\xi,\eta,\zeta_{k+1/2}\right)d\xi d\eta. \quad (8)$$

In this FV method, the flow information is stored in the cell averages $\overline{\overline{\overline{(\frac{Q}{J})}}}_{ijk}$. As the control volumes are identical with the grid cells, it belongs to the cell-centered scheme [32] (for node-centered scheme, dual control volumes are needed to be constructed). In order to update the flow variables in time, we must reconstruct the fluxes from the cell averages. To describe the process of reconstruction, we concentrate on the flux $\hat{E}_{i+1/2,j,k}$, and $\hat{F}_{i,j+1/2,k}$, $\hat{G}_{i,j,k+1/2}$ can be obtained in an analogous manner.

The spatial integrals are approximated by a numerical quadrature. Using the N-point Gaussian quadrature, the flux $\hat{E}_{i+1/2,j,k}$ can be expressed as

$$\hat{E}_{i+1/2,j,k} = \frac{1}{\Delta\eta\Delta\zeta}\sum_{\alpha=1}^{N}\sum_{\beta=1}^{N}\hat{E}\left(\xi_{i+1/2},\eta_\alpha,\zeta_\beta\right)K_\alpha K_\beta, \quad (9)$$

where η_α and ζ_β represent the positions of Gaussian integration points in different directions; K_α and K_β are the coefficients of weight and N is the number of Gaussian points that decides the order of accuracy. To achieve upwinding in the numerical method, fluxes at the Gaussian points should be determined by an approximate Riemann solver based on the left and right states obtained from the reconstruction. Therefore, Eq. (9) can be further expressed as

$$\hat{E}_{i+1/2,j,k} = \frac{1}{\Delta\eta\Delta\zeta}\sum_{\alpha=1}^{N}\sum_{\beta=1}^{N}\hat{E}_{\alpha,\beta}K_\alpha K_\beta$$

$$= \frac{1}{\Delta\eta\Delta\zeta}\sum_{\alpha=1}^{N}\sum_{\beta=1}^{N}\hat{E}[Q^L(\xi_{i+1/2},\eta_\alpha,\zeta_\beta),$$

$$Q^R(\xi_{i+1/2},\eta_\alpha,\zeta_\beta)]K_\alpha K_\beta. \quad (10)$$

As the coordinate transformation is used, the metrics and Jacobian appear. Finally, the integrated expression for fluxes at the Gaussian points is

$$\hat{E}_{\alpha,\beta} = \hat{E}[Q^L(\xi_{i+1/2},\eta_\alpha,\zeta_\beta),$$

$$Q^R(\xi_{i+1/2},\eta_\alpha,\zeta_\beta),(\hat{\xi}_x,\hat{\xi}_y,\hat{\xi}_z,J)_{i+1/2,\alpha,\beta}]. \quad (11)$$

Hence, the main procedure is to obtain the left and right states Q^L, Q^R and the corresponding metrics and Jacobian at the Gaussian points.

2.1. Reconstruction for Q^L and Q^R

The FV method in the curvilinear coordinates has the advantage of using the dimension-by-dimension reconstruction method, especially for high-order schemes, which is simple and has lower CPU cost. The general idea of the method is to divide the multi-dimensional reconstruction into multiple one-dimensional ones. The detailed procedure is described as follows.

In the first step of the three-dimensional reconstruction, the cell averages $\overline{\overline{\overline{(\frac{Q}{J})}}}_{ijk}$ are used to reconstruct in the ξ coordinate direction and the left and right cell face averages in the $\eta - \zeta$ plane are obtained.

$$\overline{\overline{\left(\frac{Q}{J}\right)}}^L_{i+1/2,j,k} = \frac{1}{\Delta\eta\Delta\zeta}\int_{\eta_{j-1/2}}^{\eta_{j+1/2}}\int_{\zeta_{k-1/2}}^{\zeta_{k+1/2}}\left(\frac{Q}{J}\right)^L d\eta d\zeta,$$

$$\overline{\overline{\left(\frac{Q}{J}\right)}}^R_{i+1/2,j,k} = \frac{1}{\Delta\eta\Delta\zeta}\int_{\eta_{j-1/2}}^{\eta_{j+1/2}}\int_{\zeta_{k-1/2}}^{\zeta_{k+1/2}}\left(\frac{Q}{J}\right)^R d\eta d\zeta. \quad (12)$$

In the second step, the one-dimensional reconstruction in the η coordinate direction is performed to obtain the one-dimensional averages respect to the ζ coordinate direction,

$$\overline{\left(\frac{Q}{J}\right)}^L_{i+1/2,j_\alpha,k} = \frac{1}{\Delta\zeta}\int_{\zeta_{k-1/2}}^{\zeta_{k+1/2}}\left(\frac{Q}{J}\right)^L d\zeta,$$

$$\overline{\left(\frac{Q}{J}\right)}^R_{i+1/2,j_\alpha,k} = \frac{1}{\Delta\zeta}\int_{\zeta_{k-1/2}}^{\zeta_{k+1/2}}\left(\frac{Q}{J}\right)^R d\zeta. \quad (13)$$

Finally, the left and right states $(\frac{Q}{J})^L_{i+1/2,j_\alpha,k_\beta}$, $(\frac{Q}{J})^R_{i+1/2,j_\alpha,k_\beta}$ at the Gaussian points are obtained from the line averages. The procedure of the dimension-by-dimension reconstruction is sketched in Fig. 2. To avoid the reconstruction across discontinuities, the WENO scheme is used. Different orders of accuracy correspond to different stencils. For the 2nd-order scheme, only one Gauss point is needed and located at the center of the cell face, so that only the first one step is necessary. However, for high-order schemes, the multi-dimensional reconstruction must be used. In two space dimensions, only the first two steps are needed.

The detailed formulae are not presented here, which can be found in [24]. The main difference between this paper and the work in [24] is the use of curvilinear meshes, which changes the reconstruction variables from $\overline{\overline{\overline{Q}}}$ to $\overline{\overline{\overline{(\frac{Q}{J})}}}$. Another difference is that the metrics and Jacobian arise, and the evaluations are discussed in the next section.

2.2. Evaluation of the metrics and Jacobian

In three space dimensions, different forms can be used in the metric evaluation which are all equivalent in the analytical sense. Eq. (5) gives the expressions that are often called the cross product form. In numerical simulations, two conservative forms can be derived [33]

$$\begin{cases} \hat{\xi}_x = (y_\eta z)_\zeta - (y_\zeta z)_\eta, \hat{\xi}_y = (z_\eta x)_\zeta - (z_\zeta x)_\eta, \hat{\xi}_z = (x_\eta y)_\zeta - (x_\zeta y)_\eta, \\ \hat{\eta}_x = (y_\zeta z)_\xi - (y_\xi z)_\zeta, \hat{\eta}_y = (z_\zeta x)_\xi - (z_\xi x)_\zeta, \hat{\eta}_z = (x_\zeta y)_\xi - (x_\xi y)_\zeta, \\ \hat{\zeta}_x = (y_\xi z)_\eta - (y_\eta z)_\xi, \hat{\zeta}_y = (z_\xi x)_\eta - (z_\eta x)_\xi, \hat{\zeta}_z = (x_\xi y)_\eta - (x_\eta y)_\xi. \end{cases}$$
(14)

$$\begin{cases} \hat{\xi}_x = (yz_\zeta)_\eta - (yz_\eta)_\zeta, \hat{\xi}_y = (zx_\zeta)_\eta - (zx_\eta)_\zeta, \hat{\xi}_z = (xy_\zeta)_\eta - (xy_\eta)_\zeta, \\ \hat{\eta}_x = (yz_\xi)_\zeta - (yz_\zeta)_\xi, \hat{\eta}_y = (zx_\xi)_\zeta - (zx_\zeta)_\xi, \hat{\eta}_z = (xy_\xi)_\zeta - (xy_\zeta)_\xi, \\ \hat{\zeta}_x = (yz_\eta)_\xi - (yz_\xi)_\eta, \hat{\zeta}_y = (zx_\eta)_\xi - (zx_\xi)_\eta, \hat{\zeta}_z = (xy_\eta)_\xi - (xy_\xi)_\eta. \end{cases}$$
(15)

Moreover, the symmetric conservative form that is coordinate invariant can be expressed as [6]

$$\begin{cases} \hat{\xi}_x = \frac{1}{2}[(zy_\eta)_\zeta + (yz_\zeta)_\eta - (zy_\zeta)_\eta - (yz_\eta)_\zeta], \\ \hat{\xi}_y = \frac{1}{2}[(xz_\eta)_\zeta + (zx_\zeta)_\eta - (xz_\zeta)_\eta - (zx_\eta)_\zeta], \\ \hat{\xi}_z = \frac{1}{2}[(yx_\eta)_\zeta + (xy_\zeta)_\eta - (yx_\zeta)_\eta - (xy_\eta)_\zeta]. \end{cases}$$
(16)

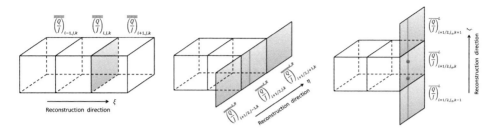

(a) Cell averages are reconstructed to obtain cell face averages (red).

(b) Cell face averages (red) are reconstructed to obtain line averages (green).

(c) Line averages (green) are reconstructed to obtain values at the Gaussian points (blue).

Fig. 2. Schematic of the dimension-by-dimension reconstruction in three space dimensions. (a) Cell averages are reconstructed to obtain cell face averages (red). (b) Cell face averages (red) are reconstructed to obtain line averages (green). (c) Line averages (green) are reconstructed to obtain values at the Gaussian points (blue). (For interpretation of the references to color in this figure legend, the reader is referred to the web version of this article.)

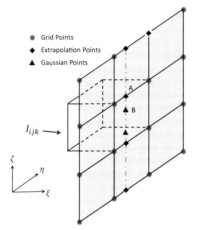

Fig. 3. Schematic of the metric evaluation in three space dimensions.

$$\begin{cases} \hat{\eta}_x = \frac{1}{2}[(zy_\zeta)_\xi + (yz_\xi)_\zeta - (zy_\xi)_\zeta - (yz_\zeta)_\xi], \\ \hat{\eta}_y = \frac{1}{2}[(xz_\zeta)_\xi + (zx_\xi)_\zeta - (xz_\xi)_\zeta - (zx_\zeta)_\xi], \\ \hat{\eta}_z = \frac{1}{2}[(yx_\zeta)_\xi + (xy_\xi)_\zeta - (yx_\xi)_\zeta - (xy_\zeta)_\xi]. \end{cases} \quad (17)$$

$$\begin{cases} \hat{\zeta}_x = \frac{1}{2}[(zy_\xi)_\eta + (yz_\eta)_\xi - (zy_\eta)_\xi - (yz_\xi)_\eta], \\ \hat{\zeta}_y = \frac{1}{2}[(xz_\xi)_\eta + (zx_\eta)_\xi - (xz_\eta)_\xi - (zx_\xi)_\eta], \\ \hat{\zeta}_z = \frac{1}{2}[(yx_\xi)_\eta + (xy_\eta)_\xi - (yx_\eta)_\xi - (xy_\xi)_\eta]. \end{cases} \quad (18)$$

No matter which form is used, the finite-difference approximation is needed. For example, as shown in Fig. 3, to evaluate the value of y_ζ at point B, we first need to obtain the coordinate value y at different points along the direction of ζ, which also passes through the Gaussian points. These values (such as at point A) are evaluated by polynomial interpolations. The detailed formulae are presented in Appendix A.

It should be noted that all above forms can satisfy the requirement in the numerical order, but may not satisfy the freestream preservation. The purpose of this paper is to find a general rule for approximating the metric terms. A detailed discussion will be presented in the following sections.

For the Jacobian, there also exist different forms. Both Eq. (4) and the expression recommended by SCMM can be used.

$$J^{-1} = \frac{1}{3}[(x\hat{\xi}_x + y\hat{\xi}_y + z\hat{\xi}_z)_\xi \\ + (x\hat{\eta}_x + y\hat{\eta}_y + z\hat{\eta}_z)_\eta + (x\hat{\zeta}_x + y\hat{\zeta}_y + z\hat{\zeta}_z)_\zeta]. \quad (19)$$

However, for the FV method, there is another form in the Jacobian evaluation. In Eq. (7), if Q is assumed to be constant, the equation can be expressed as

$$\overline{\overline{\left(\frac{1}{J}\right)}}_{ijk} = \frac{1}{\Delta\xi\Delta\eta\Delta\zeta} \int_{\xi_{i-1/2}}^{\xi_{i+1/2}} \int_{\eta_{j-1/2}}^{\eta_{j+1/2}} \int_{\zeta_{k-1/2}}^{\zeta_{k+1/2}} \left(\frac{1}{J}\right) d\xi d\eta d\zeta. \quad (20)$$

Based on the integral transform theorem, we have

$$\int_{\xi_{i-1/2}}^{\xi_{i+1/2}} \int_{\eta_{j-1/2}}^{\eta_{j+1/2}} \int_{\zeta_{k-1/2}}^{\zeta_{k+1/2}} \left(\frac{1}{J}\right) d\xi d\eta d\zeta \\ = \int_{x_{i-1/2}}^{x_{i+1/2}} \int_{y_{j-1/2}}^{y_{j+1/2}} \int_{z_{k-1/2}}^{z_{k+1/2}} dxdydz = V_{ijk}. \quad (21)$$

where V_{ijk} is the volume in the physical space corresponding to the control volume I_{ijk}. Therefore the cell averages of $\left(\frac{1}{J}\right)$ equal to the ratio of the volume in the physical space to that in the computational space. When the cell averages are known, values at the Gaussian points can be obtained by the same reconstruction algorithm discussed in Section 2.1 and hence we will obtain the left and right Jacobians at each single point. Actually, it will be proved that only in this way can the freestream be preserved in the FV method.

3. Freestream preservation in two space dimensions

The problem of freestream preservation in the FV method is first studied in two space dimensions, and the governing equaitons, metrics and Jacobian can be expressed as

$$\frac{d}{dt}\overline{\overline{\left(\frac{Q}{J}\right)}}_{ij} + \frac{1}{\Delta\xi}\left(\hat{E}_{i+1/2,j} - \hat{E}_{i-1/2,j}\right) + \frac{1}{\Delta\eta}\left(\hat{F}_{i,j+1/2} - \hat{F}_{i,j-1/2}\right) = 0,$$

(22)

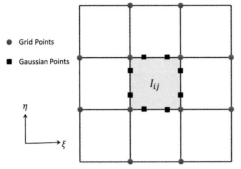

Fig. 4. Schematic of metric evaluation in two space dimensions.

$$\hat{E}_{i+1/2,j} = \frac{1}{\Delta \eta} \int_{\eta_{j-1/2}}^{\eta_{j+1/2}} \hat{E}(\xi_{i+1/2}, \eta) d\eta,$$

$$\hat{F}_{i,j+1/2} = \frac{1}{\Delta \xi} \int_{\xi_{i-1/2}}^{\xi_{i+1/2}} \hat{F}(\xi, \eta_{j+1/2}) d\xi, \tag{23}$$

$$\hat{E} = \hat{\xi}_x E + \hat{\xi}_y F,$$

$$\hat{F} = \hat{\eta}_x E + \hat{\eta}_y F, \tag{24}$$

$$\begin{cases} \hat{\xi}_x = \xi_x/J = y_\eta, & \hat{\xi}_y = \xi_y/J = -x_\eta, \\ \hat{\eta}_x = \eta_x/J = -y_\xi, & \hat{\eta}_y = \eta_y/J = x_\xi. \end{cases} \tag{25}$$

$$J^{-1} = \left| \frac{\partial(x,y)}{\partial(\xi,\eta)} \right| = \begin{vmatrix} x_\xi & x_\eta \\ y_\xi & y_\eta \end{vmatrix}. \tag{26}$$

As shown in Fig. 4, the metrics are evaluated by the polynomial interpolation through a set of points in the physical space [30]. The metric identities can be established by assuming a uniform flow $Q_{uniform}$. In addition, the conservative variables at the Gaussian points are also assumed to be $Q_{uniform}$. The rationality of this assumption will be discussed later. Hence the fluxes can be expressed as

$$\tilde{E} = E(Q_{uniform}) = \text{constant}, \quad \tilde{F} = F(Q_{uniform}) = \text{constant}.$$

$$\hat{E} = (\tilde{E}\hat{\xi}_x + \tilde{F}\hat{\xi}_y), \quad \hat{F} = (\tilde{E}\hat{\eta}_x + \tilde{F}\hat{\eta}_y). \tag{27}$$

Substituting the above expressions into Eq. (22) and eliminating the derivative of time (because $\overline{\left(\frac{Q}{J}\right)}_{ij}$ is constant), the governing equations become

$$\int_{\eta_{j-1/2}}^{\eta_{j+1/2}} [\tilde{E}(\delta^\eta y)_{i+1/2} - \tilde{F}(\delta^\eta x)_{i+1/2} - \tilde{E}(\delta^\eta y)_{i-1/2}$$
$$+ \tilde{F}(\delta^\eta x)_{i-1/2}] d\eta|^N + \int_{\xi_{i-1/2}}^{\xi_{i+1/2}} [-\tilde{E}(\delta^\xi y)_{j+1/2} + \tilde{F}(\delta^\xi x)_{j+1/2}$$
$$+ \tilde{E}(\delta^\xi y)_{j-1/2} - \tilde{F}(\delta^\xi x)_{j-1/2}] d\xi|^N = 0. \tag{28}$$

where δ is the difference operator. For different Gaussian points in one cell, the same stencil is used, which means that the metrics are obtained using the same polynomial. For convenience, the Gaussian quadrature is not explicitly expanded and is represented by the superscript N. Following the assumption of a uniform flow, Eq. (28) can be further simplified into

$$\tilde{E} \left[\int_{\eta_{j-1/2}}^{\eta_{j+1/2}} (\delta^\eta y)_{i+1/2} d\eta \bigg|^N - \int_{\eta_{j-1/2}}^{\eta_{j+1/2}} (\delta^\eta y)_{i-1/2} d\eta \bigg|^N \right.$$
$$\left. - \int_{\xi_{i-1/2}}^{\xi_{i+1/2}} (\delta^\xi y)_{j+1/2} d\xi \bigg|^N + \int_{\xi_{i-1/2}}^{\xi_{i+1/2}} (\delta^\xi y)_{j-1/2} d\xi \bigg|^N \right]$$

$$+ \tilde{F} \left[- \int_{\eta_{j-1/2}}^{\eta_{j+1/2}} (\delta^\eta x)_{i+1/2} d\eta \bigg|^N + \int_{\eta_{j-1/2}}^{\eta_{j+1/2}} (\delta^\eta x)_{i-1/2} d\eta \bigg|^N \right.$$
$$\left. + \int_{\xi_{i-1/2}}^{\xi_{i+1/2}} (\delta^\xi x)_{j+1/2} d\xi \bigg|^N - \int_{\xi_{i-1/2}}^{\xi_{i+1/2}} (\delta^\xi x)_{j-1/2} d\xi \bigg|^N \right] = 0. \tag{29}$$

It is easy to find that the following equations must be satisfied numerically in order to preserve the uniform flow

$$\int_{\eta_{j-1/2}}^{\eta_{j+1/2}} (\delta^\eta y)_{i+1/2} d\eta \bigg|^N - \int_{\eta_{j-1/2}}^{\eta_{j+1/2}} (\delta^\eta y)_{i-1/2} d\eta \bigg|^N$$
$$- \int_{\xi_{i-1/2}}^{\xi_{i+1/2}} (\delta^\xi y)_{j+1/2} d\xi \bigg|^N + \int_{\xi_{i-1/2}}^{\xi_{i+1/2}} (\delta^\xi y)_{j-1/2} d\xi \bigg|^N = 0, \tag{30}$$

$$- \int_{\eta_{j-1/2}}^{\eta_{j+1/2}} (\delta^\eta x)_{i+1/2} d\eta \bigg|^N + \int_{\eta_{j-1/2}}^{\eta_{j+1/2}} (\delta^\eta x)_{i-1/2} d\eta \bigg|^N$$
$$+ \int_{\xi_{i-1/2}}^{\xi_{i+1/2}} (\delta^\xi x)_{j+1/2} d\xi \bigg|^N - \int_{\xi_{i-1/2}}^{\xi_{i+1/2}} (\delta^\xi x)_{j-1/2} d\xi \bigg|^N = 0. \tag{31}$$

Eqs. (30) and (31) are similar to those presented in [34], where only metrics are involved and the identities hold when all the derivatives and integrals are computed analytically. Eqs. (30) and (31) are the metric identities in the FV method in curvilinear coordinates, which should be satisfied discretely. To obtain the conditions of holding the metric identities, we first prove the following theorem:

Theorem 1. *The necessary and sufficient condition for satisfying the metric identities (Eqs. (30) and (31)) in the two-dimensional FV method in curvilinear coordinates is*

$$\int_{\alpha_{j-1/2}}^{\alpha_{j+1/2}} (\delta^\alpha k)_i d\alpha \bigg|^N = k_{i,j+1/2} - k_{i,j-1/2}, \tag{32}$$

where k is the Cartesian coordinate x or y and α represents the curvilinear coordinate ξ or η.

Proof. The sufficient condition: Eqs. (30) and (31) are obviously true by using Eq. (32).

The necessary condition: As evaluated through a set of points, the metrics are actually the linear combination of the coordinate values of the mesh points. Firstly, we prove that in order to satisfy the metric identities, the unique form of $\int_{\alpha_{j-1/2}}^{\alpha_{j+1/2}} (\delta^\alpha k)_i d\alpha|^N$ is

$$\int_{\alpha_{j-1/2}}^{\alpha_{j+1/2}} (\delta^\alpha k)_i d\alpha \bigg|^N = G(k_{i,j+1/2}, k_{i,j-1/2}). \tag{33}$$

The method of proof by contradiction is used here. Assume Eqs. (30) and (31) hold and a quadrature expression that is not in the form of Eq. (33) exists. Without loss of generality, the first term in Eq. (30) is taken. As the polynomials are used in the metric evaluation, $(\delta^\eta y)_{i+1/2}$ can be expressed as

$$(\delta^\eta y)_{i+1/2} = f(y_{i+1/2, j-1/2-n}, \ldots, y_{i+1/2, j-1/2},$$
$$y_{i+1/2, j+1/2}, \ldots, y_{i+1/2, j+1/2+m}), \tag{34}$$

where n and m control the length of stencils. It is assumed that the quadrature of $(\delta^\eta y)_{i+1/2}$ is not in the form of Eq. (33), so some l exists which satisfies $j - n \leq l \leq j + m, l \neq j - 1, j$ and makes Eq. (30) hold. Hence $\int_{\eta_{j-1/2}}^{\eta_{j+1/2}} (\delta^\eta y)_{i+1/2} d\eta|^N$ can be expressed as

$$\int_{\eta_{j-1/2}}^{\eta_{j+1/2}} (\delta^\eta y)_{i+1/2} d\eta \bigg|^N = g(y_{i+1/2, j-1/2}, y_{i+1/2, j+1/2}, y_{i+1/2, j+1/2+l}).$$

$$\tag{35}$$

It can be found that $y_{i+1/2,j+1/2+l}$ will not be included in other three terms in Eq. (30), as shown in Fig. 4. If Eq. (30) holds in current state, it fails when the value of $y_{i+1/2,j+1/2+l}$ changes in mesh generations. In fact, the mesh points in the physical space are not fixed, so that for satisfying the metric identities, the coefficient of $y_{i+1/2,j+1/2+l}$ must be 0. Therefore the assumption (Eq. (35)) is false.

The above proof indicates that $\int_{\alpha_{j-1/2}}^{\alpha_{j+1/2}}(\delta^{\alpha}k)_i d\alpha\big|^N$ must be in the form of Eq. (33). Moreover, the metrics are the linear combination of coordinate values, so Eq. (33) can be simplified into

$$\int_{\alpha_{j-1/2}}^{\alpha_{j+1/2}}(\delta^{\alpha}k)_i d\alpha\bigg|^N = C_1 k_{i,j+1/2} - C_2 k_{i,j-1/2}. \tag{36}$$

Then the special condition of letting the numerical approximations approach the analytical ones is used and it is easy to obtain $C_1 = C_2 = 1$. Therefore the proposition is proved. □

In fact, Eq. (32) has the same effect as the commutativity of operators for the freestream preservation in the FD method, which is discussed in [6,13,34]. However, the detail expressions are different as a quadrature operator exits in Eq. (32), while only finite-difference operators are involved in the FD method. Moreover, the direction of the operators in each method is different. Except for the above discussion, it is also known that if the integral in Eq. (32) is computed analytically, the following equality holds (because the values of k at $(i, j+1/2)$ and $(i, j-1/2)$ are used in the polynomial)

$$\int_{\alpha_{j-1/2}}^{\alpha_{j+1/2}}(\delta^{\alpha}k)_i d\alpha = k_{i,j+1/2} - k_{i,j-1/2}. \tag{37}$$

Therefore the condition of the freestream preservation can be expressed as

$$\int_{\alpha_{j-1/2}}^{\alpha_{j+1/2}}(\delta^{\alpha}k)_i d\alpha\bigg|^N = \int_{\alpha_{j-1/2}}^{\alpha_{j+1/2}}(\delta^{\alpha}k)_i d\alpha, \tag{38}$$

which means that the numerical quadrature should equal to the analytical integral. So that the condition expressed by Eq. (32) can be viewed as the discrete version of the Newton–Leibniz formula. In the FV method, as stated above, the metrics are obtained by the polynomial approximations and the integral is computed by the Gaussian quadrature. For the $(2R-1)$th-order Gaussian quadrature, we have the following expression:

$$\int_{-1}^{1} f(x)dx = \sum_{i=1}^{R} C_i f(x_i). \tag{39}$$

If the order of the polynomial $f(x)$ is not higher than $(2R-1)$, the above expression is satisfied exactly. Hence in Eq. (38), the order of the polynomial used to fit $\delta^{\alpha}k$ is required to be lower than $2R$. Therefore we obtain

Condition 1. For guaranteeing the freestream preservation in the FV method in curvilinear coordinates, the order of the polynomials in evaluating metrics should not be higher than the order of the Gaussian quadrature.

For the two-dimensional FV method in curvilinear coordinates, the above discussion is just one of the conditions necessary for the freestream preservation, since another assumption that the reconstruction variables at the Gaussian points equal to those of the uniform flow is used. This assumption is certainly true when the variables used in the reconstruction are the primitive or conservative variables. However, it can be seen from Eq. (22) that variables used in the reconstruction are $\overline{\left(\frac{Q}{J}\right)}$, so that after the reconstruction only $\left(\frac{Q}{J}\right)_{Re}^{L/R}$ will be obtained (the superscript L and R represent the left and right states respectively). In order to obtain the conservative variables $Q^{L/R}$, the Jacobian J must be known and the following expressions should be satisfied at any Gaussian point,

$$Q^L = J\left(\frac{Q}{J}\right)_{Re}^L = Q_{uniform}, \quad Q^R = J\left(\frac{Q}{J}\right)_{Re}^R = Q_{uniform}. \tag{40}$$

In most cases, the reconstruction stencils are different, leading to different left and right states. Hence it indicates that two different values of Jacobian are also needed to satisfy Eq. (40). For this requirement, it is the method mentioned in Section 2.2 that should be used to obtain the Jacobian. That is

Condition 2. For guaranteeing the freestream preservation in the FV method in curvilinear coordinates, Jacobians at the Gaussian points are needed and should be evaluated by the same reconstruction algorithm to obtain the left and right states.

The effectiveness of this condition is explained as follows: in two space dimensions, Eq. (7) can be expressed as

$$\overline{\left(\frac{Q}{J}\right)}_{ij} = \frac{1}{\Delta\xi\Delta\eta}\int_{\xi_{i-1/2}}^{\xi_{i+1/2}}\int_{\eta_{j-1/2}}^{\eta_{j+1/2}}\left(\frac{Q}{J}\right)d\xi d\eta. \tag{41}$$

In the uniform flow, $Q_{uniform}$ has no effect on the process of reconstruction and the reconstruction results at the Gaussian points can be written as

$$\left(\frac{Q}{J}\right)_{Re}^{L/R} = Q_{uniform}\left(\frac{1}{J}\right)_{Re}^{L/R}. \tag{42}$$

Moreover, Eq. (41) can be simplified into

$$\overline{\left(\frac{1}{J}\right)}_{ij} = \frac{1}{\Delta\xi\Delta\eta}\int_{\xi_{i-1/2}}^{\xi_{i+1/2}}\int_{\eta_{j-1/2}}^{\eta_{j+1/2}}\left(\frac{1}{J}\right)d\xi d\eta$$
$$= \frac{1}{\Delta\xi\Delta\eta}\int_{x_{i-1/2}}^{x_{i+1/2}}\int_{y_{j-1/2}}^{y_{j+1/2}}dxdy = \frac{1}{\Delta\xi\Delta\eta}S_{ij}. \tag{43}$$

where S_{ij} is the area in the physical space. Eq. (43) is the two-dimensional form of Eq. (20). As $\overline{\left(\frac{1}{J}\right)}_{ij}$ is known, the same reconstruction algorithm can be used to obtain the Jacobians at the Gaussian points with the left and right states $\left(\frac{1}{J}\right)_{Re}^{L/R}$ which equal to those in Eq. (42). Therefore Eq. (40) is satisfied by

$$Q^{L/R} = \frac{1}{\left(\frac{1}{J}\right)_{Re}^{L/R}}\left(\frac{Q}{J}\right)_{Re}^{L/R} = Q_{uniform}. \tag{44}$$

4. Freestream preservation in three space dimensions

In this section, the discussions about the freestream preservation are extended to three space dimensions. As the Jacobian can be obtained in the same manner with that in two space dimensions, we focus on the metric evaluations.

Firstly we still assume a uniform flow $Q_{uniform}$. As the Jacobian is obtained by the reconstruction algorithm, the conservative values at the Gaussian points are also $Q_{uniform}$. Therefore the fluxes can be expressed as

$$\tilde{E} \equiv E(Q_{uniform}) = \text{constant},$$
$$\tilde{F} \equiv F(Q_{uniform}) = \text{constant},$$
$$\tilde{G} \equiv G(Q_{uniform}) = \text{constant}. \tag{45}$$

Substituting Eq. (45) into Eq. (6) and eliminating the derivative of time, we have

$$\int_{\eta_{j-1/2}}^{\eta_{j+1/2}} \int_{\zeta_{k-1/2}}^{\zeta_{k+1/2}} (\bar{E}\hat{\xi}_x + \bar{F}\hat{\xi}_y + \bar{G}\hat{\xi}_z)_{i+1/2} d\eta d\zeta \Big|^N$$
$$- \int_{\eta_{j-1/2}}^{\eta_{j+1/2}} \int_{\zeta_{k-1/2}}^{\zeta_{k+1/2}} (\bar{E}\hat{\xi}_x + \bar{F}\hat{\xi}_y + \bar{G}\hat{\xi}_z)_{i-1/2} d\eta d\zeta \Big|^N$$
$$+ \int_{\xi_{i-1/2}}^{\xi_{i+1/2}} \int_{\zeta_{k-1/2}}^{\zeta_{k+1/2}} (\bar{E}\hat{\eta}_x + \bar{F}\hat{\eta}_y + \bar{G}\hat{\eta}_z)_{j+1/2} d\xi d\zeta \Big|^N$$
$$- \int_{\xi_{i-1/2}}^{\xi_{i+1/2}} \int_{\zeta_{k-1/2}}^{\zeta_{k+1/2}} (\bar{E}\hat{\eta}_x + \bar{F}\hat{\eta}_y + \bar{G}\hat{\eta}_z)_{j-1/2} d\xi d\zeta \Big|^N$$
$$+ \int_{\xi_{i-1/2}}^{\xi_{i+1/2}} \int_{\eta_{j-1/2}}^{\eta_{j+1/2}} (\bar{E}\hat{\zeta}_x + \bar{F}\hat{\zeta}_y + \bar{G}\hat{\zeta}_z)_{k+1/2} d\xi d\eta \Big|^N$$
$$- \int_{\xi_{i-1/2}}^{\xi_{i+1/2}} \int_{\eta_{j-1/2}}^{\eta_{j+1/2}} (\bar{E}\hat{\zeta}_x + \bar{F}\hat{\zeta}_y + \bar{G}\hat{\zeta}_z)_{k-1/2} d\xi d\eta \Big|^N = 0. \quad (46)$$

As $\bar{E}, \bar{F}$ and $\bar{G}$ are all constant, the following equations can be derived

$$\int_{\eta_{j-1/2}}^{\eta_{j+1/2}} \int_{\zeta_{k-1/2}}^{\zeta_{k+1/2}} (\hat{\xi}_x)_{i+1/2} d\eta d\zeta \Big|^N - \int_{\eta_{j-1/2}}^{\eta_{j+1/2}} \int_{\zeta_{k-1/2}}^{\zeta_{k+1/2}} (\hat{\xi}_x)_{i-1/2} d\eta d\zeta \Big|^N$$
$$+ \int_{\xi_{i-1/2}}^{\xi_{i+1/2}} \int_{\zeta_{k-1/2}}^{\zeta_{k+1/2}} (\hat{\eta}_x)_{j+1/2} d\xi d\zeta \Big|^N - \int_{\xi_{i-1/2}}^{\xi_{i+1/2}} \int_{\zeta_{k-1/2}}^{\zeta_{k+1/2}} (\hat{\eta}_x)_{j-1/2} d\xi d\zeta \Big|^N$$
$$+ \int_{\xi_{i-1/2}}^{\xi_{i+1/2}} \int_{\eta_{j-1/2}}^{\eta_{j+1/2}} (\hat{\zeta}_x)_{k+1/2} d\xi d\eta \Big|^N - \int_{\xi_{i-1/2}}^{\xi_{i+1/2}} \int_{\eta_{j-1/2}}^{\eta_{j+1/2}} (\hat{\zeta}_x)_{k-1/2} d\xi d\eta \Big|^N$$
$$= 0, \quad (47)$$

$$\int_{\eta_{j-1/2}}^{\eta_{j+1/2}} \int_{\zeta_{k-1/2}}^{\zeta_{k+1/2}} (\hat{\xi}_y)_{i+1/2} d\eta d\zeta \Big|^N - \int_{\eta_{j-1/2}}^{\eta_{j+1/2}} \int_{\zeta_{k-1/2}}^{\zeta_{k+1/2}} (\hat{\xi}_y)_{i-1/2} d\eta d\zeta \Big|^N$$
$$+ \int_{\xi_{i-1/2}}^{\xi_{i+1/2}} \int_{\zeta_{k-1/2}}^{\zeta_{k+1/2}} (\hat{\eta}_y)_{j+1/2} d\xi d\zeta \Big|^N - \int_{\xi_{i-1/2}}^{\xi_{i+1/2}} \int_{\zeta_{k-1/2}}^{\zeta_{k+1/2}} (\hat{\eta}_y)_{j-1/2} d\xi d\zeta \Big|^N$$
$$+ \int_{\xi_{i-1/2}}^{\xi_{i+1/2}} \int_{\eta_{j-1/2}}^{\eta_{j+1/2}} (\hat{\zeta}_y)_{k+1/2} d\xi d\eta \Big|^N - \int_{\xi_{i-1/2}}^{\xi_{i+1/2}} \int_{\eta_{j-1/2}}^{\eta_{j+1/2}} (\hat{\zeta}_y)_{k-1/2} d\xi d\eta \Big|^N$$
$$= 0, \quad (48)$$

$$\int_{\eta_{j-1/2}}^{\eta_{j+1/2}} \int_{\zeta_{k-1/2}}^{\zeta_{k+1/2}} (\hat{\xi}_z)_{i+1/2} d\eta d\zeta \Big|^N - \int_{\eta_{j-1/2}}^{\eta_{j+1/2}} \int_{\zeta_{k-1/2}}^{\zeta_{k+1/2}} (\hat{\xi}_z)_{i-1/2} d\eta d\zeta \Big|^N$$
$$+ \int_{\xi_{i-1/2}}^{\xi_{i+1/2}} \int_{\zeta_{k-1/2}}^{\zeta_{k+1/2}} (\hat{\eta}_z)_{j+1/2} d\xi d\zeta \Big|^N - \int_{\xi_{i-1/2}}^{\xi_{i+1/2}} \int_{\zeta_{k-1/2}}^{\zeta_{k+1/2}} (\hat{\eta}_z)_{j-1/2} d\xi d\zeta \Big|^N$$
$$+ \int_{\xi_{i-1/2}}^{\xi_{i+1/2}} \int_{\eta_{j-1/2}}^{\eta_{j+1/2}} (\hat{\zeta}_z)_{k+1/2} d\xi d\eta \Big|^N - \int_{\xi_{i-1/2}}^{\xi_{i+1/2}} \int_{\eta_{j-1/2}}^{\eta_{j+1/2}} (\zeta_z)_{k-1/2} d\xi d\eta \Big|^N$$
$$= 0, \quad (49)$$

which are the corresponding metric identities. The conditions for the freestream preservation in three space dimensions are to satisfy the above equations discretely.

Considering the discussions in two space dimensions, it is still assumed that the order of the polynomials in evaluating the metrics is not higher than the order of the Gaussian quadrature. Take the first term in Eq. (47) as an example. If the metric $\hat{\xi}_x$ is computed by the conservative form in Eq. (14), it can be simplified into

$$\int_{\eta_{j-1/2}}^{\eta_{j+1/2}} \int_{\zeta_{k-1/2}}^{\zeta_{k+1/2}} (\hat{\xi}_x)_{i+1/2} d\eta d\zeta \Big|^N$$
$$= \int_{\eta_{j-1/2}}^{\eta_{j+1/2}} [(z\delta^\eta y)_{i+1/2,k+1/2} - (z\delta^\eta y)_{i+1/2,k-1/2}] d\eta \Big|^N$$
$$- \int_{\zeta_{k-1/2}}^{\zeta_{k+1/2}} [(z\delta^\zeta y)_{i+1/2,j+1/2} - (z\delta^\zeta y)_{i+1/2,j-1/2}] d\zeta \Big|^N. \quad (50)$$

The other terms can be obtained in the same way and all terms will be cancelled out when substituted into in Eq. (47). Hence Eqs. (47)–(49) can be satisfied, which guarantees the freestream preservation. As shown in Section 2.2, three different forms can be used in the metric evaluations. It is easy to validate that another conservative form Eq. (15) and the symmetric conservative form can also hold the metric identities (in fact, the symmetric conservative form is just the average of two conservative ones). However, the cross product form cannot satisfy these identities. This is because the Gaussian quadrature of the product of two difference operators cannot be further simplified. Taking the 2nd-order scheme as an example, Eq. (47) can be expressed as

$$(\hat{\xi}_x)_{i+1/2,j,k} - (\hat{\xi}_x)_{i-1/2,j,k} + (\hat{\eta}_x)_{i,j+1/2,k}$$
$$- (\hat{\eta}_x)_{i,j-1/2,k} + (\hat{\zeta}_x)_{i,j,k+1/2} - (\hat{\zeta}_x)_{i,j,k-1/2} = 0, \quad (51)$$

which is the same with the metric identities in the FD method [34]. It has been pointed out by Deng et al. in [34] that the cross product form cannot satisfy the above expression and the symmetric conservative or conservative forms must be used.

Therefore, except for the conditions proposed in Section 3, we obtain the third condition for the freestream preservation in the FV method. That is

Condition 3. For guaranteeing the freestream preservation in the FV method in curvilinear coordinates, the metrics should be evaluated in the conservative forms.

Actually, the metrics in two space dimensions can also be regarded as the conservative form, which indicates that the requirement always exists for the freestream preservation, merely not manifested in some special condition.

5. Numerical results

In this section, numerical tests are presented to validate the freestream preservation in the FV method. The metrics and Jacobian are evaluated as described in Sections 3 and 4. For a detailed comparison, different methods that violate the conditions of the freestream preservation are also used: for the metrics, the order of the polynomials may set to be higher than the Gaussian quadrature or the cross product form is used. For the Jacobian, we just replace the left and right values at the Gaussian points by their average for convenience. In all numerical simulations, the 3rd-order TVD Runge–Kutta method is adopted in the time discretization.

5.1. Freestream

The freestream preservation test is first performed in two and three space dimensions on randomized meshes. The mesh number is 30 × 30 and 30 × 30 × 30, respectively. The original meshes are the uniform Cartesian ones and the randomized ones are generated by adding a random quantity to the mesh coordinate values. The random quantity is controlled by a Fortran function, but will not exceed 20% magnitude of the local mesh spacing. Other randomized meshes used in this paper are obtained in the same manner, unless otherwise specified. The randomized mesh in three space dimensions is shown in Fig. 5.

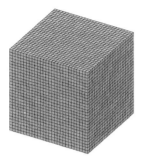

Fig. 5. The three-dimensional randomized mesh for the freestream preservation test.

Table 1
L_1 errors of v in the freestream preservation test on randomized meshes.

	FP-all	FP-metrics	FP-Jacobian1	FP-Jacobian3
2D	5.02E−16	3.53E−02	1.41E−06	–
3D	2.12E−16	5.17E−02	2.67E−06	1.06E−05

For a full comparison for the conditions of the freestream preservation, different numerical methods in evaluations of the metrics and Jacobian are used: we first make all the conditions satisfied and denote the method as FP-all. In the second method, the requirement in the metrics is satisfied, but the Jacobian is not computed following the Condition 2, which is denoted as FP-Metrics. We set another two methods denoted as FP-Jacobian1 and FP-Jacobian3, which correspond to the Condition 1 and Condition 3 are not satisfied respectively. As the Condition 3 can be automatically satisfied in two space dimensions, the result will not be presented.

In this test case, the initial conditions in two and three dimensions are set to $[\rho, u, v, p] = [1, 1, 0, 1]$ and $[\rho, u, v, w, p] = [1, 1, 0, 0, 1]$, and the periodic boundary conditions are applied in all coordinate directions. The L_1 errors of the cell average of v are collected after 100 steps and listed in Table 1. The results show that the conditions proposed in this paper are efficient in the freestream preservation. When both the metrics and Jacobian satisfy the conditions, the errors are close to machine zero for double-precision computations. However, if not satisfied, larger errors are introduced. In comparison of different methods, it is found that the evaluation of the Jacobian affects the numerical results more than that of the metrics in the freestream preservation in the FV method.

5.2. Two-dimensional channel flow

This test case is an isentropic flow governed by the Euler equations and was first discussed by Casper et al. [22] in comparison of the FD and FV methods. In this paper, the C1 geometry is used. In this case, the connections of the middle to the outer sections are continuous in ξ to only one derivative [22]. The shape of the middle section is determined by

$$\begin{cases} y_1(x) = -0.45 - 0.1 \times \left(\frac{20}{9}\right)^2 x^2 + 0.05 \left(\frac{20}{9}\right)^4 x^4, \\ y_2(x) = 0.45 + 0.1 \times \left(\frac{20}{9}\right)^2 x^2 - 0.05 \left(\frac{20}{9}\right)^4 x^4. \end{cases} \tag{52}$$

In this test case, meshes are generated by the analytical method, as described in [22]. Fig. 6 illustrates the configuration

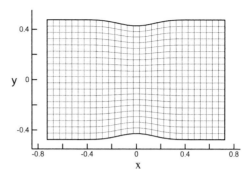

Fig. 6. The configuration and mesh used in the two-dimensional channel flow.

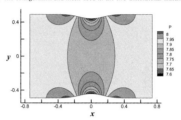

(a) The distribution of the pressure

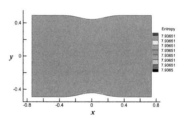

(b) The distribution of the entropy

Fig. 7. Distributions of the pressure and entropy in the two-dimensional channel flow using Method-A. (colored picture attached at the end of the book)

and a mesh of 30 × 20. Boundary conditions include the subsonic inflow, outflow and a slip wall [35].

To validate the effect of the freestream preservation in the FV method, the metrics and Jacobian are evaluated in four different numerical methods: Method-A, evaluations of both the metrics and Jacobian satisfy the freestream preservation conditions; Method-B, neither satisfies the conditions; Method-C, only the Jacobian satisfies; Method-D, only the metrics satisfy. The results of the pressure and entropy in Method-A are shown in Fig. 7. It can be seen that the flowfield is nearly isentropic and only some perturbations arise near the wall. For further comparisons, the L_1 entropy errors in four methods are listed in Table 2, and the results show an obvious increase of the entropy errors when the freestream preservation is not satisfied.

Table 2
L_1 entropy errors in the two-dimensional channel flow.

Mesh	Method-A	Method-B	Method-C	Method-D
15 × 10	2.01E−04	4.64E−03	4.12E−04	1.17E−03
30 × 20	1.97E−05	2.22E−04	4.41E−05	9.44E−05
45 × 30	5.28E−06	5.02E−05	1.28E−05	2.37E−05
60 × 40	2.18E−06	1.99E−05	5.44E−06	9.57E−06
90 × 60	6.44E−07	5.91E−06	1.64E−06	2.84E−06

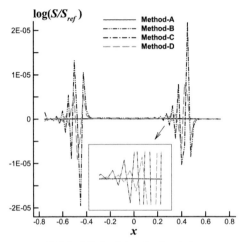

Fig. 8. The distribution of entropy alone the centerline.

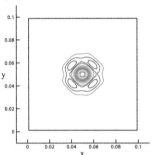

Fig. 9. The vorticity magnitude distribution of the vortex transport problem on a Cartesian mesh. (colored picture attached at the end of the book)

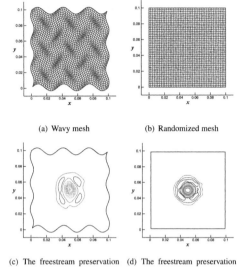

(a) Wavy mesh (b) Randomized mesh

(c) The freestream preservation is satisfied on the wavy mesh (d) The freestream preservation is satisfied on the randomized mesh

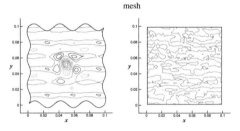

(e) The freestream preservation is not satisfied on the wavy mesh (f) The freestream preservation is not satisfied on the randomized mesh

Fig. 10. Nonuniform meshes and vorticity magnitude distributions of the results of the vortex transport problem with 16 contours.

(colored picture attached at the end of the book)

The distribution of the entropy is presented along the centerline of a 60 × 40 mesh. Fig. 8 exhibits that if the freestream preservation is satisfied, the entropy keeps constant; if not, oscillations appear at the connections of the channel, which proves the importance of the freestream preservation in the FV method.

5.3. Vortex transport by uniform flow

This test case is chosen from the benchmark cases in the international workshop on high-order CFD methods and widely used to test a high-order method's capability to preserve vorticity in an unsteady inviscid flow [36]. The initial condition is the superposition of a uniform flow with pressure p_∞, temperature T_∞, Mach number Ma and a vortical movement of characteristic radius R and strength β with the center at (X_c, Y_c). The distribution of the velocity can be expressed as,

$$\begin{cases} u_0 = U_\infty - (U_\infty \beta) \dfrac{y - Y_c}{R} \exp\left(-\dfrac{r^2}{2}\right), \\ v_0 = (U_\infty \beta) \dfrac{x - X_c}{R} \exp\left(-\dfrac{r^2}{2}\right), \end{cases} \tag{53}$$

where $r = \frac{\sqrt{(x-X_c)^2+(y-Y_c)^2}}{R}$ and U_∞ is the speed of the unperturbed flow. The computations of the pressure, temperature and density

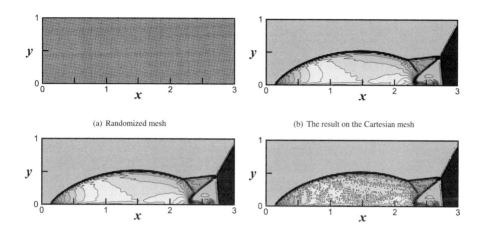

(a) Randomized mesh (b) The result on the Cartesian mesh

(c) The freestream preservation is satisfied on the randomized mesh (d) The freestream preservation is not satisfied on the randomized mesh

Fig. 11. The mesh and distributions of the density in the double Mach reflection problem with 40 contours. (colored picture attached at the end of the book)

can be found in [36]. In this paper, the computational domain is $(x, y) \in [0, L_x] \times [0, L_y]$ and $L_x = 0.1$, $L_y = 0.1$. The center of vortex is at $X_c = 0.05$, $Y_c = 0.05$. Mach number $Ma = 0.05$ is chosen with $\beta = 0.02$ and $R = 0.005$.

The numerical result on a uniform Cartesian mesh is first presented in Fig. 9, in which the vortex is well preserved after a period. In order to examine the effect of the freestream preservation, the wavy mesh and the randomized mesh are adopted in this paper. Fig. 10(a) and (b) shows the meshes and the mesh number is set to 45×45.

Fig. 10 (c)–(f) presents the results on nonuniform meshes when the freestream preservation is satisfied or not. It can be seen that on the wavy mesh, the severe nonuniformity generates significant numerical dissipation, however, the shape of the vortex is preserved as the freestream preservation is satisfied. In contrary, if not satisfied, perturbations arise in the flowfield. The effect of the freestream preservation is also examined on the randomized mesh. The results show that when the freestream preservation is satisfied, the initial vortex is well resolved, while if not satisfied, the degradation is severe and the vortex cannot be captured at all. Through the numerical results above, the importance of the freestream preservation is highlighted, especially for meshes of poor quality.

5.4. Double mach reflection

The double mach reflection problem is designed to test the capability of capturing shock in a numerical scheme [37]. The computational domain in this case is $[0, 4] \times [0, 1]$. The reflecting position at the bottom wall is $x = 1/6$; the initial right-moving shock starts from $(1/6, 0)$ and makes an angle of $60°$ with the x-axis. For the top boundary, the exact shock-moving condition is implemented and the results are obtained at $t = 0.2$.

Generally the double mach reflection problem is calculated on uniform Cartesian meshes, however, in order to examine the effect of the freestream preservation, a randomized mesh is used in this paper and the mesh number is set to 240×60, as shown in Fig. 11(a). Fig. 11(b)–(d) presents the results in the Cartesian and randomized meshes. It can be clearly seen that when the freestream is numerically preserved, the density contours are nearly the same in both Cartesian and randomized meshes. However, if it is not preserved, large numerical errors are introduced and severe oscillations arise after the shock. In this case, the effect of the freestream preservation on providing a satisfactory resolution of flowfield is pronounced as well.

5.5. Supersonic flow past a cylinder

In this case, the FV method is used to simulate the supersonic flow past a cylinder. In this paper, a randomized mesh is adopted in numerical calculation, and the mesh is generated as described in [9].

$$x = [R_x - (R_x - 1)\eta'] \cos[\theta(2\xi' - 1)],$$
$$y = [R_y - (R_y - 1)\eta'] \sin[\theta(2\xi' - 1)],$$
$$\xi' = \frac{\xi - 1}{j_{\max} - 1},$$
$$\eta' = \frac{\eta - 1}{k_{\max} - 1},$$
$$\xi = j + Random_j,$$
$$\eta = k + Random_k,$$
$$\sqrt{Random_j^2 + Random_k^2} = \begin{cases} 0.1 & \text{inner points,} \\ 0 & \text{boundary points,} \end{cases}$$
$$j = 1, 2, \ldots, j_{\max},$$
$$k = 1, 2, \ldots, k_{\max}, \tag{54}$$

where the parameters are $j_{\max} = 81$, $k_{\max} = 61$, $\theta = 5\pi/12$, $R_x = 3$ and $R_y = 6$. The original mesh is obtained by setting the random quantity to zero. The problem is initiated by a Mach 2 flow and the reflective boundary is imposed at the surface. The mesh number is 81×61, as shown in Fig. 12(a).

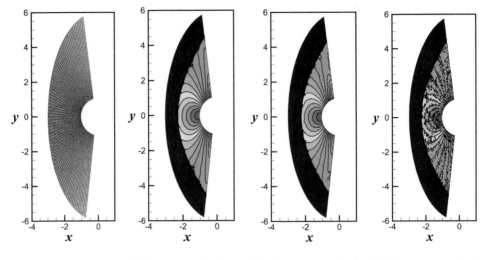

(a) Randomized mesh   (b) The result on the original mesh   (c) The freestream preservation is satisfied on the randomized mesh   (d) The freestream preservation is not satisfied on the randomized mesh

Fig. 12. The mesh and distributions of the pressure in the supersonic flow past a cylinder problem with 20 contours. (colored picture attached at the end of the book)

The result on the original mesh is first presented in Fig. 12(b) for comparisons. Fig. 12(c) and (d) shows the results on the randomized mesh when the freestream preservation is satisfied or not respectively. From the distributions of the pressure, it is found that spurious spatial oscillations arise when the freestream preservation is not satisfied, while the solution is much improved when the freestream is preserved.

6. Conclusions

For any numerical method in the curvilinear coordinates, the problem of the freestream preservation needs to be considered, otherwise numerical errors will contaminate the results. This problem is widely studied in the FD method, while discussions are seldom seen in the FV method. In the present paper, we propose the conditions for the freestream preservation in the FV method. These conditions include:

1. The order of the polynomials in evaluating metrics should not be higher than the order of accuracy of the Gaussian quadrature.
2. Jacobians at the Gaussian points are needed and should be evaluated by the same reconstruction algorithm to obtain the left and right states.
3. The metrics should be computed in the conservative forms (we consider the symmetric conservative form and the metrics in two space dimensions to be special conservative forms), but not in the cross product form.

When the above conditions are satisfied, the FV method in the curvilinear coordinates can guarantee the freestream preservation. Test cases are presented to illustrate the effect of these conditions on numerical results, which validate our conclusions as well.

It is worth noting that these conditions are different from those in the FD method and the second condition is unique. Interestingly, the conclusions in this paper are similar to those obtained by Kopriva in [3] and Abe in [38], although the numerical schemes are different. In three space dimensions, different conservative forms in metrics can all satisfy the freestream preservation. In the present study, we do not make any comparison and discuss which one is better. This work is left for further study.

Acknowledgments

This work was supported by the Basic Research Foundation of National University of Defense Technology (No. ZDYYJ-CYJ20140101). We are very grateful to the referees of this paper for their valuable suggestions to improve the quality of this paper.

Appendix A. Formulae of the metric evaluation in the FV method

In this appendix, the detailed formulae in evaluating the metrics in the FV method are presented.

In two space dimensions, the metrics are evaluated by the differece operators along the direction of ξ or η, as shown in Fig. 4. Because the formulae in different directions are the same, we take the computation of x_ξ along ξ as an example. In this paper, both two and three Gaussian points are used. For satisfying the freestream preservation, four mesh points (third order) are used in the metric evaluation for two Gaussian points and six points (fifth order) for three Gaussian points. In the numerical tests, eight points (seventh order) are also used to violate the freestream preservation. All the related formulae are given in this appendix.

For convenience, the metrics are assumed to locate in $[\xi_{i-1/2}, \xi_{i+1/2}]$ and the index j along the direction of η is neglected. For the case of two Gaussian points, the locations of the Gaussian points are $p_1^2 = \xi_{i-1/2} + \frac{1}{2}(1 - \frac{1}{\sqrt{3}})\Delta\xi$ and $p_2^2 = \xi_{i-1/2} +$

$\frac{1}{2}(1 + \frac{1}{\sqrt{3}}\Delta\xi)$. For the case of three Gaussian points, the locations are $p_1^3 = \xi_{i-1/2} + \frac{1}{2}(1 - \frac{\sqrt{15}}{5}\Delta\xi)$. $p_2^3 = \xi_{i-1/2} + \frac{1}{2}\Delta\xi$ and $p_3^3 = \xi_{i-1/2} + \frac{1}{2}(1 + \frac{\sqrt{15}}{5}\Delta\xi)$. For two Gaussian points, the 3rd-order metrics are obtained by

$$(x_\xi)_{p_1^2} = \frac{1}{12\Delta\xi}[-\sqrt{3}x_{i-3/2} + (-12 + \sqrt{3})x_{i-1/2} + (12 + \sqrt{3})x_{i+1/2} - \sqrt{3}x_{i+3/2}] \quad (55)$$

$$(x_\xi)_{p_2^2} = \frac{1}{12\Delta\xi}[\sqrt{3}x_{i-3/2} - (12 + \sqrt{3})x_{i-1/2} + (12 - \sqrt{3})x_{i+1/2} + \sqrt{3}x_{i+3/2}] \quad (56)$$

For three Gaussian points, the 5th-order metrics are obtained by

$$(x_\xi)_{p_1^3} = \frac{1}{2400\Delta\xi}\left[(9 + 22\sqrt{15})x_{i-5/2} + (-125 - 186\sqrt{15})x_{i-3/2} + (-2070 + 164\sqrt{15})x_{i-1/2} + (2070 + 164\sqrt{15})x_{i+1/2} + (125 - 186\sqrt{15})x_{i+3/2} + (-9 + 22\sqrt{15})x_{i+5/2}\right] \quad (57)$$

$$(x_\xi)_{p_2^3} = \frac{1}{1920\Delta\xi}[-9x_{i-5/2} + 125x_{i-3/2} - 2250x_{i-1/2} + 2250x_{i+1/2} - 125x_{i+3/2} + 9x_{i+5/2}] \quad (58)$$

$$(x_\xi)_{p_3^3} = \frac{1}{2400\Delta\xi}[(9 - 22\sqrt{15})x_{i-5/2} + (-125 + 186\sqrt{15})x_{i-3/2} + (-2070 - 164\sqrt{15})x_{i-1/2} + (2070 - 164\sqrt{15})x_{i+1/2} + (125 + 186\sqrt{15})x_{i+3/2} + (-9 - 22\sqrt{15})x_{i+5/2}] \quad (59)$$

The 7th-order metrics are obtained by

$$(x_\xi)_{p_1^3} = \frac{1}{5040000\Delta\xi}[-(2808 + 9541\sqrt{15})x_{i-7/2} + (38556 + 93905\sqrt{15})x_{i-5/2} - (321468 + 476469\sqrt{15})x_{i-3/2} + (-4248720 + 392105\sqrt{15})x_{i-1/2} + (4248720 + 392105\sqrt{15})x_{i+1/2} + (321468 - 476469\sqrt{15})x_{i+3/2} - (38556 - 93905\sqrt{15})x_{i+5/2} + (2808 - 9541\sqrt{15})x_{i+7/2}] \quad (60)$$

$$(x_\xi)_{p_2^3} = \frac{1}{107520\Delta\xi}[75x_{i-7/2} - 1029x_{i-5/2} + 8575x_{i-3/2} - 128625x_{i-1/2} + 128625x_{i+1/2} - 8575x_{i+3/2} + 1029x_{i+5/2} - 75x_{i+7/2}] \quad (61)$$

$$(x_\xi)_{p_3^3} = \frac{1}{5040000\Delta\xi}[-(2808 - 9541\sqrt{15})x_{i-7/2} + (38556 - 93905\sqrt{15})x_{i-5/2} - (321468 - 476469\sqrt{15})x_{i-3/2} + (-4248720 + 392105\sqrt{15})x_{i-1/2} + (4248720 - 392105\sqrt{15})x_{i+1/2} + (321468 + 476469\sqrt{15})x_{i+3/2} - (38556 + 93905\sqrt{15})x_{i+5/2} + (2808 + 9541\sqrt{15})x_{i+7/2}] \quad (62)$$

In three space dimensions, nearly the same procedure can be used in the metric evaluation. The main difference is that the interpolation should be implemented before the numerical difference. As shown in Fig. 3, we present the formulae in evaluating the coordinate value y at point A. Here, the Gaussian points are located in $[\eta_{j-1/2}, \eta_{j+1/2}]$. In this paper, six mesh points (fifth order) are used in the interpolation. For two Gaussian points, the interpolation formulae are

$$(y)_{p_1^2} = \frac{1}{25920}[(195 + 13\sqrt{3})y_{j-5/2} - (1665 + 185\sqrt{3})y_{j-3/2} + (14430 + 4810\sqrt{3})y_{j-1/2} + (14430 - 4810\sqrt{3})y_{j+1/2} - (1665 - 185\sqrt{3})y_{j+3/2} + (195 - 13\sqrt{3})y_{j+5/2}]. \quad (63)$$

$$(y)_{p_2^2} = \frac{1}{25920}[(195 - 13\sqrt{3})y_{j-5/2} - (1665 - 185\sqrt{3})y_{j-3/2} + (14430 - 4810\sqrt{3})y_{j-1/2} + (14430 + 4810\sqrt{3})y_{j+1/2} - (1665 + 185\sqrt{3})y_{j+3/2} + (195 + 13\sqrt{3})y_{j+5/2}]. \quad (64)$$

For three Gaussian points, the interpolation formulae are

$$(y)_{p_1^3} = \frac{1}{120000}[(525 + 21\sqrt{15})y_{j-5/2} - (4575 + 305\sqrt{15})y_{j-3/2} + (64050 + 12810\sqrt{15})y_{j-1/2} + (64050 - 12810\sqrt{15})y_{j+1/2} - (4575 - 305\sqrt{15})y_{j+3/2} + (525 - 21\sqrt{15})y_{j+5/2}]. \quad (65)$$

$$(y)_{p_2^3} = \frac{1}{256}[3y_{j-5/2} - 25y_{j-3/2} + 150y_{j-1/2} + 150y_{j+1/2} - 25y_{j+3/2} + 3y_{j+5/2}], \quad (66)$$

$$(y)_{p_3^3} = \frac{1}{120000}[(525 - 21\sqrt{15})y_{j-5/2} - (4575 - 305\sqrt{15})y_{j-3/2} + (64050 - 12810\sqrt{15})y_{j-1/2} + (64050 + 12810\sqrt{15})y_{j+1/2} - (4575 + 305\sqrt{15})y_{j+3/2} + (525 + 21\sqrt{15})y_{j+5/2}]. \quad (67)$$

The difference operators used in three space dimensions are the same with those in two space dimensions. Hence we do not repeat them anymore.

References

[1] Slotnick J., Khodadoust A., Alonso J., Darmofal D., Gropp W., Lurie E., et al. CFD vision 2030 study: a path to revolutionary computational aerosciences. NASA/CR-2014-218178.
[2] Vinokur M. An analysis of finite-difference and finite-volume formulations of conservation laws. J Comput Phys 1989;81:1–52.
[3] Kopriva DA. Metric identities and the discontinuous spectral element method on curvilinear meshes. J Sci Comput 2006;26(3):301–27.
[4] Cai X, Ladeinde F. Performance of WENO scheme in generalized curvilinear coordinate systems. In: Proceedings of AIAA 2008-36; 2008.
[5] Visbal R, Gaitonde D. On the use of higher-order finite-difference schemes on curvilinear and deforming meshes. J Comput Phys 2002;181:155–85.
[6] Vinokur M, Yee H. Extension of efficient low dissipation high-order schemes for 3D curvilinear moving grids. Front Comput Fluid Dyn 2002:129–64.
[7] Nonomura T, Iizuka N, Fujii K. Freestream and vortex preservation properties of high-order WENO and WCNS on curvilinear grids. Comput Fluids 2010;39;197–214.
[8] Deng X, Zhang H. Developing high-order weighted compact nonlinear schemes. J Comput Phys 2000;165:22–44.
[9] Nonomura T, Terakado D, Abe Y, Fujii K. A new technique for freestream preservation of finite-difference WENO on curvilinear grid. Comput Fluids 2015;107:242–55.
[10] Abe Y, Iizuka N, Nonomura T, Fujii K. Conservative metric evaluation for high-order finite difference schemes with the GCL identities on moving and deforming grids. J Comput Phys 2013;232:14–21.
[11] Deng X, Min Y, Mao M, Liu H, Tu G, Zhang H. Further study on geometric conservation law and application to high-order finite difference schemes with stationary grids. J Comput Phys 2013;239:90–111.

[12] Deng X, Zhu H, Min Y, Liu H, Mao M, Wang G, et al. Symmetric conservative metric method: a link between high order finite-difference and finite-volume schemes for flow computations around complex geometries. In: Proceedings of international conference on computational fluid dynamics, ICCFD8, 2014-0005; 2014.
[13] Abe Y, Nonomura T, Iizuka N, KozoFujii. Geometric interpretations and spatial symmetry property of metrics in the conservative form for high-order finite-difference schemes on moving and deforming grids. J Comput Phys 2014;260:163-203.
[14] Deng X, Jiang Y, Mao M, Liu H, Li S, Tu G. A family of hybrid cell-edge and cell-node dissipative compact schemes satisfying geometric conservation law. Comput Fluids 2015;116:29-45.
[15] Deng X, Jiang Y, Mao M, Liu H, Tu G. Developing hybrid cell-edge and cell-node dissipative compact scheme for complex geometry flows. Sci China Technol Sci 2013;56:2361-9.
[16] Calhoun DA, Helzel C, LeVeque RJ. Logically rectangular grids and finite volume methods for PDEs in circular and spherical domains. SIAM Rev 2008;50(4):723-52.
[17] Abgrall R. On essentially non-oscillatory schemes on unstructured meshes analysis and implementation. J Comput Phys 1994;114:45-58.
[18] Sonar T. On the construction of essentially non-oscillatory finite volume approximations to hyperbolic conservation laws on general triangulations: polynomial recovery, accuracy and stencil selection. Comput Methods Appl Mech Eng 1997;140:157-81.
[19] Tsoutsanis P, Titarev V, Drikakis D. WENO schemes on arbitrary mixed-element unstructured meshes in three space dimensions. J Comput Phys 2011;230:1585-601.
[20] Shu CW. High-order finite difference and finite volume WENO schemes and discontinuous Galerkin methods for CFD. Int J Comput Fluid Dyn 2003;17(2):107-18.
[21] Lele SK. Compact finite difference schemes with spectral-like resolution. J Comput Phys 1992;103:16-42.
[22] Casper J, Shu CW, Atkins H. Comparision of two formulations for high-order accurate essentially nonoscillatory schemes. AIAA J 1994;32(10):1970-7.
[23] Ekaterinaris JA. High-order accurate, lownumerical diffusion methods for aerodynamics. Prog Aerosp Sci 2005;41:192-300.
[24] Titarev V, Toro E. Finite-volume WENO schemes for three-dimensional conservation laws. J Comput Phys 2004;201:238-60.
[25] Haselbacher A. A WENO reconstruction algorithm for unstructured grids based on explicit stencil construction. In: Proceedings of AIAA 2005-879; 2005.
[26] Kok J. A high-order low-dispersion symmetry-preserving finite-volume method for compressible flow on curvilinear grids. J Comput Phys 2009;228:6811-32.
[27] Talukdar P, Steven M, Issendorff F, Trimis D. Finite volume method in 3-D curvilinear coordinates with multiblocking procedure for radiative transport problems. Int J Heat Mass Transf 2005;48:4657-66.
[28] Gallerano F, Cannata G, Tamburrino M. Upwind WENO scheme for shallow water equations in contravariant formulation. Comput Fluids 2012;62:1-12.
[29] Grimm-Strele H, Kupka F, Muthsam H. Curvilinear grids for WENO methods in astrophysical simulations.. Comput Phys Commun 2014;185:764-76.
[30] Casper J, Atkins KL. A finite-volume high-order ENO scheme for two-dimensional hyperbolic system. J Comput Phys 1993;106:62-76.
[31] Colella P, Dorr M, Hittinger J, Martin D. High-order, finite-volume methods in mapped coordinates. J Comput Phys 2011;230:2952-76.
[32] Blazek J. Computational fluid dynamics: principles and applications. Elsevier Science Ltd.; 2001.
[33] Thomas PD, Lombard CK. Geometric conservation law and its application toflow computations on moving grids. AIAA J. 1979;17(10):1030-7.
[34] Deng X, Mao M, Tu G, Liu H, Zhang H. Geometric conservation law and application to high-order finite difference schemes with stationary grids. J Comput Phys 2011;230:1100-15.
[35] Atkins H, Casper J. Nonreflective boundary conditions for high-order methods. AIAA J 1994;32(3):512-18.
[36] Wang ZJ, Fidkowski K, Abgrall R, Bassi F, Caraeni D, Cary A, et al. High-order CFD methods: current status and perspective. Int J Numer Meth Fluids 2013;72:811-45.
[37] Woodward P, Colella P. The numerical simulation of two-dimensional fluid flow with strong shocks. J Comput Phys 1984;54:115-73.
[38] Abe Y, Haga T, Nonomura T, Fujii K. On the freestream preservation of high-order conservative flux-reconstruction schemes. J Comput Phys 2015;281:28-54.

Reevaluation of high-order finite difference and finite volume algorithms with freestream preservation satisfied

Yidao Dong [a,*], Xiaogang Deng [a], Dan Xu [a], Guangxue Wang [b]

[a] College of Aerospace Science and Engineering, National University of Defense Technology, Changsha, Hunan 410073, PR China
[b] School of Physics, Sun Yat-sen University, Guangzhou 510275, PR China

ARTICLE INFO

Article history:
Received 11 January 2017
Revised 11 July 2017
Accepted 21 July 2017
Available online 22 July 2017

Keywords:
High-order schemes
Finite difference algorithms
Finite volume algorithms
Freestream preservation

ABSTRACT

High-order finite difference and finite volume algorithms based on the coordinate transformation, which satisfy the property of freestream preservation are reevaluated in this paper. The intent here is to modify the conclusion drawn by Casper et al. [28], who claimed that the finite volume implementation was less sensitive to derivative discontinuities. Therefore, for problems with complex geometries, it might pay to use the finite volume algorithm. In the present work, all the cases from Casper et al. are simulated with two advanced algorithms combined with weighting techniques and the importance of the freestream preservation is demonstrated through the comparison. It is concluded that the finite difference algorithm with the freestream preservation satisfied performs as well as the finite volume algorithm and the time consumption of high-order finite difference algorithms is remarkably lower than that of finite volume algorithms in multiple dimensions.

© 2017 Elsevier Ltd. All rights reserved.

1. Introduction

Computational Fluid Dynamics, abbreviated as CFD, essentially aims at exploiting the universal law in the numerically discrete space by discretizing the continuous space. Similar to other sciences, CFD has undergone a process from simplicity to sophistication. In the last two decades, remarkable development has been obtained in the field of CFD owing to the thorough elevation of high performance computation. Among which, the understanding of numerical algorithms is becoming much deeper. During this period, high-order schemes arise to be the focus because of the low dissipation and excellent dispersion characteristics, as well as high computational efficiency comparatively. Nonetheless, the application of high-order schemes to complex configurations still remains as a challenge, thus making it a new trend in CFD [1].

In the development of high-order schemes, great advancements in both finite difference (FD) and finite volume (FV) methods have been achieved accordingly. As for FD methods, on one hand, Shu and Osher [2,3] extended the reconstruction operator of ENO/WENO schemes from cell averages to pointwise numerical fluxes and developed finite difference ENO/WENO schemes. Lele [4] proposed the compact finite difference schemes with spectral-like resolution. To meet the requirements of aeroacoustic simulations, Tam and Webb [5] proposed dispersion relation preservation (DRP) schemes and enhanced the resolution of short waves by decreasing the order of accuracy. Deng et al. [6,7] developed a series of weighted compact nonlinear schemes (WCNS), capable of capturing shock waves and preserving high order of accuracy in smooth regions. Currently, in order to further decrease the dissipation, a family of hybrid schemes [8] were formulated, providing a switch mechanism of numerical schemes depending on the properties of flowfields, and the fundamental ingredient to this switch lies in the delicate design of discontinuity detectors. On the other hand, for FV methods, the formulation of high-order schemes may date back to the reconstruction procedure of ENO/WENO. Jay et al. [9] was among the first to apply high-order ENO schemes to structured grids and discussed the genuinely multi-dimensional reconstruction comprehensively in the framework of FV methods. Abgrall et al. [10], Harten and Chakravarthy [11], Sonar [12] managed to apply ENO schemes to unstructured grids. Friedrich [13] attempted to apply WENO schemes to unstructured grids, but optimization was not realized in his work, which was accomplished by Hu et al. [14]. However, limited by computational resources, all the above-mentioned high-order schemes in general are formulated and applied in two space dimensions. Titarev and Toro [15] extended the high-order FV methods to three space dimensions on structured grids and discussed the influence of different approximate Riemann solvers to numerical results. Tsoutsanis et al. [16] studied the application of high-order FV methods on three dimensional mixed grids. Recently, Groth et al. [17] proposed a kind of CENO schemes based on smooth indicators and applied this scheme to both structured and unstructured grids. Although multiple choices of stencil were not involved, this scheme still possessed the ENO-like properties.

It is obvious that there are some differences as well as similarities between FD and FV methods since they always refer to each other [18,19] in the process of development. To explore relations of these two methods, some researchers have conducted a series of comparison. Thomas and Lombard [20] discussed the relationship between FD and FV methods from the concept of geometric conservation law (GCL) on dynamic grids. Vinokur [21] analysed the basic discretization forms of these two methods with second order of accuracy and concluded that metrics of FD methods standed for the surface area, which linked FD and FV methods geometrically. Leonard [22] compared these two methods through the analysis of truncation errors of convective terms in one space dimension. Liu [23] made a comparison based on vortex-stream function methods and MAC methods and summarized that FV methods were superior to FD methods, but it has to be pointed out that governing equations employed there were different. Botte et al. [24] carried out a detailed comparison including the order of accuracy, conservation properties and execution time and discussed the influence of boundary conditions. Recently, based on WCNS schemes, Deng et al. proved that second order FD methods were equal to second order FV methods [25] provided that metrics and the Jacobian were treated properly, such as using the symmetric conservative metric methods (SCMM). Moreover, under high order circumstances, FD methods could be reconstructed from the weighted summation of second order FV methods [26]. Abe et al. [27] also presented the geometric interpretations and spatial symmetry property of the metrics for the high-order FD methods independently.

Among all these comparisons, the conclusion from Casper et al. [28] in 1994 based on ENO schemes was widely accepted and referenced [29–31]. It is concluded that FD methods were sensitive to the mesh quality, such as the derivative discontinuities, and the design order of accuracy could be degraded. Therefore, for complex configurations, FV methods were preferred. Also, in some practical applications, FV methods were identified to behave better than FD methods [32–34].

However, at the time Casper et al. [28] conducting the comparison, weighting techniques were not well established. (Actually, this technique was mentioned and discussed in the same paper, but not implemented.) What's more, after continual efforts and explorations in the last two decades, some breakthroughs have been made for FD methods. For example, the property of freestream preservation, considered to be influential to the numerical characteristics of FD methods, has been studied by various researchers [25,35–39], along with some convincing and impressive examples. Inspired by such kind of research, it is necessary to reevaluate the numerical performance of high order FD and FV methods combined with the latest progress. For this reason, in the present work, we would follow the framework provided by Casper et al. and simulate the same cases for comparison. The implementations for both FD and FV methods adopted in this paper are well established and possess some essential features therein, like the freestream preservation.

The rest of the paper is organized as follows. In Section 2, high order discretization algorithms for FD and FV methods are provided, along with formulations to satisfy the freestream preservation for these two methods. Based on these methods, in Section 3, cases from Casper et al. are simulated and systematic comparisons are made correspondingly. Besides, much attention is paid to distinguish the effect of freestream preservation conditions for numerical calculations. In Section 4, computation efficiency in three space dimensions is practically measured instead of approximate estimation by Casper et al. [28]. Conclusions are summarized in Section 5.

2. Governing equations

Considering that both one dimensional and two dimensional cases are included in the original paper [28], the governing equations are presented first and then, specific issues related to the freestream preservation are discussed.

2.1. Governing equations in one space dimension

In one space dimension, the differential form of Euler equations could be expressed as,

$$\frac{\partial Q}{\partial t} + \frac{\partial E}{\partial x} = 0, \qquad (1)$$

where Q and E are conservative variables and numerical fluxes,

$$Q = \begin{bmatrix} \rho \\ \rho u \\ \rho e \end{bmatrix}, E = \begin{bmatrix} \rho u \\ \rho u^2 + p \\ (\rho e + p)u \end{bmatrix}. \qquad (2)$$

For non-uniform grids, three different methods are available for the calculation, including the FD method, the FV method based on the coordinate transformation (FV-TR) and the FV method based on the physical reconstruction (FV-PR). For the FD method, after the coordinate transformation, Eq. (1) is recasted as,

$$x_\xi \frac{\partial Q}{\partial t} + \frac{\partial E}{\partial \xi} = 0. \qquad (3)$$

To discretize this equation, fifth-order WCNS scheme is employed here [7]. The metric x_ξ and the spatial derivative are calculated through sixth-order central difference,

$$E'_j = \frac{75}{64h}(E_{j+1/2} - E_{j-1/2}) - \frac{25}{384h}(E_{j+3/2} - E_{j-3/2}) + \frac{3}{640h}(E_{j+5/2} - E_{j-5/2}). \qquad (4)$$

where h is the interval and $E_{j+1/2}$ denotes the numerical flux at cell-edges. Numerical fluxes are evaluated via approximate Riemann solvers.

$$E_{j+1/2} = E\left(Q^L_{j+1/2}, Q^R_{j+1/2}\right), \qquad (5)$$

in which $Q^L_{j+1/2}$, $Q^R_{j+1/2}$ are obtained with high order interpolations. In smooth regions, linear interpolation schemes are preferred; while for discontinuities, weighted non-linear interpolations are required.

For the FV method based on the coordinate transformation, the transformed governing equations are integrated in the interval $[\xi_{j-1/2}, \xi_{j+1/2}]$.

$$\frac{\partial}{\partial t} \int_{\xi_{j-1/2}}^{\xi_{j+1/2}} (x_\xi Q) d\xi + E_{j+1/2} - E_{j-1/2} = 0. \qquad (6)$$

According to the formula of integration, it is straightforward to derive that

$$\int_{\xi_{j-1/2}}^{\xi_{j+1/2}} (x_\xi Q) d\xi = \int_{x_{j-1/2}}^{x_{j+1/2}} Q dx = \Delta \xi \overline{(x_\xi Q)}. \qquad (7)$$

where $\overline{(x_\xi Q)}$ is the cell average in the interval $[\xi_{j-1/2}, \xi_{j+1/2}]$. Therefore, $Q^L_{j+1/2}$, $Q^R_{j+1/2}$ should be reconstructed from $\overline{(x_\xi Q)}$, followed by the calculation of $E_{j+1/2}$.

For the FV method based on the physical reconstruction, Eq. (1) is integrated directly,

$$\frac{\partial}{\partial t} \int_{x_{j-1/2}}^{x_{j+1/2}} Q dx + E_{x_{j+1/2}} - E_{x_{j-1/2}} = 0. \qquad (8)$$

where $Q_{j+1/2}^L$ and $Q_{j+1/2}^R$ are reconstructed from the integration of conservative variables in the interval $[x_{j-1/2}, x_{j+1/2}]$. Since the grid distribution is non-uniform, the reconstruction formulae at cell-edges are different.

For these three above-mentioned methods, it is straightforward to extend them to multiple dimensions. However, the calculation complexity is burdensome for the high order FV-PR method, so it is not practical to apply this method to multi-dimensional cases. In this paper, only the high order FD and FV methods based on the coordinate transformation are considered in multiple space dimensions.

2.2. Governing equations in two space dimensions

In two space dimensions, hyperbolic conservative systems for FD methods are directly extended from that of one space dimension,

$$\frac{\partial Q}{\partial t} + \frac{\partial E}{\partial x} + \frac{\partial F}{\partial y} = 0. \tag{9}$$

After the coordinate transformations, governing equations in curvilinear coordinates can be derived,

$$\frac{\partial}{\partial t}\left(\frac{Q}{J}\right) + \frac{\partial \hat{E}}{\partial \xi} + \frac{\partial \hat{F}}{\partial \eta} = 0. \tag{10}$$

where

$$\hat{E} = \frac{1}{J}(E\xi_x + F\xi_y), \ \hat{F} = \frac{1}{J}(E\eta_x + F\eta_y). \tag{11}$$

J is the Jacobian and $\xi_x, \xi_y, \eta_x, \eta_y$ are metrics. For stationary grids, following relations are satisfied,

$$\begin{bmatrix} \xi_x & \xi_y \\ \eta_x & \eta_y \end{bmatrix} = \begin{bmatrix} x_\xi & x_\eta \\ y_\xi & y_\eta \end{bmatrix}^{-1}, J^{-1} = \left|\frac{\partial(x,y)}{\partial(\xi,\eta)}\right|. \tag{12}$$

Numerical fluxes calculations are still evaluated via approximate Riemann solvers and first order derivative calculations along ξ and η direction are the same as that of one space dimension.

In two space dimensions, governing equations of the FV-TR method are derived from integrating Eq. (10) in the computational domain $[\xi_{i-1/2}, \xi_{i+1/2}] \times [\eta_{j-1/2}, \eta_{j+1/2}]$,

$$\frac{d}{dt}\overline{\left(\frac{Q}{J}\right)}_{ij} + \frac{1}{\Delta \xi}\left(\hat{E}_{i+1/2,j} - \hat{E}_{i-1/2,j}\right) + \frac{1}{\Delta \eta}\left(\hat{F}_{i,j+1/2} - \hat{F}_{i,j-1/2}\right) = 0. \tag{13}$$

where

$$\overline{\left(\frac{Q}{J}\right)}_{ij} = \frac{1}{\Delta\xi\Delta\eta}\int_{\eta_{j-1/2}}^{\eta_{j+1/2}}\int_{\xi_{i-1/2}}^{\xi_{i+1/2}}\left(\frac{Q}{J}\right)d\xi d\eta. \tag{14}$$

$$\hat{E}_{i+1/2,j} = \frac{1}{\Delta\eta}\int_{\eta_{j-1/2}}^{\eta_{j+1/2}}\hat{E}(\xi_{i+1/2}, \eta)d\eta, $$
$$\hat{F}_{i,j+1/2} = \frac{1}{\Delta\xi}\int_{\xi_{i-1/2}}^{\xi_{i+1/2}}\hat{F}(\xi, \eta_{j+1/2})d\xi. \tag{15}$$

To realize the design order of accuracy, integrals in Eq. (15) should be calculated through the Gaussian quadrature with multiple Gaussian points. Numerical fluxes in Gaussian points require reconstructions along the direction of ξ and η independently. For more details, it is recommended to refer to [15].

2.3. Freestream preservation

In numerical calculations, the freestream preservation should be satisfied; otherwise excessive numerical errors would be introduced. For the verification of the freestream preservation, one immediate way is to substitute the freestream condition into discretized equations to examine whether it is preserved. Based on this method, the detailed analysis to different discretization methods in Sections 2.1 and 2.2 is presented in this section.

Obviously, for FD and FV-PR methods in one space dimension, freestream preservation can be satisfied. However, for FV-TR methods, reconstructed variables are $\overline{(x_\xi Q)}$. To evaluate numerical fluxes, corresponding metrics need to be divided to obtain conservative variables. In order to acquire numerical fluxes consistent with the uniform flow, the condition that should be satisfied is,

$$\left(x_\xi Q_{uniform}\right)_R / x_\xi = Q_{uniform}, \tag{16}$$

where subscript R corresponds to reconstructed variables while $Q_{uniform}$ represents numerical fluxes under freestream conditions. It is not difficult to derive that to satisfy Eq. (16), reconstruction operators for $\overline{(x_\xi Q)}$ and the metrics should be the same [40].

With regard to numerical calculations in two space dimensions, there also exists the issue of the freestream preservation. Specifically, as for WCNS employed here, metrics and the Jacobian need to be evaluated based on SCMM [25],

$$\hat{\xi}_x = \xi_x/J = y_\eta = \delta_\eta y, \ \hat{\xi}_y = \xi_y/J = -x_\eta = -\delta_\eta x, $$
$$\hat{\eta}_x = \eta_x/J = -y_\xi = -\delta_\xi y, \ \hat{\eta}_y = \eta_y/J = x_\xi = \delta_\xi x, \tag{17}$$

$$J = \frac{1}{2}\left[(yx_\xi)_\eta + (xy_\eta)_\xi - (xy_\xi)_\eta - (yx_\eta)_\xi\right]$$
$$= \frac{1}{2}\left[\delta_\eta(y\delta_\xi x) + \delta_\xi(x\delta_\eta y) - \delta_\xi(y\delta_\eta x) - \delta_\eta(x\delta_\xi y)\right], \tag{18}$$

where δ stands for the difference operator. Based on SCMM, difference operators for calculations of metrics and flux derivatives should be the same. It should be noted that in multiple dimensions, the freestream preservation is hard to be fulfilled for standard WENO schemes [41]. Nonomura et al. [42] demonstrated that the standart finite-difference WENO could not satisfy the freestream preservation, but the WCNS [7] could by using the conservative metrics. To remedy the shortage of WENO, Nonomura et al. [36] proposed to divide the standard finite-difference WENO into two parts, including the consistent central difference part and the numerical dissipation part. With different treatments for these two parts, the order of accuracy is maintained and the freestream is perfectly preserved. Recently, another numerical strategy has been proposed by Sun et al. [43] to ensure freestream preservation properties of the WENO schemes on stationary curvilinear grids. The essential idea of this approach is to offset the geometrically induced errors by proper discretization of the metric invariants. Still, in this paper, WCNS schemes are adopted for the FD discretization. For FV methods, freestream preservation conditions are somewhat complicated. It is demonstrated that the order of the polynomials in evaluating metrics should not be higher that of the Gaussian quadrature. In addition, the Jacobian at each Gaussian point needs to be calculated by the same reconstruction algorithm, similar to the calculation in one space dimension. For more details on the freestream preservation of FV methods, it is advisable to refer to [40].

3. Cases recalculations

In order to gain some further insights into FD and FV methods of high order of accuracy , in this section, cases from the original paper [28] would be recalculated. The biggest difference lies in that the freestream preservation is considered here. Therefore, for FD methods, fifth-order WCNS scheme is chosen instead of WENO scheme. For all these calculations, the effect of the freestream preservation to numerical results is compared comprehensively.

3.1. One-dimensional rarefaction wave

This case describes the movement of rarefaction waves governed by Euler equations in one space dimension. Before the cal-

Table 1
Initial left and right states for one-dimensional rarefaction wave.

Variables	Left state (L)	Right state (R)
Density	0.896	1.015
Velocity	1.358	1.516
Pressure	1.0	1.191

Table 2
Order of accuracy and error measurements for the FD method.

Grid	C^1 grid		C^6 grid	
120	7.837E−05	–	7.380E−05	–
240	4.565E−06	4.10	4.235E−06	4.12
480	5.824E−08	6.29	5.228E−08	6.34
960	1.625E−09	5.16	1.755E−09	4.90

Table 3
Order of accuracy and error measurements for the FV-TR method.

Grid	C^1 grid		C^6 grid	
120	8.291E−05	–	8.073E−05	–
240	4.440E−06	4.22	3.879E−06	4.38
480	5.120E−08	6.44	4.803E−08	6.34
960	1.943E−09	4.72	2.118E−09	4.50

Table 4
Order of accuracy and error measurements for the FV-PR method.

Grid	C^1 grid		C^6 grid	
120	9.221E−05	–	8.610E−05	–
240	4.747E−06	4.28	4.053E−06	4.41
480	5.459E−08	6.44	4.516E−08	6.49
960	1.885E−09	4.86	2.086E−09	4.44

culation, it has to be clarified that non-uniform grids in physical space need to be mapped from uniform grids so as to examine the influence of the grid distribution. In addition, numerical results are also influenced by the continuity of grids. For that purpose, uniform space is divided into several regions corresponding to different transformations. Based on the continuity, two family of grids, C^1 and C^6, are generated. C^1 means that the first-order derivative of the transformation is continuous, while for C^6, up to the sixth-order derivative is continuous. For C^1 grids, the transformation is given in Eq. (19),

$$x(\xi) = \begin{cases} \frac{9\pi}{20}(\xi+4) - 4\sin\left(\frac{9\pi}{20}\right), & \xi \in [-6,-4] \\ \sin\left[\frac{9\pi}{20}(\xi+4)\right] - 4\sin\left(\frac{9\pi}{20}\right), & \xi \in [-4,-3] \\ \sin\left[\frac{9\pi}{20}(\xi+2)\right] - 2\sin\left(\frac{9\pi}{20}\right), & \xi \in [-3,-1] \\ \sin\left(\frac{9\pi}{20}\xi\right), & \xi \in [-1,1] \\ \sin\left[\frac{9\pi}{20}(\xi-2)\right] + 2\sin\left(\frac{9\pi}{20}\right), & \xi \in [1,3] \\ \sin\left[\frac{9\pi}{20}(\xi-4)\right] + 4\sin\left(\frac{9\pi}{20}\right), & \xi \in [3,4] \\ \frac{9\pi}{20}(\xi-4) + 4\sin\left(\frac{9\pi}{20}\right), & \xi \in [4,6] \end{cases}$$

(19)

It is simple to verify that second-order derivative of this transformation is discontinuous in the location of $\xi = \pm 1, \pm 3$. For C^6 grids, a similar piecewise transformation is defined as,

$$x(\xi) = \begin{cases} \alpha(\xi+4) - 4\left(\alpha - \frac{5}{16}\beta\right), & \xi \in [-6,-4] \\ \left(\alpha - \frac{5}{16}\beta\right)\xi + \frac{\beta}{32}\left[\frac{15}{\pi}\sin(\pi\xi) \right. \\ \left. -\frac{3}{2\pi}\sin(2\pi\xi) + \frac{1}{3\pi}\sin(3\pi\xi)\right], & \xi \in [-4,4] \\ \alpha(\xi-4) + 4\left(\alpha - \frac{5}{16}\beta\right), & \xi \in [4,6] \end{cases}$$

(20)

Obviously, the sixth-order derivative of this transformation is continuous. Parameters α and β are determined so that the ratio of the maximum to minimum values of x_ξ are identical for the two grids and the physical distance between $x(-4)$ and $x(4)$ is the same for both.

For right-travelling rarefaction waves, initial conditions for left and right state are denoted by subscript L and R. According to the definition of the strength of rarefaction waves in [28], a mean temperature change of $\pm 5\%$ across the wave and a mean Mach number $\bar{M} = 1.134334$ is required. Initial values of the Riemann problem are listed in Table 1. In this case, the initial solution consists of an isentropic expansion smoothly distributed on $-6.5 \leqslant x \leqslant 5.5$. For $t > 0$, the rarefaction wave moves to the right until $t = 4.0$. Density error is measured with respect to the L_1 norm in a region to the left of the rarefaction wave at $t = 4.0$. In this paper, the interval $2.0 \leq x \leq 3.0$ is chosen and linear interpolation schemes are employed here.

To verify the influence of freestream preservation conditions, the density contour with the evaluation of metrics not satisfying Eq. (16) is given in Fig. 1. It can be seen that larger error occurs in the region of grid discontinuities because of freestream preservation conditions violation. For the C^1 grid, errors introduced are more profound. Fig. 2 illustrates results of FD, FV-TR and FV-PR methods with linear schemes. Since freestream preservation conditions are fulfilled, smooth density contour is obtained for both C^1 and C^6 grids. Moreover, for FV methods, there exists some twists in the region of strong grid discontinuity. However, for FD methods, this problem is smeared and avoided. Errors and the convergence order of accuracy for these three methods are listed in Tables 2–4. To summarize, for all these methods on both grids, design order of accuracy is realized and the grid discontinuity has nearly has no influence with the freestream preservation satisfied. To be more clear, errors are plotted in Fig. 3. Obviously, for this case, differences arising from numerical algorithms and the grid discontinuity are evident.

It is reasonable to conclude that for the rarefaction wave case in one space dimension, the adaptivity of FD and FV methods to grid discontinuity is identical. No obvious adverse effects are observed for both FD and FV methods. As a matter of fact, results in this paper are comparable to that of Casper et al., who claimed that the high-order convergence of the FD method was a fortuitous result inherent to this particular test case. Here, it has been clarified that when the freestream preservation is satisfied, this result is not fortuitous but natural.

3.2. Two-dimensional channel flow

This test case is a subsonic isentropic channel flow of varying areas governed by the Euler equations. Symmetric bumps are distributed in upper and lower walls. Mach number at the inflow is 0.3. For the geometry considered, the length-to-height ratio is $L/H = 1.5$. Each constant-area section near the inlet and outlet has the length $L/5$; at the throat, the constriction is 10% of H. Grids in the physical space are mapped from a rectangle computational domain $\{0 \leq \xi \leq L\} \times \{0 \leq \eta \leq H\}$. In the constant-area section,

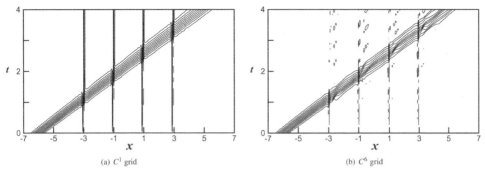

Fig. 1. Density distribution for the FV-TR method without satisfying the freestream preservation.

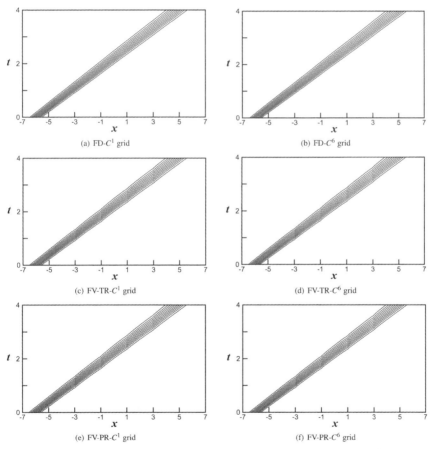

Fig. 2. Density distribution for different numerical algorithms satisfying the freestream preservation on C^1 and C^6 grids.

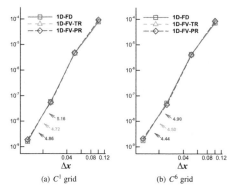

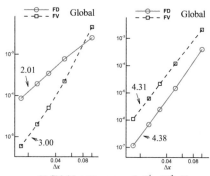

Fig. 3. Error distributions for three different methods.

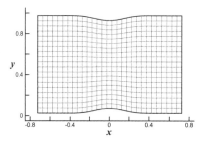

Fig. 4. The configuration and grid used in the two-dimensional channel flow.

the identity maps are employed. While for the varying-area section, the transformation is given as follows,

$$\begin{cases} x = \xi, \\ y = \left(1 - \frac{\eta}{H}\right) y_1(\xi) + \frac{\eta}{H} y_2(\xi), \end{cases} \tag{21}$$

where $y_1(\xi)$ and $y_2(\xi)$ are, respectively, the equations for the lower and upper walls.

In order to study the influence of the geometry continuity to FD and FV methods, two different functions are utilized to approximate the bump geometry,

$$\begin{cases} y_1(\xi) = 0.05 - 0.1 \times \left(\frac{20}{9}\right)^2 \xi^2 + 0.05 \left(\frac{20}{9}\right)^4 \xi^4 \\ y_2(\xi) = 0.95 + 0.1 \times \left(\frac{20}{9}\right)^2 \xi^2 - 0.05 \left(\frac{20}{9}\right)^4 \xi^4 \end{cases} \tag{22}$$

$$\begin{cases} y_1(\xi) = 0.05 \sin^4\left(\frac{10}{9}\pi\xi + \frac{\pi}{2}\right) \\ y_2(\xi) = 1 - 0.05 \sin^4\left(\frac{10}{9}\pi\xi + \frac{\pi}{2}\right) \end{cases} \tag{23}$$

which correspond to C^1 and C^3 grids, with continuous first-order and third-order derivative respectively. The geometry and grids are illustrated in Fig. 4. Boundary conditions include the subsonic inflow, outflow and a slip wall. Details on boundary condition treatments are described in [44]. To evaluate the numerical results, the entropy error is adopted here. According to Casper et al. [28], errors are computed in three different regions: the global error computed over the entire computational domain, the wall error com-

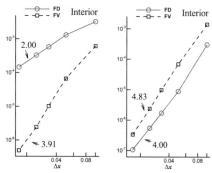

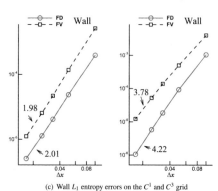

Fig. 5. L_1 entropy errors for the two-dimensional channel flow without satisfying the freestream preservation.

puted only at the points along one wall, and the interior error computed on $\{L/4 < \xi < 3L/4\} \cup \{H/4 < \eta < 3H/4\}$.

Considering that numerical results are influenced by freestream preservation conditions, in the beginning, results with freestream preservation conditions satisfied and unsatisfied are compared. Fig. 5 shows results of FD and FV methods without satisfying the freestream preservation. It can be seen that errors calculated in this paper are similar to that of Casper et al. [28]. For the C^1 grid, only second order of accuracy is achieved for FD methods; while for FV methods, global and interior accuracy are higher than that of FD methods despite that the wall accuracy is still second order. For the C^3 grid, the same order of accuracy is obtained from FD and FV methods. With respect to the absolute errors, the FD method is superior to the FV method. Concluded from these results, the FD method is sensitive to the grid discontinuity and the convergence order of accuracy is degraded when the freestream preservation is not satisfied.

Fig. 6 presents results with the freestream preservation satisfied. It is surprising that remarkable difference arises on the C^1 grid for the FD method compared with Fig. 5. Specifically, design high order of accuracy is achieved for the FD method when the freestream preservation is fulfilled and the absolute error is lower than that of the FV method. In other words, geometry nonsmoothness does not make any obvious difference for different algorithms. Results calculated with FD and FV methods on both C^1 and C^3 grids are comparable.

For more convenient comparison, results from four different algorithms on the C^1 grid are plotted in Fig. 7, where the legend FP and noFP correspond to the freestream preservation satisfied and unsatisfied. It is shown that the FD method is strongly influenced by the freestream preservation; while for the FV method, the difference is not that obvious. At this point, some interesting findings are uncovered. For the work of Casper et al. [28], the issue of the freestream preservation is not considered for the FD method, thus leading to the incomplete conclusion that the FV method was superior to the FD method, regardless of whether the freestream preservation was taken into account for the FV method. The reason for this conclusion lies in that properties of the FD method in the discretized space is not fully understood. With SCMM [25] employed, i.e., the freestream preservation satisfied, it is evident that results from the FD method is somewhat better that of the FV method even for the grid with poor continuity.

Fig. 8 exhibits the distribution of the entropy errors along the centreline with four algorithms described above. If the freestream preservation is not fulfilled, obvious oscillatory behaviours are observed in the connections of grid discontinuity, for both the FD and FV methods. Moreover, the oscillation is much severe for the FD method, similar to that of Casper et al. [28]. However, if the freestream preservation is fulfilled, the entropy error is approximately the same along the centreline. From the close-up, it is observed that small perturbations still exist for the FV method. Explanations for this phenomenon require further research.

Through the calculation of this two dimension channel flow, the importance of the freestream preservation is demonstrated and strengthened once more. As for the conclusion of Casper et al. [28], claiming that FD methods could not be applied to complex configurations, is worthwhile deliberating. Results in this paper have shown that even for grids with poor continuity, similar order of accuracy and error estimations can be obtained for FD methods.

3.3. Oblique Sod's problem

The final test under consideration is oblique Sod's problem, which is generally calculated in one space dimension. Here, Sod's problem will be solved in two space dimensions so that the planar waves produced will propagate at various angles of incidence

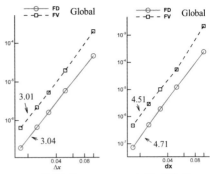

(a) Global L_1 entropy errors on the C^1 and C^3 grid

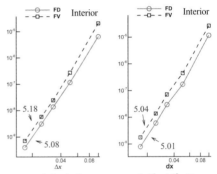

(b) Interior L_1 entropy errors on the C^1 and C^3 grid

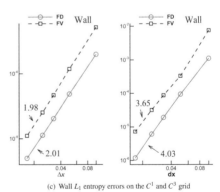

(c) Wall L_1 entropy errors on the C^1 and C^3 grid

Fig. 6. L_1 entropy errors for the two-dimensional channel flow satisfying the freestream preservation.

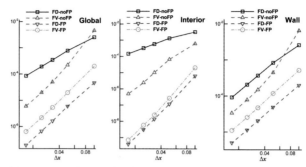

Fig. 7. L_1 entropy errors for the two-dimensional channel flow with four different algorithms on the C^1 grid.

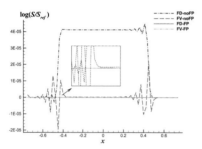

Fig. 8. The distribution of entropy errors along the centreline on the C^1 grid.

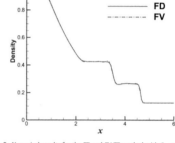

Fig. 9. Numerical results for the FD and FV-TR method with $\theta = \arctan 1$.

with respect to a rectangular grid. The intent here is not only to inspect the qualitative resolution of the oblique waves with different numerical algorithms, but also quantify the manner in which each algorithm detects an oblique wave that is not normal to the mesh.

In numerical calculations, the discontinuity is a straight line that makes an angle θ with the x axis. The computational domain is defined on the rectangle $[0, L] \times [0, H]$, with a length-to-height ratio $L/H = 6$. Specific values of L and H are related to the angle of incidence θ, defined as $L_\theta = L/\sin\theta$, $H_\theta = H/\sin\theta$. This scaling achieves the same grid resolution normal to the wave propagation on a given mesh at some fixed time for all choices of θ. The mesh scale in each direction is identical on a 96×16 grid.

Initial conditions correspond to a Riemann problem with left and right states as follows,

$$Q^L = \begin{bmatrix} 8 \\ 0 \\ 0 \\ 10 \end{bmatrix}, Q^R = \begin{bmatrix} 1 \\ 0 \\ 0 \\ 1 \end{bmatrix}. \quad (24)$$

Exact solutions to this Riemann problem are wave structures consisting of a rarefaction wave, a contact discontinuity and a shock wave from left to right. At $t = 0$, the initial discontinuity is positioned at $(x, y) = (3L/8, 0)$ and inclined at the angle θ. In finite time, waves would not propagate to the right boundary. Therefore, initial conditions are maintained in the left and right boundaries. The angles of inclination are chosen so that $\tan\theta$ is an integer and that the upper and lower boundary conditions can be determined in a "shifted-periodic" manner. In particular, the test angles are $\theta = \arctan 1$, $\arctan 2$, $\arctan 4$, in addition to the one-dimensional problem $\theta = \pi/2$.

Fig. 9 illustrates results of $\theta = \arctan 1$ from the FD and FV methods. There is no obvious difference for these two methods, except that some overshoots exist in the location of the contact discontinuity for the FV method. It should be noted that the reason for overshoots is more likely connected to the choice of reconstruction variables. For the FV method, high order of accuracy is achieved via the reconstruction of conservative variables, while for the FD method, primitive variables are used. Characteristic variables, assumed to perform better in numerical calculations, are not employed in this paper for the reconstruction or interpolation.

On a more quantitative level, the differences of density between the oblique cases and the one-dimensional case have been measured and plotted in Fig. 10. Since the cartesian grid is used here, the issue of the freestream preservation does not exist any more. The focus of this case concentrates on the influence of the new weighting techniques adopted. It can be seen that large differences arise in the connection of different wave structures. More detailed observations show that errors of the FV method are slightly larger than that of the FD method. Despite the fact that errors are due to some combined factors including choices of reconstruction variables, weighting techniques and the distribution of initial values, it is not arbitrary to conclude that the FD method is somewhat superior to the FV method from numerical results presented above.

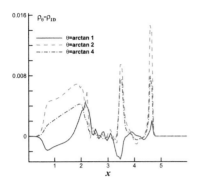

(a) Density deviation from one dimension for the FD method

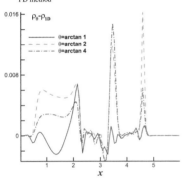

(b) Density deviation from one dimension for the FV-TR method

Fig. 10. Density deviation from one dimension with $\theta = \arctan 1$, $\theta = \arctan 2$, $\theta = \arctan 4$.

4. Cost comparison

For any numerical algorithm to be applied in the simulation of complex configurations, time consumption is a major concern. Theoretically, the cost of FD and FV methods is comparable in one space dimension regardless of the order of accuracy. Besides, in multiple dimensions, these two methods can be made equally cost effective with the implementation of first or second order schemes. However, time consumption of high-order FV methods are remarkably higher than that of FD methods in two or three space dimensions. The reason lies in that for FD methods, the high-order multidimensional reconstruction is a straightforward extension of the one-dimensional reconstruction; while for FV methods, the multidimensional reconstruction is implemented as a product of one-dimensional operators because the reconstruction is based on the cell averages. Moreover, for high order FV methods, multiple Gaussian quadrature points are required in the calculation of numerical fluxes on each cell interface.

In the original paper by Casper et al. [28], the cost comparison is conducted based on the Euler equations in two space dimensions, with the order of accuracy ranging from second to fifth, but the cost comparison in three space dimensions is just evalu-

Table 5
Time consumption statistics of Euler equations with fifth-order algorithms.

Dimensions	Algorithms	Time(s)/DoF
1D	FD	9.06E−07
	FV-TR	1.19E−06
	FV-PR	1.52E−06
2D	FD	2.10E−06
	FV-TR	1.05E−05
3D	FD	6.69E−06
	FV-TR	5.59E−05

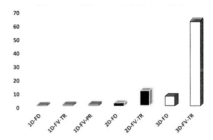

Fig. 11. Time consumption comparison of Euler equations with fifth-order algorithms.

ated. In this section, the time consumption of fifth-order FD and FV-TR methods are compared in the calculation of Euler equations, including one dimensional and multidimensional problems. To decrease the randomness of results, all the calculations are repeated for ten times. In Table 5, the average time consumption of one iteration for a single grid point is provided. To demonstrate the influence of algorithms and dimensions more clearly, normalized results based on the time consumption of FD methods in one space dimension are plotted. It can be seen that, for FD methods, the increasing factor is in the range of [2, 3]. In [28], the time consumption of FD methods in solving the Euler equations in k spatial dimensions on a grid of N points was evaluated with the formula $C_{FD} = C_1(k+2)kN^k$, in which $(k+2)$ represents the number of governing equations, and C_1 denotes the cost of solving one equation at a point in one space dimension. This formula predicts an increasing factor of 8/3 from one space dimension to two space dimensions. The corresponding factor for the extension from two to three space dimensions is 15/8. These evaluations are consistent with quantitative results tested here. For FV-TR methods, the increasing factor is in the range of [5, 10].

In addition, for the same dimension, the cost of FV-TR methods are significantly higher than that of FD methods. Specifically, in two and three space dimensions, the cost ratio of FV-TR and FD methods is 5 and 10 respectively. For one space dimension, the cost of FV-PR methods is the highest, followed by FV-TR methods, slightly higher than that of FD methods.

5. Concluding remarks

Two advanced numerical algorithms with weighting techniques introduced are systematically evaluated in this paper following the framework outlined by Casper et al. [28]. For the one-dimensional rarefaction wave and the two-dimensional channel flow, it is demonstrated that without satisfying the freestream preservation, the finite difference algorithm is sensitive to derivative discontinuities when compared with the finite volume algorithm, and this conclusion is consistent with that of Casper et al. [28]. However, with the freestream preservation satisfied, both of these two algorithms perform well and are comparable to each other in terms of accuracy, resolution of waves and sensitivity to grid irregularities. For the oblique Sod's problem, the issue of the freestream preservation does not exist since the cartesian grid is used. What is more important for this case is the influence of weighting techniques. It can be seen that there is no big difference for the FD and the FV algorithm. Besides, the time consumption in three space dimensions is measured in this paper, confirming the conclusion that in multiple dimensions, the high order finite difference algorithm is much more efficient, thus providing a great advantage over the finite volume algorithm in the simulation of practical engineering applications. It should be noted that the conclusion drawn in the present work is limited to the computation on structured, curvilinear grid. As for the application to unstructured grid, the finite volume scheme is available while the finite difference scheme is not.

Acknowledgments

This work was supported by the Basic Research Foundation of National University of Defense Technology (No. ZDYYJCYJ20140101). The authors want to acknowledge the detailed and constructive suggestions of the reviewers, which significantly improved the quality of this work.

References

[1] Slotnick J, Khodadoust A, Alonso J, Darmofal D, Gropp W, Lurie E, et al. CFD vision 2030 study: a path to revolutionary computational aerosciences. NASA/CR-2014-218178.
[2] Shu CW, Osher S. Efficient implementation of essentially non-oscillatory shock-capturing schemes. J Comput Phys 1988;77:439–71.
[3] Shu CW, Osher S. Efficient implementation of essentially non-oscillatory shock-capturing schemes. J Comput Phys 1989;83:32–78.
[4] Lele SK. Compact finite difference schemes with spectral-like resolution. J Comput Phys 1992;103:16–42.
[5] Tam CKW, Webb JC. Dispersion-relation-preserving finite difference schemes for computational acoustics. J Comput Phys 1993;107:262–81.
[6] Deng X, Maekawa H. Compact high-order accurate nonlinear schemes. J Comput Phys 1997;130. 77–91
[7] Deng X, Zhang H. Developing high-order weighted compact nonlinear schemes. J Comput Phys 2000;165:22–44.
[8] Sun Z, Luo L, Ren Y, Zhang S. A sixth order hybrid finite difference scheme based on the minimized dispersion and controllable dissipation technique. J Comput Phys 2014;270:238–54.
[9] Casper J, Atkins KL. A finite-volume high-order ENO scheme for two-dimensional hyperbolic systems. J Comput Phys 1993;106:62–76.
[10] Abgrall R. On essentially non-oscillatory schemes on unstructured meshes analysis and implementation. J Comput Phys 1994;114:45–58.
[11] Harten A, Chakravarthy S. Multi-dimensional ENO schemes for general geometries. ICASE Report; 1991. 91–76
[12] Sonar T. On the construction of essentially non-oscillatory finite volume approximations to hyperbolic conservation laws on general triangulations: polynomial recovery, accuracy and stencil selection. Comput Methods Appl Mech Eng 1997;140:157–81.
[13] Friedrichs O. Weighted essentially non-oscillatory schemes for the interpolation of mean values on unstructured grids. J Comput Phys 1998;144:194–212.
[14] Hu C, Shu C-W. Weighted essentially non-oscillatory schemes on triangular meshes. J Comput Phys 1999;150:97–127.
[15] Titarev V, Toro E. Finite-volume WENO schemes for three-dimensional conservation laws. J Comput Phys 2004;201:238–60.
[16] Tsoutsanis P, Titarev VA, Drikakis D. WENO schemes on arbitrary mixed-element unstructured meshes in three space dimensions. J Comput Phys 2011;230:1585–601.
[17] McDonald SD, Charest MRJ, Groth CPT. High-order CENO finite-volume schemes for multi-block unstructured mesh. AIAA 2011–3854; 2011.
[18] Selmin V. The node-centred finite volume approach: bridge between finite differences and finite elements. Comput Methods Appl Mech Eng 1993;102:107–38.
[19] Zhu Z, Zeng L, Liu K, Fu Y. A finite-difference scheme combining with finite-volume technique by using general quadrilateral cells. Int J Nonlinear Sci Numer Simul 2014;15:279–87.
[20] Thomas PD, Lombard CK. The geometric conservation law-a link between finite-difference and finite-volume methods of flow computation on moving grids. AIAA 78–1208; 1978.
[21] Vinokur M. An analysis of finite-difference and finite-volume formulations of conservation laws. J Comput Phys 1989;81:1–52.
[22] Leonard BP. Comparison of truncation error of finite difference and finite-volume formulations of convection terms. Appl Math Model 1994;18:46–50.
[23] Liu R, Wang D, Zhang X, Li W, Yu B. Comparison study on the performances of finite volume method and finite difference method. J Appl Math 2013. Article ID 596218
[24] Botte GG, Ritter JA, White RE. Comparison of finite difference and control volume methods for solving differential equations. Comput Chem Eng 2000;24:2633–54.
[25] Deng X, Min Y, Mao M, Liu H, Tu G, Zhang H. Further study on geometric conservation law and application to high-order finite difference schemes with stationary grids. J Comput Phys 2013;239:90–111.
[26] Deng X, Zhu H, Min Y, Liu H, Mao M, Wang G, et al. Symmetric conservative metric method: a link between high order finite-difference and finite-volume schemes for flow computations around complex geometries IC-CFD8-2014-0005; 2014.
[27] Abe Y, Nonomura T, Iizuka N, Fujii K. Geometric interpretations and spatial symmetry property of metrics in the conservative form for high-order finite-difference schemes on moving and deforming grids. J Comput Phys 2014;260:163–203.
[28] Casper J, Shu CW, Atkins H. Comparison of two formulations for high-order accurate essentially nonoscillatory schemes. AIAA J 1994;32:1970–7.
[29] Visbal MR, Gaitonde DV. High-order-accurate methods for complex unsteady subsonic flows. AIAA J 1999;37:1231–9.
[30] Shu CW. High-order finite difference and finite volume WENO schemes and discontinuous Galerkin methods for CFD. Int J Comut Fluid Dyn 2003;17:107–18.
[31] Ekaterinaris JA. High-order accurate, lownumerical diffusion methods for aerodynamics. Prog Aerosp Sci 2005;41:192–300.
[32] Pereira JMC, Kobayashi MH, Pereira JCF. A fourth-order-accurate finite volume compact method for the incompressible Navier-Stokes solutions. J Comput Phys 2001;167:217–43.
[33] Piller M, Stalio E. Compact finite volume schemes on boundary-fitted grids. J Comput Phys 2008;227:4736–62.
[34] Grimm-Strele H, Kupka F, Muthsam H. Curvilinear grids for WENO methods in astrophysical simulations.. Comput Phys Commun 2014;185:764–76.
[35] Deng X, Mao M, Tu G, Liu H, Zhang H. Geometric conservation law and application to high-order finite difference schemes with stationary grids. J Comput Phys 2011;230:1100–15.
[36] Nonomura T, Terakado D, Abe Y, Fujii K. A new technique for freestream preservation of finite-difference WENO on curvilinear grid. Comput Fluids 2015;107:242–55.
[37] Abe Y, Iizuka N, Nonomura T, Fujii K. Conservative metric evaluation for high-order finite difference schemes with the GCL identities on moving and deforming grids. J Comput Phys 2013;232:14–21.
[38] Visbal R, Gaitonde D. On the use of higher-order finite-difference schemes on curvilinear and deforming meshes. J Comput Phys 2002;181:155–85.
[39] Vinokur M, Yee H. Extension of efficient low dissipation high-order schemes for 3D curvilinear moving grids. Front Comput Fluid Dyn 2002;129–64.
[40] Xu D, Deng X, Chen Y, Dong Y, Wang G. On the freestream preservation of finite volume method in curvilinear coordinates. Comput Fluids 2016;129:20–32.
[41] Cai X, Ladeinde F. Performance of WENO scheme in generalized curvilinear coordinate systems. AIAA 2008-36; 2008.
[42] Nonomura T, Iizuka N, Fujii K. Freestream and vortex preservation properties of high-order WENO and WCNS on curvilinear grids. Comput Fluids 2010;39:197–214.
[43] Zhu Y, Sun Z, Ren Y, Hu Y, Zhang S. A numerical strategy for freestream preservation of the high order weighted essentially non-oscillatory schemes on stationary curvilinear grids. J Sci Comput 2017:1–28.
[44] Atkins H, Casper J. Nonreflective boundary conditions for high-order methods. AIAA J 1994;32:512–18.

高精度格式推广/应用技术研究

Extending Weighted Compact Nonlinear Schemes to Complex Grids with Characteristic-Based Interface Conditions

Xiaogang Deng,* Meiliang Mao,† Guohua Tu,‡ Yifeng Zhang,§ and Hanxin Zhang¶
State Key Laboratory of Aerodynamics, China Aerodynamics Research and Development Center, 621000 Mianyang, People's Republic of China

DOI: 10.2514/1.J050285

There are still some challenges, such as grid quality, numerical stability, and boundary schemes, in the practical application of high-order finite difference schemes for complex configurations. This study presents some improved strategies that indicate potential engineering applications of high-order schemes. The formally fifth-order weighted compact nonlinear scheme developed by the authors is implemented on point-matched multiblock structured grids, which are generated over complex configurations to ensure the grid quality of each component block. The information transmission between neighboring blocks is carried out by new characteristic-based interface conditions that directly exchange the spatial derivatives on each side of an interface by means of a characteristic-based projection to keep the high-order accuracy and high resolution of a spatial difference scheme. The high-order scheme combined with the interface conditions is shown to be asymptotically stable. The engineering-oriented applications of the high-order strategy are demonstrated by solving several two- and three-dimensional problems with complex grid systems.

Nomenclature

a	=	speed of sound
c	=	chord length
E	=	nondimensional total energy
F, G, H	=	fluxes in Cartesian coordinates
$\bar{F}, \bar{G}, \bar{H}$	=	fluxes in curvilinear coordinates
h	=	grid size
J	=	Jacobian of coordinate transformation, $\partial(\xi, \eta, \zeta)/\partial(x, y, z)$
L	=	left eigenvector
Ma	=	Mach number
P_{QV_C}	=	transformation matrix, $\partial Q/\partial V_C$
p	=	nondimensional pressure
Q	=	conservative variable vector
q	=	heat transfer rate
Re	=	Reynolds number
RHS	=	right-hand-side terms
T	=	nondimensional temperature
T_∞	=	inflow temperature, K
t	=	nondimensional time
u, v, w	=	nondimensional velocity components
V_C	=	characteristic variable vector
x, y, z	=	Cartesian coordinates
α	=	angle of attack, deg
γ	=	special heat ratio, 1.4
λ	=	eigenvalue
μ	=	nondimensional dynamic viscosity
ξ, η, ζ	=	curvilinear coordinates
$\xi_x, \bar{\xi}_x, \eta_x, \bar{\eta}_x$	=	metrics (grid derivatives), $\bar{\xi}_x = J^{-1}\xi_x, \ldots$
ρ	=	nondimensional density

Subscripts

b	=	interface boundary between two blocks
I	=	inviscid-related quantity
L	=	left
R	=	right
V	=	viscous-related quantity
w	=	wall
0	=	stagnation condition
∞	=	freestream condition

Received 2 November 2009; revision received 27 April 2010; accepted for publication 22 June 2010. Copyright © 2010 by State Key Laboratory of Aerodynamics. Published by the American Institute of Aeronautics and Astronautics, Inc., with permission. Copies of this paper may be made for personal or internal use, on condition that the copier pay the $10.00 per-copy fee to the Copyright Clearance Center, Inc., 222 Rosewood Drive, Danvers, MA 01923; include the code 0001-1452/10 and $10.00 in correspondence with the CCC.
*Professor, Head of Laboratory, P.O. Box 211; xgdeng@skla.cardc.cn.
†Professor, P.O. Box 211; mlmao@skla.cardc.cn.
‡Assistant Researcher, P.O. Box 211; ghtu@skla.cardc.cn.
§Assistant Researcher, P.O. Box 211; yfzhang@skla.cardc.cn.
¶Academician of Chinese Academy of Sciences, National Laboratory for Computational Fluid Dynamics; hxzhang@skla.cardc.cn.

I. Introduction

DESPITE the continuous progress in computational fluid dynamics (CFD) community, challenges still remain in the accurate numerical simulation of a broad spectrum of complex phenomena included in the direct-numerical simulation and large-eddy simulation of turbulence, aeroacoustics, fluid/structure interactions, and electromagnetics [1]. Although low-order schemes are widely used for engineering applications, they are insufficient for many viscosity-dominant domains, such as boundary-layer flows, vortex flows, shock/boundary-layer interactions, heat flux transfers, etc. An effective approach to overcome the obstacle of accurate numerical simulation is to employ high-order methods. A comprehensive review was given by Ekaterinaris [2] for high-order methods. Compact schemes with spectral-like resolution properties are more convenient to use than spectral and pseudospectral schemes and are easier to handle, especially when nontrivial geometries are involved [3]. However, central algorithms are intrinsically nondissipative and cannot prevent odd–even decoupling, which gives rise to high-frequency oscillations even in smooth regions. Reducing or removing such oscillations requires the introduction of dissipation terms. Upwind or upwind-based compact schemes with their dissipative properties are more stable than central schemes. Deng et al. [4] have proposed a type of one-parameter linear dissipative compact schemes, which was derived as a model to damp out the dispersive and parasite errors in the high-wave-number regions. Filters can also be applied to prevent numerical oscillations, such as the one proposed by Visbal and Gaitonde [5]. Some other methods resort to limiters, such as the compact-TVD schemes [6] and the characteristic-based shock-capturing compact scheme [7]. Nevertheless, in the transonic and supersonic flow regions when dealing with flows involving shock waves, one must use a numerical scheme that can both represent small scale structures with the minimum of numerical dissipation and capture discontinuities with the robustness that is common to Godunov-type methods. To achieve these dual objectives, Deng et al. [4] have first developed compact nonlinear schemes with adaptive interpolations that can capture shock waves well [8]. Furthermore, weighted compact nonlinear schemes (WCNSs) have been derived by Deng and Zhang [9] and Deng [10]. The WCNS-E-5 [10], a typical explicit scheme of WCNS, has been successfully applied to a wide range of flow simulations so far to show its flexibility and robustness [11,12]. Nonomura et al. [13] showed that WCNS-E-5 is superior to the fifth-order weighted essentially nonoscillatory (WENO) scheme in freestream and vortex preservation on curvilinear grids. Recently, we proved that WCNSs can ensure the geometric conservation laws, while WENO schemes are difficult [12].

Traditionally, high-order schemes require high-quality grids. In practical flow computations with unsuitable structured grids, numerical instability is frequently observed around singular points with metric discontinuity [14]. However, it is difficult to generate a high-quality structured single-block grid system for a complex configuration. As one approach to solve complex flows, overset grid strategy for high-order compact schemes is used by Visbal and Gaitonde [5], Gaitonde and Visbal [15], Delfs [16], Sherer et al. [17,18], and other researchers. They employed point-to-point overlap grids or generalized overset grids with high-order Lagrange or B-spline interpolation methods. As interpolations in overset grids are generally necessary to allow the grid blocks to communicate with each other, numerical instability and loss of global accuracy will trouble the usage of high-order discretization formulas on overset grids. Furthermore, the grid configurations and numerical procedures for overset grid approach are generally complicated and not easy to implement for three-dimensional complex problems. Point-matched multiblock structured grids, or patched grids for more general means, are adopted alternative approaches for complex geometries. Multiblock structured grid technique makes it possible to run high-order finite difference schemes on each individual block, and the information transmission between neighboring blocks and the propagation throughout the flowfield can be realized by some kinds of interface conditions. Rai [19] made use of a flux interpolation to construct coupled conditions on midnode with Beam–Warming and Osher schemes. Lerat and Wu [20] adopted local flux construction to establish conservative and unconditionally stable interface conditions. Kim and Lee [21] and Sumi et al. [14,22,23] employed characteristic interface conditions (CIC) or generalized characteristic interface conditions (GCIC). CIC/GCIC uses the inviscid characteristic relations straightforwardly derived from the flow transport equations and had been demonstrated excellent performance in practice [14].

However, because it is necessary to convert the governing equations to a characteristic form to compute the characteristic wave amplitude, the application of CIC or GCIC is not convenient. In this paper, we derive a new interface approach that directly exchanges the spatial derivatives (computed on each isolated block individually) on each side of an interface by means of a characteristic-based projection. Then it is no longer necessary to convert the governing equations to characteristic forms or to compute the characteristic wave-amplitude vector or other intermediate variables. The WCNS-E-5 [9–11] coupled with the new interface conditions is applied to solve some benchmark problems with complicated grid systems, such as the NLR7301 two-element airfoil, 30P-30N three-element airfoil (McDonnell Douglas 30P-30N landing configuration), and DLR-F6 (a wing-body civil aircraft configuration). The numerical results demonstrate that the present method can supply very smooth flowfields on complex multiblock structured grids.

The organization of this paper is as follows. Section II introduces the governing equations and the newly developed characteristic-based interface conditions. The high-order WCNS-E-5 scheme is introduced in Sec. III, and the asymptotic stability analysis of WCNS-E-5 with the interface conditions, together with a grid convergence test for the Burgers equation, is also given in Sec. III. In Sec. IV, the whole high-order difference strategy for complex structured grid systems is tested on several two- and three-dimensional benchmark problems. The conclusions are summarized in Sec. V.

II. Governing Equations and Characteristic-Based Interface Conditions

A. Governing Equations

The nondimensional strong conservative Navier–Stokes/Euler equations are selected and converted into curvilinear coordinates by introducing the transformation $(x, y, z) \rightarrow (\xi, \eta, \zeta)$ [17]:

$$\begin{cases} \frac{\partial \hat{Q}}{\partial t} = \text{RHS}_I + s \cdot \text{RHS}_V \\ \text{RHS}_I = -\frac{\partial \hat{F}}{\partial \xi} - \frac{\partial \hat{G}}{\partial \eta} - \frac{\partial \hat{H}}{\partial \zeta} \\ \text{RHS}_V = \frac{1}{Re}\left(\frac{\partial \hat{F}_V}{\partial \xi} + \frac{\partial \hat{G}_V}{\partial \eta} + \frac{\partial \hat{H}_V}{\partial \zeta}\right) \end{cases} \quad (1)$$

where $s = 1$ for Navier–Stokes equations, and $s = 0$ for Euler equations. $\hat{Q} = Q/J$, and $Q = [\rho, \rho u, \rho v, \rho w, \rho E]^T$ is the conservative variable vector. Here, $J = \partial(\xi, \eta, \zeta)/\partial(x, y, z)$ is the Jacobian of coordinate transformation.

B. Characteristic-Based Interface Conditions

Characteristic theory has been used for the development of boundary conditions for many years. Barth [24] proposed a characteristic projection that can be conveniently applied to flow variables when constructing boundary fluxes. Thompson [25] and Poinsot and Lele [26] proposed characteristic boundary conditions for Euler and Navier–Stokes systems, respectively. Okong'o and Bellan [27] devised characteristic boundary conditions for multicomponent real-gas flows by following the local one-dimensional inviscid (LODI) relations of [26]. Kim and Lee [28] devised generalized characteristic boundary conditions in which the transverse and viscous terms are treated as source terms in addition to the LODI relations. The basic idea behind the characteristic boundary conditions of [26–28] is to split the convective terms in the boundary-normal direction into several waves with different characteristic velocities and then express unknown incoming waves as a function of known outgoing waves. These excellent boundary treatments are frequently applied to many areas, such as large-eddy simulations, direct numerical simulations, and computational aeroacoustics. Those studies enlightened us to devise the following characteristic-based interface conditions that satisfy the discrete governing equations.

To advance the characteristic analysis, the following transformation matrices are defined in terms of conservative variables Q and characteristic variables V_C:

$$P_{QV_C} = \frac{\partial Q}{\partial V_c} \qquad \delta Q = \frac{\partial Q}{\partial V_C}\delta V_c \quad (2)$$

Hereafter, only the ξ direction is taken into account without loss of generality. The P_{QV_C} and $P_{QV_C}^{-1}$ can be found in many fluid studies, such as [28].

Equation (1) can be rearranged as

$$\frac{J^{-1}\partial Q}{\partial t} + J^{-1}\xi_x \frac{\partial F}{\partial \xi} + J^{-1}\xi_y \frac{\partial G}{\partial \xi} + J^{-1}\xi_z \frac{\partial H}{\partial \xi} = S_c \quad (3)$$

where

$$S_c = s \cdot \text{RHS}_V - \left[F\frac{\partial J^{-1}\xi_x}{\partial \xi} + G\frac{\partial J^{-1}\xi_y}{\partial \xi} + H\frac{\partial J^{-1}\xi_z}{\partial \xi} \right.$$
$$\left. + \frac{\partial J^{-1}\xi_t Q}{\partial \xi} + \frac{\partial \hat{G}}{\partial \eta} + \frac{\partial \hat{H}}{\partial \zeta} \right] \quad (4)$$

Let

$$A = \xi_x \frac{\partial F}{\partial Q} + \xi_y \frac{\partial G}{\partial Q} + \xi_z \frac{\partial H}{\partial Q} \quad (5)$$

Then P_{QV_c} is the right characteristic matrix of A, and $P_{QV_c}^{-1} A P_{QV_c} = \Lambda$. Here,

$$\Lambda = \mathrm{diag}(\lambda_1, \lambda_2, \lambda_3, \lambda_4, \lambda_5) = \mathrm{diag}(\hat{u}, \hat{u}, \hat{u}, \hat{u} + \hat{a}, \hat{u} - \hat{a})$$

$\hat{u} = \xi_x u + \xi_y v + \xi_z w$, and $\hat{a} = a\sqrt{\xi_x^2 + \xi_y^2 + \xi_z^2}$.

Let $\delta V_c = P_{QV_c}^{-1} \delta Q$; the last three terms on the left-hand side of Eq. (3) can be rewritten as

$$J^{-1}\xi_x \frac{\partial F}{\partial \xi} + J^{-1}\xi_y \frac{\partial G}{\partial \xi} + J^{-1}\xi_z \frac{\partial H}{\partial \xi} = J^{-1}A\frac{\partial Q}{\partial \xi} = J^{-1}A\frac{\partial Q}{\partial V_c}\frac{\partial V_c}{\partial \xi}$$
$$= J^{-1}AP_{QV_c}\frac{\partial V_c}{\partial \xi} = J^{-1}P_{QV_c}P_{QV_c}^{-1}AP_{QV_c}\frac{\partial V_c}{\partial \xi} = J^{-1}P_{QV_c}\Lambda\frac{\partial V_c}{\partial \xi} \quad (6)$$

Inserting Eq. (6) into Eq. (3), we get

$$\frac{\partial Q}{\partial t} = JS_c - P_{QV_c}\Lambda\frac{\partial V_c}{\partial \xi} \quad (7)$$

Assume that the interface shown in Fig. 1 is along the ξ direction with value ξ_b. Because the conservative variables and their time derivatives at the upstream limit and those at the downstream limit on the block interface are strictly matched, the following physical conditions are correct naturally:

$$Q(t, \xi_b, \eta, \zeta)|_L = Q(t, \xi_b, \eta, \zeta)|_R,$$
$$\left.\frac{\partial Q(t, \xi_b, \eta, \zeta)}{\partial t}\right|_L = \left.\frac{\partial Q(t, \xi_b, \eta, \zeta)}{\partial t}\right|_R \quad (8)$$

Then

$$\left(JS_c - P_{QV_c}\Lambda\frac{\partial V_c}{\partial \xi}\right)\bigg|_L = \left(JS_c - P_{QV_c}\Lambda\frac{\partial V_c}{\partial \xi}\right)\bigg|_R \quad (9)$$

Let $P_{QV_c}^{-1} = [L_1 \ L_2 \ L_3 \ L_4 \ L_5]^T$. Equation (9) is left-multiplied by $L_i|_L$ ($i = 1, 2, \dots, 5$):

$$\left[L_i JS_c - \left(\lambda_i \frac{\partial V_c}{\partial \xi}\right)\right]\bigg|_L = L_i|_L \left[JS_c - \left(P_{QV_c}\Lambda\frac{\partial V_c}{\partial \xi}\right)\right]\bigg|_R$$
$$\rightarrow \left(\lambda_i \frac{\partial V_c}{\partial \xi}\right)\bigg|_L = L_i|_L \left(P_{QV_c}\Lambda\frac{\partial V_c}{\partial \xi}\right)\bigg|_R + L_i|_L[(JS_c)|_L - (JS_c)|_R]$$
(10)

Similarly,

Fig. 2 Sketch of the CBIC.

$$\left(\lambda_i \frac{\partial V_c}{\partial \xi}\right)\bigg|_R = L_i|_R \left(P_{QV_c}\Lambda\frac{\partial V_c}{\partial \xi}\right)\bigg|_L + L_i|_R[(JS_c)|_R - (JS_c)|_L] \quad (11)$$

For the block on the left side of the interface, from Eq. (7), one can get

$$L_i|_L \frac{\partial Q}{\partial t}\bigg|_L = \left[JL_i S_c - L_i P_{QV_c}\Lambda\frac{\partial V_c}{\partial \xi}\right]\bigg|_L = \left[JL_i S_c - \lambda_i \frac{\partial V_c}{\partial \xi}\right]\bigg|_L \quad (12)$$

Let a positive eigenvalue denote that the characteristic wave propagates from the left to the right. For positive eigenvalues,

$$L_i|_L \frac{\partial Q}{\partial t}\bigg|_L = \frac{1 + \mathrm{sign}(\lambda_i)}{2}\left[L_i JS_c - \lambda_i \frac{\partial V_c}{\partial \xi}\right]\bigg|_L \quad (13)$$

For negative eigenvalues,

$$L_i|_L \frac{\partial Q}{\partial t}\bigg|_L = \frac{1 - \mathrm{sign}(\lambda_i)}{2}\left[L_i JS_c - \lambda_i \frac{\partial V_c}{\partial \xi}\right]\bigg|_L \quad (14)$$

The negative eigenvalues indicate that the characteristic information shall be calculated in according to the right-hand flowfield of the interface. Substitute Eq. (10) into Eq. (14):

$$L_i|_L \frac{\partial Q}{\partial t}\bigg|_L = \frac{1 - \mathrm{sign}(\lambda_i)}{2}(L_i JS_c)|_L$$
$$- \frac{1 - \mathrm{sign}(\lambda_i)}{2}\left[L_i|_L\left(P_{QV_c}\Lambda\frac{\partial V_c}{\partial \xi}\right)\bigg|_R + L_i|_L[(JS_c)|_L - (JS_c)|_R]\right]$$
$$\rightarrow L_i|_L \frac{\partial Q}{\partial t}\bigg|_L = \frac{1 - \mathrm{sign}(\lambda_i)}{2}L_i|_L\left[(JS_c)|_R - \left(P_{QV_c}\Lambda\frac{\partial V_c}{\partial \xi}\right)\bigg|_R\right] \quad (15)$$

Combine Eq. (13) and (15) together:

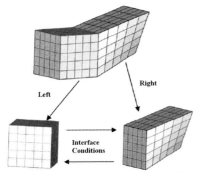

Fig. 1 Two point-matched blocks with interface conditions.

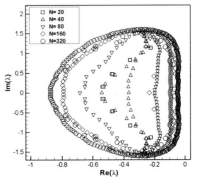

Fig. 3 Eigenvalue spectra of the WCNS-E-5 with the CBIC, $m = N/2$.

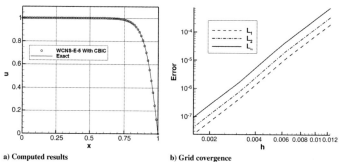

a) Computed results   b) Grid covergence

Fig. 4 Results for Eq. (34).

$$P_{QV_c}^{-1}|_L \frac{\partial Q}{\partial t}\bigg|_L = \mathrm{diag}\left(\frac{1+\mathrm{sign}(\lambda_i)}{2}\right) P_{QV_c}^{-1}|_L \left[JS_c - P_{QV_c}\left(\Lambda\frac{\partial V_c}{\partial \xi}\right)\right]\bigg|_L$$

$$+ \mathrm{diag}\left(\frac{1-\mathrm{sign}(\lambda_i)}{2}\right) P_{QV_c}^{-1}|_L \left[(JS_c)|_R - P_{QV_c}|_R\left(\Lambda\frac{\partial V_c}{\partial \xi}\right)\right]\bigg|_R$$

$$\rightarrow \frac{\partial Q}{\partial t}\bigg|_L = P_{QV_c}|_L \mathrm{diag}\left(\frac{1+\mathrm{sign}(\lambda_i)}{2}\right) P_{QV_c}^{-1}|_L \left[JS_c - P_{QV_c}\Lambda\frac{\partial V_c}{\partial \xi}\right]\bigg|_L$$

$$+ P_{QV_c}|_L \mathrm{diag}\left(\frac{1-\mathrm{sign}(\lambda_i)}{2}\right) P_{QV_c}^{-1}|_L \left[JS_c - P_{QV_c}\Lambda\frac{\partial V_c}{\partial \xi}\right]\bigg|_R \quad (16)$$

Define

$$A_s^- = P_{QV_c}\mathrm{diag}\left(\frac{1-\mathrm{sign}(\lambda_i)}{2}\right) P_{QV_c}^{-1}$$

$$A_s^+ = P_{QV_c}\mathrm{diag}\left(\frac{1+\mathrm{sign}(\lambda_i)}{2}\right) P_{QV_c}^{-1} \quad (17)$$

Then $A_s^+ + A_s^- = I$, and Eq. (16) can be expressed as

$$\frac{\partial Q}{\partial t}\bigg|_L = (A_s^+)|_L \left[JS_c - P_{QV_c}\Lambda\frac{\partial V_c}{\partial \xi}\right]\bigg|_L$$

$$+ (A_s^-)|_L \left[JS_c - P_{QV_c}\Lambda\frac{\partial V_c}{\partial \xi}\right]\bigg|_R \quad (18)$$

From Eqs. (4) and (6), one can get

$$S_c - J^{-1}P_{QV_c}\Lambda\frac{\partial V_c}{\partial \xi} = -\left[F\frac{\partial J^{-1}\xi_x}{\partial \xi} + G\frac{\partial J^{-1}\xi_y}{\partial \xi}\right.$$

$$+ H\frac{\partial J^{-1}\xi_z}{\partial \xi} + \frac{\partial J^{-1}\xi_t Q}{\partial \eta} + \frac{\partial \bar{G}}{\partial \eta} + \frac{\partial \bar{H}}{\partial \zeta}\right]$$

$$- \left(J^{-1}\xi_x\frac{\partial F}{\partial \xi} + J^{-1}\xi_y\frac{\partial G}{\partial \xi} + J^{-1}\xi_z\frac{\partial H}{\partial \xi}\right) + s\cdot\mathrm{RHS}_V$$

$$= -\left(\frac{\partial F}{\partial \xi} + \frac{\partial G}{\partial \eta} + \frac{\partial H}{\partial \zeta}\right) + s\cdot\mathrm{RHS}_V$$

$$= \mathrm{RHS}_I + s\cdot\mathrm{RHS}_V \quad (19)$$

Let

$$\mathrm{RHS} = J\left(s\cdot\mathrm{RHS}_V + \mathrm{RHS}_I\right) \quad (20)$$

Substituting Eqs. (19) and (20) into Eq. (18), one can get the following characteristic-based interface conditions (CBICs):

$$\frac{\partial Q}{\partial t}\bigg|_L = \mathrm{RHS}^{\mathrm{new}} = (A_s^+)|_L (\mathrm{RHS})|_L + (A_s^-)|_L (\mathrm{RHS})|_R \quad (21)$$

Similarly,

$$\frac{\partial Q}{\partial t}\bigg|_R = \mathrm{RHS}^{\mathrm{new}} = (A_s^+)|_R (\mathrm{RHS})|_R + (A_s^-)|_R (\mathrm{RHS})|_L \quad (22)$$

The interface conditions for the viscous terms are simple, as follows:

$$\mathrm{RHS}_V^* = \tfrac{1}{2}(\mathrm{RHS}_V|_L + \mathrm{RHS}_V|_R) \quad (23)$$

After applying Eq. (23), the algorithm for the viscous terms becomes symmetrical.

At the end, the time integration can be easily carried out for the points on the interface by the same producer as those for the inner points. The CBICs are more convenient to use than the original CIC/GCIC [21–23], as there is no longer a need to convert the governing equations to characteristic forms or to compute characteristic wave-amplitude vectors or other intermediate variables.

To eliminate numerical round-off errors, in addition to the CBIC, a simple averaging procedure is employed for the points on an interface:

$$\frac{\partial Q^*}{\partial t} = \frac{1}{2}\left(\frac{\partial Q}{\partial t}\bigg|_L + \frac{\partial Q}{\partial t}\bigg|_R\right) \quad (24)$$

$$Q^* = \frac{(Q_L + Q_R)}{2} \quad (25)$$

Table 1 Accuracy order of WCNS-E-5 with CBIC for Burgers equation

Grid size	L_1 error	Accuracy order	L_2 error	Accuracy order	L_∞ error	Accuracy order
0.0125000	1.92E − 04	———	3.45E − 04	———	7.56E − 04	———
0.0062500	1.11E − 05	4.11	2.00E − 05	4.11	4.42E − 05	4.09
0.0031250	4.29E − 07	4.69	7.75E − 07	4.69	1.71E − 06	4.69
0.0015625	2.32E − 08	4.21	4.19E − 08	4.21	9.28E − 08	4.21

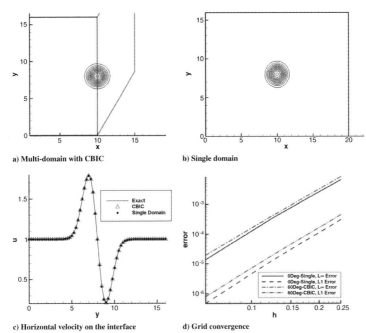

Fig. 5   Performance of the CBIC on meshes with an abrupt change of slope at the interface.

III. Implementation of CBIC for WCNS-E-5 Scheme on Multiblock Grids

A. WCNS-E-5 Scheme

The WCNS-E-5 scheme [9,10] is a formally fifth-order nonlinear scheme for inviscid terms. Let us first consider the discretization of the inviscid flux derivative along the ξ direction. The discretization for other inviscid flux derivatives can be computed by similar procedures.

The WCNS-E-5 scheme can be expressed as

$$\frac{\partial \hat{F}_i}{\partial \xi} = \frac{75}{64h}(\hat{F}_{i+1/2} - \hat{F}_{i-1/2}) - \frac{25}{384h}(\hat{F}_{i+3/2} - \hat{F}_{i-3/2}) + \frac{3}{640h}(\hat{F}_{i+5/2} - \hat{F}_{i-5/2}) \quad (26)$$

where h is the grid size, and

$$\hat{F}_{i+1/2} = \hat{F}(U^L_{i+1/2}, U^R_{i+1/2}, \bar{\xi}_{x,i+1/2}, \bar{\xi}_{y,i+1/2}, \bar{\xi}_{z,i+1/2}) \quad (27)$$

Here, Eq. (27) is computed by Steger–Warming flux-splitting method in the present work. WCNSs can be implemented on various types of variables [9], and U are adopted here to denote conservative variables, primitive variables, or characteristic variables. In this paper, we chose primitive variables. Then $U^L_{i+1/2}$ and $U^R_{i+1/2}$ are the left and right primitive variables on cell edges. The high-order nonlinear weighted interpolation derived by Deng and Zhang [9] is applied here. The idea is that each of the chosen stencils is assigned a weight factor, which determines its contribution to the final approximation of the midnode value. The weights are designed in such a way that in the smooth region they approach the optimal weights to achieve fifth-order accuracy, whereas in the regions near the discontinuities, the weight of the stencil, which contains the discontinuities, is assigned nearly zero. Therefore, the weighted interpolations can prevent numerical oscillations around discontinuities.

Nonomura et al. [13] have given some detailed study on the freestream and vortex preservation properties of WCNS and WENO. It is worth noting that the main differences between WCNS and WENO are follows:

For WCNS, high-order nonlinear *interpolations* are adopted to get the cell-edge variables $U^{WCNS,L/R}_{j+1/2}$:

$$\begin{cases} U^{WCNS,L}_{i+1/2} = \mathcal{W}^{WCNS,L}(U_{i-r+1},\ldots,U_{i+r}) + \mathcal{O}(h^{2r-1}) \\ U^{WCNS,R}_{i+1/2} = \mathcal{W}^{WCNS,R}(U_{i-r+2},\ldots,U_{i+r}) + \mathcal{O}(h^{2r-1}) \\ \hat{F}^{WCNS}_{i+1/2} = \hat{F}(U^{WCNS,L}_{i+1/2}, U^{WCNS,R}_{i+1/2}, \bar{\xi}_{x,i+1/2}, \bar{\xi}_{y,i+1/2}, \bar{\xi}_{z,i+1/2}) \\ \kappa \frac{\partial \hat{F}_{j-1}}{\partial \xi} + \frac{\partial \hat{F}_j}{\partial \xi} + \kappa \frac{\partial \hat{F}_{j+1}}{\partial \xi} = \frac{1}{h}\sum_k \alpha_k \hat{F}^{WCNS}_{i+1/2+k} \end{cases} \quad (28)$$

where $\mathcal{W}^{WCNS,L/R}$ denotes the weighted interpolating procedure, as presented in [9], and κ and α_k are coefficients. When $\kappa = 0$,

Table 2 Accuracy order of WCNS-E-5 with CBIC for skewed grids and smooth grids

	60-deg-skewness multidomain with CBIC				Single domain (without skewness)			
h	L_∞	Accuracy order	L_1	Accuracy order	L_∞	Accuracy order	L_1	Accuracy order
0.25	8.57E−3	——	4.50E−4	——	6.66E−3	——	3.07E−4	——
0.125	4.33E−4	4.31	2.00E−5	4.49	3.60E−4	4.21	1.33E−5	4.28
0.0625	1.93E−5	4.49	7.93E−7	4.66	1.38E−5	4.70	4.74E−7	4.79

$\alpha_0 = -\alpha_{-1} = 75/64$, $\alpha_1 = -\alpha_{-2} = -25/384$, $\alpha_2 = -\alpha_{-3} = 3/640$, and the last equation in Eq. (28) is equal to Eq. (26).

The nonlinear interpolations based on primitive variables are

$$U_{i+1/2}^L = U_i + \frac{1}{2} h f_{Li}^* + \frac{1}{8} h^2 s_{Li}^*$$

$$U_{i-1/2}^R = U_i - \frac{1}{2} h f_{Ri}^* + \frac{1}{8} h^2 s_{Ri}^* \qquad f_{Li}^* = \sum_{k=1}^{3} \omega_{Lk} f_i^k$$

$$s_{Li}^* = \sum_{k=1}^{3} \omega_{Lk} s_i^k \qquad f_{Ri}^* = \sum_{k=1}^{3} \omega_{Rk} f_i^k \qquad s_{Ri}^* = \sum_{k=1}^{3} \omega_{Rk} s_i^k$$

$$f_i^1 = \frac{1}{2h}[U_{i-2} - 4U_{i-1} + 3U_i] \qquad s_i^1 = \frac{1}{h^2}[U_{i-2} - 2U_{i-1} + U_i]$$

$$f_i^2 = \frac{1}{2h}[(U)_{i+1} - (U)_{i-1}] \qquad s_i^2 = \frac{1}{h^2}[U_{i-1} - 2U_i + U_{i+1}]$$

$$f_i^3 = \frac{1}{2h}[-3U_i + 4U_{i+1} - U_{i+2}] \qquad s_i^3 = \frac{1}{h^2}[U_i - 2U_{i+1} + U_{i+2}]$$

$$\omega_{Lk} = \frac{\beta_{Lk}}{\sum_{m=1}^{3} \beta_{Lm}} \qquad \omega_{Rk} = \frac{\beta_{Rk}}{\sum_{m=1}^{3} \beta_{Rm}} \qquad \beta_{Lk} = \frac{C_{Lk}}{(\varepsilon + IS_k)^2}$$

$$\beta_{Rk} = \frac{C_{Rk}}{(\varepsilon + IS_k)^2} \qquad IS_k = (hf_i^k)^2 + (h^2 s_i^k)^2$$

$$C_{L1} = C_{R3} = \frac{1}{16} \qquad C_{L2} = C_{R2} = \frac{10}{16} \qquad C_{L3} = C_{R1} = \frac{5}{16}$$

where $C_{L1} \sim C_{L3}$ and $C_{R1} \sim C_{R3}$ are the optimized coefficients to ensure fifth-order interpolations as follows:

$$\begin{cases} U_{i+1/2}^{L,op} = \frac{1}{128}(3U_{i-2} - 20U_{i-1} + 90U_i + 60U_{i+1} - 5U_{i+2}) + \mathcal{O}(h^5) \\ U_{i+1/2}^{R,op} = \frac{1}{128}(-5U_{i-1} + 60U_i + 90U_{i+1} - 20U_{i+2} + 3U_{i+3}) + \mathcal{O}(h^5) \end{cases} \tag{29}$$

For WENO, nonlinear *reconstructions* of fluxes are adopted to get $\tilde{F}_{j+1/2}^{\text{WENO}}$:

$$\begin{cases} \tilde{F}_{i+1/2}^{\text{WENO},+} = \mathcal{W}^{\text{WENO},+}(\hat{F}_{i-r+1},\dots,\hat{F}_{i+r-1}) \\ \tilde{F}_{i+1/2}^{\text{WENO},-} = \mathcal{W}^{\text{WENO},-}(\hat{F}_{i-r+2},\dots,\hat{F}_{i+r}) \\ \tilde{F}_{i+1/2}^{\text{WENO}} = \tilde{F}_{i+1/2}^{\text{WENO},+} + \tilde{F}_{i+1/2}^{\text{WENO},-} \\ \frac{\partial F^{\text{WENO}}}{\partial \xi} = \frac{1}{h}(\tilde{F}_{i+1/2}^{\text{WENO}} - \tilde{F}_{i-1/2}^{\text{WENO}}) + \mathcal{O}(h^{2r-1}) \end{cases} \tag{30}$$

where $\mathcal{W}^{\text{WENO},\pm}$ denotes the weighted reconstructing procedure, as presented in [29].

The procedures of the CBICs for WCNS-E-5 are as follows:

1) Get the RHS in each individual block. For example, the RHS_I in Eq. (20) can be calculated by WCNS-E-5. If the governing equations contain viscous terms, RHS_V are also computed, and Eq. (23) is applied to the RHS_V on interfaces.

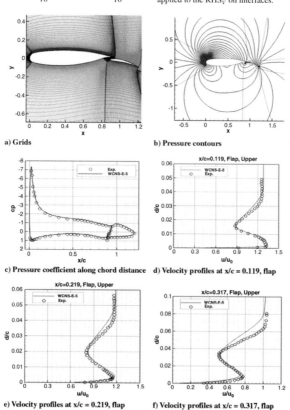

Fig. 6 Grids and results of NLR7301 two-element airfoil. (colored picture attached at the end of the book)

2) Modify the RHS of an interface by Eq. (21) for the left-hand-side block or by Eq. (22) for the right-hand-side block.
3) Eliminate round-off errors by Eq. (24).
4) Obtain Q by a time-integration method such as the Runge–Kutta method.
5) If Q belong to an interface, they are modified by Eq. (25) to eliminate round-off errors.

For the viscous terms, the fourth-order central explicit scheme [10] is adopted here, and the three-step Runge–Kutta method or the lower/upper symmetric Gauss–Seidel method is adopted for temporal integration. In addition, the geometric conservation law is very important for high-order difference schemes, even when the grid is stationary [12]. A conservation metric method that can ensure the discrete geometric conservation law for any high-order-accuracy schemes is given in [12]. This method is employed in this paper to compute the metrics such as $\tilde{\xi}_x$.

B. Asymptotic Stability of the WCNS-E-5 Scheme with the CBIC

Consider the following linear hyperbolic system:

$$\frac{\partial u}{\partial t} + a \frac{\partial u}{\partial x} = 0 \quad (31)$$

where $a > 0$ is a constant speed. The domain is uniformity discretized into N nodes with grid size h. Supposing that the field is decomposed into two parts at point m (see Fig. 2), connecting by the CBIC, the semidiscretized form of Eq. (31) can be written as

$$\frac{d\mathbf{u}}{dt} = \frac{a}{h}\mathbf{D}\mathbf{u} + \mathbf{g}(t) \quad (32)$$

where $\mathbf{u} = [u_1, \ldots, u_m, u_{m+1}, \ldots, u_N]^T$, $\mathbf{g}(t)$ corresponds to the initial and boundary conditions (without losing generality, we set $\mathbf{g}(t) = 0$), and $\mathbf{D}$ is the difference matrix. Here, the CBICs are

$$\frac{du_m}{dt} = \text{RHS}_L = -c \frac{\partial u_m}{\partial x}\Big|_L$$

The CBICs are contained in $\mathbf{D}$ at the connecting point m as follows:

$$\mathbf{D} = \{d_{ij}\} = \begin{cases} d_{i,j}^L & \text{if } 1 \leq i \leq m \text{ and } 1 \leq j \leq m \\ d_{i-m,j-m}^R & \text{if } m+1 \leq i \leq N \text{ and } m \leq j \leq N \\ 0 & \text{else} \end{cases} \quad (33)$$

where $d_{i,j}^L$ and $d_{i,j}^R$ ($i = 1, 2, \ldots$) are the components of $\mathbf{D}_L$ and $\mathbf{D}_R$, respectively. $\mathbf{D}_L = \mathbf{A}_L \mathbf{B}_L$, and $\mathbf{D}_R = \mathbf{A}_R \mathbf{B}_R$. Here, $\mathbf{A}_L$, $\mathbf{A}_R$, $\mathbf{B}_L$, and $\mathbf{B}_R$ are specified by the optimized WCNS-E-5 scheme and the boundary schemes. For example,

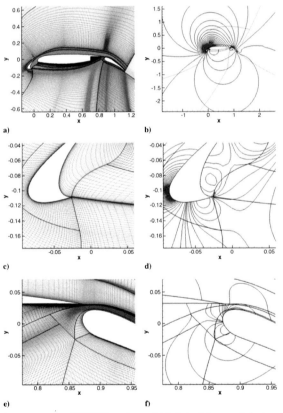

Fig. 7 **Grids and pressure contours of 30P-30N three-element airfoil.** (colored picture attached at the end of the book)

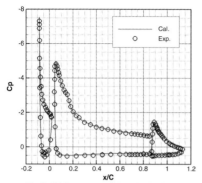

Fig. 8 Wall pressure coefficient distributions.

$$A_L = -\frac{1}{24}\begin{bmatrix} -23 & 21 & 3 & -1 & & & & & \\ & -22 & 17 & 9 & -5 & 1 & & & \\ & 1 & -27 & 27 & -1 & & & & \\ & -\frac{9}{80} & \frac{25}{16} & -\frac{225}{8} & \frac{225}{8} & -\frac{25}{16} & \frac{9}{80} & & \\ & & \ddots & \ddots & \ddots & \ddots & \ddots & \ddots & \\ & & & -\frac{9}{80} & \frac{25}{16} & -\frac{225}{8} & \frac{225}{8} & -\frac{25}{16} & \frac{9}{80} \\ & & & & 1 & -27 & 27 & -1 & \\ & & & & & -1 & 5 & -9 & -17 & 22 \\ & & & & & & 1 & -3 & -21 & 23 \end{bmatrix}_{(0 \sim m) \times (0 \sim m+1)}$$

$$B_L = \frac{1}{16}\begin{bmatrix} 35 & -35 & 21 & -5 & & & & \\ 5 & 15 & -5 & 1 & & & & \\ -1 & 9 & 9 & -1 & & & & \\ \frac{3}{8} & -\frac{20}{8} & \frac{90}{8} & \frac{60}{8} & -\frac{5}{8} & & & \\ & \ddots & \ddots & \ddots & \ddots & \ddots & & \\ & & \frac{3}{8} & -\frac{20}{8} & \frac{90}{8} & \frac{60}{8} & -\frac{5}{8} & \\ & & -1 & 9 & 9 & -1 & \\ & & & 1 & -5 & 15 & 5 \\ & & & -5 & 21 & -35 & 35 \end{bmatrix}_{(0 \sim m+1) \times (0 \sim m)}$$

The asymptotic stability condition for the semidiscretized equation is that all eigenvalues of the matrix **D** have no positive real parts. Figure 3 shows the eigenvalue spectra of **D**. It can be seen that the WCNS-E-5 scheme with the CBIC is asymptotically stable.

C. Accuracy Analysis in Solving the Burgers Equation

To show the accuracy of the present method for nonlinear viscous problems, the following Burgers equation is chosen as a model problem:

$$\frac{\partial u}{\partial t} + \frac{\partial f}{\partial x} = \mu \frac{\partial^2 u}{\partial x^2} \qquad (34)$$

where $f = u^2/2$, $\mu = 0.05$, and $x \in [0, 1]$. The boundary conditions is $u(t, 0) = 1$ and $u(t, 1) = 0$. The analytical solution for this problem is

$$u(x) = \bar{u}\frac{1 - \exp[(x-1)\bar{u}/\mu]}{1 + \exp[(x-1)\bar{u}/\mu]}, \qquad \frac{\bar{u} - 1}{\bar{u} + 1} = \exp[-\bar{u}/\mu] \qquad (35)$$

The computed domain is split into two sections. The left section is $x \in [0, 0.75]$, and the right section is $x \in [0.75, 1]$. The point at $x = 0.75$ is collected by the CBIC. Figure 4a shows the result of $h = 1/80$. The grid convergence result is shown in Fig. 4b. The L_1,

L_2, and L_∞ errors and the corresponding orders of accuracy are shown in Table 1. It is obvious that the present method is fourth-order-accurate.

IV. Applications and Discussions

To validate the CBIC and to show the potential ability of high-order scheme in solving engineering problems, the following two- and three-dimensional benchmark cases are simulated. These are a vortex convection problem, subsonic flows over NLR7301 two-element airfoil, subsonic flows over 30N-30P three-element airfoil, and transonic flows over DLR-F6.

A. Vortex Convection Across an Interface

This case is chosen to validate the CBIC. The governing equations are the nondimensional Euler equations. The flow is set to be two-dimensional with $(u, v, p, T) = (1.0, 0, 1, 1)$. The initial condition is imposed by prescribing a vortex, centered at the location (x_c, y_c), and satisfying the following relations:

$$\begin{cases} \delta u = \frac{\varepsilon}{2\pi} e^{(1-r^2)/2}(y_c - y), & \delta v = \frac{\varepsilon}{2\pi} e^{(1-r^2)/2}(x - x_c) \\ \delta T = -\frac{(\gamma-1)\varepsilon^2}{8\gamma\pi^2} e^{1-r^2}, & \delta S = 0 \\ r^2 = (x - x_c)^2 + (y - y_c)^2 \end{cases} \qquad (36)$$

where S is the entropy, and e is the natural number. The computational domains are shown in Fig. 5a. An interface is imposed at $x = 10$ with an abrupt change of the slope on its two sides. The left-side domain is $x \in [0, 10]$ and $y \in [0, 16]$ with the 41×65 orthogonal mesh, while the right-side domain is 60 deg away from orthogonality with 41×65 points.

Starting from $t = 0$ with initial position $x_c = 5$ and $y_c = 8$, the computation is carried out until the vortex core arrived at the interface ($t = 5$). Figures 5a and 5b show the pressure contours of the multidomain field and a single-domain field, respectively. The single domain is a uniform 81×65 grid without the interface or the abrupt change of slope. The distribution of u along the interface is shown in Fig. 5c. It is obvious that, due to the effectiveness of the CBIC, excellent agreement between the result computed on the skewed grid and the result computed on the single domain is obtained, and both results are very close to the theoretical distribution. The grid convergence for L_∞ errors and L_1 errors of ρ are shown in Fig. 5d. The errors as well as the order of accuracy are listed in Table 2. Because one-side schemes are necessary for the RHS of the interface,

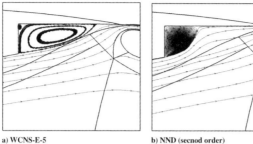

a) WCNS-E-5   b) NND (secnod order)

Fig. 9 Streamlines.

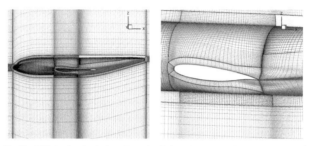

Fig. 10 Grids for the wing-body configuration. (colored picture attached at the end of the book)

the errors of the multiblock grids (with an abrupt skewness) are greater than that of the single-block grids (without skewness). However, it can be found that the CBIC can preserve the formal accuracy of WCNS-E-5.

B. NLR7301 Two-Element Airfoil

The Reynolds-averaged Navier–Stokes equations are used as the governing equations. The Spalart–Allmaras one-equation turbulence model [30] is applied hereafter without the trip function of f_{t1} and f_{t2}. In this paper, the turbulence model is discretized by the second-order nonoscillatory and non-free-parameter dissipative (NND) scheme [31] for simplicity.

Comprehensive experimental results about the airfoil can be found in [32]. Case 2 of [32] is chosen in this paper. The incoming flow conditions are as follows. Mach number is 0.185, Reynolds number is 2.51×10^6, angle of attack is 6.0 deg, flap deflection angle is 20.0 deg, gap width is 1.3%, and overlap width is 5.3%.

The grid system is shown in Fig. 6a, and the computed pressure contours are shown in Fig. 6b. It can be seen that the result is very smooth without any spurious oscillations, despite there being many interfaces. Figure 6c shows the pressure coefficient along the nondimensional chord distance. Good agreement between the WCNS-E-5 result and the experimental result is acquired. Figures 6d–6f show the boundary velocity profiles at three different chord locations, where c denotes the chord length of the main foil or the flap, and x denotes the chord distance between a measured position and the forefront point of an element foil. Figures 6d–6f indicate qualitative agreements between the WCNS-E-5 results and the experimental results. The experimental lift coefficient and drag coefficient are 2.366 and 0.0225, respectively. The computed results are 2.461 and 0.0276, correspondingly. The difference between the experimental results and the computational results is about 4% for lift coefficient and 23% for drag coefficient. The main purpose of this paper is to introduce the CBIC as well as to show the ability of the WCNS-E-5 in solving complex configuration problems. The

discrepancies between the numerical results and the experimental results indicate that, in addition to high-order schemes, many other aspects such as far-field boundary, wall boundary, Euler flux, and transition still remain to affect the accuracy of CFD.

C. 30P-30N Three-Element Airfoil

This is a classic case to check the stability of a scheme, because the flow is complex with many shear layers. The flow is supposed to be full turbulence, and the inflow conditions are $Ma = 0.2$, $Re = 9 \times 10^6$, and $T_\infty = 300$ K.

Figure 7 contains the grids and pressure contours. Some enlarged parts near the complex connecting points are also shown in Figs. 7c–7f. For quantitative validation of the accuracy of the present high-order strategy, the pressure coefficient distributions on the wall are given in Fig. 8. The computational results are in good agreement with the experimental results. The streamlines near the main foil tail are shown in Fig. 9: Fig. 9a shows WCNS-E-5, and Fig. 9b shows the second-order NND scheme [31]. The WCNS-E-5 scheme successfully resolves the small secondary vortex in the corner of the main foil, while the second-order scheme fails.

D. DLR-F6 with FX2B Wing-Body Fairing

From 2001 to 2009, the AIAA Applied Aerodynamics Technical Committee sponsored four Drag Prediction Workshops (DPW), with the aim of assessing the state-of-the-art current CFD solvers at predicting absolute and incremental drag changes on generic transonic transport aircraft configurations.**

We conduct the present study in according to DPW-III. The DLR-F6 wing-body configuration with the FX2B fairing is chosen to show the potential applications of the WCNS-E-5 scheme for general aircraft configurations. The flow conditions are $Ma = 0.75$,

---

**Data available online at http://aaac.larc.nasa.gov/tsab/cfdlarc/aiaa-dpw [retrieved 20 September 2010].

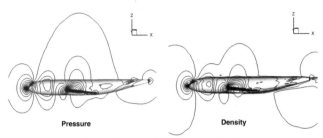

Fig. 11 Pressure and density contours ($\alpha = 0°$).

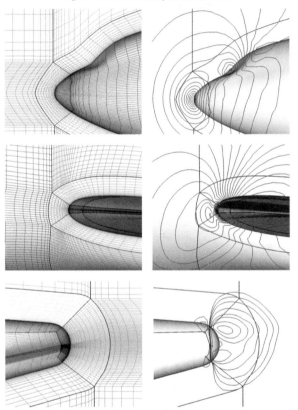

Fig. 12 Grids and pressure contours at the nose (top), the wing-body junction (middle), and the tail (bottom). $\alpha = 0°$.
(colored picture attached at the end of the book)

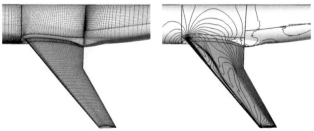

Fig. 13 Grids and pressure contours on the leeward of the wing ($\alpha = 0°$).
(colored picture attached at the end of the book)

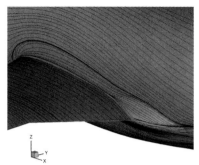

Fig. 14 Surface streamlines near the wing-body junction ($\alpha = 0°$).

$Re = 5 \times 10^6$, $T_\infty = 322.22$ K, and $\alpha = 0°$. The reference quantities are half-model reference area $S/2 = 72{,}700$ mm^2, mean aerodynamic chord $c_{\text{ref}} = 141.22$ mm, projected half-span $b/2 = 585.647$ mm, and aspect ratio AR $= 9.5$.

The grid number is 4.28 million, which is a coarse one when compared with the grids of DPW-III. The grid system is shown in Fig. 10. The computed pressure contours and density contours are shown in Fig. 11. The close-up views near the nose, the wing-body junction, and the tail, where the grid topology is complex, are shown in Fig. 12. The surface grids and pressure contours of the wing are shown in Fig. 13. The surface streamlines are shown in Fig. 14. When $\alpha = 0°$, the computed lift coefficient is 0.4763, and drag coefficient is 0.0294. When $C_l = 0.5$, we get $C_d = 0.0304$.

V. Conclusions

WCNSs are a kind of high-order nonlinear scheme and have been used for a large range of flow simulations. We successfully applied the fifth-order WCNS (WCNS-E-5) to complex grid systems. The characteristic-based interface conditions (CBIC), which are very convenient for practical applications, are proposed in order to fulfill the high-order schemes on complex grids. With the CBIC, a high-order strategy for complex point-matched multiblock grids is presented. The high-order WCNS-E-5 scheme with the CBIC was proved to be asymptotically stable in linear system with uniform grids. Validation of the CBIC with WCNS-E-5 is given by solving the Burgers equation and a vortex convection problem. Numerical results show that the CBIC can keep the high-order accuracy of WCNS-E-5. For two- and three-dimensional engineering-oriented problems, the strategy performs excellently, despite the grids being very complex with multiple connecting points. Further investigations will be carried on in the aspects such as wall boundary treatments and transition, which may degrade the performance of the high-order WCNS processes for engineering-oriented applications.

Acknowledgments

This study was supported by the project of National Natural Science Foundation of China (grant nos. 10621062 and 11072259) and National Basic Research Program of China (grant no. 2009CB723800). The authors would like to thank Laiping Zhang, Shuhai Zhang, Guangxue Wang, Xin Liu, and Junwu Hong for their contributions and useful suggestions.

References

[1] Wang, Z. J., "High-Order Methods for The Euler and Navier–Stokes Equations on Unstructured Grids," *Progress in Aerospace Sciences*, Vol. 43, 2007, pp. 1–41.
doi:10.1016/j.paerosci.2007.05.001
[2] Ekaterinaris, J. A., "High-Order Accurate, Low Numerical Diffusion Methods for Aerodynamics," *Progress in Aerospace Sciences*, Vol. 41, 2005, pp. 192–300.
doi:10.1016/j.paerosci.2005.03.003
[3] Lele, S. K., "Compact Finite Difference Schemes with Spectral-Like Resolution," *Journal of Computational Physics*, Vol. 103, 1992, pp. 16–42.
doi:10.1016/0021-9991(92)90324-R
[4] Deng, X., Maekawa, H., and Shen, Q., "A Class of High Order Dissipative Compact Schemes," AIAA Paper 96-1972, 1996.
[5] Visbal, M. R., and Gaitonde, D. V., "High-Order Accurate Methods for Complex Unsteady Subsonic Flows," *AIAA Journal*, Vol. 37, No. 10, 1999, pp. 1231–1239.
doi:10.2514/2.591
[6] Tu, G-H., Yuan. X- J., Xia, Z-Q, and Hu, Z., "A Class of Compact Upwind TVD Difference Schemes," *Journal of Applied Mathematics and Mechanics*, Vol. 27, No. 6, 2006, pp. 765–772.
doi:10.1007/s10483-006-0607-1
[7] Tu, G-H., and Yuan, X-J., "A Characteristic-Based Shock-Capturing Scheme for Hyperbolic Problems," *Journal of Computational Physics*, Vol. 225, 2007, pp. 2083–2097.
doi:10.1016/j.jcp.2007.03.007
[8] Deng, X., and Maekawa, H., "Compact High-Order Accurate Nonlinear Schemes," *Journal of Computational Physics*, Vol. 130, 1997, pp. 77–91.
doi:10.1006/jcph.1996.5553
[9] Deng, X., and Zhang, H., "Developing High-Order Accurate Nonlinear Schemes," *Journal of Computational Physics*, Vol. 165, 2000, pp. 22–44.
doi:10.1006/jcph.2000.6594
[10] Deng, X., "High-Order Accurate Dissipative Weighted Compact Nonlinear Schemes," *Science in China, Series A (Mathematics, Physics, Astronomy and Technological Sciences)*, Vol. 45, No. 3, 2002, pp. 356–370.
doi:10.1360/02ys9037
[11] Liu, X., Deng, X., and Mao, M. L., "High-Order Behaviors of Weighted Compact Fifth-Order Nonlinear Schemes," *AIAA Journal*, Vol. 45, No. 8, 2007, pp. 2093–2097.
doi:10.2514/1.23797
[12] Deng, X., Mao, M., Tu, G., Zhang, Y., and Zhang, H., "High-Order and High-Accurate CFD Methods and Their Applications," 8th Asian Computational Fluid Dynamics Conference, Hong Kong, Paper ACFD0126-T002-A-001, 10–14 Jan. 2010.
[13] Nonomura, T., Iizuka, N., and Fujii, K., "Freestream and Vortex Preservation Properties of High-Order WENO and WCNS on Curvilinear Grids," *Computers and Fluids*, Vol. 39, 2010, pp. 197–214.
doi:10.1016/j.compfluid.2009.08.005
[14] Sumi, T., Kurotaki, T., and Hiyama, J., "Generalized Characteristic Interface Conditions with High-Order Interpolation Method," AIAA Paper 2008-752, 2008.
[15] Gaitonde, D. V., and Visbal, M. R., "Padé-Type High-Order Boundary Filters for the Navier–Stokes Equations," *AIAA Journal*, Vol. 38, No. 11, 2000, pp. 2103–2112.
doi:10.2514/2.872
[16] Delfs, J. W., "Sound Generation From Gust-Airfoil Interaction Using CAA-Chimera Method," AIAA Paper 2001-2136, 2001.
[17] Sherer, S. E., Gordnier, R. E., and Visbal, M. R., "Computational Study of a UCAV Configuration Using a High-Order Overset-Grid Algorithm," AIAA Paper 2008-626, 2008.
[18] Sherer, S. E., and Scott, J. N., "High-Order Finite-Difference Methods on General Overset Grids," *Journal of Computational Physics*, Vol. 210, 2005, pp. 459–496.
doi:10.1016/j.jcp.2005.04.017
[19] Rai, M. M., "A Relaxation Approach to Patched Grid Calculations with the Euler Equations," *Journal of Computational Physics*, Vol. 66, 1986, pp. 99–131.
doi:10.1016/0021-9991(86)90056-2
[20] Lerat, A., and Wu, Z. N., "Stable Conservative Multidomain Treatments for Implicit Euler Equations," *Journal of Computational Physics*, Vol. 123, 1996, pp. 45–64.
doi:10.1006/jcph.1996.0004
[21] Kim, J. W., and Lee, D. J., "Characteristic Interface Conditions for Multi-Block High-Order Computation on Singular Structured Grid," AIAA Paper 2003-3122, 2003.
[22] Sumi, T., Kurotaki, T., and Hiyama, J., "Generalized Characteristic Interface Conditions for Accurate Multi-Block Computation," AIAA Paper 2006-1272, 2006.
[23] Sumi, T., Kurotaki, T., and Hiyama, J., "Practical Multi-Block Computation with Generalized Characteristic Interface Conditions Around Complex Geometry," AIAA Paper 2007-4471, 2007.
[24] Barth, T. J., "Aspects of Unstructured Grids and Finite-Volume Solvers

[25] Thompson, K. W., "Time Dependent Boundary Conditions for Hypersonic System, II," *Journal of Computational Physics*, Vol. 89, 1990, pp. 439–461.
doi:10.1016/0021-9991(90)90152-Q

[26] Poinsot, T. J., and Lele, S. K., "Boundary Conditions for Direction Simulations of Compressible Viscous Flows," *Journal of Computational Physics*, Vol. 101, 1992, pp. 104–129.
doi:10.1016/0021-9991(92)90046-2

[27] Okong'o, N., and Bellan, J., "Consistent Boundary Conditions for Multicomponent Real Gas Mixtures Based on Characteristic Waves," *Journal of Computational Physics*, Vol. 176, 2002, pp. 330–344.
doi:10.1006/jcph.2002.6990

[28] Kim, J. W., and Lee, D. J., "Generalized Characteristic Boundary Conditions for Computational Aeroacoustics, Part 2," *AIAA Journal*, Vol. 42, No. 1, 2004, pp. 47–55.
doi:10.2514/1.9029

[29] Jiang, G.-S., and Shu, C.-W., "Efficient Implementation of Weighted ENO Schemes," *Journal of Computational Physics*, Vol. 126, 1996, pp. 202–228.
doi:10.1006/jcph.1996.0130

[30] Spalart, P. R., and Allmaras, S. R., "A One-Equation Turbulence Model for Aerodynamic Flows," AIAA Paper 92-0439, 1992.

[31] Zhang, H., and Zhuang, F., "NND Schemes and Their Applications to Numerical Simulation of Two- and Three-Dimensional Flows," *Advances in Applied Mechanics*, Vol. 29, 1991, pp. 193–256.
doi:10.1016/S0065-2156(08)70165-0

[32] Van den, B., and Oskam, B., "Boundary Layer Measurements on a Two-Dimensional Wing with Flap and a Comparison with Calculations," National Aerospace Laboratory/NLR, TR 79009U, Emmeloord, The Netherlands, 1979.

High-Order Behaviors of Weighted Compact Fifth-Order Nonlinear Schemes

Liu Xin,* Deng Xiaogang,[†] and Mao Meiliang[‡]
China Aerodynamics Research and Development Center, 621000 Sichuan, People's Republic of China

DOI: 10.2514/1.23797

I. Introduction

THE typical theory for the finite difference method is that partial difference equations (PDEs) are discretized in space using a Taylor series expansion and solving it for the variables at discrete points. The derivatives are written as functions of variables on neighboring points. The approximations are said to be high-order accurate when the power p of the leading truncate error $O(h^p)$ is greater than two. Because of the advantages of high-flow structure resolution and nice discontinuity capturing precision, high-order accurate schemes are playing important roles in computational fluid dynamics (CFD).

Nonlinear high-order schemes for computing turbulence and aeroacoustic flows, which contain shock, have received more and more attention since the last decade. Although compact schemes have been derived by some researchers like Lele [1] and Gaitonde and Shang [2], they were poorly used in capturing the shock owing to their linear property in construction. Cockburn and Shu [3] proposed fourth-order compact nonlinear schemes based on a total variation diminishing (TVD) and total variation bounded (TVB) concept. Hu et al. [4] developed a fifth-order upwind compact scheme to solve the acoustic phenomena arising from shock-vortex interactions. According to the principles from the physical consideration, Zhang et al. [5] constructed a mixing method, which degenerated into the second-order nonoscillatory, containing no free parameters and dissipative (NND) scheme in the shock region and retained high-order accuracy elsewhere. An appropriate weight technique treated in high-order scheme construction may also differentiate the vortex in the smooth region and shocks robustly. Levy et al. [6] applied fourth-order central weighted essentially nonoscillatory (WENO) to hypersonic flows. Shu [7] used fifth-order WENO for shock problems.

Deng [8] and other researchers have developed a variety of compact high-order accurate schemes. A type of one-parameter linear dissipative compact scheme (DCS) was derived as to damp out the dispersive and parasite errors in the high-wave-number regions when central compact schemes were used. Although the linear compact schemes often caused numerical oscillations in the vicinity of discontinuities, a fourth-order cell-centered compact nonlinear finite difference scheme (CNS) [9] was then derived. An adaptive interpolation of variables at cell edges was designed, which automatically jumped to local one as discontinuities were encountered. It was a way to make the overall compact scheme capture discontinuities in a nonoscillatory manner. Afterward, using the weighted technique of Jiang and Shu [10] for interpolation at cell edge, fourth- and fifth-order weighted compact nonlinear schemes (WCNS) were developed [11,12]. They were more efficient than CNS, for two tridiagonal inversions were cancelled and no logical algorithms were needed. To eliminate the unphysical oscillations in the vortex field near the shock wave when inviscid flow was computed, Deng constructed high-order dissipative weighted compact nonlinear schemes (DWCNS) [13], which improved the ability of restraining nonphysical oscillation in the vorticity field. For viscous flow computations, as proposed in [13], WCNS-5 and WCNS-E-5 would better be used.

This Note shows the high-order features of WCNS including WCNS-E-5 and WCNS-5 with such typical high-order schemes as the explicit upwind biased fifth-order scheme (EUW-5) [14], Padé [1], and other special schemes like WENOs [6,7], and the explicit fifth-order shock-fitting upwind scheme [15]. We do the Fourier analysis to discuss WCNS-E-5 and WCNS-5 with EUW-5 and Padé. Owing to nontridiagonal inversion, WCNS-E-5 acts more efficiently than WCNS-5 during computations. Therefore, in Sec. IV, we use WCNS-E-5 to solve the multidimensional inviscid/viscous flows. WCNS-E-5 obtains several numerical results to show the good performances with those by WENOs and the explicit fifth-order shock-fitting upwind scheme.

II. Construction of Weighted Compact Fifth-Order Nonlinear Schemes

Let $U_i = U(\xi_i, t)$ ($\xi_i = ih$) be the numerical solution of the hyperbolic conservation laws, at every node ξ_i, the model equation can be semidiscretized as

$$\left(\frac{\partial U}{\partial t}\right)_i = -E'_i \qquad (1)$$

here E'_i represents the approximation to the spatial derivative. Based on the cell-centered finite difference compact schemes, E'_i can be computed as the following forms:
1) WCNS-5,

$$\frac{9}{62}E'_{i-1} + E'_i + \frac{9}{62}E'_{i+1} = \frac{63}{62h}(\tilde{E}_{i+1/2} - \tilde{E}_{i-1/2})$$
$$+ \frac{17}{186h}(\tilde{E}_{i+3/2} - \tilde{E}_{i-3/2}) \qquad (2)$$

2) WCNS-E-5,

$$E'_i = \frac{75}{64h}(\tilde{E}_{i+1/2} - \tilde{E}_{i-1/2}) - \frac{25}{384h}(\tilde{E}_{i+3/2} - \tilde{E}_{i-3/2})$$
$$+ \frac{3}{640h}(\tilde{E}_{i+5/2} - \tilde{E}_{i-5/2}) \qquad (3)$$

Received 29 April 2006; revision received 17 April 2007; accepted for publication 18 April 2007. Copyright © 2007 by the American Institute of Aeronautics and Astronautics, Inc. All rights reserved. Copies of this paper may be made for personal or internal use, on condition that the copier pay the $10.00 per-copy fee to the Copyright Clearance Center, Inc., 222 Rosewood Drive, Danvers, MA 01923; include the code 0001-1452/07 $10.00 in correspondence with the CCC.

*Associate Professor, High Speed Aerodynamics Institute; liuxing_76@sina.com.
[†]Professor, Computational Fluid Dynamics Institute; xgdeng@my-public.sc.cninfo.net. Member AIAA.
[‡]Professor, Computational Fluid Dynamics Institute; mml219@163.com.

In Eqs. (2) and (3), $\tilde{E}_{i\pm 1/2} = E(\tilde{U}_{i\pm 1/2})$ denotes the numerical flux at the cell edges. The weighted technique is used in the $\tilde{U}_{i\pm 1/2}$ interpolations. The idea is that each of the chosen stencils is assigned a weight factor, which determines its contribution to the final approximation of the cell-edge value. The weights are designed in such a way that in the smooth region they approach the optimal weights to achieve fifth-order accuracy and require nontridiagonal inversion, whereas in the regions near the discontinuities, the weight of the stencil, which contains the discontinuities, is assigned nearly zero. Therefore the weighted interpolations are prevented from crossing discontinuities, whereafter the third-order interpolations are enforced in these regions. For more details, see, [11–13].

The boundary schemes have an important effect on the solution of the whole flowfield. As shown by Gustaffson [16], for a pth order interior scheme, the accuracy of the boundary schemes can be $(p-1)$th-order accurate without reducing the global accuracy of the interior scheme. For Eqs. (2) and (3), several boundary and near boundary schemes are derived in [12,13], in which more detailed discretization and interpolation of viscous flux for Navier–Stokes equations are also developed.

III. Fourier Analysis of Weighted Compact Fifth-Order Nonlinear Schemes

The dissipative and dispersive properties of WCNS by the Fourier analysis are further discussed in this section. Consider the semidiscrete form of the scaled linear hyperbolic equation

$$\frac{\partial u_i}{\partial t} + cu'_i = 0 \quad (4)$$

with a periodic function

$$u \equiv \exp(\mathrm{Im}\,\omega x) \quad (5)$$

where $c > 0$ is a constant speed. The difference error of first derivative is approximated by the comparison of the Fourier coefficients $(u'_k)_{\mathrm{fd}}$ with the exact u'_k. Defining the modified wave number $\omega^* = \omega_r^* + \mathrm{Im}\,\omega_i^*$, then Eq. (5) can be differenced as

$$(u')_{\mathrm{fd}} = \mathrm{Im}\,\omega^* u \quad (6)$$

The accurate solution of Eq. (4) with the initial condition of Eq. (5) can be obtained by substituting $u(x,t) = A(t)\exp(\mathrm{Im}\,\omega x)$ into Eq. (4)

$$u_i = \exp(c\omega_i^* t)\cdot \exp\left[\mathrm{Im}\,\omega\left(x_i - c\frac{\omega_r^*}{\omega}t\right)\right] \quad (7)$$

The cell-centered schemes for the u'_i have the form

$$\kappa u'_{i-1} + u'_i + \kappa u'_{i+1} = \frac{a}{h}(\tilde{u}_{i+1/2} - \tilde{u}_{i-1/2}) + \frac{b}{h}(\tilde{u}_{i+3/2} - \tilde{u}_{i-3/2}) + \frac{c}{h}(\tilde{u}_{i+5/2} - \tilde{u}_{i-5/2}) \quad (8)$$

Substituting Eqs. (5) and (6) into Eq. (8) and using the optimal interpolation

$$\tilde{u}_{i+1/2} = u_i + \frac{1}{128}(3u_{i-2} - 20u_{i-1} - 38u_i + 60u_{i+1} - 5u_{i+2})$$

we get the modified wave number function $\omega^* h = \omega^*(\omega h)h$.

Figure 1 shows the real and imaginary parts of $\omega^* h$ of WCNS-5 and WCNS-E-5 compared with EUW-5 and fourth-order Padé schemes. It can be observed that $\omega_i h < 0$ for all schemes but Padé (whose $\omega_i h$ equals zero), which just corresponds to the dissipative property. In terms of the modified wave numbers, WCNS-5 and WCNS-E-5 have almost the same dissipative and dispersive features as each other, and they both are superior to EUW-5.

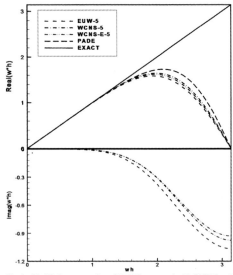

Fig. 1 Modified wave number of WCNS compared with EUW-5 and Padé.

Now define an error tolerance ε as

$$\left|\frac{\omega^*(\omega h)h - \omega h}{\omega h}\right| \leq \varepsilon$$

Within the specified error tolerance, the range of well-resolved waves can be determined. We then use the fraction $e_1 \equiv \omega_f/\pi$ to represent as a measure of the resolving efficiency of a scheme, where ω_f is the shortest well-resolved wave. Table 1 shows the resolving efficiency of schemes for three different values of the error tolerance, i.e., $\varepsilon = 0.1, 0.01$, and 0.001. It is obvious that WCNS-5 and WCNS-E-5 have little difference from each other, and they stay close to the exact differentiation over a wider range of wave numbers than the other two schemes, when $\varepsilon = 0.01$ and 0.001.

In the multidimensional equation computation, the phase error is characterized by the form of anisotropic phase speed. The phase speeds of different wave numbers in a two-dimensional problem are shown as [1]

$$\omega^{**} = \frac{\omega^*(\omega h, \theta)h}{\omega h} = \frac{\cos\theta\omega^*(\omega h\cos\theta)h + \sin\theta\omega^*(\omega h\sin\theta)h}{\omega h}$$

where θ is the angle between the propagation direction and the axis. The anisotropy is displayed in Fig. 2 for the second-order center scheme, WCNS-5, WCNS-E-5, EUW-5, and Padé. The curves are plotted for

$$\frac{\omega h}{\pi} = \frac{1}{50}, \frac{5}{50}, \cdots, \frac{45}{50}, \frac{50}{50}$$

The outermost curves are circles corresponding to the small ωh. For these waves, the propagation is isotropic. The innermost curves

Table 1 Resolving efficiency of schemes

Scheme	$\varepsilon = 0.1$	$\varepsilon = 0.01$	$\varepsilon = 0.001$
EUW-5	0.545	0.355	0.235
WCNS-5	0.570	0.370	0.250
WCNS-E-5	0.560	0.365	0.245
Padé	0.595	0.360	0.205

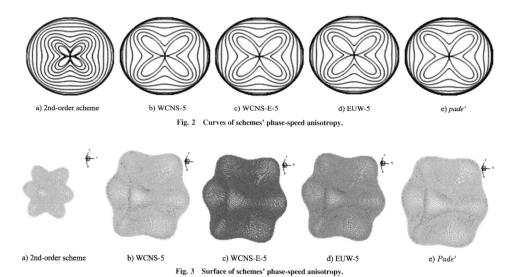

Fig. 2 Curves of schemes' phase-speed anisotropy.

Fig. 3 Surface of schemes' phase-speed anisotropy.

correspond to the shortest waves resolved on the mesh. Among the schemes, the innermost curve of the second-order scheme is the smallest. The anisotropic features of the high-order schemes are little different from one another, and especially the behaviors of WCNS-5 are very close to those of WCNS-E-5.

For the three-dimensional phase speed, the following formula is derived as

$$\omega^{***} = \frac{\cos\theta\cos\varphi\omega^*(\omega h\cos\theta\cos\varphi)h + \cos\theta\sin\varphi\omega^*(\omega h\cos\theta\sin\varphi)h + \sin\theta\omega^*(\omega h\sin\theta)h}{\omega h}$$

Here θ is the angle between the propagation direction and the $x-y$ plane, and φ means the angle caused by the wave propagation in the $x-y$ plane with x axis. Subfigures in Fig. 3 show the anisotropic characteristics of the schemes when $\omega h = \pi$. Similar to those in two-dimensional flows, the resolved shorter waves propagate in the anisotropic way with the least error along ± 45 deg angles. The larger the surface configuration is, the narrower range of short wave the phase error is limited to.

By the Fourier analysis, the fifth-order WCNS (WCNS-E-5 and WCNS-5) behave as good as EUW-5 and Padé, and they act as almost the same characteristics as each other. However compared with WCNS-5, WCNS-E-5 requires nontridiagonal inversion during computation, which improves the computational efficiency. Therefore we apply WCNS-E-5 to the following complicated flows.

IV. Numerical Tests

Several transient flows are numerically simulated to compare WCNS-E-5 with WENOs [6,7] and the explicit fifth-order shock-fitting upwind scheme [15]. They are divided into two groups: inviscid flow and viscous flow. Roe's flux difference scheme [17] is used in the inviscid flow simulation and Steger–Warming vector flux splitting scheme [18] in the viscous flow simulation. The temporal term is discretized with the third-order TVD Runge–Kutta method [19].

Example 1. The two-dimensional Riemann problem for the Euler equations of gas dynamics is solved with the initial conditions shown in Fig. 4. In this computation, we simulate the results of interactions between different waves including rarefaction wave (from 2 to 1), shock wave (from 4 to 2), and contact surfaces (from 3 to 4 and from 3 to 1). We use the grid points of 200×200. Figure 5 shows the computed contours by WCNS-E-5 and those [6] by WENO when time equals 0.2. WENO with 400×400 indicates some oscillation within 2 and 4, whereas WCNS-E-5 does not. Furthermore WCNS-E-5 uses much fewer grids than WENO.

Example 2. The second example describes the interaction problem between a vortex and a stationary shock. It is a good model for sound waves that are generated when turbulence interacts with shock waves with shock structures in a jet plume resulting in broadband noise. The stationary Mach 1.1 shock is positioned at $x = 0.5$ and normal to the x axis. Its left state is $(\rho, u, v, p) = (1, M\sqrt{\gamma}, 0, 1)$. A small vortex

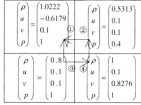

Fig. 4 Initial conditions for two-dimensional Riemann problem.

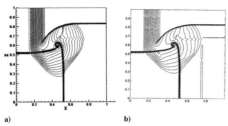

Fig. 5 Density contours a) WCNS-E-5 with 200 × 200, and b) WENO with 400 × 400 [6].

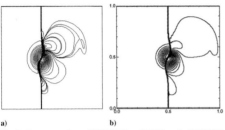

Fig. 6 Pressure contours a) WCNS-E-5, and b) fifth-order WENO [7].

centers at $(x_c, y_c) = (0.25, 0.5)$ and is superposed to flow left to the shock. We describe the vortex as a perturbation to the velocity, temperature, and entropy of the mean flow and denote it by $(\delta u, \delta v) = \varepsilon \iota e^{\alpha(1-\iota^2)}(\sin\theta, -\cos\theta)$, $\delta T = -[(\gamma-1)\varepsilon^2/4\alpha\gamma]e^{2\alpha(1-\iota^2)}$, and $\delta S = 0$, where $\iota = r/r_c$ and $r^2 = (x-x_c)^2 + (y-y_c)^2$. Here ε indicates the strength of the vortex, α controls the decay rate of the vortex, and r_c is the critical radius for which the vortex has the maximum strength. In our test, we choose $\varepsilon = 0.35$, $\alpha = 0.204$, and $r_c = 0.05$. Figure 6 shows the results of the pressure contours at $t = 0.2$ by WCNS-E-5 and those [7] by Shu's WENO. When the vortex core passes through the shock, the bifurcation is being created, which causes shock to be split into two triple points connected with a Mach stem. That the vortex remains a well-defined circular shape shows the two fifth-order schemes have accurately represented the shock-vortex interaction, except for the small oscillation caused near the shock by WENO.

Example 3. We use WCNS-E-5 to study the hypersonic boundary-layer receptivity over a parabola in this example. It is helpful to understand the physical mechanisms of the transition process depending on the character of small environmental perturbations, which often causes boundary-layer instability, amplification, and interaction of different instability modes to laminar flow breakdown. After obtaining the steady solution, we impose perturbations to the basic flow in the freestream. The freestream is $M_\infty = 15$, $T_\infty^* = 192.989K$, $Re_\infty = 6026.55$, and the perturbations are $[u', v', p', \rho']_\infty^T = [|u'|, |v'|, |p'|, |\rho'|]_\infty^T e^{Imk[x-(1+M_\infty^{-1})t]}$. Figure 7 shows the contours of the instantaneous velocity perturbation v', the Fourier amplitude, and the phase angle after the unsteady flow has been computed for 30 period time. The v' contours indicate the instability waves developing in the boundary layer because of the dominant instability waves near the surface. It is also seen very clearly in the Fourier amplitude and phase angle contours that two instability waves, named the first mode wave and the second mode wave, respectively, are generated. On the surface, the first mode is generated near the leading edge ($x < 0.2$), whereas the second mode is generated downstream. Figure 8 shows the solutions [15] by Zhong with his fifth-order linear scheme for the same test. Zhong's wave fields provide the region between the bow shock and the body surface. However our results provide a larger region where there is a bit of oscillation behind the bow shock. After comparing the wave structures between the bow shock and the surface, especially in the boundary layer, we know the solutions by Zhong's scheme and our WCNS-E-5 are well conformed to each other.

From the instantaneous entropy perturbation distributions in Fig. 9, we can see that the decay of the first mode wave and the growth of the second mode wave at about $x = 0.2$ are divided by a sudden change on the surface. The perturbation changes dramatically between the two modes. The change points captured by Zhong's scheme and ours are almost at the same location.

V. Conclusions

The Fourier analysis and applications of WCNS are studied to show their high-order characteristics. By the Fourier analysis, the features of WCNS, in contrast with EUW-5 and Padé schemes, are discussed in terms of dissipative and dispersive errors, the resolving

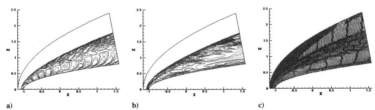

Fig. 7 Unsteady contours by WCNS-E-5: a) instantaneous velocity, b) Fourier amplitude, and c) Fourier phase angle.

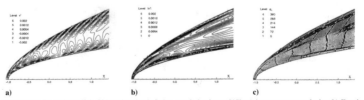

Fig. 8 Unsteady contours by Zhong's explicit fifth-order shock-fitting upwind scheme [15]: a) instantaneous velocity, b) Fourier amplitude, and c) Fourier phase angle.

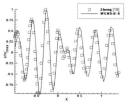

Fig. 9 Instantaneous entropy on surface.

efficiency and the multidimensional phase-speed anisotropy in the smooth region. The analytical results show that WCNS have better characteristics than EUW-5 in the parasite errors and they also seem more resolving efficient than EUW-5 and Padé. It may be seen that for the high-order schemes, the difference of anisotropic errors is little among one another, and they are altogether limited to a narrow range of short waves.

The interpolations of variables at cell-edges are the dominant factor for the accuracy of WCNS, which is of fifth-order accuracy in the smooth region and third-order accuracy in the vicinity of discontinuities. Besides its behavior similar to WCNS-5, WCNS-E-5 is more efficient because of nontridiagonal inversion. After computing several multidimensional Euler and Navier–Stokes equations, WCNS-E-5 shows that it captures discontinuities robustly. Owing to the quite low numerical error, WCNS-E-5 differentiates the flow structure clearly and accurately. Moreover, WCNS-E-5 needs fewer grids and obtains less unphysical oscillation nearby the shock than some high-order WENO schemes.

Acknowledgments

This study was supported by China National Natural Science Foundation (No. 10225208 and No. 10321002). The first author thanks Zhang Hanxin for his very helpful advice.

References

[1] Lele, S., "Compact Finite Difference Schemes with Spectral-Like Resolution," *Journal of Computational Physics*, Vol. 103, No. 1, 1992, pp. 16–42.

[2] Gaitonde, D., and Shang, J. S., "Optimized Compact-Difference-Based Finite Volume Schemes For Linear Wave Phenomena," *Journal of Computational Physics*, Vol. 138, No. 2, 1997, pp. 617–643.

[3] Cockburn, B., and Shu, C. W., "Nonlinearly Stable Compact Schemes for Shock Calculations," *SIAM Journal on Numerical Analysis*, Vol. 31, No. 3, 1994, pp. 607–630.

[4] Hu, G. Q., Fu, D. X., and Ma, Y. W., "Numerical Study of Sound Generated by Shock-Vortex Interactions," *Acta Mechanica Sinica*, Vol. 33, No. 6, 2001, pp. 721–728.

[5] Zhang, H. X., Li, Q., and Zhuang, F. G., "On the Construction of High Order Accuracy Difference Schemes," *ACTA Aerodynamica Sinica*, Vol. 16, No. 1, 1998, pp. 14–23.

[6] Levy, D., Puppo, G., and Russo, G., "Fourth Order Central WENO Scheme for Multi-Dimensional Hypersonic Systems of Conservation Laws," *SIAM Journal on Scientific Computing*, Vol. 24, No. 2, 2002, pp. 480–506.

[7] Shu, C. W., "Essentially Non-Oscillatory and Weighted Essentially Non-Schemes for Hyperbolic Conservation Laws," NASA CR-97-206253, 1997.

[8] Deng, X. G., Maekawa, H., and Shen, Q., "Class of High Order Dissipative Compact Schemes," AIAA Paper 96-1972, 1996.

[9] Deng, X. G., and Maekawa, H., "Uniform Fourth-Order Compact Scheme for Discontinuities Capturing," AIAA Paper 96-1974, 1996.

[10] Jiang, G., and Shu, C. W., "Efficient Implementation of Weighted ENO Schemes," *Journal of Computational Physics*, Vol. 126, No. 1, 1996, pp. 202–228.

[11] Deng, X. G., and Mao, M. L., "Weighted Compact High-Order Nonlinear Schemes for the Euler Equations," AIAA Paper 97-1941, 1997.

[12] Deng, X. G., and Zhang, H. X., "Developing High-Order Accurate Nonlinear Schemes," *Journal of Computational Physics*, Vol. 165, No. 1, 2000, pp. 22–44.

[13] Deng, X. G., "High-Order Accurate Dissipative Weighted Compact Nonlinear Schemes," *Science in China, Series A*, Vol. 31, No. 12, 2001, pp. 1104–1117.

[14] Rai, M. M., and Moin, P., "Direct Simulations of Turbulent Flow Using Finite-Difference Schemes," *Journal of Computational Physics*, Vol. 96, No. 1, 1991, pp. 15–53.

[15] Zhong X., "Direct Numerical Simulation of Hypersonic Boundary Layer Transition over Blunt Leading Edges, Part 2: Receptivity to Sound," AIAA Paper 97-0756, 1997.

[16] Gustaffson, B., "Convergence Rate for Difference Approximations to Mixed Initial Boundary Value Problems," *Mathematics of Computation*, Vol. 49, No. 2, 1975, pp. 396–406.

[17] Leer, B. V., Thomas, J. L., and Roe, P. L., "Comparison of Numerical Flux Formulas for the Euler and Navier–Stokes Equations," AIAA Paper 87-1104, 1987.

[18] Steger, J. L., and Warming, R. F., "Flux Vector Splitting of the Inviscid Gasdynamic Equations with Application to Finite-Difference Methods," *Journal of Computational Physics*, Vol. 40, No. 2, 1981, pp. 263–293.

[19] Shu, C. W., and Osher, S., "Efficient Implementation of Essentially Non-Oscillatory Shock-Capturing Schemes 2," *Journal of Computational Physics*, Vol. 77, No. 2, 1988, pp. 439–471.

Large eddy simulation on curvilinear meshes using seventh-order dissipative compact scheme

Yi Jiang [a,c,*], Meiliang Mao [c], Xiaogang Deng [b], Huayong Liu [a]

[a] State Key Laboratory of Aerodynamics, China Aerodynamics Research and Development Center, P.O. Box 211, Mianyang 621000, PR China
[b] National University of Defense Technology, Changsha, Hunan 410073, PR China
[c] Computational Aerodynamics Institute, China Aerodynamics Research and Development Center, P.O. Box 211, Mianyang 621000, PR China

ARTICLE INFO

Article history:
Received 15 January 2014
Received in revised form 24 May 2014
Accepted 9 August 2014
Available online 27 August 2014

Keywords:
Large eddy simulation (LES)
High-order implicit large eddy simulation (HILES)
High-order hybrid cell-edge and cell-node dissipative compact scheme (HDCS)
Geometric conservation law (GCL)
Curvilinear mesh

ABSTRACT

This work investigates the performance of the high-order implicit large eddy simulation (HILES) on curvilinear meshes. The HILES is developed based on a seventh-order hybrid cell-edge and cell-node dissipative compact scheme (HDCS-E8T7) satisfying the surface conservation law (SCL). Efficiency of implicit subgrid-scale model is tested by three-dimensional Taylor–Green vortex case. According to the test of flow over a cylinder, the influence of the SCL errors has been investigated on curvilinear mesh. Then stall phenomena of thin airfoil NACA64A006 have been simulated by the HILES. The slope of lift curve, the maximum lift and the stall angle are successfully predicted. Moreover, the lift characteristic seems to be satisfactorily captured even after the stall angle. The solutions demonstrate the potential of HILES for simulating complex turbulent flow.

© 2014 Elsevier Ltd. All rights reserved.

1. Introduction

The need for predictive simulation methods for turbulent flows has led to a significant interest in large-eddy simulations (LES) in recent years. Without LES that captures unsteady behavior of the turbulent flows, accurate result may not be obtained [1]. For example, it is difficult for Reynolds-averaged Navier–Stokes simulations (RANS) model to estimate stall characteristic of thin-airfoil NACA64A006 [1], where the laminar flow separation occurs at the leading edge and the transition causes the turbulent reattachment. The reattachment point gradually moves rearward with increasing angles of attack [2]. The small vortexes shed from the leading edge, which produces strong unsteadiness in the flow. It is difficult for RANS simulations to resolve this feature, and this is the main reason that RANS simulations do not give satisfactory results for this case. According to the limitation of RANS methods, Fujii [1] proposed that LES should be employed for the prediction of thin-airfoil stall characteristics. Furthermore, LES has been successfully applied for the simulations of stall phenomena [3,4].

As well known, high-order scheme has the advantage over low-order scheme for the simulation of turbulent flow containing unsteady vortex shedding. However, the application of high-order compact schemes still has some challenges, such as robustness and grid-quality sensitivity [5,6]. This deficiency can be largely removed by the researches of the Geometric Conservation Law (GCL) [7–11]. The GCL contains surface conservation law (SCL) and volume conservation law (VCL). The VCL has been widely studied for time-dependent grids, while the SCL is merely discussed for finite difference schemes. Recent research [7] shows that the GCL is very important for the application of finite difference schemes on curvilinear grids. If the SCL has not been satisfied, numerical instabilities and even computing collapse may appear on complex curvilinear grids during numerical simulation. In order to fulfill the SCL for high-order finite difference schemes, a conservative metric method (CMM) is derived by Deng et al. [7]. Not long after, a symmetrical conservative metric method (SCMM) [12], which can evidently increase the numerical accuracy on irregular grids, is proposed based on the CMM. According to the principle of satisfying the SCMM, a seventh-order hybrid cell-edge and cell-node dissipative compact scheme (HDCS-E8T7) has been proposed for complex geometry [13]. The HDCS-E8T7 has inherent dissipation to dissipate unresolvable wavenumbers, therefore filtering is not needed. The properties of HDCS-E8T7 scheme have been systemically analyzed by Deng et. al. [14].

Based on the HDCS-E8T7, a new high-order implicit large eddy simulation (HILES) is developed following the concept of monotone integrated LES (MILES) [15], i.e. the effects of explicit LES models are imitated by the truncation error of the discretization scheme itself. Although the conception of HILES is similar to that of MILES, HDCS-E8T7 is a new seventh-order compact scheme having inherent dissipation, and the HILES based on HDCS-E8T7 can eliminate the SCL errors, which may contaminate flowfield of the HILES. Moreover, if we use the seventh-order compact scheme for LES without subgrid-scale (SGS) model, i.e., HILES, the accuracy of the flowfield obtained will be seventh-order. However, if subgrid model, for instance, the Smgorinsky subgrid model is added, the accuracy of the LES results will be degenerated to the undesirable second order. If a carefully designed high-order subgrid model is absent, high-order scheme without model may be better than with second-order subgrid models for LES [16].

In this paper, we will investigate the performance of the HILES on curvilinear meshes. Based on the test of flow over a cylinder, the influence phenomena of the SCL errors on the HILES has been shown. Then the stall phenomena of thin airfoil NACA64A006 have been simulated by the HILES. In the next section, the governing equations are given. In Section 3, we will give the numerical method comprising three main components: the spatial discretization, the grid metric calculation and the time-integration method. The numerical tests are presented in Section 4.

2. Governing equations

Three dimensional Navier–Stokes equations in computational coordinates may be written as

$$\frac{\partial \tilde{U}}{\partial t} + \frac{\partial \tilde{E}}{\partial \xi} + \frac{\partial \tilde{F}}{\partial \eta} + \frac{\partial \tilde{G}}{\partial \zeta} = \frac{1}{R_e}\left(\frac{\partial \tilde{E}_v}{\partial \xi} + \frac{\partial \tilde{F}_v}{\partial \eta} + \frac{\partial \tilde{G}_v}{\partial \zeta}\right), \quad (1)$$

where,

$\tilde{U} = U/J,$
$\tilde{E} = (\xi_t U + \xi_x E + \xi_y F + \xi_z G)/J, \quad \tilde{E}_v = (\xi_x E_v + \xi_y F_v + \xi_z G_v)/J,$
$\tilde{F} = (\eta_t U + \eta_x E + \eta_y F + \eta_z G)/J, \quad \tilde{F}_v = (\eta_x E_v + \eta_y F_v + \eta_z G_v)/J,$
$\tilde{G} = (\zeta_t U + \zeta_x E + \zeta_y F + \zeta_z G)/J, \quad \tilde{G}_v = (\zeta_x E_v + \zeta_y F_v + \zeta_z G_v)/J,$

and

$U = [\rho, \rho u, \rho v, \rho w, \rho e]^T,$
$E = [\rho u, \rho u^2 + p, \rho v u, \rho w u, (\rho e + p)u]^T,$
$F = [\rho v, \rho u v, \rho v^2 + p, \rho w v, (\rho e + p)v]^T,$
$G = [\rho w, \rho u w, \rho v w, \rho w^2 + p, (\rho e + p)w]^T,$
$E_v = [0, \tau_{xx}, \tau_{xy}, \tau_{xz}, u\tau_{xx} + v\tau_{xy} + w\tau_{xz} + \dot{q}_x]^T,$
$F_v = [0, \tau_{yx}, \tau_{yy}, \tau_{yz}, u\tau_{yx} + v\tau_{yy} + w\tau_{yz} + \dot{q}_y]^T,$
$G_v = [0, \tau_{zx}, \tau_{zy}, \tau_{zz}, u\tau_{zx} + v\tau_{zy} + w\tau_{zz} + \dot{q}_z]^T,$

with viscous stress terms written as

$\tau_{ij} = \mu\left(\frac{\partial u_i}{\partial x_j} + \frac{\partial u_j}{\partial x_i} - \frac{2}{3}\delta_{ij}\frac{\partial u_l}{\partial x_l}\right),$

and heat transfer terms

$\dot{q}_x = \frac{1}{(\gamma-1)P_r M_\infty^2}\mu\frac{\partial T}{\partial x},$
$\dot{q}_y = \frac{1}{(\gamma-1)P_r M_\infty^2}\mu\frac{\partial T}{\partial y},$
$\dot{q}_z = \frac{1}{(\gamma-1)P_r M_\infty^2}\mu\frac{\partial T}{\partial z},$

where γ is the ratio of specific heats, and the viscous coefficient μ can be calculated by the Sutherland's law. The equation of state and the energy function are

$$p = \frac{1}{\gamma M_\infty^2}\rho T, \quad e = \frac{p}{(\gamma-1)\rho} + \frac{u^2 + v^2 + w^2}{2}.$$

In the above u, v and w are the velocity components in x, y and z directions, respectively, p is the pressure, ρ is the density and T is temperature. The non-dimensional variables are defined as $\rho = \rho^*/\rho_\infty^*$, $(u, v, w) = (u, v, w)^*/V_\infty^*$, $T = T^*/T_\infty^*$, $p = p^*/\rho_\infty^* V_\infty^{*2}$, $\mu = \mu^*/\mu_\infty^*$, respectively, and $M_\infty = u_\infty^*/\sqrt{\gamma R T_\infty^*}$, $R_e = \rho_\infty^* u_\infty^* r^*/\mu_\infty^*$, $P_r = \mu_\infty^* C_p/\kappa_\infty^*$ are the Mach number, Reynolds number and Prandtl number, r^* is the characteristic length. J is the Jacobi of grid transformation. $\xi_t, \xi_x, \xi_y, \xi_z, \eta_t, \eta_x, \eta_y, \eta_z, \zeta_t, \zeta_x, \zeta_y, \zeta_z$ are grid derivatives. The grid metric derivatives have the conservative form as

$$\begin{cases}\tilde{\xi}_x = \xi_x/J = (y_\eta z)_\zeta - (y_\zeta z)_\eta, \tilde{\xi}_y = \xi_y/J = (z_\eta x)_\zeta - (z_\zeta x)_\eta, \tilde{\xi}_z = \xi_z/J = (x_\eta y)_\zeta - (x_\zeta y)_\eta,\\ \tilde{\eta}_x = \eta_x/J = (y_\zeta z)_\xi - (y_\xi z)_\zeta, \tilde{\eta}_y = \eta_y/J = (z_\zeta x)_\xi - (z_\xi x)_\zeta, \tilde{\eta}_z = \eta_z/J = (x_\zeta y)_\xi - (x_\xi y)_\zeta,\\ \tilde{\zeta}_x = \zeta_x/J = (y_\xi z)_\eta - (y_\eta z)_\xi, \tilde{\zeta}_y = \zeta_y/J = (z_\xi x)_\eta - (z_\eta x)_\xi, \tilde{\zeta}_z = \zeta_z/J = (x_\xi y)_\eta - (x_\eta y)_\xi.\end{cases} \quad (2)$$

3. Numerical method

3.1. Spatial discretization

A seventh-order finite difference scheme HDCS-E8T7 is employed to discretize the equations (1). Considering discretization of the inviscid terms,

$$\frac{\partial \tilde{U}}{\partial t} + \frac{\partial \tilde{E}}{\partial \xi} + \frac{\partial \tilde{F}}{\partial \eta} + \frac{\partial \tilde{G}}{\partial \zeta} = 0, \quad (3)$$

and theirs semi-discrete approximations,

$$\frac{\partial \tilde{U}}{\partial t} = -\delta_1^\xi \tilde{E} - \delta_1^\eta \tilde{F} - \delta_1^\zeta \tilde{G}. \quad (4)$$

The discretizations $\delta_1^\xi, \delta_1^\eta$ and δ_1^ζ are the same, thus we only give the discretization in ξ direction. The δ_1^ξ of the HDCS-E8T7 is

$$\delta_1^\xi \tilde{E}_j = \frac{256}{175h}\left(\tilde{E}_{j+1/2} - \tilde{E}_{j-1/2}\right) - \frac{1}{4h}\left(\tilde{E}_{j+1} - \tilde{E}_{j-1}\right) + \frac{1}{100h}\left(\tilde{E}_{j+2} - \tilde{E}_{j-2}\right) - \frac{1}{2100h}\left(\tilde{E}_{j+3} - \tilde{E}_{j-3}\right), \quad (5)$$

where, $\tilde{E}_{j+1/2} = \tilde{E}\left(\hat{U}_{j\pm 1/2}, \tilde{\xi}_{x,j\pm 1/2}, \tilde{\xi}_{y,j\pm 1/2}, \tilde{\xi}_{z,j\pm 1/2}\right)$ and $\tilde{E}_{j+m} = \tilde{E}\left(U_{j\pm m}, \tilde{\xi}_{x,j\pm m}, \tilde{\xi}_{y,j\pm m}, \tilde{\xi}_{z,j\pm m}\right)$ are the fluxes at cell edges and at cell nodes, respectively. The numerical fluxes $\hat{E}_{j\pm 1/2}$ may be evaluated by the variables at cell-edges,

$$\hat{E}_{j\pm 1/2} = \tilde{E}\left(\hat{U}_{j\pm 1/2}^L, \hat{U}_{j\pm 1/2}^R, \tilde{\xi}_{x,j\pm 1/2}, \tilde{\xi}_{y,j\pm 1/2}, \tilde{\xi}_{z,j\pm 1/2}\right). \quad (6)$$

where, $\hat{U}_{j\pm 1/2}^L, \hat{U}_{j\pm 1/2}^R$ are variables at cell-edge.

$$\frac{5}{14}(1-\alpha)\hat{U}_{j-1/2}^L + \hat{U}_{j+1/2}^L + \frac{5}{14}(1+\alpha)\hat{U}_{j+3/2}^L$$
$$= \frac{25}{32}(U_{j+1} + U_j) + \frac{5}{64}(U_{j+2} + U_{j-1}) - \frac{1}{448}(U_{j+3} + U_{j-2})$$
$$+ \alpha\left[\frac{25}{64}(U_{j+1} - U_j) + \frac{15}{128}(U_{j+2} - U_{j-1}) - \frac{5}{896}(U_{j+3} - U_{j-2})\right]. \quad (7)$$

where $\alpha < 0$ is the dissipative parameter employed to control the dissipation of the HDCS-E8T7. The corresponding $\hat{U}_{j+1/2}^R$ can be obtained easily by setting $\alpha > 0$. Fig. 1 plots the modified wavenumber of the HDCS-E8T7 with different dissipative parameters. It may be noticed that the resolution of the HDCS-E8T7 is spectral-like.

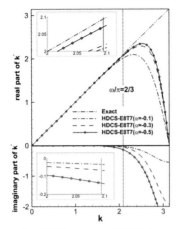

Fig. 1. Modified wavenumber of the HDCS-E8T7.

Dissipative parameter α has no influence on the accuracy of HDCS-E8T7 scheme, however, it has effect on resolution of the scheme. According to the concept of Dispersion-Relation-Preserving (DRP) [17], HDCS-E8T7 scheme has been optimized by adjusted the value of α. The optimized dissipative parameter is equal to 0.3 [14], which will be adopted in this paper.

Seven-point stencil is used by the HDCS-E8T7, thus three kinds of boundary and near-boundary schemes are required for the left-side boundary nodes $j = 1, 2, 3$ as well as for the right-side boundary nodes $j = N, N-1, N-2$. The sixth-order scheme is used for the near boundary $j = 3$ and $j = N - 2$,

$$\widetilde{E}'_j = \frac{64}{45h}(\widehat{E}_{j+1/2} - \widehat{E}_{j-1/2}) - \frac{2}{9h}(\widetilde{E}_{j+1} - \widetilde{E}_{j-1}) + \frac{1}{180h}(\widetilde{E}_{j+2} - \widetilde{E}_{j-2}). \quad (8)$$

The fourth-order scheme is employed for the near boundary $j = 2$ and $j = N - 1$,

$$\widetilde{E}'_j = \frac{4}{3h}(\widehat{E}_{j+1/2} - \widehat{E}_{j-1/2}) - \frac{1}{6h}(\widetilde{E}_{j+1} - \widetilde{E}_{j-1}). \quad (9)$$

The fourth-order boundary schemes for $j = 1$ and $j = N$ are

$$\widetilde{E}'_1 = \frac{1}{6h}(48\widehat{E}_{3/2} + 16\widehat{E}_{5/2} - 25\widetilde{E}_1 - 36\widetilde{E}_2 - 3\widetilde{E}_3),$$

$$\widetilde{E}'_N = \frac{-1}{6h}(48\widehat{E}_{N-1/2} + 16\widehat{E}_{N-3/2} - 25\widetilde{E}_N - 36\widetilde{E}_{N-1} - 3\widetilde{E}_{N-2}). \quad (10)$$

For boundary schemes, the fifth-order interpolations for $j = 3/2$ and $j = N - 1/2$ are

$$\widehat{U}^L_{3/2} = \frac{1}{128}(35U_1 + 140U_2 - 70U_3 + 28U_4 - 5U_5),$$

$$\widehat{U}^L_{N-1/2} = \frac{1}{128}(35U_N + 140U_{N-1} - 70U_{N-2} + 28U_{N-3} - 5U_{N-4}), \quad (11)$$

synchronously, the fifth-order dissipative compact interpolations for $j = 5/2$ and $j = N - 3/2$ are given by

$$\frac{3}{10}(1-\alpha)\widehat{U}^L_{j-1/2} + \widehat{U}^L_{j+1/2} + \frac{3}{10}(1+\alpha)\widehat{U}^L_{j+3/2}$$

$$= \frac{3}{4}(U_{j+1} + U_j) + \frac{1}{20}(U_{j+2} + U_{j-1})$$

$$+ \alpha\left[\frac{3}{8}(U_{j+1} - U_j) + \frac{3}{40}(U_{j+2} - U_{j-1})\right]. \quad (12)$$

For the computation of the viscous terms, the primitive variables, u, v, w, T, are first differentiated to form the components of the stress tensor and the heat flux vector. The viscous flux derivatives are then computed by a second application of the HDCS-E8T7.

3.2. Grid metric calculation

The conservative grid metric derivatives (2) have the discretization form as

$$\begin{cases} \tilde{\xi}^x_x = \delta^\xi_{II}((\delta^\eta_{III}y)z) - \delta^\xi_{II}((\delta^\zeta_{III}y)y), \tilde{\xi}^y_y = \delta^\xi_{II}((\delta^\eta_{III}z)x) - \delta^\xi_{II}((\delta^\zeta_{III}z)x), \tilde{\xi}^z_z = \delta^\xi_{II}((\delta^\eta_{III}x)y) - \delta^\xi_{II}((\delta^\zeta_{III}x)y), \\ \tilde{\eta}^x_x = \delta^\eta_{II}((\delta^\zeta_{III}y)z) - \delta^\eta_{II}((\delta^\xi_{III}y)z), \tilde{\eta}^y_y = \delta^\eta_{II}((\delta^\zeta_{III}z)x) - \delta^\eta_{II}((\delta^\xi_{III}z)x), \tilde{\eta}^z_z = \delta^\eta_{II}((\delta^\zeta_{III}x)y) - \delta^\eta_{II}((\delta^\xi_{III}x)y), \\ \tilde{\zeta}^x_x = \delta^\zeta_{II}((\delta^\xi_{III}y)z) - \delta^\zeta_{II}((\delta^\eta_{III}y)z), \tilde{\zeta}^y_y = \delta^\zeta_{II}((\delta^\xi_{III}z)x) - \delta^\zeta_{II}((\delta^\eta_{III}z)x), \tilde{\zeta}^z_z = \delta^\zeta_{II}((\delta^\xi_{III}x)y) - \delta^\zeta_{II}((\delta^\eta_{III}x)y). \end{cases} \quad (13)$$

where $\delta^\xi_{II}, \delta^\eta_{II}, \delta^\zeta_{II}$ and $\delta^\xi_{III}, \delta^\eta_{III}, \delta^\zeta_{III}$ are numerical derivative operators used for the metric calculations in ξ, η and ζ coordinate direction. For the applications on curvilinear meshes, finite difference scheme should satisfy the SCL which means $I_x = I_y = I_z = 0$ [7],

$$\begin{cases} I_x = \delta^\xi_I\left(\tilde{\xi}_x\right) + \delta^\eta_I\left(\tilde{\eta}_x\right) + \delta^\zeta_I\left(\tilde{\zeta}_x\right), \\ I_y = \delta^\xi_I\left(\tilde{\xi}_y\right) + \delta^\eta_I\left(\tilde{\eta}_y\right) + \delta^\zeta_I\left(\tilde{\zeta}_y\right), \\ I_z = \delta^\xi_I\left(\tilde{\xi}_z\right) + \delta^\eta_I\left(\tilde{\eta}_z\right) + \delta^\zeta_I\left(\tilde{\zeta}_z\right). \end{cases} \quad (14)$$

In order to fulfill the SCL, the grid metrics should be calculated with a conservative form by the same scheme used for flux derivative calculations, i.e. $\delta_I = \delta_{II}$, to implement the CMM [7]. It has been proved in Ref. [7] that the CMM can be easily applied in high-order schemes if the difference operator δ_I of flux derivatives is not split, while it is difficult to be applied in the schemes where the δ_I is split into two upwind operators as δ^+_I and δ^-_I. Although the inner-level difference operator δ_{III} in the conservative metrics has no effect on the GCL, the constraint $\delta_{II} = \delta_{III}$ is recommended by Deng et al. [7]. Not long after, the constraint $\delta_{II} = \delta_{III}$ is explained from geometry viewpoint, and a symmetrical conservative metric method (SCMM) [12], which can evidently increase the numerical accuracy on irregular grids, *is proposed*. To eliminate the SCL errors on curvilinear meshes, here we calculate the grid metrics with a conservative form by the same scheme used for flux derivative calculations, i.e. $\delta_I = \delta_{II} = \delta_{III}$. Then explicit eighth-order central interpolation is employed for δ_{II} and δ_{III} at interior nodes,

$$\widehat{U}_{j+1/2} = \frac{1}{2048}(1225(U_{j+1} + U_j) - 245(U_{j+2} + U_{j-1}) + 49(U_{j+3} + U_{j-2}) - 5(U_{j+4} + U_{j-3})). \quad (15)$$

At boundary points, the interpolations for δ_{II} and δ_{III} are

$$\widehat{U}_{3/2} = \frac{1}{128}(35U_1 + 140U_2 - 70U_3 + 28U_4 - 5U_5),$$

$$\widehat{U}_{5/2} = \frac{1}{128}(-5U_1 + 60U_2 + 90U_3 - 20U_4 + 3U_5),$$

$$\widehat{U}_{7/2} = \frac{1}{256}(3U_1 - 25U_2 + 150U_3 + 150U_4 - 25U_5 + 3U_6),$$

$$\widehat{U}_{N-1/2} = \frac{1}{128}(35U_N + 140U_{N-1} - 70U_{N-2} + 28U_{N-3} - 5U_{N-4}),$$

$$\widehat{U}_{N-3/2} = \frac{1}{128}(-5U_N + 60U_{N-1} + 90U_{N-2} - 20U_{N-3} + 3U_{N-4}),$$

$$\widehat{U}_{N-5/2} = \frac{1}{256}(3U_N - 25U_{N-1} + 150U_{N-2} + 150U_{N-3} - 25U_{N-4} + 3U_{N-5}). \quad (16)$$

3.3. Time Integration

In the following numerical tests, highly stretched meshes are employed for the HILES studies of wall-bounded flows. The stability constraint of explicit time-marching methods is too restrictive and the use of an implicit approach becomes necessary, then dual time stepping scheme [18] with Newton-like subiterations [19] is employed. With R denoting the residual, the governing equation is

$$\frac{\partial \tilde{U}}{\partial t} = R = \frac{1}{R_e}\left(\frac{\partial \tilde{E}_v}{\partial \xi} + \frac{\partial \tilde{F}_v}{\partial \eta} + \frac{\partial \tilde{G}_v}{\partial \zeta}\right) - \frac{\partial \tilde{E}}{\partial \xi} - \frac{\partial \tilde{F}}{\partial \eta} - \frac{\partial \tilde{G}}{\partial \zeta}. \quad (17)$$

The second-order dual time stepping scheme may be written as

$$\left[J^{-1}\left(\frac{1}{\Delta\tau} + \frac{3}{2\Delta t}\right)I + M(U^p)\right]\Delta U^{p+1}$$
$$= -\left[J^{-1}\frac{3U^p - 4U^n + U^{n-1}}{2\Delta t} + R(U^p)\right], \quad (18)$$

where $M(U) = \partial R(U)/\partial U$, and τ is pseudo-time. For the first subiteration, $p = 1, U^p = U^n$, and as $p \to \infty, U^p \to U^{n+1}$. The spatial derivatives in the implicit operators are represented using standard second-order approximations whereas seventh-order discretizations are employed for the residual.

4. Numerical results

4.1. Transition and turbulence decay in the three-dimensional Taylor–Green vortex

For the assessment of HILES, we consider the three-dimensional Taylor–Green vortex (TGV) case which has been used as fundamental prototype for vortex stretching and consequent production of small-scale eddies to test LES methods [20]. The TGV initial configuration involves triple-periodic boundary conditions enforced on a cubical domain with box side length 2π. The computational domain is chosen to be $(x,y,z) \in [-\pi,\pi] \times [-\pi,\pi] \times [-\pi,\pi]$. The flow is initialized with the solenoidal velocity components,

$$u = V_\infty \sin(x)\cos(y)\cos(z),$$
$$v = -V_\infty \cos(x)\sin(y)\cos(z), \quad (19)$$
$$w = 0,$$

and pressure given by a solution of the Poisson equation for the above given velocity field is

$$p = p_\infty + \frac{1}{16}\rho_\infty V_\infty^2[2 + \cos(2z)][\cos(2x) + \cos(2y)]. \quad (20)$$

Since the present code solves the compressible form of the Navier–Stokes equations, a low Mach number, $M_\infty = 0.1$, is specified to ensure the solutions obtained for the velocity and pressure fields are indeed very close to those obtained assuming an incompressible flow. The considered Reynolds number in this test is $R_e = 400$.

Three meshes are used to solve this problem, the grid dimensions are $65 \times 65 \times 65$ (coarse), $97 \times 97 \times 97$ (medium), $129 \times 129 \times 129$ (fine), respectively. Mesh spacing is constant in the every direction. Third-order Runge–Kutta scheme [21] is used for the time integration. The physical duration of the computation is set to $t_{final} = 10$ (non-dimensional time).

In Fig. 2, we compare the evolution in time of the mean kinetic energy dissipation rate ε between the HILES and the DNS of Brachet et al. [22]. ε is written as

$$\varepsilon = -\frac{de(t)}{dt}, \quad (21)$$

where $e(t) = 1/2\langle u^2 + v^2 + w^2\rangle$ and $\langle \cdot \rangle$ denotes mean (volumetric average). In this figure, the solutions of HILES are approach to that of DNS when the mesh size is increased, and results of medium grid and fine grid are in good agreement with the previous DNS results.

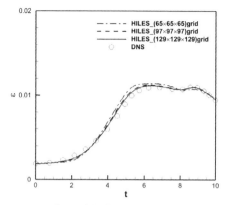

Fig. 2. Evolution of energy dissipation rate.

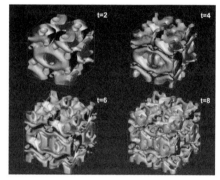

Fig. 3. Iso-surfaces of the second invariant of the velocity gradient tensor colored by streamwise velocity. (colored picture attached at the end of the book)

Fig. 3 shows the evolution of TGV flow dynamics based on instantaneous visualizations of the second invariant of the velocity gradient tensor Q from the present TGV simulations with medium grid. The second invariant Q is

$$Q = (\Omega_{ij}\Omega_{ij} + S_{ij}S_{ij})/2, \quad (22)$$

where $\Omega_{ij} = (u_{i,j} - u_{j,i})/2$ and $S_{ij} = (u_{i,j} + u_{j,i})/2$ are the antisymmetric and the symmetric components of the curl of the velocity respectively. The TGV evolution is laminar and strongly anisotropic at the early times, then energy is transferred to larger wave numbers by vortex stretching, and the flow eventually becomes fully developed turbulent. In the final steps, the small scales are nearly isotropic and exhibit an $k^{-5/3}$ inertial range of the kinetic energy spectrum. To show this process clearly, the kinetic energy spectrum $e_k(k,t)$ from HILES simulation with medium grid at different times is plotted in Fig. 4. The energy spectrum $e_k(k,t)$ is computed by the velocity field in Fourier space, and

$$e(t) = \int e_k(k,t)dk. \quad (23)$$

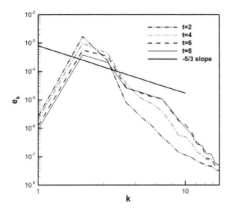

Fig. 4. Turbulence kinetic energy spectra at different time.

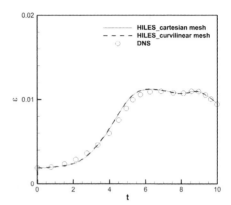

Fig. 6. Evolution of energy dissipation rate on curvilinear mesh comparing with that on cartesian mesh and the DNS solution.

The illustrated slope in Fig. 4 indicates that the HILES can reliably capture the turbulence spectrum. Moreover, the laminar-turbulent transition is resolved by the HILES naturally. For most eddy-viscosity models, some modifications are required to catch this transition [20].

In order to show the ability of the HILES in handling complex mesh, the TGV problem is solved by a curvilinear mesh which is generated from the medium mesh used above. To obtain this new curvilinear mesh, the medium mesh is modified only in the domain $(x,y,z) \in [-\pi/2, \pi/2] \times [-\pi/2, \pi/2] \times [-\pi/2, \pi/2]$ according to the equations,

$$\begin{cases} x = -\pi/2 + \Delta x[(i-1) + 0.5\sin(4\pi y_0) \times \sin(4\pi z_0)], \\ y = -\pi/2 + \Delta y[(j-1) + 0.5\sin(4\pi x_0) \times \sin(4\pi z_0)], \\ z = -\pi/2 + \Delta z[(k-1) + 0.5\sin(4\pi x_0) \times \sin(4\pi y_0)], \end{cases} \quad (24)$$

where $x_0 = \Delta x(i-1), y_0 = \Delta y(j-1), z_0 = \Delta z(k-1), i = 1, \ldots, 48, j = 1, \ldots, 48, j = 1, \ldots, 48, \Delta x = \Delta y = \Delta z = \pi/48$. This 3D curvilinear mesh is sketchily shown in Fig. 5. Comparing with the calculation from the solutions of HILES on medium cartesian mesh, Fig. 6 plots the mean kinetic energy dissipation rate ε on the new curvilinear mesh. It is clear that the slight differences between these two solutions can be rarely observed. This phenomenon demonstrates that the HILES has potential ability to simulate the complex flow problem on complex geometry.

4.2. Flow past a circular cylinder

The flow past a circular cylinder at Mach number of 0.2 and Reynolds number based on cylinder diameter, D, of $Re_D = 3900$ is simulated by HILES. The large eddy simulation for this problem has been performed extensively by different research groups [23–25]. The nondimensional computational domain size is $(90 \times 60 \times \pi)$ in (x, y, z). The mesh size is $180 \times 180 \times 45$ in steamwise, normal and spanwise directions respectively. The 2D grid system is shown in Fig. 7. Grid points are clustered near the cylinder surface, with a minimum normal spacing of 0.001. Due to a H-type mesh is employed for the HILES, the gird size here is smaller than that of Rizzetta et al. [24], where a $199 \times 197 \times 53$ O-type grid is used for sixth-order central compact scheme to perform implicit LES (ILES) of the same case. At the outer boundary, far-field boundary conditions based on the Local One-Dimensional Inviscid (LODI) approximation [26] and sponge technique [27] are prescribed. The no-slip condition is invoked on the cylinder surface, together with fifth-order accurate approximations for an adiabatic wall and zero normal pressure gradient. At the spanwise boundaries, periodicity is applied.

According to the study of Deng et al.[7], the SCL is satisfied automatically on uniform meshes. However, the SCL may be dissatisfied on curvilinear meshes. If the SCL has not been satisfied, numerical instabilities and even computing collapse may appear on complex curvilinear grids during numerical simulation. Thus we will investigate the effect of SCL errors in current simulations. To show the influence of the SCL errors, we design a new scheme HDCS-E8T7A, which is same as HDCS-E8T7 except that the grid metric derivatives are calculated by combining the conservation grid metric derivatives and traditional grid metric derivatives. Traditional grid metric derivatives are given by

$$\begin{cases} \bar{\xi}_x^T = \delta_y^\eta(y)\delta_z^\zeta(z) - \delta_z^\eta(z)\delta_y^\zeta(y), \bar{\xi}_y^T = \delta_z^\eta(z)\delta_x^\zeta(x) - \delta_x^\eta(x)\delta_z^\zeta(z), \bar{\xi}_z^T = \delta_x^\eta(x)\delta_y^\zeta(y) - \delta_y^\eta(y)\delta_x^\zeta(x), \\ \bar{\eta}_x^T = \delta_y^\zeta(y)\delta_z^\xi(z) - \delta_z^\zeta(z)\delta_y^\xi(y), \bar{\eta}_y^T = \delta_z^\zeta(z)\delta_x^\xi(x) - \delta_x^\zeta(x)\delta_z^\xi(z), \bar{\eta}_z^T = \delta_x^\zeta(x)\delta_y^\xi(y) - \delta_y^\zeta(y)\delta_x^\xi(x), \\ \bar{\zeta}_x^T = \delta_y^\xi(y)\delta_z^\eta(z) - \delta_z^\xi(z)\delta_y^\eta(y), \bar{\zeta}_y^T = \delta_z^\xi(z)\delta_x^\eta(x) - \delta_x^\xi(x)\delta_z^\eta(z), \bar{\zeta}_z^T = \delta_x^\xi(x)\delta_y^\eta(y) - \delta_y^\xi(y)\delta_x^\eta(x). \end{cases} \quad (25)$$

Fig. 5. Curvilinear mesh sketch for the TGV problem.

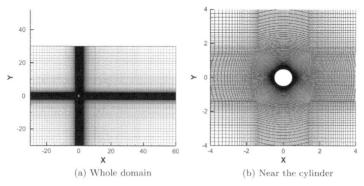

(a) Whole domain (b) Near the cylinder

Fig. 7. The grid system of the cylinder test.

Table 1
The maximal errors of I_x and I_y for the HDCS-E8T7 and HDCS-E8T7A.

	HDCS-E8T7	HDCS-E8T7A Weight σ							
		10^{-2}	10^{-1}	0.2	0.4	0.6	0.8	1.0
max(I_x)	4.83E−14	3.35E−6	3.35E−5	6.70E−5	1.34E−4	2.01E−4	2.68E−4	3.35E−4
max(I_y)	7.08E−14	3.36E−6	3.36E−5	6.72E−5	1.34E−4	2.02E−4	2.69E−4	3.36E−4

For instance, one of the grid metric derivatives is written as

$$\tilde{\xi}_x^A = \sigma \tilde{\xi}_x^T + (1-\sigma)\tilde{\xi}_x^C, \qquad (26)$$

where $0 \leq \sigma \leq 1$ is the weight coefficient, $\tilde{\xi}_x^A$ is the grid metric derivative of HDCS-E8T7A, $\tilde{\xi}_x^T$ and $\tilde{\xi}_x^C$ are traditional and conservation grid metric derivatives, respectively. If $\sigma = 0$, the HDCS-E8T7 satisfying the SCL is recovered by the HDCS-E8T7A. However, the SCL is not satisfied if $\sigma \neq 0$. The maximal SCL errors of the HDCS-E8T7 and HDCS-E8T7A are listed in Table 1, it can be seen that the SCL error is highest when $\sigma = 1$. As shown in Fig. 8, the largest SCL error of the HDCS-E8T7A with $\sigma = 1$ presents at the singular point.

The uniform initial flowfield is used to start the simulation and the transition period is up to dimensionless time $t = 100$. Then the statistical results are calculated from $t = 100$ to $t = 200$. The time step size in the computations is $\Delta t = 0.01$. Comparing with the solution of implicit LES (ILES) [24] and MILES [23], mean surface pressure from the HILES computation appears in Fig. 9. Agreement between the simulation of HILES satisfying the SCL and the measurement [28] is reasonable, particularly in the upstream region, and the result of HILES satisfying the SCL lies closer to the experimental date on the aft cylinder surface. For the result of HILES dissatisfying the SCL, however, oscillations are observed at the location near $\theta = 130°$ where the singular point exists. The fluctuating pressure on the surface is also shown in Fig. 9. Due to the SCL errors, the difference between solutions satisfying the SCL and dissatisfying the SCL is significant. In the solution dissatisfying the SCL, there is a second peak at the location near $\theta = 130°$.

Fig. 10 plots mean streamwise velocity field calculated by the HILES with HDCS-E8T7 and HDCS-E8T7A. Highly smooth flowfield is obtained by the HILES with HDCS-E8T7. However, when the

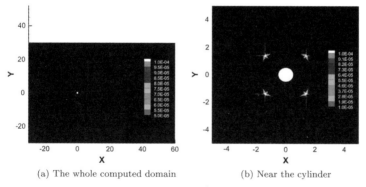

(a) The whole computed domain (b) Near the cylinder

Fig. 8. The SCL error ($|I_x| + |I_y|$) of HDCS-E8T7A with $\sigma = 1$. (colored picture attached at the end of the book)

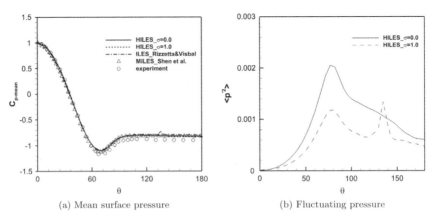

(a) Mean surface pressure

(b) Fluctuating pressure

Fig. 9. Mean and fluctuating pressure on the cylinder surface.

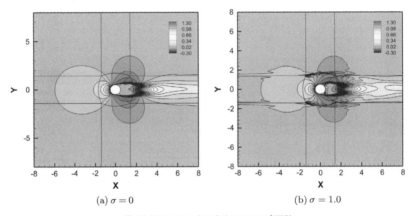

(a) $\sigma = 0$

(b) $\sigma = 1.0$

Fig. 10. Mean streamwise velocity contours of HILES.
(colored picture attached at the end of the book)

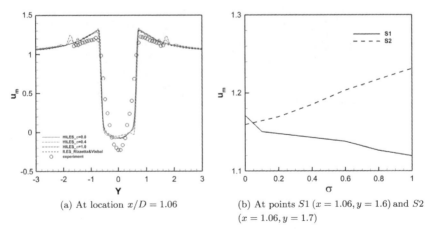

(a) At location $x/D = 1.06$

(b) At points $S1$ ($x = 1.06, y = 1.6$) and $S2$ ($x = 1.06, y = 1.7$)

Fig. 11. Spanwise averaged mean streamwise velocity distributions near the cylinder.

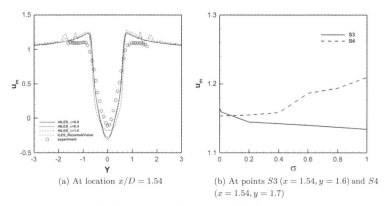

Fig. 12. Spanwise averaged mean streamwise velocity distributions.

(a) At location $x/D = 1.54$

(b) At points $S3$ ($x = 1.54, y = 1.6$) and $S4$ ($x = 1.54, y = 1.7$)

HDCS-E8T7A with $\sigma = 1$ is employed for the computation, the field has been contaminated near the singular point where the SCL error is the largest. It is clear that HILES can eliminate the SCL errors which may contaminate the flowfield. To quantitatively show the influence of the SCL errors, we plot vertical distributions of the mean streamwise velocity component at the streamwise location $x/D = 1.06$ in Fig. 11. Following the SCL errors are enlarged, it is clear that unphysical oscillation is getting more visible. This phenomenon is also demonstrated by the velocities at two points $S_1(x = 1.06, y = 1.6)$ and $S_2(x = 1.06, y = 1.7)$, as shown in Fig. 11. In Fig. 12, similar oscillation is observed at streamwise location $x/D = 1.54$ as well as two points $S_3(x = 1.54, y = 1.6)$ and $S_4(x = 1.54, y = 1.7)$ near the singular point. The overall results satisfying the SCL are very similar to the numerical results of Rizzetta et al. [24]. However, there are some differences between the results of the simulations and those of the experiment of Lourenco and Shih [28]. Those differences are most likely due to that the particular experiment suffered from some external disturbances that

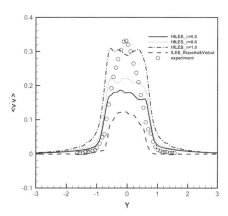

Fig. 14. Lateral Reynolds stress at location $x/D = 1.54$.

contributed to an earlier transition in the separating shear layers [25].

The streamwise and lateral Reynolds stress calculated by the HILES are given in Figs. 13 and 14. Comparing with the experimental data [28], it may be seen that both HILES and ILES significantly under-predict the Reynolds stresses. Due to the SCL errors, the unphysical oscillations may magnify the Reynolds stress. The instantaneous vorticity contours are plotted in Fig. 15. Illustrated in the figure is the fine-scale details calculated by HILES with HDCS-E8T7, particularly in the near wake. However, it is clear that the vorticity may be significantly contaminated by the SCL errors.

4.3. Stall phenomena of thin airfoil NACA64A006

This computation is chosen to check the ability of the HILES for simulating the stall phenomena of the thin airfoil NACA64A006, which is failed to be successfully simulated by RANS model. The airfoil is the same as the experimental model used by McCullough et al. [29]. According to the experiment, freestream Mach number

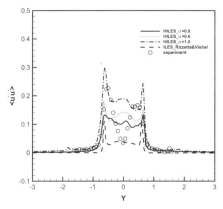

Fig. 13. Streamwise Reynolds stress at location $x/D = 1.06$.

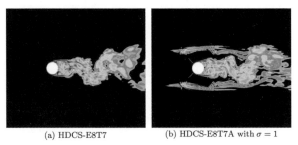

(a) HDCS-E8T7   (b) HDCS-E8T7A with $\sigma = 1$

Fig. 15. The instantaneous vorticity contours. (colored picture attached at the end of the book)

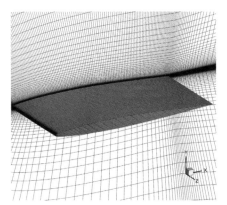

Fig. 16. Computational grid of the airfoil NACA64a006.

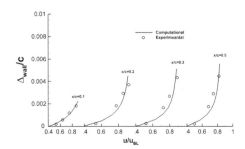

Fig. 18. Boundary-layer velocity profiles developed along the suction surface.

is set to be $M_\infty = 0.17$, chord length is $c = 1.524$ m, the Reyonlds number based on mean aerodynamic chord c is 5.8×10^6, and the angle of attack ranges from $4°$ to $11°$.

A C-type mesh is employed for this computation as shown in Fig. 16. The x, y and z directions correspond to the steamwise, normal and spanwise directions respectively. For the large eddy simulation, the computational domain extends to $0.5c$ in the spanwise, and the mesh size is $341 \times 121 \times 37$. The grid size here is similar to that of Kawai et al. [2], where the LES/RANS hybrid methodology with sixth-order central compact scheme is used to simulate this stall phenomenon. Although the required mesh may be relatively coarser for hybrid methodology than that for LES, our mesh resolution is refined near the wall, then this relatively coarse mesh is suitable for the HILES to capture the stall phenomenon.

At the outer boundary, far-field boundary conditions based on the LODI approximation [26] are prescribed. The no-slip condition is invoked on the cylinder surface, together with fifth-order accurate approximations for an adiabatic wall and zero normal pressure gradient. At the spanwise boundaries, periodicity is applied. The

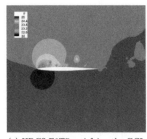 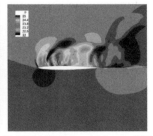

(a) HDCS-E8T7 satisfying the GCL (b) HDCS-E8T7 dissatisfying the GCL

Fig. 17. Instantaneous pressure contours of HILES with HDCS-E8T7 and HDCS-E8T7A. (colored picture attached at the end of the book)

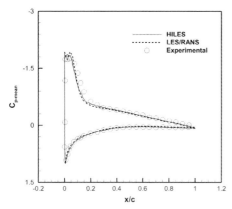

Fig. 19. Time-averaged pressure coefficient distributions over airfoil, $\alpha = 5.5°$.

dimensionless time step size is $\Delta t = 0.001$, computations are advanced for total 25,000 time steps, and flow statistics are collected for 10,000 time steps. For periodic boundaries are used in the spanwise, flow statistics are averaged in this direction.

Although this mesh does not contain singular points, the role of the GCL is also very important since the mesh is highly stretched and curved. If the traditional grid metric derivatives (25) are employed for this calculation, the GCL is dissatisfied and computing collapse will appear during numerical simulation. To illustrate this importance, we continue the calculation from the solution of the HDCS-E8T7 satisfying the GCL at 9° angle of attack, and employ the HDCS-E8T7A with $\sigma = 1$. The computing is collapse after 1300 iterations. Fig. 17 plots the instantaneous pressure contours calculated by the HDCS-E8T7A on the plane $z = 0$ at 100 iteration step. It is clear that unphysical information dominates the flowfield obtained by the HDCS-E8T7A.

The boundary layer should be computed correctly to simulate a separation phenomenon, thus we will investigate the ability of the HILES for the simulation of boundary-layer profile. Fig. 18 compares the computational time-averaged boundary-layer velocity profiles with the corresponding experimental data on the suction surface ($x/c = 0.1, 0.2, 0.3, 0.5$) at 4° angle of attack. It may be seen that the HILES can successfully predict the growth and shape of the boundary-layer profiles at all of the locations. Fig. 19 shows the time-averaged pressure distributions over the airfoil at 5.5° angle of attack. Comparing with the experimental data [29,30] and the numerical solution of LES/RANS hybrid methodology with sixth-order central compact scheme [2], it may be noted that numerical result of the HILES is reliable.

Fig. 20 shows the computed time-averaged lift characteristics. Time-averaged pressure contours at 9°, 10° and 11° angles of attack are also included in this figure. Good agreement between the experimental data [29,30] and the numerical solutions has been observed. It is demonstrated that the HILES has successfully predicted the slope of lift curve, the maximum lift, and the stall angle. Moreover, the lift characteristic seems to be satisfactorily captured even after the stall angle.

The flowfield around NACA64A006 airfoil has thin-airfoil stall characteristics. In this flowfield, the laminar flow separation occurs at the leading edge and the transition causes the turbulent reattachment. According to the numerical research [2], reliable simulation of separation bubble is very important for the successful prediction of stall phenomenon. Fig. 21 shows the time-averaged Mach number distributions and velocity vectors near the leading edge obtained by the HILES. It is obvious that separation bubble is successfully resolved. Such a flowfield is also observed in the experiment by Fitzgerald et al. [31] using laser Doppler velocimetry. Fig. 22 shows the instantaneous pressure distributions over airfoil surface and iso-surface of the second invariant of the velocity gradient tensor at 5.5° angle of attack. Separation from the leading edge and unsteady vortex shedding caused by the instability of the shear layer are clearly observed in this figure. The small vortexes shed from the leading edge, which produces strong unsteadiness in the flow.

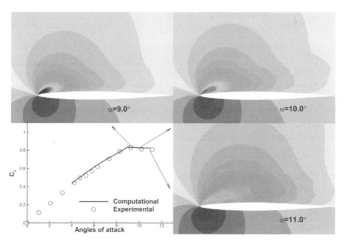

Fig. 20. The time-averaged lift coefficient vs angles of attack and pressure contours at 9°, 10° and 11° angles of attack.

(colored picture attached at the end of the book)

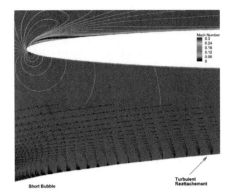

Fig. 21. Time-averaged Mach number distribution and velocity vectors at 5.5° angle of attack (colored picture attached at the end of the book)

Fig. 22. The instantaneous pressure distribution over airfoil surface and iso-surface of the second invariant of the velocity gradient tensor at 5.5° angle of attack.

(colored picture attached at the end of the book)

5. Conclusion

The performance of HILES has been investigated on curvilinear meshes. The capability of HILES is tested by three-dimensional Taylor–Green vortex case, where the kinetic energy dissipation, the kinetic energy spectrum and the laminar-turbulent transition are reliably captured. Moreover, the potential ability of the HILES handling complex geometry is demonstrated by solving TGV problem on a 3D curvilinear mesh.

The HILES is developed based on HDCS-E8T7 satisfying the SCL, and the influence of the SCL errors is shown in the test of flow past a circular cylinder. Highly smooth flowfield is obtained by the HILES with HDCS-E8T7. However, when the HDCS-E8T7A is employed for the computation, the field has been contaminated near the singular point where the SCL error is the largest. Comparing with experimental and other computational data, it is obvious that unphysical oscillation is getting more visible following the SCL error is enlarged. If the SCL errors are eliminated, the results of HILES are reasonable.

According to the numerical solutions of the thin airfoil NACA64A006, the HILES has successfully predicted the boundary-layer profiles over the suction surface, the slope of lift curve, the maximum lift, and the stall angle. Furthermore, separation from the leading edge and unsteady vortex shedding caused by the instability of the shear layer are clearly resolved.

Based on the above investigation, HILES is a suitable method for simulating the unsteady turbulence flow on curvilinear meshes. In our next paper, we will give more applications of HILES on complex geometry to show its ability of simulating complex flow.

Acknowledgments

This study was supported by the project of National Basic Research Program of China (Grant No. 2009CB723800), National Natural Science Foundation of China (Grant Nos. 11072259 and 11202226) and the foundation of State Key Laboratory of Aerodynamics (Grant No. JBKY11030902 and JBKY11010100).

References

[1] Fujii K. Progress and future prospects of CFD in aerospace – wind tunnel and beyond. Progr Aerosp Sci 2005;41:455–70.
[2] Kawai S, Fujii K. Analysis and prediction of thin-airfoil stall phenomena with hybrid turbulence methodology. AIAA J 2005;43(5):953–61.
[3] Mellen PC, Frohlich J, Rodi W. Lessons from LESFOIL project on large eddy simulation of flow around an airfoil. AIAA J 2003;41(4):573–81.
[4] Mary I, Sagant P. Large eddy simulation of flow around an airfoil near stall. AIAA J 2002;40(6):1139–45.
[5] Wang ZJ. High-order methods for the Euler and Navier–Stokes equations on unstructured grids. Progr Aerosp Sci 2007;43:1–41.
[6] Ekaterinaris JA. High-order accurate, low numerical diffusion methods for aerodynamics. Progr Aerosp Sci 2005;41(3–4):192–300.
[7] Deng XG, Mao ML, Tu GH, Zhang HX. Geometric conservation law and applications to high-order finite difference schemes with stationary grids. J Comput Phys 2011;230:1100–15.
[8] Visbal RM, Gaitonde DV. On the use of higher-order finite-difference schemes on curvilinear and deforming meshes. J Comput Phys 2002;181:155–85.
[9] Nonomura T, Iizuka N, Fujii K. Freestream and vortex preservation properties of high-order WENO and WCNS on curvilinear grids. Comput Fluids 2010;39:197–214.
[10] Thomas PD, Lombard CK. Geometric conservation law and its application to flow computations on moving grids. AIAA J 1979;17(10):1030–7.
[11] Pulliam TH, Steger JL. On implicit finite-difference simulations of three-dimensional flow. AIAA Paper 78-10; 1978.
[12] Deng Xiaogang, Min Yaobing, Mao Meiliang, Liu Huayong, Tu Guohua. Further studies on geometric conservation law and applications to high-order finite difference schemes with stationary grids. J Comput Phys 2013;239:90–111.
[13] Deng Xiaogang, Jiang Yi, Mao Meiliang, Liu Huayong, Tu Guohua. Developing hybrid cell-edge and cell-node dissipative compact scheme for complex geometry flows. Sci China Technol Sci 2013;56:2361–9.
[14] Deng Xiaogang, Jiang Yi, Mao Meiliang, Liu Huayong, Li Song, Tu Guohua. A family of hybrid cell-edge and cell-node dissipative compact schemes satisfying geometric conservation law. Commun Comput Phys; 2014 [submitted for publications].
[15] Boris JP, Grinstein FF, Oran ES, Kolbe RL. New insights into large eddy simulation. Fluid Dyn Res 1992;10:199.
[16] Xie Peng, Liu Chaoqun. Weighted compact and non-compact scheme for shock tube and shock entropy interaction. AIAA 2007-509.
[17] Tam CKW, Webb JC. Dispersion-relation-preserving finite difference schemes for computational acoustics. J Comput Phys 1993;107:262–81.
[18] Hsu John M, Jameson Antony. An implicit–explicit hybrid scheme for calculating complex unsteady flows, AIAA 2002-0714.
[19] Gordnier RE, Visbal MR. Numerical simulation of delta-wing roll. AIAA Paper 93-0554; January 1993].
[20] Hickel S, Adams NA, Domaradzki JA. An adaptive local deconvolution method for implicit LES. J Comput Phys 2006;213:413–36.
[21] Gottlieb S, Shu C-W. Total variation diminishing Runge–Kutta schemes. Math Comput 1998;67:73–85.
[22] Brachet M, Meiron D, Orszag S, Nickel B, Morf R, Frisch U. Small-scale structure of the Taylor Green vortex. J Fluid Mech 1983;130:411–52.
[23] Shen Yiqing, Zha Gecheng. Large eddy simulation using a new set of sixth order schemes for compressible viscous terms. J Comput Phys 2010;229:8296–312.
[24] Rizzetta DP, Visbal MR, Blaisdell GA. A time-implicit high-order compact differencing and filtering scheme for large-eddy simulation. Int J Numer Methods Fluids 2003;42:665–93.
[25] Kravchenko AG, Moin P. Numerical studies of flow over a circular cylinder at Re D = 3900. Phys Fluids 1992;12:403–17.
[26] Poinsot T, Lele SK. Boundary conditions for direct simulations of compressible viscous flows. J Comput Phys 1992;101:104–29.
[27] Bodony Daniel J. Analysis of sponge zones for computational fluid mechanics. J Comput Phys 2006;212:681–702.
[28] Lourenco LM, Shih C. Characteristics of the plane turbulent near wake of a circular cylinder. A particle image velocimetry study (private communication).
[29] McCullough GB, Gault DE. Examples of three representative types of airfoil-section stall at low speed. NACA TN2502; September 1951.
[30] McCullough GB, Gault, DE. Boundary-layer and stalling characteristics of the NACA64A006 airfoil section. NACATN1923; August 1949.
[31] Fitzgerald JE, Mueller JT. Measurements in separation bubble on an airfoil using laser velocimetry. AIAA J 1990;28(4):584–92.

Developing a hybrid flux function suitable for hypersonic flow simulation with high-order methods

Dongfang Wang[1], Xiaogang Deng[1,*,†], Guangxue Wang[2] and Yidao Dong[1]

[1] *College of Aerospace Science and Engineering, National University of Defense Technology, Changsha, Hunan 410073, China*
[2] *Computational Aerodynamics Institute, China Aerodynamics Research and Development Center, Mianyang, Sichuan 621000, China*

SUMMARY

In this paper, we develop a new hybrid Euler flux function based on Roe's flux difference scheme, which is free from shock instability and still preserves the accuracy and efficiency of Roe's flux scheme. For computational cost, only 5% extra CPU time is required compared with Roe's FDS. In hypersonic flow simulation with high-order methods, the hybrid flux function would automatically switch to the Rusanov flux function near shock waves to improve the robustness, and in smooth regions, Roe's FDS would be recovered so that the advantages of high-order methods can be maintained. Multidimensional dissipation is introduced to eliminate the adverse effects caused by flux function switching and further enhance the robustness of shock-capturing, especially when the shock waves are not aligned with grids. A series of tests shows that this new hybrid flux function with a high-order weighted compact nonlinear scheme is not only robust for shock-capturing but also accurate for hypersonic heat transfer prediction. Copyright © 2015 John Wiley & Sons, Ltd.

Received 28 April 2015; Revised 7 September 2015; Accepted 11 October 2015

KEY WORDS: Riemann solver; hybrid method; Roe's FDS; shock instability; high-order methods; hypersonic heating

1. INTRODUCTION

Over the past decade, significant advancement has been made in the field of hypersonic technology, among which, the accurate and efficient prediction of the aerothermodynamic environment is essential and urgent for the analysis and design of hypersonic vehicles, presenting a great challenge for the development of computational fluid dynamics. Hypersonic heat transfer prediction requires high-fidelity simulation of the boundary layer, laminar-turbulent transition and shock wave/boundary layer interaction. To meet these demands, it is necessary to develop more accurate and robust computational fluid dynamics algorithms [1, 2].

Recently, great attention has been paid to high-order numerical methods for their potential in simulating complex flow physics with comparatively lower computational cost. Because of these features, high-order methods are now widely used in large eddy simulations, direct numerical simulations and computational aeroacoustics [3, 4]. However, further applications to hypersonic flow are limited for their poor capability of capturing shock waves robustly. Aside from the improvement and modification of high-order numerical algorithms, our research indicates that adopting an appropriate flux function can significantly improve the robustness. Therefore, this paper aims to construct a new flux function suitable for hypersonic flow simulation with high-order methods.

Wang Dongfang, Deng Xiaogang, Wang Guangxue, Dong Yidao. Developing a hybrid flux function suitable for hypersonic flow simulation with high-order methods . International Journal for Numerical Methods in Fluids, 2016, 81(5): 309-327.

A flux function should possess the following two crucial properties [5] for hypersonic heat transfer prediction: (i) shock stability/robustness (i.e., free from both one-dimensional and multidimensional anomalies) and (ii) the ability to resolve the boundary layer (and hence, the temperature gradient near the wall).

The shock stability of flux functions will be affected by spatial schemes. For the traditional second-order monotone upstream-centered scheme for conservation laws (MUSCL), advection upstream splitting method (AUSM) family flux functions are shown to be sensitive to limiters [6]. For high-order methods, solutions are at higher risk of shock instability [7]. Tissera [8] used the Harten-Lax-van Leer flux function and weighted essentially non-oscillatory scheme for the simulation of the standard model HB-2 at a high Mach number of 17.8. Serious numerical oscillations are observed in the results. Our numerical tests also demonstrate that many accredited robust flux functions may encounter oscillations and even fail in hypersonic simulation with high-order methods. This may be caused by the comparatively low dissipation of high-order methods. Therefore, it is necessary to construct a robust flux function suitable for shock wave capturing with high-order methods.

Lax–Friedrichs flux splitting is frequently used with high-order WENO schemes to simulate flows with strong shock waves, and smooth results are obtained [9, 10]. We have evaluated the ability of the Lax–Friedrichs flux function and Rusanov flux function to capture shock waves with high-order methods. Numerical results show that these two flux functions are as good as Steger-Warming and Van Leer flux-vector splitting and can capture shock waves within 4 cells without oscillations. However, these two flux functions are too dissipative to accurately simulate the boundary layer, so they are not suitable for hypersonic heat transfer prediction. In terms of dissipation characteristics, robustly capturing shock waves and accurately simulating the boundary layer are contradictory, as a flux function with constant dissipation characteristics can generally not meet these two requirements simultaneously. Therefore, hybrid flux functions are often employed in hypersonic simulation, which can be adaptively adjusted according to the flow field properties, such as AUSM family flux functions [11], rotated flux functions [12] and the modified Steger-Warming flux function [13].

Nishikawa [12] constructed rotated Roe–Rusanov (RR)/Roe-HLL flux functions based on Ren's [14] rotated Riemann solver approach. We have evaluated their abilities with a fifth-order weighted compact nonlinear scheme, and found that the rotated RR flux function can robustly capture shock waves, but the rotated Roe-HLL flux function may encounter oscillations, both of them can accurately simulate the boundary layer. Upon applying the rotated RR flux function to steady flow simulation, the normal vectors of the rotated RR flux function should be fixed after an order of magnitude reduction of the residuals [12], otherwise, full convergence could not be obtained. This correction method may increase the computational complexity and incur an additional memory cost and is not applicable in unsteady flow simulation.

To solve the convergence problem of the rotated RR flux function, we adopt Quirk's [15] adaptive strategy to replace the rotated Riemann solver approach and develop a new hybrid flux function, a combination of Roe's flux function and the Rusanov flux function. The flux function used in the cell face is determined by the pressure gradient around the cell. The Rusanov flux function is adopted in cell faces near large pressure gradients (such as shock waves), and Roe's flux function is adopted in smooth regions. The switch between flux functions is achieved automatically through a pressure-based switch function instead of logical statements, such that the efficiency of this hybrid flux function is improved. Multidimensional effects, corresponding to multidimensional dissipation (MD) are considered and coupled in the switch function. MD, on the one hand, can enhance the robustness of the hybrid flux function in capturing shock waves, and on the other hand, it can eliminate the adverse effects of flux function switching. In the hybrid flux function, Roe's flux function will not handle strong shock waves, so we can use an entropy correction with smaller dissipation effect. We use an entropy correction based on a flow field, which tends to a small value in smooth regions to improve the resolution of the boundary layer and the accuracy of the heat transfer prediction.

This paper is organized as follows: in the next section, the governing equations and numerical methods are described briefly. The construction of the hybrid Roe–Rusanov flux function, including the switch function and entropy correction, is detailed and analyzed in Section 3. In Section 4, a series of typical tests are presented to assess the effectiveness of this new hybrid Roe–Rusanov flux function in shock capturing and boundary layer resolution. In Section 5, hypersonic flows over a blunt cone and double ellipsoid are simulated, and surface heat transfer profiles are compared with experimental results. Concluding remarks are summarized in Section 6.

2. COMPUTATIONAL METHOD

The non-dimensional strong conservative Navier–Stokes/Euler equations are selected and converted into curvilinear coordinates by introducing the transformation $(x, y, z, t) \to (\xi, \eta, \zeta, \tau)$:

$$\frac{\partial \hat{Q}}{\partial \tau} + \frac{\partial \hat{E}}{\partial \xi} + \frac{\partial \hat{F}}{\partial \eta} + \frac{\partial \hat{G}}{\partial \zeta} = \frac{s}{Re} \left(\frac{\partial \hat{E}_v}{\partial \xi} + \frac{\partial \hat{F}_v}{\partial \eta} + \frac{\partial \hat{G}_v}{\partial \zeta} \right), \tag{1}$$

where $s = 1$ for the Navier–Stokes equations, and $s = 0$ for the Euler equations. Details of Equation (1) can be found in [16]. To ensure the robustness and accuracy of the high-order finite difference schemes on low-quality grids [17], the symmetrical conservative metric method [18] is adopted for the computation of the metrics and Jacobian.

Convective and viscous flux derivatives are discretized by the WCNS-E6 [19] scheme of the sixth order. First, let us consider the discretization of the convective flux derivative along the ξ direction

$$\frac{\partial \hat{E}}{\partial \xi} = \frac{75}{64 \Delta \xi} \left(\hat{E}_{i+\frac{1}{2}} - \hat{E}_{i-\frac{1}{2}} \right) - \frac{25}{384 \Delta \xi} \left(\hat{E}_{i+\frac{3}{2}} - \hat{E}_{i-\frac{3}{2}} \right) + \frac{3}{640 \Delta \xi} \left(\hat{E}_{i+\frac{5}{2}} - \hat{E}_{i-\frac{5}{2}} \right), \tag{2}$$

where $\hat{E}_{i+\frac{1}{2}}$ is the numerical flux at the cell edge:

$$\hat{E}_{i+\frac{1}{2}} = \hat{E} \left(Q^L_{i+\frac{1}{2}}, Q^R_{i+\frac{1}{2}}, \hat{\xi}_{x,i+\frac{1}{2}}, \hat{\xi}_{y,i+\frac{1}{2}}, \hat{\xi}_{z,i+\frac{1}{2}} \right). \tag{3}$$

Both the flux vector splitting schemes and flux difference schemes can be used in Equation (3) for the flux evaluation. $Q^L_{i+\frac{1}{2}}$ and $Q^R_{i+\frac{1}{2}}$ correspond to the left and the right variables at the cell edges, and they are computed from variables at the cell-center through a fifth-order weighted nonlinear interpolation method [19]. As for the variables interpolated, different choices can be made, such as conservative variables, primitive variables, or characteristic variables. Without loss of generality, we choose primitive variables for simple operation. "WCNS-E6E5" is adopted here to denote this sixth-order difference scheme (Equation (2)) combined with the fifth-order weighted nonlinear interpolation. Variables at the cell edges for viscous fluxes are computed through a sixth-order central interpolation.

For time integration, the explicit third-order total variation diminishing Runge–Kutta method [10] and implicit lower-upper symmetric Gauss–Seidel method [20] are employed for unsteady and steady flows, respectively.

3. HYBRID ROE–RUSANOV RIEMANN SOLVER

To illustrate the construction of the hybrid Roe–Rusanov flux function, the computation of the numerical flux $\hat{E}_{i+\frac{1}{2}}$ at the cell edge along the ξ direction is presented in this section.

3.1. Base Riemann solvers

When we employ Roe's approximate Riemann solver [21] in Equation (3), the numerical flux at the cell edge is given by

$$\hat{E}^{Roe}_{i+\frac{1}{2}} = \frac{1}{2} \left[\hat{E} \left(Q^L_{i+\frac{1}{2}} \right) + \hat{E} \left(Q^R_{i+\frac{1}{2}} \right) \right] + \frac{1}{2} \tilde{R} |\tilde{\Lambda}| \tilde{L} \left(Q^L_{i+\frac{1}{2}} - Q^R_{i+\frac{1}{2}} \right), \tag{4}$$

where $\tilde{A} = \partial \hat{E} / \partial Q$ is the flux Jacobian computed by Roe-averaged variables, $\tilde{R}$ and $\tilde{L}$ correspond to the right and left eigenvector matrixes of $\tilde{A}$, respectively ($\tilde{L} = \tilde{R}^{-1}$), and $|\tilde{\Lambda}|$ is the eigenvalue matrix defined by

$$|\hat{\Lambda}| = diag\left(|\tilde{V}^n - \tilde{a}|, |\tilde{V}^n|, |\tilde{V}^n|, |\tilde{V}^n|, |\tilde{V}^n + \tilde{a}|\right), \tag{5}$$

where $\tilde{a}$ is the Roe-averaged sound speed, and $\tilde{V}^n$ is the Roe-averaged speed along the face-normal direction.

Similarly, the numerical flux at the cell edge evaluated by the Rusanov solver [22] in Equation (3) can be expressed as follows:

$$\hat{E}^{Rusanov}_{i+\frac{1}{2}} = \frac{1}{2}\left[\hat{E}\left(Q^L_{i+\frac{1}{2}}\right) + \hat{E}\left(Q^R_{i+\frac{1}{2}}\right)\right] + \frac{1}{2}\tilde{S}\left(Q^L_{i+\frac{1}{2}} - Q^R_{i+\frac{1}{2}}\right), \tag{6}$$

$$\tilde{S} = \left(|\tilde{V}^n| + \tilde{a}\right)\mathbf{I}. \tag{7}$$

According to Liou's analysis [23], the dissipation coefficients of the mass flux for Roe's flux function are in the form of

$$\mathcal{D}^{(\rho)}_{Roe} = |\tilde{V}^n|, \qquad \mathcal{D}^{(p)}_{Roe} = \left(|\tilde{V}^n - \tilde{a}| + |\tilde{V}^n + \tilde{a}| - 2|\tilde{V}^n|\right)/2\tilde{a}. \tag{8}$$

Likewise, the dissipation coefficients of the mass flux for the Rusanov flux function are

$$\mathcal{D}^{(\rho)}_{Rusanov} = |\tilde{V}^n| + |\tilde{a}|, \qquad \mathcal{D}^{(p)}_{Rusanov} = 0. \tag{9}$$

According to Liou's lemma [23], the Rusanov flux function is a shock-stable scheme, but it is incapable of accurately computing the contact discontinuity, thus leading to a weak resolution of the boundary layer.

3.2. Hybrid method

Roe's flux function is an approximate Riemann solver taking into account all waves in the exact Riemann solver, whereas for the Rusanov flux function, only the wave corresponding to the maximum eigenvalue is considered. The difference between these two flux functions lies in the eigenvalue matrix of the dissipation part. Therefore, it is easy enough for these two flux functions to switch through the manipulation of the eigenvalues. Based on the discussions previously, we construct a new hybrid flux function:

$$\hat{E}^{Roe-Rusanov}_{i+\frac{1}{2}} = \frac{1}{2}\left[\hat{E}\left(Q^L_{i+\frac{1}{2}}\right) + \hat{E}\left(Q^R_{i+\frac{1}{2}}\right)\right] + \frac{1}{2}\tilde{R}\Lambda^*\tilde{L}\cdot\left(Q^L_{i+\frac{1}{2}} - Q^R_{i+\frac{1}{2}}\right). \tag{10}$$

In smooth regions, $\Lambda^* = |\hat{\Lambda}|$ (Equation (5)) and Roe's flux function is recovered. In the vicinity of shock waves, $\Lambda^* = \tilde{S}$ (Equation (7)) and the hybrid flux function will switch to the Rusanov flux function. Construction of the switch function ω is similar to that used by Scalabrin [24]. ω is 1 in smooth regions and 0 elsewhere. The eigenvalue matrix of the hybrid flux function can be expressed as

$$\Lambda^* = \omega|\tilde{\Lambda}| + (1 - \omega)\tilde{S}, \tag{11}$$

$$\omega = 0.5 - 0.5\,\mathrm{sign}\,(\nabla p - \alpha)\left(1 - \exp\left(-100|\nabla p - \alpha|\right)\right), \tag{12}$$

$$\nabla p = \frac{|p^L - p^R|}{\min\left(p^L, p^R\right)}. \tag{13}$$

where α is a threshold parameter related to the shock strength. The switch functions with different α are plotted in Figure 1. For many problems, a sensible threshold can be specified a priori. In Ref [24], α is chosen to be 0.3. For forward step flow in Section 4, we found no difference in the results when α is in the range of $0.15 \sim 0.7$. In this paper, α is set to be 0.15 to ensure that the Rusanov flux function is active near a weak shock.

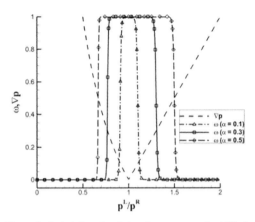

Figure 1. Switch function (ω) and pressure gradient (∇p).

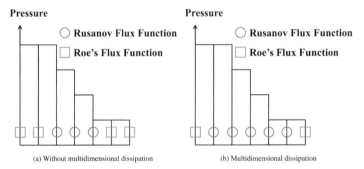

Figure 2. Flux function switch near shock wave.

Multidimensional effects on the pressure gradient should be considered in practice. For a three-dimensional structured mesh, without loss of generality, the pressure gradient along the ξ direction can be computed as follows:

$$\nabla p_{i+\frac{1}{2},j,k} = \max\left(\nabla p_{i,j,k}, \nabla p_{i+1,j,k}\right), \tag{14}$$

$$\nabla p_{i,j,k} = \max\left(\nabla p^*_{i+\frac{1}{2},j,k}, \nabla p^*_{i-\frac{1}{2},j,k}, \nabla p^*_{i,j+\frac{1}{2},k}, \nabla p^*_{i,j-\frac{1}{2},k}, \nabla p^*_{i,j,k+\frac{1}{2}}, \nabla p^*_{i,j,k-\frac{1}{2}}\right), \tag{15}$$

$$\nabla p^*_{i+\frac{1}{2},j,k} = \frac{|p_{i,j,k} - p_{i+1,j,k}|}{\min\left(p_{i,j,k}, p_{i+1,j,k}\right)}. \tag{16}$$

The aforementioned algorithms are somewhat like adding multidimensional dissipation [25], which can obviously strengthen the robustness of the hybrid Roe–Rusanov flux function. Furthermore, this algorithm ensures that the switch of flux functions occurs in only smooth regions to avoid nonphysical numerical oscillations, improving the convergence characteristics of the hybrid flux function. Flux functions switching near shocks is shown in Figure 2. The histogram bars represent the cells in the computational grid, and the heights of the histogram bars represent the pressure levels. The shock is located in the place between bars 3 and 4.

It can be seen that Roe's flux function is used only in smooth regions when α is set to a small value, so we use an entropy correction that can minimize the dissipation effect. Detailed formulas of entropy correction are given as follows [25]

$$|\Lambda|_* = \begin{cases} |\Lambda| & \text{if } |\Lambda| \geq \eta \\ \frac{1}{2}\left(\frac{|\Lambda|^2}{\eta} + \eta\right) & \text{otherwise} \end{cases}, \quad (17)$$

$$\eta(Q^L, Q^R) = |V_L^n - V_R^n| + |a_L - a_R|, \quad (18)$$

where $V_{L/R}^n$ is the speed along face-normal direction. In this way, nonphysical expansion shocks can be eliminated, whereas entropy corrections in the smooth regions are small enough to maintain the high resolution of Roe's flux function, especially for boundary layer resolution and heat transfer prediction.

4. NUMERICAL TESTS

In this section, we use a series of classic test cases to evaluate the capabilities of the hybrid Roe–Rusanov flux function with a fifth-order WCNS-E6E5 to capture the shock waves and simulate the boundary layer.

First, a hypersonic inviscid flow over a cylinder is simulated to evaluate the ability of the hybrid Roe–Rusanov flux function to capture strong shock waves, and a comparison is made with other flux functions. Then, three classic inviscid flows, forward step flow, double Mach reflection, and shock diffraction are tested. The distribution of the switch function is plotted to analyze the adaptive features of the hybrid flux function. Finally, the incompressible flat-plate boundary layer is computed, and the results are compared with the Blasius solution.

4.1. Mach 17 hypersonic inviscid flow over a cylinder

Hypersonic inviscid flow over a cylinder is a well-known severe test case to examine the catastrophic carbuncle phenomenon of some Riemann solvers. Two sets of grids, grids aligned with shock and grids not aligned with shock, have been designed, as shown in Figure 3. It has been shown that for grids not aligned with shock, spurious oscillations would occur behind shock waves, posing a great difficulty for numerical calculations [6, 12, 26]. For comparison, various flux functions are employed here: Van Leer [27, 28], AUSMPW+ [29] and rotated RR [12]. For the rotated RR flux function, the normal vector of the flux function is updated at every time step.

For grids aligned with shock, the pressure contours, pressure distribution along the stagnation streamline, and convergence history of the residue are shown in Figures 4, 5 and 6. All of these four flux functions perform well and are free from oscillations and carbuncle phenomena. One could observe that the hybrid/rotated Roe–Rusanov flux function obtains a somewhat more dissipative solution than do the Van Leer and AUSMPW+ flux functions. All residues converge to 10^{-12}, except that of the rotated RR flux function, which stagnates after four orders of magnitude reduction.

For grids not aligned with shock, the AUSMPW+ flux function fails. The results of the other three flux functions are illustrated in Figure 7. Slight oscillations near shock at the outflow boundary are observed for the Van Leer flux function, whereas the pressure contour line is comparatively smooth for the hybrid/rotated Roe–Rusanov flux function. The residue and aerodynamic force convergence histories demonstrate that the hybrid Roe–Rusanov flux function performs better than the rotated RR flux function, as shown in Figures 8 and 9.

The effects of MD in the switch function are also considered. For the grid aligned with shock, the results without MD are very similar to those with MD, as shown in Figure 4, but the residue stagnates after a three order of magnitude reduction, as shown in Figure 6. For grids not aligned with shock, near shock waves at the outflow boundary slight oscillations are observed in the results without MD, as shown in Figure 7.

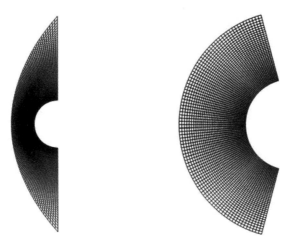

Figure 3. Computational grids: Left, grids aligned with shock; Right, grids not aligned with shock.

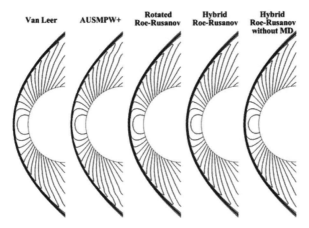

Figure 4. Pressure contours for grids aligned with shock.

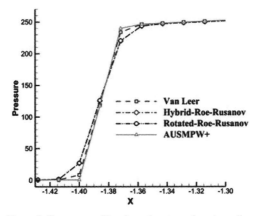

Figure 5. Pressure profiles along the stagnation streamline.

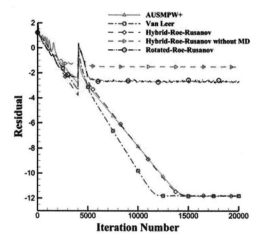

Figure 6. Convergence histories for grids aligned with shock.

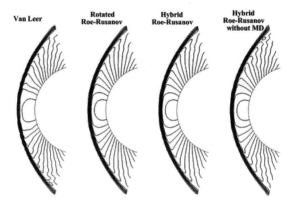

Figure 7. Pressure contours for grids not aligned with shock.

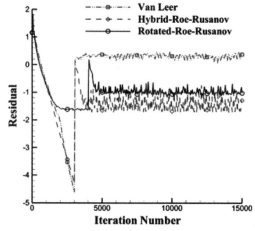

Figure 8. Convergence histories for grids not aligned with shock.

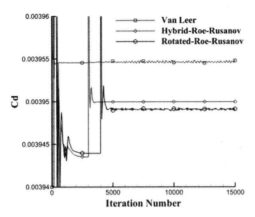

Figure 9. Drag coefficient convergence histories for grids not aligned with shock.

4.2. Forward facing step problem

Woodward and Colella conducted a detailed research on this case to evaluate the numerical methods in simulation of fluid flows with strong shocks [30]. The wind tunnel is 1-length-unit wide and 3-length-units long. The step is 0.2-length-units high and located 0.6 length units from the left-hand end of the tunnel. The problem is initialized by a right-moving Mach 3 flow. Reflective boundary conditions are imposed at the wind tunnel walls and step surface. Inflow and outflow boundary conditions are applied at the left-hand and right-hand sides, respectively. For the treatment of the singularity at the corner of the step, the method suggested in [30] is adopted, which is based on the assumption of a nearly steady flow in the region near the corner. For this case, a grid with 300×100 grid points and two flux functions, hybrid Roe–Rusanov and AUSMPW+ are employed.

To evaluate the effects of α in the hybrid Roe–Rusanov flux function, we computed the forward step flow with $\alpha = 0.15 \sim 0.7$. The density contours and distribution of the switch function at $t = 4$ are shown in Figure 10. The black regions indicate the switch function is 0, where the Rusanov flux function is active. No obvious differences in the density contours are observed. However, the switch function changes greatly with α, and we can see that the Rusanov flux function will not be active near reflected shock when $\alpha = 0.7$. For $\alpha = 0.15$, the Rusanov flux function is active near all the shocks and expansion fans around the corner. Therefore, we recommend that reader adjust α according to the shock strength in actual problems. Density contours at nondimensional time intervals of 1 up to 4 are shown in Figure 11. Both of these two flux functions have good resolutions for shock reflection and interaction as well as contact discontinuity, but the hybrid Roe–Rusanov flux function gives a smoother solution than the AUSMPW+ flux function.

4.3. Double Mach reflection problem

This double Mach reflection problem was first studied by Woodward and Colella [30] and later by many others [31]. It is well known that Roe's flux function produces a spurious triple point (also known as the kinked Mach stem) in this problem. The setups of the initial and boundary conditions are based on Ref [32]. As shown in Figure 12, the shock wave, inclined at 60° with respect to the x-axis and propagated to the right at $Ma = 10$, against a region of still air (state $[\rho, u, v, p]_B^T = [1.4, 0, 0, 1]^T$). $[\rho, u, v, p]_A^T = [8, 8.25\cos\frac{\pi}{6}, -8.25\sin\frac{\pi}{6}, 116.5]^T$ corresponds to the state trailing behind the right-moving shock wave. The boundary conditions correspond to the exact shockwave motion at the upper boundary ($y = 1$). More details can be found in [32]. The computational domain is $[0, 4] \times [0, 1]$, and the mesh size is chosen to be 800×200. The test was computed with the AUSMPW+, Van Leer, and hybrid Roe–Rusanov flux functions up to nondimensional time $t = 0.20$.

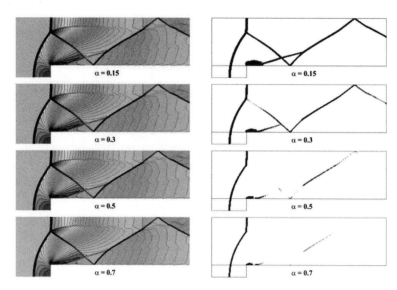

Figure 10. Density contours and switch function at $t = 4$.
(colored picture attached at the end of the book)

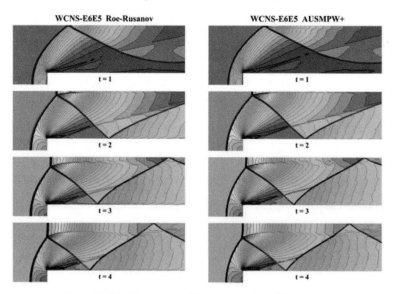

Figure 11. Density contours of flow past a step at different time.
(colored picture attached at the end of the book)

The results of the density contours are shown in Figure 13. It can be observed that the flux functions all have a good resolution of shock waves and contact discontinuity, without suffering from a kinked Mach stem. As for the simulation of the jet flow structure, the hybrid Roe–Rusanov and AUSMPW+ flux functions produce better results than the Van Leer flux function, but the hybrid Roe–Rusanov flux function gives a smoother solution than the AUSMPW+ flux function. The distribution of the switch function is illustrated in Figure 14, indicating that the Rusanov flux function would be activated near the shock waves.

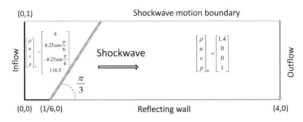

Figure 12. Initial and boundary conditions of double Mach reflection.

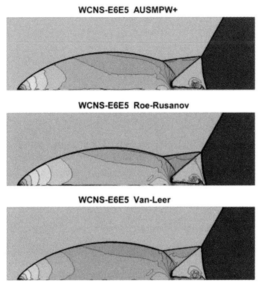

Figure 13. Density contours of double Mach reflection.
(colored picture attached at the end of the book)

Figure 14. Switch function of double Mach reflection.

4.4. Shock diffraction

The shock diffraction problem is another severe test case. Many Godunov-type flux functions are known to fail because a series of complex flow structures are included, such as shock reflection, diffraction, and interaction. In addition, instability is likely to appear for the primary shock wave, and expansion shock waves violating entropy conditions would possibly be produced. The setup of computation parameters is based on Ref [12]. As shown in Figure 15, a normal shock wave in the tunnel moves toward right at $Ma = 5.09$ and then will diffract around a 90° corner. We compute only the right part of this problem. The computational domain is a unit square $[0, 1] \times [0, 1]$, and the mesh size is set to be 400×400. The computations were performed up to nondimensional time $t = 0.18$. For the high-order method, the Van Leer and AUSMPW+ flux functions fail in this simulation, whereas the hybrid Roe–Rusanov flux function performs well. The result of the

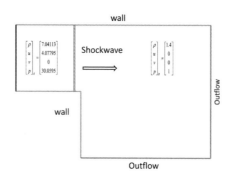

Figure 15. Initial and boundary conditions of shock diffraction.

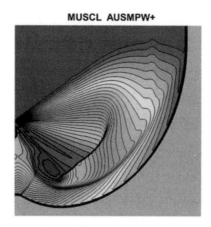

Figure 16. Density contours of shock diffraction. (colored picture attached at the end of the book)

second-order MUSCL with AUSMPW+ flux function is shown here in Figure 16, and the result of the high-order method with the hybrid Roe–Rusanov flux function is shown in Figure 17. No shock instability or expansion shock waves are observed in either of the results. Despite the fact that the Rusanov flux function is more dissipative than the AUSMPW+ flux function, the high-order result is more accurate than that of the second-order MUSCL. The distribution of the switch function is illustrated in Figure 18, indicating that Rusanov flux function would be activated near the shock waves and expansion fan around the corner.

4.5. Subsonic flat plate laminar flow

In the previous numerical examples, the shock capturing capabilities of the hybrid Roe–Rusanov flux function are evaluated, but the resolution in smooth regions has not been assessed. For this purpose, a subsonic flat plate laminar flow is simulated here. The computational grid and boundary conditions are shown in Figure 19. The hybrid Roe–Rusanov flux function and Roe's flux function are employed. Flow conditions are $Ma = 0.2$, $T_\infty = 300K$, $Re_L = 2 \times 10^5$, $L = 2$m. Figure 20 shows the comparison of the streamwise velocity profile, normal velocity profile (at $x = 1$m), and wall skin friction with the Blasius solution [33]. As expected, the hybrid Roe–Rusanov flux function gives an almost identical result to that of Roe's flux function. In this test case, we also make a comparison of the cost. In a calculation of 2000 time steps, the hybrid Roe–Rusanov flux function costs only 5% more CPU time than Roe's flux function.

The results of the aforementioned test cases show that the hybrid Roe–Rusanov flux function with WCNS-E6E5 is not only robust for complex flows with a strong shock and shock/shock interaction

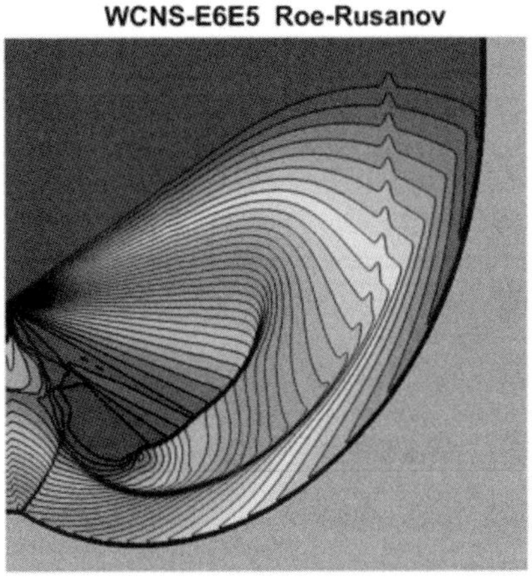

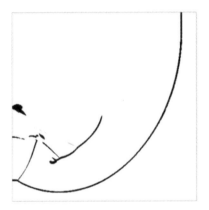

Figure 17. Density contours of shock diffraction. (colored picture attached at the end of the book)

Figure 18. Switch function of shock diffraction.

but is also accurate for the resolution of contact discontinuity and the boundary layer. The distribution of the switch function indicates that the hybrid flux function would switch to the Rusanov flux function in regions near shock waves, just in accordance with the design requirements. However, in forward facing step and shock diffraction problems, we observe that the Rusanov flux function was activated near the strong expansion regions around the corner. It is hard to distinguish shock and strong expansion only according to the pressure gradient. Some other parameter should be introduced to solve this problem, making the switch function more complex.

5. APPLICATION

To evaluate the capabilities of the hybrid Roe–Rusanov flux function in hypersonic heat transfer prediction, hypersonic flows over a blunt cone and double ellipsoid are simulated, and wall heat transfer profiles are compared with experimental data.

5.1. Hypersonic flow over a blunt cone

Cleary's experimental heat-transfer data for a 15-deg blunt cone [34] have been used in the past for comparisons with computational techniques. In this experiment, tests were performed to determine

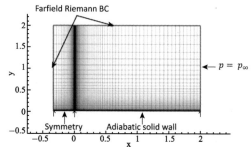

Figure 19. Computational grid and boundary conditions for flat plate boundary calculations.

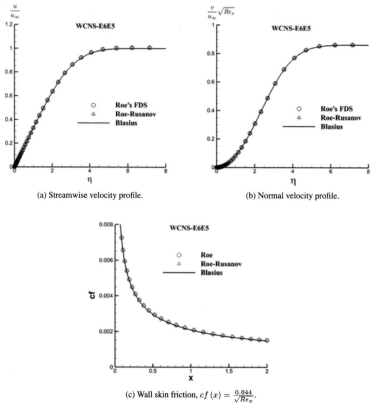

(a) Streamwise velocity profile.

(b) Normal velocity profile.

(c) Wall skin friction, $cf(x) = \frac{0.044}{\sqrt{Re_x}}$.

Figure 20. Boundary layer calculation on a flat plate: $\eta = \frac{y}{x}\sqrt{Re_x}$, $Re_x = \frac{\rho_\infty u_\infty x}{\mu_\infty}$.

the effects of the angle of attack and nose bluntness on laminar heating rates. The cone nose radius varied from 0 to 1.1 inch, and the angle of attack ranged from 0 to 20 degrees. A cone at a 20-degrees angle of attack with a 1.1 inch nose radius is selected for the comparison. In this case, the ability of the hybrid flux function to predict the wall heat transfer rate is explored.

The following flow conditions are adopted [34]: $Ma = 10.6$, $T_\infty = 47.3$ K, $Re = 3.937 \times 10^6 \text{m}^{-1}$, $T_{wall} = 294.44 \text{K}$. Computational grids are $81 \times 181 \times 121$ (streamwise, circumferential, and normal direction), as shown in Figure 21. The cell Reynolds number based on the minimum cell size is set to 10.

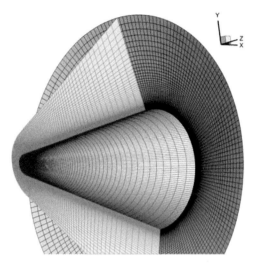

Figure 21. Computational grids for blunt cone.
(colored picture attached at the end of the book)

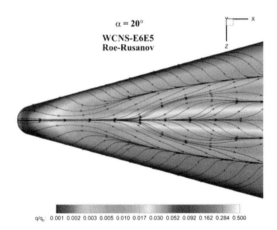

Figure 22. Upper-surface streamlines and wall heat transfer rate contours.
(colored picture attached at the end of the book)

Figure 22, a plot of upper surface streamlines and the computed heat transfer rate nondimensionalized by a theoretical stagnation heat transfer rate, reveals the line of separation and an increase in leeside heating because of vortex reattachment. It is clear that the solution well identifies the second separation existing in real flow.

Comparisons of heat transfer rate profiles are presented in the streamwise and circumferential directions, as shown in Figures 23 and 24. The circumferential angle ϕ equal to 0 and 180 degrees corresponds to the upper and lower symmetry planes, respectively, and $\phi = 90$ degrees represents the side plane. In general, the computed heat transfer rate is in good agreement with the experiments and shows the same trends found in the experimental data. The highest heating values are found on the windward symmetry plane. Proceeding to the leeward surface, the values decrease to a minimum and then rise in the area of the vortex reattachment. In the circumferential direction, the minimum heating occurs almost at $\phi = 30$, within the region of separated flow.

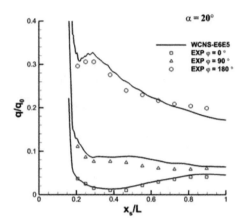

Figure 23. Streamwise heating distributions on blunt cone.

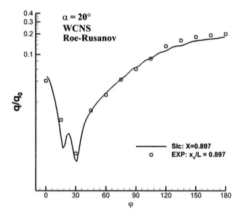

Figure 24. Circumferential heating distributions on blunt cone at $x/L = 0.897$.

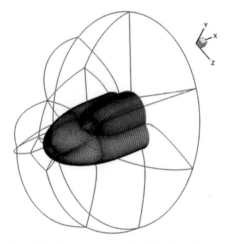

Figure 25. Computational grids of double-ellipsoid.

5.2. Hypersonic flow over a double-ellipsoid

The three-dimensional double-ellipsoid configuration is typical in reentry vehicles for experimental and computational investigations. It always possesses complicated flow structures such as shock/shock interaction, shock/boundary-layer interaction, and separation flow.

The setup of freestream conditions comes from the FD-14A wind tunnel experiment conducted at the China Aerodynamics Research and Development Center. The detailed parameters are $Ma = 10.02$, $Re = 2.2 \times 10^6 \text{m}^{-1}$, $P_0 = 6.9 \text{MPa}$, $T_0 = 1457 \text{K}$, and $\alpha = 0°$. The double-ellipsoid test model has a main semi-axis 158mm, and the cylindrical part in the rear is 57mm long. Two sets of grids are generated for the computation, as shown in Figure 25, and the cell Reynolds number based on the minimum cell size is 4.4 and 11, respectively.

The pressure contours in the symmetry plane, the wall heat transfer rate contours and the streamlines are shown in Figure 26. The pressure contours in the symmetry plane show a strong expansion downstream of the stagnation point of the base ellipsoid, as well as the interaction of the embedded shock with the bow shock. The streamlines clearly reveals the main separation and second separation, which are caused by the upper ellipsoid.

A comparison of the heat transfer rate profiles of the symmetry plane and circumferential directions at $x = 120$ are presented in Figure 27. The circumferential angle ϕ means the same as in Section 5.1. The wall heat transfer results of the two sets of grids are almost the same,

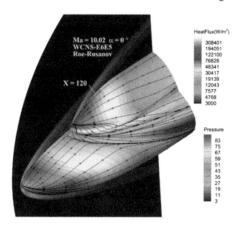

Figure 26. Wall heat transfer rate contours, streamlines, and pressure contours in the symmetry plane.
(colored picture attached at the end of the book)

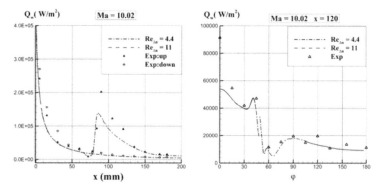

Figure 27. Wall heat transfer rate profiles with experimental data (left: streamwise direction in symmetry plane; right: circumferential direction at $x = 120$).

indicating that grid-independent solutions are obtained. In general, the computed heat transfer rate on the windward symmetry plane is in good agreement with the experimental data. However, the maximal value of the heat transfer of upper ellipsoid in the leeward symmetry plane is much smaller than the experimental data. This may be caused by the transition of the leeward flow after separation. The trends of the heat transfer rate in the circumferential directions at $x = 120$ is also in good agreement with the experimental data.

6. CONCLUSIONS

In this paper, we developed a very robust hybrid Riemann solver that is suitable for high-order methods and hypersonic heat transfer prediction. An adaptive strategy is adopted to combine the Roe solver and Rusanov solver with a switch function based on a pressure gradient. Near shock waves, the hybrid flux function would automatically switch to the Rusanov flux function, whereas in smooth regions, Roe's flux function is recovered. Multidimensional effects, corresponding to multidimensional dissipation, are also considered in the switch function, thus strengthening the robustness of the hybrid flux function to capture shocks and ensuring that the flux function switches in smooth regions before or after the shock waves to minimize their influence.

A series of numerical tests indicates that this new hybrid flux function can robustly simulate hypersonic flow characteristics and provide an accurate prediction of surface heat transfer. In addition, a high resolution in the smooth regions, like the boundary layer, is obtained. Compared with Roe's FDS, only 5 percent extra CPU time is needed. Further investigations will be focused on the simulation of practical engineering configurations and prediction of aerothermodynamic characteristics.

ACKNOWLEDGEMENTS

This study was supported by the foundation of the National University of Defense Technology (ZDYYJ-CYJ20140101). The first author thanks Prof. Ming Zeng for her valuable comments. The authors would like to thank Dr. Dan Xu and Dr. Yaming Chen for their useful suggestions. We also thank the Research Center of Supercomputing Application of the National University of Defense Technology for the support provided with computational resources.

REFERENCES

1. Schmisseur JD. Hypersonics into the 21st century: a perspective on afosr-sponsored research in aerothermodynamics. *Progress in Aerospace Sciences* 2015; **72**:3–16.
2. Knight D, Longo J, Drikakis D, Gaitonde D, Lani A, Nompelis I, Reimann B, Walpot L. Assessment of cfd capability for prediction of hypersonic shock interactions. *Progress in Aerospace Sciences* 2012; **48**:8–26.
3. Wang ZJ, Fidkowski K, Abgrall R, Bassi F, Caraeni D, Cary A, Deconinck H, Hartmann R, Hillewaert K, Huynh HT, *et al.* High-order cfd methods: current status and perspective. *International Journal for Numerical Methods in Fluids* 2013; **72**(8):811–845.
4. Ekaterinaris JA. High-order accurate, low numerical diffusion methods for aerodynamics. *Progress in Aerospace Sciences* 2005; **41**(3):192–300.
5. Kitamura K, Shima E, Nakamura Y, Roe PL. Evaluation of Euler fluxes for hypersonic heating computations. *AIAA Journal* 2010; **48**(4):763–776.
6. Kitamura K. A further survey of shock capturing methods on hypersonic heating issues. *AIAA Paper* 2013; **2698**.
7. Tu G, Zhao X, Mao M, Chen J, Deng X, Liu H. Evaluation of Euler fluxes by a high-order cfd scheme: shock instability. *International Journal of Computational Fluid Dynamics* 2014; **28**(5):171–186.
8. Tissera S. Assessment of High-Resolution Methods in Hypersonic Real-Gas Flows, Cranfield University, 2010.
9. Kotov DV, Yee HC, Panesi M, Prabhu DK, Wray AA. Computational challenges for simulations related to the nasa electric arc shock tube (east) experiments. *Journal of Computational Physics* 2014; **269**:215–233.
10. Shu C-W, Osher S. Efficient implementation of essentially non-oscillatory shock-capturing schemes. *Journal of Computational Physics* 1988; **77**(2):439–471.
11. Liou MS. A sequel to ausm, part ii: Ausm+-up for all speeds. *Journal of Computational Physics* 2006; **214**(1):137–170.
12. Nishikawa H, Kitamura K. Very simple, carbuncle-free, boundary-layer-resolving, rotated-hybrid Riemann solvers. *Journal of Computational Physics* 2008; **227**(4):2560–2581.

13. Druguet M-C, Candler GV, Nompelis I. Effects of numerics on Navier–Stokes computations of hypersonic double-cone flows. *AIAA Journal* 2005; **43**(3):616–623.
14. Ren Y-X. A robust shock-capturing scheme based on rotated Riemann solvers. *Computers & Fluids* 2003; **32**(10):1379–1403.
15. Quirk JJ. *A Contribution to the Great Riemann Solver Debate*. Springer: Berlin Heidelberg, 1997.
16. Deng X, Mao M, Tu G, Liu H, Zhang H. Geometric conservation law and applications to high-order finite difference schemes with stationary grids. *Journal of Computational Physics* 2011; **230**(4):1100–1115.
17. Xiaogang D, Zhu H, Min Y, Liu H, Meiliang M, Guangxue W, Hanxin Z. Symmetric conservative metric method: a link between high order finite-difference and finite-volume schemes for flow computations around complex geometries. *Eight International Conference on Computational Fluid Dynamics (ICCFD8)*, Chengdu, China, 2014. ICCFD8-2014-0005.
18. Deng X, Min Y, Mao M, Liu H, Tu G, Zhang H. Further studies on geometric conservation law and applications to high-order finite difference schemes with stationary grids. *Journal of Computational Physics* 2013; **239**:90–111.
19. Deng X, Zhang H. Developing high-order weighted compact nonlinear schemes. *Journal of Computational Physics* 2000; **165**(1):22–44.
20. Yoon S, Jameson A. Lower-upper symmetric-Gauss–Seidel method for the Euler and Navier–Stokes equations. *AIAA Journal* 1988; **26**(9):1025–1026.
21. Roe PL. Approximate Riemann solvers, parameter vectors, and difference schemes. *Journal of Computational Physics* 1981; **43**(2):357–372.
22. Rusanov VV. Calculation of interaction of non-steady shock waves with obstacles. *Journal of Computational and Mathematical Physics USSR* 1961; **1**:267–279.
23. Liou M-S. Mass flux schemes and connection to shock instability. *Journal of Computational Physics* 2000; **160**(2):623–648.
24. Scalabrin LC, Boyd ID. Development of an unstructured Navier–Stokes solver for hypersonic nonequilibrium aerothermodynamics. *AIAA Paper* 2005; **5203**:2005.
25. Sanders R, Morano E, Druguet M-C. Multidimensional dissipation for upwind schemes: stability and applications to gas dynamics. *Journal of Computational Physics* 1998; **145**(2):511–537.
26. Kitamura K, Shima E. Towards shock-stable and accurate hypersonic heating computations: A new pressure flux for ausm-family schemes. *Journal of Computational Physics* 2013; **245**:62–83.
27. Van Leer B. Flux-vector splitting for the Euler equations. In *Eighth International Conference on Numerical Methods in Fluid Dynamics*. Springer: Aachen, West Germany, 1982; 507–512.
28. Hanel D, Schwane R. An implicit flux-vector splitting scheme for the computation of viscous hypersonic flow. *AIAA Paper* 1989; **274**:1989.
29. Kim KH, Kim C, Rho O-H. Methods for the accurate computations of hypersonic flows: I. ausmpw+ scheme. *Journal of Computational Physics* 2001; **174**(1):38–80.
30. Woodward P, Colella P. The numerical simulation of two-dimensional fluid flow with strong shocks. *Journal of Computational Physics* 1984; **54**(1):115–173.
31. Kim Ss, Kim C, Rho O-H, Hong SK. Cures for the shock instability: development of a shock-stable Roe scheme. *Journal of Computational Physics* 2003; **185**(2):342–374.
32. Gerolymos GA, Sénéchal D, Vallet I. Very-high-order weno schemes. *Journal of Computational Physics* 2009; **228**(23):8481–8524.
33. Schlichting H, Gersten K. *Boundary Layer Theory*. Springer-Verlag: Berlin Heidelberg, 2000.
34. Cleary JW. Effects of angle of attack and bluntness on Laminar heating-rate distributions of a 1S cone at a mach number of 10.6. NASA TN D-5450, Ames Research Center, Molfett Field, California, 1969.

Efficiency benchmarking of seventh-order tri-diagonal weighted compact nonlinear scheme on curvilinear mesh

Shengye Wang[a], Xiaogang Deng[a], Guangxue Wang[b], Dan Xu[a] and Dongfang Wang[a]

[a]College of Aerospace Science and Engineering, National University of Defense Technology, Changsha, Hunan, P.R. China; [b]Computational Aerodynamics Institute, China Aerodynamics Research and Development Center, Mianyang, Sichuan, P.R. China

ABSTRACT

Efficiency improvements of high-order weighted compact nonlinear scheme (WCNS) are verified using a series of benchmark cases, proposed at the International Workshop on High-Order CFD Methods. A seventh-order tri-diagonal compact one of WCNSs (WCNS-E8T7), constructed in recent years, is investigated as a basic scheme, and compared to a typical fifth-order explicit WCNS (WCNS-E6E5) and a traditional second-order TVD scheme MUSCL. Among these tests, a symmetrical conservative metric method (SCMM) is adopted to ensure the accuracy and robustness of WCNSs when solving cases in the curvilinear coordinates. The computational efficiency of schemes is evaluated based on a non-dimensional cost in achieving the same level of accuracy. Related results show that WCNS-E8T7 has a better performance than WCNS-E6E5 with the same interpolation stencils. Moreover, the opinion that high-order methods can obtain higher computational efficiency than second-order methods is demonstrated on the cases ranging from academic problems to real-life computations.

ARTICLE HISTORY
Received 28 April 2016
Accepted 2 October 2016

KEYWORDS
high-order methods; finite difference scheme; geometric conservation law; CFD

1. Introduction

For the past two decades, high-order computational fluid dynamics (CFD) methods have received considerable attention because of their potential in delivering higher accuracy with lower cost than first/second-order methods. Many types of high-order methods have been developed to deal with a diverse of problems (Wang 2007; Cheng and Shu 2009; Deng et al. 2012; Svärd and Nordstrem 2014). However, they have not gained much popularity for real-life CFD computations: partly because the impression that high-order methods are expensive is widely held and partly because high-order methods are not as robust as various second-order methods commonly used in production codes.

The impression that high-order methods are expensive is perhaps generated when obtaining a converged steady solution with the high-order method takes much longer than a second-order method on a given mesh. However, the fair way to compare efficiency should be to look at the cost to achieve the same level of accuracy or given the same CPU time, what error is produced (Wang et al. 2013). In order to dispel the myth, Shu (2003) gave an example of the evolution of a two-dimensional periodic vortex with a shock, and indicated that second-order MUSCL-type TVD scheme has a larger CPU cost than fifth-order WENO scheme to reach the same resolution. Pirozzoli (2006) defined a error metric for shock-capturing schemes using the approximate dispersion relation (ADR) and indicated that the high-order scheme WENO is more cost-effective than second-order TVD scheme when relatively strict error tolerances are placed for a one-dimensional problem. Although proven to be efficient for many academic problems, high-order methods still need to be evaluated for a range of real-life CFD cases. Wang et al. (2013) summarised the manner to evaluate the efficiency of high-order and second-order methods and the results from the participants at the First International Workshop on High-order CFD Methods[1] (HiOCFD1). It was indicated that high-order methods are able to demonstrate better performance than the second-order methods on the basis of error versus cost for the problems such as the bump, the NACA0012 airfoil and so on. But for compressible turbulent flows especially involving shocks, high-order methods need more comparisons. Weide, Giangaspero, and Svärd (2015) presented a series of results of the benchmark cases at the Second International Workshop on High-Order CFD Methods[2] (HiOCFD2) to demonstrate the effectiveness of the SBP/SAT approach in a quantitative way. Very good results were obtained for many cases such as the flow over smooth bump, the vortex transport, the Taylor–Green vortex and so on. It was demonstrated that the efficiency can be greatly enhanced by employing higher-order accurate numerical schemes for some smooth flows. But for some real-life cases using RANS turbulence models or with shocks, they were not presented.

For simulations of compressible flows in real-life CFD problems, the desired numerical scheme should be free of spurious numerical oscillations across discontinuities and obtain higher-order accuracy in smooth flow regions in an efficient manner (Brehm et al. 2015). A popular family of this desired scheme is high-order nonlinear schemes such as WENO by Jiang and Shu (1996) and WCNS by Deng and Zhang (2000). However too large built-in dissipation of original versions causes that the efficiency is still not enough. A lot of efforts have been devoted to overcoming the deficiencies of traditional high-order nonlinear schemes. One method is to decrease the bias away from the linear scheme. Henrick, Aslam, and Powers (2005) proposed a mapping function that pushes the nonlinear weights smoothly to the optimal values by a regularisation process. Taylor, Wu, and Martin (2007) developed a relatively smoothness indicator. The nonlinear weights are switched on directly when the ratios of smoothness indicators exceed a threshold value. Nonomura et al. (2011) proposed a localisation method to make the nonlinear weight computation localised only in non-smooth region and demonstrated that it is more efficient than Henrick's and Taylor's methods adopted in WCNS. Another method is to include the contribution of a downwind sub-stencil to form a symmetric stencil, such as the sixth-order adaptive central-upwind WENO (WENO-CU6) by Hu, Wang, and Adams (2010). Brehm et al. (2015) assessed a series of types of WENO and indicated that WENO-CU6 is more efficient and accurate than the others for inviscid flows. Liu et al. (2015) extended the idea of WENO-CU6 and proposed a new type of less dissipative WCNS (WCNS-HW6) with hybrid weighted upwind-biased and central interpolation. Wong and Lele (2016) combined relative limiters with WCNS-HW6 and demonstrated that this new type of WCNS (WCNS-HW6-LD) has better performance to capture discontinuities and preserve highly fluctuating flow features than WCNS-CU6-M2 based on the idea of Hu et al. and original WCNS. The third method is using compact reconstructions to replace explicit ones such as Compact-Reconstruction WENO by Ghosh and Baeder (2012). Liu et al. (2014) developed a seventh-order nonlinear compact interpolation, which can improve the order of accuracy for smooth solutions with the same stencil width as the traditional fifth-order one, and yields lower dissipation and dispersion errors. This new type of interpolation has been assessed in some basic flows, but for real-life CFD computations, it needs more works. In this paper, the WCNS with eighth-order explicit differencing and seventh-order tri-diagonal interpolation is called WCNS-E8T7. It will be carried out to compare efficiency with the traditional fifth-order scheme WCNS-E6E5 (Deng et al. 2005) and second-order TVD scheme MUSCL (van Leer 1979).

The poor robustness is an obstacle to using high-order methods, especially high-order finite difference methods (HiOFDMs) for real-life CFD computations. HiOFDMs have the advantage of simplicity and lower CPU cost for multi-dimensional problems (Shu 2003). But for practical fluid problems, FDMs usually require a grid transformation from the Cartesian coordinates to the curvilinear coordinates. It has been proved that a geometric conservation law (GCL) should be satisfied as a precondition for this transformation (Pulliam and Steger 1978; Thomas and Lombard 1979; Zhang et al. 1993), otherwise some negative effects, such as numerical oscillations and violation of free-stream conservation, may appear. Recent years, a number of researches on GCL have been carried out to improve the robustness of HiOFDMs used in complex grids. Visbal and Gaitonde (2002) studied the metric evaluation errors for the high-order central type FDMs and found the errors could be largely decreased by means of the conservative forms of metric derivatives. The conservative form was extended to the symmetric conservative one by Vinokur and Yee (2002). Nonomura, Iizuka, and Fujii (2010) successfully applied Visbal and Gaitonde's technique to WCNS and showed the excellent free-stream and vortex preservation properties of WCNS. Deng et al. (2011b, 2013, 2014) carefully studied the GCL and developed the symmetrical conservative metric method (SCMM) that not only fulfilled the requirements of the free-stream preservation, but also presented the geometric meanings of the metrics and Jacobian, establishing a link between the FVM and FDM. Abe, Iizuka, Nonomura, and Fujii (2013) and Abe et al. (2014) developed symmetric conservative forms on moving and deforming grids and also presented the geometric interpretations of the metrics for the HiOFDMs . On the basis of the above researches, HiOFDMs such as WCNS with SCMM can be applied to solve the problems with complex geometries more robustly.

The purpose of this paper is to evaluate the performance of WCNSs mainly the WCNS-E8T7 with SCMM. The test cases also stem from HiOCFD in order to at least partially dispel some myths regarding HiOFDMs. A benchmark code (TauBench) is run on each particular machine to provide a normalisation such that the results presented in this paper are referable for other research groups during the development of CFD code. This paper is organised as follows. In Section 2, the high-order weighted compact nonlinear schemes (WCNSs) and symmetrical conservative metric method (SCMM) are introduced briefly. In Section 3, the general three-dimensional code containing WCNS-E8T7, WCNS-E6E5 and MUSCL is described. In Section 4, the computing processes and results on each of benchmark cases are presented in detail. Finally, conclusions with a summary of the results and future works are given in Section 5.

2. A brief introduction of WCNS with SCMM

The weighted compact nonlinear scheme (WCNS) was first derived by Deng and Zhang (2000). Afterward, a series of WCNSs were developed by Deng et al. (2005), Nonomura, Iizuka, and Fujii (2007), Zhang, Jiang, and Shu (2008), and Deng et al. (2011a), and so on. In this paper, the version with sixth-order explicit differencing and fifth-order explicit interpolation (WCNS-E6E5) and the version with eighth-order explicit differencing and seventh-order tri-diagonal interpolation (WCNS-E8T7) are carried out. The WCNS-E6E5 (Deng et al. 2005) has the advantages of simplicity and high efficiency compared with the original version (Deng and Zhang 2000) and has been successfully applied to a wide range of flow simulations (Liu and Deng 2007; Deng et al. 2010; Deng, Wang, and Tu 2011). The WCNS-E8T7 (Liu et al. 2014) was developed in recent years. It can yield lower dissipation and dispersion errors with the same stencil width as WCNS-E6E5 and is going to be applied to complicated flow problems.

In this section, a grid transformation from Cartesian coordinates to curvilinear coordinates for the three-dimensional Navier-Stokes equations will be shown first. Then, the procedure to solve the three-dimensional Navier-Stokes equations using WCNSs will be presented. Finally, the importance of GCL will be pointed out and the SCMM will be given.

2.1. Governing equation

In Cartesian coordinates, the three-dimensional nondimensional Navier-Stokes equations are

$$\frac{\partial Q}{\partial t} + \frac{\partial E}{\partial x} + \frac{\partial F}{\partial y} + \frac{\partial G}{\partial z} = \frac{1}{Re}\left(\frac{\partial E_v}{\partial x} + \frac{\partial F_v}{\partial y} + \frac{\partial G_v}{\partial z}\right), \quad (1)$$

where

$$Q = [\rho, \rho u, \rho v, \rho w, \rho e]^T,$$
$$E = [\rho u, \rho u^2 + p, \rho vu, \rho wu, (\rho e + p)u]^T,$$
$$F = [\rho v, \rho uv, \rho v^2 + p, \rho wv, (\rho e + p)v]^T, \quad (2)$$
$$G = [\rho w, \rho uw, \rho vw, \rho w^2 + p, (\rho e + p)w]^T,$$
$$E_v = [0, \tau_{xx}, \tau_{xy}, \tau_{xz}, u\tau_{xx} + v\tau_{xy} + w\tau_{xz} + \dot{q}_x]^T,$$
$$F_v = [0, \tau_{yx}, \tau_{yy}, \tau_{yz}, u\tau_{yx} + v\tau_{yy} + w\tau_{yz} + \dot{q}_y]^T,$$
$$G_v = [0, \tau_{zx}, \tau_{zy}, \tau_{zz}, u\tau_{zx} + v\tau_{zy} + w\tau_{zz} + \dot{q}_z]^T,$$

with viscous stress terms written as

$$\tau_{ij} = \mu\left(\frac{\partial u_i}{\partial x_j} + \frac{\partial u_j}{\partial x_i} - \frac{2}{3}\delta_{ij}\frac{\partial u_k}{\partial x_k}\right), \quad (3)$$

and heat transfer terms written as

$$\dot{q}_x = \frac{1}{(\gamma-1)P_r Ma_\infty^2}\mu\frac{\partial T}{\partial x},$$
$$\dot{q}_y = \frac{1}{(\gamma-1)P_r Ma_\infty^2}\mu\frac{\partial T}{\partial y}, \quad (4)$$
$$\dot{q}_z = \frac{1}{(\gamma-1)P_r Ma_\infty^2}\mu\frac{\partial T}{\partial z},$$

where γ is the ratio of specific heats, and the viscous coefficient μ can be calculated by the Sutherland's law. u, v and w are the velocity components in x, y and z directions, respectively, p is pressure, ρ is density and T is temperature.

Using the transformation from Cartesian coordinates to curvilinear coordinates, i.e.,

$$x = x(\xi, \eta, \zeta), y = y(\xi, \eta, \zeta), z = z(\xi, \eta, \zeta), \quad (5)$$

Equation (1) becomes

$$\frac{\partial \hat{Q}}{\partial t} + \frac{\partial \hat{E}}{\partial \xi} + \frac{\partial \hat{F}}{\partial \eta} + \frac{\partial \hat{G}}{\partial \zeta} = \frac{1}{Re}\left(\frac{\partial \hat{E}_v}{\partial \xi} + \frac{\partial \hat{F}_v}{\partial \eta} + \frac{\partial \hat{G}_v}{\partial \zeta}\right), \quad (6)$$

where

$$\hat{Q} = JQ,$$
$$\hat{E} = \tilde{\xi}_x E + \tilde{\xi}_y F + \tilde{\xi}_z G, \hat{E}_v = \tilde{\xi}_x E_v + \tilde{\xi}_y F_v + \tilde{\xi}_z G_v,$$
$$\hat{F} = \tilde{\eta}_x E + \tilde{\eta}_y F + \tilde{\eta}_z G, \hat{F}_v = \tilde{\eta}_x E_v + \tilde{\eta}_y F_v + \tilde{\eta}_z G_v,$$
$$\hat{G} = \tilde{\zeta}_x E + \tilde{\zeta}_y F + \tilde{\zeta}_z G, \hat{G}_v = \tilde{\zeta}_x E_v + \tilde{\zeta}_y F_v + \tilde{\zeta}_z G_v.$$
$$(7)$$

Here, J is the Jacobian of the grid transformation and $\tilde{\xi}_x$, $\tilde{\xi}_y$, $\tilde{\xi}_z$, $\tilde{\eta}_x$, $\tilde{\eta}_y$, $\tilde{\eta}_z$, $\tilde{\zeta}_x$, $\tilde{\zeta}_y$ and $\tilde{\zeta}_z$ are the grid metrics, which will be presented in Section 2.3. Note that only stationary grids are considered in this paper.

2.2. High-order scheme: WCNS

The WCNS procedure consists of three components: (i) cell-edge to cell-node central flux differencing, (ii) flux evaluation at the cell-edge, and (iii) cell-node to cell-edge weighted nonlinear interpolation of flow variables.

Considering discretisation of the inviscid terms in Equation (6), the explicit sixth-order central flux differencing is adopted for WCNS-E6E5,

$$\frac{\partial \hat{E}_i}{\partial \xi} = \frac{75}{64\Delta\xi}\left(\hat{E}_{i+1/2} - \hat{E}_{i-1/2}\right) - \frac{25}{384\Delta\xi}\left(\hat{E}_{i+3/2} - \hat{E}_{i-3/2}\right)$$
$$+ \frac{3}{640\Delta\xi}\left(\hat{E}_{i+5/2} - \hat{E}_{i-5/2}\right), \quad (8)$$

and the explicit eighth-order central flux differencing is adopted for WCNS-E8T7,

$$\frac{\partial \hat{E}_i}{\partial \xi} = \frac{1225}{1024\Delta\xi}\left(\hat{E}_{i+1/2} - \hat{E}_{i-1/2}\right)$$
$$-\frac{245}{3072\Delta\xi}\left(\hat{E}_{i+3/2} - \hat{E}_{i-3/2}\right)$$
$$+\frac{49}{5120\Delta\xi}\left(\hat{E}_{i+5/2} - \hat{E}_{i-5/2}\right)$$
$$-\frac{5}{7168\Delta\xi}\left(\hat{E}_{i+7/2} - \hat{E}_{i-7/2}\right), \quad (9)$$

where

$$\hat{E}_{i+1/2} = \hat{E}\big[Q^L_{i+1/2}, Q^R_{i+1/2}, (\tilde{\xi}_x)_{i+1/2}, (\tilde{\xi}_y)_{i+1/2},$$
$$(\tilde{\xi}_z)_{i+1/2}\big]. \quad (10)$$

Here, only the discretisation in ξ direction is given, and the discretisation for other direction can be computed by similar procedures. Equation (10) is the flux at cell edges which can be evaluated by various flux schemes. In this paper, Roe's flux difference scheme (Roe 1981) is used. $Q^L_{i+1/2}$, $Q^R_{i+1/2}$ are the left and right cell-edge variables value, which are calculated by weighted nonlinear interpolation methods. In this paper, we will only describe how to compute $Q^L_{i+1/2}$.

For WCNS-E6E5, an explicit fifth-order weighted nonlinear interpolation is adopted,

$$Q^L_{j+1/2} = \begin{pmatrix} \frac{3}{8}\omega_0 \\ \frac{1}{8}(-10\omega_0 - \omega_1) \\ \frac{1}{8}(15\omega_0 + 6\omega_1 + 3\omega_2) \\ \frac{1}{8}(3\omega_1 + 6\omega_2) \\ -\frac{1}{8}\omega_2 \end{pmatrix}^T \begin{pmatrix} Q_{j-2} \\ Q_{j-1} \\ Q_j \\ Q_{j+1} \\ Q_{j+2} \end{pmatrix} \quad (11)$$

where ω_k, $k = 0, 1, 2$ are nonlinear weights, which are defined in such a way that in smooth regions they approach the optimal weights to achieve the optimal order of accuracy, while in regions near discontinuities, the sub-stencils which contain discontinuities are assigned nearly zero weights,

$$\omega_k = \frac{\alpha_k}{\sum_{l=0}^{2}\alpha_l}, \quad \alpha_k = C_k\left(1 + \frac{\tau_5}{\beta_k + \varepsilon}\right), \quad (12)$$

where $\tau_5 = |\beta_0 - \beta_2|$, which is proposed by Borges et al. (2008). $C_0 = 1/16$, $C_1 = 5/8$ and $C_2 = 5/16$ are the optimal weights. $\varepsilon = 10^{-6}$ is a small number to avoid the denominator to be zero and β_k are the smoothness indicators. The evaluation of the smoothness indicators was presented in Deng et al. (2005).

For WCNS-E8T7, the seventh-order weighted compact nonlinear interpolation is derived as

$$\begin{pmatrix} \frac{1}{2}\omega_5 \\ \omega_0 + \omega_1 + \omega_2 \\ \frac{3}{14}\omega_5 \end{pmatrix}^T \begin{pmatrix} Q^L_{j-1/2} \\ Q^L_{j+1/2} \\ Q^L_{j+3/2} \end{pmatrix}$$
$$= \begin{pmatrix} \frac{3}{8}\omega_0 - \frac{25}{896}\omega_5 \\ \frac{1}{32}(-40\omega_0 - 4\omega_1 + 9\omega_5) \\ \frac{3}{64}(40\omega_0 + 16\omega_1 + 8\omega_2 + 5\omega_5) \\ \frac{1}{8}(3\omega_1 + 6\omega_2 + \frac{5}{4}\omega_5) \\ \frac{1}{8}(-\omega_2 + \frac{9}{16}\omega_5) \end{pmatrix}^T \begin{pmatrix} Q_{j-2} \\ Q_{j-1} \\ Q_j \\ Q_{j+1} \\ Q_{j+2} \end{pmatrix}. \quad (13)$$

The weights ω_k are given by

$$\omega_k = D_k\left(1 + \frac{\tau_7}{\beta_k + \varepsilon}\right), \quad k = 0, 1, 2, 5 \quad (14)$$

where

$$\tau_7 = \left|\beta_5 - \frac{\beta_0 + 4\beta_1 + \beta_2}{6}\right|\left(\frac{\beta_0 - \beta_2}{\beta_0 + \beta_2 + \varepsilon}\right)^2. \quad (15)$$

Here β_0, β_1 and β_2 are the smoothness indicators, which are the same as those of WCNS-E6E5. β_5 is designed to control the compact part, which was presented in Liu et al. (2014). When there are discontinuities in the whole stencil, β_5 will be much greater than min $\{\beta_0, \beta_1, \beta_2\}$ and the compact part will be almost switched off. The interpolation degenerates to the explicit interpolation of WCNS-E6E5. $D_0 = 1/16$, $D_1 = 5/8$, $D_2 = 5/16$ and $D_5 = 1$ are the optimal weights to recover seventh-order interpolation.

At boundary points, two boundary and near-boundary schemes are required for WCNS-E6E5 and three for WCNS-E8T7. Here only left-side boundary is discussed and the right-side boundary can be obtained by the same manner. At the boundary nodes, the fourth-order differencings for WCNS-E6E5 are,

$$\hat{E}'_1 = \frac{1}{\Delta\xi}\left(-\frac{11}{12}\hat{E}_{1/2} + \frac{17}{24}\hat{E}_{3/2} + \frac{3}{8}\hat{E}_{5/2}\right.$$
$$\left. - \frac{5}{24}\hat{E}_{7/2} + \frac{1}{24}\hat{E}_{9/2}\right),$$
$$\hat{E}'_2 = \frac{1}{\Delta\xi}\left(\frac{1}{24}\hat{E}_{1/2} - \frac{9}{8}\hat{E}_{3/2}\right.$$
$$\left. + \frac{9}{8}\hat{E}_{5/2} - \frac{1}{24}\hat{E}_{7/2}\right), \quad (16)$$

and the sixth-order differencings for WCNS-E8T7 are,

$$\hat{E}'_1 = \frac{1}{\Delta\xi}\left(-\frac{1019}{1440}\hat{E}_{1/2} - \frac{179}{72}\hat{E}_{3/2} - \frac{153}{16}\hat{E}_{5/2}\right.$$
$$\left. + \frac{389}{45}\hat{E}_2 - \frac{103}{72}\hat{E}_{7/2} + \frac{247}{45}\hat{E}_3 + \frac{77}{1440}\hat{E}_{9/2}\right),$$

$$\hat{E}'_2 = \frac{1}{\Delta \xi} \left(-\frac{1}{30}\hat{E}_{1/2} + \frac{3}{10}\hat{E}_1 - \frac{3}{2}\hat{E}_{3/2} + \frac{3}{2}\hat{E}_{5/2} \right.$$
$$\left. -\frac{3}{10}\hat{E}_3 + \frac{1}{30}\hat{E}_{7/2} \right),$$
$$\hat{E}'_3 = \frac{1}{\Delta \xi} \left[\frac{64}{45} \left(\hat{E}_{7/2} - \hat{E}_{5/2} \right) - \frac{2}{9} \left(\hat{E}_4 - \hat{E}_2 \right) \right.$$
$$\left. + \frac{1}{180} \left(\hat{E}_5 - \hat{E}_1 \right) \right]. \tag{17}$$

Besides numerical differencing, the corresponding interpolations adopted at the boundary and near-boundary cell-edges for WCNS-E6E5 are

$$Q_{1/2} = \frac{1}{128} (315Q_1 - 420Q_2 + 378Q_3 - 180Q_4 + 35Q_5),$$
$$Q_{3/2} = \frac{1}{128} (35Q_1 + 140Q_2 - 70Q_3 + 28Q_4 - 5Q_5), \tag{18}$$

and for WCNS-E8T7 are

$$Q_{1/2} = \frac{1}{256}(693Q_1 - 1155Q_2 + 1386Q_3 - 990Q_4$$
$$+ 385Q_5 - 63Q_6),$$
$$Q_{3/2} = \frac{1}{256}(63Q_1 + 315Q_2 - 210Q_3 + 126Q_4$$
$$- 45Q_5 + 7Q_6),$$
$$Q_{5/2} = \frac{1}{256}(-7Q_1 + 105Q_2 + 210Q_3 - 70Q_4$$
$$+ 21Q_5 - 3Q_6). \tag{19}$$

For the computation of the viscous terms, the sixth-order central scheme is also adopted for WCNS-E6E5,

$$\frac{\partial \hat{E}_{v,i}}{\partial \xi} = \frac{75}{64\Delta \xi} \left(\hat{E}_{v,i+1/2} - \hat{E}_{v,i-1/2} \right)$$
$$- \frac{25}{384\Delta \xi} \left(\hat{E}_{v,i+3/2} - \hat{E}_{v,i-3/2} \right)$$
$$+ \frac{3}{640\Delta \xi} \left(\hat{E}_{v,i+5/2} - \hat{E}_{v,i-5/2} \right), \tag{20}$$

and the eighth-order central scheme is adopted for WCNS-E8T7,

$$\frac{\partial \hat{E}_{v,i}}{\partial \xi} = \frac{1225}{1024\Delta \xi} \left(\hat{E}_{v,i+1/2} - \hat{E}_{v,i-1/2} \right)$$
$$- \frac{245}{3072\Delta \xi} \left(\hat{E}_{v,i+3/2} - \hat{E}_{v,i-3/2} \right)$$
$$+ \frac{49}{5120\Delta \xi} \left(\hat{E}_{v,i+5/2} - \hat{E}_{v,i-5/2} \right)$$
$$- \frac{5}{7168\Delta \xi} \left(\hat{E}_{v,i+7/2} - \hat{E}_{v,i-7/2} \right), \tag{21}$$

where $\hat{E}_{v,i\pm1/2}$ are the viscous fluxes at cell edges. The detailed computation procedure is described in Liu et al. (2011).

The simulations of the steady turbulence flows are based on RANS equations. It is necessary to point out that the modelling equations such as $\tilde{v}$ equation in Spalart-Allmaras turbulence model (Spalart and Rumsey 2007) are discretised by the same high-order schemes. In this paper, the second-order scheme MUSCL is also used for comparison. The value of κ is set to be 1/3 and the van Leer limiter is selected.

2.3. Symmetrical conservative metric method (SCMM)

As outlined previously, if the geometric conservation law (GCL) is not satisfied, it will cause large errors and even result in numerical instabilities for HiOFDMs (Visbal and Gaitonde 2002; Nonomura, Iizuka, and Fujii 2010; Deng et al. 2011b). Typically in uniform flow where Q, E, F and G are constant and E_v, F_v and G_v are zero, Equation (6) can be written as

$$\frac{\partial \hat{Q}}{\partial t} = -\left(I_x E + I_y F + I_z G \right), \tag{22}$$

where

$$\begin{cases} I_x = \delta^1_\xi \left(\tilde{\xi}_x \right) + \delta^1_\eta \left(\tilde{\eta}_x \right) + \delta^1_\zeta \left(\tilde{\zeta}_x \right), \\ I_y = \delta^1_\xi \left(\tilde{\xi}_y \right) + \delta^1_\eta \left(\tilde{\eta}_y \right) + \delta^1_\zeta \left(\tilde{\zeta}_y \right), \\ I_z = \delta^1_\xi \left(\tilde{\xi}_z \right) + \delta^1_\eta \left(\tilde{\eta}_z \right) + \delta^1_\zeta \left(\tilde{\zeta}_z \right). \end{cases} \tag{23}$$

δ^1_ξ, δ^1_η and δ^1_ζ are numerical difference operators in ξ, η and ζ coordinate directions. In order to cancel the right-hand side of Equation (22), the GCL should be satisfied, namely $I_x = I_y = I_z = 0$ (Zhang et al. 1993).

To satisfy GCL, Deng et al. (2011b) have developed the symmetrical conservative metric method (SCMM) and proven that high-order WCNS based on SCMM can be applied to robustly solve problems with complex geometries. SCMM was presented in Deng et al. (2013) which satisfies the following three conditions:

(i) the symmetric conservative form of metrics is adopted,

$$\tilde{\xi}_x = \frac{1}{2} \left[\delta^2_\zeta \left(z\delta^3_\eta y - y\delta^3_\eta z \right) + \delta^2_\eta \left(y\delta^3_\zeta z - z\delta^3_\zeta y \right) \right],$$
$$\tilde{\xi}_y = \frac{1}{2} \left[\delta^2_\zeta \left(x\delta^3_\eta z - z\delta^3_\eta x \right) + \delta^2_\eta \left(z\delta^3_\zeta x - x\delta^3_\zeta z \right) \right],$$
$$\tilde{\xi}_z = \frac{1}{2} \left[\delta^2_\zeta \left(y\delta^3_\eta x - x\delta^3_\eta y \right) + \delta^2_\eta \left(x\delta^3_\zeta y - y\delta^3_\zeta x \right) \right],$$

$$\tilde{\eta}_x = \frac{1}{2}\left[\delta_\xi^2\left(z\delta_\zeta^3 y - y\delta_\zeta^3 z\right) + \delta_\zeta^2\left(y\delta_\xi^3 z - z\delta_\xi^3 y\right)\right],$$

$$\tilde{\eta}_y = \frac{1}{2}\left[\delta_\xi^2\left(x\delta_\zeta^3 z - z\delta_\zeta^3 x\right) + \delta_\zeta^2\left(z\delta_\xi^3 x - x\delta_\xi^3 z\right)\right],$$

$$\tilde{\eta}_z = \frac{1}{2}\left[\delta_\xi^2\left(y\delta_\zeta^3 x - x\delta_\zeta^3 y\right) + \delta_\zeta^2\left(x\delta_\xi^3 y - y\delta_\xi^3 x\right)\right],$$

$$\tilde{\zeta}_x = \frac{1}{2}\left[\delta_\eta^2\left(z\delta_\xi^3 y - y\delta_\xi^3 z\right) + \delta_\xi^2\left(y\delta_\eta^3 z - z\delta_\eta^3 y\right)\right],$$

$$\tilde{\zeta}_y = \frac{1}{2}\left[\delta_\eta^2\left(x\delta_\xi^3 z - z\delta_\xi^3 x\right) + \delta_\xi^2\left(z\delta_\eta^3 x - x\delta_\eta^3 z\right)\right],$$

$$\tilde{\zeta}_z = \frac{1}{2}\left[\delta_\eta^2\left(y\delta_\xi^3 x - x\delta_\xi^3 y\right) + \delta_\xi^2\left(x\delta_\eta^3 y - y\delta_\eta^3 x\right)\right],$$

(24)

where $\delta_\xi^2, \delta_\eta^2, \delta_\zeta^2$ and $\delta_\xi^3, \delta_\eta^3, \delta_\zeta^3$ are numerical derivative operators used for the metric calculations in ξ, η and ζ coordinate directions,

(ii) the symmetric conservative form of Jacobian is adopted,

$$J = \frac{1}{3}\Big\{\left[(x\tilde{\xi}_x) + (y\tilde{\xi}_y) + (z\tilde{\xi}_z)\right]_\xi \\ + \left[(x\tilde{\eta}_x) + (y\tilde{\eta}_y) + (z\tilde{\eta}_z)\right]_\eta \\ + \left[(x\tilde{\zeta}_x) + (y\tilde{\zeta}_y) + (z\tilde{\zeta}_z)\right]_\zeta\Big\}, \quad (25)$$

(iii) difference operators δ^1, δ^2 and δ^3 are equal in the same direction,

$$\delta_\xi^1 = \delta_\xi^2 = \delta_\xi^3, \delta_\eta^1 = \delta_\eta^2 = \delta_\eta^3, \delta_\zeta^1 = \delta_\zeta^2 = \delta_\zeta^3. \quad (26)$$

Finally, it should be pointed out that the methods of calculating the cell-edge flow-field variables have no effects on the GCL, so that any upwind-based linear or nonlinear interpolation methods, such as that adopted in WCNS and MUSCL, may be used to calculate $Q_{i+1/2}^L$, $Q_{i+1/2}^R$.

3. Code description

A general three-dimensional (3D) code, which can handle multi-block grids and run on parallel platforms, is used in this paper. A series of spatial discretisation schemes including WCNS-E8T7, WCNS-E6E5 and MUSCL are integrated in the code. For explicit high-order scheme WCNS-E6E5, a series of efforts have been devoted to improving the parallel performance. Deng et al. (2010) proposed a characteristic-based interface condition (CBIC) to fulfill high-order multi-block computing by directly exchanging the spatial derivatives on each side of an interface. Based on the CBIC, the high-order WCNS-E6E5 scheme was proven to be asymptotically stable in linear system with uniform grids and more convenient for parallel applications. Xu et al. (2014) optimised the code based on WCNS for 3D multi-block structured grids onto the supercomputer TianHe-1A and improved the parallel performance of WCNS-E6E5 by some novel techniques. For compact high-order scheme WCNS-E8T7, it requires a solution to the tri-diagonal system at each time integration step. In order to break the global dependence of compact methods and inherit the efficient and robust parallel performance of WCNS-E6E5, a explicit seventh-order interpolation (E7) (Nonomura, Iizuka, and Fujii 2007) is adopted at block interfaces in this paper. Chao, Haselbacher, and Balachandar (2009) developed this idea based on hybrid compact-WENO several years ago and found that numerical errors increase slightly as the number of blocks increases and the maximum allowable time steps increase with the number of blocks.

For all steady test cases, the Lower-Upper Symmetric Gauss-Seidel (LU-SGS) method is adopted for time iterative. In order to ensure that the accuracy of spatial scheme is not to be affected by the time scheme, the machine zero is required. For the unsteady test case (the vortex transport), the total variation diminishing (TVD) third-order Runge-Kutta (RK3) scheme is selected. Finally it should be pointed out that for two-dimensional (2D) test cases, if one cell (i.e. two grid points) is set in the z dimension, the code would be changed to the 2D version automatically.

4. Case summary

In this section, the test cases computed by WCNSs and MUSCL are presented in each of subsections. These benchmark cases are proposed at the International Workshop on High-Order CFD Methods, including the flow over smooth bump, the vortex transport, the laminar boundary layer on flat plate, the flow around delta wing, the transonic flow over airfoil and the transonic flow over wing-body configuration. All of the results are obtained on the TianHe supercomputer of the National University of Defense Technology (NUDT) which is equipped with Inter Xeon E5-2692 CPUs. It should be pointed out that the code only runs on CPU nodes.

Mathematically, a numerical method is said to be pth-order if the solution error e is proportional to the mesh size h to the power p, that is $e \propto h^p$. The mesh size h can be computed by

$$h = 1/(DoFs)^{1/\dim} \quad (27)$$

where dim is the physical dimension (1, 2 or 3) and DoFs are degrees of freedom (Wang 2007). For WCNS or MUSCL with SCMM, each of cells has one DoF.

Efficiency is the important aspect of verification, but the cost is somewhat difficult to quantify. In this paper,

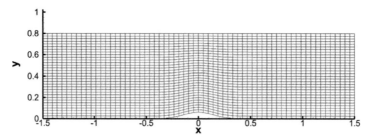

Figure 1. Coarse (64 × 32) grid for the bump.

the TauBench code recommended in the Workshop is adopted to measure computer performance and nondimensionalise cost. The work unit is used to quantify nondimensional cost. Moreover, it equivalent to the CPU time taken to run the Taubench code for 10 steps with 250,000 DoFs. For TianHe supercomputer of NUDT, it runs the TauBench code led to a CPU time of 8.5s on one core.

4.1. Inviscid flow through a channel with a smooth bump

This test case is the internal flow with a high-order curved boundary representation governed by the Euler equations. The computational domain is given by $-1.5 \leq x \leq 1.5, y = 0.8$ (upper boundary) and $y = 0.0625\exp(-25x^2)$ (lower boundary). The inflow Mach number is 0.5 at zero angle of attack. The lower boundary is a slip adiabatic wall, and the upper boundary is a symmetry boundary. For the left boundary and the right boundary, subsonic inflow boundary condition and pressure exit boundary condition are employed, respectively.

The solutions are computed on five structured grids containing 32 × 16, 64 × 32, 128 × 64, 256 × 128 and 512 × 256 cells. The grid point distributions on all boundaries of the finest grid are uniform, and the coarser grids are obtained by deleting every other grid line from the fine grid. The coarse (64 × 32) grid is shown in Figure 1.

The simulations are initiated from the freestream throughout the domain, and the LU-SGS method is adopted for the time iterations. A typical convergence history for WCNSs and MUSCL on the extra-coarse (32 × 16) grid is shown in Figure 2. The machine zero is achieved for WCNSs and MUSCL, but we identify that the reduction of the global residual to 10^{-10} relative to freestream conditions is the mark of iterative convergence[3].

As the exact solution for this case is isentropic, the deviation from the exact entropy can be used as a

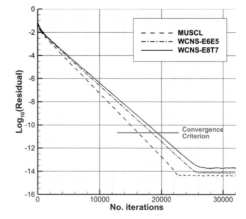

Figure 2. Convergence history for WCNSs and MUSCL on the extra-coarse (32 × 16) grid for the bump.

measure of the error, i.e.

$$\text{error} = \frac{s - s^{\text{exact}}}{s^{\text{exact}}}, \quad s = \frac{p}{\rho^\gamma}. \quad (28)$$

The contour plot of this error for WCNS-E8T7 on the finest (512 × 256) grid is shown in Figure 3. It can be seen that the large entropy error is located at the top and downstream of the smooth bump. The L^2 norm of this quantity is defined as

$$\text{Error}_{L^2} = \left[\frac{\sum_{i=1}^N \text{error}_i^2 |J_i|}{\sum_{i=1}^N |J_i|} \right]^{1/2} \quad (29)$$

where J_i is Jacobian on each cells, which is computed by Equation (25).

Table 1 shows the convergence rate on five grids for WCNS-E6E5 and WCNS-E8T7. It can be seen that the global convergence rates of WCNS-E6E5 are close to 5

Table 1. L^2 norm of the entropy error, actual order of accuracy and work units of WCNSs for the bump.

Grid	WCNS-E6E5			WCNS-E8T7		
	L^2-norm entropy error	Convergence rate	Work units	L^2-norm entropy error	Convergence rate	Work units
32 × 16	2.27E−04	− −	5.37E+00	4.84E−05	− −	7.52E+00
64 × 32	1.20E−05	4.25	8.76E+01	1.16E−06	5.38	1.18E+02
128 × 64	4.45E−07	4.75	7.36E+02	7.52E−09	7.27	1.03E+03
256 × 128	1.47E−08	4.93	6.30E+03	5.31E−11	7.15	8.82E+03
512 × 256	4.66E−10	4.98	5.51E+04	7.29E−13	6.19	7.60E+04

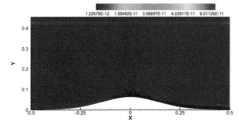

Figure 3. Contour plot of the entropy error for WCNS-E8T7 on the extra-fine (512 × 256) grid.
(colored picture attached at the end of the book)

while the global convergence rates of WCNS-E8T7 are close to 7. To address the question of efficiency, both the computational errors against the grid size and work units are shown in Figure 4. It is evident that the error decreases with the mesh size to the power as the order of the scheme. For the efficiency, Figure 4(b) shows that the computational cost of WCNSs is less than that of MUSCL achieving the same error, namely high-order methods can obtain a numerical solution more efficiently than low-order ones, given the same error threshold. Moreover, comparing the results between WCNS-E6E5 and WCNS-E8T7, it is evident that using compact interpolation with the same stencils can not only improve the order of accuracy but also the computational efficiency.

4.2. Vortex transport by uniform flow

This test case consists of a 2D vortex transported by a uniform flow. It is governed by the Euler equations and aims at characterising the solver's ability to preserve vorticity in an inviscid flow. The computational domain is rectangular, with $(x, y) \in [0, L_x] \times [0, L_y]$. The initial configuration of the vortex, centered in (X_c, Y_c) and superimposed onto the uniform flow, is given by the following equations:

$$u_0 = U_\infty + \delta u, v_0 = \delta v, \rho_0 = \rho_\infty \left(\frac{T_0}{T_\infty}\right)^{1/(\gamma-1)},$$

$$T_0 = T_\infty - \delta T \qquad (30)$$

where

$$\delta u = -(\beta U_\infty)\frac{y - Y_c}{R}e^{-r^2/2}, \delta v = (\beta U_\infty)\frac{y - X_c}{R}e^{-r^2/2},$$

$$\delta T = \frac{0.5}{C_p}(\beta U_\infty)^2 e^{-r^2}, \qquad (31)$$

and

$$C_p = \frac{\gamma}{\gamma - 1}R_{\text{gas}}, r = \sqrt{(x - X_c)^2 + (y - Y_c)^2}/R. \qquad (32)$$

The variable R is the vortex characteristic radius and β is strength. Furthermore, $\gamma = 1.4$ is the constant specific heat ratio; $R_{\text{gas}} = 287.87\text{J}/(\text{kg} \cdot \text{K})$ is the gas constant; and $U_\infty = Ma_\infty\sqrt{\gamma R_{\text{gas}} T_\infty}$ and $\rho_\infty = p_\infty/(T_\infty R_{\text{gas}})$ are the velocity and density of the unperturbed flow, respectively. In this paper, the slow vortex configuration with $Ma_\infty = 0.05$, $\beta = 1/50$ and $R = 0.005$ is computed. Moreover, $L_x = L_y = 0.1$, $X_c = Y_c = 0.05\text{m}$, $p_\infty = 10^5 \text{Pa}$ and $T_\infty = 300\text{K}$ are set.

The flow is periodic in both directions and the period T is defined as $T = L_x/U_\infty$. The solution is advanced in time with the TVD RK3 scheme up to 50 periods. The time step $\Delta t = 0.00001T$ is dictated by the requirement that reducing the time step did not change the results significantly on all grids. Four uniform grids are generated, which contains 32^2, 64^2, 128^2 and 256^2 cells. Since the vortex should be transported without distortion, the initial solution can be used to assess the accuracy of the algorithm. In this paper, the u velocity is chosen to assess the accuracy and the error is defined as

$$\text{error}_i = u_i^{\text{final}} - u_i^{\text{initial}}. \qquad (33)$$

Figure 5 shows the L^2 error of u velocity as a function of h and work units. The second-order scheme MUSCL does not show the design accuracy; in fact, the error is almost flat. This indicates that the asymptotic range has not been reached; therefore, the second-order scheme

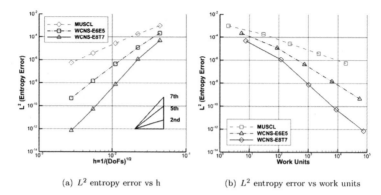

(a) L^2 entropy error vs h (b) L^2 entropy error vs work units

Figure 4. Entropy error of WCNSs and MUSCL for the bump. (a) L^2 entropy error vs. h. (b) L^2 entropy error vs. work units.

requires much finer grids. Table 2 shows the details of L^2 norm of the u velocity error and the convergence rate obtained by WCNS-E6E5 and WCNS-E8T7. It confirms that the expected order of accuracy is obtained for WCNSs.

To address the question of efficiency, Figure 6 shows the comparison for u velocity among MUSCL on 256^2 grid, WCNS-E6E5 on 128^2 grid, WCNS-E8T7 on 64^2 grid and initial solution. For the WCNS-E8T7, the result shows a good agreement with the initial solution with 1.70×10^4 work units. For the WCNS-E6E5, it needs 5.09×10^4 work units to reach the same solution at least. However, for the MUSCL, it still does not provide a satisfactory solution on the finest 256^2 grid with 1.03×10^5 work units. Clearly, WCNS-E8T7 is more efficiency than WCNS-E6E5 and MUSCL because of

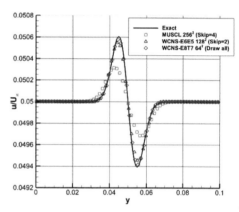

Figure 6. The u velocity along $x/L_x = 0.5$.

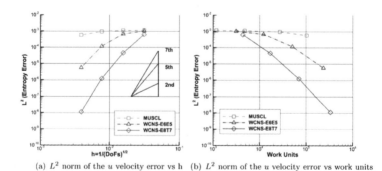

(a) L^2 norm of the u velocity error vs h (b) L^2 norm of the u velocity error vs work units

Figure 5. The u error of WCNSs and MUSCL for the vortex. (a) L^2 norm of the u velocity error vs. h. (b) L^2 norm of the u velocity error vs. work units.

Table 2. L^2 norm of the u velocity error, convergence rate and work units of WCNSs for the vortex transport.

Grid	WCNS-E6E5			WCNS-E8T7		
	L^2-norm u velocity error	Convergence rate	Work units	L^2-norm entropy error	Convergence rate	Work units
32 × 32	1.04E-03	--	3.11E+03	6.24E-04	--	4.34E+03
64 × 64	6.74E-04	0.62	1.22E+04	4.59E-05	3.77	1.70E+04
128 × 128	1.17E-04	2.53	5.09E+04	1.23E-06	5.22	7.04E+04
256 × 256	5.73E-06	4.35	2.37E+05	1.13E-08	6.77	3.34E+05

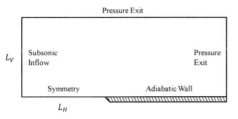

Figure 7. Computational domain for the flat plate boundary layer.

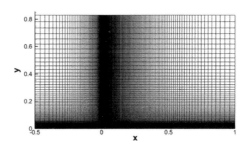

Figure 8. Fine (176 × 144) grid for the flat plate.

Table 3. Drag coefficient obtained on the coarse, medium and fine grids with WCNSs.

h	WCNS-E6E5	WCNS-E8T7
2.51E-02	0.001352600	0.001337343
1.26E-02	0.001332397	0.001330854
6.28E-03	0.001330725	0.001330630

low dissipation, which will be very important for large eddy simulation (LES) and direct numerical simulation (DNS).

4.3. Laminar boundary layer on a flat plate

This test case is a subsonic, laminar flow over a flat plate governed by the Navier-Stokes equations. In Wang et al. (2013), the flow conditions are set to be $Ma_\infty = 0.5$, $\alpha = 0°$ and $Re_L = 10^6$. While in this paper, the freestream Mach number is replaced to be $M_\infty = 0.2$ for the comparison between the numerical solution and the Blasius solution (Kundu, Cohen, and Dowling 2013). The plate length L is assumed to be 1, and the computational domain has two other length scales L_H and L_V, as shown in Figure 7. A sensitivity study of the drag against the domain parameters has been done, and the domain parameters are set to be $L_H = 0.5$, $L_V = 1.0$.

The solution is computed on three structured grids (coarse, medium and fine) containing 44 × 36, 88 × 72 and 176 × 144 cells. For the fine grid (Figure 8), the point distribution in the parallel direction is created using hyperbolic tangent stretching, and in the vertical direction is created using the equal proportion stretching (the height of the first cell at the wall is $\Delta_1 y = 1.0 \times 10^{-5}$ and growth rate is 1.05). For the coarser grids, they are obtained by deleting every other grid line from the finer grid.

For this case, the LU-SGS method is adopted for the time iterations, and all grids are computed on one core. The convergence criterion is a reduction of the global residual to 10^{-10} relative to freestream conditions.

The computed drag coefficients (C_d) on three different grids are shown in Table 3. Due to the lack of the exact solution, a Richardson extrapolated (Roache 1998) value obtain by WCNS-E8T7 is adopted as a reference solution ($C_{d,\text{ref}} = 0.001330622$) to evaluate the C_d errors. Figure 9 presents the C_d errors against the grid size and work unit. The convergence rate of WCNS-E8T7 is slightly lower than the design accuracy. In contrast with the performance in the bump (Section 4.1), where the convergence rate of WCNS-E8T7 on coarse grid only reaches 5.38, it is clear that this is caused by the insufficience of the grid size. For the computational efficiency, looking at the cost to obtain a numerical solution similarly closed to reference value, it is clear that the WCNS-E8T7 gets the best performance.

This problem aims at testing the capacity of high-order methods in simulating viscous boundary layers. The boundary layer profiles computed by WCNSs and

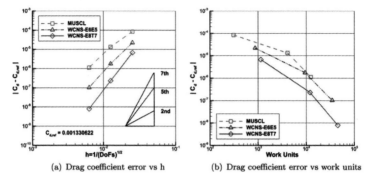

(a) Drag coefficient error vs h
(b) Drag coefficient error vs work units

Figure 9. Drag coefficient error for the flat plate. (a) Drag coefficient error vs. h. (b) Drag coefficient error vs. work units.

MUSCL on coarse grid are shown in Figure 10. For streamwise velocity profiles, all of the numerical solutions show a good agreement with the Blasius profile. However, for the normal velocity, which is a small value and difficult to predict, a poor agreement is obtained for MUSCL. To address the question of efficiency, Figure 11 presents the v velocity profile on different grids. For the second-order scheme MUSCL, it does not provide a good resolution until fine grid with 12,493 work units. For the fifth-order scheme WCNS-E6E5, it takes 9382 work units (on medium grid) to reach the same resolution as the WCNS-E8T7 on coarse grid (only with 1141 work units). By this comparison, it is clear that the good resolution for v velocity profile on the flat plate requires high-order scheme.

4.4. Laminar flow around a delta wing

This test case is a steady, laminar flow at high angle of attack, around a delta wing with sharp edges. It is governed by the 3D Navier-Stokes equations. The geometry of the delta wing is depicted in Leicht and Hartmann (2010) and a free stream Mach number $Ma_\infty = 0.3$ is considered. The Reynolds number based on the root chord of the wing is $Re_c = 4000$ and the angle of attack is $\alpha = 12.5°$. The solution is computed on three structured grids and the fine grid contains 18,717,440 cells. Figure 12 shows the surface mesh of the coarse grid which is obtained from the fine grid by applying a regular coarsing twice. It can be seen that a Y-shape conjunction is adopted. The main purpose is to avoid the generation of odd axes on the top of wing.

The simulations are initiated from the freestream throughout the domain and the no-slip adiabatic wall condition is set on the wing surface. Coarse, medium and fine grids are computed on 8 processors, 48 processors and 96 processors, respectively, and the LU-SGS method

Table 4. Lift and drag coefficients obtained on three different grids with WCNSs for the delta wing.

	WCNS-E6E5		WCNS-E8T7	
h	Cd	Cl	Cd	Cl
1.51E-02	0.171457	0.362144	0.174759	0.374886
7.53E-03	0.167855	0.354515	0.167751	0.354502
3.77E-03	0.167666	0.354114	0.167650	0.354092

is adopted for the time iterations. In order to nondimensionalize the computational costs, the TauBench code is run on 8 cores with 1.9s, on 48 cores with 0.61s and on 96 cores with 0.47s.

This problem is a steady flow and the steady-state solutions are obtained when the global residual reduces to 10^{-10} relative to freestream conditions. Figure 13 shows the streamlines and slices of Mach number contours along and behind the delta wing. It is clear that the flow is dominated by the complex vortices. As the flow passes the leading edge, it rolls up and creates a large vortex structure which is convected far behind the wing, at the same time, near the leading edge a smaller secondary vortex appears.

The computed lift and drag coefficients on the three different grids with WCNSs are presented in Table 4. Figures 14 and 15 show the C_d errors and C_l errors against the grid size and work unit, respectively. The reference solutions are obtained by the Richardson extrapolation with WCNS-E8T7, which are $C_{d,\,ref} = 0.167649$ and $C_{l,\,ref} = 0.354085$. The design accuracy is obtained for both WCNS-E8T7 and WCNS-E6E5. However on the coarse grid, the numerical solutions obtained by WCNS-E8T7 are worse than those of WCNS-E6E5, which leads to a slightly lower efficiency.

To compare the performance of predicting vortex-dominated flows between high-order schemes and

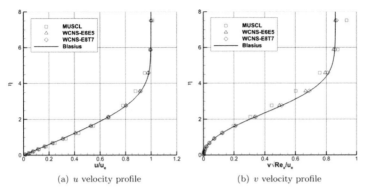

Figure 10. Boundary layer profiles on coarse (44 × 36) grid for laminar flat-plate flow. (a) *u* velocity profile. (b) *v* velocity profile.

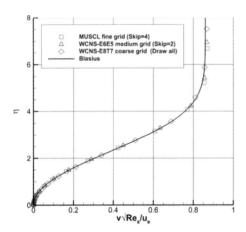

Figure 11. The *v* velocity profile on different grids for laminar flat-plate flow.

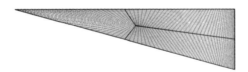

Figure 12. Surface mesh of coarse grid for the delta wing.

second-order scheme, graphs of isoline of the Mach number at $x = 1.75c$ (the apex of the wing is located at the origin of the x-coordinate) are displayed in Figures 16 and 17. Figure 16 shows the results on the medium gird. It is clear that the structure of secondary vortex can not be resolved clearly using second-order scheme MUSCL. Moreover, the resolution of WCNS-E6E5 is slightly lower

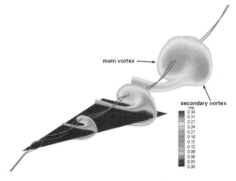

Figure 13. Streamlines and slices of Mach number contours along and behind the delta wing.

(colored picture attached at the end of the book)

than that of WCNS-E8T7. On the fine grid, shown in Figure 17, the resolution of MUSCL just reaches the same level as that of WCNS-E8T7 on medium grid. Comparing the work units of MUSCL on fine grid and WCNS-E8T7 on medium grid (2.16×10^7 vs. 1.90×10^6), it is evident that high-order scheme is more efficient in resolving the vortex-dominated flows.

4.5. Steady turbulent transonic flow over an airfoil

This test case is a 2D turbulent flow under transonic conditions with weak shock-boundary layer interaction effects. It is the RAE2822 airfoil Case 9 (Cook, McDonald, and Firmin 1979), and the original flow conditions in the wind tunnel experiment are $Ma_\infty = 0.730$, $\alpha = 3.19°$ and $Re_c = 6.5 \times 10^6$. However, to take into account the wind tunnel corrections for the comparison with experimental

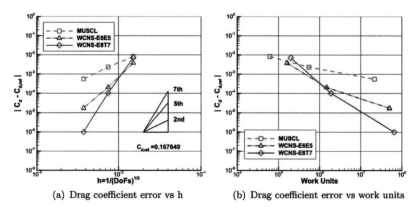

Figure 14. Drag coefficient error for the delta wing. (a) Drag coefficient error vs. h. (b) Drag coefficient error vs. work units.

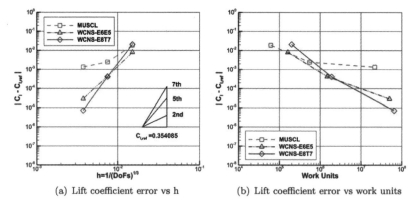

Figure 15. Lift coefficient error for the delta wing. (a) Lift coefficient error vs. h. (b) Lift coefficient error vs. work units.

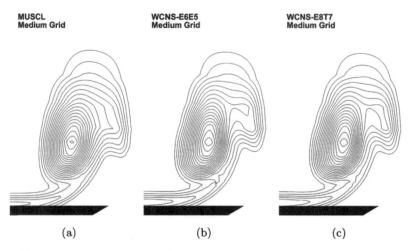

Figure 16. Isolines of the Mach number at $x = 1.75c$ on medium grid.

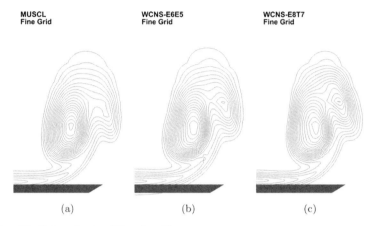

Figure 17. Isolines of the Mach number at $x = 1.75c$ on fine grid.

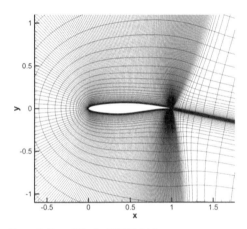

Figure 18. Fine grid for the RAE2822 airfoil.

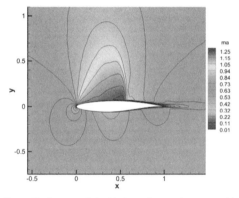

Figure 19. Contours of the Mach number on the coarse grid obtained by WCNS-E8T7 for the RAE2822 airfoil.
(colored picture attached at the end of the book)

data, the computations in this paper are made with corrected flow conditions, namely $Ma_\infty = 0.734$, $\alpha = 2.79°$ and $Re_c = 6.5 \times 10^6$.

The flow is governed by the RANS equations, and the Spalart-Allmaras (SA) turbulence model is adopted. Three structure C grids are generated, which contains 10,384, 23,364 and 52,569 cells, respectively. The mesh sizes of coarser grids are only 1.5 times that of the finer grids, ensuring that the height of the first cell at the wall of the coarse grid are less than 1 ($\Delta_1 y^+ < 1$). The fine grid is shown in Figure 18.

For this test case, the LU-SGS method is adopted for the time iterations. Coarse, medium and fine grids are computed on 2 processors, 4 processors and 8 processors, and the CPU time running the TauBench is 4.4s on 2 cores, 2.8s on 4 cores and 1.9s on 8 cores, respectively. The reduction of the global residual to 10^{-8} relative to freestream conditions is the mark of convergence. Figure 19 shows the contours of the Mach number on the coarse grid obtained by WCNS-E8T7, which shows the excellent capture of the shock.

The computed lift and drag coefficients on the three different grids are presented in Table 5. Same as the previous cases, the reference solutions ($C_{d,\,\text{ref}} = 0.0186527$ and $C_{l,\,\text{ref}} = 0.792486$) are obtained by the Richardson extrapolation with WCNS-E8T7. The C_d errors and C_l errors against the grid size and work units are plotted in Figures 20 and 21, respectively. From these plots, there

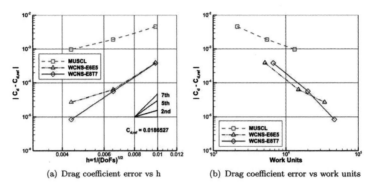

Figure 20. Drag coefficient error for the RAE2822 airfoil. (a) Drag coefficient error vs. h. (b) Drag coefficient error vs. work units.

Table 5. Lift and drag coefficients obtained on three different grids with WCNSs for the RAE2822 airfoil.

h	WCNS-E6E5		WCNS-E8T7	
	Cd	Cl	Cd	Cl
9.81E-03	0.019041	0.77719	0.018271	0.77109
6.54E-03	0.018716	0.78986	0.018596	0.78963
4.36E-03	0.018680	0.79469	0.018644	0.79210

is little efficiency improvement for WCNS-E8T7 compared with WCNS-E6E5. Moreover for the convergence rate, both WCNS-E6E5 and WCNS-E8T7 are lower than the fifth-order accuracy. This is caused by the weighted nonlinear interpolation method in regions near shock. As outlined in Section 2.2, the weighted nonlinear interpolation methods are adopted for high-order schemes WCNSs. Take WCNS-E6E5 as an example, the weights are designed in such a way that in smooth region they approach the optimal weights to achieve fifth-order of accuracy, while in regions near discontinuities, the stencils which contain discontinuities are assigned nearly zero weights, such that third-order interpolations are achieved in these regions. Therefore no definitive conclusion of whether the designed accuracy is obtained can be drawn in this case.

To compare the performance of predicting weak shock-boundary layer interaction effects among WCNSs and MUSCL, graphs of the pressure distribution are displayed in Figure 22. In the comparison between the numerical solution and experimental data, a good agreement is obtained for WCNSs and MUSCL in smooth fields. However, near the shock wave, it is clear that the resolution of MUSCL on medium grid just reaches the same level as that of WCNS-E6E5 on coarse grid. Comparing the work units (WCNS-E6E5 on coarse grid: 5.16 $\times 10^3$ vs. MUSCL on medium grid: 5.51 $\times 10^3$), it is indicated that high-order WCNS-E6E5 is more efficient than

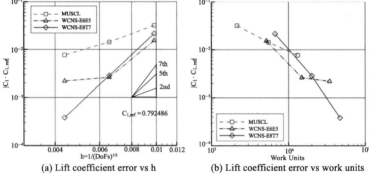

Figure 21. Lift coefficient error for the RAE2822 airfoil. (a) Lift coefficient error vs. h. (b) Lift coefficient error vs. work units.

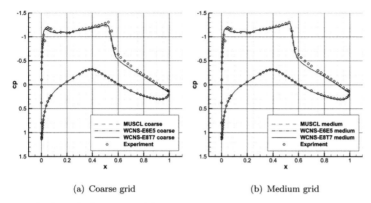

Figure 22. The pressure distribution for the RAE2822 airfoil.

MUSCL. Moreover, comparing the resolution of shock between WCNS-E6E5 and WCNS-E8T7, there is no obvious difference. This is caused by the fact that the interpolation of WCNS-E8T7 would degenerate to the explicit interpolation same as WCNS-E6E5 when there are discontinuities in the whole stencil. In summary, the computational efficiency is not obviously improved for WCNS-E8T7 in this test case.

4.6. Wing-body configuration in DPW-5

This test case is a wing-body configuration under transonic cruise conditions. The flow is assumed to be steady-state and fully turbulent. Computations are to be performed in a target lift mode, i.e. given a specified lift coefficient $C_L = 0.500(\pm 0.001)$ and the corresponding angle of attack has to be determined. This configuration is a Common Research Model (CRM) (Vassberg et al. 2008) and was extensively studied with state-of-the-art CFD codes in the 'fifth drag prediction workshop (DPW-5)[4]. In DPW-5, second-order results were fully displayed while high-order methods were used rarely. Therefore, the objective of this simulation is to compare mesh-converged drag and moment coefficient values between high-order and second-order methods.

This study corresponds to the first test case of DPW-5, and the flow conditions are $Ma_\infty = 0.85$ and $Re = 5 \times 10^6$ based on reference chord c_{ref} =275.80 (inch). The moment reference centre is $x_{ref} = 1325.90$ (inch), $y_{ref} = 177.95$ (inch) and the reference area (half model) for coefficient computations is $A = 297360$ (inch)^2. A series of grids has been provided at the website of DPW-5. In this simulation, three of the structured multi-block grids with O-O topology (Vassberg 2011) are used,

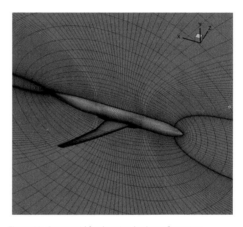

Figure 23. Coarse grid for the wing-body configuration.

which are called tiny, coarse and medium grid and contain 638,976, 2,156,544 and 5,111,808 cells, respectively. Figure 23 shows the coarse grid.

The simulations are initiated from the freestream with the angle of attack $\alpha = 2.1°$ and the LU-SGS method is adopted for the time iterations. Tiny, coarse and medium grids are computed on 12, 36 and 48 processors, and the CPU time running the TauBench is 0.87s on 12 cores, 0.65s on 36 cores and 0.61s on 48 cores respectively. In order to obtain the target lift, the method of α-bisection is adopted. The reduction of the global residual to 10^{-8} relative to initiated conditions is the mark of convergence in each step of the bisection. Moreover, the computational costs are only relative to the last step of the bisection. The lift convergence history on tiny grid with WCNS-E8T7 is shown in Figure 24. Figure 25 presents the contours of

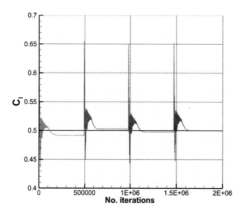

Figure 24. The lift convergence history on coarse grid with WCNS-E8T7. (colored picture attached at the end of the book)

Figure 25. Contours of the pressure coefficient on the medium grid obtained by WCNS-E8T7 for the wing-body configuration.

(colored picture attached at the end of the book)

the pressure coefficient on the medium grid obtained by WCNS-E8T7, which demonstrates that the WCNS-E8T7 based on SCMM could be applied to such a real-life case with complex transonic flows.

The computed drag coefficients against the grid size and work units are plotted in Figure 26, where the mesh-converged values are obtain by the Richardson extrapolation with different schemes. The WCNS-E8T7 shows the best grid convergence in all of the schemes. However for the mesh-converged drag coefficient values, there are large variation in WCNSs and MUSCL. This phenomenon is mainly caused by the difference of spatial discretisation schemes, which also appeared in the DPW-5 and HiOCFD2. For problem of the computational efficiency, it is difficult to obtain a reference value. However, looking at the cost to achieve an error of five drag counts between computed value and mesh-converged value, MUSCL requires 3.55×10^5 work units on coarse grid, while WCNS-E8T7 only requires 2.74×10^5 work units on tiny grid. To an extent, it is indicated that high-order method is more efficient than second-order scheme in such a real-life cases. Figure 27 shows the mesh convergence for the corresponding angle of attack and moment coefficients. WCNS-E8T7 also shows a best moment convergence.

5. Conclusion

In this paper, the seventh-order tri-diagonal compact WCNS (WCNS-E8T7) have been applied to several test cases proposed at the International Workshop on High-Order CFD Methods. Good results have been obtained, especially for the bump and the vortex. For these inviscid cases, WCNS-E8T7 has achieved the order of accuracy closed to 7 with the same stencil width as the WCNS-E6E5. Moreover, the improved the accuracy of WCNS-E8T7 allows the lower costs than WCNS-E6E5 to achieve results of the same error level. In addition, to disapprove the impression that high-order method is expensive, the traditional second-order scheme MUSCL has been used.

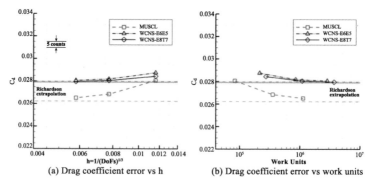

Figure 26. The computed drag coefficients against the grid size and work units. (a) Drag coefficient error vs. h. (b) Drag coefficient error vs. work units.

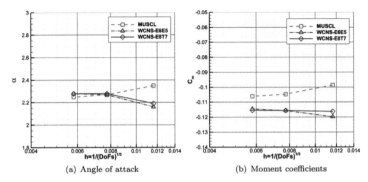

Figure 27. The mesh convergence for the corresponding angle of attack and moment coefficients. (a) Angle of attack. (b) Moment coefficients.

By means of comparing the computational cost to achieve the same level of accuracy, it is confirmed that high-order methods have the advantage in efficiency.

For steady laminar flows (the flat plate and the delta wing), the Richardson extrapolated values obtained by WCNS-E8T7 have been adopted as the reference solutions. With the similar performance in inviscid cases, the higher-order scheme has obtained the numerical solutions with lower costs than the lower-order one, similarly close to reference values. Furthermore, in accurately simulating viscous boundary layers on plate and predicting the secondary vortex in the wake of delta wing, it is also indicated that the WCNS-E8T7 is more efficient.

For turbulent flows, it is difficult to evaluate the performance of these schemes with a reasonable standard. In the RAE2822 airfoil, the reference solutions still have been obtained by the Richardson extrapolation with WCNS-E8T7. Both WCNS-E6E5 and WCNS-E8T7 have obtained a lower convergence accuracy than 5, because of the weighted nonlinear interpolation methods used in the region near shock. Furthermore, it has not been found obvious efficiency improvement for WCNS-E8T7 compared with WCNS-E6E5 in this case. However, comparing the performance of predicting weak shock, high-order WCNSs are still more efficient than MUSCL. In the wing-body configuration, WCNS-E8T7 shows the best grid convergence in all of the considered schemes. Moreover, it leads to a low cost when achieve an error of five drag counts between computed value and mesh-converged value.

Efficiency improvements of WCNS-E8T7 have been verified in the inviscid and laminar test cases. It will be very important for large eddy simulation (LES) and direct numerical simulation (DNS). However, for RANS cases, the efficiency improvement is not obvious. Therefore, the fifth-order explicit scheme WCNS-E6E5 is still a good choice for engineering calculation based on RANS models at present. Future works include the thorough investigations of the efficiency behaviour for RANS test cases and the applications to LES for unsteady turbulence cases. The sub-grid-scale (SGS) models are currently being implemented, such that the unsteady turbulence cases can also be considered.

Acknowledgements

The authors would like to thank the Research Center of Supercomputing Application of the National University of Defense Technology for the support provided with computational resources.

Disclosure statement

No potential conflict of interest was reported by the authors.

Funding

Foundation of the National University of Defense Technology [grant number ZDYYJCYJ20140101].

Notes

1. Data available online at http://zjwang.com/hiocfd.html
2. Data available online at http://www.dlr.de/as/desktopdefault.aspx/tabid-128/268-read-35550
3. The machine zero can be achieved for all steady flows in this paper. In order to avoid the duplication, the convergence history in the following test cases will not be given.
4. Data available online at http://aaac.larc.nasa.gov/tsab/cfdlarc/aiaa-dpw/

References

Abe, Yoshiaki, Nobuyuki Iizuka, Taku Nonomura, and Kozo Fujii. 2013. "Conservative Metric Evaluation for High-Order Finite Difference Schemes with the GCL Identities on Moving and Deforming Grids." *Journal of Computational Physics* 232: 14–21.

Abe, Yoshiaki, Taku Nonomura, Nobuyuki Iizuka, and Kozo Fujii. 2014. "Geometric Interpretations and Spatial Symmetry Property of Metrics in the Conservative form for High-Order Finite-Difference Schemes on Moving and Deforming Grids." *Journal of Computational Physics* 260: 163–203.

Borges, R., M. Carmona, B. Costa, and W. S. Don. 2008. "An Improved Weighted Essentially Non-Oscillatory Scheme for Hyperbolic Conservation Laws." *Journal of Computational Physics* 227(6): 3191–3211.

Brehm, Christoph, Michael F. Barad, Jeffrey A. Housman, and Cetin C. Kiris. 2015. "A Comparison of Higher-Order Finite-Difference Shock Capturing Schemes." *Computers and Fluids* 122: 184–208.

Chao, J., A. Haselbacher, and S. Balachandar. 2009. "A Massively Parallel Compact Multi-block Hybrid Compact-WENO Scheme for Compressible Flows." *Journal of Computational Physics* 228: 7473–7491.

Cheng, Juan, and C. Shu. 2009. "High Order Schemes for CFD: A Review." *Chinese Journal of Computational Physics* 26(5): 633–655.

Cook, P., M. McDonald, and M. Firmin. 1979. "Aerofoil RAE 2822-Pressure Distributions, and Boundary Layer and Wake Measurements, Experimental Data Base for Computer Program Assessment." In Advanced Guidance for Alliance Research and Development, part of NATO Science & Technology Organization, AGARD Report AR-138, A6: 1–77.

Deng, Xiaogang, Xin Liu, Meiliang Mao, and Hanxin Zhang. 2005. "Investigation on Weighted Compact Fifth-order Nonlinear Scheme and Applications to Complex Flow." Paper presented at the 17th AIAA Computational Fluid Dynamics Conference, AIAA 2005-5246, June 6–9, Toronto, ON.

Deng, Xiaogang, Meiliang Mao, Yi Jiang, and Huayong Liu. 2011a. "New High-Order Hybrid Cell-Edge and Cell-Node Weighted Compact Nonlinear Schemes." Paper presented at the 20th AIAA Computational Fluid Dynamics Conference, AIAA 2011-3857, June 27–30, Honolulu, HI.

Deng, Xiaogang, Meiliang Mao, Guohua Tu, Huayong Liu, and Hanxin Zhang. 2011b. "Geometric Conservation Law and Application to High-Order Finite Difference Schemes with Stationary Grids." *Journal of Computational Physics* 230: 1100–1115.

Deng, Xiaogang, Meiliang Mao, Guohua Tu, Hanxin Zhang, and Yifeng Zhang. 2012. "High-Order and High Accurate CFD Methods and Their Applications for Complex Grid Problems." *Communications in Computational Physics* 11: 1081–1102.

Deng, Xiaogang, Meiliang Mao, Guohua Tu, Yifeng Zhang, and Hanxin Zhang. 2010. "Extending Weighted Compact Nonlinear Schemes to Complex Grids with Characteristic-Based Interface Conditions." *AIAA Journal* 48(12): 2840–2851.

Deng, Xiaogang, Yaobing Min, Meiliang Mao, Huayong Liu, Guohua Tu, and Hanxin Zhang. 2013. "Further Study on Geometric Conservation Law and Application to High-Order Finite Difference Schemes with Stationary Grids." *Journal of Computational Physics* 239: 90–111.

Deng, Xiaogang, Guangxue Wang, and Guohua Tu. 2011. "Applications of High-Order Weighted Compact Nonlinear Scheme for Complex Transonic Flows." In *49th AIAA Aerospace Sciences Meeting including the New Horizons Forum and Aerospace Exposition*, 4–7 January 2011, Orlando, FL, AIAA 2011-364.

Deng, Xiaogang, and Hanxin Zhang. 2000. "Developing High-Order Weighted Compact Nonlinear Schemes." *Journal of Computational Physics* 165: 22–44.

Deng, Xiaogang, Huajun Zhu, Yaobing Min, Huayong Liu, Meiliang Mao, Guangxue Wang, and Hanxin Zhang. 2014. "Symmetric Conservative Metric Method: A Link Between High Order Finite-Difference and Finite-Volume Schemes for Flow Computations Around Complex Geometries." Paper presented at the 8th International Conference on Computational Fluid Dynamics (ICCFD8), ICCFD8 2014-0005, July, Chengdu, China.

Ghosh, D., and J.D. Baeder. 2012. "Compact Reconstruction Schemes with Weighted ENO Limiting for Hyperbolic Conservation Laws." *SIAM Journal on Scientific Computing* 34(3): A1678–A1706.

Henrick, A. k., T. D. Aslam, and J. M. Powers. 2005. "Mapped Weighted Essentially Non-oscillatory Schemes: Achieving Optimal Order Near Critical Points." *Journal of Computational Physics* 207: 542–576.

Hu, X. Y., Q. Wang, and N. Adams. 2010. "An Adaptive Central-Upwind Weighted Essentially Non-oscillatory Scheme." *Journal of Computational Physics* 229: 8952–8965.

Jiang, G., and C. Shu. 1996. "Efficient Implementation of Weighted ENO Schemes." *Journal of Computational Physics* 126: 202–228.

Kundu, Pijush, Ira Cohen, and David Dowling. 2013. *Fluid Mechanics*. 5thed. Singapore: Elsevier.

Leicht, Tobias, and Ralf Hartmann. 2010. "Error Estimation and Anisotropic Mesh Refinement for 3d Laminar Aerodynamic Flow Simulations." *Journal of Computational Physics* 229: 7344–7360.

Liu, Huayong, Xiaogang Deng, Meiliang Mao, and Guangxue Wang. 2011. "High Order Nonlinear Schemes for Viscous Terms and the Application to Complex Configuration Problems." 49th AIAA Aerospace Sciences Meeting Including the New Horizons Forum and Aerospace Exposition, 4–7 January 2011, Orlando, FL, AIAA 2011-369.

Liu, H., Y. Ma, Z. Yan, M. Mao, and X. Deng. 2014. "A Shock-Capturing Methodology Based on High Order Compact Interpolation." In Paper presented at the 8th International Conference on Computational Fluid Dynamics (ICCFD8), ICCFD8 2014-0082, July, Chengdu, China.

Liu, Xin, and Xiaogang Deng. 2007. "Application of High-Order Accurate Algorithm to Hypersonic Viscous Flows for Calculating Heat Transfer Distributions." 45th AIAA Aerospace Sciences Meeting and Exhibit, 8-11 January 2007, Reno, NV, AIAA 2007-691.

Liu, X., S. Zhang, H. Zhang, and C.-W. Shu. 2015. "A New Class of Central Compact Schemes with Spectral-Like Resolution II: Hybrid Weighted Nonlinear Schemes." *Journal of Computational Physics* 284: 133–154.

Nonomura, T., N. Iizuka, and K. Fujii. 2007. "Increasing Order of Accuracy of Weighted Compact Nonlinear Scheme." Paper presented at the 45th AIAA Aerospace Sciences Meeting and Exhibit, AIAA 2007-893, January 8–11, Reno, NV.

Nonomura, T., N. Iizuka, and K. Fujii. 2010. "Freestream and Vortex Preservation Properties of High-Order WENO and WCNS on Curvilinear Grids." *Computers & Fluids* 39: 197–214.

Nonomura, Taku, Weipeng Li, Yoshinori Goto, and Kozo Fujii. 2011. "Efficiency Improvements of Seventh-Order Weighted Compact Nonlinear Scheme." Computational Fluid Dynamics Journal 18 (2): 180–186.

Pirozzoli, Sergio. 2006. "On the Spectral Properties of Shock-Capturing Schemes." *Journal of Computational Physics* 219: 489–497.

Pulliam, T., and J. Steger. 1978. "On Implicit Finite-Difference Simulations of Three-Dimensional Flow." In Paper presented at the 16th AIAA Aerospace Sciences Meeting, AIAA 78-10, January 16–18, Huntsville, AL.

Roache, J. 1998. *Verification and Validation in Computational Science and Engineering*. Albuquerque, NM: Hermosa.

Roe, P. L. 1981. "Approximate Riemann Solvers, Parameter Vectors and Difference Schemes." *Journal of Computational Physics* 43: 357–372.

Shu, Chi Wang. 2003. "High-Order Finite Difference and Finite Volume WENO Schemes and Discontinuous Galerkin Methods for CFD." *International Journal of Computational Fluid Dynamics* 17(2): 107–118.

Spalart, P., and C. L. Rumsey. 2007. "Effective Inflow Conditions for Turbulence Models in Aerodynamic Calculations." *AIAA Journal* 45(10): 2544–2553.

Svärd, Magnus, and Jan Nordstrem. 2014. "Review of Summation-by-Parts Schemes for Initial ˝Cboundary-Value Problems." *Journal of Computational Physics* 268: 17–38.

Taylor, E. M., M. Wu, and M. P. Martin. 2007. "Optimization of Nonlinear Error for Weighted Essentially Non-Oscillatory Methods in Direct Numerical Simulations of Compressible Turbulence." *Journal of Computational Physics* 223: 384–397.

Thomas, P. D., and C. K. Lombard. 1979. "Geometric Conservation Law and Its Application to Flow Computations on Moving Grids." *AIAA Journal* 17(10): 1030–1037.

van Leer, B. 1979. "Towards the Ultimate Conservative Difference Scheme V : A Second-Order Sequal to Godunov's Methods." *Journal of Computational Physics* 32: 101–136.

Vassberg, John C. 2011. "A Unified Baseline Grid about the Common Research Model Wing-Body for the Fifth AIAA CFD Drag Prediction Workshop." Paper presented at the 29th AIAA Applied Aerodynamics Conference, AIAA 2011-3508, June 27–30, Honolulu, HI.

Vassberg, J. C., M. A. DeHaan, S. M. Rivers, and R. A. Wahls. 2008. "Development of a Common Research Model for Applied CFD Validation Studies." In *AIAA Paper 2008-6919, AIAA Applied Aerodynamics Conference*, Honolulu, HI, August.

Vinokur, M., and H. C. Yee. 2002. "Extension of Efficient Low Dissipation High-Order Schemes for 3D Curvilinear Moving Grids." Frontiers of Computational Fluid Dynnamics 2002: 129–164.

Visbal, R. M., and D. V. Gaitonde. 2002. "On the Use of Higher-Order Finite-Difference Schemes on Curvilinear and Deforming Meshes." *Journal of Computational Physics* 181: 155–185.

Wang, Z. J. 2007. "High-Order Methods for the Euler and Navier-Stokes Equations on Unstructured Grids." *Progress in Aerospace Sciences* 43: 1–41.

Wang, Z. J., Krzysztof Fidkowski, Rmi Abgrall, Francesco Bassi, Doru Caraeni, Andrew Cary, Herman Deconinck, et al. 2013. "High-Order CFD Methods: Current Status and Perspective." *International Journal for Numerical Methods in Fluids* 72: 811–845.

Weide, E., G. Giangaspero, and M. Svärd. 2015. "Efficient Benchmarking of an Energy Stable High-Order Finite Difference Discretization." *AAIA Journal* 001: 1–10.

Wong, Man Long, and Sanjiva K. Lele. 2016. "Improved Weighted Compact Nonlinear Scheme for Flows with Shocks and Material Interfaces: Algorithm and Assessment." Paper presented at the 54th AIAA Aerospace Sciences Meeting, AIAA 2016-1807, January 4–8, San Diego, CA.

Xu, Chuanfu, Xiaogang Deng, Lilun Zhang, Jianbin Fang, Guangxue Wang, Wei Cao Yi Jiang, Yonggang Che, et al. 2014. "Collaborating CPU and GPU for Large-Scale High-Order CFD Simulations with Complex Grids on the TianHe-1A Supercomputer." *Journal of Computational Physics* 278: 275–294.

Zhang, H., M. Reggio, J. Y. Trpanier, and R. Camarero. 1993. "Discrete form of the GCL for Moving Meshes and Its Implementation in CFD Schemes." *Computers and Fluids* 22(1): 9–23.

Zhang, S., S. Jiang, and C. -W. Shu. 2008. "Development of Nonlinear Weighted Compact Schemes with Increasingly Higher Order Accuracy." *Journal of Computational Physics* 227(15): 7294–7321.

Developing a new mesh deformation technique based on support vector machine

Xiang Gao[a], Yidao Dong[b], Chuanfu Xu[a], Min Xiong[a], Zhenghua Wang[a,c] and Xiaogang Deng[a,b]

[a]College of Computer, National University of Defense Technology, Changsha, Hunan, China; [b]College of Aerospace Science and Engineering, National University of Defense Technology, Changsha, Hunan, China; [c]State Key Laboratory of Aerodynamics, Mianyang, Sichuan, China

ABSTRACT
Mesh deformation technique is widely applied in numerical simulations involving moving boundaries, and the deforming capability and efficiency is the key of it. In this paper, we present a new point-by-point mesh deformation method based on the support vector machine. This proposed method, to certain extent, is similar to the radial basis function (RBF) interpolation method with data reduction, but the new approach selects key boundary points automatically without specifying an initial set, and the function coefficients are obtained by solving a simple quadratic programming problem. Therefore, it is more efficient than the RBF method. Typical 2D/3D applications and realistic unsteady flow over a pitching airfoil are simulated to demonstrate the capability of the new method. With proper setting, the quality of the deformed meshes after using the new method is comparable to that of the RBF method, and the performance is improved by up to 7 ×.

ARTICLE HISTORY
Received 27 November 2016
Accepted 29 April 2017

KEYWORDS
Mesh deformation; support vector machine; radial basis function; control point; mesh quality metric

1. Introduction

Unsteady numerical simulations like the aeroelastic computation and the aerodynamic shape optimisation usually involve moving boundaries. To be able to perform the flow calculations accurately and reliably, an efficient method is needed to conform the computational mesh to the new deformed domain. The popular way to solve this kind of relatively small deformation problem is using the so-called mesh deformation approach, which just moves the position of grid points without changing their topological relations.

For structured meshes, the transfinite interpolation method interpolates the displacements of boundary points into inner points along with the grid lines, which is fast and accurate (Wang and Przekwas 1994). For unstructured meshes, due to the uncertain connectivity of each grid point, the transfinite interpolation method is no longer suitable. Two types of strategies are developed for unstructured mesh movement. The first employs the connectivity information of the grid points, by analogising the edges as springs (Batina 1991) or solid body elasticity (Lynch and ONeill 1980). Computational costs for these methods are huge because a system of equations involving all the mesh points is needed to be solved. The other point-by-point strategy moves each grid point independently based on its coordinates and distance to the boundary, and is easily implemented in parallel. In recent years, two main point-by-point interpolation methods have been applied for mesh deformations. The inverse distance weighting (IDW) interpolation method computes the displacements of interior points explicitly, but needs different power parameters for different kinds of movement (Witteveen and Bijl 2009; Stadler et al. 2011; Luke, Collins, and Blades 2012). The radial basis function (RBF) interpolation method is more flexible and robust to deal with large deformations (De Boer, Der Schoot, and Bijl 2007; Liu, Guo, and Liu 2012; Sheng and Allen 2013). However, the motion requires the solution of a system of equations whose size equals to the number of boundary points. Moreover, this system of equations leads to a dense matrix system, which is difficult to solve using standard iterative techniques. Although some greedy algorithms can be employed to reduce the selection of boundary points, solving this system can still be expensive in terms of the computation time in large-scale three-dimensional simulations.

In this paper, we present a new mesh deformation approach based on the support vector machine (SVM). The final form of this machine learning algorithm is similar to the RBF method. Our approach contains two steps. The first step is to compute a fitting function formed by a summation of kernel functions over boundary points, but the SVM method selects key boundary points automatically, and the function coefficients are obtained by solving a simple quadratic programming problem instead of a dense linear system of equations. The second step is to calculate the displacements of inner points individually

---

Gao Xiang, Dong Yidao, Xu Chuanfu, Xiong Min, Wang Zhenghua, Deng Xiaogang. Developing a new mesh deformation technique based on support vector machine . International Journal of Computational Fluid Dynamics, 2017, 31(10): 1-12.

according to the trained formula. Since it avoids computationally intensive operations and preserves the parallel ability, this approach is more efficient than the RBF interpolation method. Typical 2D/3D numerical examples are tested with the SVM and RBF methods to demonstrate that the new method is more efficient and gives similar results of the deformed mesh. Furthermore, compressible flow over an oscillatory and transient pitching airfoil is simulated using the new mesh motion approach. The calculated results compare well with the experimental data. These provide a definite proof about the applicability of the new method. The proposed method could be used in real CFD analysis like aeroelastic computation or optimisation process.

The remainder of the paper is organised as follows. Section 2 briefly describes the RBF interpolation method, Section 3 presents the methodology of the SVM regression applying to the mesh morphing. In Section 4, typical cases are tested to demonstrate the deforming capability and efficiency of the proposed method. Finally, we conclude in Section 5.

2. RBF interpolation and data reduction

The interpolation function of the RBF method, describing the displacement of the entire computational domain in each coordinate direction, usually takes the form as follows. Typically, a linear polynomial $p(\mathbf{x})$ is added to the function to recover rigid body translations exactly (De Boer, Der Schoot, and Bijl 2007).

$$g^{(k)}(\mathbf{x}) = \sum_{i=1}^{N_b} \alpha_i^{(k)} \phi\left(\|\mathbf{x} - \mathbf{x}_i\|\right), k \in \{x, y, z\}, \quad (1)$$

where $g(\mathbf{x})$ is the function to be evaluated at location $\mathbf{x}$ and gives the offset of the grid points, k stands for x, y, z directions. ϕ is the basis function with respect to the Euclidean distance $\|\cdot\|$ (in this work, the thin plate spline (TPS) function, $x^2\ln(x)$, is used). $\mathbf{x}_i$ are the known centres (i.e. mesh boundary points) and N_b denotes the number of boundary points. The coefficients α_i are determined by recovering known values of the offset at the boundary in the following form:

$$d_i^{(k)} = g^{(k)}(\mathbf{x}_i). \quad (2)$$

For all boundary points, the matrix form can be written in the following fashion:

$$\mathbf{Y}^{(k)} = \mathbf{M}\alpha^{(k)}, \quad (3)$$

where

$$\mathbf{Y} = \begin{pmatrix} d_1 \\ d_2 \\ \vdots \\ d_{N_b} \end{pmatrix}, \alpha = \begin{pmatrix} \alpha_1 \\ \alpha_2 \\ \vdots \\ \alpha_{N_b} \end{pmatrix} \quad (4)$$

and

$$\mathbf{M} = \begin{pmatrix} \phi_{11} & \phi_{12} & \cdots & \phi_{1N_b} \\ \phi_{21} & \phi_{22} & \cdots & \phi_{2N_b} \\ \vdots & \vdots & \ddots & \vdots \\ \phi_{N_b1} & \phi_{N_b2} & \cdots & \phi_{N_bN_b} \end{pmatrix}, \phi_{ij} = \phi(\|\mathbf{x}_i - \mathbf{x}_j\|). \quad (5)$$

In general, Equation (3) leads to a dense matrix system, which is very expensive to be solved using standard iterative techniques. Therefore, LU factorisation is needed to obtain the coefficients:

$$\alpha^{(k)} = \mathbf{M}^{-1}\mathbf{Y}^{(k)}. \quad (6)$$

Once the interpolation function is obtained, the displacement of internal points can be calculated directly using their coordinate values.

In order to speed up the RBF method, the greedy point selection algorithm is proposed in literature (Sheng and Allen 2013; Rendall and Allen 2009, 2010). Its main idea is to reduce the number of boundary points selected as RBF centres and control the interpolation error. These selected points are also called as *control points*. The typical procedure may be described as follows: first choose a subset of boundary points to compute the interpolation function; then calculate the displacement error of other boundary points; if the largest displacement error is smaller than a given criterion, use this function as the final interpolation function; otherwise, add one or more boundary points with the largest displacement error to the subset list and repeat the procedure again. The control points for each coordinate direction may or may not keep the same and it depends on the error control criterion.

In this paper, we applied the real surface displacement method (Sheng and Allen 2013) as the RBF method with greedy point selection (RBF-Greedy), using the same control points for each direction. Here, we can rewrite the algorithm for one direction as follows:

$$\begin{cases} d_i = g(\mathbf{x}_i) & i \in \{N_b'\} \\ |d_i - g(\mathbf{x}_i)| \leq \varepsilon & i \in \{N_b\} - \{N_b'\} \end{cases} \quad (7)$$

where N_b' denotes the number of selected points, $\{N_b'\}$ denotes the selected subset of boundary points, $\{N_b\}$

denotes the set of all boundary points, and ε is the maximum displacement error. Formula (7) means finding N_b' points that accurately satisfy the fitting function, and keeping the error in an acceptable range for other boundary points. Furthermore, we could relax the constraints and make the discrepancy between the fitting function and the known displacements under a predefined threshold for all the boundary points.

3. Support vector machine regression

The SVM was originally developed by Vapnik based on statistical learning theory at the AT&T Bell Laboratories since 1979 (Vapnik, Golowich, and Smola 1997; Burges 1998; Vapnik 2013). In recent three decades, SVMs have been successfully applied to a wide variety of problems such as face detection, time series and financial prediction, and approximation of complex engineering analysing (Li et al. 2012, 2015). There are two main categories of SVM based on classification and regression. For the purpose of mesh deformation, SVM regression can be utilised to predict the movements of mesh nodes. The idea of SVM regression is to calculate a linear function in a high-dimensional feature space mapping from the input space via a nonlinear kernel function. The resulting function produced by SVM regression only depends on a subset of the training data, because it ignores the training data that lie within a threshold ε to the prediction function (Smola and Scholkopf 2004; Basak, Pal, and Patranabis 2007).

Supposing the training data $\{(\mathbf{x}_1, y_1), \ldots, (\mathbf{x}_n, y_n)\} \subset \chi \times \Re$, where χ denotes the space of input data (for 3D space, is $\Re^3$). The case of linear function $f(\mathbf{x})$ can be described in the form as

$$f(\mathbf{x}) = \langle \omega, \mathbf{x} \rangle + b \quad \text{with } \omega \in \chi, b \in \Re, \quad (8)$$

where $\langle \cdot, \cdot \rangle$ denotes the dot product in χ. The goal of SVM regression is to find a function f that has at most ε deviation from y_i for all the training data, and at the same time make the function as flat as possible. Flatness means to minimise the Euclidean norm $\|\omega\|^2$, hence this can be written as the following convex optimisation problem:

$$\min \frac{1}{2}\|\omega\|^2 \quad \text{subject to } \begin{cases} y_i - \langle \omega, \mathbf{x}_i \rangle - b \leq \varepsilon \\ \langle \omega, \mathbf{x}_i \rangle + b - y_i \leq \varepsilon \end{cases} \quad (9)$$

This problem is solvable if f actually exists and approximates all data pairs $(\mathbf{x}_i, y_i)$ with ε precision. Figure 1 graphically demonstrates the regression solution. In the mesh deformation, we assume all training data lie in/on the ε-insensitive tube around f without exception.

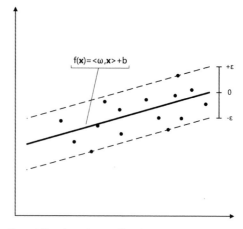

Figure 1. The schematic view of linear SVM regression.

The aforementioned convex optimisation problem, more specifically, the quadratic programming (QP) problem, could be dualised by utilising Lagrange multipliers which make it efficient to solve and easy to extend to nonlinear cases. The Lagrangian function can be written as

$$\mathcal{L} = \frac{1}{2}\|\omega\|^2 - \sum_{i=1}^n \alpha_i \left(\varepsilon - y_i + \langle \omega, \mathbf{x}_i \rangle + b\right)$$
$$- \sum_{i=1}^n \alpha_i^* \left(\varepsilon + y_i - \langle \omega, \mathbf{x}_i \rangle - b\right), \quad (10)$$

where dual variables $\alpha_i, \alpha_i^* \geq 0$. Therefore, it can be deduced that

$$\min \frac{1}{2}\|\omega\|^2 = \min(\max_{\alpha_i, \alpha_i^* \geq 0} \mathcal{L}). \quad (11)$$

When the Karush-Kuhn-Tucker (KKT) conditions (Smola and Scholkopf 2004) are satisfied, the order of min and max in (11) can be exchanged without changing the solution, i.e.

$$\min \frac{1}{2}\|\omega\|^2 = \max_{\alpha_i, \alpha_i^* \geq 0} (\min_{\omega, b} \mathcal{L}). \quad (12)$$

This means minimise the function $\mathcal{L}$ first with respect to the primal variables (ω, b), so the partial derivatives should be zero.

$$\frac{\partial \mathcal{L}}{\partial \omega} = 0 \Rightarrow \omega = \sum_{i=1}^n (\alpha_i - \alpha_i^*)\mathbf{x}_i, \quad (13)$$

$$\frac{\partial \mathcal{L}}{\partial b} = 0 \Rightarrow \sum_{i=1}^{n}(\alpha_i - \alpha_i^*) = 0. \quad (14)$$

Hence, substituting (13) and (14) into (10), the dual optimisation problem takes the form as

$$\max \left\{ -\frac{1}{2}\sum_{i,j=1}^{n}(\alpha_i - \alpha_i^*)(\alpha_j - \alpha_j^*)\langle \mathbf{x}_i, \mathbf{x}_j \rangle \right.$$
$$\left. -\varepsilon \sum_{i=1}^{n}(\alpha_i + \alpha_i^*) + \sum_{i=1}^{n} y_i(\alpha_i - \alpha_i^*) \right\}.$$

subject to $\alpha_i, \alpha_i^* \geq 0$ and $\sum_{i=1}^{n}(\alpha_i - \alpha_i^*) = 0$ (15)

Substituting (13) into linear function (8), the function can be rewritten as follows:

$$f(\mathbf{x}) = \sum_{i=1}^{n}(\alpha_i - \alpha_i^*)\langle \mathbf{x}_i, \mathbf{x} \rangle + b. \quad (16)$$

The coefficients α_i, α_i^* can be efficiently calculated by solving the special dual problem (15). Because the maximum of $\mathcal{L}$ is equal to $\frac{1}{2}\|\omega\|^2$ in (11), computation of b can using the following formula:

$$\begin{aligned} \alpha_i \left(\varepsilon - y_i + \langle \omega, \mathbf{x}_i \rangle + b\right) &= 0, \\ \alpha_i^* \left(\varepsilon + y_i - \langle \omega, \mathbf{x}_i \rangle - b\right) &= 0. \end{aligned} \quad (17)$$

Therefore, for all samples inside the ε-tube, i.e. $|f(\mathbf{x}_i) - y_i| < \varepsilon$, dual variables α_i, α_i^* have to be zero. Moreover, there can never be both nonzero dual variables which deduce $\varepsilon = 0$. Hence, b could be computed using one of the following formulae:

$$\begin{aligned} b &= y_i - \langle \omega, \mathbf{x}_i \rangle - \varepsilon & \text{for } \alpha_i > 0, \\ b &= y_i - \langle \omega, \mathbf{x}_i \rangle + \varepsilon & \text{for } \alpha_i^* > 0. \end{aligned} \quad (18)$$

Consequently, only a few samples are needed to describe ω. The input data that come with nonzero coefficients are called *support vectors*, which are similar to the control points of the RBF method.

SVM can be easily extended to nonlinear cases by mapping the input space into higher dimensional feature space, by using the kernel function $k(\mathbf{x}_i, \mathbf{x})$ defined as a linear dot product of the nonlinear mapping (Basak, Pal, and Patranabis 2007), then the regression function becomes

$$f(\mathbf{x}) = \sum_{i=1}^{n}(\alpha_i - \alpha_i^*)k(\mathbf{x}_i, \mathbf{x}) + b. \quad (19)$$

In this paper, we applied the Gaussian kernel function to map the input data. In fact, it is one of the most widely used radial basis functions (Scholkopf et al. 1997; Burges 1998):

$$k(\mathbf{x}_i, \mathbf{x}) = e^{-\frac{\|\mathbf{x}_i - \mathbf{x}\|^2}{2\sigma^2}}, \sigma > 0, \quad (20)$$

where σ is the kernel width parameter that controls the radial scope of the function. With the increase of σ, the faster the Gaussian function decays. When σ tends to infinity, the function degenerates into a constant. On the contrary, when σ sets smaller, it could be mapped to a space with much higher dimensions than the original space, and there will be serious over-fitting problem. In the mesh deformation, we could set the kernel width close to the size of the deformable part of the shape (e.g. the span of the airfoil), so the displacements of the control points mainly affect the appropriate local domain.

Finally, the efficient sequential minimal optimisation (SMO) algorithm (Platt 1998) can be applied to address the dual quadratic programming problem (Formula (15)). SMO breaks this large QP problem into a series of smallest possible QP problems by selecting two variables α_i, α_j in each iteration and treating other variables as constant so that $\alpha_i + \alpha_j = $ const, and then these small QP problems are solved analytically without the matrix computation. Therefore, coefficients of the regression function, $\alpha, \alpha*$, could be found more quickly than solving a linear system of equations.

In the mesh deformation, positions of boundary points and their known displacements of the next step are used as training samples. After training the SVMs (one coordinate direction for an SVM), the displacements of internal points can be computed individually just like the RBF method. With the use of SVM, we successfully relax Equation (7) to $|y_i - f(\mathbf{x}_i)| \leq \varepsilon$ for all the boundary points. It should be noticed that there is no need to pre-scale the data to [0, 1] range as what SVM usually does, because the dimension of each component of the grid coordinates is the same, and the normalisation may degrade the data accuracy.

4. Numerical examples and discussions

The new mesh deformation method is tested with three numerical examples to demonstrate its deforming capability and efficiency. Typical cases involving rotation and translation of a rectangle, 3D bending wing and real flow over oscillatory pitching airfoil are simulated in this work. The mesh quality and efficiency of the deformations are compared with the RBF method which has been widely applied. For the purpose of simplicity, the TPS function is chosen as the radial basis function, which has global support and does not need support radius. De Boer, Der Schoot, and Bijl (2007) has proved it is one of the best functions for the RBF method. All examples are tested on a high-performance cluster node with Intel Xeon E5-2692 CPU at 2.20 GHz, and 63 GB memory.

In this work, we use the size–skew metric f_{ss} as the mesh quality factor for each mesh element (Knupp 2003; De Boer, Der Schoot, and Bijl 2007), and it takes the form as

$$f_{ss} = \sqrt{f_{size}} f_{skew}, \quad (21)$$

where f_{size} and f_{skew} are the relative size and skew metrics, respectively. The range of these measure factors are all between 0 and 1. The size–skew metric is 0 when the element is degenerate or has negative area/volume. While the mesh quality factor reaches the maximum of 1, the area/volume of the element is not changed after deformation and the deformed element attains the ideal node configuration. The minimum and average f_{ss} values of the mesh are calculated to represent its mesh quality.

4.1. Rotation and translation of a rectangle

The first test case consists of mesh movement due to severe rotation and translation of a rectangle in a rather small square domain, which is previously implemented by De Boer, Der Schoot, and Bijl (2007). The mesh nodes on the rectangle follow its movement, while the nodes on the outer boundary are fixed. The square domain has dimensions of 25×25. Within this domain, a rectangle with dimensions of 5×1 is embedded, as illustrated in Figure 2. The computational mesh has 850 inner points, 156 boundary points and 928 triangular elements. The rectangle is translated down and to the left 5 in each direction and is rotated $60°$ counter-clockwise around the centre of the rectangle. The mesh deformation is performed

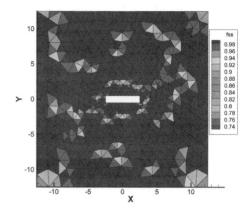

Figure 2. The original mesh of rectangle rotation and translation.
(colored picture attached at the end of the book)

in multiple steps controlled by two sine functions with different amplitudes. The size–skew metrics of the initial mesh are presented in Figure 2.

To compare the SVM and the RBF method, the mesh deformation is performed in a minimum of 1 step and a maximum of 64 steps. The maximum displacement error of RBF-Greedy method and the threshold ε of SVM method are set as 5×10^{-3}, and the Gaussian kernel width parameter of the SVM method is $\sigma = 5$, which equals the length of the rectangle. To ensure the success of the deformation, the threshold ε should be less than the minimum mesh element size divided by the number of deforming steps, i.e. $\varepsilon < \frac{\text{minimum size}}{\text{number of steps}}$, and with larger ε, the deforming ability of the SVM method is better.

The minimum mesh quality value of f_{ss} after movement is shown in Figure 3(a). It can be seen that for both SVM and RBF methods, the minimum value of f_{ss} increases when more intermediate steps are taken. The final mesh quality deformed by the RBF method with/without greedy selection algorithm keeps constant

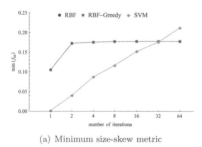
(a) Minimum size-skew metric

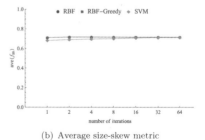
(b) Average size-skew metric

Figure 3. The minimum and average size–skew metric of the final mesh.

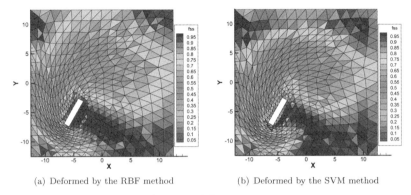

(a) Deformed by the RBF method (b) Deformed by the SVM method

Figure 4. Final meshes using the RBF and SVM method after 64 steps. (colored picture attached at the end of the book)

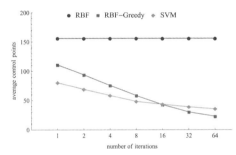

Figure 5. Average control points of different number of iterations.

while the number of intermediate steps reaches 8, but the SVM method needs more than 32 steps to reach comparable results. However, when the number of the intermediate steps reaches 64, the minimum mesh cell quality is even better than that of the RBF method. This means the SVM method only available where the deformation is relatively small for each step. Figure 3(b) gives the average value of the mesh quality metric, the average quality deformed by the SVM method is close to that of the RBF.

The final meshes after deformation with the RBF and SVM methods are presented in Figure 4, respectively. The results illustrate that after severe rotation and translation, the boundary meshes still move together with the rectangle, and their mesh quality has been preserved well during the deformation. This case demonstrates that the SVM method gives similar results compared with the RBF method using adequate intermediate steps.

Figure 5 shows the average number of control points for different number of iterations. The RBF method uses all the boundary points as control points, so its average number of control points keeps constant. The RBF-Greedy method selects boundary point one by one until the displacement error condition is satisfied, and once a new point is added as the control point, the RBF interpolation procedure will repeat again. In contrast, once the dual problem of the SVM method is solved, the control points (i.e. the so-called support vectors) of the SVM method is obtained in one time. Moreover, the SVM method does not need to select an initial set of boundary points as control points. As the number of iterations increases, the average control points of each step for both methods decrease, which indicates that smaller deformation needs fewer points to construct its interpolation (regression) function. The SVM method requires nearly the same number of control points compared with the RBF-Greedy algorithm in Figure 5. Figure 6 shows the control points selected by the RBF-Greedy and SVM method for the first deforming step, which preforms the deformation in 64 steps. The SVM method selects different support vectors to predict the displacement of interior points for their x- and y-components. In general, the difference between the point sets selected by the two methods is not large, but those selected by the SVM are more regular and symmetric. We can see the points selected by the SVM method in Figure 6 are symmetrical about $x = 0$ and $y = 0$, respectively.

The CPU runtime of different methods with increasing iteration steps is illustrated in Figure 7. When the number of iterations is less than 32, the time cost of the RBF with greedy selection is even higher than that of the original RBF because the mesh points are relatively few. Solving the interpolation equations directly is more efficient than the RBF-Greedy method with one-by-one selection and recomputing. As the intermediate steps increase over 32, the number of control points decreases, resulting in the reduction of the greedy selection time cost. Figure 7

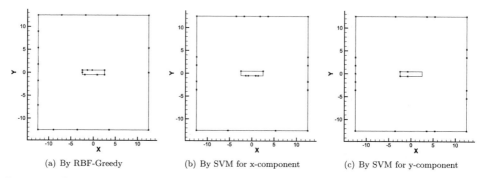

(a) By RBF-Greedy (b) By SVM for x-component (c) By SVM for y-component

Figure 6. Control points selected by the RBF-Greedy and SVM method for the first deforming step.

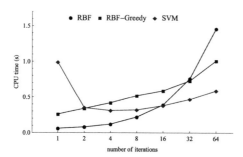

Figure 7. CPU runtime of each method with different iteration numbers.

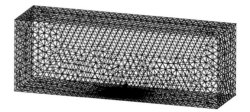

Figure 8. The computation domain of the initial ONERA M6 mesh.

clearly demonstrates that the SVM method is more efficient than the two RBF methods. When the deformation is performed in 64 steps, the CPU cost of the RBF method is 2.5 times than that of the SVM method, while the cost of the RBF-Greedy method is 1.7 times than that of the SVM method. With the deforming steps increase, the advantage of the SVM approach will be more obvious.

Overall, with enough intermediate steps, the rotation and translation of a rectangle can maintain good mesh quality using the SVM method, and the CPU runtime is lower compared with the RBF method. In real CFD computation such as the aeroelastic computation, the number of intermediate steps depends on not only the deformation algorithm, but also the flow fields, and which makes the mesh moves gently and slowly in each step.

4.2. 3D bending wing

The example is presented to demonstrate the deformation capability of the SVM method for 3D complex boundary movement. This case is constructed by artificially bending the ONERA M6 wing, which is previously used by Liu, Guo, and Liu (2012). A periodic vertical movement is prescribed at the wing tip, and the motion can be formulated as follows:

$$h_{tip} = h_0 sin(2\pi t),$$
$$h(z) = (z/z_{tip})^2 h_{tip}, \tag{22}$$

where h_{tip} is the displacement of the wing tip in y direction, z_{tip} is the distance from the wing tip to the root. The wing root is fixed, and a quadratic variation of the bending motion is prescribed between the wing tip and the root, z is the distance from the point on the wing surface to the symmetric plane at the wing root, and $h(z)$ is the vertical displacement of the point. In this case, the bending magnitude h_0 equals c, where c is the chord length of the wing.

The mesh has 281,032 elements, 39,496 internal points and 14,992 boundary (surface) points. The size of the computational domain, as shown in Figure 8, is $23L \times 6L \times 8L$, and L is the span of the ONERA M6 wing. The number of boundary points is so large that huge memory storage is required to store the coefficients matrix (Formula (5)), and unacceptable runtime is needed to

 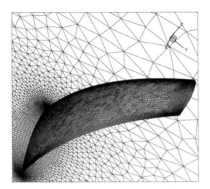

(a) The wing mesh at initial (b) The wing mesh at the maximum displacement

Figure 9. The mesh configuration of the wing at initial and the maximum displacement.

obtain the results directly using the RBF interpolation, therefore we only use the RBF-Greedy method for comparison.

The test performs three deformation cycles in 300 steps, the maximum displacement error of the RBF-Greedy method and the parameter ε of the SVM method are set as 10^{-5}, and the Gaussian kernel width parameter σ of the SVM method approximately equals the span of the wing L, about 1.2. The wing mesh configuration at initial and the maximum displacement after deformation are presented in Figure 9. Figure 10(a) shows the minimum tetrahedron mesh quality after each step; at the maximum displacement of a cycle, the minimum cell quality deformed by the SVM method is a little bit lower than that of the RBF, but the quality is still good enough to perform aeroelastic computation. The average size–skew metric of tetrahedron mesh is illustrated in Figure 10(b), the difference between the SVM and RBF methods is ignorable, which demonstrates that the SVM method is also capable for 3D mesh deformation.

Table 1. Total and average results of the wing bending.

Method	Total time (s)	Average time (s)	Average control points
RBF-Greedy	80,728.71	269.10	540.44
SVM	11,417.50	38.06	381.58

Like the RBF method, the SVM method performs the prediction in three times, one direction a time. The greedy selection algorithm of the RBF method used in this paper selects control points once and applies in three interpolation functions, but the SVM method selects control points for each direction respectively. In this case, boundary points only move in y direction, so the displacement of inner points only has nonzero value in y component applying the SVM/RBF method. The number of support vectors in x and z direction is zero while using the SVM method, and the average number of control points in each deforming step compared with the RBF-Greedy

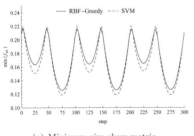 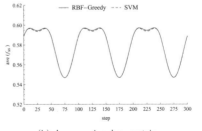

(a) Minimum size-skew metric (b) Average size-skew metric

Figure 10. The minimum and average size–skew metric in each deforming step.

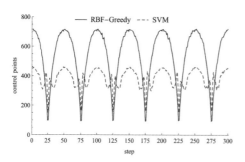

Figure 11. The number of control points in each deforming step.

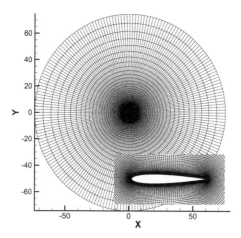

Figure 12. Initial grid about the NACA0012 airfoil with a detailed view.

Table 2. Deformation cost of the NACA0012 airfoil pitching.

Method	Deformation time (s)	Average control points of each step
RBF-Greedy	2050.04	141.61
SVM	1680.85	290.87

method is given in Figure 11. Near the maximum displacement of the wing deformation, the number of control points of both methods decreases to their minimum value. In general, the SVM method uses about one time more point information than the RBF in y direction, but the RBF method wastes control points in the other two directions, therefore resulting in larger average number. Because the RBF method uses the same control points for different directions, the coefficients matrix (Equation (6)) of the three equations are the same, so we can inverse the matrix only one time.

Table 1 shows the total time and the average control points of each step of the mesh deformation; the SVM method is 7.1 times faster than the RBF-Greedy method. The average control points of the SVM method is 2.55% of all the boundary points, and that of the RBF-Greedy method is 3.60%.

4.3. Flow over pitching airfoil

Until now, only the mesh quality of the deformed meshes is investigated without applying to real physical problem. In this section, two calculations of inviscid flow around a pitching NACA0012 airfoil are performed. These results are obtained using the quadrilateral grid shown in Figure 12 as the initial grid. The mesh has 40,400 inner points, 800 boundary points and 20,400 elements, and extends 75 times chord length from the airfoil

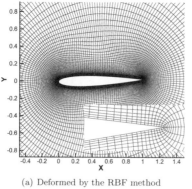 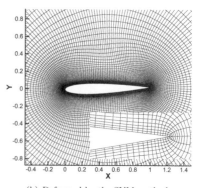

(a) Deformed by the RBF method (b) Deformed by the SVM method

Figure 13. The deformed NACA0012 mesh using the RBF and SVM method.

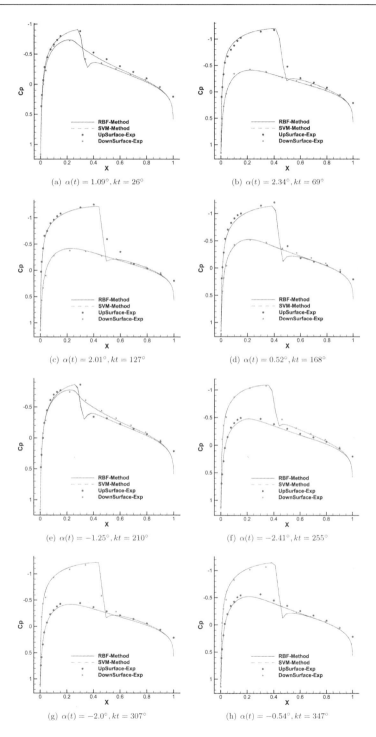

Figure 14. Comparison of instantaneous pressure distributions for the NACA0012 airfoil pitching.

with a circular outer boundary. The unsteady calculations are performed for the airfoil pitching harmonically about the quarter chord with the following formula:

$$\alpha = \alpha_0 + \alpha_m \sin(kt), \tag{23}$$

where the amplitude of $\alpha_m = 2.51°$, the reduced frequency based on semichord of $k = 0.0814$ at $M_\infty = 0.755$ and $\alpha_0 = 0.016°$. After transforming the above dimensionless formula into dimensional, one oscillatory cycle takes about 0.147 s.

The flow solver uses the linear least-squares reconstruction technique, the ALE formulation of HLLC flux function (Daude et al. 2014) and implicit dual timestepping approach with the LU-SGS method (Zhang and Wang 2004). The flow starts with the steady results at $\alpha = \alpha_0$, and the step size is chosen as $\Delta t = 4 \times 10^{-4}$ s (about 1/368 cycle). In the first calculation, we use the SVM method to deform the mesh at every physical time step, and for comparison the RBF-Greedy method is applied in the other calculation. Figure 13 shows a close-up of the airfoil mesh at the oscillatory amplitude after using the RBF and SVM methods, respectively. In both cases, three cycles of motion are computed to obtain a periodic solution.

The maximum displacement error of the RBF-Greedy method and the parameter ε of the SVM method are set as 10^{-7}, and the Gaussian kernel coefficient σ of the SVM method is 1, which equals the chordlength of the airfoil. The calculated pressure coefficient distributions at different angle of attacks in time during the third cycle of motion are presented in Figure 14, compared with the experimental data in Landon (1982). Because the meshes deformed by the RBF and SVM methods are both good enough, the resulting pressure distributions over the airfoil of the RBF and SVM methods are almost identical. There is a shock wave on the upper surface of the airfoil during the first part of the cycle, and during the latter part of the cycle, a shock forms along the lower surface. In general, the two sets of calculated results compare well with the experimental data.

Table 2 shows the deformation time and average number of control points for each method. The average control points of the SVM and RBF-Greedy methods are 36.4% and 17.7% of all the boundary points in this case, respectively, but the SVM method is more efficient than the RBF.

5. Conclusions

In this paper, a novel point-by-point unstructured mesh deformation algorithm based on the support vector machine is developed. This machine learning method is similar to the RBF method with point selection algorithm, but further relaxing the constraints by making the discrepancy between the fitting function and the known displacements under a predefined threshold for all the boundary points. The proposed approach solves a special quadratic programming problem instead of some linear systems of equations, which is more efficient. Moreover, the new method selects control points in the programming problem solving automatically instead of adding mesh points to the initial set of control points iteratively, so it avoids the problem of the initial set selection.

The proposed method is tested with the Gaussian kernel function for three typical examples. The rotation and translation of a 2D rectangle case is shown that with enough deforming steps the SVM method produces meshes of similar quality as the RBF method. The 3D bending wing and pitching airfoil cases demonstrate that the SVM method is capable for 3D mesh deformation and real CFD computation. Compared with the RBF method, the CPU time cost of the SVM method is much lower, especially in large-scale 3D cases.

With proper setting of the parameters σ and ε, the quality of the meshes after deformation is high enough to perform accurate flow calculations like the aeroelastic computation and the aerodynamic shape optimisation. In general, the Gaussian kernel width parameter σ can set close to the length of the deformable part of the object, and ε is depended on the minimum mesh cell size and the approximate number of the computation steps.

Further exploration is needed to investigate the influence of different kernel functions and their parameters setting for the mesh movement, and to solve the problem in several intermediate steps to make it possible not to select control points every time; a linear polynomial could be added to the regression function to improve the deforming capability.

Disclosure statement

No potential conflict of interest was reported by the authors.

Funding

This paper was supported by the National Natural Science Foundation of China [grant number 11502296], [grant number 61561146395]; the Foundation of National University of Defense Technology [grant number ZDYYJCYJ20140101]; the Open Research Program of China State Key Laboratory of Aerodynamics [grant number SKLA20160104]; and the Defense Industrial Technology Development Program [grant number C1520110002].

References

Basak, Debasish, Srimanta Pal, and Dipak Chandra Patranabis. 2007. "Support Vector Regression." *International Journal of Neural Information Processing Letters and Reviews* 11 (10): 203–224.

Batina, John T. 1991. "Unsteady Euler Algorithm with Unstructured Dynamic Mesh for Complex-Aircraft Aerodynamic Analysis." *AIAA Journal* 29 (3): 327–333.

Burges, Christopher J. C. 1998. "A Tutorial on Support Vector Machines for Pattern Recognition." *Data Mining and Knowledge Discovery* 2 (2): 121–167.

Daude, Frederic, Pascal Galon, Zhenlan Gao, and Etienne Blaud. 2014. "Numerical Experiments Using a HLLC-Type Scheme with ALE Formulation for Compressible Two-Phase Flows Five-Equation Models with Phase Transition." *Computers & Fluids* 94: 112–138.

De Boer, A., M. S. Van Der Schoot, and Hester Bijl. 2007. "Mesh Deformation Based on Radial Basis Function Interpolation." *Computers & Structures* 85: 784–795.

Knupp, Patrick M. 2003. "Algebraic Mesh Quality Metrics for Unstructured Initial Meshes." *Finite Elements in Analysis and Design* 39 (3): 217–241.

Landon, R. H. 1982. *NACA0012 Oscillating and Transient Pitching, Compendium of Unsteady Aerodynamic Measurements*. Tech. Rep. Data Set3 AGARD Report R2702. London: Technical Editing and Reproduction Ltd.

Li, Xiang, Huaimin Wang, Bin Gu, and Charles X Ling. 2015. "Data Sparseness in Linear SVM." In *International Conference on Artificial Intelligence*, 3628–3634.

Li, Kuan, Jianping Yin, Zhi Lu, and Xiangfei Kong. 2012. "Multiclass Boosting SVM Using Different Texture Features in HEp-2 Cell Staining Pattern Classification." In *International Conference on Pattern Recognition*, 170–173. IEEE.

Liu, Yu, Zheng Guo, and Jun Liu. 2012. "RBFs-MSA Hybrid Method for Mesh Deformation." *Chinese Journal of Aeronautics* 25 (4): 500–507.

Luke, Edward A., Eric Collins, and Eric Blades. 2012. "A Fast Mesh Deformation Method Using Explicit Interpolation." *Journal of Computational Physics* 231 (2): 586–601.

Lynch, D. R., and K. ONeill. 1980. "Elastic Grid Deformation for Moving Boundary Problems in Two Space Dimensions." *Finite Elements in Water Resources* 2: 111–120.

Platt, John C. 1998. "Sequential Minimal Optimization: A Fast Algorithm for Training Support Vector Machines." In *Technical Report MST-TR-98-14*. Microsoft Research.

Rendall, Thomas, and Christian B. Allen. 2009. "Efficient Mesh Motion Using Radial Basis Functions with Data Reduction Algorithms." *Journal of Computational Physics* 228 (17): 6231–6249.

Rendall, Thomas, and Christian B. Allen. 2010. "Reduced Surface Point Selection Options for Efficient Mesh Deformation Using Radial Basis Functions." *Journal of Computational Physics* 229 (8): 2810–2820.

Scholkopf, Bernhard, Kahkay Sung, Christopher J. C. Burges, Federico Girosi, Partha Niyogi, T. Poggio, and Vladimir Vapnik. 1997. "Comparing Support Vector Machines with Gaussian Kernels to Radial Basis Function Classifiers." *IEEE Transactions on Signal Processing* 45 (11): 2758–2765.

Sheng, Chunhua, and Christian B. Allen. 2013. "Efficient Mesh Deformation Using Radial Basis Functions on Unstructured Meshes." *AIAA Journal* 51 (3): 707–720.

Smola, Alexander J., and Bernhard Scholkopf. 2004. "A Tutorial on Support Vector Regression." *Statistics and Computing* 14 (3): 199–222.

Stadler, Domen, Franc Kosel, Damjan Celic, and Andrej Lipej. 2011. "Mesh Deformation Based on Artificial Neural Networks." *International Journal of Computational Fluid Dynamics* 25 (8): 439–448.

Vapnik, Vladimir. 2013. *The Nature of Statistical Learning Theory*. Dordrecht: Springer Science & Business Media.

Vapnik, Vladimir, Steven E. Golowich, and Alexander J. Smola. 1997. "Support Vector Method for Function Approximation, Regression Estimation and Signal Processing." *Advances in Neural Information Processing Systems* 9: 281–287.

Wang, Z. J., and A. J. Przekwas. 1994. *Unsteady Flow Computation Using Moving Grid with Mesh Enrichment*. Tech. Rep. AIAA-94-0285. AIAA.

Witteveen, Jeroen, and Hester Bijl. 2009. "Explicit Mesh Deformation Using Inverse Distance Weighting Interpolation." In *19th AIAA Computational Fluid Dynamics, Fluid Dynamics and Co-located Conferences* 3996. AIAA.

Zhang, L. P., and Z. J. Wang. 2004. "A Block LU-SGS Implicit Dual Time-Stepping Algorithm for Hybrid Dynamic Meshes." *Computers & Fluids* 33 (7): 891–916.

Large eddy simulation on curvilinear meshes using seventh-order dissipative compact scheme

Yi Jiang [a,c,*], Meiliang Mao [c], Xiaogang Deng [b], Huayong Liu [a]

[a] State Key Laboratory of Aerodynamics, China Aerodynamics Research and Development Center, P.O. Box 211, Mianyang 621000, PR China
[b] National University of Defense Technology, Changsha, Hunan 410073, PR China
[c] Computational Aerodynamics Institute, China Aerodynamics Research and Development Center, P.O. Box 211, Mianyang 621000, PR China

ARTICLE INFO

Article history:
Received 15 January 2014
Received in revised form 24 May 2014
Accepted 9 August 2014
Available online 27 August 2014

Keywords:
Large eddy simulation (LES)
High-order implicit large eddy simulation (HILES)
High-order hybrid cell-edge and cell-node dissipative compact scheme (HDCS)
Geometric conservation law (GCL)
Curvilinear mesh

ABSTRACT

This work investigates the performance of the high-order implicit large eddy simulation (HILES) on curvilinear meshes. The HILES is developed based on a seventh-order hybrid cell-edge and cell-node dissipative compact scheme (HDCS-E8T7) satisfying the surface conservation law (SCL). Efficiency of implicit subgrid-scale model is tested by three-dimensional Taylor–Green vortex case. According to the test of flow over a cylinder, the influence of the SCL errors has been investigated on curvilinear mesh. Then stall phenomena of thin airfoil NACA64A006 have been simulated by the HILES. The slope of lift curve, the maximum lift and the stall angle are successfully predicted. Moreover, the lift characteristic seems to be satisfactorily captured even after the stall angle. The solutions demonstrate the potential of HILES for simulating complex turbulent flow.

© 2014 Elsevier Ltd. All rights reserved.

1. Introduction

The need for predictive simulation methods for turbulent flows has led to a significant interest in large-eddy simulations (LES) in recent years. Without LES that captures unsteady behavior of the turbulent flows, accurate result may not be obtained [1]. For example, it is difficult for Reynolds-averaged Navier–Stokes simulations (RANS) model to estimate stall characteristic of thin-airfoil NACA64A006 [1], where the laminar flow separation occurs at the leading edge and the transition causes the turbulent reattachment. The reattachment point gradually moves rearward with increasing angles of attack [2]. The small vortexes shed from the leading edge, which produces strong unsteadiness in the flow. It is difficult for RANS simulations to resolve this feature, and this is the main reason that RANS simulations do not give satisfactory results for this case. According to the limitation of RANS methods, Fujii [1] proposed that LES should be employed for the prediction of thin-airfoil stall characteristics. Furthermore, LES has been successfully applied for the simulations of stall phenomena [3,4].

As well known, high-order scheme has the advantage over low-order scheme for the simulation of turbulent flow containing unsteady vortex shedding. However, the application of high-order compact schemes still has some challenges, such as robustness and grid-quality sensitivity [5,6]. This deficiency can be largely removed by the researches of the Geometric Conservation Law (GCL) [7–11]. The GCL contains surface conservation law (SCL) and volume conservation law (VCL). The VCL has been widely studied for time-dependent grids, while the SCL is merely discussed for finite difference schemes. Recent research [7] shows that the GCL is very important for the application of finite difference schemes on curvilinear grids. If the SCL has not been satisfied, numerical instabilities and even computing collapse may appear on complex curvilinear grids during numerical simulation. In order to fulfill the SCL for high-order finite difference schemes, a conservative metric method (CMM) is derived by Deng et al. [7]. Not long after, a symmetrical conservative metric method (SCMM) [12], which can evidently increase the numerical accuracy on irregular grids, is proposed based on the CMM. According to the principle of satisfying the SCMM, a seventh-order hybrid cell-edge and cell-node dissipative compact scheme (HDCS-E8T7) has been proposed for complex geometry [13]. The HDCS-E8T7 has inherent dissipation to dissipate unresolvable wavenumbers, therefore the filtering is not needed. The properties of HDCS-E8T7 scheme have been systemically analyzed by Deng et. al. [14].

Based on the HDCS-E8T7, a new high-order implicit large eddy simulation (HILES) is developed following the concept of monotone integrated LES (MILES) [15], i.e. the effects of explicit LES models are imitated by the truncation error of the discretization scheme itself. Although the conception of HILES is similar to that of MILES, HDCS-E8T7 is a new seventh-order compact scheme having inherent dissipation, and the HILES based on HDCS-E8T7 can eliminate the SCL errors, which may contaminate flowfield of the HILES. Moreover, if we use the seventh-order compact scheme for LES without subgrid-scale (SGS) model, i.e., HILES, the accuracy of the flowfield obtained will be seventh-order. However, if subgrid model, for instance, the Smgorinsky subgrid model is added, the accuracy of the LES results will be degenerated to the undesirable second order. If a carefully designed high-order subgrid model is absent, high-order scheme without model may be better than with second-order subgrid models for LES [16].

In this paper, we will investigate the performance of the HILES on curvilinear meshes. Based on the test of flow over a cylinder, the influence of the SCL errors on the HILES has been shown. Then the stall phenomena of thin airfoil NACA64A006 have been simulated by the HILES. In the next section, the governing equations are given. In Section 3, we will give the numerical method comprising three main components: the spatial discretization, the grid metric calculation and the time-integration method. The numerical tests are presented in Section 4.

2. Governing equations

Three dimensional Navier–Stokes equations in computational coordinates may be written as

$$\frac{\partial \widetilde{U}}{\partial t} + \frac{\partial \widetilde{E}}{\partial \xi} + \frac{\partial \widetilde{F}}{\partial \eta} + \frac{\partial \widetilde{G}}{\partial \zeta} = \frac{1}{R_e}\left(\frac{\partial \widetilde{E}_v}{\partial \xi} + \frac{\partial \widetilde{F}_v}{\partial \eta} + \frac{\partial \widetilde{G}_v}{\partial \zeta}\right), \quad (1)$$

where,

$\widetilde{U} = U/J$,
$\widetilde{E} = (\xi_t U + \xi_x E + \xi_y F + \xi_z G)/J$, $\widetilde{E}_v = (\xi_x E_v + \xi_y F_v + \xi_z G_v)/J$,
$\widetilde{F} = (\eta_t U + \eta_x E + \eta_y F + \eta_z G)/J$, $\widetilde{F}_v = (\eta_x E_v + \eta_y F_v + \eta_z G_v)/J$,
$\widetilde{G} = (\zeta_t U + \zeta_x E + \zeta_y F + \zeta_z G)/J$, $\widetilde{G}_v = (\zeta_x E_v + \zeta_y F_v + \zeta_z G_v)/J$,

and

$U = [\rho, \rho u, \rho v, \rho w, \rho e]^T$,
$E = [\rho u, \rho u^2 + p, \rho v u, \rho w u, (\rho e + p)u]^T$,
$F = [\rho v, \rho u v, \rho v^2 + p, \rho w v, (\rho e + p)v]^T$,
$G = [\rho w, \rho u w, \rho v w, \rho w^2 + p, (\rho e + p)w]^T$,
$E_v = [0, \tau_{xx}, \tau_{xy}, \tau_{xz}, u\tau_{xx} + v\tau_{xy} + w\tau_{xz} + \dot{q}_x]^T$,
$F_v = [0, \tau_{yx}, \tau_{yy}, \tau_{yz}, u\tau_{yx} + v\tau_{yy} + w\tau_{yz} + \dot{q}_y]^T$,
$G_v = [0, \tau_{zx}, \tau_{zy}, \tau_{zz}, u\tau_{zx} + v\tau_{zy} + w\tau_{zz} + \dot{q}_z]^T$,

with viscous stress terms written as

$$\tau_{ij} = \mu\left(\frac{\partial u_i}{\partial x_j} + \frac{\partial u_j}{\partial x_i} - \frac{2}{3}\delta_{ij}\frac{\partial u_l}{\partial x_l}\right),$$

and heat transfer terms

$$\dot{q}_x = \frac{1}{(\gamma - 1)P_r M_\infty^2}\mu\frac{\partial T}{\partial x},$$
$$\dot{q}_y = \frac{1}{(\gamma - 1)P_r M_\infty^2}\mu\frac{\partial T}{\partial y},$$
$$\dot{q}_z = \frac{1}{(\gamma - 1)P_r M_\infty^2}\mu\frac{\partial T}{\partial z},$$

where γ is the ratio of specific heats, and the viscous coefficient μ can be calculated by the Sutherland's law. The equation of state and the energy function are

$$p = \frac{1}{\gamma M_\infty^2}\rho T, e = \frac{p}{(\gamma - 1)\rho} + \frac{u^2 + v^2 + w^2}{2}.$$

In the above u, v and w are the velocity components in x, y and z directions, respectively, p is the pressure, ρ is the density and T is temperature. The non-dimensional variables are defined as $\rho = \rho^*/\rho_\infty^*$, $(u, v, w) = (u, v, w)^*/V_\infty^*$, $T = T^*/T_\infty^*$, $p = p^*/\rho_\infty^* V_\infty^{*2}$, $\mu = \mu^*/\mu_\infty^*$ respectively, and $M_\infty = u_\infty/\sqrt{\gamma R T_\infty^*}$, $R_e = \rho_\infty^* u_\infty^* r^*/\mu_\infty^*$, $P_r = \mu_\infty^* C_p/\kappa_\infty^*$ are the Mach number, Reynolds number and Prandtl number, r^* is the characteristic length. J is the Jacobi of grid transformation. $\xi_t, \xi_x, \xi_y, \xi_z, \eta_t, \eta_x, \eta_y, \eta_z, \zeta_t, \zeta_x, \zeta_y, \zeta_z$ are grid derivatives. The grid metric derivatives have the conservative form as

$$\begin{cases} \widetilde{\xi}_x = \xi_x/J = (y_\eta z)_\zeta - (y_\zeta z)_\eta, \widetilde{\xi}_y = \xi_y/J = (z_\eta x)_\zeta - (z_\zeta x)_\eta, \widetilde{\xi}_z = \xi_z/J = (x_\eta y)_\zeta - (x_\zeta y)_\eta, \\ \widetilde{\eta}_x = \eta_x/J = (y_\zeta z)_\xi - (y_\xi z)_\zeta, \widetilde{\eta}_y = \eta_y/J = (z_\zeta x)_\xi - (z_\xi x)_\zeta, \widetilde{\eta}_z = \eta_z/J = (x_\zeta y)_\xi - (x_\xi y)_\zeta, \\ \widetilde{\zeta}_x = \zeta_x/J = (y_\xi z)_\eta - (y_\eta z)_\xi, \widetilde{\zeta}_y = \zeta_y/J = (z_\xi x)_\eta - (z_\eta x)_\xi, \widetilde{\zeta}_z = \zeta_z/J = (x_\xi y)_\eta - (x_\eta y)_\xi. \end{cases} \quad (2)$$

3. Numerical method

3.1. Spatial discretization

A seventh-order finite difference scheme HDCS-E8T7 is employed to discretize the equations (1). Considering discretization of the inviscid terms,

$$\frac{\partial \widetilde{U}}{\partial t} + \frac{\partial \widetilde{E}}{\partial \xi} + \frac{\partial \widetilde{F}}{\partial \eta} + \frac{\partial \widetilde{G}}{\partial \zeta} = 0, \quad (3)$$

and theirs semi-discrete approximations,

$$\frac{\partial \widetilde{U}}{\partial t} = -\delta_i^\xi \widetilde{E} - \delta_i^\eta \widetilde{F} - \delta_i^\zeta \widetilde{G}. \quad (4)$$

The discretizations $\delta_i^\xi, \delta_i^\eta$ and δ_i^ζ are the same, thus we only give the discretization in ξ direction. The δ_i^ξ of the HDCS-E8T7 is

$$\delta_i^\xi \widetilde{E}_j = \frac{256}{175h}\left(\widehat{E}_{j+1/2} - \widehat{E}_{j-1/2}\right) - \frac{1}{4h}\left(\widetilde{E}_{j+1} - \widetilde{E}_{j-1}\right)$$
$$+ \frac{1}{100h}\left(\widetilde{E}_{j+2} - \widetilde{E}_{j-2}\right) - \frac{1}{2100h}\left(\widetilde{E}_{j+3} - \widetilde{E}_{j-3}\right), \quad (5)$$

where, $\widehat{E}_{j\pm 1/2} = \widetilde{E}\left(\widehat{U}_{j\pm 1/2}, \widetilde{\xi}_{x,j\pm 1/2}, \widetilde{\xi}_{y,j\pm 1/2}, \widetilde{\xi}_{z,j\pm 1/2}\right)$ and $\widetilde{E}_{j+m} = \widetilde{E}\left(U_{j+m}, \widetilde{\xi}_{x,j\pm m}, \widetilde{\xi}_{y,j\pm m}, \widetilde{\xi}_{z,j\pm m}\right)$ are the fluxes at cell edges and at cell nodes, respectively. The numerical fluxes $\widehat{E}_{j+1/2}$ may be evaluated by the variables at cell-edges,

$$\widehat{E}_{j\pm 1/2} = \widetilde{E}\left(\widehat{U}^L_{j\pm 1/2}, \widehat{U}^R_{j\pm 1/2}, \widetilde{\xi}_{x,j\pm 1/2}, \widetilde{\xi}_{y,j\pm 1/2}, \widetilde{\xi}_{z,j\pm 1/2}\right), \quad (6)$$

where, $\widehat{U}^L_{j\pm 1/2}, \widehat{U}^R_{j\pm 1/2}$ are variables at cell-edge.

$$\frac{5}{14}(1-\alpha)\widehat{U}^L_{j-1/2} + \widehat{U}^L_{j+1/2} + \frac{5}{14}(1+\alpha)\widehat{U}^L_{j+3/2}$$
$$= \frac{25}{32}(U_{j+1} + U_j) + \frac{5}{64}(U_{j+2} + U_{j-1}) - \frac{1}{448}(U_{j+3} + U_{j-2})$$
$$+ \alpha\left[\frac{25}{64}(U_{j+1} - U_j) + \frac{15}{128}(U_{j+2} - U_{j-1}) - \frac{5}{896}(U_{j+3} - U_{j-2})\right], \quad (7)$$

where $\alpha < 0$ is the dissipative parameter employed to control the dissipation of the HDCS-E8T7. The corresponding $\widehat{U}^R_{j+1/2}$ can be obtained easily by setting $\alpha > 0$. Fig. 1 plots the modified wavenumber of the HDCS-E8T7 with different dissipative parameters. It may be noticed that the resolution of the HDCS-E8T7 is spectral-like.

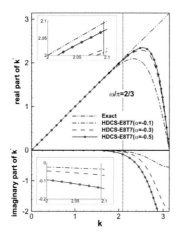

Fig. 1. Modified wavenumber of the HDCS-E8T7.

Dissipative parameter α has no influence on the accuracy of HDCS-E8T7 scheme, however, it has effect on resolution of the scheme. According to the concept of Dispersion-Relation-Preserving (DRP) [17], HDCS-E8T7 scheme has been optimized by adjusted the value of α. The optimized dissipative parameter is equal to 0.3 [14], which will be adopted in this paper.

Seven-point stencil is used by the HDCS-E8T7, thus three kinds of boundary and near-boundary schemes are required for the left-side boundary nodes $j = 1, 2, 3$ as well as for the right-side boundary nodes $j = N, N - 1, N - 2$. The sixth-order scheme is used for the near boundary $j = 3$ and $j = N - 2$,

$$\tilde{E}'_j = \frac{64}{45h}(\hat{E}_{j+1/2} - \hat{E}_{j-1/2}) - \frac{2}{9h}(\tilde{E}_{j+1} - \tilde{E}_{j-1}) + \frac{1}{180h}(\tilde{E}_{j+2} - \tilde{E}_{j-2}). \quad (8)$$

The fourth-order scheme is employed for the near boundary $j = 2$ and $j = N - 1$,

$$\tilde{E}'_j = \frac{4}{3h}(\hat{E}_{j+1/2} - \hat{E}_{j-1/2}) - \frac{1}{6h}(\tilde{E}_{j+1} - \tilde{E}_{j-1}). \quad (9)$$

The fourth-order boundary schemes for $j = 1$ and $j = N$ are

$$\tilde{E}'_1 = \frac{1}{6h}(48\hat{E}_{3/2} + 16\hat{E}_{5/2} - 25\tilde{E}_1 - 36\tilde{E}_2 - 3\tilde{E}_3),$$

$$\tilde{E}'_N = \frac{-1}{6h}(48\hat{E}_{N-1/2} + 16\hat{E}_{N-3/2} - 25\tilde{E}_N - 36\tilde{E}_{N-1} - 3\tilde{E}_{N-2}). \quad (10)$$

For boundary schemes, the fifth-order interpolations for $j = 3/2$ and $j = N - 1/2$ are

$$\hat{U}^L_{3/2} = \frac{1}{128}(35U_1 + 140U_2 - 70U_3 + 28U_4 - 5U_5),$$

$$\hat{U}^L_{N-1/2} = \frac{1}{128}(35U_N + 140U_{N-1} - 70U_{N-2} + 28U_{N-3} - 5U_{N-4}). \quad (11)$$

synchronously, the fifth-order dissipative compact interpolations for $j = 5/2$ and $j = N - 3/2$ are given by

$$\frac{3}{10}(1 - \alpha)\hat{U}^L_{j-1/2} + \hat{U}^L_{j+1/2} + \frac{3}{10}(1 + \alpha)\hat{U}^L_{j+3/2}$$

$$= \frac{3}{4}(U_{j+1} + U_j) + \frac{1}{20}(U_{j+2} + U_{j-1})$$

$$+ \alpha\left[\frac{3}{8}(U_{j+1} - U_j) + \frac{3}{40}(U_{j+2} - U_{j-1})\right]. \quad (12)$$

For the computation of the viscous terms, the primitive variables, u, v, w, T, are first differentiated to form the components of the stress tensor and the heat flux vector. The viscous flux derivatives are then computed by a second application of the HDCS-E8T7.

3.2. Grid metric calculation

The conservative grid metric derivatives (2) have the discretization form as

$$\begin{cases}
\tilde{\xi}^c_x = \delta^\eta_\text{II}((\delta^\zeta_\text{III}y)z) - \delta^\zeta_\text{II}((\delta^\eta_\text{III}y)z), \tilde{\xi}^c_y = \delta^\zeta_\text{II}((\delta^\eta_\text{III}z)x) - \delta^\eta_\text{II}((\delta^\zeta_\text{III}z)x), \tilde{\xi}^c_z = \delta^\eta_\text{II}((\delta^\zeta_\text{III}x)y) - \delta^\zeta_\text{II}((\delta^\eta_\text{III}x)y), \\
\tilde{\eta}^c_x = \delta^\zeta_\text{II}((\delta^\xi_\text{III}y)z) - \delta^\xi_\text{II}((\delta^\zeta_\text{III}y)z), \tilde{\eta}^c_y = \delta^\xi_\text{II}((\delta^\zeta_\text{III}z)x) - \delta^\zeta_\text{II}((\delta^\xi_\text{III}z)x), \tilde{\eta}^c_z = \delta^\xi_\text{II}((\delta^\zeta_\text{III}x)y) - \delta^\zeta_\text{II}((\delta^\xi_\text{III}x)y), \\
\tilde{\zeta}^c_x = \delta^\xi_\text{II}((\delta^\eta_\text{III}y)z) - \delta^\eta_\text{II}((\delta^\xi_\text{III}y)z), \tilde{\zeta}^c_y = \delta^\eta_\text{II}((\delta^\xi_\text{III}z)x) - \delta^\xi_\text{II}((\delta^\eta_\text{III}z)x), \tilde{\zeta}^c_z = \delta^\xi_\text{II}((\delta^\eta_\text{III}x)y) - \delta^\eta_\text{II}((\delta^\xi_\text{III}x)y).
\end{cases} \quad (13)$$

where $\delta^\xi_\text{II}, \delta^\eta_\text{II}, \delta^\zeta_\text{II}$ and $\delta^\xi_\text{III}, \delta^\eta_\text{III}, \delta^\zeta_\text{III}$ are numerical derivative operators used for the metric calculations in ξ, η and ζ coordinate direction. For the applications on curvilinear meshes, finite difference scheme should satisfy the SCL which means $I_x = I_y = I_z = 0$ [7],

$$\begin{cases}
I_x = \delta^\xi_1\left(\tilde{\xi}_x\right) + \delta^\eta_1(\tilde{\eta}_x) + \delta^\zeta_1\left(\tilde{\zeta}_x\right), \\
I_y = \delta^\xi_1\left(\tilde{\xi}_y\right) + \delta^\eta_1\left(\tilde{\eta}_y\right) + \delta^\zeta_1\left(\tilde{\zeta}_y\right), \\
I_z = \delta^\xi_1\left(\tilde{\xi}_z\right) + \delta^\eta_1(\tilde{\eta}_z) + \delta^\zeta_1\left(\tilde{\zeta}_z\right).
\end{cases} \quad (14)$$

In order to fulfill the SCL, the grid metrics should be calculated with a conservative form by the same scheme used for flux derivative calculations, i.e. $\delta_\text{I} = \delta_\text{II}$, to implement the CMM [7]. It has been proved in Ref. [7] that the CMM can be easily applied in high-order schemes if the difference operator δ_I of flux derivatives is not split, while it is difficult to be applied in the schemes where the δ_I is split into two upwind operators as δ^+_I and δ^-_I. Although the inner-level difference operator δ_III in the conservative metrics has no effect on the GCL, the constraint $\delta_\text{II} = \delta_\text{III}$ is recommended by Deng et al. [7]. Not long after, the constraint $\delta_\text{II} = \delta_\text{III}$ is explained from geometry viewpoint, and a symmetrical conservative metric method (SCMM) [12], which can evidently increase the numerical accuracy on irregular grids, is proposed. To eliminate the SCL errors on curvilinear meshes, here we calculate the grid metrics with a conservative form by the same scheme used for flux derivative calculations, i.e. $\delta_\text{I} = \delta_\text{II} = \delta_\text{III}$. Then explicit eighth-order central interpolation is employed for δ_II and δ_III at interior nodes,

$$\hat{U}_{j+1/2} = \frac{1}{2048}(1225(U_{j+1} + U_j) - 245(U_{j+2} + U_{j-1})$$

$$+ 49(U_{j+3} + U_{j-2}) - 5(U_{j+4} + U_{j-3})). \quad (15)$$

At boundary points, the interpolations for δ_II and δ_III are

$$\hat{U}_{3/2} = \frac{1}{128}(35U_1 + 140U_2 - 70U_3 + 28U_4 - 5U_5),$$

$$\hat{U}_{5/2} = \frac{1}{128}(-5U_1 + 60U_2 + 90U_3 - 20U_4 + 3U_5),$$

$$\hat{U}_{7/2} = \frac{1}{256}(3U_1 - 25U_2 + 150U_3 + 150U_4 - 25U_5 + 3U_6),$$

$$\hat{U}_{N-1/2} = \frac{1}{128}(35U_N + 140U_{N-1} - 70U_{N-2} + 28U_{N-3} - 5U_{N-4}),$$

$$\hat{U}_{N-3/2} = \frac{1}{128}(-5U_N + 60U_{N-1} + 90U_{N-2} - 20U_{N-3} + 3U_{N-4}),$$

$$\hat{U}_{N-5/2} = \frac{1}{256}(3U_N - 25U_{N-1} + 150U_{N-2} + 150U_{N-3}$$

$$- 25U_{N-4} + 3U_{N-5}). \quad (16)$$

3.3. Time Integration

In the following numerical tests, highly stretched meshes are employed for the HILES studies of wall-bounded flows. The stability constraint of explicit time-marching methods is too restrictive and the use of an implicit approach becomes necessary, then dual time stepping scheme [18] with Newton-like subiterations [19] is employed. With R denoting the residual, the governing equation is

$$\frac{\partial \widetilde{U}}{\partial t} = R = \frac{1}{R_e}\left(\frac{\partial \widetilde{E}_v}{\partial \xi} + \frac{\partial \widetilde{F}_v}{\partial \eta} + \frac{\partial \widetilde{G}_v}{\partial \zeta}\right) - \frac{\partial \widetilde{E}}{\partial \xi} - \frac{\partial \widetilde{F}}{\partial \eta} - \frac{\partial \widetilde{G}}{\partial \zeta}. \quad (17)$$

The second-order dual time stepping scheme may be written as

$$\left[J^{-1}\left(\frac{1}{\Delta\tau} + \frac{3}{2\Delta t}\right)I + M(U^p)\right]\Delta U^{p+1}$$
$$= -\left[J^{-1}\frac{3U^p - 4U^n + U^{n-1}}{2\Delta t} + R(U^p)\right], \quad (18)$$

where $M(U) = \partial R(U)/\partial U$, and τ is pseudo-time. For the first subiteration, $p = 1, U^p = U^n$, and as $p \to \infty, U^p \to U^{n+1}$. The spatial derivatives in the implicit operators are represented using standard second-order approximations whereas seventh-order discretizations are employed for the residual.

4. Numerical results

4.1. Transition and turbulence decay in the three-dimensional Taylor–Green vortex

For the assessment of HILES, we consider the three-dimensional Taylor–Green vortex (TGV) case which has been used as fundamental prototype for vortex stretching and consequent production of small-scale eddies to test LES methods [20]. The TGV initial configuration involves triple-periodic boundary conditions enforced on a cubical domain with box side length 2π. The computational domain is chosen to be $(x, y, z) \in [-\pi, \pi] \times [-\pi, \pi] \times [-\pi, \pi]$. The flow is initialized with the solenoidal velocity components,

$$u = V_\infty \sin(x)\cos(y)\cos(z),$$
$$v = -V_\infty \cos(x)\sin(y)\cos(z), \quad (19)$$
$$w = 0,$$

and pressure given by a solution of the Poisson equation for the above given velocity field is

$$p = p_\infty + \frac{1}{16}\rho_\infty V_\infty^2 [2 + \cos(2z)][\cos(2x) + \cos(2y)]. \quad (20)$$

Since the present code solves the compressible form of the Navier–Stokes equations, a low Mach number, $M_\infty = 0.1$, is specified to ensure the solutions obtained for the velocity and pressure fields are indeed very close to those obtained assuming an incompressible flow. The considered Reynolds number in this test is $R_e = 400$.

Three meshes are used to solve this problem, the grid dimensions are $65 \times 65 \times 65$ (coarse), $97 \times 97 \times 97$ (medium), $129 \times 129 \times 129$ (fine), respectively. Mesh spacing is constant in the every direction. Third-order Runge–Kutta scheme [21] is used for the time integration. The physical duration of the computation is set to $t_{final} = 10$ (non-dimensional time).

In Fig. 2, we compare the evolution in time of the mean kinetic energy dissipation rate ε between the HILES and the DNS of Brachet et al. [22]. ε is written as

$$\varepsilon = -\frac{de(t)}{dt}, \quad (21)$$

where $e(t) = 1/2\langle u^2 + v^2 + w^2\rangle$ and $\langle \cdot \rangle$ denotes mean (volumetric average). In this figure, the solutions of HILES are approach to that of DNS when the mesh size is increased, and results of medium grid and fine grid are in good agreement with the previous DNS results.

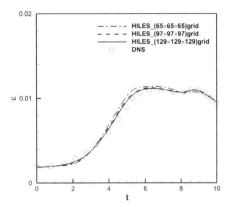

Fig. 2. Evolution of energy dissipation rate.

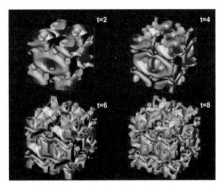

Fig. 3. Iso-surfaces of the second invariant of the velocity gradient tensor colored by streamwise velocity. (colored picture attached at the end of the book)

Fig. 3 shows the evolution of TGV flow dynamics based on instantaneous visualizations of the second invariant of the velocity gradient tensor Q from the present TGV simulations with medium grid. The second invariant Q is

$$Q = (\Omega_{ij}\Omega_{ij} + S_{ij}S_{ij})/2, \quad (22)$$

where $\Omega_{ij} = (u_{i,j} - u_{j,i})/2$ and $S_{ij} = (u_{i,j} + u_{j,i})/2$ are the antisymmetric and the symmetric components of the curl of the velocity respectively. The TGV evolution is laminar and strongly anisotropic at the early times, then energy is transferred to larger wave numbers by vortex stretching, and the flow eventually becomes fully developed turbulent. In the final steps, the small scales are nearly isotropic and exhibit an $k^{-5/3}$ inertial range of the kinetic energy spectrum. To show this process clearly, the kinetic energy spectrum $e_k(k, t)$ from HILES simulation with medium grid at different times is plotted in Fig. 4. The energy spectrum $e_k(k, t)$ is computed by the velocity field in Fourier space, and

$$e(t) = \int e_k(k, t) dk. \quad (23)$$

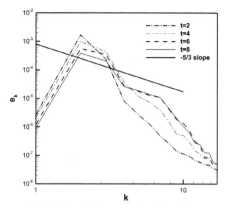

Fig. 4. Turbulence kinetic energy spectra at different time.

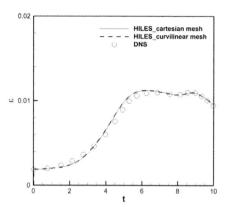

Fig. 6. Evolution of energy dissipation rate on curvilinear mesh comparing with that on cartesian mesh and the DNS solution.

The illustrated slope in Fig. 4 indicates that the HILES can reliably capture the turbulence spectrum. Moreover, the laminar-turbulent transition is resolved by the HILES naturally. For most eddy-viscosity models, some modifications are required to catch this transition [20].

In order to show the ability of the HILES in handling complex mesh, the TGV problem is solved by a curvilinear mesh which is generated from the medium mesh used above. To obtain this new curvilinear mesh, the medium mesh is modified only in the domain $(x,y,z) \in [-\pi/2, \pi/2] \times [-\pi/2, \pi/2] \times [-\pi/2, \pi/2]$ according to the equations,

$$\begin{cases} x = -\pi/2 + \Delta x[(i-1) + 0.5sin(4\pi y_0) \times sin(4\pi z_0)], \\ y = -\pi/2 + \Delta y[(j-1) + 0.5sin(4\pi x_0) \times sin(4\pi z_0)], \\ z = -\pi/2 + \Delta z[(k-1) + 0.5sin(4\pi x_0) \times sin(4\pi y_0)], \end{cases} \quad (24)$$

where $x_0 = \Delta x(i-1), y_0 = \Delta y(j-1), z_0 = \Delta z(k-1), i = 1,\ldots,48, j = 1,\ldots,48, j = 1,\ldots,48, \Delta x = \Delta y = \Delta z = \pi/48$. This 3D curvilinear mesh is sketchily shown in Fig. 5. Comparing with the calculation from the solutions of HILES on medium cartesian mesh, Fig. 6 plots the mean kinetic energy dissipation rate ε on the new curvilinear

Fig. 5. Curvilinear mesh sketch for the TGV problem.

mesh. It is clear that the slight differences between these two solutions can be rarely observed. This phenomenon demonstrates that the HILES has potential ability to simulate the complex flow problem on complex geometry.

4.2. Flow past a circular cylinder

The flow past a circular cylinder at Mach number of 0.2 and Reynolds number based on cylinder diameter, D, of $Re_D = 3900$ is simulated by HILES. The large eddy simulation for this problem has been performed extensively by different research groups [23–25]. The nondimensional computational domain size is $(90 \times 60 \times \pi)$ in (x,y,z). The mesh size is $180 \times 180 \times 45$ in steamwise, normal and spanwise directions respectively. The 2D grid system is shown in Fig. 7. Grid points are clustered near the cylinder surface, with a minimum normal spacing of 0.001. Due to a H-type mesh is employed for the HILES, the gird size here is smaller than that of Rizzetta et al. [24], where a $199 \times 197 \times 53$ O-type grid is used for sixth-order central compact scheme to perform implicit LES (ILES) of the same case. At the outer boundary, far-field boundary conditions based on the Local One-Dimensional Inviscid (LODI) approximation [26] and sponge technique [27] are prescribed. The no-slip condition is invoked on the cylinder surface, together with fifth-order accurate approximations for an adiabatic wall and zero normal pressure gradient. At the spanwise boundaries, periodicity is applied.

According to the study of Deng et al.[7], the SCL is satisfied automatically on uniform meshes. However, the SCL may be dissatisfied on curvilinear meshes. If the SCL has not been satisfied, numerical instabilities and even computing collapse may appear on complex curvilinear grids during numerical simulation. Thus we will investigate the effect of SCL errors in current simulations. To show the influence of the SCL errors, we design a new scheme HDCS-E8T7A, which is same as HDCS-E8T7 except that the grid metric derivatives are calculated by combining the conservation grid metric derivatives and traditional grid metric derivatives. Traditional grid metric derivatives are given by

$$\begin{cases} \tilde{\xi}_x^T = \delta_1^{\eta}(y)\delta_1^{\zeta}(z) - \delta_1^{\eta}(z)\delta_1^{\zeta}(y), \tilde{\xi}_y^T = \delta_1^{\eta}(z)\delta_1^{\zeta}(x) - \delta_1^{\eta}(x)\delta_1^{\zeta}(z), \tilde{\xi}_z^T = \delta_1^{\eta}(x)\delta_1^{\zeta}(y) - \delta_1^{\eta}(y)\delta_1^{\zeta}(x), \\ \tilde{\eta}_x^T = \delta_1^{\zeta}(y)\delta_1^{\xi}(z) - \delta_1^{\zeta}(z)\delta_1^{\xi}(y), \tilde{\eta}_y^T = \delta_1^{\zeta}(z)\delta_1^{\xi}(x) - \delta_1^{\zeta}(x)\delta_1^{\xi}(z), \tilde{\eta}_z^T = \delta_1^{\zeta}(x)\delta_1^{\xi}(y) - \delta_1^{\zeta}(y)\delta_1^{\xi}(x), \\ \tilde{\zeta}_x^T = \delta_1^{\xi}(y)\delta_1^{\eta}(z) - \delta_1^{\xi}(z)\delta_1^{\eta}(y), \tilde{\zeta}_y^T = \delta_1^{\xi}(z)\delta_1^{\eta}(x) - \delta_1^{\xi}(x)\delta_1^{\eta}(z), \tilde{\zeta}_z^T = \delta_1^{\xi}(x)\delta_1^{\eta}(y) - \delta_1^{\xi}(y)\delta_1^{\eta}(x). \end{cases}$$

(25)

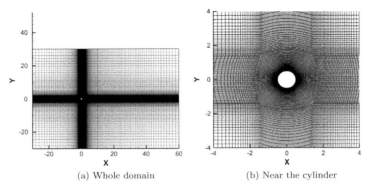

(a) Whole domain   (b) Near the cylinder

Fig. 7. The grid system of the cylinder test.

Table 1
The maximal errors of I_x and I_y for the HDCS-E8T7 and HDCS-E8T7A.

	HDCS-E8T7	HDCS-E8T7A Weight σ						
		10^{-2}	10^{-1}	0.2	0.4	0.6	0.8	1.0
max(I_x)	4.83E − 14	3.35E − 6	3.35E − 5	6.70E − 5	1.34E − 4	2.01E − 4	2.68E − 4	3.35E − 4
max(I_y)	7.08E − 14	3.36E − 6	3.36E − 5	6.72E − 5	1.34E − 4	2.02E − 4	2.69E − 4	3.36E − 4

For instance, one of the grid metric derivatives is written as

$$\tilde{\xi}_x^A = \sigma \tilde{\xi}_x^T + (1-\sigma)\tilde{\xi}_x^c, \quad (26)$$

where $0 \leqslant \sigma \leqslant 1$ is the weight coefficient, $\tilde{\xi}_x^A$ is the grid metric derivative of HDCS-E8T7A, $\tilde{\xi}_x^T$ and $\tilde{\xi}_x^c$ are traditional and conservation grid metric derivatives, respectively. If $\sigma = 0$, the HDCS-E8T7 satisfying the SCL is recovered by the HDCS-E8T7A. However, the SCL is not satisfied if $\sigma \neq 0$. The maximal SCL errors of the HDCS-E8T7 and HDCS-E8T7A are listed in Table 1, it can be seen that the SCL error is highest when $\sigma = 1$. As shown in Fig. 8, the largest SCL error of the HDCS-E8T7A with $\sigma = 1$ presents at the singular point.

The uniform initial flowfield is used to start the simulation and the transition period is up to dimensionless time $t = 100$. Then the statistical results are calculated from $t = 100$ to $t = 200$. The time step size in the computations is $\Delta t = 0.01$. Comparing with the solution of implicit LES (ILES) [24] and MILES [23], mean surface pressure from the HILES computation appears in Fig. 9. Agreement between the simulation of HILES satisfying the SCL and the measurement [28] is reasonable, particularly in the upstream region, and the result of HILES satisfying the SCL lies closer to the experimental date on the aft cylinder surface. For the result of HILES dissatisfying the SCL, however, oscillations are observed at the location near $\theta = 130°$ where the singular point exists. The fluctuating pressure on the surface is also shown in Fig. 9. Due to the SCL errors, the difference between solutions satisfying the SCL and dissatisfying the SCL is significant. In the solution dissatisfying the SCL, there is a second peak at the location near $\theta = 130°$.

Fig. 10 plots mean streamwise velocity field calculated by the HILES with HDCS-E8T7 and HDCS-E8T7A. Highly smooth flowfield is obtained by the HILES with HDCS-E8T7. However, when the

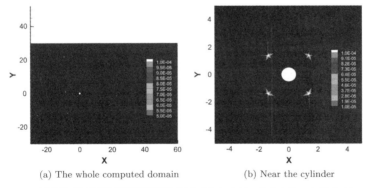

(a) The whole computed domain   (b) Near the cylinder

Fig. 8. The SCL error ($|I_x| + |I_y|$) of HDCS-E8T7A with $\sigma = 1$. (colored picture attached at the end of the book)

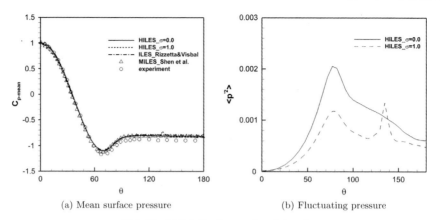

(a) Mean surface pressure (b) Fluctuating pressure

Fig. 9. Mean and fluctuating pressure on the cylinder surface.

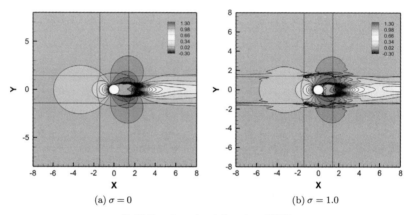

(a) $\sigma = 0$ (b) $\sigma = 1.0$

Fig. 10. Mean streamwise velocity contours of HILES.
(colored picture attached at the end of the book)

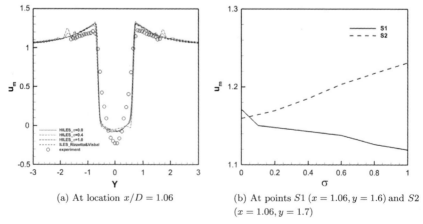

(a) At location $x/D = 1.06$ (b) At points $S1$ ($x = 1.06, y = 1.6$) and $S2$ ($x = 1.06, y = 1.7$)

Fig. 11. Spanwise averaged mean streamwise velocity distributions near the cylinder.

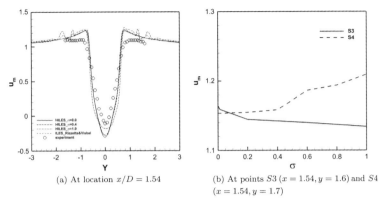

(a) At location $x/D = 1.54$

(b) At points $S3$ ($x = 1.54, y = 1.6$) and $S4$ ($x = 1.54, y = 1.7$)

Fig. 12. Spanwise averaged mean streamwise velocity distributions.

HDCS-E8T7A with $\sigma = 1$ is employed for the computation, the field has been contaminated near the singular point where the SCL error is the largest. It is clear that HILES can eliminate the SCL errors which may contaminate the flowfield. To quantitatively show the influence of the SCL errors, we plot vertical distributions of the mean streamwise velocity component at the streamwise location $x/D = 1.06$ in Fig. 11. Following the SCL errors are enlarged, it is clear that unphysical oscillation is getting more visible. This phenomenon is also demonstrated by the velocities at two points $S_1(x = 1.06, y = 1.6)$ and $S_2(x = 1.06, y = 1.7)$, as shown in Fig. 11. In Fig. 12, similar oscillation is observed at streamwise location $x/D = 1.54$ as well as two points $S_3(x = 1.54, y = 1.6)$ and $S_4(x = 1.54, y = 1.7)$ near the singular point. The overall results satisfying the SCL are very similar to the numerical results of Rizzetta et al. [24]. However, there are some differences between the results of the simulations and those of the experiment of Lourenco and Shih [28]. Those differences are most likely due to that the particular experiment suffered from some external disturbances that

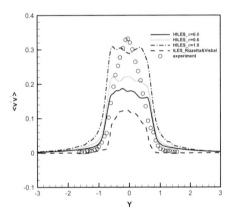

Fig. 14. Lateral Reynolds stress at location $x/D = 1.54$.

contributed to an earlier transition in the separating shear layers [25].

The streamwise and lateral Reynolds stress calculated by the HILES are given in Figs. 13 and 14. Comparing with the experimental data [28], it may be seen that both HILES and ILES significantly under-predict the Reynolds stresses. Due to the SCL errors, the unphysical oscillations may magnify the Reynolds stress. The instantaneous vorticity contours are plotted in Fig. 15. Illustrated in the figure is the fine-scale details calculated by HILES with HDCS-E8T7, particularly in the near wake. However, it is clear that the vorticity may be significantly contaminated by the SCL errors.

4.3. Stall phenomena of thin airfoil NACA64A006

This computation is chosen to check the ability of the HILES for simulating the stall phenomena of the thin airfoil NACA64A006, which is failed to be successfully simulated by RANS model. The airfoil is the same as the experimental model used by McCullough et al. [29]. According to the experiment, freestream Mach number

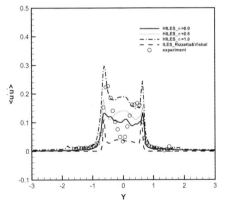

Fig. 13. Streamwise Reynolds stress at location $x/D = 1.06$.

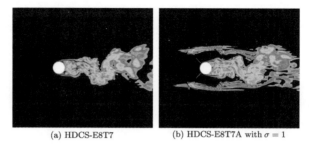

(a) HDCS-E8T7 (b) HDCS-E8T7A with $\sigma = 1$

Fig. 15. The instantaneous vorticity contours. (colored picture attached at the end of the book)

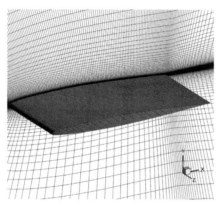

Fig. 16. Computational grid of the airfoil NACA64a006.

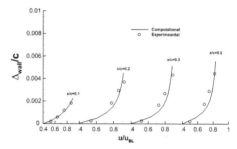

Fig. 18. Boundary-layer velocity profiles developed along the suction surface.

is set to be $M_\infty = 0.17$, chord length is $c = 1.524$ m, the Reyonlds number based on mean aerodynamic chord c is 5.8×10^6, and the angle of attack ranges from $4°$ to $11°$.

A C-type mesh is employed for this computation as shown in Fig. 16. The x, y and z directions correspond to the steamwise, normal and spanwise directions respectively. For the large eddy simulation, the computational domain extends to $0.5c$ in the spanwise, and the mesh size is $341 \times 121 \times 37$. The grid size here is similar to that of Kawai et al. [2], where the LES/RANS hybrid methodology with sixth-order central compact scheme is used to simulate this stall phenomenon. Although the required mesh may be relatively coarser for hybrid methodology than that for LES, our mesh resolution is refined near the wall, then this relatively coarse mesh is suitable for the HILES to capture the stall phenomenon.

At the outer boundary, far-field boundary conditions based on the LODI approximation [26] are prescribed. The no-slip condition is invoked on the cylinder surface, together with fifth-order accurate approximations for an adiabatic wall and zero normal pressure gradient. At the spanwise boundaries, periodicity is applied. The

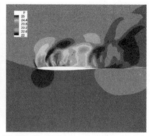

(a) HDCS-E8T7 satisfying the GCL (b) HDCS-E8T7 dissatisfying the GCL

Fig. 17. Instantaneous pressure contours of HILES with HDCS-E8T7 and HDCS-E8T7A. (colored picture attached at the end of the book)

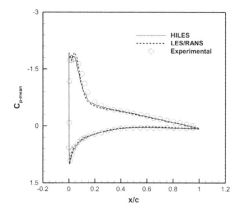

Fig. 19. Time-averaged pressure coefficient distributions over airfoil, $\alpha = 5.5°$.

dimensionless time step size is $\Delta t = 0.001$, computations are advanced for total 25,000 time steps, and flow statistics are collected for 10,000 time steps. For periodic boundaries are used in the spanwise, flow statistics are averaged in this direction.

Although this mesh does not contain singular points, the role of the GCL is also very important since the mesh is highly stretched and curved. If the traditional grid metric derivatives (25) are employed for this calculation, the GCL is dissatisfied and computing collapse will appear during numerical simulation. To illustrate this importance, we continue the calculation from the solution of the HDCS-E8T7 satisfying the GCL at 9° angle of attack, and employ the HDCS-E8T7A with $\sigma = 1$. The computing is collapse after 1300 iterations. Fig. 17 plots the instantaneous pressure contours calculated by the HDCS-E8T7A on the plane $z = 0$ at 100 iteration step. It is clear that unphysical information dominates the flowfield obtained by the HDCS-E8T7A.

The boundary layer should be computed correctly to simulate a separation phenomenon, thus we will investigate the ability of the HILES for the simulation of boundary-layer profile. Fig. 18 compares the computational time-averaged boundary-layer velocity profiles with the corresponding experimental data on the suction surface ($x/c = 0.1, 0.2, 0.3, 0.5$) at 4° angle of attack. It may be seen that the HILES can successfully predict the growth and shape of the boundary-layer profiles at all of the locations. Fig. 19 shows the time-averaged pressure distributions over the airfoil at 5.5° angle of attack. Comparing with the experimental data [29,30] and the numerical solution of LES/RANS hybrid methodology with sixth-order central compact scheme [2], it may be noted that numerical result of the HILES is reliable.

Fig. 20 shows the computed time-averaged lift characteristics. Time-averaged pressure contours at 9°, 10° and 11° angles of attack are also included in this figure. Good agreement between the experimental data [29,30] and the numerical solutions has been observed. It is demonstrated that the HILES has successfully predicted the slope of lift curve, the maximum lift, and the stall angle. Moreover, the lift characteristic seems to be satisfactorily captured even after the stall angle.

The flowfield around NACA64A006 airfoil has thin-airfoil stall characteristics. In this flowfield, the laminar flow separation occurs at the leading edge and the transition causes the turbulent reattachment. According to the numerical research [2], reliable simulation of separation bubble is very important for the successful prediction of stall phenomenon. Fig. 21 shows the time-averaged Mach number distributions and velocity vectors near the leading edge obtained by the HILES. It is obvious that separation bubble is successfully resolved. Such a flowfield is also observed in the experiment by Fitzgerald et al. [31] using laser Doppler velocimetry. Fig. 22 shows the instantaneous pressure distributions over airfoil surface and iso-surface of the second invariant of the velocity gradient tensor at 5.5° angle of attack. Separation from the leading edge and unsteady vortex shedding caused by the instability of the shear layer are clearly observed in this figure. The small vortexes shed from the leading edge, which produces strong unsteadiness in the flow.

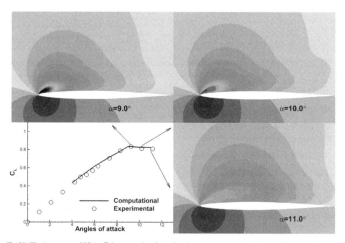

Fig. 20. The time-averaged lift coefficient vs angles of attack and pressure contours at 9°, 10° and 11° angles of attack.
(colored picture attached at the end of the book)

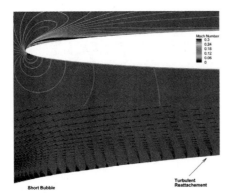

Fig. 21. Time-averaged Mach number distribution and velocity vectors at 5.5° angle of attack. (colored picture attached at the end of the book)

Fig. 22. The instantaneous pressure distribution over airfoil surface and iso-surface of the second invariant of the velocity gradient tensor at 5.5° angle of attack.

(colored picture attached at the end of the book)

5. Conclusion

The performance of HILES has been investigated on curvilinear meshes. The capability of HILES is tested by three-dimensional Taylor–Green vortex case, where the kinetic energy dissipation, the kinetic energy spectrum and the laminar-turbulent transition are reliably captured. Moreover, the potential ability of the HILES handling complex geometry is demonstrated by solving TGV problem on a 3D curvilinear mesh.

The HILES is developed based on HDCS-E8T7 satisfying the SCL, and the influence of the SCL errors is shown in the test of flow past a circular cylinder. Highly smooth flowfield is obtained by the HILES with HDCS-E8T7. However, when the HDCS-E8T7A is employed for the computation, the field has been contaminated near the singular point where the SCL error is the largest. Comparing with experimental and other computational data, it is obvious that unphysical oscillation is getting more visible following the SCL error is enlarged. If the SCL errors are eliminated, the results of HILES are reasonable.

According to the numerical solutions of the thin airfoil NACA64A006, the HILES has successfully predicted the boundary-layer profiles over the suction surface, the slope of lift curve, the maximum lift, and the stall angle. Furthermore, separation from the leading edge and unsteady vortex shedding caused by the instability of the shear layer are clearly resolved.

Based on the above investigation, HILES is a suitable method for simulating the unsteady turbulence flow on curvilinear meshes. In our next paper, we will give more applications of HILES on complex geometry to show its ability of simulating complex flow.

Acknowledgments

This study was supported by the project of National Basic Research Program of China (Grant No. 2009CB723800), National Natural Science Foundation of China (Grant Nos. 11072259 and 11202226) and the foundation of State Key Laboratory of Aerodynamics (Grant No. JBKY11030902 and JBKY11010100).

References

[1] Fujii K. Progress and future prospects of CFD in aerospace – wind tunnel and beyond. Progr Aerosp Sci 2005;41:455–70.
[2] Kawai S, Fujii K. Analysis and prediction of thin-airfoil stall phenomena with hybrid turbulence methodology. AIAA J 2005;43(5):953–61.
[3] Mellen PC, Frohlich J, Rodi W. Lessons from LESFOIL project on large eddy simulation of flow around an airfoil. AIAA J 2003;41(4):573–81.
[4] Mary I, Sagant P. Large eddy simulation of flow around an airfoil near stall. AIAA J 2002;40(6):1139–45.
[5] Wang ZJ. High-order methods for the Euler and Navier–Stokes equations on unstructured grids. Progr Aerosp Sci 2007;43:1–41.
[6] Ekaterinaris JA. High-order accurate, low numerical diffusion methods for aerodynamics. Progr Aerosp Sci 2005;41(3–4):192–300.
[7] Deng XG, Mao ML, Tu GH, Zhang HX. Geometric conservation law and applications to high-order finite difference schemes with stationary grids. J Comput Phys 2011;230:1100–15.
[8] Visbal RM, Gaitonde DV. On the use of higher-order finite-difference schemes on curvilinear and deforming meshes. J Comput Phys 2002;181:155–85.
[9] Nonomura T, Iizuka N, Fujii K. Freestream and vortex preservation properties of high-order WENO and WCNS on curvilinear grids. Comput Fluids 2010;39:197–214.
[10] Thomas PD, Lombard CK. Geometric conservation law and its application to flow computations on moving grids. AIAA J 1979;17(10):1030–7.
[11] Pulliam TH, Steger JL. On implicit finite-difference simulations of three-dimensional flow. AIAA Paper 78-10; 1978.
[12] Deng Xiaogang, Min Yaobing, Mao Meiliang, Liu Huayong, Tu Guohua. Further studies on geometric conservation law and applications to high-order finite difference schemes with stationary grids. J Comput Phys 2013;239:90–111.
[13] Deng Xiaogang, Jiang Yi, Mao Meiliang, Liu Huayong, Tu Guohua. Developing hybrid cell-edge and cell-node dissipative compact scheme for complex geometry flows. Sci China Technol Sci 2013;56:2361–9.
[14] Deng Xiaogang, Jiang Yi, Mao Meiliang, Liu Huayong, Li Song, Tu Guohua. A family of hybrid cell-edge and cell-node dissipative compact schemes satisfying geometric conservation law. Commun Comput Phys; 2014 [submitted for publications].
[15] Boris JP, Grinstein FF, Oran ES, Kolbe RL. New insights into large eddy simulation. Fluid Dyn Res 1992;10:199.
[16] Xie Peng, Liu Chaoqun. Weighted compact and non-compact scheme for shock tube and shock entropy interaction. AIAA 2007-509.
[17] Tam CKW, Webb JC. Dispersion-relation-preserving finite difference schemes for computational acoustics. J Comput Phys 1993;107:262–81.
[18] Hsu John M, Jameson Antony. An implicit–explicit hybrid scheme for calculating complex unsteady flows. AIAA 2002-0714.
[19] Gordnier RE, Visbal MR. Numerical simulation of delta-wing roll. AIAA Paper 93-0554; January 1993].
[20] Hickel S, Adams NA, Domaradzki JA. An adaptive local deconvolution method for implicit LES. J Comput Phys 2006;213:413–36.
[21] Gottlieb S, Shu C-W. Total variation diminishing Runge–Kutta schemes. Math Comput 1998;67:73–85.
[22] Brachet M, Meiron D, Orszag S, Nickel B, Morf R, Frisch U. Small-scale structure of the Taylor–Green vortex. J Fluid Mech 1983;130:411–52.
[23] Shen Yiqing, Zha Gecheng. Large eddy simulation using a new set of sixth order schemes for compressible viscous terms. J Comput Phys 2010;229:8296–312.
[24] Rizzetta DP, Visbal MR, Blaisdell GA. A time-implicit high-order compact differencing and filtering scheme for large-eddy simulation. Int J Numer Methods Fluids 2003;42:665–93.
[25] Kravchenko AG, Moin P. Numerical studies of flow over a circular cylinder at Re D = 3900. Phys Fluids 2000;12:403–17.
[26] Poinsot T, Lele SK. Boundary conditions for direct simulations of compressible viscous flows. J Comput Phys 1992;101:104–29.
[27] Bodony Daniel J. Analysis of sponge zones for computational fluid mechanics. J Comput Phys 2006;212:681–702.
[28] Lourenco LM, Shih C. Characteristics of the plane turbulent near wake of a circular cylinder. A particle image velocimetry study (private communication).
[29] McCullough GB, Gault DE. Examples of three representative types of airfoil-section stall at low speed. NACA TN2502; September 1951.
[30] McCullough GB, Gault, DE. Boundary-layer and stalling characteristics of the NACA64A006 airfoil section. NACATN1923; August 1949.
[31] Fitzgerald JE, Mueller JT. Measurements in separation bubble on an airfoil using laser velocimetry. AIAA J 1990;28(4):584–92.

Effect of Nonuniform Grids on High-Order Finite Difference Method

Dan Xu[1,*], Xiaogang Deng[1], Yaming Chen[2], Guangxue Wang[3] and Yidao Dong[1]

[1] *College of Aerospace Science and Engineering, National University of Defense Technology, Changsha, Hunan 410073, China*
[2] *College of Science, National University of Defense Technology, Changsha, Hunan 410073, China*
[3] *School of Physics, Sun Yat-sen University, Guangzhou, Guangdong 510006, China*

Received 1 March 2016; Accepted (in revised version) 20 September 2016

Abstract. The finite difference (FD) method is popular in the computational fluid dynamics and widely used in various flow simulations. Most of the FD schemes are developed on the uniform Cartesian grids; however, the use of nonuniform or curvilinear grids is inevitable for adapting to the complex configurations and the coordinate transformation is usually adopted. Therefore the question that whether the characteristics of the numerical schemes evaluated on the uniform grids can be preserved on the nonuniform grids arises, which is seldom discussed. Based on the one-dimensional wave equation, this paper systematically studies the characteristics of the high-order FD schemes on nonuniform grids, including the order of accuracy, resolution characteristics and the numerical stability. Especially, the Fourier analysis involving the metrics is presented for the first time and the relation between the resolution of numerical schemes and the stretching ratio of grids is discussed. Analysis shows that for smooth varying grids, these characteristics can be generally preserved after the coordinate transformation. Numerical tests also validate our conclusions.

AMS subject classifications: 65N06, 65N22

Key words: Finite difference method, nonuniform grids, coordinate transformation, Fourier analysis.

1 Introduction

The FD method is historically old and plays an important role in the computational fluid dynamics [1]. In recent 20 years, due to the efficiency and simplicity, various high-order schemes based on the FD method have been proposed and widely used in direct numerical simulations (DNS), computational aeroacoustics (CAA) and large eddy simulations (LES), in which the high resolution is needed. At present, the high-order FD method has been successfully applied in the simulations of incompressible, compressible and hypersonic flows [2–4] and several other practical applications [5,6].

In the studies of high-order FD method, different discretization techniques for the spatial derivative are developed. The common one is the explicit scheme which is directly derived from the Taylor series expansion. For steadiness, the numerical dissipation should be introduced by different ways, such as using upwind schemes. However, a shortage of such schemes is that a long stencil is needed to achieve the desired order of accuracy, which makes the boundary schemes difficult to design [7]. To reduce the stencil width, the compact scheme becomes another choice. Compact FD schemes with spectral-like resolution are first systematically studied by Lele [8] and gain a quick development. Recently, Rizzetta et al. [5] carried out a high-order compact scheme with compact filter, which has been demonstrated to produce accurate and stable results in large eddy simulations. A family of hybrid dissipative compact schemes is proposed by Deng et al. [9] and suitable for simulations in aeroacoustics. To make compact schemes possess the shock-capturing capabilities, many efforts have been devoted [10–12], which were well summarized by Shen and Zha [13]. However, as stated by Tam [14], the Taylor series truncation cannot be used to quantity the wave propagation errors which are dominant in CAA and this issue results in the development of optimized schemes, where the order of accuracy is lowered to reduce errors over a range of wavenumbers [15]. One of the classical optimized schemes is the dispersion-relation-preserving (DRP) scheme developed by Tam and Webb [16], which is capable to accurately resolve harmonic components with few points-per-wavelength [17].

In most cases, the development of the FD schemes is based on the uniform Cartesian grids. However, in the simulations of practical problems, the use of the nonuniform or curvilinear grids is inevitable for adapting to the complex configurations. For solving this issue, some schemes are especially designed. Gamet et al. [18] modified the original compact scheme to approximate the first derivative on the nonuniform meshes. Cheong and Lee [19] developed the GODRP finite difference scheme to locally preserve the same dispersion relation as the original partial differential equations on the nonuniform mesh. Moreover, Zhong et al. [20,21] used the polynomial interpolation to derive arbitrary high-order compact schemes on nonuniform grids, which have been adopted for simulations of hypersonic boundary-layer stability and transition. Although these schemes can be directly applied on the nonuniform grids, the distribution of grids is still relatively simple (for example, the grids are only stretched along each direction of the Cartesian grids), making the schemes difficult to extend to practical conditions. Another method of dealing with the nonuniform or curvilinear grids in the FD schemes is to employ the coordinate transformation (or named Jacobian transformation), which is the most popular way. Using this method, the original schemes can be applied in the computational space where the grid is uniform Cartesian one, but as a result, the metrics and Jacobian are involved in the governing equations, which should be carefully treated [22, 23].

At this moment, a problem arises naturally that when the coordinate transformation is introduced, whether the characteristics of the FD schemes that are evaluated on the uniform grids can be preserved, especially for the high-order schemes. The characteristics include the order of accuracy, resolution characteristics and the numerical stability. These characteristics are important for any numerical scheme and this problem is critical as we expect to apply the FD schemes to a wide range of flow simulations. Actually, many researchers have noticed this problem and carried out some studies. Gamet et al. [18] compared the calculation results based on the coordinate transformation with those using the nonuniform compact schemes on the random mesh. Visbal and Gaitonde [24] carried out detailed studies on the errors introduced by the mesh nonuniformities in finite-difference formulations. Chung and Tucker [25] compared the resolution characteristics by the Fourier analysis on uniform and nonuniform grids, but the detail expressions were not given. In this paper, we present a systematic analysis of the characteristics of the FD schemes in computing the first order spatial derivative on nonuniform grids, when the coordinate transformation is used. The nonuniformity involves many factors such as the grid size, stretching ratio and smoothness, which is a very complex problem. In this study, we restrict our attention to the smooth varying grids and the non-smooth case is left for further study. For the order of accuracy, we find that the designed order can be preserved, as the metrics are evaluated in the same high-order schemes. The resolution characteristics are studied by the Fourier analysis and we derive the dispersion and dissipation errors for the FD schemes when the metrics are included, which is not discussed before. The results show that the resolution characteristics on the nonuniform grids can be preserved as well, as long as the stretching ratio is controlled. The asymptotic stability is also analyzed by computing the eigenvalues of the matrices obtained by the spatial discretization. By comparison, it is found that the coordinate transformation may degrade the stability of the FD schemes when the stretch of grid is stronger at the boundary points, which puts forward higher requirements to the boundary schemes.

The rest of this paper is organized as follows. In the next section, we first present the one-dimensional wave equation after the coordinate transformation. Based on this equation, the order of accuracy, resolution characteristics and the asymptotic stability of the FD schemes on nonuniform grids are analyzed in Section 2.2 to Section 2.4. Numerical tests are presented in Section 3 to validate the analyses. Section 4 concludes this paper.

2 Analysis of the characteristics of the FD schemes on nonuniform grids

2.1 One-dimensional wave equation

In this section, the characteristics of the FD schemes when the coordinate transformation is used are analyzed in detail. These characteristics include the order of accuracy, resolution characteristics and the numerical stability. For convenience, we take the one-dimensional wave equation into consideration,

$$\frac{\partial f}{\partial t}+a\frac{\partial f}{\partial x}=0. \tag{2.1}$$

Without loss of generality, we set a to be 1, so that the analytical solution of Eq. (2.1) is

$$f=g(x-t), \tag{2.2}$$

where the form of function g is determined by the initial conditions. As classified by Whitham [26], the solution represents a hypersonic wave with the constant phase speed and group velocity.

In numerical simulations, the first order spatial derivative is approximated by the numerical difference. In this paper, we discuss the effect of the coordinate transformation on the nonuniform grids, so Eq. (2.1) is transformed into

$$\frac{\partial f}{\partial t}+\xi_x\frac{\partial f}{\partial \xi}=0, \tag{2.3}$$

where ξ_x is the metric and satisfies $\xi_x=1/x_\xi$. As the coordinate transformation only influences the spatial derivative, the time integration is not discretized and keeps the analytical form in this study. Therefore, only the semi-discrete equation is used,

$$\frac{\partial f}{\partial t}+\left(\xi_x\delta^\xi f\right)_n=0, \tag{2.4}$$

where δ is the difference operator and the subscript n represents the position of the corresponding grid point.

2.2 The order of accuracy

For Eq. (2.4), the truncation errors are introduced by both the numerical difference and the evaluation of the metrics. After the transformation, the FD schemes in computing $\delta^\xi f$ are the same with those on uniform grids. In this paper, we adopt two FD schemes (COM6 and WCNS) that are developed by Lele [8] and Deng [11] respectively.

The compact scheme used by Lele is 6th-order and can be expressed as

$$\frac{1}{3}f'_{i-1}+f'_i+\frac{1}{3}f'_{i+1}=\frac{14}{9}\frac{f_{i+1}-f_{i-1}}{2\Delta\xi}+\frac{1}{9}\frac{f_{i+2}-f_{i-2}}{4\Delta\xi}, \tag{2.5}$$

where f' represents $\partial f/\partial \xi$ and $\Delta\xi$ is the interval in the computational space. In practice, the classical 4th-order Padé scheme and 3th-order compact relation are used for the non-periodic boundary problems [18] at points 2 and 1. The boundary formulations are

$$\frac{1}{4}f'_1+f'_2+\frac{1}{4}f'_3=\frac{3}{4\Delta\xi}(f_3-f_1), \tag{2.6a}$$

$$f'_1+2f'_2=\frac{1}{\Delta\xi}\left(-\frac{5}{2}f_1+2f_2+\frac{1}{2}f_3\right). \tag{2.6b}$$

Moreover, the explicit WCNS scheme based on the values at cell-edges is in the following form,

$$f'_i = \frac{75}{64\Delta\xi}(f_{i+1/2}-f_{i-1/2}) - \frac{25}{384\Delta\xi}(f_{i+3/2}-f_{i-3/2}) + \frac{3}{640\Delta\xi}(f_{i+5/2}-f_{i-5/2}), \qquad (2.7)$$

which is also 6th-order. The values at cell-edges can be obtained by the interpolation and in consideration of the characteristic direction of Eq. (2.1), a 5th-order upwind interpolation is used,

$$f_{i+1/2} = \frac{3}{128}f_{i-2} - \frac{5}{32}f_{i-1} + \frac{45}{65}f_i + \frac{15}{32}f_{i+1} - \frac{5}{128}f_{i+2}. \qquad (2.8)$$

The boundary schemes are discussed in [27]. The 5th-order interpolations at cell-edges 1/2 and 3/2 can be expressed as

$$f_{1/2} = \frac{1}{128}(315f_1 - 420f_2 + 378f_3 - 180f_4 + 35f_5), \qquad (2.9a)$$

$$f_{3/2} = \frac{1}{128}(35f_1 + 140f_2 - 70f_3 + 28f_4 - 5f_5). \qquad (2.9b)$$

The difference schemes at points 1 and 2 are

$$f'_1 = \frac{1}{24}(-22f_{1/2} + 17f_{3/2} + 9f_{5/2} - 5f_{7/2} + f_{9/2}), \qquad (2.10a)$$

$$f'_2 = \frac{1}{24}(f_{1/2} - 27f_{3/2} + 27f_{5/2} - f_{7/2}), \qquad (2.10b)$$

which are both 4th-order.

Besides the spatial derivative, the metrics should also be evaluated by the high-order numerical scheme to keep the overall order of accuracy. When the coordinate transformation is adopted in the FD schemes, another issue of the freestream preservation arises, which has been fully studied by many authors [22,23,28,29]. However, for Eq. (2.4), this issue is negligible and the metrics can be obtained in the same manner with the spatial derivative.

To study the effect of the coordinate transformation on the order of accuracy, Eq. (2.4) is solved. Numerical errors and the accuracy order are presented. The computational domain is $[-1,1]$ and the non-periodic boundary condition is specified at the left boundary. Two analytical solutions are used in this paper,

$$f(x,t) = \sin(-\pi x + \pi t - \pi), \qquad (2.11a)$$

$$f(x,t) = \sin\left[\pi(x-t) - \frac{\sin[\pi(x-t)]}{\pi}\right]. \qquad (2.11b)$$

Both the uniform and nonuniform grids are adopted and the stretching function proposed by Gamet [18] is used,

$$x_i = \eta_i + \frac{C-1}{\pi}\sin[\pi(\eta_i+1)] \quad \text{with} \quad -1 \le \eta_i = 2\frac{i-1}{N-1} - 1 \le +1, \qquad (2.12)$$

Table 1: Numerical errors and the accuracy order on Eq. (2.11a).

	WCNS				COM6			
	Uniform grid		Nonuniform grid		Uniform grid		Nonuniform grid	
20	5.44088E-05	—	8.07521E-04	—	1.21289E-03	—	8.65392E-03	—
40	1.79230E-06	4.924	3.07065E-05	4.717	7.01554E-05	4.112	5.63927E-04	3.940
60	2.84135E-07	4.542	4.47558E-06	4.750	1.34682E-05	4.070	1.09116E-04	4.051
80	6.12175E-08	5.336	1.15817E-06	4.699	4.19476E-06	4.055	3.40422E-05	4.049
120	1.12671E-08	4.174	1.56420E-07	4.938	8.15535E-07	4.039	6.61649E-06	4.040

Table 2: Numerical errors and the accuracy order on Eq. (2.11b).

	WCNS				COM6			
	Uniform grid		Nonuniform grid		Uniform grid		Nonuniform grid	
20	5.44422E-04	—	2.45131E-03	—	2.93717E-03	—	1.56838E-02	—
40	1.72302E-05	4.982	9.05211E-05	4.759	1.55150E-04	4.242	1.03129E-03	3.992
60	2.36988E-06	4.893	1.33190E-05	4.726	2.91068E-05	4.127	1.19184E-04	5.322
80	5.58360E-07	5.025	3.34101E-06	4.807	9.00768E-06	4.077	5.92262E-05	2.431
120	7.57268E-08	4.927	4.81152E-07	4.779	1.74570E-06	4.047	1.14501E-05	4.053

where C is a constant and taken as $C=1.5$. The time advancement is accomplished through a 3th-order Runge-Kutta scheme and the time step is chosen to be small enough to make the errors in time negligible. Tables 1 and 2 present the errors and the accuracy order at $t=1$. It can be seen that although the absolute errors on the nonuniform grids are larger, nearly the same accuracy order is achieved on both the uniform and nonuniform grids. This shows that the designed accuracy order of the scheme can be preserved on nonuniform grids as the coordinate transformation is utilized in the FD method, as long as all terms in the equations are discretized in the corresponding high-order scheme.

2.3 Analysis of the resolution characteristics

The resolution characteristics of the FD schemes are analysed by the Fourier analysis, which is extensively described by Vichnevetsky and Bowles [30] and has become a classical technique for comparing differencing schemes [8]. However, most of the analyses are implemented on the uniform grids [31] or directly on the nonuniform grids [21]. The case that the coordinate transformation is involved is seldom studied. In this section, the detail formulae of the Fourier analysis of Eq. (2.3) are presented and the comparisons with the results on the uniform grids are made.

Following the discussions in [25], the mapping function between the uniform gird in the computational space ζ and the nonuniform grid in the physical space x is assumed to be

$$x=\frac{\sinh(\gamma\zeta)}{\sinh(\gamma)}, \qquad (2.13)$$

where γ is the control parameter and is set to be 3 in this paper. As the mapping function is known, we can use the analytical metrics in the discussion. For the compact scheme in

Eq. (2.5), the difference operator δ can be expressed in the matrix form,

$$\begin{bmatrix} \cdots & & & & \\ \alpha & 1 & \alpha & & \\ & \alpha & 1 & \alpha & \\ & & \alpha & 1 & \alpha \\ & & & & \cdots \end{bmatrix} \begin{bmatrix} \vdots \\ f'_{n-1} \\ f'_n \\ f'_{n+1} \\ \vdots \end{bmatrix} \quad (2.14)$$

$$= \begin{bmatrix} \cdots & & & & & \\ -\dfrac{a}{2\Delta\xi} & 0 & \dfrac{a}{2\Delta\xi} & \dfrac{b}{4\Delta\xi} & & \\ -\dfrac{b}{4\Delta\xi} & -\dfrac{a}{2\Delta\xi} & 0 & \dfrac{a}{2\Delta\xi} & \dfrac{b}{4\Delta\xi} & \\ & -\dfrac{b}{4\Delta\xi} & -\dfrac{a}{2\Delta\xi} & 0 & \dfrac{a}{2\Delta\xi} & \\ & & & & \cdots & \end{bmatrix} \begin{bmatrix} \vdots \\ f_{n-1} \\ f_n \\ f_{n+1} \\ \vdots \end{bmatrix}, \quad (2.15)$$

where α, a and b are the coefficients in the FD scheme. The matrices at the LHS and RHS are defined as A_1 and A_2 respectively. Hence the semi-discrete equation Eq. (2.4) becomes

$$\frac{\partial}{\partial t} \begin{bmatrix} \vdots \\ f_{n-1} \\ f_n \\ f_{n+1} \\ \vdots \end{bmatrix} + \begin{bmatrix} \cdots & & & & \\ 0 & (\xi_x)_{n-1} & 0 & 0 & 0 \\ 0 & 0 & (\xi_x)_n & 0 & 0 \\ 0 & 0 & 0 & (\xi_x)_{n+1} & 0 \\ & & & & \cdots \end{bmatrix} A_1^{-1} A_2 \begin{bmatrix} \vdots \\ f_{n-1} \\ f_n \\ f_{n+1} \\ \vdots \end{bmatrix} = 0, \quad (2.16)$$

which can be further expressed as

$$\frac{\partial}{\partial t} A_1 \begin{bmatrix} \cdots & & & & \\ 0 & (\xi_x)_{n-1} & 0 & 0 & 0 \\ 0 & 0 & (\xi_x)_n & 0 & 0 \\ 0 & 0 & 0 & (\xi_x)_{n+1} & 0 \\ & & & & \cdots \end{bmatrix}^{-1} \begin{bmatrix} \vdots \\ f_{n-1} \\ f_n \\ f_{n+1} \\ \vdots \end{bmatrix} + A_2 \begin{bmatrix} \vdots \\ f_{n-1} \\ f_n \\ f_{n+1} \\ \vdots \end{bmatrix} = 0. \quad (2.17)$$

Taking any row in the above matrix, the semi-discrete equation can be written in another form,

$$\alpha \frac{1}{(\xi_x)_{n-1}} \frac{\partial f_{n-1}}{\partial t} + \frac{1}{(\xi_x)_n} \frac{\partial f_n}{\partial t} + \alpha \frac{1}{(\xi_x)_{n+1}} \frac{\partial f_{n+1}}{\partial t} + a \frac{f_{n+1} - f_{n-1}}{2\Delta\xi} + b \frac{f_{n+2} - f_{n-2}}{4\Delta\xi} = 0. \quad (2.18)$$

As $\xi_x = 1/x_\xi$, the above equation can be further simplified into

$$\alpha (x_\xi)_{n-1} \frac{\partial f_{n-1}}{\partial t} + (x_\xi)_n \frac{\partial f_n}{\partial t} + \alpha (x_\xi)_{n+1} \frac{\partial f_{n+1}}{\partial t} + a \frac{f_{n+1} - f_{n-1}}{2\Delta\xi} + b \frac{f_{n+2} - f_{n-2}}{4\Delta\xi} = 0, \quad (2.19)$$

which will be used in the Fourier analysis.

The essence of the Fourier analysis is to analyze the accuracy of the finite difference approximation of a partial differential equation [30]. Therefore the most straightforward method is to compare the analytical and approximate solutions. For Eq. (2.1), the analytical sinusoidal solution is in the form of

$$f(x,t) = v(0)e^{i\omega(x-t)}, \tag{2.20}$$

where ω is the space frequency (for the analytical solution of the wave equation, the time frequency Ω equals to ω) and $v(0)$ is defined by the initial condition. In the same manner, the approximate solution of the partial differential equation that is determined by Eq. (2.19) can be assumed in the following form,

$$p(\xi,t) = v(t)e^{i\omega\xi}, \tag{2.21}$$

where $v(t) = \tilde{v}(0)e^{\hat{A}(\omega)t}$ and $\tilde{v}(0)$ is also defined by the initial condition. As the approximate solution is obtained in the computational space, the independent variable in the function p is ξ, while ξ is a function of x. It is easy to verify that Eq. (2.21) is the solution of Eq. (2.19), as long as $\hat{A}(\omega)$ satisfies

$$\hat{A}(\omega) = -\frac{\frac{b}{4\Delta\xi}e^{i\omega\xi_{n+2}} + \frac{a}{2\Delta\xi}e^{i\omega\xi_{n+1}} - \frac{a}{2\Delta\xi}e^{i\omega\xi_{n-1}} - \frac{b}{4\Delta\xi}e^{i\omega\xi_{n-2}}}{\alpha(x_\xi)_{n-1}e^{i\omega\xi_{n-1}} + (x_\xi)_n e^{i\omega\xi_n} + \alpha(x_\xi)_{n+1}e^{i\omega\xi_{n+1}}}. \tag{2.22}$$

Substituting it into Eq. (2.21) leads to

$$p(\xi,t) = \tilde{v}(0)e^{\hat{A}(\omega)t}e^{i\omega\xi} = \tilde{v}(0)e^{Re[\hat{A}(\omega)]t}e^{i(\omega\xi + Im[\hat{A}(\omega)]t)}. \tag{2.23}$$

$Re[\hat{A}(\omega)]$ and $Im[\hat{A}(\omega)]$ represent the real part and the imaginary part of $\hat{A}(\omega)$, which correspond to the dissipative and dispersive errors respectively. $\hat{A}(\omega)$ can be further simplified into

$$\hat{A}(\omega) = -\frac{\left[\frac{b}{2\Delta\xi}\sin(2\omega\Delta\xi) + \frac{a}{\Delta\xi}\sin(\omega\Delta\xi)\right]i}{\left[\alpha(x_\xi)_{n-1}\cos(\omega\Delta\xi) + (x_\xi)_n + \alpha(x_\xi)_{n+1}\cos(\omega\Delta\xi)\right] + \left[\alpha(x_\xi)_{n+1}\sin(\omega\Delta\xi) - \alpha(x_\xi)_{n-1}\sin(\omega\Delta\xi)\right]i}, \tag{2.24}$$

from which it is found that it is the metrics that make sense in the Fourier analysis, not the coordinates of grid points. As the distribution of the grid points in space ξ is uniform, x_ξ reflects the stretching ratio of the grid points in the physical space, which is widely used and controlled in grid generations. Therefore the effect of the nonuniform grids on the resolution characteristics is evaluated by the stretching ratio in this paper, instead of the coordinates of grid points.

Based on above equations, the dissipation errors of the FD schemes are evaluated by the real part of $\hat{A}(\omega)$, which is zero for the analytical solution. The dispersive errors are related to the imaginary part of $\hat{A}(\omega)$. In order to express the dispersive errors in an intuitive way, we first calculate the phase velocity of Eq. (2.23),

$$c^* = -(x_\xi)_n \frac{\text{Im}[\hat{A}(\omega)]}{\omega}. \tag{2.25}$$

As the phase velocity is different as the spatial position changes, the value of the metric in Eq. (2.25) is obtained at the calculation grid point n. The exact phase velocity of the wave equation is $c^*=1$, so that the dispersive errors can be evaluated by comparing the phase velocity, which is often depicted in the figure of time frequency Ω and space frequency ω [30],

$$\Omega = \omega c^*(\omega). \tag{2.26}$$

In Eq. (2.26), the analytical metrics are used and the influence of the nonuniform grids are evaluated by the stretching ratio, which is defined as

$$r = \frac{x_{n+1} - x_n}{x_n - x_{n-1}}. \tag{2.27}$$

For Eq. (2.13), the computational domain in the physical space is [1,2] and the stretching ratio is controlled by adding or removing grid points. The results on the uniform grids can be easily obtained by setting the metrics to be constant. Fig. 1 shows the dissipative and dispersive errors with different stretching ratios. Moreover, results on the uniform grids and the exact solution are presented at the same time. It can be observed from Fig. 1 that although the symmetry schemes are designed to be non-dissipative on the uniform grids, the dissipative errors are non-zero on the nonuniform grids. As the stretching ratio increases, more dissipative errors are introduced in the numerical scheme. Dispersive errors are also greatly influenced by the stretching ratio: if it is small, the difference of dispersive errors between the uniform and nonuniform grids is indistinguishable; however, if amplified, the dispersive errors on the nonuniform grids experience an obvious increase, and degrade the resolution characteristics.

The above discussions are based on the compact scheme (COM6). For the explicit one, it is just a special case of the compact scheme with $\alpha = 0$. Substitute it into Eqs. (2.24) and (2.25) and it is easy to verify that the dissipative and dispersive errors are the same with those on the uniform grids. This is because the metric used in Eq. (2.25) is obtained at the calculation grid point n, which will be divided out by the metric in Eq. (2.24). Therefore $\hat{A}(\omega)$ becomes a pure imaginary number and brings in no dissipation error. However, different from the consistent resolution characteristics on the uniform grids, the dispersive errors vary in space on the nonuniform grids. To further study the effect of the stretching ratio on the explicit scheme, we present the dispersive errors at the half points on the left and right of the calculation point. Accordingly, the metrics in Eq. (2.25) are replaced by $(x_\xi)_{n-1/2}$ and $(x_\xi)_{n+1/2}$. For convenience, we use a 4th-order explicit scheme with $\alpha = 0$, $a = 4/3$ and $b = -1/3$. The dispersion errors for different stretching

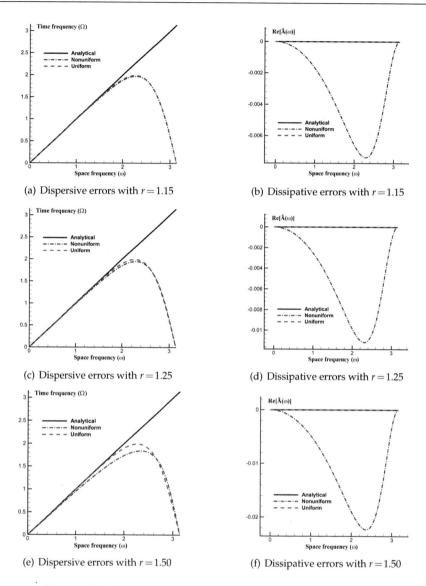

Figure 1: The dispersive and dissipative errors for different stretching ratios.

ratios are shown in Fig. 2. Similar with the results in the compact scheme, the dispersion errors are enlarged rapidly and become unacceptable as the stretching ratio increases.

From the above discussions, it is found that the resolution characteristics of the FD

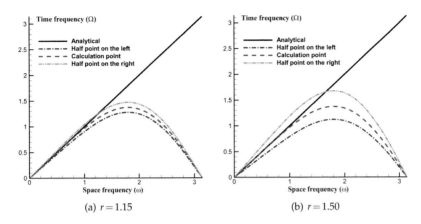

Figure 2: The dispersive errors for different stretching ratios at the calculation point and the half points on its left and right.

schemes have a close relation with the stretching ratio of the grids. The characteristics can be preserved as long as the stretching ratio is controlled, no matter for the compact or explicit schemes. Actually, in the practical process of grid generations, the stretching ratio is often required to be less than 1.25. The current study also validates the requirement from the resolution characteristics point of view.

2.4 Asymptotic stability analysis

The asymptotic stability is an important characteristic of the numerical schemes and it decides whether a scheme can be used for solving Eq. (2.1). Carpenter et al. [32] started an early discussion about this issue. Based on the semi-discrete equation, the first-order spatial derivative at all grid points including interior and the boundary can be written as

$$[M_1]\{f'\}=[M_2]\{f\}. \tag{2.28}$$

Substituting the above equation into Eq. (2.1) leads to

$$\left\{\frac{df}{dt}\right\}=[M]\{f\}+\{g(t)\}, \tag{2.29}$$

where $[M]=-[M_1]^{-1}[M_2]$, and $g(t)$ represents the physical boundary conditions imposed on the boundary grid point. If the coordinate transformation is used, the metrics are introduced and Eq. (2.29) becomes

$$\left\{\frac{df}{dt}\right\}=[D]^{-1}[M]\{f\}+\{g'(t)\}, \tag{2.30}$$

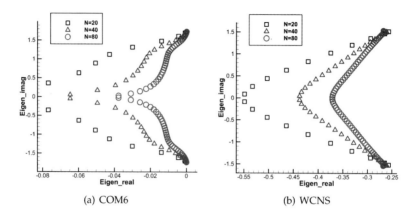

(a) COM6 (b) WCNS

Figure 3: The eigenvalue spectra of COM6 and WCNS schemes on the uniform grid.

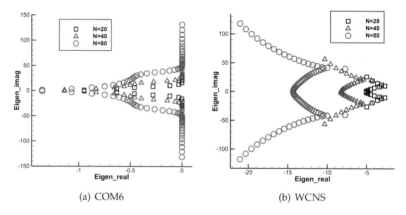

(a) COM6 (b) WCNS

Figure 4: The eigenvalue spectra of COM6 and WCNS schemes on the nonuniform grid defined by Eq. (2.12).

where $[D]$ is a diagonal matrix such that

$$diag[D] = [(\xi_x)_i], \quad i=1,\cdots,N. \tag{2.31}$$

For guaranteeing the stability of the overall FD schemes, it is required that the eigenvalues of Eq. (2.30) lie in the left half of the complex plane. In this section, we first present the asymptotic stability analysis on the uniform grids. The eigenvalue spectra of COM6 and WCNS schemes is shown in Fig. 3. It can be seen that all the eigenvalues are in the left half and both schemes are steady. For the nonuniform grids, the coordinate transformation is introduced and the characteristic of the eigenvalues has a close relation with the grid distribution. Fig. 4 presents the eigenvalue spectra on the nonuniform grid defined by Eq. (2.12). On this grid, both schemes are steady, which is also demonstrated

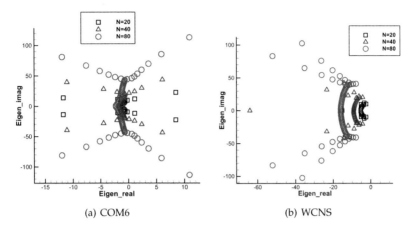

Figure 5: The eigenvalue spectra of COM6 and WCNS schemes on the nonuniform grid defined by Eq. (2.32).

by the practical calculations in Section 2.2. In order to further study the behavior of the eigenvalues on the nonuniform grids, another spacing function with strong stretch is used [33],

$$x_i = \frac{\sin^{-1}[-\alpha\cos(\pi i/N)]}{\sin^{-1}\alpha}, \quad i=0,\cdots,N, \tag{2.32}$$

where α controls the grid points from a Chebyshev grid ($\alpha \to 0$) to a uniform grid ($\alpha = 0$). In this paper, α is chosen to be 0.5 and Fig. 5 presents the eigenvalue spectra. Fig. 5 shows that the WCNS scheme is still steady on this grid, but the real part of the eigenvalues gets close to zero. On the other hand, the COM6 scheme becomes unsteady as some eigenvalues appear in the right half of the complex plane. This is because the stretch is stronger at the boundary points where the boundary schemes are implemented.

Actually, the design of steady boundary schemes is a challenge for the high-order schemes in the FD method. The above discussions indicate that the nonuniform grids may aggravate this problem, especially when the strong stretch exists at boundary points. However this condition is usually encountered in solving the Euler or Navier-Stokes equations. Therefore the development of steady boundary schemes in the high-order FD method is critical. Furthermore, it should be noted that the asymptotic stability analysis is based on the linear wave equation, which cannot fully represent the characteristics of the numerical schemes in solving nonlinear equations. This may leave us the possibility to obtain the high-order and steady boundary schemes.

From the above analysis in Sections 2.2 to 2.4, we find that the characteristics of the FD schemes can be generally preserved on the nonuniform grids when the coordinate transformation is used. However, the discussions are based on the smooth varying nonuniform grids. The cases of non-smooth or random grids are out of the scope of this paper.

3 Numerical results

In this section, numerical tests are presented to validate the analyses in this paper. In consideration of the good numerical stability, the 5th-order explicit WCNS scheme is used. Although the second spatial derivative corresponding to the viscous terms is not discussed, a test case governed by the Navier-Stokes equations is also included. The numerical schemes for the viscous terms can be found in [34] for detail. In multi-dimensional problems, the metrics and Jacobian are evaluated by the symmetric conservative metric method (SCMM) developed by Deng [23].

3.1 Two-dimensional channel flow

The two-dimensional channel flow is an isentropic flow governed by the Euler equations, which was first discussed by Casper et al. [35]. In this paper, the C3 geometry is used, and the shape of the middle section is determined by

$$\begin{cases} y_1(x) = -0.5 + 0.05\sin^4\left(\frac{10}{9}\pi x + \frac{\pi}{2}\right), \\ y_2(x) = 0.5 - 0.05\sin^4\left(\frac{10}{9}\pi x + \frac{\pi}{2}\right), \end{cases} \quad (3.1)$$

where $y_1(x)$ and $y_2(x)$ represent the lower and upper walls respectively. Other parts of the walls are alone the direction of x and defined by

$$\begin{cases} y_1(x) = -0.5, \\ y_2(x) = 0.5. \end{cases} \quad (3.2)$$

The grid is generated by a mapping from a rectangle space (ξ,η) to the physical point (x,y),

$$\begin{cases} x = \xi, \\ y = \left(1 - \frac{\eta}{H}\right)y_1(\xi) + \frac{\eta}{H}y_2(\xi). \end{cases} \quad (3.3)$$

Moreover, the boundary conditions used in this test case are presented in [36].

To demonstrate that the FD schemes after the coordinate transformation can preserve the original characteristics, four different grids are used in the numerical calculations, which are labeled as Uniform, Chebyshev, Chebyshev-0.5 and Exp. The Uniform grid does not mean the grid in the physical space is uniform, but the gird in the rectangle space (ξ,η) is uniform and the other grids are defined in the same manner. The Chebyshev represents a Chebyshev grid alone η and the Chebyshev-0.5 defines the grid by Eq. (2.32) with $\alpha = 0.5$. The grid labeled as Exp is generated by the following mapping function,

$$\eta(\varphi) = \frac{A(\beta + 2\gamma) + 2\gamma - \beta}{(2\gamma + 1)(1 + A)}, \quad A = \left(\frac{\beta + 1}{\beta - 1}\right)^{(1 - \varphi - \gamma)/(\gamma - 1)}, \quad (3.4)$$

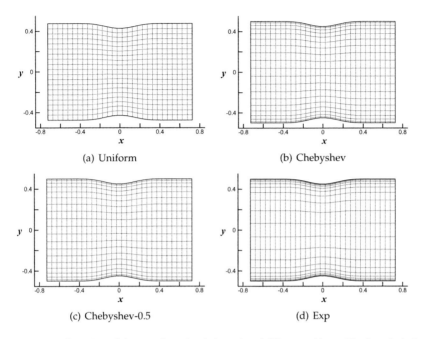

Figure 6: The configuration of the two-dimensional channel and different grids used in the calculations.

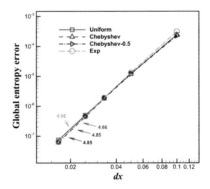

Figure 7: Global entropy errors in the two-dimensional channel flow on different grids.

where $\gamma = 0.5$ and $\beta = 1.01015$ are constant [37]. Fig. 6 illustrates the configuration of the channel and different grids.

The global L_1 entropy errors on different grids are listed in Table 3 and also plotted in Fig. 7. It can be seen that the entropy errors agree well with each other and the

Table 3: Global L_1 entropy errors on different grids.

Grid	Uniform	Chebyshev	Chebyshev-0.5	Exp
15×10	2.4369E-04	2.3247E-04	2.2515E-04	3.0725E-04
30×20	1.2189E-05	1.1906E-05	1.2034E-05	1.3227E-05
45×30	1.8928E-06	1.7744E-06	1.8408E-06	1.8853E-06
60×40	4.9728E-07	4.3888E-07	4.6006E-07	4.6299E-07
90×60	7.5057E-08	6.1315E-08	6.4355E-08	6.4433E-08

distribution of the grid points does not influence the numerical results much. The calculated order of accuracy is presented in Fig. 7 as well, and all the values are close to 5.0. It indicates that the designed accuracy order can be preserved when the coordinate transformation is used and this technique is indeed usable in the practical simulations.

3.2 Two-dimensional isentropic vortex

This problem is widely used in testing a high-order CFD method's capacity to preserve vorticity in an unsteady inviscid flow [38]. The initial flow is the superposition of a uniform flow and a vortical movement. As periodic boundary conditions are adopted, the analytical solution after several periods euqals to the initial state. Therefore the errors introduced by the numerical schemes can be easily evaluated. The distribution of the velocity can be expressed as

$$\begin{cases} u_0 = U_\infty - (U_\infty \beta) \dfrac{y-Y_c}{R} \exp\left(-\dfrac{r^2}{2}\right), \\ v_0 = (U_\infty \beta) \dfrac{x-X_c}{R} \exp\left(-\dfrac{r^2}{2}\right), \end{cases} \quad (3.5)$$

where $r = \dfrac{\sqrt{(x-X_c)^2+(y-Y_c)^2}}{R}$ and U_∞ is the unperturbed velocity. The calculation of the pressure, temperature and density can be found in [38]. In this paper, the computational domain is $(x,y)\in[0,L_x]\times[0,L_y]$ and $L_x=0.1$, $L_y=0.1$. The center of the vortex is at $X_c=0.05$, $Y_c=0.05$ and the initial pressure and temperature are $p_\infty=10^5$Pa and $T_\infty=300$K. Mach number $Ma=0.05$ is chosen with $\beta=0.02$ and $R=0.005$. Because the vortex is transported by a uniform flow, we define the period T as $T=L_x/U_\infty$ and the errors of the velocity u are collected after 50 periods.

Two kinds of girds are used in this test case, which are the uniform grid and the wavy grid respectively. The wavy grid is generated by the following formula [24],

$$x_{i,j} = x_{\min} + \Delta x_o \left[(i-1) + \sin\left(\dfrac{6\pi(j-1)\Delta y_o}{L_y}\right)\right], \quad (3.6\text{a})$$

$$y_{i,j} = y_{\min} + \Delta y_o \left[(j-1) + 2\sin\left(\dfrac{6\pi(i-1)\Delta x_o}{L_x}\right)\right], \quad (3.6\text{b})$$

$$(3.6\text{c})$$

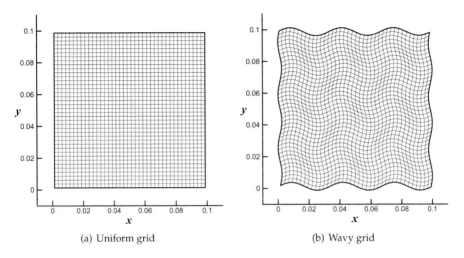

Figure 8: The uniform and wavy grids used in the problem of two-dimensional isentropic vortex.

$$\Delta x_o = \frac{L_x}{IL-1}, \quad \Delta y_o = \frac{L_y}{JL-1}, \quad 1 \leq i \leq IL, \quad 1 \leq j \leq JL, \quad (3.6d)$$

where IL and JL denote the number of points in the ξ and η directions and $x_{min} = y_{min} = 0$. Both girds are shown in Fig. 8. Table 4 presents the errors of u and the accuracy order with different grid numbers. It can be seen that the designed order of the scheme is both achieved. The abnormality of the order between the first two girds is due to relatively stronger dissipation on coarse grids, which causes problems in the error statistics. Fig. 9 shows the numerical results with the grid number 120×120. In Fig. 9(a), the value of u alone the vertical centerline is plotted. The numerical result on the wavy grid deviates the analytical solution more than that on the uniform grid. This indicates that the scheme introduces more dissipation errors on the wavy grid. Figs. 9(b) and 9(c) compare the vorticity magnitude contours. Some nonphysical wiggles are observed on the wavy grid, due to the dispersive errors. As we know, in this test case, the best results are obtained on the uniform Cartesian grids. However, from the above comparisons, we find the

Table 4: Numerical errors and accuracy order of u with different grid number at $t=50T$.

Grid	Uniform grid		Wavy grid	
40×40	3.59861E-04	—	3.83602E-04	—
80×80	1.19638E-04	1.589	2.30041E-04	0.738
120×120	3.18750E-05	3.262	6.06152E-05	3.289
160×160	9.21134E-06	4.315	1.44412E-05	4.986
240×240	1.30105E-06	4.827	1.68295E-06	4.779

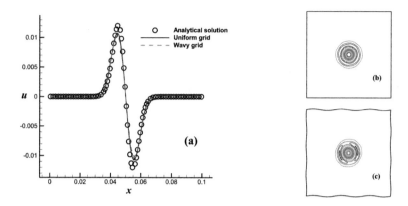

Figure 9: Numerical results on the uniform and wavy grids with grid number 120×120. (a) u alone the vertical centerline. (b) Vorticity magnitude contours on uniform grid. (c) Vorticity magnitude contours on wavy grid.
(colored picture attached at the end of the book)

results on the wavy mesh are equally acceptable. The flowfield is well preserved when the coordinate transformation is used in the FD method.

3.3 Laminar flow around a delta wing

This test case is designed in the EU project ADIGMA [40]. With a free stream Mach number $Ma=0.5$, Reynolds number $Re=4000$ and at an angle of $\alpha=12.5°$, the delta wing is considered at the laminar conditions. As the flow passes the leading edge, it rolls up and creates a large vortex and a secondary vortex which are convected far behind the wing.

The geometry of delta wing [39] and a coarse grid are presented in Fig. 10. In this paper, three different grids labeled as coarse, medium and fine are used in simulations and the grid number is increased by the ratio of 8. Because of the complex structure of the grids, the coordinate transformation should be used. The solution is initialized by a uniform flow and the no-slip adiabatic wall boundary condition is imposed. For comparison, both the 2nd-order and 5th-order schemes are adopted in this paper and all the calculations can converge to the machine zero. Table 5 collects the lift and drag coeffi-

Table 5: Lift and drag coefficients of the delta wing on three grids.

	Coarse	Medium	Fine	Referance [41]
Lift coefficient (5th order)	0.362	0.355	0.352	0.347
Drag coefficient (5th order)	0.1715	0.1700	0.1675	0.1658
Lift coefficient (2th order)	0.373	0.357	0.351	0.347
Drag coefficient (2th order)	0.1759	0.1700	0.1682	0.1658

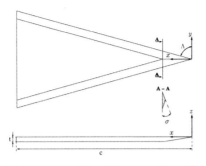
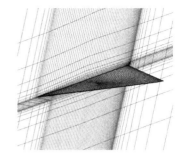

(a) Geometry of the delta wing [39] with $\Lambda=75°$, $\sigma=60°$, $t/c=0.024$.

(b) The constructed grid used in the simulation.

Figure 10: The configuration and grid of the delta wing.

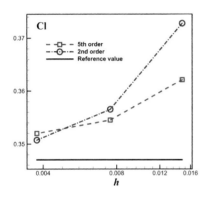
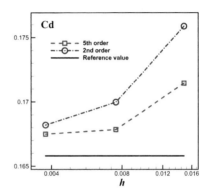

Figure 11: Lift (left) and drag (right) coefficients of using 5th-order and 2nd-order schemes with different grid sizes.

cients on three grids. For reference, the values of extrapolating the results obtained with a higher order DG method [41] are included. Fig. 11 shows the lift and drag coefficients with different grid sizes. It can be seen that as the grid is refined, the lift and drag coefficients both converge to the reference values, but the high-order scheme has a more rapid rate. The benefit of using a high-order scheme in the numerical calculations is obvious, as the vortex structure over the wing can be better captured. This test case demonstrates the schemes used in this paper can well resolve the features of the flowfield and can be applied in the simulations of practical problems.

In addition, the Mach number contours at $x=2.6c$ for both the 2nd-order and 5th-order schemes are presented in Fig. 12. From the comparison, it is found that the secondary vortex can be well captured using the 5th-order scheme, while the 2nd-order scheme has a much worse resolution. The advantage of the high-order scheme is ob-

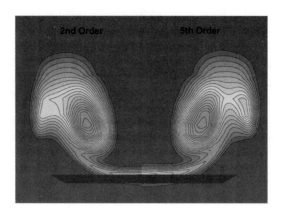

Figure 12: Comparison of the Mach number contours of the 2nd-order and 5th-order schemes at $x=2.6c$.
(colored picture attached at the end of the book)

vious in this test case. In [25], the authors pointed out that as the grid nonuniformity increased, if the coordinate transformation was used, the benefit of using higher-order schemes would almost disappear. However, from the results obtained in this paper, we consider this conclusion may be worth discussing.

4 Conclusions

In this paper, systematical studies of the characteristics of the FD schemes on nonuniform grids when the coordinate transformation is introduced are carried out. Focused on the smooth varying grids, detail analyses are presented and can be summarized as follows:

(i) For the order of accuracy, the designed order of the schemes can be preserved on the nonuniform grids. The accuracy deterioration in higher-order FD schemes described in [25] is not found.

(ii) Formulae of the Fourier analysis involving the coordinate transformation are presented for the first time. Based on the current method, it is found that the resolution characteristics of the FD schemes on nonuniform grids have a close relation with the stretching ratio of grids. The resolution characteristics can be preserved as well, as long as the stretching ratio is reasonably controlled.

(iii) Asymptotic stability analysis shows that the coordinate transformation may degrade the stability of some FD schemes, which is closely related to the boundary schemes. This issue deserves more attention.

In general, the characteristics of the FD schemes can be preserved after the coordinate transformation, which is also demonstrated by numerical tests.

In the current paper, we only analyze the behaviors of first spatial derivative that represents the calculations of the convective terms. In fact, the viscous terms are equally important and worth studying. The effect of non-smooth girds on the FD schemes is a big and complex topic. Quantitative analysis is seldom seen. These issues are of great importance in the development of the FD method and we expect further studies.

Acknowledgments

This work was supported by the Basic Research Foundation of National University of Defense Technology (No. ZDYYJCYJ20140101). We are very grateful to the referees of this paper for their valuable suggestions to improve the quality of this paper.

References

[1] M. VINOKUR, *An analysis of finite-difference and finite-volume formulations of conservation laws*, J. Comput. Phys., 81 (1989), pp. 1–52.

[2] D. J. GARMANN, *Compact finite-differencing and filtering procedure applied to the incompressible navier-stokes equations*, AIAA J., 51(9) (2013).

[3] X. ZHANG AND C. W. SHU, *Positivity-preserving high order finite difference WENO schemes for compressible Euler equations*, J. Comput. Phys., 231 (2012), pp. 2245–2258.

[4] O. MARXEN, G. IACCARINO AND T. E. MAGIN, *Direct numerical simulations of hypersonic boundary-layer transition with finite-rate chemistry*, J. Fluid Mech., 755 (2014), pp. 35–49.

[5] D. P. RIZZETTA, M. R. VISBAL AND P. E. MORGAN, *A high-order compact finite-difference scheme for large-eddy simulation of active flow control*, Progress Aerospace Sci., 44 (2008), pp. 397–426.

[6] Y. MOR-YOSSEF, *Unconditionally stable time marching scheme for reynolds stress models*, J. Comput. Phys., 276 (2014), pp. 635–664.

[7] J. A. EKATERINARIS, *High-order accurate, low numerical diffusion methods for aerodynamics*, Progress Aerospace Sci., 41 (2005), pp. 192–300.

[8] S. K. LELE, *Compact finite difference schemes with spectral-like resolution*, J. Comput. Phys., 103 (1992), pp. 16–42.

[9] X. DENG, Y. JIANG, M. MAO, H. LIU, S. LI AND G. TU, *A family of hybrid cell-edge and cell-node dissipative compact schemes satisfying geometric conservation law*, Comput. Fluids, 116 (2015), pp. 29–45.

[10] B. COCKBURN AND C. W. SHU, *Nonlinearly stable copact schemes for shock calculations*, SIAM J. Numer. Anal., 31(3) (1994), pp. 607–627.

[11] X. DENG AND H. MAEKAWA, *Compact high-order accurate nonlinear schemes*, J. Comput. Phys., 130 (1997), pp. 77–91.

[12] Y.-X. REN, M. LIU AND H. ZHANG, *A characteristic-wise hybrid compact-WENO scheme for solving hyperbolic conservation laws*, J. Comput. Phys., 192 (2003), pp. 365–386.

[13] Y. SHEN AND G. ZHA, *Generalized finite compact difference scheme for shock/complex flowfield interaction*, J. Comput. Phys., 230 (2011), pp. 4419–4436.

[14] C. K. W. TAM, *Computational Aeroacoustics: A Wave Number Approach*, Cambridge University Press, 2012.

[15] D. W. ZINGG, S. D. RANGO, M. NEMEC AND T. H. PULLIAM, *Comparison of several spatial discretizations for the Navier–Stokes equations*, J. Comput. Phys., 160 (2000), pp. 683–704.
[16] C. K. W. TAM AND J. C. WEBB, *Dispersion-relation-preserving finite difference schemes for computational acoustics*, J. Comput. Phys., 107 (1993), pp. 262–281.
[17] S. PIROZZOLI, *Performance analysis and optimization of finite-difference schemes for wave propagation problems*, J. Comput. Phys., 222 (2007), pp. 809–831.
[18] L. GAMET, F. DUCROS, F. NICOUD AND T. POINSOT, *Compact finite difference schemes on non-uniform meshes, application to direct numerical simulations of compressible flows*, Int. J. Numer. Meth. Fluids, 29 (1999), pp. 159–191.
[19] C. CHEONG AND S. LEE, *Grid-optimized dispersion-relation-preserving schemes on general geometries for computational aeroacoustics*, J. Comput. Phys., 174 (2001), pp. 248–276.
[20] X. ZHONG AND M. TATINENI, *High-order non-uniform grid schemes for numerical simulation of hypersonic boundary-layer stability and transition*, J. Comput. Phys., 190 (2003), pp. 419–458.
[21] R. K. SHUKLA AND X. ZHONG, *Derivation of high-order compact finite difference schemes for non-uniform grid using polynomial interpolation*, J. Comput. Phys., 204 (2005), pp. 404–429.
[22] X. DENG, M. MAO, G. TU, H. LIU AND H. ZHANG, *Geometric conservation law and application to high-order finite difference schemes with stationary grids*, J. Comput. Phys., 230 (2011), pp. 1100–1115.
[23] X. DENG, Y. MIN, M. MAO, H. LIU, G. TU AND H. ZHANG, *Further study on geometric conservation law and application to high-order finite difference schemes with stationary grids*, J. Comput. Phys., 239 (2013), pp. 90–111.
[24] R. VISBAL AND D. GAITONDE, *On the use of higher-order finite-difference schemes on curvilinear and deforming meshes*, J. Comput. Phys., 181 (2002), pp. 155–185.
[25] Y. M. CHUNG AND P. G. TUCKER, *Accuracy of higher-order finite difference schemes on nonuniform grids*, AIAA J., 41(8) (2003).
[26] G. B. WHITHAM, Linear and Nonlinear Waves, John Wiley & Sons, 1974.
[27] X. DENG AND H. ZHANG, *Developing high-order weighted compact nonlinear schemes*, J. Comput. Phys., 165 (2000), pp. 22–44.
[28] M. VINOKUR AND H. YEE, *Extension of efficient low dissipation high-order schemes for 3D curvilinear moving grids*, Front. Comput. Fluid Dyn., (2002), pp. 129–164.
[29] T. NONOMURA, N. IIZUKA AND K. FUJII, *Freestream and vortex preservation properties of high-order WENO and WCNS on curvilinear grids*, Comput. Fluids, 39 (2010), pp. 197–214.
[30] R. VICHNEVETSKY AND J. B. BOWLES, Fourier Analysis of Numerical Approximations of Hyperbolic Equations, SIAM Philadelphia, 1982.
[31] X. LIU, X. DENG AND M. MAO, *High-order behaviors of weighted compact fifth-order nonlinear schemes*, AIAA J., 45(8) (2007).
[32] M. H. CARPENTER, D. GOTTLIEB AND S. ABARBANEL, *The stability of numerical boundary treatments for compact high-order finite-difference schemes*, J. Comput. Phys., 108 (1993), pp. 272–295.
[33] D. KOSLOFF AND H. TAL-EZER, *A modified Chebyshev pseudo-spectral method with an $\mathcal{O}(1/n)$ time step restriction*, J. Comput. Phys., 104 (1993), pp. 457–469.
[34] H. LIU, X. DENG, M. MAO AND G. WANG, *High order nonlinear schemes for viscous terms and the application to complex configuration problems*, AIAA, (2011), 2011-369.
[35] J. CASPER, C. W. SHU AND H. ATKINS, *Comparision of two formulations for high-order accurate essentially nonoscillatory schemes*, AIAA J., 32(10) (1994).
[36] H. ATKINS AND J. CASPER, *Nonreflective boundary conditions for high-order methods*, AIAA J., 32(3) (1994).

[37] G. O. ROBERTS, *Computational meshes for the boundary problems*, in: Second International Conference Numerical Methods Fluid Dynamics, Lecture Notes in Physics (Springer-Verlag, New York, 1971).

[38] Z. J. WANG, K. FIDKOWSKI, R. ABGRALL, F. BASSI, D. CARAENI, A. CARY, H. DECONINCK, R. HARTMANN, K. HILLEWAERT, H. T. HUYNH, N. KROLL, G. MAY, P. O. PERSSON, B. VAN LEER AND M. VISBAL, *High-order CFD methods: current status and perspective*, Int. J. Numer. Meth. Fluids, 72 (2013), pp. 811–845.

[39] R. ABGRALL AND D. D. SANTIS, *Linear and non-linear high order accurate residual distribution schemes for the discretization of the steady compressible Navier-Stokes equations*, J. Comput. Phys., 283 (2015), pp. 329–359.

[40] N. KROLL, *Adigma–a European project on the development of adaptive higher-order variational methods for aerospace applications*, AIAA, (2009), 2009-176.

[41] T. LEICHT AND R. HARTMANN, *Error estimation and anisotropic mesh refinement for 3D laminar aerodynamic flow simulations*, J. Comput. Phys., 229 (2010), pp. 7344–7360.

Extending Seventh-Order Dissipative Compact Scheme Satisfying Geometric Conservation Law to Large Eddy Simulation on Curvilinear Grids

Yi Jiang[1,3,*], Meiliang Mao[3], Xiaogang Deng[2] and Huayong Liu[1]

[1] *State Key Laboratory of Aerodynamics, China Aerodynamics Research and Development Center, Mianyang 621000, China*
[2] *National University of Defense Technology, Changsha 410073, China*
[3] *Computational Aerodynamics Institute, China Aerodynamics Research and Development Center, Mianyang 621000, China*

Received 14 November 2013; Accepted (in revised version) 23 October 2014

Abstract. Seventh-order hybrid cell-edge and cell-node dissipative compact scheme (HDCS-E8T7) is extended to a new implicit large eddy simulation named HILES on stretched and curvilinear meshes. Although the conception of HILES is similar to that of monotone integrated LES (MILES), i.e., truncation error of the discretization scheme itself is employed to model the effects of unresolved scales, HDCS-E8T7 is a new high-order finite difference scheme, which can eliminate the surface conservation law (SCL) errors and has inherent dissipation. The capability of HILES is tested by solving several benchmark cases. In the case of flow past a circular cylinder, the solutions of HILES fulfilling the SCL have good agreement with the corresponding experiment data, however, the flowfield is gradually contaminated when the SCL error is enlarged. With the help of fulling the SCL, ability of HILES for handling complex geometry has been enhanced. The numerical solutions of flow over delta wing demonstrate the potential of HILES in simulating turbulent flow on complex configuration.

AMS subject classifications: 76Fxx, 76Gxx

Key words: Implicit large eddy simulation, high-order hybrid cell-edge and cell-node dissipative compact scheme (HDCS), geometric conservation law (GCL), complex geometry.

1 Introduction

Large Eddy Simulation (LES) is a promising approach for engineering problems with lower cost than direct numerical simulation (DNS) and higher accuracy than Reynolds averaged NavierStokes (RANS) models. In large eddy simulation of turbulent flows, the small-scale structures are left unresolved and should be accounted by a subgrid-scale (S-GS) turbulence model. As well known, there is complex interaction between SGS model and truncation errors of numerical methods [1–4]. This complex interaction leads to that it is difficult to quantify and control discretization errors in LES methods. However, this interference can also be beneficial. Kawamura et al. [5] indicated that the truncation error of a linear upwind scheme in some cases may function as implicit SGS model, and this is the conception of monotone integrated LES (MILES), which is proposed by Boris et al. [6]. Instead of an explicit computation of the SGS stress, the truncation error of the discretization scheme itself is employed to model the effects of unresolved scales. For MILES, the resolved scales are connected with the unresolved scales properly [7,8]. In 2007, a theoretical connection between explicit LES models and the implicit modeling of MILES was derived by Margolin et al. [8] using modified equation analysis (MEA). In particular, they have developed a structural explanation of why some numerical methods work well as implicit subgrid models whereas others are inadequate. Following the idea of MILES, ILES is developed by Visbal et al. [9]. Nowadays, MILES and ILES are widely accepted and applied [10–12]. Although the conception of ILES and MILES is similar, there are two main differences between MILES and ILES. Firstly, the descretization schemes of ILES are the fourth-order and sixth-order central compact scheme proposed by Lele [13]. Secondly, numerical dissipation of ILES is introduced by high-order filtering, not by the schemes themselves.

Applications of LES to increasingly complex configurations of engineering interest is motivated by the need to provide more realistic characterizations of complex flows. On the other hand, the necessity of high-order scheme for LES of turbulent flows has been recognized by many researchers [1–3]. Then high-order schemes with ability handling complex geometry are attractive methods for the LES. Finite difference schemes are widely used for their relative simpleness and flexibility. However, applications of high-order finite difference schemes are still challenged by complex meshes. When the numerical simulation is performed by these schemes on complex mesh, there may be some challenges, such as robustness and grid-quality sensitivity [14,15]. Fortunately, this deficiency can be largely removed by the researches of the Geometric Conservation Law (GCL) [16–20]. The GCL contains surface conservation law (SCL) and volume conservation law (VCL). The VCL has been widely studied for time-dependent grids, while the SCL is merely discussed for finite difference schemes. If the SCL has not been satisfied, numerical instabilities and even computing collapse may appear on complex curvilinear grids during numerical simulation. In order to fulfill the SCL for high-order finite difference schemes, a conservative metric method (CMM) is derived by Deng et al. [16]. The CMM is achieved by computing grid metric derivatives through a conservative form with the same scheme applied to fluxes. According to the principle of satisfying the CMM, a seventh-order hybrid cell-edge and cell-node dissipative compact scheme (HDCS-E8T7) has been proposed for complex geometry [21]. The HDCS-E8T7 has inherent dissipation to dissipate unresolvable wavenumbers, therefore the filtering is not needed. The properties of HDCS-E8T7 scheme have been systemically analyzed by Deng et al. [22].

In this paper, we extend HDCS-E8T7 scheme to a new implicit large eddy simulation named HILES on stretched and curvilinear meshes. Although the conception of HILES is similar to that of ILES and MILES, HDCS-E8T7 for HILES is a new seventh-order dissipative compact scheme which can eliminate the SCL errors and has inherent dissipation. If the SCL errors are eliminated, the ability of the HILES in handling complex geometry is enhanced. Moreover, if the HDCS-E8T7 is used for LES without model (HILES), we will obtain seventh-order of accuracy. However, if subgrid model, for instance, the Smgorinsky subgrid model is added, the LES results will be degenerated to second order of accuracy, which is undesirable. If a carefully designed high-order subgrid model is absent, high-order scheme without model may be better than with second-order subgrid models for LES [23]. In the next section, the governing equations are given. In Section 3, numerical methods for the HILES are given. Numerical tests are presented in Section 4.

2 Governing equations

Three dimensional Navier-Stokes equations in computational coordinates may be written as

$$\frac{\partial \widetilde{U}}{\partial t}+\frac{\partial \widetilde{E}}{\partial \xi}+\frac{\partial \widetilde{F}}{\partial \eta}+\frac{\partial \widetilde{G}}{\partial \zeta}=\frac{1}{R_e}\left(\frac{\partial \widetilde{E}_v}{\partial \xi}+\frac{\partial \widetilde{F}_v}{\partial \eta}+\frac{\partial \widetilde{G}_v}{\partial \zeta}\right)+\frac{\vec{S}}{J}, \quad (2.1)$$

where $\vec{S}$ is a source term, and it is non-zero only for the computation of turbulent channel flow (Subsection 4.1),

$$\widetilde{U}=U/J,$$
$$\widetilde{E}=(\xi_t U+\xi_x E+\xi_y F+\xi_z G)/J, \qquad \widetilde{E}_v=(\xi_x E_v+\xi_y F_v+\xi_z G_v)/J,$$
$$\widetilde{F}=(\eta_t U+\eta_x E+\eta_y F+\eta_z G)/J, \qquad \widetilde{F}_v=(\eta_x E_v+\eta_y F_v+\eta_z G_v)/J,$$
$$\widetilde{G}=(\varsigma_t U+\zeta_x E+\zeta_y F+\zeta_z G)/J, \qquad \widetilde{G}_v=(\zeta_x E_v+\zeta_y F_v+\zeta_z G_v)/J,$$

and

$$U=[\rho,\rho u,\rho v,\rho w,\rho e]^T, \qquad E=[\rho u,\rho u^2+p,\rho vu,\rho wu,(\rho e+p)u]^T,$$
$$F=[\rho v,\rho uv,\rho v^2+p,\rho wv,(\rho e+p)v]^T, \qquad G=[\rho w,\rho uw,\rho vw,\rho w^2+p,(\rho e+p)w]^T,$$
$$E_v=[\,0,\ \tau_{xx},\ \tau_{xy},\ \tau_{xz},\ u\tau_{xx}+v\tau_{xy}+w\tau_{xz}+\dot{q}_x\,]^T,$$
$$F_v=[\,0,\ \tau_{yx},\ \tau_{yy},\ \tau_{yz},\ u\tau_{yx}+v\tau_{yy}+w\tau_{yz}+\dot{q}_y\,]^T,$$
$$G_v=[\,0,\ \tau_{zx},\ \tau_{zy},\ \tau_{zz},\ u\tau_{zx}+v\tau_{zy}+w\tau_{zz}+\dot{q}_z\,]^T,$$

with viscous stress terms written as

$$\tau_{ij}=\mu\left(\frac{\partial u_i}{\partial x_j}+\frac{\partial u_i}{\partial x_j}-\frac{2}{3}\delta_{ij}\frac{\partial u_l}{\partial x_l}\right),$$

and heat transfer terms

$$\dot{q}_x = \frac{1}{(\gamma-1)P_r M_\infty^2}\mu\frac{\partial T}{\partial x}, \quad \dot{q}_y = \frac{1}{(\gamma-1)P_r M_\infty^2}\mu\frac{\partial T}{\partial y}, \quad \dot{q}_z = \frac{1}{(\gamma-1)P_r M_\infty^2}\mu\frac{\partial T}{\partial z},$$

where γ is the ratio of specific heats, and the viscous coefficient μ can be calculated by the Sutherland's law. The equation of state and the energy function are

$$p = \frac{1}{\gamma M^2}\rho T, \quad e = \frac{p}{(\gamma-1)\rho} + \frac{u^2+v^2+w^2}{2}.$$

In the above u, v and w are the velocity components in x, y and z directions, respectively, p is the pressure, ρ is the density and T is temperature. The non-dimensional variables are defined as $\rho = \rho^*/\rho_\infty^*$, $(u,v,w) = (u,v,w)^*/V_\infty^*$, $T = T^*/T_\infty^*$, $p = p^*/\rho_\infty^* V_\infty^{*2}$, $\mu = \mu^*/\mu_\infty^*$, respectively, and $M_\infty = u_\infty/\sqrt{\gamma R T_\infty^*}$, $R_e = \rho_\infty^* u_\infty^* r^*/\mu_\infty^*$, $P_r = \mu_\infty^* C_p/\kappa_\infty^*$ are the Mach number, Reynolds number and Prandtl number, r^* is the characteristic length. J is the Jacobi of grid transformation, $\xi_t, \xi_x, \xi_y, \xi_z, \eta_t, \eta_x, \eta_y, \eta_z, \zeta_t, \zeta_x, \zeta_y, \zeta_z$ are grid derivatives.

3 Numerical methods

3.1 Spatial discretization

Seventh-order finite difference scheme HDCS-E8T7 is employed for HILES to discretize the Eq. (2.1). Considering discretization of the inviscid terms,

$$\frac{\partial \tilde{U}}{\partial t} + \frac{\partial \tilde{E}}{\partial \xi} + \frac{\partial \tilde{F}}{\partial \eta} + \frac{\partial \tilde{G}}{\partial \zeta} = 0, \tag{3.1}$$

and theirs semi-discrete approximation,

$$\frac{\partial \tilde{U}}{\partial t} = -\delta_I^\xi \tilde{E} - \delta_I^\eta \tilde{F} - \delta_I^\zeta \tilde{G}. \tag{3.2}$$

The discretization δ_I^ξ, δ_I^η and δ_I^ζ are the same, thus we only give the discretization in ξ direction. The δ_I^ξ of the HDCS-E8T7 is

$$\delta_I^\xi \tilde{E}_j = \frac{256}{175h}(\hat{E}_{j+1/2} - \hat{E}_{j-1/2}) - \frac{1}{4h}(\tilde{E}_{j+1} - \tilde{E}_{j-1}) \\ + \frac{1}{100h}(\tilde{E}_{j+2} - \tilde{E}_{j-2}) - \frac{1}{2100h}(\tilde{E}_{j+3} - \tilde{E}_{j-3}), \tag{3.3}$$

where, $\hat{E}_{j\pm 1/2} = \tilde{E}(\hat{U}_{j\pm 1/2}, \hat{\xi}_{x,j\pm 1/2}, \hat{\xi}_{y,j\pm 1/2}, \hat{\xi}_{z,j\pm 1/2})$ and $\tilde{E}_{j+m} = \tilde{E}(U_{j+m}, \hat{\xi}_{x,j\pm m}, \hat{\xi}_{y,j\pm m}, \hat{\xi}_{z,j\pm m})$ are the fluxes at cell edges and at cell nodes, respectively. The numerical flux $\hat{E}_{j\pm 1/2}$ is evaluated by the variables at cell-edges,

$$\hat{E}_{j\pm 1/2} = \tilde{E}(\hat{U}_{j\pm 1/2}^L, \hat{U}_{j\pm 1/2}^R, \hat{\xi}_{x,j\pm 1/2}, \hat{\xi}_{y,j\pm 1/2}, \hat{\xi}_{z,j\pm 1/2}), \tag{3.4}$$

where $\hat{U}^L_{j\pm1/2}$, $\hat{U}^R_{j\pm1/2}$ are variables at cell-edge

$$\frac{5}{14}(1-\alpha)\hat{U}^L_{j-1/2}+\hat{U}^L_{j+1/2}+\frac{5}{14}(1+\alpha)\hat{U}^L_{j+3/2}$$
$$=\frac{25}{32}(U_{j+1}+U_j)+\frac{5}{64}(U_{j+2}+U_{j-1})-\frac{1}{448}(U_{j+3}+U_{j-2})$$
$$+\alpha\left[\frac{25}{64}(U_{j+1}-U_j)+\frac{15}{128}(U_{j+2}-U_{j-1})-\frac{5}{896}(U_{j+3}-U_{j-2})\right], \quad (3.5)$$

where $\alpha<0$ is the dissipative parameter to control the dissipation of the HDCS-E8T7. The corresponding $\hat{U}^R_{j+1/2}$ can be obtained easily by setting $\alpha>0$. Dissipative parameter α has no influence on the accuracy of HDCS-E8T7 scheme, however, it has effect on resolution of the scheme. According to the concept of Dispersion-Relation-Preserving (DRP) [24], HDCS-E8T7 scheme has been optimized by adjusting the value of α. The optimized dissipative parameter is equal to 0.3 [22], which will be adopted in this paper.

Seven-point stencil is used by the HDCS-E8T7, thus three kinds of boundary and near-boundary schemes are required for the left-side boundary nodes $j=1,2,3$ as well as for the right-side boundary nodes $j=N,N-1,N-2$. The sixth-order scheme is used for the near boundary $j=3$ and $j=N-2$,

$$\widetilde{E}'_j=\frac{64}{45h}(\hat{E}_{j+1/2}-\hat{E}_{j-1/2})-\frac{2}{9h}(\widetilde{E}_{j+1}-\widetilde{E}_{j-1})+\frac{1}{180h}(\widetilde{E}_{j+2}-\widetilde{E}_{j-2}). \quad (3.6)$$

The fourth-order scheme is employed for the near boundary $j=2$ and $j=N-1$,

$$\widetilde{E}'_j=\frac{4}{3h}(\hat{E}_{j+1/2}-\hat{E}_{j-1/2})-\frac{1}{6h}(\widetilde{E}_{j+1}-\widetilde{E}_{j-1}). \quad (3.7)$$

The fourth-order boundary schemes for $j=1$ and $j=N$ are

$$\widetilde{E}'_1=\frac{1}{6h}(48\hat{E}_{3/2}+16\hat{E}_{5/2}-25\widetilde{E}_1-36\widetilde{E}_2-3\widetilde{E}_3), \quad (3.8a)$$

$$\widetilde{E}'_N=\frac{-1}{6h}(48\hat{E}_{N-1/2}+16\hat{E}_{N-3/2}-25\widetilde{E}_N-36\widetilde{E}_{N-1}-3\widetilde{E}_{N-2}). \quad (3.8b)$$

For boundary schemes, the fifth-order interpolations for $j=3/2$ and $j=N-1/2$ are

$$\hat{U}^L_{3/2}=\frac{1}{128}(35U_1+140U_2-70U_3+28U_4-5U_5), \quad (3.9a)$$

$$\hat{U}^L_{N-1/2}=\frac{1}{128}(35U_N+140U_{N-1}-70U_{N-2}+28U_{N-3}-5U_{N-4}), \quad (3.9b)$$

synchronously, the fifth-order dissipative compact interpolations for $j=5/2$ and $j=N-3/2$ are given by

$$\frac{3}{10}(1-\alpha)\hat{U}^L_{j-1/2}+\hat{U}^L_{j+1/2}+\frac{3}{10}(1+\alpha)\hat{U}^L_{j+3/2}$$
$$=\frac{3}{4}(U_{j+1}+U_j)+\frac{1}{20}(U_{j+2}+U_{j-1})+\alpha\left[\frac{3}{8}(U_{j+1}-U_j)+\frac{3}{40}(U_{j+2}-U_{j-1})\right]. \quad (3.10)$$

For the computation of the viscous terms, the primitive variables, u, v, w, T, are first differentiated to form the components of the stress tensor and the heat flux vector. The viscous flux derivatives are then computed by a second application of the HDCS-E8T7.

3.2 Time integration

In the following numerical tests, highly stretched meshes are employed for the HILES studies of wall-bounded flows. The stability constraint of explicit time-marching methods is too restrictive and the use of an implicit approach becomes necessary, then dual time stepping scheme [25] with Newton-like subiterations [26] is employed. With R denoting the residual, the governing equation is

$$\frac{\partial \widetilde{U}}{\partial t} = R = \frac{\theta}{R_e}\left(\frac{\partial \widetilde{E}_v}{\partial \xi}+\frac{\partial \widetilde{F}_v}{\partial \eta}+\frac{\partial \widetilde{G}_v}{\partial \zeta}\right)+\frac{\vec{S}}{J}-\frac{\partial \widetilde{E}}{\partial \xi}-\frac{\partial \widetilde{F}}{\partial \eta}-\frac{\partial \widetilde{G}}{\partial \zeta}. \tag{3.11}$$

The second-order dual time stepping scheme may be written as

$$\left[J^{-1}\left(\frac{1}{\Delta \tau}+\frac{3}{2\Delta t}\right)I+M(U^p)\right]\Delta U^{p+1}=-\left[J^{-1}\frac{3U^p-4U^n+U^{n-1}}{2\Delta t}+R(U^p)\right], \tag{3.12}$$

where $M(U)=\partial R(U)/\partial U$, and τ is pseudo-time. For the first subiteration, $p=1$, $U^p=U^n$, and as $p\rightarrow\infty$, $U^p\rightarrow U^{n+1}$. The spatial derivatives in the implicit operators are represented using standard second-order approximations whereas seventh-order discretizations are employed for the residual.

3.3 The SCL on curvilinear meshes

According to the study of Deng et al. [16], the SCL is satisfied on uniform meshes. However, the SCL may be dissatisfied on curvilinear meshes. If the SCL has not been satisfied, numerical instabilities and even computing collapse may appear on complex curvilinear grids during numerical simulation. In this subsection, we will discuss the SCL on curvilinear meshes. The grid metric derivatives of governing equations (2.1) have the conservative form as

$$\begin{cases}\tilde{\xi}_x=\delta_{\text{II}}^{\zeta}((\delta_{\text{III}}^{\eta}y)z)-\delta_{\text{II}}^{\eta}((\delta_{\text{III}}^{\zeta}y)z), & \tilde{\xi}_y=\delta_{\text{II}}^{\eta}((\delta_{\text{III}}^{\zeta}z)x)-\delta_{\text{II}}^{\zeta}((\delta_{\text{III}}^{\eta}z)x), & \tilde{\xi}_z=\delta_{\text{II}}^{\zeta}((\delta_{\text{III}}^{\eta}x)y)-\delta_{\text{II}}^{\eta}((\delta_{\text{III}}^{\zeta}x)y),\\ \tilde{\eta}_x=\delta_{\text{II}}^{\zeta}((\delta_{\text{III}}^{\eta}y)z)-\delta_{\text{II}}^{\eta}((\delta_{\text{III}}^{\zeta}y)z), & \tilde{\eta}_y=\delta_{\text{II}}^{\eta}((\delta_{\text{III}}^{\zeta}z)x)-\delta_{\text{II}}^{\zeta}((\delta_{\text{III}}^{\eta}z)x), & \tilde{\eta}_z=\delta_{\text{II}}^{\zeta}((\delta_{\text{III}}^{\eta}x)y)-\delta_{\text{II}}^{\eta}((\delta_{\text{III}}^{\zeta}x)y),\\ \tilde{\zeta}_x=\delta_{\text{II}}^{\eta}((\delta_{\text{III}}^{\zeta}y)z)-\delta_{\text{II}}^{\zeta}((\delta_{\text{III}}^{\eta}y)z), & \tilde{\zeta}_y=\delta_{\text{II}}^{\zeta}((\delta_{\text{III}}^{\eta}z)x)-\delta_{\text{II}}^{\eta}((\delta_{\text{III}}^{\zeta}z)x), & \tilde{\zeta}_z=\delta_{\text{II}}^{\eta}((\delta_{\text{III}}^{\zeta}x)y)-\delta_{\text{II}}^{\zeta}((\delta_{\text{III}}^{\eta}x)y),\end{cases} \tag{3.13}$$

where $\delta_{\text{II}}^{\xi}, \delta_{\text{II}}^{\eta}, \delta_{\text{II}}^{\zeta}$ and $\delta_{\text{III}}^{\xi}, \delta_{\text{III}}^{\eta}, \delta_{\text{III}}^{\zeta}$ are numerical derivative operators used for the metric calculations in ξ, η and ζ coordinate direction, respectively. For the applications on curvilinear meshes, finite difference scheme should satisfy the SCL which means $I_x=I_y=I_z=0$ [16],

$$\begin{cases}I_x=\delta_I^{\xi}(\tilde{\xi}_x)+\delta_I^{\eta}(\tilde{\eta}_x)+\delta_I^{\zeta}(\tilde{\zeta}_x),\\ I_y=\delta_I^{\xi}(\tilde{\xi}_y)+\delta_I^{\eta}(\tilde{\eta}_y)+\delta_I^{\zeta}(\tilde{\zeta}_y),\\ I_z=\delta_I^{\xi}(\tilde{\xi}_z)+\delta_I^{\eta}(\tilde{\eta}_z)+\delta_I^{\zeta}(\tilde{\zeta}_z).\end{cases} \tag{3.14}$$

In order to fulfill the SCL, the grid metrics should be calculated with a conservative form by the same schemes used for flux derivative calculations, i.e., $\delta_\mathrm{I}=\delta_\mathrm{II}$, to implement the CMM [16]. It has been proved in [16] that the CMM can be easily applied in high-order schemes if the difference operator δ_I of flux derivatives is not split, while it is difficult to be applied in the schemes where the δ_I is split into two upwind operators as δ_I^+ and δ_I^-. Although the inner-level difference operators δ_III in the conservative metrics have no effect on the GCL, the constraint $\delta_\mathrm{III}=\delta_\mathrm{II}$ is recommended by Deng et al. [16]. Not long after, the constraint $\delta_\mathrm{III}=\delta_\mathrm{II}$ is explained from geometry viewpoint, and a symmetrical conservative metric method (SCMM) [27], which can evidently increase the numerical accuracy on irregular grids is proposed. To eliminate the SCL errors on curvilinear mesh, here we calculate the grid metrics with a conservative form by the same schemes used for flux derivative calculations, i.e., $\delta_\mathrm{I}=\delta_\mathrm{II}=\delta_\mathrm{III}$, to implement the SCMM [27]. Then explicit eighth-order central interpolation is employed for δ_II and δ_III at interior nodes,

$$\hat{U}_{j+1/2}=\frac{1}{2048}(1225(U_{j+1}+U_j)-245(U_{j+2}+U_{j-1})+49(U_{j+3}+U_{j-2})\\-5(U_{j+4}+U_{j-3})). \tag{3.15}$$

At boundary points, the interpolations for δ_II and δ_III are

$$\hat{U}_{3/2}=\frac{1}{128}(35U_1+140U_2-70U_3+28U_4-5U_5), \tag{3.16a}$$

$$\hat{U}_{5/2}=\frac{1}{128}(-5U_1+60U_2+90U_3-20U_4+3U_5), \tag{3.16b}$$

$$\hat{U}_{7/2}=\frac{1}{256}(3U_1-25U_2+150U_3+150U_4-25U_5+3U_6), \tag{3.16c}$$

$$\hat{U}_{N-1/2}=\frac{1}{128}(35U_N+140U_{N-1}-70U_{N-2}+28U_{N-3}-5U_{N-4}), \tag{3.16d}$$

$$\hat{U}_{N-3/2}=\frac{1}{128}(-5U_N+60U_{N-1}+90U_{N-2}-20U_{N-3}+3U_{N-4}), \tag{3.16e}$$

$$\hat{U}_{N-5/2}=\frac{1}{256}(3U_N-25U_{N-1}+150U_{N-2}+150U_{N-3}-25U_{N-4}+3U_{N-5}). \tag{3.16f}$$

3.4 Interface scheme for multi-block structured mesh

It is generally known that structured single-block grid system is difficult to be generated. for complex configuration, thus multi-block structured grid technique may be required for the applications on complex geometry. Moreover, the grid lines may be abruptly bent owing to the body configurations. A simple example of these meshes is illustrated in Fig. 1. Sudden slope changes of the grid line occur at the particular points which are represented by the circle dots in Fig. 1. It is clear that the grid metrics of the left-side and right-side unequal to each other at these points, then a singularity problem presents [28]. In order to avoid the grid singularity, we adopt the idea of Kim et al. [28], where the domain is decomposed into two blocks along the singular line. After the decomposition,

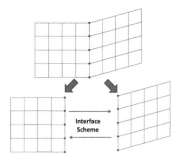

Figure 1: Decomposing a computational domain along the singular line.

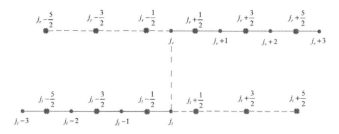

(a) Computational grids for the singular point divided

(b) Computational grids for the singular point united

Figure 2: Sketch of computational grids near the singular point.

special interface schemes will be constructed for the HDCS-E8T7 scheme at interface and near-interface nodes. As illustrated in Fig. 1, the singular point is divided into two virtual points. For the left-hand and right-hand of grid metrics, the interface scheme is expected to be a one-side difference whose stencil does not cross the block interface. Also, the GCL should be satisfied by the interface scheme to eliminate the GCL errors, i.e., $\delta_{\mathrm{I}} = \delta_{\mathrm{II}} = \delta_{\mathrm{III}}$. The construction of the interface schemes is divided into two steps.

Firstly, the interpolations are formulated for δ_{I}, δ_{II} and δ_{III}. To avoid the grid singularity, one-side interpolations are derived. As illustrated in Fig. 2(a), the singular point j is divided into two virtual points j_l and j_r with different values of the grid metrics, although they have the same location. The grid metrics of j_l and j_r are calculated independently at the singular point j of the left and right blocks, respectively. One-side high-order interpolations at j_l and j_r are given by

$$\hat{U}_{jl-5/2} = \frac{1}{256}(3U_{jl} - 25U_{jl-1} + 150U_{jl-2} + 150U_{jl-3} - 25U_{jl-4} + 3U_{jl-5}),$$

$$\hat{U}_{jl-3/2} = \frac{1}{128}(-5U_{jl} + 60U_{jl-1} + 90U_{jl-2} - 20U_{jl-3} + 3U_{jl-4}),$$

$$\hat{U}_{jl-1/2} = \frac{1}{128}(35U_{jl} + 140U_{jl-1} - 70U_{jl-2} + 28U_{jl-3} - 5U_{jl-4}),$$

$$\hat{U}_{jl+1/2} = \frac{1}{128}(315U_{jl} - 420U_{jl-1} + 378U_{jl-2} - 180U_{jl-3} + 35U_{jl-4}),$$

$$\hat{U}_{jl+3/2} = \frac{1}{128}(1155U_{jl} - 2772U_{jl-1} + 2970U_{jl-2} - 1540U_{jl-3} + 315U_{jl-4}),$$

$$\hat{U}_{jl+5/2} = \frac{1}{128}(3003U_{jl} - 8580U_{jl-1} + 10010U_{jl-2} - 5460U_{jl-3} + 1155U_{jl-4}),$$

and

$$\hat{U}_{jr-5/2} = \frac{1}{128}(3003U_{jr} - 8580U_{jr+1} + 10010U_{jr+2} - 5460U_{jr+3} + 1155U_{jr+4}),$$

$$\hat{U}_{jr-3/2} = \frac{1}{128}(1155U_{jr} - 2772U_{jr+1} + 2970U_{jr+2} - 1540U_{jr+3} + 315U_{jr+4}),$$

$$\hat{U}_{jr-1/2} = \frac{1}{128}(315U_{jr} - 420U_{jr+1} + 378U_{jr+2} - 180U_{jr+3} + 35U_{jr+4}),$$

$$\hat{U}_{jr+1/2} = \frac{1}{128}(35U_{jr} + 140U_{jr+1} - 70U_{jr+2} + 28U_{jr+3} - 5U_{jr+4}),$$

$$\hat{U}_{jr+3/2} = \frac{1}{128}(35U_{jr} + 140U_{jr+1} - 70U_{jr+2} + 28U_{jr+3} - 5U_{jr+4}),$$

$$\hat{U}_{jr+5/2} = \frac{1}{256}(3U_{jr} - 25U_{jr+1} + 150U_{jr+2} + 150U_{jr+3} - 25U_{jr+4} + 3U_{jr+5}).$$

Secondly, central type δ_I based on cell-edge is employed to construct one-side difference scheme for the calculations of grid metrics. We unite the two virtual points divided from the singular point as shown in Fig. 2(b), and a sixth-order central type δ_I, which is equal to δ_{II} and δ_{III}, at the interface point j is written as

$$\widetilde{E}'_j = \frac{2250}{1920h}(\hat{E}_{j+1/2} - \hat{E}_{j-1/2}) - \frac{125}{1920h}(\hat{E}_{j+3/2} - \hat{E}_{j-3/2}) + \frac{9}{1920h}(\hat{E}_{j+5/2} - \hat{E}_{j-5/2}). \quad (3.17)$$

For the calculations of grid metrics, Eq. (3.17) will be used in the left and right block independently. This interface scheme will be applied in the region $[j-2, j+2]$ for that seven-point stencil is used by the HDCS-E8T7. The Characteristic-Based Interface Conditions (CBIC) [29] is employed to communicate the information between the left and right blocks. After time integration, the variables on the interface need to be refined by averaging the left and right side values, which is described as

$$U_L^* = U_R^* = (U_L + U_R)/2, \quad (3.18)$$

where the asterisk denotes the refined values. The accuracy and efficiency of this interface scheme have been demonstrated by Jiang et al. [30].

4 Numerical tests

4.1 Turbulent channel flow

Turbulent channel flow problem is commonly used to test LES methods. For the situation considered here, the Reynolds number based on channel height, $Re_h = 6600$ ($Re_\tau = 180$), corresponds to the incompressible DNS of Kim et al. [31]. Since the present code solves the compressible form of the Navier-Stokes equations, a low Mach number, $M_\infty = 0.1$, is specified.

The nondimensional computational domain size is $(2\pi \times 1 \times \pi)$ in (x,y,z). Three meshes are used to solve this problem, the grid dimensions are $41 \times 41 \times 41$ (coarse), $61 \times 61 \times 61$ (medium) and $81 \times 81 \times 81$ (fine), respectively. Mesh spacing is constant in the x and z directions, and geometrically stretched in y from the walls, where $\Delta y = 0.001$. Periodic boundary conditions are performed for all variables in the streamwise and spanwise directions. At the channel walls, the no-slip condition is satisfied, along with a constant surface temperature, and a vanishing normal pressure gradient. An artificial forcing mechanism is needed to mimic an imposed streamwise pressure gradient and to maintain a fixed mass flow rate [32]. In this study, the artificial forcing algorithm suggested by Rizzetta et al. [32] is employed for the source term appearing in Eq. (2.1). The vector source term $\vec{S}$ is defined by:

$$\vec{S} = [0, s_1, s_2, s_3, s_i u_i]. \tag{4.1}$$

Considering the steady two-dimensional limiting form of the momentum equation in the streamwise direction and integrating over the channel height, the source term may be written as

$$s_1 = -\left[Re \int \left(\frac{1}{J}\right) d\eta\right]^{-1} \left\{\left[\left(\frac{\mu \eta_y^2}{J}\right) \frac{\partial u}{\partial \eta}\right]_{\eta = \eta_{\max}} - \left[\left(\frac{\mu \eta_y^2}{J}\right) \frac{\partial u}{\partial \eta}\right]_{\eta = 1}\right\}, \tag{4.2a}$$

$$s_2 = s_3 = 0, \tag{4.2b}$$

where the terms in square brackets imply averaging over the lower or upper channel walls. The streamwise velocity is initialized by a Poiseuille parabolic streamwise velocity profile, and random fluctuations are imposed on the three velocity components [33]. The computations are advanced with a time step of 0.001 for total 120,000 time steps, and flow statistics are collected for 40,000 time steps for each case. For periodic boundaries are used in the streamwise and spanwise directions, flow statistics are averaged over $x-z$ planes in this test.

Comparing with the DNS and LES results, the mean streamwise velocity profile solved by HILES is shown in Fig. 3. The LES results are calculated by Jungsoo et al. [34], where

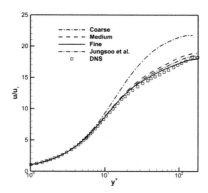

Figure 3: The mean streamwise profile solved by HILES comparing with DNS and LES results.

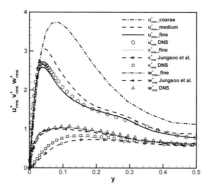

Figure 4: The fluctuating velocity profiles solved by HILES comparing with DNS and LES results.

sixth-order compact finite difference scheme, dynamic Smagorinsky SGS model and a $80 \times 80 \times 81$ mesh in the domain $(2\pi \times 1 \times 2\pi/3)$ are used. In this figure, HILES results of medium grid and fine grid are in good agreement with the previous DNS results [31], especially in the viscous sublayer ($y^+ < 5$). However there is slight overprediction of the HILES in the region $y^+ > 40$. This difference is most likely due to the combined influences of the mesh resolution and the other numerical factors, similar difference is observed in the LES studies of Jungsoo et al. [34] and Lenormand et al. [33]. Comparing with these LES studies, our HILES predictions are in as good, if not better, agreement with the DNS results.

Fig. 4 plots root mean square (RMS) values normalized by friction velocity for the fluctuating velocity solved by HILES. It can be seen that streamwise velocity fluctuations calculated by the HILES approach to that of DNS when the mesh size is increased. A similar trend is observed for the other two velocity fluctuation components (not shown).

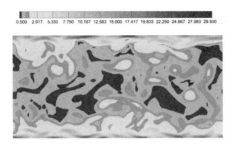

Figure 5: Instantaneous vorticity magnitude in the (y,z)-plane (fine grid).
(colored picture attached at the end of the book)

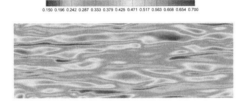

Figure 6: Instantaneous streamwise velocity in the (x,z)-plane at $y=0.03$ (fine grid).
(colored picture attached at the end of the book)

The profile of streamwise velocity fluctuations calculated on the fine grid is in good agreement with the DNS results, and the location of the peak is also well predicted. The spanwise and normal velocity fluctuations are also in good agreement with the DNS results, and they lie closer to the DNS data than LES results of Jungsoo et al. [34]. To examine the instantaneous structures of fully developed turbulent channel flow, the vorticity magnitude near the walls is shown in Fig. 5, and streamwise velocity streaks are shown in Fig. 6.

4.2 Flow past a circular cylinder

The flow past a circular cylinder at Mach number of 0.2 and Reynolds number based on cylinder diameter, D, of $Re_D = 3900$ is simulated by HILES. The large eddy simulation for this problem has been performed extensively by different research groups [10, 32, 35]. We investigate the effect of SCL errors in current simulation. The nondimensional computational domain size is $(90 \times 60 \times \pi)$ in (x,y,z). The mesh size is $180 \times 180 \times 45$ in steamwise, normal and spanwise direction respectively. The 2D grid system is shown in Fig. 7. Grid points are clustered near the cylinder surface, with a minimum normal spacing of 0.001.

At the outer boundary, far-field boundary conditions based on the LODI approximation [36] and sponge technique [37] are prescribed. The no-slip condition is invoked on the cylinder surface, together with fifth-order accurate approximations for an adiabat-

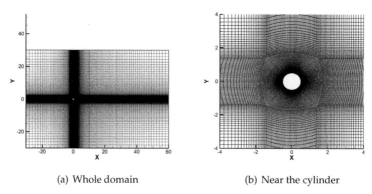

(a) Whole domain (b) Near the cylinder

Figure 7: The grid system of the cylinder test.

ic wall and zero normal pressure gradient. At the spanwise boundaries, periodicity is applied.

To show the influence of the SCL errors, we design a new scheme HDCS-E8T7A, which is same as HDCS-E8T7 except that the grid metric derivatives are calculated by combining the conservation grid metric derivatives and traditional grid metric derivatives. Traditional grid metric derivatives are given by

$$\begin{cases} \tilde{\xi}_x = \delta_I^\eta(y)\delta_I^\zeta(z) - \delta_I^\eta(z)\delta_I^\zeta(y), & \tilde{\xi}_y = \delta_I^\eta(z)\delta_I^\zeta(x) - \delta_I^\eta(x)\delta_I^\zeta(z), & \tilde{\xi}_z = \delta_I^\eta(x)\delta_I^\zeta(y) - \delta_I^\eta(y)\delta_I^\zeta(x), \\ \tilde{\eta}_x = \delta_I^\zeta(y)\delta_I^\xi(z) - \delta_I^\zeta(z)\delta_I^\xi(y), & \tilde{\eta}_y = \delta_I^\zeta(z)\delta_I^\xi(x) - \delta_I^\zeta(x)\delta_I^\xi(z), & \tilde{\eta}_z = \delta_I^\zeta(x)\delta_I^\xi(y) - \delta_I^\zeta(y)\delta_I^\xi(x), \\ \tilde{\zeta}_x = \delta_I^\xi(y)\delta_I^\eta(z) - \delta_I^\xi(z)\delta_I^\eta(y), & \tilde{\zeta}_y = \delta_I^\xi(z)\delta_I^\eta(x) - \delta_I^\xi(x)\delta_I^\eta(z), & \tilde{\zeta}_z = \delta_I^\xi(x)\delta_I^\eta(y) - \delta_I^\xi(y)\delta_I^\eta(x). \end{cases} \quad (4.3)$$

For instance, one of the grid metric derivatives is written as

$$\tilde{\xi}_x^A = \sigma \tilde{\xi}_x^T + (1-\sigma)\tilde{\xi}_x^C, \quad (4.4)$$

where $0 \leq \sigma \leq 1$ is the weight coefficient, $\tilde{\xi}_x^A$ is the grid metric derivative of HDCS-E8T7A, $\tilde{\xi}_x^T$ and $\tilde{\xi}_x^C$ are traditional and conservation grid metric derivatives, respectively. If $\sigma = 0$, the HDCS-E8T7 satisfying the SCL is recovered by the HDCS-E8T7A. However, the SCL is not satisfied if $\sigma \neq 0$. The maximal SCL errors of the HDCS-E8T7 and HDCS-E8T7A are listed in Table 1, it can be seen that the SCL error is highest when $\sigma = 1$. As shown in Fig. 8, the largest SCL error of the HDCS-E8T7A with $\sigma = 1$ presents at the singular point.

Table 1: The maximal errors of I_x and I_y for the HDCS-E8T7 and HDCS-E8T7A.

	HDCS-E8T7	HDCS-E8T7A weight σ						
		10^{-2}	10^{-1}	0.2	0.4	0.6	0.8	1.0
max(I_x)	4.83E−14	3.35E−6	3.35E−5	6.70E−5	1.34E−4	2.01E−4	2.68E−4	3.35E−4
max(I_y)	7.08E−14	3.36E−6	3.36E−5	6.72E−5	1.34E−4	2.02E−4	2.69E−4	3.36E−4

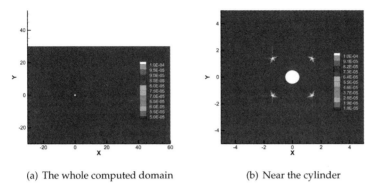

(a) The whole computed domain (b) Near the cylinder

Figure 8: The SCL error ($|I_x|+|I_y|$) of HDCS-E8T7A with $\sigma=1$.
(colored picture attached at the end of the book)

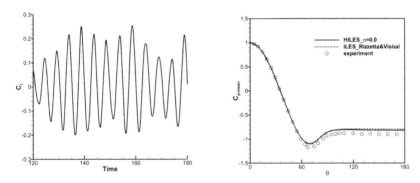

Figure 9: The time history of the cylinder lift coefficient.

Figure 10: Mean surface pressure.

The uniform initial flowfield is used to start the simulation and the transition period is up to dimensionless time $t=100$. Then the statistical results are calculated from $t=100$ to $t=200$. The time step size in the computations is $\Delta t=0.01$. For the HILES with HDCS-E8T7, the time history of the cylinder lift coefficient is shown in Fig. 9. Comparing with the solution of ILES [32], mean surface pressure from the HILES computation appears in Fig. 10. Agreement between the simulation of HILES and the measurement [38] is reasonable, particularly in the upstream region. Corresponding mean wake centerline streamwise velocity distributions are displayed in Fig. 11.

Fig. 12 plots mean streamwise velocity field calculated by the HILES with HDCS-E8T7 and HDCS-E8T7A. It shows that flowfield is slightly contaminated when the SCL error is low ($\sigma=10^{-2}$). However, when σ is enlarged from 10^{-2} to 0.2, the field is getting worse near the singular point where the SCL error is the largest. As shown in Fig. 13,

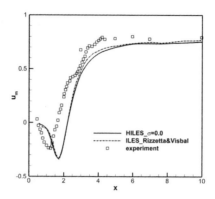

Figure 11: Mean wake centerline velocity.

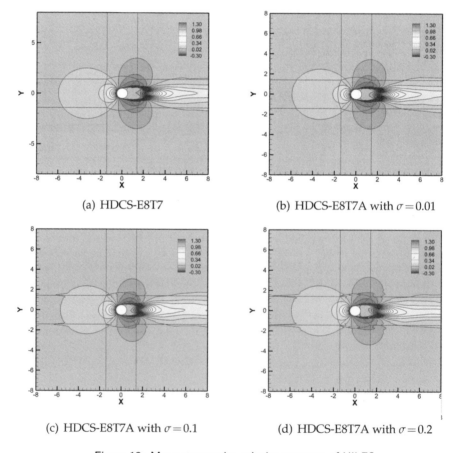

(a) HDCS-E8T7

(b) HDCS-E8T7A with $\sigma=0.01$

(c) HDCS-E8T7A with $\sigma=0.1$

(d) HDCS-E8T7A with $\sigma=0.2$

Figure 12: Mean streamwise velocity contours of HILES.
(colored picture attached at the end of the book)

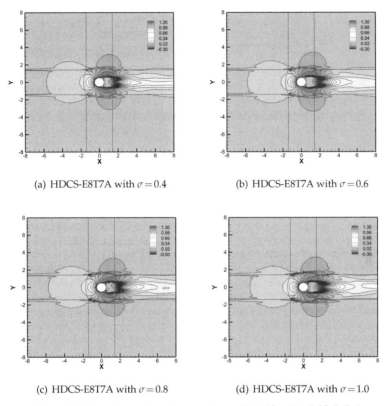

(a) HDCS-E8T7A with $\sigma=0.4$
(b) HDCS-E8T7A with $\sigma=0.6$
(c) HDCS-E8T7A with $\sigma=0.8$
(d) HDCS-E8T7A with $\sigma=1.0$

Figure 13: Mean streamwise velocity contours of HILES with HDCS-E8T7A.
(colored picture attached at the end of the book)

the contamination is even obvious when σ is larger than 0.2, and the worst area is still near the singular point. It is clear that HILES can eliminate the SCL errors which may contaminate the flowfield. To quantitatively show influence of the SCL errors, we plot vertical distributions of the mean streamwise velocity component for the streamwise location $x/D=1.06$ in Fig. 14. It is clear that unphysical oscillation is getting more visible following the SCL error is enlarged. In Fig. 15, similar oscillation is observed at streamwise location $x/D=1.54$ near the singular point where the SCL error is the largest. The overall results satisfying the SCL are very similar to the numerical results of Rizzetta et al. [32].

The streamwise and lateral Reynolds stress calculated by HILES are given in Figs. 16 and 17, respectively. It may be seen that both HILES and ILES significantly under-predict the Reynolds stresses. Due to the SCL errors, the unphysical oscillations may magnify the Reynolds stress. The instantaneous vorticity contours are plotted in Fig. 18. Illustrated in the figure is the fine-scale details, particularly in the near wake.

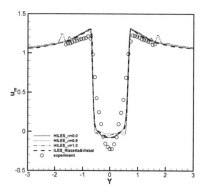

Figure 14: Spanwise averaged mean streamwise velocity distributions at location $x/D=1.06$.

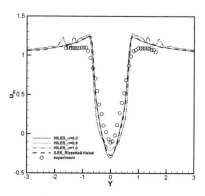

Figure 15: Spanwise averaged mean streamwise velocity distributions at location $x/D=1.54$.

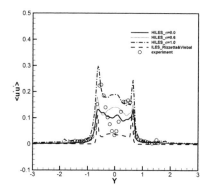

Figure 16: Streamwise Reynolds stress at location $x/D=1.06$.

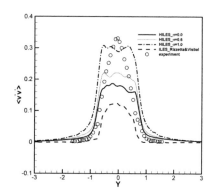

Figure 17: Lateral Reynolds stress at location $x/D=1.54$.

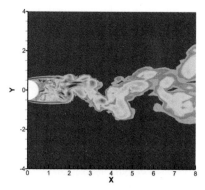

Figure 18: The instantaneous vorticity contours. (colored picture attached at the end of the book)

4.3 Flow past a delta wing configuration

We continue the test of HILES on more complex geometry. In this computation, delta wing geometry is the same as the experimental model with a fully rounded edge used by Verhaageen [39]. Computations are performed for the same conditions as in the experiment [39], where the incoming velocity is 20m/s, chord length is $c=0.12$m, the Reyonlds number based on mean aerodynamic chord c is 2.0×10^5, and angle of attack is $15°$.

Fig. 19 shows the computational grid, two different grids are used. The coarser mesh contains approximately 10.0 million grid points, where the spacing on the wing is $\Delta y=0.01$ (normalized by the reference length $L_{ref}=0.001$m) in the normal direction. A refined mesh having approximately 86.8 million grid points is also developed to significantly decrease the mesh spacing in all three directions, where the normal spacing on the wing is halved to $\Delta y=0.005$. The computations are advanced with a time step of 5×10^{-6}s for total 20,000 time steps, and flow statistics are collected for 10,000 time steps.

For the instantaneous flow, Fig. 20 demonstrates the impact of mesh refinement on

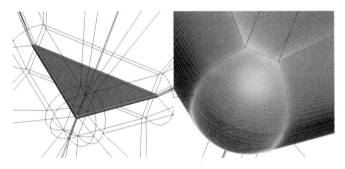

Figure 19: Delta wing grid structure (colored picture attached at the end of the book)

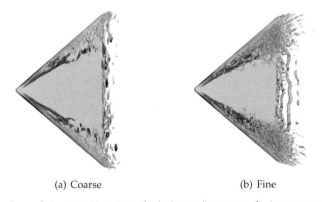

(a) Coarse   (b) Fine

Figure 20: Iso-surfaces of the second invariant of velocity gradient tensor for instantaneous vortex structure. (colored picture attached at the end of the book)

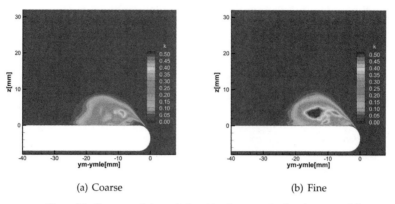

(a) Coarse (b) Fine

Figure 21: Contours of the turbulent kinetic energy in the plane $x_m=0.5$.
(colored picture attached at the end of the book)

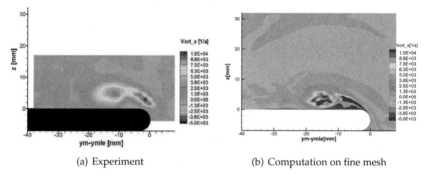

(a) Experiment (b) Computation on fine mesh

Figure 22: Time-averaged axial-vorticity distribution, $x_m=0.5$.
(colored picture attached at the end of the book)

the global structure of the vortex. In this figure, iso-surfaces of the second invariant of the velocity gradient tensor are colored by density contours. The second invariant Q is

$$Q = (\Omega_{ij}\Omega_{ij} + S_{ij}S_{ij})/2, \tag{4.5}$$

where $\Omega_{ij}=(u_{i,j}-u_{j,i})/2$ and $S_{ij}=(u_{i,j}+u_{j,i})/2$. Due to the fine mesh, we may see more details about the vortex structure in shear layer that separates from the leading edge and rolls up to form the vortex. The effect of grid refinement on the computed flowfield can be further understood by examining the structure in plane normal to the leading edge. Fig. 21 shows the turbulent kinetic energy $k=\langle \rho(u'2+v'2+w'2)\rangle$ in $x_m=0.5$, which denotes the plane cross the leading edge at 50% root chord. It is clear that higher levels of turbulent kinetic energy is obtained by the fine mesh.

Comparing with the experimental PIV measurements, Fig. 22 shows the mean axial vorticity distribution calculated by fine mesh in the measurement plane $x_m=0.5$.

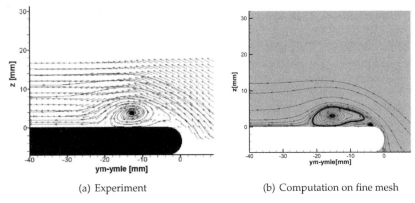

(a) Experiment (b) Computation on fine mesh

Figure 23: Time-averaged velocity-vector distribution, $x_m = 0.5$.

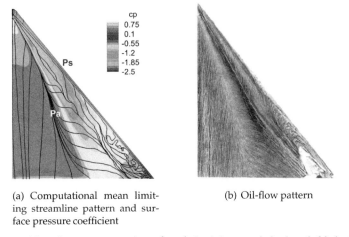

(a) Computational mean limiting streamline pattern and surface pressure coefficient

(b) Oil-flow pattern

Figure 24: Mean flow patterns on the surface. (colored picture attached at the end of the book)

Good qualitative agreement is seen between the computation and the experiment with all salient features in the experiment being reproduced in the computation. The mean velocity-vector on the plane $x_m = 0.5$, Fig. 23, also shows good agreement between computation and experiment. Considering the mean surface flowfield, the pressure coefficient calculated on fine mesh is shown in Fig. 24, moreover, the mean limiting streamline pattern is compared with the experimental oil-flow pattern. Primary separation (Ps) occurs at the leading edge of the wing with the primary attachment line (Pa) located outboard of the symmetry plane. The surface streamline pattern in the outboard region of the delta wing is getting more complex as the trailing edge is approached.

5 Conclusions

Seventh-order HDCS-E8T7 scheme is extended to HILES on stretched and curvilinear meshes in this paper. The HDCS-E8T7 scheme is a new high-order finite difference scheme which can eliminate the SCL errors and has inherent dissipation concurrently. Due to the inherent dissipation, numerical dissipation of HILES is introduced by HDCS-E8T7 itself. Moreover, ability of HILES for handling complex geometry has been enhanced with the help of eliminating SCL errors. To enable the use of HILES on complex geometry, a special sixth-order interface scheme is employed to avoid grid singularity.

In the test of turbulent channel flow, the solutions of HILES approach to that of DNS when the mesh size is increased. Moreover, our HILES predictions are in as good, if not better, agreement with the DNS results comparing with the LES results calculated by Jungsoo et al. [34], in which sixth-order compact finite difference scheme and dynamic Smagorinsky SGS model are used. The influence of the SCL errors is shown in the test of flow past a circular cylinder. The flowfield is slightly contaminated if the SCL error is low. However, when the SCL error is enlarged, the field is getting worse near the singular point where the SCL error is the largest. Comparing with experimental and other computational data, it is obvious that unphysical oscillation of the mean streamwise velocity component is getting more visible following the SCL error is enlarged. Due to the SCL errors, the unphysical oscillations may magnify the Reynolds stress. Furthermore, the potential of HILES in simulating turbulent flow on complex configuration is shown by the numerical solutions of flow over detla wing.

Acknowledgments

This study was supported by the project of National Basic Research Program of China (Grant No. 2009CB723800), National Natural Science Foundation of China (Grant Nos. 11072259 and 11202226) and the foundation of State Key Laboratory of Aerodynamics (Grant Nos. JBKY11030902 and JBKY11010100). The authors would like to thank Dr. Guohua Tu of China Aerodynamics Research & Development Center for his contributions and useful discussions.

References

[1] S. GHOSAL, *An analysis of numerical errors in large-eddy simulations of turbulence*, J. Comput. Phys., 125 (1996), pp. 187–206

[2] A. G. KRAVCHENKO AND P. MOIN, *On the effect of numerical errors in large eddy simulations of turbulent flows*, J. Comput. Phys., 131(2) (1997), pp. 310–322.

[3] FOTINI KATOPODES CHOW AND PARVIZ MOIN, *A further study of numerical errors in large-eddy simulations*, J. Comput. Phys., 184 (2003), pp. 366–380

[4] NOMA PARK, JUNG YUL YOO AND HAECHEON CHOI, *Discretization errors in large eddy simulation: on the suitability of centered and upwind-biased compact difference schemes*, J. Comput. Phys., 198 (2004), pp. 580–616.
[5] T. KAWAMURA AND K. KUWAHARA, *Computation of high Reynolds number flow around a circular cylinder with surface roughness*, AIAA-paper, 84-0340.
[6] J. P. BORIS, F. F. GRINSTEIN, E. S. ORAN AND R. L. KOLBE, *New insights into large eddy simulation*, Fluid Dyn. Res., 10 (1992), pp. 199–228.
[7] J. P. BORIS, *More for LES: a brief historical perspective of MILES*, in: F. F. Grinstein, L. G. Margolin, W. J. Rider (Eds.), Implicit Large Eddy Simulation, Cambridge University Press, 2007, pp. 9–38.
[8] L. G. MARGOLIN AND W. J. RIDER, *Numerical regularization: The numerical analysis of implicit subgrid models*, in: F. F. Grinstein, L. G. Margolin, W. J. Rider (Eds.), Implicit Large Eddy Simulation, Cambridge University Press, 2007, pp. 195–221.
[9] M. R. VISBAL AND D. P. RIZZETTA, *Large-eddy simulation on curvilinear grids using compact differencing and filtering schemes*, J. Fluids Eng., 124(4) (2002), PP. 836–847.
[10] YIQING SHEN AND GECHENG ZHA, *Large eddy simulation using a new set of sixth order schemes for compressible viscous terms*, J. Comput. Phys., 229 (2010), pp. 8296–8312.
[11] F. VUILLOT, N. LUPOGLAZOFF AND M. HUET, *Effect of chevrons on double stream jet noise from hybrid CAA computations*, 49th AIAA Aerospace Sciences Meeting including the New Horizons Forum and Aerospace Exposition 4-7 January 2011, Orlando, Florida.
[12] DONALD P. RIZZETTA, MIGUEL R. VISBAL AND PHILIP E. MORGAN, *A high-order compact finite-difference scheme for large-eddy simulation of active flow control*, Progress Aerospace Sci., 44 (2008), pp. 397–426.
[13] S. K. LELE, *Compact finite difference schemes with spectral-like resolution*, J. Comput. Phys., 103 (1992), pp. 16–42.
[14] Z. J. WANG, *High-order methods for the Euler and Navier-Stokes equations on unstructured grids*, Progress Aerospace Sci., 43 (2007), pp. 1–41.
[15] J. A. EKATERINARIS, *High-order accurate, low numerical diffusion methods for aerodynamics*, Progress Aerospace Sci., 41(3-4) (2005), pp. 192–300.
[16] X. G. DENG, M. L. MAO, G. H. TU, H. Y. LIU AND H. X. ZHANG, *Geometric conservation law and applications to high-order finite difference schemes with stationary grids*, J. Comput. Phys., 230 (2011), pp. 1100–1115.
[17] R. M. VISBAL AND D. V. GAITONDE, *On the use of higher-order finite-difference schemes on curvilinear and deforming meshes*, J. Comput. Phys., 181 (2002), pp. 155–185.
[18] T. NONOMURA, N. IIZUKA AND K. FUJII, *Freestream and vortex preservation properties of high-order WENO and WCNS on curvilinear grids*, Comput. Fluids, 39 (2010), pp. 197–214.
[19] P. D. THOMAS AND C. K. LOMBARD, *Geometric conservation law and its application to flow computations on moving grids*, AIAA J., 17(10) (1979), pp. 1030–1037.
[20] T. H. PULLIAM AND J. L. STEGER, *On implicit finite-difference simulations of three-dimensional flow*, AIAA Paper 78-10, 1978.
[21] XIAOGANG DENG, YI JIANG, MEILIANG MAO, HUAYONG LIU AND GUOHUA TU, *Developing hybrid cell-edge and cell-node dissipative compact scheme for complex geometry flows*, Sci. China Tech. Sci., 56 (2013), pp. 2361–2369.
[22] XIAOGANG DENG, YI JIANG, MEILIANG MAO, HUAYONG LIU, SONG LI AND GUOHUA TU, *A family of hybrid cell-edge and cell-node dissipative compact schemes satisfying geometric conservation law*, submitted to Computers & Fluids.
[23] PENG XIE AND CHAOQUN LIU, *Weighted compact and non-compact scheme for shock tube and*

shock entropy interaction, AIAA, 2007-509.

[24] C. K. W. TAM AND J. C. WEBB, *Dispersion-relation-preserving finite difference schemes for computational acoustics,* J. Comput. Phys., 107 (1993), pp. 262–281.

[25] JOHN M. HSU AND ANTONY JAMESON, *An Implicit-explicit hybrid scheme for calculating complex unsteady flows,* AIAA, 2002-0714.

[26] R. E. GORDNIER AND M. R. VISBAL, *Numerical simulation of delta-wing roll,* AIAA Paper 93-0554, January 1993.

[27] XIAOGANG DENG, YAOBING MIN, MEILIANG MAO, HUAYONG LIU, GUOHUA TU AND HANXIN ZHANG, *Further studies on geometric conservation law and applications to high-order finite difference schemes with stationary grids,* J. Comput. Phys., 239 (2013), pp. 90–111.

[28] J. W. KIM AND D. J. LEE, *Characteristic interface conditions for multiblock high-order computation on singular structured grid,* AIAA J., 41(2) (2003), pp. 2341–2348.

[29] X. G. DENG, M. L. MAO, G. H. TU, Y. F. ZHANG AND H. X. ZHANG, *Extending weighted compact nonlinear schemes to complex grids with characteristic-based interface conditions,* AIAA J., 48(12) (2010), pp. 2840–2851.

[30] YI JIANG, MEILIANG MAO, XIAOGANG DENG, HUAYONG LIU, SONG LI AND ZHENGUO YAN, *Extending seventh-order hybrid cell-edge and cell-node dissipative compact scheme to complex grids,* The 4th Asian Symposium on Computational Heat Transfer and Fluid Flow, Hong Kong, 3-6 June 2013.

[31] JOHN KIM, PARVIZ MOIN AND ROBERT MOSER, *Turbulence statistics in fully developed channel flow at low Reynolds number,* J. Fluid Mech. 177 (1987), pp. 133–166.

[32] D. P. RIZZETTA, M. R. VISBAL AND G. A. BLAISDELL, *A time-implicit high-order compact differencing and filtering scheme for large-eddy simulation,* Int. J. Numer. Meth. Fluids, 42 (2003), pp. 665–693.

[33] E. LENORMAND, P. SAGAUT AND L. TA PHUOC, *Large eddy simulation of subsonic and supersonic channel flow at moderate Reynolds number,* Int. J. Numer. Meth. Fluids, 32 (2000), pp. 369–406.

[34] JUNGSOO SUH, STEVEN H. FRANKEL, LUC MONGEAU AND MICHAEL W. PLESNIAK, *Compressible large eddy simulations of wall-bounded turbulent flows using a semi-implicit numerical scheme for low Mach number aeroacoustics,* J. Comput. Phys., 215 (2006), pp. 526–551.

[35] G. KRAVCHENKO AND P. MOIN, *Numerical studies of flow over a circular cylinder at $Re=3900$,* Phys. Fluids, 12 (2000), pp. 403–417.

[36] T. POINSOT AND S. T. LELE, *Boundary conditions for direct simulations of compressible viscous flows,* J. Comput. Phys., 101 (1992), pp. 104–129.

[37] DANIEL J. BODONY, *Analysis of sponge zones for computational fluid mechanics,* J. Comput. Phys., 212 (2006), pp. 681–702.

[38] A. G. KRAVCHENKO AND P. MOIN, *B-spline methods and zonal grids for numerical simulations of turbulent flows,* Flow Physics and Computation Division, Department of Mechanical Engineering, Stanford University, Report No. TF-73, Stanford, CA, Feb. 1998.

[39] N. G. VERHAAGEN AND M. ELSAYED, *Leading-edge radius effects on 50° delta wing flow,* AIAA paper 2009-540, 47th AIAA Aerospace Sciences Meeting Including The New Horizons Forum and Aerospace Exposition 5-8 January 2009, Orlando, Florida.

Effect of Geometric Conservation Law on Improving Spatial Accuracy for Finite Difference Schemes on Two-Dimensional Nonsmooth Grids

Meiliang Mao[1,2,*], Huajun Zhu[1], Xiaogang Deng[3], Yaobing Min[1] and Huayong Liu[1]

[1] *State Key Laboratory of Aerodynamics, CARDC, Mianyang, 621000, P.R. China.*
[2] *Computational Aerodynamics Institute, CARDC, Mianyang, 621000, P.R. China.*
[3] *National University of Defense Technology, Changsha, Hunan, 410073, P.R. China.*

Communicated by Chi-Wang Shu

Received 25 June 2014; Accepted (in revised version) 6 February 2015

Abstract. It is well known that grid discontinuities have significant impact on the performance of finite difference schemes (FDSs). The geometric conservation law (GCL) is very important for FDSs on reducing numerical oscillations and ensuring free-stream preservation in curvilinear coordinate system. It is not quite clear how GCL works in finite difference method and how GCL errors affect spatial discretization errors especially in nonsmooth grids. In this paper, a method is developed to analyze the impact of grid discontinuities on the GCL errors and spatial discretization errors. A violation of GCL cause GCL errors which depend on grid smoothness, grid metrics method and finite difference operators. As a result there are more source terms in spatial discretization errors. The analysis shows that the spatial discretization accuracy on nonsufficiently smooth grids is determined by the discontinuity order of grids and can approach one higher order by following GCL. For sufficiently smooth grids, the spatial discretization accuracy is determined by the order of FDSs and FDSs satisfying the GCL can obtain smaller spatial discretization errors. Numerical tests have been done by the second-order and fourth-order FDSs to verify the theoretical results.

AMS subject classifications: 35L04, 65D25, 65D18, 65M06, 65M15

Key words: Geometric conservation law, finite difference scheme, spatial discretization error, nonsmooth grids, high-order accuracy.

1 Introduction

High-order finite difference schemes (FDSs), which can be constructed easily and have high computational efficiency, are widely used in large eddy simulations (LES) and direct numerical simulations (DNS) of turbulences and aeroacoustics [1–4]. However, in order to obtain their designed accuracy, FDSs always require sufficiently smooth grids which are very difficult or even impossible to be generated for complex geometries. Complex grids for calculations on practical geometries usually contain nonsmooth features such as slope discontinuities, skewness and stretching, and the impact of these factors on scheme performance is usually significant [5]. Therefore, it is extremely important to study the influence of grid nonsmoothness on spatial accuracy for applications of high-order FDSs to practical engineering problems involving complex geometries.

It has been announced that high-order FDSs can achieve their designed accuracy on sufficiently smooth grids but will degrade their accuracy on non-sufficiently smooth grids. Casper et al. [6] pointed out that a nonsmooth grid will adversely affect the results of finite difference ENO schemes and a sufficiently smooth grid is required to achieve the designed accuracy. Castillo et al. [7] found that the accuracy of fourth-order method degrades gradually as the smoothness of the grid degenerates. Shu [8] also pointed out that the smoothness of meshes must be comparable with the order of accuracy of finite difference WENO schemes in order to obtain a truly high-order result, and the smooth meshes for the fifth-order method mean that at least the fifth derivative of coordinate transformation is continuous. However, the impact of grid smoothness on accuracy of finite difference schemes is usually analyzed numerically but not theoretically.

It was proved that the geometric conservation law (GCL) shall be also satisfied when high-order FDSs are used in the curvilinear coordinates, otherwise some negative effects may appear [9–20], such as violation of free-stream conservation, numerical oscillation. It is true that the GCL is satisfied analytically, but the discrete GCL may be dissatisfied by unsuitable discrete algorithm even on some sufficiently smooth grids, let alone non-sufficiently smooth grids. In recent years, some numerical results showed that satisfying GCL can reduce the requirement on grid smoothness for high-order FDSs, which make it possible to use them to solve problems with complex geometries. In addition, increasing numerical evidences show that satisfying the discrete GCL can improve the time-accuracy of numerical computations and can improve stabilities of these schemes as well (see, e.g., [21–24]). Nonomura et al. [25] showed that the free-stream and vortex preservation properties of WCNS [26–28] is superior to those of the WENO scheme [29], and Deng et al. [30] found the essential reason is that conservative metric method (CMM) can be introduced to ensure WCNS satisfying the GCL. Recently, Nonomura et al. [31] also introduced a technique to make WENO satisfy the GCL and found that with this technique the resolution of vortex is much improved on wavy and random grids. In order to ensure the GCL for high-order finite difference schemes, CMM has been proposed in [30] and a special case of the CMM can be found in Visbal and Gaitonde's numerical technique [5]. Thereafter, symmetrical conservative metric method (SCMM) is derived by analyzing the geometric meaning of metrics and Jacobian [32]. However, the impact of the GCL on spatial accuracy of FDSs has not been analyzed theoretically. For instance, as shown in Refs. [5, 25, 30] correct solutions can be obtained on random grids by FDSs provided that the GCL is satisfied strictly, otherwise, wrong solutions may be yielded if the GCL is not satisfied.

Most of previous work focus on ensuring the GCL for high-order FDSs to obtain better solutions in nonsmooth grids while the reason GCL-FD(FDSs satisfying the GCL) can improve solutions is ignored and not clear up to now. To understand this reason, theoretical and numerical analysis on the accuracy of GCL-FD and NGCL-FD(FDSs which does not satisfy the GCL) are made on nonsmooth grids in the current work. In addition, we denote FDSs with a rth-order accuracy as FDr. Then the rth-order accurate GCL-FD and NGCL-FD are denoted as GCL-FDr and NGCL-FDr, respectively. The main contributions of the work go as follows:

1. Systematic methodology for analyzing numerical errors of finite difference schemes on nonsmooth grids is developed. Our main technique is to use Taylor series expansion based on single-side partial derivatives on discontinuous points.

2. GCL errors and spatial discretization errors are analyzed theoretically. GCL errors are proved to be zero for GCL-FD but be related to grid smoothness and difference operators for NGCL-FD. GCL errors can affect spatial discretization errors. GCL-FD has one order higher spatial accuracy than NGCL-FD for non-sufficiently smooth grids.

3. Numerical tests on GCL errors and spatial discretization errors are made based on grids with different orders of smoothness.

This paper is organized as follows. In Section 2, definitions of grid smoothness and some propositions are presented. In Section 3, GCL errors are analyzed theoretically. In Section 4, spatial discretization errors are analyzed theoretically. In Section 5, numerical tests are made for GCL errors and spatial discretization errors based on grids with different orders of smoothness. Finally, concluding remarks are made in Section 6.

2 Definition of grid smoothness and some propositions

Grid discontinuity will cause some difficulties in analyzing the discrete errors of FDSs theoretically. There are mainly two difficulties: The first is how to describe grid discontinuities mathematically, and the second is how to analyze the influence of grid discontinuities on spatial discretization errors of finite difference operators. In addition, the analysis becomes more complicated when the spatial dimension increases. Take three dimensions for an example, discontinuities may exist in any of the three directions and the Taylor expanding will have many cross-partial derivatives. As a start we consider two dimensional grids and discuss the case that grid coordinates can be written as a smooth function plus another nonsmooth function which can be written as a finite sum of products of one-dimensional functions. Some grids of this kind can be found later in Fig. 1. As for more general grids and three dimensional grids, we like to leave for considering in the future. Moreover, we analyze only the stationary grids in this paper. The grids considered in this paper can be transformed from Cartesian $x-y$ coordinates to curvilinear $\xi-\eta$ coordinates,

$$\begin{cases} x(\xi,\eta) = f(\xi,\eta) + \sum_{m=1}^{M} [\varphi_{m,1}(\xi)\varphi_{m,2}(\eta)], \\ y(\xi,\eta) = g(\xi,\eta) + \sum_{n=1}^{N} [\psi_{n,1}(\xi)\psi_{n,2}(\eta)], \end{cases} \xi \in [0,1], \eta \in [0,1]. \tag{2.1}$$

where f and g are sufficiently smooth functions while $\varphi_{m,l}$ and $\psi_{n,l}$ ($l=1,2$, $m=1,2,\cdots,M$, $n=1,2,\cdots,N$) are piecewise-smooth functions. Suppose computational grid has $N_\xi+1$ and $N_\eta+1$ grid points along grid lines $\xi=\xi_i$ and $\eta=\eta_j$, respectively. Coordinates of grid points are $x_{i,j}=x(\xi_i,\eta_j)$ and $y_{i,j}=y(\xi_i,\eta_j)$, where $\xi_i=(i-1)\Delta\xi$, $\eta_j=(j-1)\Delta\eta$, $\Delta\xi=1/N_\xi$, $\Delta\eta=1/N_\eta$, $i=1,2,\cdots,N_\xi+1$ and $j=1,2,\cdots,N_\eta+1$.

In the following, we give some definitions of grid smoothness and some propositions which are used in Sections 3 and 4 to derive and analyze GCL errors and spatial discretization errors.

Definition 2.1. (Local ξ-continuity) Assume $\left(\frac{\partial^q \phi}{\partial \xi^q}\right)_{i,j}^L$ and $\left(\frac{\partial^q \phi}{\partial \xi^q}\right)_{i,j}^R$ are the qth partial derivatives on the left side and the right side of the ith-point along the grid line $\eta=\eta_j$, if

$$\left(\frac{\partial^q \phi}{\partial \xi^q}\right)_{i,j}^L = \left(\frac{\partial^q \phi}{\partial \xi^q}\right)_{i,j}^R \text{ for } q \leq k-1; \quad \left(\frac{\partial^k \phi}{\partial \xi^k}\right)_{i,j}^L \neq \left(\frac{\partial^k \phi}{\partial \xi^k}\right)_{i,j}^R,$$

then ϕ is $(k-1)$th-order ξ-continuous at the point (i,j), which is the kth-order ξ-discontinuous point. Here k is called the order of ξ-discontinuity of ϕ.

Definition 2.2. (global ξ-continuity) Suppose ϕ is $n_{i,j}$th-order continuous about ξ at (i,j), and $n=\min(n_{1,1},n_{1,2},\cdots,n_{N_\xi+1,N_\eta+1})$. We say that ϕ is nth-order globally ξ-continuous.

Definition 2.3. (C^n grid) If $x(\xi,\eta)$ is n_1th-order globally ξ-continuous and n_2th-order globally η-continuous, and $y(\xi,\eta)$ is n_3th-order globally ξ-continuous and n_4th-order globally η-continuous, we call such a grid is nth-order global continuous grid or C^n grid, where $n=\min\{n_1,n_2,n_3,n_4\}$.

For sufficiently smooth function ϕ, some propositions are presented in the following.

Proposition 2.1. Assume ϕ is sufficiently continuous at ξ_i, and $\phi(\xi)$ can be expanded in Taylor series and the corresponding convergent region is $[-R,R]$. Then, for arbitrary two points ξ_M and ξ_N in this convergent region, we have

$$\sum_{q=1}^{\infty} \frac{(\xi_M-\xi_i)^q}{q!}\left(\frac{\partial^q \phi}{\partial \xi^q}\right)_i + \sum_{q=1}^{\infty} \frac{(\xi_N-\xi_M)^q}{q!}\left(\frac{\partial^q \phi}{\partial \xi^q}\right)_M = \sum_{q=1}^{\infty} \frac{(\xi_N-\xi_i)^q}{q!}\left(\frac{\partial^q \phi}{\partial \xi^q}\right)_i. \tag{2.2}$$

This proposition can be easily proved by using the following Taylor series expansion

$$\left(\frac{\partial^q \phi}{\partial \xi^q}\right)_M = \left(\frac{\partial^q \phi}{\partial \xi^q}\right)_i + \sum_{r=1}^{\infty}\left[\frac{(\xi_M-\xi_i)^r}{r!}\left(\frac{\partial^{q+r}\phi}{\partial \xi^{q+r}}\right)_i\right]$$

and Binomial Theorem.

Suppose ϕ is discontinuous at the point (i,j), the following Taylor series expansion based on single-side partial derivatives can be obtained,

$$\phi_{i+l,j}-\phi_{i,j} = \begin{cases} \sum_{q=1}^{\infty} l^q \frac{(\Delta\xi)^q}{q!}\left(\frac{\partial^q \phi}{\partial^q \xi}\right)_{i,j}^R, & l>0, \\ \sum_{q=1}^{\infty} l^q \frac{(\Delta\xi)^q}{q!}\left(\frac{\partial^q \phi}{\partial^q \xi}\right)_{i,j}^L, & l<0. \end{cases} \quad (2.3)$$

Proposition 2.2. Suppose ϕ is continuous at the point (i,j) but discontinuous at point $i_A = i+i_D$ ($i_D \neq 0$) which is between i and $i+l$, then we have

$$\phi_{i+l,j}-\phi_{i,j} = \sum_{q=1}^{\infty} l^q \frac{(\Delta\xi)^q}{q!}\left(\frac{\partial^q \phi}{\partial^q \xi}\right)_{i,j} + \text{sign}(i_D)\sum_{q=1}^{\infty}(l-i_D)^q \frac{(\Delta\xi)^q}{q!}\left(\frac{\partial^q \phi}{\partial^q \xi}\right)_{i_A,j}^{R-L}, \quad (2.4)$$

where $(\cdot)^{R-L}$ is defined as the difference between the right side and the left side values at the same point, i.e., $(\cdot)^{R-L} = (\cdot)^R - (\cdot)^L$.

Proof. For $i_D > 0$, we notice that

$$\phi_{i+l,j}-\phi_{i,j} = \phi_{i+l,j}-\phi_{i_A,j}+\phi_{i_A,j}-\phi_{i,j}$$
$$= \sum_{q=1}^{\infty}(l-i_D)^q \frac{(\Delta\xi)^q}{q!}\left(\frac{\partial^q \phi}{\partial^q \xi}\right)_{i_A,j}^R + \sum_{q=1}^{\infty}(i_D)^q \frac{(\Delta\xi)^q}{q!}\left(\frac{\partial^q \phi}{\partial^q \xi}\right)_{i,j}$$
$$= \sum_{q=1}^{\infty}(l-i_D)^q \frac{(\Delta\xi)^q}{q!}\left(\frac{\partial^q \phi}{\partial^q \xi}\right)_{i_A,j}^{R-L} + \sum_{q=1}^{\infty}(i_D)^q \frac{(\Delta\xi)^q}{q!}\left(\frac{\partial^q \phi}{\partial^q \xi}\right)_{i,j}$$
$$+ \sum_{q=1}^{\infty}(l-i_D)^q \frac{(\Delta\xi)^q}{q!}\left(\frac{\partial^q \phi}{\partial^q \xi}\right)_{i_A,j}^L. \quad (2.5)$$

In addition, from Proposition 2.1 we have

$$\sum_{q=1}^{\infty}(i_D)^q \frac{(\Delta\xi)^q}{q!}\left(\frac{\partial^q \phi}{\partial^q \xi}\right)_{i,j} + \sum_{q=1}^{\infty}(l-i_D)^q \frac{(\Delta\xi)^q}{q!}\left(\frac{\partial^q \phi}{\partial^q \xi}\right)_{i_A,j}^L = \sum_{q=1}^{\infty} l^q \frac{(\Delta\xi)^q}{q!}\left(\frac{\partial^q \phi}{\partial^q \xi}\right)_{i,j}. \quad (2.6)$$

Therefore, (2.4) can be obtained by combining (2.5) and (2.6). The result for $i_D < 0$ can be obtained by a similar deductive procedure. □

For the difference operator of the first partial derivative $\frac{\partial \phi}{\partial \xi}$,

$$\delta^{\xi}\phi_{i,j} = \frac{1}{\Delta\xi} \sum_{m=-M_1}^{M_2} b_m(\phi_{i+m,j} - \phi_{i,j}), \tag{2.7}$$

if the operator has rth-order accuracy, the relation of $\{b_m\}$ can be easily obtained by taking Taylor series expansion of $\phi_{i+m,j}$ at $\phi_{i,j}$

$$\sum_{m=-M_1}^{M_2} mb_m = 1, \quad \sum_{m=-M_1}^{M_2} m^q b_m = 0, \ 2 \leq q \leq r \text{ or } q = 0. \tag{2.8}$$

Since the operator has rth-order accuracy, i.e.

$$\delta^{\xi}\phi_{i,j} = \left(\frac{\partial \phi}{\partial \xi}\right)_{i,j} + \mathcal{O}((\Delta\xi)^r). \tag{2.9}$$

Then, we consider the discretization errors of the difference operator $\delta^{\xi}\phi$ of the first derivative ϕ_{ξ}.

Proposition 2.3. Suppose ϕ is kth-order ξ-partial discontinuous at point (i_A, j) along the grid line $\eta = \eta_j$, and is continuous on the both sides of this point. Set $i_A = i_D + i$. We have for $i_D > 0$

$$\delta^{\xi}\phi_{i,j} = \sum_{m=-M_1^{\xi}}^{M_2^{\xi}} b_m^{\xi} \sum_{q=1}^{\infty} \frac{m^q (\Delta\xi)^{q-1}}{q!} \left(\frac{\partial^q \phi}{\partial \xi^q}\right)_{i,j}$$
$$+ \sum_{m=i_D+1}^{M_2^{\xi}} b_m^{\xi} \sum_{q=k}^{\infty} \frac{(m-i_D)^q (\Delta\xi)^{q-1}}{q!} \left(\frac{\partial^q \phi}{\partial \xi^q}\right)_{i_A,j}^{R-L}; \tag{2.10a}$$

for $i_D < 0$,

$$\delta^{\xi}(\phi)_{i,j} = \sum_{m=-M_1^{\xi}}^{M_2^{\xi}} b_m^{\xi} \sum_{q=1}^{\infty} \frac{m^q (\Delta\xi)^{q-1}}{q!} \left(\frac{\partial^q \phi}{\partial \xi^q}\right)_{i,j}$$
$$- \sum_{m=-M_1^{\xi}}^{i_D} b_m^{\xi} \sum_{q=k}^{\infty} \frac{(m-i_D)^q (\Delta\xi)^{q-1}}{q!} \left(\frac{\partial^q \phi}{\partial \xi^q}\right)_{i_A,j}^{R-L}; \tag{2.10b}$$

and for $i_D = 0$,

$$\delta^{\xi}(\phi)_{i,j} = \sum_{m=-M_1^{\xi}}^{i_D} b_m^{\xi} \sum_{q=1}^{\infty} \frac{(m-i_D)^q (\Delta\xi)^{q-1}}{q!} \left(\frac{\partial^q \phi}{\partial \xi^q}\right)_{i_A,j}^{L}$$
$$+ \sum_{m=i_D+1}^{M_2^{\xi}} b_m^{\xi} \sum_{q=1}^{\infty} \frac{(m-i_D)^q (\Delta\xi)^{q-1}}{q!} \left(\frac{\partial^q \phi}{\partial \xi^q}\right)_{i_A,j}^{R}. \tag{2.10c}$$

Here the superscript in b_m^ξ denotes the coefficients of the partial derivatives in ξ direction. Proposition 2.3 can be easily proved based on single-side Taylor expanding (2.3) and Proposition 2.2.

For convenient and without loss of generality, hereinafter we assume that the computational point (i,j) locates at the left side of the discontinuous point (i_A,j) and satisfies that $i_A = i + i_D$ with $i_D > 0$.

For rth-order difference operator,

$$\sum_{m=-M_1^\xi}^{M_2^\xi} b_m^\xi \sum_{q=1}^{\infty} \frac{m^q (\Delta\xi)^{q-1}}{q!} \left(\frac{\partial^q \phi}{\partial \xi^q}\right)_{i,j} = \frac{\partial \phi}{\partial \xi} + \mathcal{O}((\Delta\xi)^r). \tag{2.11}$$

Inserting (2.11) into (2.10), we can get

$$\delta^\xi \phi_{i,j} = \left(\frac{\partial \phi}{\partial \xi}\right)_{i,j} + \sum_{m=i_D+1}^{M_2^\xi} b_m^\xi \sum_{q=k}^{\infty} \frac{(m-i_D)^q (\Delta\xi)^{q-1}}{q!} \left(\frac{\partial^q \phi}{\partial \xi^q}\right)_{i_A,j}^{R-L} + \mathcal{O}((\Delta\xi)^r). \tag{2.12}$$

Proposition 2.4. *If u is an infinitely differentiable function about x and y, and i_A is the kth-order ξ-discontinuous point of x, y. Then, the following relation holds*

$$\left(\frac{\partial^k u}{\partial \xi^k}\right)_{i_A,j}^{R-L} = \left(\frac{\partial u}{\partial y}\right)_{i_A,j} \left(\frac{\partial^k y}{\partial \xi^k}\right)_{i_A,j}^{R-L} + \left(\frac{\partial u}{\partial x}\right)_{i_A,j} \left(\frac{\partial^k x}{\partial \xi^k}\right)_{i_A,j}^{R-L}. \tag{2.13}$$

Proof. Firstly, expanding $\frac{\partial^k u}{\partial \xi^k}$ by chain rule, we notice that the terms which contain the kth partial derivatives of x and y about ξ are $\frac{\partial u}{\partial x}\frac{\partial^k x}{\partial \xi^k}$ and $\frac{\partial u}{\partial y}\frac{\partial^k y}{\partial \xi^k}$. In addition, according to the kth partial derivative discontinuity in Definition 2.1, we have

$$\left(\frac{\partial^q x}{\partial \xi^q}\right)_{i_A,j}^R = \left(\frac{\partial^q x}{\partial \xi^q}\right)_{i_A,j}^L, \quad \left(\frac{\partial^q y}{\partial \xi^q}\right)_{i_A,j}^R = \left(\frac{\partial^q y}{\partial \xi^q}\right)_{i_A,j}^L, \quad q \leq k-1,$$

$$\left(\frac{\partial^k x}{\partial \xi^k}\right)_{i_A,j}^R \neq \left(\frac{\partial^k x}{\partial \xi^k}\right)_{i_A,j}^L, \quad \left(\frac{\partial^k y}{\partial \xi^k}\right)_{i_A,j}^R \neq \left(\frac{\partial^k y}{\partial \xi^k}\right)_{i_A,j}^L.$$

Therefore, (2.13) is obtained immediately. $\square$

If u is an infinitely differentiable function about y, and i_A is the kth-order ξ-discontinuous point of y, then the relation (2.13) reduces to

$$\left(\frac{\partial^k u}{\partial \xi^k}\right)_{i_A,j}^{R-L} = \left(\frac{\partial u}{\partial y}\right)_{i_A,j} \left(\frac{\partial^k y}{\partial \xi^k}\right)_{i_A,j}^{R-L}.$$

Remark 2.1. Under the transformation in (2.1), $\frac{\partial^p \phi}{\partial \eta^p}$ ($p \geq 1$) have the same order of ξ-discontinuity as ϕ, and $\frac{\partial^q \phi}{\partial \xi^q}$ ($q \geq 1$) have the same order of η-discontinuity as ϕ. Here ϕ is the coordinate x or y.

3 GCL discretization errors

If a FDS does not satisfy the GCL, the GCL errors may cause many unpleasant problems like numerical oscillations and free-stream violation. The GCL errors of GCL-FD and NGCL-FD are analyzed in this section.

Two-dimensional transformation from Cartesian coordinates $x-y$ to curvilinear coordinates $\xi-\eta$ shall satisfy the following differential GCL

$$I_x = \frac{\partial \widehat{\xi}_x}{\partial \xi} + \frac{\partial \widehat{\eta}_x}{\partial \eta} = 0, \quad I_y = \frac{\partial \widehat{\xi}_y}{\partial \xi} + \frac{\partial \widehat{\eta}_y}{\partial \eta} = 0, \tag{3.1}$$

where

$$\begin{cases} \widehat{\xi}_x = J\xi_x = y_\eta, \\ \widehat{\xi}_y = J\xi_y = -x_\eta, \end{cases} \quad \begin{cases} \widehat{\eta}_x = J\eta_x = -y_\xi, \\ \widehat{\eta}_y = J\eta_y = x_\xi. \end{cases} \tag{3.2}$$

Here the grid metrics are

$$J = \left| \frac{\partial(x,y)}{\partial(\xi,\eta)} \right| = x_\xi y_\eta - x_\eta y_\xi. \tag{3.3}$$

According to (3.2), the discrete form of the GCL (3.1) can be written as

$$\widetilde{I}_x = \delta_1^\xi(\delta_2^\eta y) - \delta_1^\eta(\delta_2^\xi y), \quad \widetilde{I}_y = -\delta_1^\xi(\delta_2^\eta x) + \delta_1^\eta(\delta_2^\xi x), \tag{3.4}$$

where δ_1 and δ_2 are called the first and second difference operator of (3.1). Here $\widetilde{(\cdot)}$ represents the discrete form of $(\cdot)$. It is worth noting that δ_2 is used to compute the grid metrics while δ_1 is used to compute flux derivatives. In curvilinear coordinate, since flux derivatives contain grid metrics, δ_1 also operates the metrics. A general form of δ_1 and δ_2 can be written as,

$$\delta_1^\xi \psi_i = \frac{1}{\Delta \xi} \sum_{l=-L_1^\xi}^{L_2^\xi} a_l^\xi (\psi_{i+l} - \psi_i), \quad \delta_2^\xi \phi_i = \frac{1}{\Delta \xi} \sum_{m=-M_1^\xi}^{M_2^\xi} b_m^\xi (\phi_{i+m} - \phi_i), \tag{3.5a}$$

$$\delta_1^\eta \psi_j = \frac{1}{\Delta \eta} \sum_{l=-L_1^\eta}^{L_2^\eta} a_l^\eta (\psi_{j+l} - \psi_j), \quad \delta_2^\eta \phi_j = \frac{1}{\Delta \eta} \sum_{m=-M_1^\eta}^{M_2^\eta} b_m^\eta (\phi_{j+m} - \phi_j). \tag{3.5b}$$

The discrete GCL errors of GCL-FD and NGCL-FD are given in the following theorem.

Theorem 3.1. *Assume the grid is kth-order discontinuous, and x,y are sufficiently smooth in η direction but kth-order ξ-discontinuous. Suppose the finite difference operator has rth-order accuracy. Then, we have*

(i) *For GCL-FD, which satisfies the CMM condition as $\delta_1^\xi = \delta_2^\xi$ and $\delta_1^\eta = \delta_2^\eta$, then $\widetilde{I}_x = 0$ and $\widetilde{I}_y = 0$.*

(ii) *For NGCL-FD, which does not satisfy the above CMM condition, the GCL errors are*

$$(\widetilde{I}_x)_{i,j} = (\widetilde{I}_y)_{i,j} = \mathcal{O}((\Delta\xi)^{\min\{r,k-1\}}, (\Delta\eta)^r). \tag{3.6}$$

Proof. Using the results (2.10) in Proposition 2.3, we obtain

$$(\widetilde{I}_x)_{i,j} = \delta_1^\xi(\delta_2^\eta y_{i,j}) - \delta_1^\eta(\delta_2^\xi y_{i,j})$$

$$= \frac{1}{\Delta\xi} \sum_{l=-L_1^\xi}^{L_2^\xi} a_l^\xi (\delta_2^\eta y_{i+l,j} - \delta_2^\eta y_{i,j}) - \frac{1}{\Delta\eta} \sum_{l=-L_1^\eta}^{L_2^\eta} a_l^\eta (\delta_2^\xi y_{i,j+l} - \delta_2^\xi y_{i,j})$$

$$= \frac{1}{\Delta\xi} \sum_{l=-L_1^\xi}^{L_2^\xi} a_l^\xi \sum_{m=-M_1^\eta}^{M_2^\eta} b_m^\eta \sum_{p=1}^\infty \frac{m^p (\Delta\eta)^{p-1}}{p!} \left[\left(\frac{\partial^p y}{\partial \eta^p}\right)_{i+l,j} - \left(\frac{\partial^p y}{\partial \eta^p}\right)_{i,j} \right]$$

$$- \frac{1}{\Delta\eta} \sum_{l=-L_1^\eta}^{L_2^\eta} a_l^\eta \sum_{m=-M_1^\xi}^{M_2^\xi} b_m^\xi \sum_{q=1}^\infty \frac{m^q (\Delta\xi)^{q-1}}{q!} \left[\left(\frac{\partial^q y}{\partial \xi^q}\right)_{i,j+l} - \left(\frac{\partial^q y}{\partial \xi^q}\right)_{i,j} \right]$$

$$- \frac{1}{\Delta\eta} \sum_{l=-L_1^\eta}^{L_2^\eta} a_l^\eta \sum_{m=i_D+1}^{M_2^\xi} b_m^\xi \sum_{q=1}^\infty \frac{(m-i_D)^q (\Delta\xi)^{q-1}}{q!} \times \left[\left(\frac{\partial^q y}{\partial \xi^q}\right)_{i_A,j+l}^{R-L} - \left(\frac{\partial^q y}{\partial \xi^q}\right)_{i_A,j}^{R-L} \right]. \tag{3.7}$$

Then, by using Taylor series expansion in η direction and Proposition 2.3 once again, we can obtain

$$(\widetilde{I}_x)_{i,j} = \widetilde{I}_{sm} + \widetilde{I}_{nsm}, \tag{3.8}$$

where

$$\widetilde{I}_{sm} = \sum_{p=1}^\infty \sum_{q=1}^\infty \frac{(\Delta\eta)^{p-1}(\Delta\xi)^{q-1}}{p!q!} \left(\frac{\partial^{p+q} y}{\partial \eta^p \partial \xi^q}\right)_{i,j}$$

$$\times \left(\sum_{l=-L_1^\xi}^{L_2^\xi} a_l^\xi l^q \sum_{m=-M_1^\eta}^{M_2^\eta} b_m^\eta m^p - \sum_{l=-L_1^\eta}^{L_2^\eta} a_l^\eta l^p \sum_{m=-M_1^\xi}^{M_2^\xi} b_m^\xi m^q \right),$$

$$\widetilde{I}_{nsm} = \sum_{p=1}^\infty \sum_{q=1}^\infty \frac{(\Delta\xi)^{q-1}}{q!} \frac{(\Delta\eta)^{p-1}}{p!} \left(\frac{\partial^{p+q} y}{\partial \xi^q \partial \eta^p}\right)_{i_A,j}^{R-L}$$

$$\times \left(\sum_{l=i_D+1}^{L_2^\xi} a_l^\xi (l-i_D)^q \sum_{m=-M_1^\eta}^{M_2^\eta} b_m^\eta m^p - \sum_{l=-L_1^\eta}^{L_2^\eta} a_l^\eta l^p \sum_{m=i_D+1}^{M_2^\xi} b_m^\xi (m-i_D)^q \right).$$

(i) Since $\delta_1^\xi = \delta_2^\xi$ and $\delta_1^\eta = \delta_2^\eta$, both of $\widetilde{I}_{sm}$ and $\widetilde{I}_{nsm}$ are equal to zero. Therefore, we have $\widetilde{I}_x = 0$ for GCL-FD no matter the grid is smooth or not.

(ii) For NGCL-FD, based on the coefficient relations (2.8) and note that the order of discontinuity of $\frac{\partial x}{\partial \eta}$ and $\frac{\partial y}{\partial \eta}$ are the same with x and y correspondingly in ξ direction, the main error of $\tilde{I}_{sm}$ and $\tilde{I}_{nsm}$ becomes

$$\left(\sum_{m=-M_1^\eta}^{M_2^\eta} b_m^\eta m^{r+1} - \sum_{l=-L_1^\eta}^{L_2^\eta} a_l^\eta l^{r+1}\right) \frac{(\Delta\eta)^r}{(r+1)!} \left(\frac{\partial^{r+2}y}{\partial\eta^{r+1}\partial\xi}\right)_{i,j}$$
$$+ \left(\sum_{l=-L_1^\xi}^{L_2^\xi} a_l^\xi l^{r+1} - \sum_{m=-M_1^\xi}^{M_2^\xi} b_m^\xi m^{r+1}\right) \frac{(\Delta\xi)^r}{(r+1)!} \left(\frac{\partial^{r+2}y}{\partial\xi^{r+1}\partial\eta}\right)_{i,j} = \mathcal{O}((\Delta\xi)^r,(\Delta\eta)^r),$$

and

$$\left(\sum_{l=i_D+1}^{L_2^\xi} a_l^\xi (l-i_D)^k - \sum_{m=i_D+1}^{M_2^\xi} b_m^\xi (m-i_D)^k\right) \frac{(\Delta\xi)^{k-1}}{k!} \left(\frac{\partial^{1+k}y}{\partial\xi^k\partial\eta}\right)_{i_A,j}^{R-L} = \mathcal{O}((\Delta\xi)^{k-1}).$$

Thus, the GCL error is

$$(\tilde{I}_x)_{i,j} = \mathcal{O}\left((\Delta\xi)^{\min\{r,k-1\}},(\Delta\eta)^r\right). \tag{3.9}$$

By similar deductive procedure, the GCL error of I_y can be obtained by replacing y in (3.9) with x. Thus, we have $\tilde{I}_y=0$ for GCL-FD and $\tilde{I}_y=\mathcal{O}((\Delta\xi)^{\min\{r,k-1\}},(\Delta\eta)^r)$ for NGCL-FD. □

Then it is evident that the convergence order of the GCL errors for NGCL-FDr is $\min\{k-1,r\}$. From (3.6), we note that if the grid is sufficiently smooth ($k\geq r+1$), the convergence order of the GCL errors is equal to the spatial accuracy order of the finite difference operator.

Proposition 3.1. For a C^k grid (defined in 2.3) where x,y have discontinuities in both ξ and η direction, the GCL errors of NGCL-FDr are determined by the lowest order of discontinuities and the convergence order is also $\min\{k-1,r\}$.

This proposition can be proved by using Taylor series expansion and considering both the ξ-partial discontinuity and the η-partial discontinuity simultaneously. Details of the proof process are omitted in this paper to save space.

In addition, the relation between L_m norm errors and L_∞ norm errors is also derived. For sufficiently smooth grids, the convergence rates of L_m errors are the same as L_∞ errors, while for nonsmooth grids which have a finite number of discontinuities, the relation becomes $Order(L_m)=Order(L_\infty)+1/m$. The derivation process of the relation is not detailed here.

4 Spatial discretization errors

In this section, we analyze spatial discretization errors of GCL-FD and NGCL-FD based on nonsmooth grids.

Consider the following conservation law

$$\frac{\partial u}{\partial t} + \frac{\partial E(u)}{\partial x} + \frac{\partial F(u)}{\partial y} = 0. \tag{4.1}$$

By using curvilinear coordinates transformation (2.1), (4.1) becomes

$$\frac{\partial Ju}{\partial t} + \frac{\partial (E\hat{\xi}_x + F\hat{\xi}_y)}{\partial \xi} + \frac{\partial (E\hat{\eta}_x + F\hat{\eta}_y)}{\partial \eta} = 0. \tag{4.2}$$

Substituting (3.2) into (4.2), we obtain

$$\frac{\partial u}{\partial t} + \frac{A}{J} = 0, \tag{4.3}$$

where A is the flux part

$$A = \left[\frac{\partial (Ey_\eta - Fx_\eta)}{\partial \xi} + \frac{\partial (Fx_\xi - Ey_\xi)}{\partial \eta} \right]. \tag{4.4}$$

We split the flux part A into two parts, i.e.,

$$A = A_E + A_F \tag{4.5}$$

with the E-flux part

$$A_E = (Ey_\eta)_\xi - (Ey_\xi)_\eta \tag{4.6}$$

and the F-flux part

$$A_F = -\left((Fx_\eta)_\xi - (Fx_\xi)_\eta\right). \tag{4.7}$$

In the following, we will analyze the discretization errors of A and J, respectively.

4.1 Discretization errors of the flux part

Now we analyze the discretization errors of A. Firstly, the errors of A_E and A_F will be analyzed separately. Then, the discretization errors of A will be obtained by adding the errors of A_E and A_F.

4.1.1 Discretization errors of the E-flux part

By using the operators in (3.5), the discretization form of the E-flux part (4.6) reads

$$\left(\widetilde{A_E}\right)_{i,j} = \delta_1^\xi (E\delta_2^\eta y)_{i,j} - \delta_1^\eta (E\delta_2^\xi y)_{i,j} = (A_E)_{i,j} + (e_{A_E})_{i,j}. \tag{4.8}$$

At the end, the discretization errors of A_E can be evaluated as (see Appendix for details)

$$(e_{A_E})_{i,j} = e_{GCL}^E + e_{nsm1}^E + e_{nsm2}^E + \mathcal{O}\left((\Delta\xi)^{\min\{r,k\}}, (\Delta\eta)^r\right), \tag{4.9}$$

where

$$e_{GCL}^E = (\widetilde{I}_x)_{i,j} E_{i,j},$$

$$e_{nsm1}^E = \left(\frac{\partial E}{\partial y}\right)_{i,j} \left(\frac{\partial y}{\partial \eta}\right)_{i,j} \left(\frac{\partial^k y}{\partial \xi^k}\right)_{i_A,j}^{R-L} \frac{(\Delta\xi)^{k-1}}{k!} \left[\sum_{l=i_D+1}^{L_2^\xi} a_l^\xi (l-i_D)^k - \sum_{m=i_D+1}^{M_2^\xi} b_m^\xi (m-i_D)^k\right],$$

$$e_{nsm2}^E = \left(\frac{\partial E}{\partial x}\right)_{i,j} \left(\frac{\partial y}{\partial \eta}\right)_{i,j} \left(\frac{\partial^k x}{\partial \xi^k}\right)_{i_A,j}^{R-L} \frac{(\Delta\xi)^{k-1}}{k!} \sum_{l=i_D+1}^{L_2^\xi} a_l^\xi (l-i_D)^k$$

$$- \left(\frac{\partial E}{\partial x}\right)_{i,j} \left(\frac{\partial x}{\partial \eta}\right)_{i,j} \left(\frac{\partial^k y}{\partial \xi^k}\right)_{i_A,j}^{R-L} \frac{(\Delta\xi)^{k-1}}{k!} \sum_{m=i_D+1}^{M_2^\xi} b_m^\xi (m-i_D)^k.$$

If a GCL-FD is applied to the flux part (4.4), which means that the difference operators satisfy the conditions of $\delta_1^\xi = \delta_2^\xi$ and $\delta_1^\eta = \delta_2^\eta$. Then, according to Theorem 3.1(i), the GCL error is equal to zero, thus

$$e_{GCL}^E = 0.$$

Furthermore, since $\delta_1^\xi = \delta_2^\xi$, we have

$$\sum_{l=i_D+1}^{L_2^\xi} a_l^\xi (l-i_D)^k - \sum_{m=i_D+1}^{M_2^\xi} b_m^\xi (m-i_D)^k = 0. \tag{4.10}$$

Thus $e_{nsm1}^E = 0$. Moreover, from (4.10), e_{nsm2}^E can be simplified as

$$e_{nsm2}^E = \frac{(\Delta\xi)^{k-1}}{k!} \left(\frac{\partial E}{\partial x}\right)_{i,j} g_{i,j} \sum_{l=i_D+1}^{L_2^\xi} a_l^\xi (l-i_D)^k,$$

where

$$g_{i,j} = \left(\frac{\partial y}{\partial \eta}\right)_{i,j} \left(\frac{\partial^k x}{\partial \xi^k}\right)_{i_A,j}^{R-L} - \left(\frac{\partial x}{\partial \eta}\right)_{i,j} \left(\frac{\partial^k y}{\partial \xi^k}\right)_{i_A,j}^{R-L}. \tag{4.11}$$

Therefore, e_{A_E} becomes

$$(e_{A_E})_{i,j} = \mathcal{O}\left((\Delta\xi)^{\min\{r,k\}}, (\Delta\eta)^r\right) + \frac{(\Delta\xi)^{k-1}}{k!} \left(\frac{\partial E}{\partial x}\right)_{i,j} g_{i,j} \sum_{l=i_D+1}^{L_2^\xi} a_l^\xi (l-i_D)^k.$$

From (4.9) we can see that, for sufficiently smooth grid ($k \geq r+1$), the E flux part of GCL-FD does not have the error term e_{GCL}^E, which is of $\mathcal{O}((\Delta\xi)^r, (\Delta\eta)^r)$ according to Theorem 3.1(ii). Therefore, satisfying GCL may help to reduce numerical errors even for sufficiently smooth grid.

4.1.2 Discretization errors of the F-flux part

In a similar deductive procedure, the discretization errors of the F-flux part can be obtained,

$$(e_{A_F})_{i,j} = e_{GCL}^F + e_{nsm1}^F + e_{nsm2}^F + \mathcal{O}\left((\Delta\xi)^{\min\{r,k\}}, (\Delta\eta)^r\right), \tag{4.12}$$

where

$$e_{GCL}^F = (\tilde{I}_y)_{i,j} F_{i,j},$$

$$e_{nsm1}^F = -\frac{(\Delta\xi)^{k-1}}{k!}\left[\sum_{l=i_D+1}^{L_2^\xi} a_l^\xi (l-i_D)^k - \sum_{m=i_D+1}^{M_2^\xi} b_m^\xi (m-i_D)^k\right]\left(\frac{\partial F}{\partial x}\right)_{i,j}\left(\frac{\partial x}{\partial \eta}\right)_{i,j}\left(\frac{\partial^k x}{\partial \xi^k}\right)_{i_A,j}^{R-L},$$

$$e_{nsm2}^F = -\frac{(\Delta\xi)^{k-1}}{k!}\sum_{l=i_D+1}^{L_2^\xi} a_l^\xi (l-i_D)^k \left(\frac{\partial F}{\partial y}\right)_{i,j}\left(\frac{\partial x}{\partial \eta}\right)_{i,j}\left(\frac{\partial^k y}{\partial \xi^k}\right)_{i_A,j}^{R-L}$$

$$+\frac{(\Delta\xi)^{k-1}}{k!}\sum_{m=i_D+1}^{M_2^\xi} b_m^\xi (m-i_D)^k \left(\frac{\partial F}{\partial y}\right)_{i,j}\left(\frac{\partial y}{\partial \eta}\right)_{i,j}\left(\frac{\partial^k x}{\partial \xi^k}\right)_{i_A,j}^{R-L}.$$

If GCL-FD is applied to the flux part (4.4), namely $\delta_1^\xi = \delta_2^\xi$ and $\delta_1^\eta = \delta_2^\eta$, we have

$$(e_{A_F})_{i,j} = \mathcal{O}\left((\Delta\xi)^{\min\{r,k\}}, (\Delta\eta)^r\right) + \frac{(\Delta\xi)^{k-1}}{k!}\left(\frac{\partial F}{\partial y}\right)_{i,j} g_{i,j} \sum_{l=i_D+1}^{L_2^\xi} a_l^\xi (l-i_D)^k.$$

For sufficiently smooth grid ($k \geq r+1$), the F-flux part of GCL-FD does not have the error term e_{GCL}^F, which is of $\mathcal{O}((\Delta\xi)^r, (\Delta\eta)^r)$.

4.1.3 Discretization errors of the flux part

The discretization errors of A can be obtained by adding the errors of A_E and A_F, i.e.,

$$(e_A)_{i,j} = (e_{A_E})_{i,j} + (e_{A_F})_{i,j}, \tag{4.13}$$

where $(e_{A_E})_{i,j}$ and $(e_{A_F})_{i,j}$ are given in (4.9) and (4.12), respectively.

If $\delta_1^\xi = \delta_2^\xi$, then we have

$$(e_A)_{i,j} = \mathcal{O}\left((\Delta\xi)^{\min\{r,k\}}, (\Delta\eta)^r\right)$$

$$+\frac{(\Delta\xi)^{k-1}}{k!}\left(\left(\frac{\partial E}{\partial x}\right)_{i,j} + \left(\frac{\partial F}{\partial y}\right)_{i,j}\right) g_{i,j} \sum_{l=i_D+1}^{L_2^\xi} a_l^\xi (l-i_D)^k. \tag{4.14}$$

For sufficiently smooth grid ($k \geq r+1$), the flux part of GCL-FD does not have the error term $e_{GCL}^E + e_{GCL}^F$, which is of $\mathcal{O}((\Delta\xi)^r, (\Delta\eta)^r)$.

We compare the error of NGCL-FD (4.13) with the error of GCL-FD (4.14). For unsteady problem $\left(\frac{\partial E}{\partial x}\right)_{i,j}+\left(\frac{\partial F}{\partial y}\right)_{i,j}\neq 0$, $(e_A)_{i,j}$ is of $\mathcal{O}((\Delta\xi)^{\min\{r,k-1\}},(\Delta\eta)^r)$ for both NGCL-FD and GCL-FD, but the latter one has less error terms. For steady problem we have

$$(e_A)_{i,j}=\begin{cases} \mathcal{O}\left((\Delta\xi)^{\min\{r,k-1\}},(\Delta\eta)^r\right), & \text{for NGCL-FD,} \\ \mathcal{O}\left((\Delta\xi)^{\min\{r,k\}},(\Delta\eta)^r\right), & \text{for GCL-FD.} \end{cases} \quad (4.15)$$

4.2 Discretization errors of Jacobian

The Jacobian J defined in (3.3) has two basic equivalent conservation forms,

$$J_0=(xy_\eta)_\xi-(xy_\xi)_\eta, \quad J_1=(yx_\xi)_\eta-(yx_\eta)_\xi.$$

In fact, we have the following general form of the Jacobian,

$$J_a=(1-a)J_0+aJ_1, \quad (4.16)$$

where $0\leq a\leq 1$. The discretization form of J_0 is

$$\left(\widetilde{J_0}\right)_{i,j}=\delta_1^\xi(x\delta_2^\eta y)_{i,j}-\delta_1^\eta(x\delta_2^\xi y)_{i,j}=(J_0)_{i,j}+(e_{J_0})_{i,j}.$$

Replacing E in (4.9) with x and noticing that $\frac{\partial y}{\partial x}=0$, $\frac{\partial x}{\partial x}=1$, $\frac{\partial x}{\partial y}=0$ and $\frac{\partial y}{\partial y}=1$, we obtain

$$(e_{J_0})_{i,j}=(\widetilde{I}_x)_{i,j}x_{i,j}+\mathcal{O}\left((\Delta\xi)^{\min\{r,k\}},(\Delta\eta)^r\right)$$

$$+\frac{(\Delta\xi)^{k-1}}{k!}\sum_{l=i_D+1}^{L_2^\xi}a_l^\xi(l-i_D)^k\left(\frac{\partial y}{\partial \eta}\right)_{i,j}\left(\frac{\partial^k x}{\partial \xi^k}\right)_{i_A,j}^{R-L}$$

$$-\frac{(\Delta\xi)^{k-1}}{k!}\sum_{m=i_D+1}^{M_2^\xi}b_m^\xi(m-i_D)^k\left(\frac{\partial x}{\partial \eta}\right)_{i,j}\left(\frac{\partial^k y}{\partial \xi^k}\right)_{i_A,j}^{R-L}. \quad (4.17)$$

And the discretization form of J_1 is

$$\widetilde{J_1}=\delta_1^\eta(y\delta_2^\xi x)-\delta_1^\xi(y\delta_2^\eta x)=(J_1)_{i,j}+(e_{J_1})_{i,j}.$$

Similarly, e_{J_1} can be obtained by replacing F in (4.12) with y,

$$(e_{J_1})_{i,j}=-\left\{-(\widetilde{I}_y)_{i,j}y_{i,j}+\mathcal{O}\left((\Delta\xi)^{\min\{r,k\}},(\Delta\eta)^r\right)\right.$$

$$+\frac{(\Delta\xi)^{k-1}}{k!}\sum_{l=i_D+1}^{L_2^\xi}a_l^\xi(l-i_D)^k\left(\frac{\partial x}{\partial \eta}\right)_{i,j}\left(\frac{\partial^k y}{\partial \xi^k}\right)_{i_A,j}^{R-L}$$

$$\left.-\frac{(\Delta\xi)^{k-1}}{k!}\sum_{m=i_D+1}^{M_2^\xi}b_m^\xi(m-i_D)^k\left(\frac{\partial y}{\partial \eta}\right)_{i,j}\left(\frac{\partial^k x}{\partial \xi^k}\right)_{i_A,j}^{R-L}\right\}. \quad (4.18)$$

From (4.16) we know that

$$(e_{J_a})_{i,j} = (1-a)(e_{J_0})_{i,j} + a(e_{J_1})_{i,j}. \tag{4.19}$$

In particular, when $\delta_1^{\xi} = \delta_2^{\xi}$, we have

$$(e_{J_a})_{i,j} = \mathcal{O}\left((\Delta\xi)^{\min\{r,k\}}, (\Delta\eta)^r\right) + \frac{(\Delta\xi)^{k-1}}{k!} g_{i,j} \sum_{l=i_D+1}^{L_2^{\xi}} a_l^{\xi}(l-i_D)^k, \tag{4.20}$$

where $g_{i,j}$ is given in (4.11). In fact, the main discretization errors of different Jacobians J_a ($0 \le a \le 1$) are all the same for the case $\delta_1^{\xi} = \delta_2^{\xi}$, but are different for the case $\delta_1^{\xi} \ne \delta_2^{\xi}$.

4.3 Spatial discretization errors

The errors of the flux part and Jacobian can be written as

$$\widetilde{A} = A + e_A, \quad \widetilde{J} = J + e_J.$$

Hence, the spatial discretization errors reads

$$e = \frac{\widetilde{A}}{\widetilde{J}} - \frac{A}{J} = \frac{A + e_A}{J + e_J} - \frac{A}{J} = \frac{Je_A - Ae_J}{J(J + e_J)}. \tag{4.21}$$

Firstly, we consider the unsteady problem $A \ne 0$. From (4.13) and (4.19), it is easy to observe that $Je_A - Ae_J$ is of $\mathcal{O}\left((\Delta\xi)^{\min\{r,k-1\}}, (\Delta\eta)^r\right)$ for NGCL-FD. Since GCL-FD satisfies CMM condition $\delta_1^{\xi} = \delta_2^{\xi}$ and $\delta_1^{\eta} = \delta_2^{\eta}$, from (4.14) and (4.20) we have

$$Je_A - Ae_J = \frac{(\Delta\xi)^{k-1}}{k!} J_{i,j} \left[\left(\frac{\partial E}{\partial x}\right)_{i,j} + \left(\frac{\partial F}{\partial y}\right)_{i,j}\right] g_{i,j} \sum_{l=i_D+1}^{L_2^{\xi}} a_l^{\xi}(l-i_D)^k$$
$$- \frac{(\Delta\xi)^{k-1}}{k!} \left[\frac{\partial(Ey_\eta - Fx_\eta)}{\partial \xi} + \frac{\partial(Fx_\xi - Ey_\xi)}{\partial \eta}\right]_{i,j} g_{i,j} \sum_{l=i_D+1}^{L_2^{\xi}} a_l^{\xi}(l-i_D)^k$$
$$+ \mathcal{O}\left((\Delta\xi)^{\min\{r,k\}}, (\Delta\eta)^r\right)$$
$$= \mathcal{O}\left((\Delta\xi)^{\min\{r,k\}}, (\Delta\eta)^r\right).$$

Hence, for GCL-FD we get the error estimate

$$e_{i,j} = \mathcal{O}\left((\Delta\xi)^{\min\{r,k\}}, (\Delta\eta)^r\right). \tag{4.22}$$

Therefore, we have the following error estimate

$$e_{i,j} = \begin{cases} \mathcal{O}\left((\Delta\xi)^{\min\{r,k-1\}}, (\Delta\eta)^r\right), & \text{for NGCL-FD,} \\ \mathcal{O}\left((\Delta\xi)^{\min\{r,k\}}, (\Delta\eta)^r\right), & \text{for GCL-FD.} \end{cases} \quad (4.23)$$

For steady problem, i.e., $A=0$, we have $e = e_A/(J+e_J)$ from (4.21). In this case, the spatial discretization errors are determined by the discretization errors of the flux part e_A and thus the result (4.23) also holds according to (4.15). Therefore, the result (4.23) holds for both steady and unsteady problems.

The above theoretical result can be concluded as the following theorem.

Theorem 4.1. *For C^{k-1} grid under the hypothesis of Theorem 3.1, the spatial discretization errors by (r)th-order finite difference schemes have the following relations:*

(i) *For nonsmooth grid ($k \leq r$), the spatial discretization accuracy approaches (k)th-order by GCL-FDr while only $(k-1)$th-order by NGCL-FDr. Thus, GCL-FDr can obtain one-order higher spatial discretization accuracy than NGCL-FDr.*

(ii) *For sufficiently smooth grid ($k > r$), the spatial discretization error term $((\widetilde{I}_x)_{i,j} E_{i,j} + (\widetilde{I}_y)_{i,j} F_{i,j})/J_{i,j}$, which is of $\mathcal{O}((\Delta\xi)^r, (\Delta\eta)^r)$, vanishes by GCL-FDr and maybe not by NGCL-FDr. Thus, GCL-FDr tends to obtain smaller spatial discretization errors than NGCL-FDr.*

For example, for a grid with $k=2$ (C^1 grid), GCL-FD2 can obtain second-order accuracy while NGCL-FD2 has only first-order accuracy.

Proposition 4.1. *For C^{k-1} grid in Definition 2.3, where x, y have discontinuities in both ξ and η direction, the spatial discretization error is determined by the lowest order of discontinuities and the results in Theorem 4.1 also hold.*

5 Numerical investigation

In this section, GCL errors and spatial discretization errors are tested for GCL-FD and NGCL-FD. For numerical test, we consider second-order and fourth-order finite difference operators. In Section 5.1, we make tests on grids with discontinuities in one direction to verify the results in Theorem 3.1 and Theorem 4.1. Then we choose one kind of grids with discontinuities in two directions for test in Section 5.2 to verify the results in Proposition 3.1 and Proposition 4.1.

Consider second-order and fourth-order FD operators. The general form of second-order FD operator reads

$$\delta^\xi E_{i,j} = \frac{\alpha}{2\Delta\xi}(E_{i+1,j} - E_{i-1,j}) + \frac{1-\alpha}{4\Delta\xi}(E_{i+2,j} - E_{i-2,j}), \quad (5.1)$$

which is second-order for $\alpha \neq \frac{4}{3}$ and is fourth-order for $\alpha = \frac{4}{3}$. And the general form of fourth-order FD operator is

$$\delta^{\xi} E_{i,j} = \frac{\alpha}{2\Delta\xi}(E_{i+1,j} - E_{i-1,j}) + \frac{9-8\alpha}{20\Delta\xi}(E_{i+2,j} - E_{i-2,j}) + \frac{3\alpha-4}{30\Delta\xi}(E_{i+3,j} - E_{i-3,j}), \quad (5.2)$$

which is fourth-order for $\alpha \neq \frac{3}{2}$ and is sixth-order for $\alpha = \frac{3}{2}$. We take the operators δ_1 and δ_2 in (3.5) as the form (5.1) for second-order procedure (FD2) and the form (5.2) for fourth-order procedure (FD4). In addition, the parameter α in the operators are denoted as α_1 and α_2 for δ_1 and δ_2, respectively. For FD2 with the form (5.1), GCL-FD2 is obtained by taking the same parameter $\alpha_1 = \alpha_2 = \frac{2}{3}$ for δ_1 and δ_2, thus $\delta_1 = \delta_2$. As for NGCL-FD2, $\alpha_1 = \frac{2}{3}$ and $\alpha_2 = \frac{1}{3}$ are taken for δ_1 and δ_2, respectively, thus $\delta_1 \neq \delta_2$ and the GCL may not be hold. Similarly, for FD4 with (5.2), GCL-FD4 is obtained by taking $\alpha_1 = \alpha_2 = \frac{2}{3}$ and NGCL-FD4 by taking $\alpha_1 = \frac{2}{3}$, $\alpha_2 = \frac{1}{3}$.

5.1 Tests on grids with discontinuities in one direction

The generation of five kinds of grids with different smoothness in one direction is given in Subsection 5.1.1, and GCL errors and spatial discretization errors are given in Subsections 5.1.2 and 5.1.3, respectively. In Subsection 5.1.4, the vortex evolution problem is solved to demonstrate the errors and orders of accuracy at a fixed final time. Periodic boundary conditions are used.

5.1.1 Design of grids with different orders of smoothness

In order to test the impact of grid smoothness on GCL errors and numerical errors, grids with different orders of smoothness are generated. The grids are generated via the following coordinates transformation

$$x_{i,j} = \xi_i, \quad y_{i,j} = \eta_j + \psi_1(\xi_i)\psi_2(\eta_j), \quad \xi_i \in [-0.5, 0.5], \quad \eta_j \in [-0.5, 0.5], \quad (5.3)$$

where

$$\psi_1(\xi_i) = \begin{cases} \sin^k\left(\frac{5\pi}{3}\xi_i + \frac{\pi}{2}\right), & |\xi_i| < 0.3, \\ 0, & |\xi_i| \geq 0.3, \end{cases} \quad \psi_2(\eta_j) = A\sin^2\left(\pi\eta_j + \frac{\pi}{2}\right),$$

where A is amplitude of grid irregularity, $\xi_i = (i-1)\Delta\xi - 0.5$ and $\eta_j = (j-1)\Delta\eta - 0.5$. In the following, we choose $A = 0.1$. The grids with different discontinuity properties at $x = \pm 0.3$ can be produced by taking different k for $y_{i,j}$ in (5.3). These grids have the following properties:

(i) Satisfying periodic boundary conditions. The grid derivatives $x_\xi, x_\eta, y_\xi, y_\eta$ and the Jacobian J are periodic in both ξ and η directions.

(ii) The parameter k determines the order of discontinuity, or more exactly, y and $\frac{\partial^p y}{\partial \eta^p}$ have kth ξ-partial discontinuities in finite points, which means that $\frac{\partial^{p+k} y}{\partial \xi^k \partial \eta^p}$ is discontinuous for $p \geq 0$ at $\xi = \pm 0.3$. In addition, x is infinite differentiable in both ξ and η directions,

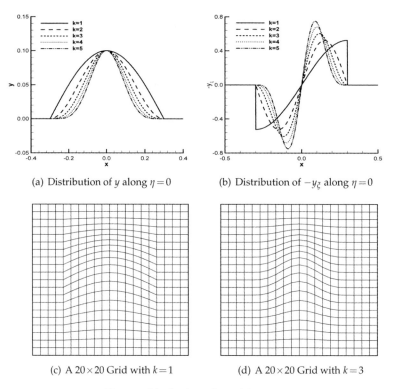

Figure 1: Distributions of y and $\hat{\eta}_x = -y_\xi$.

which means that $\partial^{p+q}x/\partial\xi^q\partial\eta^p$ is continuous for $p \geq 0$ and $q \geq 0$. Moreover, since $x_\eta = 0$ is satisfied everywhere, the Jacobian J is infinite differentiable in both ξ and η directions.

Figs. 1(a) and 1(b) show the distributions of $y_{i,j}$ and $\hat{\eta}_x = -y_\xi$ along the grid line $\eta = 0$ for $k=1,2,3,4,5$. We can see that the continuity property of y and y_ξ will change when the parameter k changes. For grids with the parameter k, the derivative $\frac{\partial^{k-1}y}{\partial\xi^{k-1}}$ is continuous in the whole line while $\frac{\partial^k y}{\partial\xi^k}$ is discontinuous at $x = \pm 0.3$. Figs. 1(c) and 1(d) shows two 20×20 grids with $k=1$ and $k=3$, correspondingly.

The theoretical analysis has shown that GCL-FDr cannot obtain their designed accuracy order unless the grid is at least $(r-1)$th-order smooth, and NGCL-FDr requires that the grid is at least rth-order smooth. In the next two subsections, the above findings will be verified by numerical tests based on FD2 and FD4. In order to cover the grids from nonsmooth grids to sufficiently smooth grids for the two kinds of schemes, five kinds of grids with $k=1,2,3,4,5$ are generated. Table 1 shows the grid discontinuity properties and the scheme requirements of GCL-FDr and NGCL-FDr.

Table 1: Relations between discontinuity orders of grids and the scheme requirements.

k	Grid smoothness at $x=\pm 0.3$ grid derivative $\partial^q y/\partial \xi^q$ continuous	discontinuous	Satisfy scheme requirement or not FD2 GCL	NGCL	FD4 GCL	NGCL
1	y	$\partial y/\partial \xi$	No	No	No	No
2	$\partial y/\partial \xi$	$\partial^2 y/\partial \xi^2$	Yes	No	No	No
3	$\partial^2 y/\partial \xi^2$	$\partial^3 y/\partial \xi^3$	Yes	Yes	No	No
4	$\partial^3 y/\partial \xi^3$	$\partial^4 y/\partial \xi^4$	Yes	Yes	Yes	No
5	$\partial^4 y/\partial \xi^4$	$\partial^5 y/\partial \xi^5$	Yes	Yes	Yes	Yes

5.1.2 Numerical test for GCL errors

In this subsection, we test the GCL errors of NGCL-FD for five kinds of grids with different smoothness of $k=1,2,3,4,5$ given in Subsection 5.1.1. The L_1, L_2 and L_∞ GCL errors on these grids are given in Tables 2 and 3. The results are further summarized in Table 4

Table 2: GCL errors $|\tilde{I}_x|$ for NGCL-FD2 on five kinds of grids.

NGCL	k	$N_\xi \times N_\eta$	L_1 error	order	L_2 error	order	L_∞ error	order
		300×300	1.17E-03	—	7.91E-03	—	6.85E-02	—
		500×500	7.01E-04	1.01	6.13E-03	0.50	6.85E-02	0.00
	1	800×800	4.38E-04	1.00	4.85E-03	0.50	6.85E-02	0.00
		1200×1200	2.91E-04	1.00	3.96E-03	0.50	6.85E-02	0.00
		2000×2000	1.75E-04	1.00	3.07E-03	0.50	6.85E-02	0.00
		300×300	9.24E-05	—	3.55E-04	—	4.19E-03	—
		500×500	3.33E-05	2.00	1.63E-04	1.52	2.51E-03	1.00
	2	800×800	1.30E-05	2.00	8.01E-05	1.51	1.57E-03	1.00
		1200×1200	5.79E-06	2.00	4.35E-05	1.51	1.05E-03	1.00
		2000×2000	2.08E-06	2.00	2.01E-05	1.51	6.28E-04	1.00
		300×300	1.15E-04	—	1.83E-04	—	4.97E-04	—
		500×500	4.15E-05	2.00	6.59E-05	2.00	1.80E-04	1.99
FD2	3	800×800	1.62E-05	2.00	2.58E-05	2.00	7.04E-05	2.00
		1200×1200	7.20E-06	2.00	1.15E-05	2.00	3.13E-05	2.00
		2000×2000	2.59E-06	2.00	4.13E-06	2.00	1.13E-05	2.00
		300×300	1.52E-04	—	2.49E-04	—	7.63E-04	—
		500×500	5.48E-05	2.00	8.98E-05	2.00	2.75E-04	2.00
	4	800×800	2.14E-05	2.00	3.51E-05	2.00	1.07E-04	2.00
		1200×1200	9.53E-06	2.00	1.56E-05	2.00	4.77E-05	2.00
		2000×2000	3.43E-06	2.00	5.62E-06	2.00	1.72E-05	2.00
		300×300	1.90E-04	—	3.30E-04	—	1.11E-03	—
		500×500	6.85E-05	2.00	1.19E-04	2.00	4.00E-04	2.00
	5	800×800	2.68E-05	2.00	4.65E-05	2.00	1.56E-04	2.00
		1200×1200	1.19E-05	2.00	2.07E-05	2.00	6.96E-05	2.00
		2000×2000	4.28E-06	2.00	7.44E-06	2.00	2.50E-05	2.00

Table 3: GCL errors $|\tilde{I}_x|$ for NGCL-FD4 on five kinds of grids.

NGCL	k	$N_\xi \times N_\eta$	L_1 error	order	L_2 error	order	L_∞ error	order
FD4	1	300×300	9.31E-04	–	5.48E-03	–	5.48E-02	–
		500×500	5.59E-04	1.00	4.25E-03	0.50	5.48E-02	0.00
		800×800	3.49E-04	1.00	3.36E-03	0.50	5.48E-02	0.00
		1200×1200	2.33E-04	1.00	2.74E-03	0.50	5.48E-02	0.00
		2000×2000	1.40E-04	1.00	2.12E-03	0.50	5.48E-02	0.00
	2	300×300	2.44E-05	–	1.54E-04	–	1.44E-03	–
		500×500	8.78E-06	2.00	7.15E-05	1.50	8.62E-04	1.00
		800×800	3.43E-06	2.00	3.53E-05	1.50	5.38E-04	1.00
		1200×1200	1.52E-06	2.00	1.92E-05	1.50	3.59E-04	1.00
		2000×2000	5.48E-07	2.00	8.94E-06	1.50	2.15E-04	1.00
	3	300×300	9.60E-07	–	5.46E-06	–	4.86E-05	–
		500×500	1.95E-07	3.12	1.52E-06	2.51	1.74E-05	2.02
		800×800	4.59E-08	3.08	4.68E-07	2.50	6.76E-06	2.01
		1200×1200	1.33E-08	3.05	1.70E-07	2.50	3.00E-06	2.00
		2000×2000	2.83E-09	3.03	4.73E-08	2.50	1.08E-06	2.00
	4	300×300	3.72E-07	–	6.85E-07	–	5.56E-06	–
		500×500	4.84E-08	3.99	1.02E-07	3.73	1.20E-06	3.00
		800×800	7.40E-09	4.00	1.81E-08	3.67	2.94E-07	3.00
		1200×1200	1.46E-09	4.00	4.16E-09	3.63	8.71E-08	3.00
		2000×2000	1.90E-10	4.00	6.64E-10	3.59	1.88E-08	3.00
	5	300×300	4.98E-07	–	8.17E-07	–	2.60E-06	–
		500×500	6.47E-08	4.00	1.06E-07	4.00	3.37E-07	4.00
		800×800	9.87E-09	4.00	1.62E-08	4.00	5.15E-08	4.00
		1200×1200	1.95E-09	4.00	3.20E-09	4.00	1.02E-08	4.00
		2000×2000	2.53E-10	4.00	4.15E-10	4.00	1.32E-09	4.00

to show the relation between the convergence rate of GCL errors, the grid discontinuity order k, and the accuracy order r of difference operators.

Firstly, we have a close look at the relation between grid smoothness and the convergent order of GCL errors. For the grids with $k=1$, we can see from Tables 2 and 3 that the order of L_1, L_2 and L_∞ GCL errors are 1, 0.5 and 0, respectively, for both NGCL-FD2 and NGCL-FD4. Similarly, for the other four kinds of grids, the L_1, L_2 and L_∞ GCL errors of NGCL-FDr ($r=2$ or $r=4$) converge in the rates of $\min(k,r)$, $\min(k-0.5,r)$ and $\min(k-1,r)$, respectively. The L_∞ errors agree with the theoretical results (3.9).

Secondly, It is worth noting that the GCL errors of the fourth-order scheme are generally lower than that of the second-order scheme, although the GCL errors of the two schemes show similar convergence rate on non-sufficiently smooth grids. Taking the grids with $k=2$ for an example, although the convergence rates of NGCL-FD2 and NGCL-FD4 are the same, the latter one produces smaller GCL errors than the former.

Thirdly, grids with $k=r+1$ (C^r grid) are sufficiently smooth for NGCL-FDr from the

Table 4: The correlation among the convergent rate of GCL errors $|\tilde{I}_x|$, the grid continuity order k, and the accuracy order r of difference operators.

	grid property		$|\tilde{I}_x|$					
			NGCL-FD2 ($r=2$)			NGCL-FD4 ($r=4$)		
k	grid coordinate order of continuity	grid derivative order of continuity	L_1	L_2	L_∞	L_1	L_2	L_∞
1	1st-order	zero-order	1.0	0.5	0.0	1.0	0.5	0.0
2	2nd-order	1st-order	2.0	1.5	1.0	2.0	1.5	1.0
3	3th-order	2nd-order	**2.0**	**2.0**	**2.0**	3.0	2.5	2.0
4	4th-order	3th-order	2.0	2.0	2.0	4.0	3.5	3.0
5	5th-order	4th-order	2.0	2.0	2.0	**4.0**	**4.0**	**4.0**

aspect of GCL errors. For example, the grids with $k=3$ are smooth enough for NGCL-FD2 to obtain second-order accuracy from the aspect of all the L_m ($m=1,2,\cdots,\infty$) GCL errors, and the grids with $k=5$ are smooth enough for the NGCL-FD4.

5.1.3 Numerical test for spatial discretization errors

In this subsection, spatial discretization errors of GCL-FD and NGCL-FD are tested on the five kinds of grids given in Subsection 5.1.1. Since y is discontinuous in η direction, we check the spatial discretization errors of $\frac{\partial E}{\partial x} = \frac{A_E}{J}$ and choose $E(x,y) = \sin(2\pi x)\sin(2\pi y)$.

The discretization errors and the accuracy orders for FD4 is listed in Tables 5 and 6. Firstly, when the grids are first-order discontinuous ($k=1$), the orders of L_1, L_2 and L_∞ errors are 2, 1.5 and 1 for GCL-FD4, but 1, 0.5 and 0 for NGCL-FD4. The L_∞ errors of NGCL-FD4 can not converge to zero as refining the grids because the discrete version of the spatial terms in Eq. (4.2) is inconsistent with Eq. (4.1) due to the strong non-smoothness of the grids. Secondly, it can be concluded that the orders of L_1, L_2 and L_∞ errors are $\min(k+1,r)$, $\min(k+0.5,r)$ and $\min(k,r)$ for GCL-FDr, but $\min(k,r)$, $\min(k-0.5,r)$ and $\min(k-1,r)$ for NGCL-FDr. In order to facilitate the analysis, numerical spatial errors by NGCL-FD4 and GCL-FD4 are also plotted together as shown in Fig. 2. This result indicates that for nonsmooth grids ($k \leq r$), the GCL-FDr has one order higher spatial accuracy than NGCL-FDr. Thirdly, to obtain the designed accuracy orders, the grids with $k=r$ (C^{r-1} grid) are smooth enough for GCL-FDr schemes while the grids with $k=r+1$ (C^r grid) are smooth enough for NGCL-FDr schemes. Fourthly, we can see from the results of the case $k=5$ that satisfying the GCL can still reduce the discretization errors for sufficiently smooth grids, which agrees with Theorem 4.1.

FD2 is also tested and the results are not listed here. It can be found that FD2 also have the above four properties. Table 7 gives the relation between numerical errors, the grid smoothness, and the accuracy orders of difference operator for FD2 and FD4. In addition, although GCL-FD2 and GCL-FD4 have the same accuracy on the grids with $k=1$, the former has larger numerical errors than the latter. This is similar to the comparison between NGCL-FD2 and NGCL-FD4. The results indicate that high-order schemes might

Table 5: Numerical errors $\left|\frac{\partial E}{\partial x}^N - \frac{\partial E}{\partial x}^{exact}\right|$ for GCL-FD4 on five kinds of grids.

FD4	k	$N_\xi \times N_\eta$	L_1 error	L_1 order	L_2 error	L_2 order	L_∞ error	L_∞ order
GCL	1	300×300	5.81E-05	—	4.26E-04	—	8.40E-03	—
		500×500	2.08E-05	2.01	1.98E-04	1.50	5.08E-03	0.99
		800×800	8.08E-06	2.01	9.76E-05	1.50	3.19E-03	0.99
		1200×1200	3.59E-06	2.00	5.31E-05	1.50	2.13E-03	0.99
		2000×2000	1.29E-06	2.00	2.47E-05	1.50	1.28E-03	1.00
	2	300×300	2.11E-06	—	1.00E-05	—	1.48E-04	—
		500×500	3.86E-07	3.33	2.75E-06	2.53	5.02E-05	2.12
		800×800	8.45E-08	3.23	8.46E-07	2.51	1.89E-05	2.07
		1200×1200	2.34E-08	3.16	3.07E-07	2.50	8.34E-06	2.02
		2000×2000	4.78E-09	3.11	8.55E-08	2.50	3.00E-06	2.00
	3	300×300	1.45E-06	—	2.46E-06	—	1.09E-05	—
		500×500	1.91E-07	3.98	3.27E-07	3.95	2.45E-06	2.92
		800×800	2.93E-08	3.98	5.17E-08	3.93	6.10E-07	2.96
		1200×1200	5.81E-09	3.99	1.07E-08	3.90	1.83E-07	2.97
		2000×2000	7.56E-10	3.99	1.49E-09	3.85	3.99E-08	2.98
	4	300×300	1.92E-06	—	3.65E-06	—	1.86E-05	—
		500×500	2.49E-07	4.00	4.75E-07	4.00	2.42E-06	3.99
		800×800	3.81E-08	4.00	7.25E-08	4.00	3.69E-07	4.00
		1200×1200	7.52E-09	4.00	1.43E-08	4.00	7.30E-08	4.00
		2000×2000	9.75E-10	4.00	1.86E-09	4.00	9.46E-09	4.00
	5	300×300	2.53E-06	—	5.20E-06	—	2.88E-05	—
		500×500	3.28E-07	3.99	6.77E-07	3.99	3.74E-06	3.99
		800×800	5.02E-08	4.00	1.03E-07	4.00	5.72E-07	4.00
		1200×1200	9.91E-09	4.00	2.04E-08	4.00	1.13E-07	4.00
		2000×2000	1.28E-09	4.00	2.65E-09	4.00	1.47E-08	4.00

be superior to low-order schemes in reducing numerical errors even for grids with first-order discontinuity.

5.1.4 Vortex evolution problem

In this subsection, GCL-FD and NGCL-FD are applied to solve isentropic moving vortex problem on five kinds of grids given in Subsection 5.1.1 with initial conditions

$$\rho = \left(1 - \frac{\gamma-1}{4\alpha\gamma}\Phi^2\right)^{(\gamma-1)}, \quad u = a + \frac{y-y_c}{r_c}\Phi, \quad v = b - \frac{x-x_c}{r_c}\Phi, \quad p = \rho^\gamma,$$

where $\gamma = 1.4$ is the specific heat ratio $\alpha = 1.2$ is the parameter of the length scale of vortex decay. Here we define a function $\Phi = \varepsilon e^{\alpha(1-d^2)}$ with the vortex strength $\varepsilon = 0.2$, $d = \frac{r}{r_c}$, $r = \sqrt{(x-x_c)^2 + (y-y_c)^2}$, the vortex core width $r_c = 0.1$ and the initial center of the vortex $(x_c, y_c) = (0,0)$. $(a,b) = (1,0)$ is the moving speed of the isentropic vortex.

Table 6: Numerical errors $\left|\frac{\partial E}{\partial x}^N - \frac{\partial E}{\partial x}^{exact}\right|$ for NGCL-FD4 on five kinds of grids.

FD4	k	$N_\xi \times N_\eta$	L_1 error	L_1 order	L_2 error	L_2 order	L_∞ error	L_∞ order
		300×300	1.15E-03	—	7.37E-03	—	1.09E-01	—
		500×500	6.89E-04	1.00	5.71E-03	0.50	1.07E-01	0.04
	1	800×800	4.31E-04	1.00	4.52E-03	0.50	1.06E-01	0.02
		1200×1200	2.87E-04	1.00	3.69E-03	0.50	1.06E-01	0.01
		2000×2000	1.72E-04	1.00	2.86E-03	0.50	1.05E-01	0.01
		300×300	3.08E-05	—	2.06E-04	—	2.83E-03	—
		500×500	1.09E-05	2.03	9.61E-05	1.50	1.68E-03	1.03
	2	800×800	4.25E-06	2.01	4.75E-05	1.50	1.04E-03	1.02
		1200×1200	1.88E-06	2.00	2.59E-05	1.50	6.90E-04	1.01
		2000×2000	6.78E-07	2.00	1.20E-05	1.50	4.12E-04	1.01
		300×300	2.39E-06	—	7.62E-06	—	9.72E-05	—
		500×500	4.00E-07	3.50	2.05E-06	2.57	3.38E-05	2.07
NGCL	3	800×800	8.11E-08	3.39	6.30E-07	2.51	1.30E-05	2.03
		1200×1200	2.13E-08	3.30	2.28E-07	2.50	5.77E-06	2.01
		2000×2000	4.12E-09	3.21	6.36E-08	2.50	2.07E-06	2.01
		300×300	2.03E-06	—	3.84E-06	—	1.85E-05	—
		500×500	2.64E-07	3.99	5.03E-07	3.98	2.40E-06	4.00
	4	800×800	4.05E-08	3.99	7.79E-08	3.97	5.45E-07	3.15
		1200×1200	8.01E-09	4.00	1.56E-08	3.96	1.63E-07	2.98
		2000×2000	1.04E-09	4.00	2.09E-09	3.94	3.54E-08	2.99
		300×300	2.77E-06	—	5.53E-06	—	2.85E-05	—
		500×500	3.60E-07	3.99	7.18E-07	3.99	3.71E-06	3.99
	5	800×800	5.50E-08	4.00	1.10E-07	4.00	5.67E-07	4.00
		1200×1200	1.09E-08	4.00	2.17E-08	4.00	1.12E-07	4.00
		2000×2000	1.41E-09	4.00	2.81E-09	4.00	1.45E-08	4.00

Table 7: Accuracy of numerical errors.

k	FD2 $\left|\frac{\partial E}{\partial x}^N - \frac{\partial E}{\partial x}^{exact}\right|$						FD4					
	GCL			NGCL			GCL			NGCL		
	L_1	L_2	L_∞	L_1	L_2	L_∞	L_1	L_2	L_∞	L_1	L_2	L_∞
1	2.0	1.5	1.0	1.0	0.5	0.0	2.0	1.5	1.0	1.0	0.5	0.0
2	**2.0**	**2.0**	**2.0**	2.0	1.5	1.0	3.0	2.5	2.0	2.0	1.5	1.0
3	2.0	2.0	2.0	**2.0**	**2.0**	**2.0**	4.0	3.5	3.0	3.0	2.5	2.0
4	2.0	2.0	2.0	2.0	2.0	2.0	**4.0**	**4.0**	**4.0**	4.0	3.5	3.0
5	2.0	2.0	2.0	2.0	2.0	2.0	4.0	4.0	4.0	**4.0**	**4.0**	**4.0**

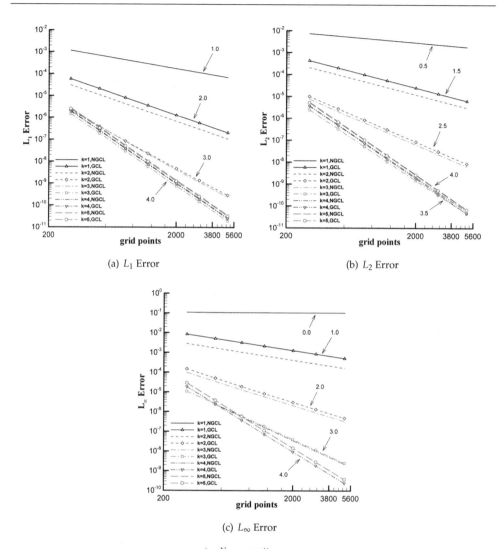

Figure 2: Numerical errors $\left|\frac{\partial E}{\partial x}^N - \frac{\partial E}{\partial x}^{exact}\right|$ with FD4 on four kinds of grids.

Firstly, the problem is solved in the area $[-0.5, 0.5] \times [-0.5, 0.5]$ till $T = 1.0$. Tables 8 and 9 gives numerical accuracy of density for GCL-FD4 and NGCL-FD4, respectively. It can be seen that GCL-FD4 performs at nearly second order accuracy on grids with $k = 1$ while NGCL-FD4 at nearly first order accuracy. Thus GCL-FD4 has one order higher accuracy than NGCL-FD4. In addition GCL-FD4 has second order higher order accuracy

Table 8: Numerical errors $|u^N - u^{exact}|$ of GCL-FD4 at time $T=1$.

FD4	k	$N_\xi \times N_\eta$	L_1 error	L_1 order	L_2 error	L_2 order	L_∞ error	L_∞ order
GCL	1	300×300	4.91E-06	—	9.51E-06	—	1.67E-04	—
		500×500	1.46E-06	2.38	3.15E-06	2.16	6.14E-05	1.96
		800×800	4.77E-07	2.37	1.11E-06	2.22	2.63E-05	1.81
		1200×1200	1.80E-07	2.40	4.42E-07	2.27	9.81E-06	2.43
		2000×2000	4.71E-08	2.63	1.28E-07	2.42	2.84E-06	2.43
	2	300×300	5.01E-07	—	1.76E-06	—	2.41E-05	—
		500×500	6.96E-08	3.86	2.32E-07	3.97	3.25E-06	3.92
		800×800	1.16E-08	3.82	3.60E-08	3.96	4.96E-07	4.00
		1200×1200	2.53E-09	3.74	7.37E-09	3.92	9.82E-08	4.00
		2000×2000	3.70E-10	3.77	1.03E-09	3.85	1.48E-08	3.70
	3	300×300	4.61E-07	—	1.75E-06	—	2.42E-05	—
		500×500	5.94E-08	4.01	2.29E-07	3.99	3.03E-06	4.07
		800×800	9.08E-09	3.99	3.50E-08	3.99	4.74E-07	3.94
		1200×1200	1.80E-09	4.00	6.92E-09	4.00	9.35E-08	4.01
		2000×2000	2.33E-10	4.00	8.97E-10	4.00	1.21E-08	4.00
	4	300×300	4.87E-07	—	1.77E-06	—	2.43E-05	—
		500×500	6.26E-08	4.02	2.31E-07	3.99	3.45E-06	3.82
		800×800	9.56E-09	4.00	3.53E-08	3.99	5.29E-07	3.99
		1200×1200	1.89E-09	4.00	6.97E-09	4.00	1.05E-07	4.00
		2000×2000	2.45E-10	4.00	9.04E-10	4.00	1.36E-08	4.00
	5	300×300	5.22E-07	—	1.79E-06	—	2.66E-05	—
		500×500	6.72E-08	4.02	2.34E-07	3.99	3.80E-06	3.81
		800×800	1.03E-08	4.00	3.57E-08	3.99	5.86E-07	3.98
		1200×1200	2.03E-09	4.00	7.06E-09	4.00	1.16E-07	4.00
		2000×2000	2.63E-10	4.00	9.16E-10	4.00	1.50E-08	4.00

than NGCL-FD4 on grids with $k=2$. Moreover, GCL-FD4 has a bit smaller error than NGCL-FD4 for $k=3$ and has nearly the same errors for $k=4,5$.

Secondly, we solve the problem till $T=2.0$ to see the time variation of orders for GCL-FD4 and NGCL-FD4 on grids with $k=1$. And numerical errors on two grids with 500×500 and 800×800 grid points are used to calculate numerical orders. From Fig. 3 we can see that GCL-FD4 has one order higher accuracy than NGCL-FD4 with respect to L_1 order and L_2 order. As for L_∞ order, both schemes has large vibration. The former one always has higher order accuracy than the latter one and the exceeding order vibrates with time. Moreover, we can see from Fig. 4 that both the errors and the area influenced by nonsmooth grid for GCL-FD4 are much smaller than that for NGCL-FD4.

Therefore, GCL-FD has superiority over NGCL-FD4 in spatial accuracy at a fixed final time. In addition, the exceeding order of GCL-FD4 over NGCL-FD4 may be different for different grids. The reason may be that the influence of local truncation errors near discontinuous points on global errors may rely on two aspects which are properties of the

Table 9: Numerical errors $|u^N - u^{exact}|$ of NGCL-FD4 at time $T=1$.

FD4	k	$N_\xi \times N_\eta$	L_1 error	L_1 order	L_2 error	L_2 order	L_∞ error	L_∞ order
NGCL	1	300×300	1.16E-04	—	1.59E-04	—	7.62E-04	—
		500×500	5.44E-05	1.49	7.41E-05	1.49	4.01E-04	1.26
		800×800	2.97E-05	1.29	4.39E-05	1.12	2.15E-04	1.33
		1200×1200	1.70E-05	1.38	2.58E-05	1.31	1.43E-04	1.00
		2000×2000	9.15E-06	1.21	1.47E-05	1.10	7.74E-05	1.20
	2	300×300	3.36E-06	—	4.42E-06	—	2.57E-05	—
		500×500	9.13E-07	2.55	1.29E-06	2.41	5.95E-06	2.87
		800×800	3.22E-07	2.22	4.66E-07	2.17	2.17E-06	2.14
		1200×1200	1.29E-07	2.26	2.00E-07	2.09	9.53E-07	2.03
		2000×2000	3.87E-08	2.35	6.22E-08	2.29	3.40E-07	2.01
	3	300×300	4.76E-07	—	1.76E-06	—	2.42E-05	—
		500×500	6.25E-08	3.97	2.29E-07	3.99	3.03E-06	4.07
		800×800	9.90E-09	3.92	3.51E-08	3.99	4.76E-07	3.94
		1200×1200	2.05E-09	3.89	6.96E-09	3.99	9.38E-08	4.00
		2000×2000	2.91E-10	3.82	9.10E-10	3.98	1.22E-08	4.00
	4	300×300	4.87E-07	—	1.77E-06	—	2.43E-05	—
		500×500	6.26E-08	4.02	2.31E-07	3.99	3.45E-06	3.82
		800×800	9.56E-09	4.00	3.53E-08	3.99	5.29E-07	3.99
		1200×1200	1.89E-09	4.00	6.97E-09	4.00	1.04E-07	4.00
		2000×2000	2.45E-10	4.00	9.04E-10	4.00	1.36E-08	4.00
	5	300×300	5.23E-07	—	1.79E-06	—	2.66E-05	—
		500×500	6.72E-08	4.02	2.34E-07	3.99	3.80E-06	3.81
		800×800	1.03E-08	4.00	3.57E-08	3.99	5.85E-07	3.98
		1200×1200	2.03E-09	4.00	7.06E-09	4.00	1.16E-07	3.99
		2000×2000	2.63E-10	4.00	9.16E-10	4.00	1.50E-08	4.00

grid (discontinuity order and discontinuity distribution) and properties of the solution (solution distribution and solution evolution).

5.2 Tests on grids with discontinuities in two directions

In this subsection, we test GCL errors and numerical errors on grids with discontinuities in two directions. The grids are generated via the following coordinates transformation

$$x_{i,j}=\xi_i, \quad y_{i,j}=\eta_j+\psi_1(\xi_i)\psi_2(\eta_j), \quad \xi_i\in[-0.5,0.5], \quad \eta_j\in[-0.5,0.5]. \quad (5.4)$$

where

$$\psi_1(\xi_i)=\begin{cases}\sin^{k_\xi}\left(\frac{5\pi}{3}\xi_i+\frac{\pi}{2}\right), & |\xi_i|<0.3,\\ 0, & |\xi_i|\geq0.3,\end{cases} \quad \psi_2(\eta_j)=\begin{cases}\sin^{k_\eta}\left(\frac{5\pi}{3}\eta_j+\frac{\pi}{2}\right), & |\eta_j|<0.3,\\ 0, & |\eta_j|\geq0.3.\end{cases}$$

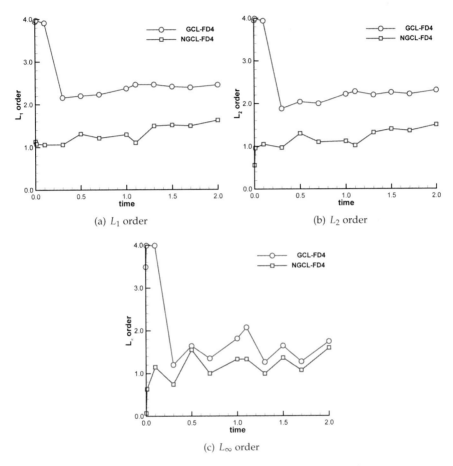

Figure 3: Time variation of numerical order for GCL-FD and NGCL-FD.

Here k_ξ and k_η denote the order of discontinuity of y in ξ and η direction, respectively. We test the case $k_\xi = 2$ and $k_\eta = 4$.

The GCL errors of NGCL-FD are summarized in Table 10. We can see that the order of L_1, L_2 and L_∞ GCL errors are 2, 1.5 and 1, respectively, for both NGCL-FD2 and NGCL-FD4. This result indicates that the GCL errors of NGCL-FD are determined by the lower order of discontinuity order $k = \min(k_\xi, k_\eta)$ and is consistent with Proposition 3.1.

Tables 11 and 12 give the spatial discretization errors of GCL-FD and NGCL-FD. It can be concluded that the orders of L_1, L_2 and L_∞ errors are $\min(k+1,r)$, $\min(k+0.5,r)$ and $\min(k,r)$ for GCL-FDr, but $\min(k,r)$, $\min(k-0.5,r)$ and $\min(k-1,r)$ for NGCL-FDr. This result agrees with Proposition 4.1.

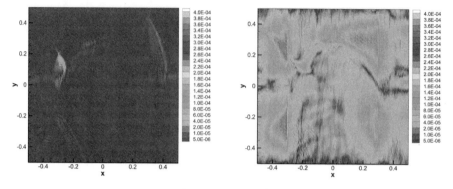

(a) Error distribution of density for GCL-FD4   (b) Error distribution of density for NGCL-FD4

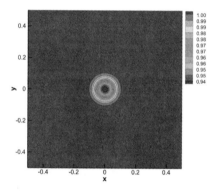

(c) Density distribution for NGCL-FD4

Figure 4: Error distribution of density for GCL-FD4 and NGCL-FD4 at time $T=2$ in solving vortex problem on 500×500 grid with $k=1$. (colored picture attached at the end of the book)

6 Concluding remarks

In the present work, the impact of grid discontinuities on the GCL errors and spatial discretization errors is analyzed theoretically and numerically to explore how GCL works in finite difference method. The GCL errors are proved to be zero for GCL-FD. However, for NGCL-FD, the GCL errors are related to grid smoothness. For C^{k-1} grids and rth-order finite difference operators, the GCL errors of NGCL-FD are approaching to zero with refining grids at $\min\{r,k-1\}$th-order convergence rate. The convergence rate of spatial discretization errors is $\min\{r,k\}$th order for GCL-FD but $\min\{r,k-1\}$th order for

Table 10: GCL errors $|\tilde{I}_x|$ for NGCL-FD2 and NGCL-FD4 on grids with $k_\xi=2$ and $k_\eta=4$.

NGCL	$N_\xi \times N_\eta$	L_1 error	order	L_2 error	order	L_∞ error	order
FD2	300×300	1.08E-04	–	4.11E-04	–	6.91E-03	–
	500×500	3.90E-05	2.00	1.85E-04	1.57	4.14E-03	1.00
	800×800	1.52E-05	2.00	8.95E-05	1.54	2.59E-03	1.00
	1200×1200	6.77E-06	2.00	4.81E-05	1.53	1.73E-03	1.00
	2000×2000	2.44E-06	2.00	2.22E-05	1.52	1.04E-03	1.00
FD4	300×300	1.64E-05	–	1.60E-04	–	2.07E-03	–
	500×500	5.87E-06	2.01	7.41E-05	1.50	1.24E-03	1.00
	800×800	2.29E-06	2.01	3.66E-05	1.50	7.77E-04	1.00
	1200×1200	1.02E-06	2.00	1.99E-05	1.50	5.18E-04	1.00
	2000×2000	3.66E-07	2.00	9.27E-06	1.50	3.11E-04	1.00

Table 11: Numerical errors $\left|\frac{\partial E}{\partial x}^N - \frac{\partial E}{\partial x}^{exact}\right|$ for GCL-FD2 and NGCL-FD2 on grids with $k_\xi=2$ and $k_\eta=4$.

FD2	$N_\xi \times N_\eta$	L_1 error	order	L_2 error	order	L_∞ error	order
GCL	300×300	2.67E-04	–	3.67E-04	–	1.22E-03	–
	500×500	9.60E-05	2.00	1.32E-04	2.00	4.40E-04	2.00
	800×800	3.75E-05	2.00	5.17E-05	2.00	1.72E-04	2.00
	1200×1200	1.67E-05	2.00	2.30E-05	2.00	7.65E-05	2.00
	2000×2000	6.01E-06	2.00	8.28E-06	2.00	2.75E-05	2.00
NGCL	300×300	3.34E-04	–	5.31E-04	–	8.92E-03	–
	500×500	1.21E-04	1.99	2.06E-04	1.85	5.40E-03	0.98
	800×800	4.72E-05	2.00	8.84E-05	1.80	3.39E-03	0.99
	1200×1200	2.10E-05	2.00	4.36E-05	1.75	2.26E-03	0.99
	2000×2000	7.56E-06	2.00	1.84E-05	1.69	1.36E-03	1.00

NGCL-FD. Then for non-sufficiently smooth grids with $k<r+1$, GCL-FD has one order higher spatial accuracy than NGCL-FD. If the grid is sufficiently smooth, namely $k \geq r+1$, both GCL-FD and NGCL-FD can obtain rth-order spatial accuracy, but the former is proved to produce smaller numerical errors than the latter. It can be concluded that satisfying the GCL helps FDSs to obtain one more order of accuracy for non-sufficiently smooth grids and reduce discretization errors for sufficiently smooth grid.

The following two issues are worth to be emphasized: (i) Satisfying the GCL can reduce the grid requirement of FDSs. GCL-FDr only needs a C^{r-1} grid while NGCL-FDr needs a C^r grid to obtain their rth-order accuracy. Thus, from the viewpoint of grid generation, satisfying the GCL can reduce the difficulties on grid generation for high-order FDSs when complex configurations are involved. (ii) Satisfying the GCL can make finite difference equation be consistent with the corresponding differential equation. For C^0 grids, GCL-FD still has first-order accuracy and thus satisfies consistent condition.

Table 12: Numerical errors $\left|\frac{\partial E}{\partial x}^N - \frac{\partial E}{\partial x}^{exact}\right|$ for GCL-FD4 and NGCL-FD4 on grids with $k_\xi = 2$ and $k_\eta = 4$.

FD4	$N_\xi \times N_\eta$	L_1 error	order	L_2 error	order	L_∞ error	order
GCL	300×300	5.61E-07	—	2.96E-06	—	7.31E-05	—
	500×500	9.14E-08	3.55	7.97E-07	2.57	2.56E-05	2.06
	800×800	1.82E-08	3.43	2.44E-07	2.52	9.91E-06	2.02
	1200×1200	4.71E-09	3.33	8.85E-08	2.51	4.39E-06	2.01
	2000×2000	9.00E-10	3.24	2.47E-08	2.50	1.57E-06	2.01
NGCL	300×300	1.21E-05	—	1.31E-04	—	4.54E-03	—
	500×500	4.27E-06	2.03	6.10E-05	1.50	2.73E-03	1.00
	800×800	1.66E-06	2.01	3.02E-05	1.50	1.71E-03	1.00
	1200×1200	7.35E-07	2.00	1.64E-05	1.50	1.14E-03	1.00
	2000×2000	2.64E-07	2.00	7.63E-06	1.50	6.83E-04	1.00

However, NGCL-FD only have zeroth-order accuracy, and the L_∞ error dose not decrease to zero because the local truncation errors of the difference equation cannot converge to zero as the grid is refined. This means that the finite difference equation of the NGCL-FD is no longer consistent with the corresponding differential equation on C^0 grids.

FDSs satisfying the GCL can obtain better solutions because the spatial discretization errors are reduced. Some researches show the numerical stability of finite difference method can also be improved by following GCL. Therefore, More attention will be paid to the influence of the GCL on the numerical stability in our future works.

Acknowledgments

This study was supported by the project of National Basic Research Program of China (Grant No. 2009CB723800), National Natural Science Foundation of China (Grant Nos. 11372342, 11072259 and 11301525) and the foundation of State Key Laboratory of Aerodynamics (Grant No. JBKY11030902). We would like to thank Dr. Guohua Tu for his valuable suggestions on improving this manuscript.

Appendix: Discretization errors of the E flux part

By using the operators in (3.5), the discretization form of the E-flux part (4.6) reads

$$\left(\widetilde{A_E}\right)_{i,j} = \frac{1}{\Delta \xi} \sum_{l=-L_1^\xi}^{L_2^\xi} a_l^\xi \left(E_{i+l,j} \delta_2^\eta y_{i+l,j} - E_{i,j} \delta_2^\eta y_{i+l,j} + E_{i,j} \delta_2^\eta y_{i+l,j} - E_{i,j} \delta_2^\eta y_{i,j} \right)$$

$$- \frac{1}{\Delta \eta} \sum_{l=-L_1^\eta}^{L_2^\eta} a_l^\eta \left(E_{i,j+l} \delta_2^\xi y_{i,j+l} - E_{i,j} \delta_2^\xi y_{i,j+l} + E_{i,j} \delta_2^\xi y_{i,j+l} - E_{i,j} \delta_2^\xi y_{i,j} \right)$$

$$= \frac{1}{\Delta\xi} \sum_{l=-L_1^\xi}^{L_2^\xi} a_l^\xi (E_{i+l,j} - E_{i,j}) \delta_2^\eta y_{i+l,j} - \frac{1}{\Delta\eta} \sum_{l=-L_1^\eta}^{L_2^\eta} a_l^\eta (E_{i,j+l} - E_{i,j}) \delta_2^\xi y_{i,j+l} + E_{i,j} (\widetilde{I}_x)_{i,j}, \quad (A.1)$$

where we have used (3.4) to get the last equality. Using (2.10), we have

$$\left(\widetilde{A_E}\right)_{i,j} = (\widetilde{I}_x)_{i,j} E_{i,j} + \frac{1}{\Delta\xi} \sum_{l=-L_1^\xi}^{L_2^\xi} a_l^\xi (E_{i+l,j} - E_{i,j}) \sum_{m=-M_1^\eta}^{M_2^\eta} b_m^\eta \sum_{p=1}^{\infty} \frac{m^p (\Delta\eta)^{p-1}}{p!} \left(\frac{\partial^p y}{\partial \eta^p}\right)_{i+l,j}$$

$$- \sum_{l=-L_1^\eta}^{L_2^\eta} a_l^\eta \sum_{p=1}^{\infty} \frac{l^p (\Delta\eta)^{p-1}}{p!} \left(\frac{\partial^p E}{\partial \eta^p}\right)_{i,j} \left\{ \sum_{m=-M_1^\xi}^{M_2^\xi} b_m^\xi \sum_{q=1}^{\infty} \frac{m^q (\Delta\xi)^{q-1}}{q!} \left(\frac{\partial^q y}{\partial \xi^q}\right)_{i,j+l} \right.$$

$$\left. + \sum_{m=i_D+1}^{M_2^\xi} b_m^\xi \sum_{q=1}^{\infty} \frac{(m-i_D)^q (\Delta\xi)^{q-1}}{q!} \left(\frac{\partial^q y}{\partial \xi^q}\right)_{i_A,j+l}^{R-L} \right\}. \quad (A.2)$$

Getting the main error terms according to the coefficient relations, we have

$$\left(\widetilde{A_E}\right)_{i,j} = \frac{1}{\Delta\xi} \sum_{l=-L_1^\xi}^{L_2^\xi} a_l^\xi (E_{i+l,j} - E_{i,j}) \left(\frac{\partial y}{\partial \eta}\right)_{i+l,j} - \left(\frac{\partial E}{\partial \eta}\right)_{i,j} \left(\frac{\partial y}{\partial \xi}\right)_{i,j} + (\widetilde{I}_x)_{i,j} E_{i,j}$$

$$- \left(\frac{\partial E}{\partial \eta}\right)_{i,j} \sum_{m=i_D+1}^{M_2^\xi} b_m^\xi \sum_{q=1}^{\infty} \frac{(m-i_D)^q (\Delta\xi)^{q-1}}{q!} \left(\frac{\partial^q y}{\partial \xi^q}\right)_{i_A,j}^{R-L} + \mathcal{O}((\Delta\xi)^r, (\Delta\eta)^r). \quad (A.3)$$

Now we analyze the first term of (A.3). By using Taylor series expansion (2.4) with discontinuous points in Proposition 2.2, we can obtain

$$\text{First term} = \frac{1}{\Delta\xi} \sum_{l=-L_1^\xi}^{i_D} a_l^\xi (E_{i+l,j} - E_{i,j}) \left(\frac{\partial y}{\partial \eta}\right)_{i+l,j} + \frac{1}{\Delta\xi} \sum_{l=i_D+1}^{L_2^\xi} a_l^\xi (E_{i+l,j} - E_{i,j}) \left(\frac{\partial y}{\partial \eta}\right)_{i+l,j}$$

$$= \sum_{l=-L_1^\xi}^{i_D} a_l^\xi \left(\sum_{q=1}^{\infty} l^q \frac{(\Delta\xi)^{q-1}}{q!} \left(\frac{\partial^q E}{\partial \xi^q}\right)_{i,j} \right) \left[\left(\frac{\partial y}{\partial \eta}\right)_{i,j} + \sum_{\tilde{q}=1}^{\infty} l^{\tilde{q}} \frac{(\Delta\xi)^{\tilde{q}}}{\tilde{q}!} \left(\frac{\partial^{1+\tilde{q}} y}{\partial \eta \partial \xi^{\tilde{q}}}\right)_{i,j} \right]$$

$$+ \sum_{l=i_D+1}^{L_2^\xi} a_l^\xi \left[\sum_{q=1}^{\infty} l^q \frac{(\Delta\xi)^{q-1}}{q!} \left(\frac{\partial^q E}{\partial \xi^q}\right)_{i,j} + \sum_{q=1}^{\infty} (l-i_D)^q \frac{(\Delta\xi)^{q-1}}{q!} \left(\frac{\partial^q E}{\partial \xi^q}\right)_{i_A,j}^{R-L} \right]$$

$$\times \left[\left(\frac{\partial y}{\partial \eta}\right)_{i,j} + \sum_{\tilde{q}=1}^{\infty} l^{\tilde{q}} \frac{\Delta\xi^{\tilde{q}}}{\tilde{q}!} \left(\frac{\partial^{1+\tilde{q}} y}{\partial \eta \partial \xi^{\tilde{q}}}\right)_{i,j} + \sum_{\tilde{q}=1}^{\infty} (l-i_D)^{\tilde{q}} \frac{\Delta\xi^{\tilde{q}}}{\tilde{q}} \left(\frac{\partial^{1+\tilde{q}} y}{\partial \eta \partial \xi^{\tilde{q}}}\right)_{i_A,j}^{R-L} \right],$$

Notice that

$$\sum_{l=-L_1^{\xi}}^{L_2^{\xi}} a_l^{\xi}\left(\sum_{q=1}^{\infty}l^q\frac{(\Delta\xi)^{q-1}}{q!}\left(\frac{\partial^q E}{\partial\xi^q}\right)_{i,j}\right)\sum_{\tilde{q}=1}^{\infty}l^{\tilde{q}}\frac{(\Delta\xi)^{\tilde{q}}}{\tilde{q}!}\left(\frac{\partial^{1+\tilde{q}} y}{\partial\eta\partial\xi^{\tilde{q}}}\right)_{i,j}$$

$$=\sum_{q=1}^{\infty}\sum_{\tilde{q}=1}^{\infty}\left(\sum_{l=-L_1^{\xi}}^{L_2^{\xi}}a_l^{\xi}l^{q+\tilde{q}}\right)\frac{(\Delta\xi)^{q+\tilde{q}-1}}{q!\tilde{q}!}\left(\frac{\partial^q E}{\partial\xi^q}\right)_{i,j}\left(\frac{\partial^{1+\tilde{q}}y}{\partial\eta\partial\xi^{\tilde{q}}}\right)_{i,j}=\mathcal{O}((\Delta\xi)^r)$$

and

$$\sum_{l=i_D+1}^{L_2^{\xi}}a_l^{\xi}\left[\sum_{q=1}^{\infty}l^q\frac{(\Delta\xi)^{q-1}}{q!}\left(\frac{\partial^q E}{\partial\xi^q}\right)_{i,j}\sum_{\tilde{q}=1}^{\infty}(l-i_D)^{\tilde{q}}\frac{\Delta\xi^{\tilde{q}}}{\tilde{q}}\left(\frac{\partial^{1+\tilde{q}}y}{\partial\eta\partial\xi^{\tilde{q}}}\right)_{i_A,j}^{R-L}\right]$$

$$=\sum_{l=i_D+1}^{L_2^{\xi}}a_l^{\xi}\left[\sum_{q=1}^{\infty}l^q\frac{(\Delta\xi)^{q-1}}{q!}\left(\frac{\partial^q E}{\partial\xi^q}\right)_{i,j}(l-i_D)^k\frac{\Delta\xi^k}{k!}\left(\frac{\partial^{1+k}y}{\partial\eta\partial\xi^k}\right)_{i_A,j}^{R-L}\right]+\mathcal{O}(\Delta\xi^{k+1})$$

$$=\frac{\Delta\xi^k}{k!}\left(\frac{\partial E}{\partial\xi}\right)_{i,j}\left(\frac{\partial^{1+k}y}{\partial\eta\partial\xi^k}\right)_{i_A,j}^{R-L}\sum_{l=i_D+1}^{L_2^{\xi}}a_l^{\xi}l(l-i_D)^k+\mathcal{O}((\Delta\xi)^{k+1}).$$

By using the relations (2.8) and getting the main terms of the errors, we obtain

$$\text{First term}=\left(\frac{\partial E}{\partial\xi}\right)_{i,j}\left(\frac{\partial y}{\partial\eta}\right)_{i,j}+\frac{\Delta\xi^k}{k!}\left(\frac{\partial E}{\partial\xi}\right)_{i,j}\left(\frac{\partial^{1+k}y}{\partial\eta\partial\xi^k}\right)_{i_A,j}^{R-L}\sum_{l=i_D+1}^{L_2^{\xi}}a_l^{\xi}l(l-i_D)^k$$

$$+\frac{(\Delta\xi)^{k-1}}{k!}\left(\frac{\partial y}{\partial\eta}\right)_{i,j}\left(\frac{\partial^k E}{\partial\xi^k}\right)_{i_A,j}^{R-L}\sum_{l=i_D+1}^{L_2^{\xi}}a_l^{\xi}(l-i_D)^k$$

$$+\frac{(\Delta\xi)^k}{k!}\left(\frac{\partial^2 y}{\partial\eta\partial\xi}\right)_{i,j}\left(\frac{\partial^k E}{\partial\xi^k}\right)_{i_A,j}^{R-L}\sum_{l=i_D+1}^{L_2^{\xi}}a_l^{\xi}(l-i_D)^k+\mathcal{O}((\Delta\xi)^r)+\mathcal{O}((\Delta\xi)^{k+1})$$

$$=\left(\frac{\partial E}{\partial\xi}\right)_{i,j}\left(\frac{\partial y}{\partial\eta}\right)_{i,j}+\mathcal{O}((\Delta\xi)^{\min\{r,k\}})+\frac{(\Delta\xi)^{k-1}}{k!}\left(\frac{\partial y}{\partial\eta}\right)_{i,j}\left(\frac{\partial^k E}{\partial\xi^k}\right)_{i_A,j}^{R-L}\sum_{l=i_D+1}^{L_2^{\xi}}a_l^{\xi}(l-i_D)^k.$$
(A.4)

Substituting (A.4) into (A.3), we can obtain the error of A_E

$$(e_{A_E})_{i,j}=\frac{(\Delta\xi)^{k-1}}{k!}\left(\frac{\partial y}{\partial\eta}\right)_{i,j}\left(\frac{\partial^k E}{\partial\xi^k}\right)_{i_A,j}^{R-L}\sum_{l=i_D+1}^{L_2^{\xi}}a_l^{\xi}(l-i_D)^k+E_{i,j}(\widetilde{I}_x)_{i,j}$$

$$-\frac{(\Delta\xi)^{k-1}}{k!}\left(\frac{\partial E}{\partial\eta}\right)_{i,j}\left(\frac{\partial^k y}{\partial\xi^k}\right)_{i_A,j}^{R-L}\sum_{m=i_D+1}^{M_2^{\xi}}b_m^{\xi}(m-i_D)^k+\mathcal{O}((\Delta\xi)^{\min\{r,k\}},(\Delta\eta)^r).$$
(A.5)

Then, (4.9) can be obtained by substituting (2.13) into (A.5) and noting that

$$\left(\frac{\partial E}{\partial y}\right)_{i_A,j} = \left(\frac{\partial E}{\partial y}\right)_{i,j} + \sum_{q=1}^{\infty} (i_D)^q \frac{(\Delta \xi)^q}{q!} \left(\frac{\partial^q}{\partial \xi^q}\left(\frac{\partial E}{\partial y}\right)\right)_{i,j}.$$

References

[1] S. D. Rango and D.W. Zingg. Aerodynamic computations using a higher-order algorithm. *AIAA paper 99-0167*, 1999.

[2] S. D. Rango and D. W. Zingg. Further investigation of a higher-order algorithm for aerodynamic computations. *AIAA paper 2000-0823*, 2000.

[3] J.A. Ekaterinaris. High-order accurate, low numerical diffusion methods for aerodynamics. *Progress in Aerospace Sciences*, 41:192–300, 2005.

[4] Z.J. Wang, Krzysztof Fidkowski, and etal. High-order cfd methods: current status and perspective. *Int. J. Numer. Meth. Fluids*, 00:1–42, 2012.

[5] M.R. Visbal and D.V. Gaitonde. On the use of higher-order finite-difference schemes on curvilinear and deforming meshes. *J. Comput. Phys.*, 181:155–185, 2002.

[6] J. Casper, C.W. Shu, and H. Atkins. Comparison of two formulations for high-order accurate essentially nonoscillatory schemes. *AIAA Journal*, 32:1970–1977, 1994.

[7] J.E. Castillo, J.M. Hyman, M.J. Shashkov, and S. Steinberg. The sensitivity and accuracy of fourth order finite-difference schemes on nonuniform grids in one dimension. *Computers Math. Applic.*, 30:41–55, 1995.

[8] C.W. Shu. High-order finite difference and finite volume weno schemes and discontinuous galerkin methods for cfd. *Int. J. Comput. Fluid Dynamics*, 17:107–118, 2003.

[9] T.H. Pulliam and J.L. Steger. On implicit finite-difference simulations of three-dimensional flow. *AIAA Paper 78-10*, 1978.

[10] P.D. Thomas and C.K. Lombard. The geometric conservation law-a link between finite difference and finite volume methods of flow computation on moving grids. *AIAA paper 78-1208*, 1978.

[11] P. D. Thomas and C. K. Lombard. Geometric conservation law and its application to flow computations on moving grids. *AIAA Journal*, 17(10):1030–1037, 1979.

[12] H. Zhang, M. Reggio, J.Y. TršŠpanier, and R. Camarero. Discrete form of the gcl for moving meshes and its implementation in cfd schemes. *Computers and Fluids*, 22 (1):9–23, 1993.

[13] M. Lesoinne and C. Farhat. Geometric conservation laws for flow problems with moving boundaries and deformable meshes, and their impact on aeroelastic computations. *Comput. Methods Appl. Mech. Eng.*, 134:71–90, 1996.

[14] H. Guillard and C. Farhat. On the significance of the geometric conservation law for flow computations on moving meshes. *Comput. Methods Appl. Mech. Eng.*, 190:1467–1482, 2000.

[15] P. Geuzaine, C. Grandmont, and C. Farhat. Design and analysis of ale schemes with provable second-order time-accuracy for inviscid and viscous flow simulations. *J. Comput. Phys., 191 (2003), pp. .*, 191:206–227, 2003.

[16] D.J. Mavriplis and Z. Yang. Construction of the discrete geometric conservation law for high-order time accurate simulations on dynamic meshes. *J. Comput. Phys.*, 213 (2):557–573, 2006.

[17] J. Sitaraman and J.D. Baeder. Field velocity approach and geometric conservation law for unsteady flow simulations. *AIAA Journal*, 44(9):2084–2094, 2006.

[18] S. Etienne, A. Garon, and D. Pelletier. Geometric conservation law and finite element methods for ale simulations of incompressible flow. *AIAA Paper 2008-733*, 2008.

[19] D.J. Mavriplis and C.R. Nastase. On the geometric conservation law for high-order discontinuous galerkin discretizations on dynamically deforming meshes. *J. Comput. Phys.*, 230:4285–4300, 2011.

[20] B. Sjogreen, H. C. Yee, and M. Vinokur. On high order finite-difference metric discretizations satisfying gcl on moving and deforming grids. Technical report, Lawrence Livermore National Laboratory, LLNL-TR-637397, 2013.

[21] C. Farhat, P. Geuzainne, and C. Grandmont. The discrete geometric conservation law and the nonlinear stability of ale schemes for the solution of flow problems on moving grids. *J. Comput. Phys.*, 174:669–694, 2001.

[22] L. Formaggia and F. Nobile. Stability analysis of second-order time accurate schemes for ale-fem. *Comput. Methods Appl. Mech. Eng.*, 193:4097–4116, 2004.

[23] K. Ou and A. Jameson. On the temporal and spatial accuracy of spectral difference method on moving deformable grids and the effect of geometry conservation law. *AIAA Paper 2010-5032*, 2010.

[24] X. Deng, Y. Jiang, M. Mao, H. Liu, and G. Tu. Developing hybrid cell-edge and cell-node dissipative compact scheme for complex geometry flows. *Science China*, 56(10):2361–2369, 2013.

[25] T. Nonomura, N. Iizuka, and K. Fujii. Freestream and vortex preservation properties of high-order weno and wcns on curvilinear grids. *Computers and Fluids*, 39:197–214, 2010.

[26] X. Deng and H. Zhang. Developing high-order weighted compact nonlinear schemes. *J. Comput. Phys.*, 165:22–44, 2000.

[27] X. Deng. High-order accurate dissipative weighted compact nonlinear schemes. *Science in China (Series A) x*, 45 (3):356–370, 2002.

[28] X. Deng, X. Liu, and M. Mao. Advances in high-order accurate weighted compact nonlinear schemes. *Advances in Mechanics*, 37 (3):417–427, 2007.

[29] G.S. Jiang and C.W. Shu. Efficient implementation of weighted eno schemes. *J. Comput. Phys.*, 126:202–228, 1996.

[30] X. Deng, M. Mao, G. Tu, H. Liu, and H. Zhang. Geometric conservation law and applications to high-order finite difference schemes with stationary grids. *J. Comput. Phys.*, 230:1100–1115, 2011.

[31] T. Nonomura, D. Terakado, Y. Abe, and K. Fujii. A new technique for finite difference weno with geometric conservation law. *AIAA Paper 2013-2569*, 2013.

[32] X. Deng, Y. Min, M. Mao, H. Liu, G. Tu, and H. Zhang. Further studies on geometric conservation law and applications to high-order finite difference schemes with stationary grids. *J. Comput. Phys.*, 239:90–111, 2013.

Further improvement of weighted compact nonlinear scheme using compact nonlinear interpolation

Zhen-Guo Yan [a,*], Huayong Liu [a], Yankai Ma [a], Meiliang Mao [a,b], Xiaogang Deng [c]

[a] State Key Laboratory of Aerodynamics, China Aerodynamics Research and Development Center, P.O. Box 211, Mianyang, Sichuan 621000, People's Republic of China
[b] Computational Aerodynamics Institute, China Aerodynamics Research and Development Center, P.O. Box 211, Mianyang, Sichuan 621000, People's Republic of China
[c] National University of Defense Technology, Changsha, Hunan 410073, People's Republic of China

ARTICLE INFO

Article history:
Received 15 January 2017
Revised 20 June 2017
Accepted 29 June 2017
Available online 4 July 2017

Keywords:
Hyperbolic conservation laws
High-order schemes
Weighted compact nonlinear scheme (WCNS)
Compact nonlinear interpolation
Nonlinear weights

ABSTRACT

To further improve the resolution of weighted compact nonlinear schemes(WCNS), a new 7th-order compact nonlinear interpolation method is proposed on the same stencil as 5th-order CRWENO scheme. Proper nonlinear weights are developed based on the Y type nonlinear weights, the properties of which are further analyzed in this paper. It is found that the Y type nonlinear weights are equivalent to the original nonlinear weights with an adaptive ϵ, which has very small value near discontinuities and very large value in smooth regions. As a result, the new nonlinear scheme has good shock capturing ability and achieves optimal order of accuracy in smooth regions. A new characteristic projection method is put forward, which largely reduces the computation costs. Simulations of Lax problem, Osher-Shu problem, double Mach reflection problem, decay of homogenous turbulence, shock turbulence interaction and turbulent channel flow show that the new scheme captures shocks without obvious oscillations, has higher resolution than other schemes tested, has obvious advantage in distinguishing turbulence from shocks and has higher efficiency.

© 2017 Elsevier Ltd. All rights reserved.

1. Introduction

High-order accurate and high-resolution schemes with discontinuity capturing ability are desired in simulating multi-scale flows which contain shock waves, such as direct numerical simulation (DNS) and large eddy simulation (LES) of high speed turbulence [1–3]. Many high-order discontinuity capturing schemes have been constructed, in which weighted nonlinear schemes(such as WENO [4], WCNS [5] and WCS [6]) have achieved great success for their superb shock capturing ability and relatively high resolution. However, these schemes are still low in resolution and too dissipative for some simulations such as compressible turbulence simulations and acoustics simulations. A lot of efforts have been devoted to overcoming this deficiency.

One feasible way is to develop better nonlinear mechanisms(such as nonlinear weights and nonlinear limiters). In 1980s, third-order essentially non-oscillatory (ENO) scheme was constructed by developing a dynamic sub-stencil choosing method [7]. In 1994, Liu et al. [8] proposed weighted ENO (WENO), which achieves higher order accuracy and higher resolution on the same stencil. Later, Jiang and Shu [4] further improved the nonlinear weights, which help WENO scheme achieve optimal 5th order in smooth regions. However, the corresponding WENO-JS scheme [4] still loses accuracy near critical points [9,10]. To solve this problem, Henrick et al. [9] put forward a Mapped WENO scheme (WENO-M), which ensures optimal order accuracy at first-order critical points. The WENO-M scheme exhibits better resolution than the WENO-JS scheme, but is more expensive. Borges and co-workers [10] solved this problem by proposing Z nonlinear weights. And the corresponding WENO-Z scheme achieves optimal order accuracy at first-order critical points [10]. Compared with the WENO-JS scheme, the WENO-Z scheme has similar computation cost, but higher resolution [10]. However, the WENO-M and WENO-Z schemes still suffer a loss in accuracy near high-order critical points. Yamaleev and Carpenter [11] proposed some limitations of ϵ, a parameter originally introduced to avoid division by zero, and optimal order accuracy is achieved near high-order critical points. Yamaleev and Carpenter's method has been widely investigated and used [12–14]. In [15], it is found that limiting ϵ is effective in improving the accuracy because the method has implicitly introduced the values of smoothness indicators into the nonlinear weights. Based on this observation, new Y type nonlinear weights are developed by explicitly considering the values of smoothness indicators. WCNS with Y type nonlinear weights has optimal accuracy and higher resolution while maintaining good shock capturing ability.

Another way is to use compact schemes instead of explicit ones to improve the resolution. Compact linear schemes have already been widely used for their excellent spectral properties [16–18]. Meanwhile, many researchers have tried to develop compact nonlinear schemes, which not only have excellent spectral properties but also have good shock capturing ability. Cockburn and Shu proposed a local nonlinear limiting method to control spurious oscillations while keeping the accuracy and proved that the scheme is total variation stable in one-space dimension for scalar conservation laws [19]. Deng et al. [20] developed CNS, which uses a dynamic sub-stencil choosing idea like ENO. Fu and Ma proposed a 5th-order upwind compact scheme by using group velocity control method, and further applied the scheme to direct numerical simulations. A weighted compact scheme(WCS) is constructed in [6], in which three compact sub-stencils are used to form a six-order compact scheme. However, WCS may lead to oscillatory solutions and is further combined with WENO scheme to achieve local dependency in shock regions [21]. By replacing explicit reconstructions with compact ones, a class of Compact-Reconstruction WENO schemes (CRWENO) is proposed in [22]. The smoothness indicators of CRWENO schemes are the same with those of the WENO scheme, and the resulting compact schemes achieve 5th-order accuracy in the same stencil as WENO. Although great progresses have been made in compact nonlinear schemes, they are still not as robust as explicit schemes in capturing shocks and they still suffer a great loss in resolution compared with their linear counterparts.

In this paper, we further improve the resolution of WCNS by developing a new compact nonlinear interpolation method. The main contributions of this paper are as follows:

- Developing a 7th-order compact nonlinear interpolation method and a new WCNS-C7 scheme on the same stencil as 5th-order WENO and 5th-order CRWENO schemes.
- Further analyzing properties of Y type nonlinear weights in an adaptive ϵ form and making comparisons with other choices of ϵ.
- Proposing a new characteristic projection method and largely improving efficiency.

The present work is based on weighted compact nonlinear scheme (WCNS) [5] which uses nonlinear weights similar to those of the WENO schemes. The WCNS can easily satisfy geometric conservation law (GCL) [23,24] and has superiority in simulations on complex grids [25,26]. Some canonical cases are used to test the shock capturing ability, high frequency wave simulating ability and turbulence simulating ability of the new WCNS scheme.

The organization of the paper is as follows. The WCNS scheme is briefly introduced in Section 2. Section 3 presents the development of the new scheme, which includes proposing new compact nonlinear scheme, further analyzing Y type nonlinear weights and efficiently implementing in Euler system. Section 4 gives numerical experiments to verify the theoretical analysis. At last, conclusions are drawn in Section 5.

2. WCNS

Hyperbolic conservation law has the following form

$$\frac{\partial u}{\partial t} + \frac{\partial f(u)}{\partial x} = 0, \quad (1)$$

where u is a conserved quantity, f describes its flux. Consider a uniform grid defined by $x_j = j\Delta x = jh, j = 0, \ldots, N$, where $\Delta x = h$ is the uniform grid spacing. The semi-discrete form of (1) yields an ordinary differential equation:

$$\frac{du_j(t)}{dt} = -F_j', \quad (2)$$

where $u_j(t)$ is a numerical approximation of $u(x_j, t)$, and F_j' is a spacial discretization of $\frac{\partial f}{\partial x}|_{x=x_j} = f_j'$. In this paper, the WCNS is used for spatial discretization.

The WCNS consists of three parts [5]: high-order flux difference scheme for flux derivatives, numerical flux for cell-edge fluxes and high-order interpolation for cell-edge variable values.

In the first part, a special case (HCS-E6) of the general hybrid cell-edge and cell-node compact scheme (HCS) [27,28] is adopted to calculate flux derivatives, which is

$$F_j' = \frac{\varphi}{h}(\tilde{F}_{j+\frac{1}{2}} - \tilde{F}_{j-\frac{1}{2}}) + \frac{192 - 175\varphi}{256h}(f_{j+1} - f_{j-1}) + \frac{-48 + 35\varphi}{320h}(f_{j+2} - f_{j-2}) + \frac{64 - 45\varphi}{3840h}(f_{j+3} - f_{j-3}). \quad (3)$$

where f_{j+m} is the flux at cell-node $j + m$.

$$\tilde{F}_{j+\frac{1}{2}} = \tilde{F}(\tilde{u}_{j+\frac{1}{2}}^+, \tilde{u}_{j+\frac{1}{2}}^-), \quad (4)$$

is the numerical flux at cell-edge $j + \frac{1}{2}$. $\tilde{u}_{j+\frac{1}{2}}^+$ and $\tilde{u}_{j+\frac{1}{2}}^-$ are cell-edge variable values calculated by upwind-like nonlinear interpolations. Generally, the HCS-E6 of (3) has 6th-order in accuracy. However, $\varphi = 256/175$ is taken in this paper, with which the HCS of (3) is 8th-order in accuracy.

In the second part, the choice of numerical flux can be flexible. Steger and Warming's numerical flux [29] is used, if not specified.

In the third part, nonlinear methods are used for cell-edge variable interpolations to capture discontinuities [5]. Only formulas for $\tilde{u}_{j+\frac{1}{2}}^+$ are given. $\tilde{u}_{j+\frac{1}{2}}^-$ can be similarly calculated since it is symmetry with $\tilde{u}_{j+\frac{1}{2}}^+$. In the following, $\pm$ will be dropped for simplicity. Fifth-order weighted nonlinear interpolation has the following form

$$\tilde{u}_{j+\frac{1}{2}} = \sum_{k=0}^{2} \omega_k \tilde{u}_{j+\frac{1}{2}}^k. \quad (5)$$

$\tilde{u}_{j+\frac{1}{2}}^k$ is the low order interpolation on the kth sub-stencil, which is

$$\tilde{u}_{j+\frac{1}{2}}^k = u_j + \left(\frac{h}{2}\right)u_j' + \frac{1}{2}\left(\frac{h}{2}\right)^2 u_j''$$
$$= u_j + \frac{1}{2}g_j^k h + \frac{1}{8}s_j^k h^2, \quad (6)$$

where $g_j^k h$ and $s_j^k h^2$ are the first and second order undivided differences on the kth sub-stencil. If $g_j^k h = u_j' h + O(h^3)$ and $s_j^k h^2 = u_j'' h^2 + O(h^3)$, (6) would be 3rd order. $g_j^k h$ and $s_j^k h^2$ are calculated by

$$g_j^0 h = \frac{1}{2}(3u_j - 4u_{j-1} + u_{j-2}) = u_j' h + O(h^3),$$
$$g_j^1 h = \frac{1}{2}(u_{j+1} - u_{j-1}) = u_j' h + O(h^3),$$
$$g_j^2 h = \frac{1}{2}(-u_{j+2} + 4u_{j+1} - 3u_j) = u_j' h + O(h^3), \quad (7)$$

$$s_j^0 h^2 = (u_j - 2u_{j-1} + u_{j-2}) = u_j'' h^2 + O(h^3),$$
$$s_j^1 h^2 = (u_{j+1} - 2u_j + u_{j-1}) = u_j'' h^2 + O(h^3),$$
$$s_j^2 h^2 = (u_{j+2} - 2u_{j+1} + u_j) = u_j'' h^2 + O(h^3), \quad (8)$$

respectively.

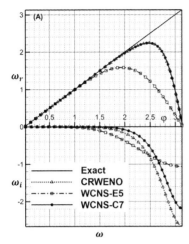

Fig. 1. Spectral properties of different schemes.

The Z nonlinear weights [10] is used for ω_k in (5), which have the following form

$$\omega_k = \frac{\alpha_k}{\sum_{l=0}^{2}\alpha_l}, \qquad (9)$$

$$\alpha_k = d_k\left(1 + \frac{\tau_5}{\beta_k + \epsilon}\right), \ \tau_5 = |\beta_2 - \beta_0|, \qquad (10)$$

where $\{d_0, d_1, d_2\} = \{1/16, 10/16, 5/16\}$ are optimal weights, and $\epsilon = 10^{-40}$ [10]. The β_k in (10) is the smoothness indicator of the kth sub-stencil. In this paper, it is calculated by

$$\beta_k = \left(g_j^k h\right)^2 + \left(s_j^k h^2\right)^2. \qquad (11)$$

3. Improvement of WCNS

3.1. Compact nonlinear interpolation

To improve the resolution of WCNS, we use compact nonlinear interpolations instead of explicit ones. On the same stencil as the explicit 5th-order WCNS(WCNS-E5), the tri-diagonal compact interpolation is 7th order in accuracy at most, which is

$$\frac{1}{2}\tilde{u}_{j-\frac{1}{2}} + \tilde{u}_{j+\frac{1}{2}} + \frac{3}{14}\tilde{u}_{j+\frac{3}{2}} = \frac{-1}{224}u_{j-2} + \frac{1}{8}u_{j-1}$$
$$+ \frac{15}{16}u_j + \frac{5}{8}u_{j+1} + \frac{1}{32}u_{j+2}. \qquad (12)$$

Using (12), the WCNS scheme is 7th-order in accuracy, which has much higher resolution as shown in Fig. 1.

To capture discontinuities, nonlinear mechanism has to be introduced into the compact interpolation. Following the idea of weighted nonlinear interpolation in (5), (12) should be divided into low order interpolations, which are coupled with nonlinear weights to capture shocks. In [6,20,22], the compact scheme is divided into several low order compact interpolations. However, it is divided in a different way, which is

$$S_0: \ \tilde{u}_{j+\frac{1}{2}}^0 = \frac{1}{8}(3u_{j-2} - 10u_{j-1} + 15u_j)$$

$$S_1: \ \tilde{u}_{j+\frac{1}{2}}^1 = \frac{1}{8}(-u_{j-1} + 6u_j + 3u_{j+1})$$

$$S_2: \ \tilde{u}_{j+\frac{1}{2}}^2 = \frac{1}{8}(3u_j + 6u_{j+1} - u_{j+2})$$

$$S_3: \ \frac{1}{2}\tilde{u}_{j-\frac{1}{2}}^3 + \frac{3}{14}\tilde{u}_{j+\frac{3}{2}}^3$$
$$= \frac{1}{128}\left(-\frac{25}{7}u_{j-2} + 36u_{j-1} + 30u_j + 20u_{j+1} + 9u_{j+2}\right) \qquad (13)$$

The compact interpolation of (12) is divided into four low order interpolations, three of which are explicit ones(S_0, S_1, S_2) and the rest one is a compact one(S_3). Actually, the interpolations of S_0, S_1, S_2 are the same as the three 3rd-order interpolations in (6).

Assigning nonlinear weights(ω_0, ω_1, ω_2 and ω_3) to interpolations of S_0, S_1, S_2, S_3, the nonlinear compact interpolation can be expressed in the following form

$$\begin{pmatrix} \frac{1}{2}\omega_3 \\ \omega_0 + \omega_1 + \omega_2 \\ \frac{3}{14}\omega_3 \end{pmatrix}^T \begin{pmatrix} \tilde{u}_{j-\frac{1}{2}} \\ \tilde{u}_{j+\frac{1}{2}} \\ \tilde{u}_{j+\frac{3}{2}} \end{pmatrix}$$
$$= \begin{pmatrix} 3/8\omega_0 \\ (-40\omega_0 - 4\omega_1)/32 \\ 3(40\omega_0 + 16\omega_1 + 8\omega_2)/64 \\ (3\omega_1 + 6\omega_2)/8 \\ (-\omega_2)/8 \end{pmatrix}^T \begin{pmatrix} u_{j-2} \\ u_{j-1} \\ u_j \\ u_{j+1} \\ u_{j+2} \end{pmatrix}$$
$$+ \omega_3 \begin{pmatrix} -25/896 \\ 9/32 \\ 15/64 \\ 5/32 \\ 9/128 \end{pmatrix}^T \begin{pmatrix} u_{j-2} \\ u_{j-1} \\ u_j \\ u_{j+1} \\ u_{j+2} \end{pmatrix}. \qquad (14)$$

Designing proper smoothness indicators is vital in capturing discontinuities. Since interpolations of S_0, S_1 and S_2 are exactly the same with the interpolations in (6), the same smoothness indicators of (11) are used for S_0, S_1 and S_2. This means (5) would be recovered if $\omega_3 = 0$. Since the interpolation of S_3 would definitely lead to oscillations if there is a discontinuity in $[j-2, j+2]$, ω_3 should be small near discontinuities. So we use the smoothness indicator of the whole stencil(β_3) for S_3, which is

$$\beta_3 = \left(g_j^3 h\right)^2 + \left(s_j^3 h^2\right)^2 + \left(t_j^3 h^3\right)^2 + \left(r_j^3 h^4\right)^2, \qquad (15)$$

$$g_j^3 h = \left(g_j^0 + 4g_j^1 + g_j^2\right)/6 = u_j' h + O(h^5),$$
$$s_j^3 h^2 = \left(-s_j^0 h^2 + 14 s_j^1 + s_j^2 h^2\right)/12 = u_j'' h^2 + O(h^5),$$
$$t_j^3 h^3 = \left(s_j^2 h^2 - s_j^0 h^2\right)/2 = u_j^{(3)} h^3 + O(h^5),$$
$$r_j^3 h^4 = 2\left(g_j^0 h - g_j^2 h\right) = u_j^{(4)} h^4 + O(h^6). \qquad (16)$$

If there is a discontinuity in $[j-2, j+2]$, β_3 would be much larger than the β_k of the most smooth sub-stencil. In consequence, ω_3 would be very small to maintain the essentially non-oscillatory property near discontinuities.

Through convergence analysis [15], it is deduced that a sufficient condition for the WCNS with (14) to obtain optimal 7th-order is

$$\omega_k = d_k + O(h^5), \qquad (17)$$

where $\{d_0, d_1, d_2, d_3\} = \{1/16, 10/16, 5/16, 1\}$. If the Z nonlinear weights, which can only satisfy $\omega_k = d_k + O(h^3)$, is used, the resulting nonlinear scheme is only 5th-order in accuracy, which has properties very similar to 5th-order CRWENO.

3.2. Y type nonlinear weights

In this paper, the Y type nonlinear weights are used because they can satisfy the condition of (17) and they exhibit advantages in turbulence simulations [15]. We also give further analysis

Table 1
Different choices of ϵ.

	ϵ_A	ϵ_B	ϵ_C	ϵ_D	ϵ_E	ϵ_F	ϵ_G
ϵ	10^{-40}	10^{-6}	10^{-2}	h^5	h^2	$10^{-8}h^2$	$10^{-6}\min\left(1, \frac{\min(\beta_k)}{\max(\beta_k,1-\min(\beta_k)+10^{-30}}\right)+10^{-99}$
Proposed in	[9]	[4]	[30]	[11]	[13]	[31]	[32]

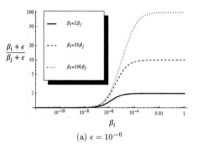
(a) $\epsilon = 10^{-6}$

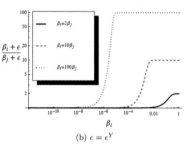
(b) $\epsilon = \epsilon^Y$

Fig. 2. Analysis of the effect of ϵ.

about the properties of Y type nonlinear weights. The Z-Y non-liner weights(Y type nonlinear weights based on the Z nonlinear weights) are

$$\alpha_k = d_k\left(1 + \frac{\phi\tau_5}{\bar{\beta} + \phi(\beta_k - \bar{\beta}) + \epsilon}\right), \quad (18)$$

where

$$\phi = \tanh\left(\frac{\bar{\beta}}{\kappa}\left(\frac{\beta_{max} + \kappa h^5}{\beta_{min} + \kappa h^5}\right)^{p_Y}\right),$$

$\bar{\beta} = (\beta_0 + \beta_1 + \beta_2)/3,$

$\beta_{max} = \max(\beta_0, \beta_1, \beta_2),$

$\beta_{min} = \min(\beta_0, \beta_1, \beta_2). \quad (19)$

Substituting (18) into (9), we get the final form of the Z-Y nonlinear weights. In this paper, $p_Y = 4$ is used.

In (19), the κ is added to make the scheme scale-invariant of flow variables [15], which is calculated by

$\kappa(\rho) = C_Y \max(\rho_0^2),$

$\kappa(p) = C_Y \max(p_0^2),$

$\kappa(v) = C_Y \max(c_0^2). \quad (20)$

Since (20) is based on the initial flow field, no repeated cost is needed. $C_Y = 200$ is used in this paper.

It can be verified that the Z-Y nonlinear weights satisfy the following error estimation [15]

$$\omega_k - d_k = O(h^{\max(5, 2n_{cp}+3)}), \quad (21)$$

where n_{cp} is the order of critical points ($n_{cp} = 0$ means non-critical points). It can be seen that the Z-Y nonlinear weights can satisfy the condition of (17) in all the smooth regions including critical points. As a result, the WCNS-C7-Z-Y scheme achieves optimal 7th order.

Through equivalent transformations, (18) can be expressed in an adaptive ϵ form, which is

$$\alpha_k = d_k\left(1 + \frac{\tau_5}{\beta_k + \epsilon^Y}\right). \quad (22)$$

In another word, the Y type nonlinear weights are equivalent to the original nonlinear weights with an adaptive ϵ^Y, which is

$$\epsilon^Y = \frac{(1-\phi)(\bar{\beta} + \epsilon_0)}{(\phi + \epsilon_0)} + \epsilon_0. \quad (23)$$

where ϵ_0 is the machine zero added to avoid division by zero, and $\epsilon_0 = 10^{-40}$ is used in this paper.

Previous research work in [9,11,13] has shown that ϵ is a very important parameter because of its great impact on some properties of the nonlinear weights. In fact, the nonlinear weights are determined by the ratios between $\beta_k + \epsilon$, namely

$$(\beta_0 + \epsilon) : (\beta_1 + \epsilon) : (\beta_2 + \epsilon), \quad (24)$$

for interpolations with 3 sub-stencils. Ratios in (24) closer to one lead to nonlinear weights closer to the corresponding linear weights. With the assumptions $\beta_i = q\beta_j = q\beta_l$, $(i, j, l = 0, 1, 2; i \neq j \neq l)$, the ratio of $(\beta_i + \epsilon) : (\beta_j + \epsilon)$ becomes a function of β_i. The curves of $(\beta_i + \epsilon) : (\beta_j + \epsilon)$ as a function of β_i with $q = 2, 10, 100$ are given in Fig. 2(a). It can be seen that ϵ serves as a threshold value here. If $\beta_k (k = 0, 1, 2)$ is much smaller than ϵ, the nonlinear effect is shut down by the ϵ despite the value of q. As a consequence, larger ϵ makes the nonlinear scheme closer to the corresponding optimal linear scheme. Since the ϵ has so dramatic influence on the nonlinear scheme, many researchers have proposed their own choices as is shown in Table 1. The ϵ_A, ϵ_B and ϵ_C are simply fixed numbers. The ϵ_D and ϵ_E, which are functions of the grid spacing, can help the YC nonlinear weights and Z nonlinear weights obtain optimal order, respectively. The ϵ_F is similar to ϵ_E but has much smaller value. The ϵ_G is adaptive based on the flow field. However, all the choices in Table 1 have to be small to capture shocks, and the relatively large ones (ϵ_C and ϵ_E) already destroy the essentially non-oscillatory property in some cases as will be shown in Section 4.

Now we analyze the properties of the newly designed ϵ^Y. Near discontinuities, $\phi \to 1.0$ [15], which leads to $\epsilon^Y \to \frac{1-1}{1}\bar{\beta} = 0$. Since $\phi \to 0$ and $(\beta_{max} + \kappa h^5)/(\beta_{min} + \kappa h^5) \to 1$ in smooth regions [15], we get $\epsilon^Y \to 1/\kappa$ by expanding ϕ near zero. The test case corresponding to Fig. 2 of [10] is used here to verify the theoretical analysis(for simplicity, κ is set to one in this section). Its initial

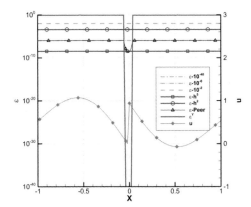

Fig. 3. Comparison of values of different ϵ.

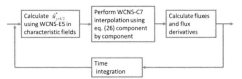

Fig. 4. Flowchart of solution procedure.

conditions are

$$u(x,0) = \begin{cases} -\sin(x\pi) - \frac{1}{2}x^3 & x \in [-1,0), \\ -\sin(x\pi) - \frac{1}{2}x^3 + 1 & x \in [0,1]. \end{cases} \quad (25)$$

Periodic boundary condition and 100 grid points($h = 2 \times 10^{-2}$) are used. Distribution of ϵ^Y is plotted in Fig. 3. Different ϵ in Table 1 are also plotted for comparison. Near the discontinuity, $\epsilon^Y = \epsilon_0$. which is very small. So the scheme has good essentially non-oscillatory property using ϵ^Y. In smooth regions, ϵ^Y is of $O(1/\kappa)$, which is much larger than the choices in Table 1. As a result, the ϵ^Y dominates the term $\beta_k + \epsilon^Y$ in most smooth regions and the nonlinear weights are much closer to the optimal ones.

Just like Fig. 2(a), curves of $(\beta_i + \epsilon^Y) : (\beta_j + \epsilon^Y)$ are given in Fig. 2(b) with the same assumptions as in Fig. 2(a). Unlike the almost constant threshold value for the ratios of $(\beta_i + \epsilon) : (\beta_j + \epsilon)$ to switch to 1 in Fig. 2(a), the threshold value varies with $q = \beta_i : \beta_j$ when ϵ^Y is used. The threshold value goes smaller as q becomes larger. If β_i/β_j is large(like the situation near strong discontinuities) there is a great suspicion of discontinuities, and ϵ^Y switches the nonlinear weights to the optimal linear ones only if β_k is very small. On the contrary, in smooth regions, β_i/β_j are much smaller than β_i/β_j near discontinuities, ϵ^Y switches the nonlinear weights to the linear ones at relatively large β_k value. As a result, much smaller nonlinear dissipation is added to smooth regions.

3.3. Efficient implementation in euler system

Using characteristic-wise nonlinear interpolation is very effective in suppressing oscillations near discontinuities in Euler system. However, the tri-diagnoal matrices of component-wise compact interpolations become a block tri-diagnoal matrix if characteristic-wise interpolations are used, which greatly increases the computation cost. According to our codes, simulations based on characteristic-wise interpolations take approximately 6 times as much CPU time as those based on component-wise interpolations, which makes the characteristic-wise compact schemes very low in efficiency.

Since β_3 is the smoothness indicator of the whole stencil, β_3 would be much larger than β_k of the most smooth sub-stencil near discontinuities, and the interpolation of S_3 would have very little contribution to the nonlinear scheme near discontinuities. In another word, the interpolations of S_0, S_1 and S_2 are the most important part to capture shocks. We try to just use the characteristic projection to interpolations of S_0, S_1 and S_2, and to avoid solving the block tri-diagnoal matrix. (14) can be expressed in another form

$$\begin{pmatrix} \frac{1}{2}\omega_3 \\ \omega_0 + \omega_1 + \omega_2 \\ \frac{3}{14}\omega_3 \end{pmatrix}^T \begin{pmatrix} \tilde{u}_{j-\frac{1}{2}}^* \\ \tilde{u}_{j+\frac{1}{2}}^* \\ \tilde{u}_{j+\frac{3}{2}}^* \end{pmatrix} = (\omega_0 + \omega_1 + \omega_2)\tilde{u}_{j+\frac{1}{2}}^*$$

$$+ \omega_3 \begin{pmatrix} -25/896 \\ 9/32 \\ 15/64 \\ 5/32 \\ 9/128 \end{pmatrix}^T \begin{pmatrix} u_{j-2} \\ u_{j-1} \\ u_j \\ u_{j+1} \\ u_{j+2} \end{pmatrix} \quad (26)$$

in which $\tilde{u}_{j+\frac{1}{2}}^*$ is exactly the same with the nonlinear interpolation of (5) if the same nonlinear weights are used. Since the $\tilde{u}_{j+\frac{1}{2}}^*$ is the most important part on the right hand side of (26) near discontinuities, we just calculate $\tilde{u}_{j+\frac{1}{2}}^*$ in characteristic fields, while the nonlinear interpolations in (26) is still calculated component by component, as can be seen from the flowchart in Fig. 4. As a result we only need to perform N explicit interpolations in characteristic fields and solve N tri-diagnoal matrices (N is the number of equations, which is 5 for 3 dimension problems) instead solving a block tri-diagnoal matrix.

Remark. The new scheme is mainly compared with the 5th-order CRWENO scheme because they share the same stencil, have similar computation cost(their costs are also similar in characteristic fields if the same characteristic projection method are used) and share similar smoothness indicators and nonlinear weights. The original 5th-order explicit schemes are also used for comparison because they are the most widely used ones.

4. Numerical experiments

In this section, the new WCNS-C7-Z-Y scheme are tested using some canonical cases. The new nonlinear weights are applied to the test cases without problem dependent parameters. In all the test cases, third-order Runge–Kutta scheme [5] is used to discrete the time derivatives, and CFL = 0.3 is used if not specified. Characteristic-wise nonlinear interpolations are adopted for cases based on Euler and Navier–Stokes equations. And sixth-order central explicit scheme is adopted to discrete the viscous terms in the Navier–Stokes equations [33]. The turbulent channel flow is the only test case that adopts parallel computing.

4.1. Convergence tests

The evolution of isotropic vortex is used to test the convergence rate of WCNS-C7-Z-Y. The mean flow conditions are $\{\rho, u, v, p\} = \{1,1,0,1\}$ and an isotropic vortex is added to the

Table 2
Convergence test based on isotropic vortex evolution.

Grid	L_∞ error	L_∞ accuracy	L_1 error	L_1 accuracy
40^2	2.89E −4	/	2.32E −6	/
80^2	2.24E −6	7.02	1.58E −8	7.20
100^2	4.48E −7	7.20	3.10E −9	7.29
150^2	2.67E −8	6.96	1.75E −10	7.09
200^2	3.54E −9	7.02	2.30E −11	7.04
300^2	2.04E −10	7.03	1.34E −12	7.02

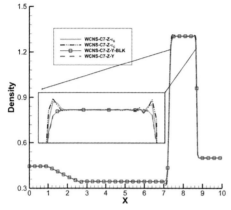

Fig. 5. Computed results of Lax problem.

mean flow. The isotropic vortex is described as

$$\begin{cases} \delta u = -\frac{\chi}{2\pi} e^{(1-r^2)/2}(y-10) \\ \delta v = \frac{\chi}{2\pi} e^{(1-r^2)/2}(x-10) \\ \delta T = -\frac{(\gamma-1)\chi^2}{8\gamma\pi^2} e^{1-r^2} \\ \delta S = 0 \end{cases} \quad (27)$$

where $r^2 = (x-10)^2 + (y-10)^2$ and $\chi = 5.0$. A very small time step is used to reduce the time discretization errors. The errors and accuracy are listed in Table 2. It can be seen that optimal 7th order accuracy is achieved for both the L_∞ error and the L_1 error. The vortex core, where the largest error(L_∞ error) occurs, is a critical point. From the L_∞ accuracy, we can conclude that the scheme also achieves optimal order near critical points.

4.2. Lax problem

This test case is to test the shock capturing ability of the new WCNS scheme. It's initial conditions are

$$(\rho, u, p) = \begin{cases} (0.445, 0.698, 3.528) & x \in [0, 5], \\ (0.5, 0, 0.571) & x \in (5, 10]. \end{cases} \quad (28)$$

The case is calculated using 200 grid points and results at $t = 1.5$ are given in Fig. 5. Essentially non-oscillatory property is achieved by the new WCNS-C7-Z-Y scheme. We also compare different characteristic projection methods in Fig. 5, where WCNS-C7-Z-Y represents the scheme based on the characteristic projection method in Section 3.3, WCNS-C7-Z-Y-BLK represents the one based on solving block tri-diagnoal matrix as the CRWENO. It can be seen that their results coincide with each other, which indicate that the new characteristic projection method is comparable to the original one in suppressing oscillations. Results with $\epsilon_C = 10^{-2}$ and

$\epsilon_E = h^2$ are also shown in Fig. 5. It is clear that they are already too large to maintain the essentially non-oscillatory property.

4.3. Osher-Shu problem

The Osher-Shu problem is a one-dimensional idealization of shock-turbulence interaction. This problem is used to test the capability to accurately capture shock waves and high frequency waves. The initial conditions are

$$(\rho, u, p) = \begin{cases} (3.87143, 2.629369, \frac{31}{3}) & x \in [-5, -4], \\ (1 + 0.2\sin(kx), 0, 1) & x \in (-4, 5]. \end{cases} \quad (29)$$

where $k = 5$. The problem is solved on 200 grid points and results at $t = 1.8$ are given in Fig. 6(a), in which the reference curve is the result of the WCNS-E5-JS [5] on 2000 grid points. It can be seen from Fig. 6(a) that the new WCNS-C7-Y scheme gets the best result. In Fig. 6(a), we also compare different characteristic projection methods as in Section 4.2. It can be seen that the result of WCNS-C7-Z-Y-BLK and the result of WCNS-C7-Z-Y coincide with each other in most regions. However, ratio between their computation costs is approximately 4.4:1 in this test.

We also test the performance of different ϵ in this test case. The results of WCNS-C7-Z with different ϵ are plotted in Fig. 6(b). The results with ϵ of Table 1 coincide with each other in most flow regions. The relatively large ones($\epsilon_C = 10^{-2}$ and $\epsilon_E = h^2$) already lead to oscillations near discontinuities. Result with $\epsilon = \epsilon^Y$ is the closest to the reference curve and no obvious oscillation appears. In Fig. 6(c), the ϵ^Y distribution is given. We can see that ϵ^Y is very small near discontinuities and is very large in smooth regions including the high frequency wave region. As a result, in smooth regions ratios between $\beta_k + \epsilon^Y$ is much closer to one than ratios between β_k and the resulting nonlinear weights are very close to the optimal weights.

4.4. Double Mach reflection problem

Double Mach reflection problem is widely used to test the performance of nonlinear schemes in capturing strong two-dimensional discontinuities. The initial conditions are

$$(\rho, u, v, p) = \begin{cases} (8.0, 7.14471, -4.125, 116.5), & x < \frac{1}{6} + \frac{y}{\sqrt{3}}, \\ (1.4, 0, 0, 1.0), & x \geq \frac{1}{6} + \frac{y}{\sqrt{3}}. \end{cases} \quad (30)$$

with a Mach 10 shock reflected from the wall with an incidence angle of $60°$ [4]. The calculations are based on the two-dimensional Euler equations in $[0, 4] \times [0, 1]$ computation region with 960×240 grid points. Global Lax–Friedrichs numerical flux is used in this problem.

Density distribution of WCNS-C7-Z-Y is shown in Fig. 7(a), it can be seen that the discontinuities are captured without obvious oscillation. The flow details near the shear layer is given in Fig. 7(b). From the vortices captured along the shear layer, it can be seen that the WCNS-C7-Z-Y has very high resolution. The distribution of ϵ^Y based on one of characteristic variables in the Y-direction is illustrated in Fig. 7(c). In consistent with theoretical analysis, ϵ^Y has large values in most smooth regions and very small value near discontinuities.

4.5. Homogeneous compressible turbulence

The WCNS-C7-Z-Y is now tested in freely decaying homogeneous compressible turbulence without shock in a periodic $[0, 2\pi] \times [0, 2\pi] \times [0, 2\pi]$ square box. The efficiency of the WCNS-C7-Z-Y is also investigated using side-by-side comparisons of different schemes. The initial conditions are random, isotropic velocity fluctuations satisfying a prescribed energy spectrum [34]

$$E(k) = Ak^4 \exp(-2k^2/k_0^2), \quad (31)$$

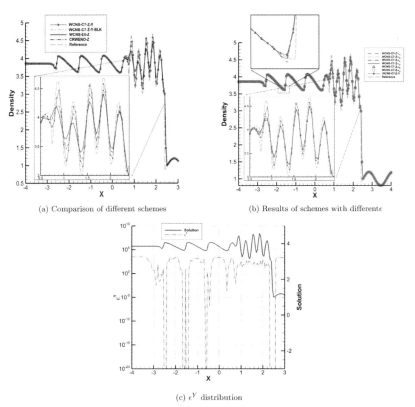

(a) Comparison of different schemes

(b) Results of schemes with different ϵ

(c) ϵ^Y distribution

Fig. 6. Results of Osher-Shu problem.

Table 3
Computation costs of different schemes.

Grids	128^3			80^3	96^3	108^3	192^3	
Schemes	WCNS-E5	WCNS-C7	CRWENO	WCNS-C7	WCNS-C7	WCNS-C7	WCNS-E5	
Weights	JS/Z	Z-Y	Z-Y	Z	Z-Y	Z-Y	Z-Y	JS/Z
Costs	1.00	1.16	1.93	7.24	0.29	0.64	0.95	5.04

where k is the wave number, k_0 is the wave number at which the spectrum peaks, and A is a constant chosen to get a specified initial kinetic energy. In this paper, we take $k_0 = 8.0$, $M_t = 0.3$ and $Re_\lambda = 72.0$. where M_t is the initial turbulence Mach number and Re_λ is the initial Taylor micro-scale Reynolds number. In the results the time is normalized by the large-eddy-turnover time(τ) [34] and kinetic energy is normalized by the initial kinetic energy.

For the simulation of WCNS-C7-Z-Y on 129^3 grid, it takes 396 steps to reach $t = 1.0$ and about 120 seconds per time step. Side-by-side comparisons are conducted to investigate the efficiency of different schemes. The kinetic energy decay curve are shown in Fig. 8(a). The WCNS-C7-Z-Y gets similar results on coarser grid. Fig. 8(b) shows the energy spectrum at $t = 1.0$. From Fig. 8(b), it can be seen that WCNS-C7-Z-Y can capture more small scale struc-

tures. The computation costs (normalized by the computation cost of the WCNS-E5-JS) is listed in Table 3. From Fig. 8 and Table 3, we can conclude that the WCNS-C7-Z-Y scheme largely improves the efficiency, because it gives better or comparable results on coarser grid. However, the computation cost of CRWENO-Z is too high that its efficiency is even lower than WCNS-E5-Z, although it can also give better results on coarser grid. Overall, the efficiency of WCNS-C7-Z-Y is improved by approximately 17.4 times compared with WCNS-E5-JS, because it preserves similar kinetic energy on 80^3 grid as the WCNS-E5-JS does on 192^3 grid. And similarly the efficiency improvement estimations of WCNS-C7-Z-Y compared with WCNS-E5-Z and CRWENO-Z are 3.4 times and 11.3 times, respectively.

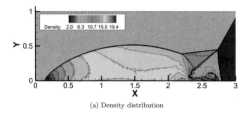

(a) Density distribution

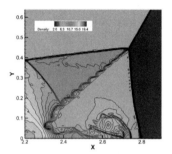

(b) Flow details near the shear layer

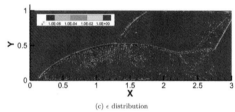

(c) ϵ distribution

Fig. 7. Results of double Mach problem.
(colored picture attached at the end of the book)

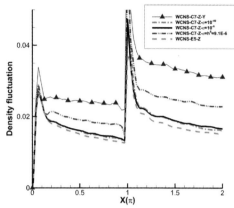

Fig. 9. Streamwise variation of density fluctuation.

4.6. Shock turbulence interaction

The interaction of an isotropic turbulent flow with a normal shock wave has been extensively studied [1,35], which can test the turbulence simulation ability at the present of shock. Details of initial conditions, boundary treatment and method to calculate statistically relevant quantities can be found in [15]. Only value of some key parameters are listed here: $M_{shock} = 2.0$(shock Mach number), $M_t = 0.1$, $k_0 = 8.0$ and $Re_\lambda = 72.0$. $96 \times 32 \times 32$ uniform grid points are used in the simulations (with $32 \times 32 \times 32$ for the sponge zone).

Results with different ϵ are given in Fig. 9. The result with $\epsilon = \epsilon^Y$ preserves the most turbulence fluctuations. Nonlinear index (NI) in [36] is used to estimate the nonlinear errors of WCNS with different ϵ. And the smaller NI is, the less nonlinear errors the scheme has. Fig. 10 shows the distributions of NI for different ϵ. Nonlinear weights with $\epsilon = \epsilon^Y$ are very close to the optimal weights through out the flow field except for the regions near the shock and near the inflow boundary. Near the shock, nonlinear effect is activated to maintain essentially non-oscillatory property,

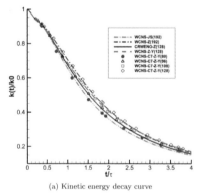

(a) Kinetic energy decay curve

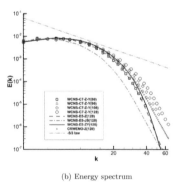

(b) Energy spectrum

Fig. 8. Comparison of kinetic energy decay curves with different schemes.

4.7. Turbulent channel flow

Turbulent channel flow is very simple from a geometrical point of view. But most of wall turbulence interactions are represented in it [37,38]. The computation domain is $[0, 2\pi] \times [0, 2] \times [0, \pi]$ in streamwise (x), wall-normal (y) and spanwise (z) direction, respectively. $80 \times 80 \times 80$ grid points are used. They are uniformly distributed in the x and z direction, while they are clustered toward walls in the y direction. Periodical boundary condition is applied in the x and the z directions, and nonslip condition is used at the walls. Details of initial conditions, boundary treatments, method to calculate the body force and method to calculate statistics can be found in [37,38]. Only values of some key parameters are listed here: the Mach number based on bulk velocity and wall temperature is 0.5 and Reynolds number based on bulk velocity, bulk density, wall viscosity and half channel width is 3000. The boundary closure method similar to the one in [39] are used to the WCNS-C7 scheme near walls. The computational domain is extended using ghost cells and 7th-order explicit interpolation (WCNS-E7) is used for the first and last cell-edges. The Symmetrical Conservative Metric Method (SCMM) is used to calculate the grid metrics [24] to satisfy GCL.

Four cores are used in the parallel computation. The CFL number is approximately 0.63. 100,000 steps are calculated in total, in which the first 50,000 steps are used to form the statistically stationary state and the following 50,000 steps are to obtain statistics by averaging the flow field. It takes approximately 2.04 s per step for WCNS-C7-Z-Y. Results of WCNS-C7-Z-Y and WCNS-E5-Z are given in Fig. 11, in which the EXP curves are experimental data of Nierdershulte et al. [38] in Fig. 11(a) and corrected experimental data of Eckelmann [38] in Fig. 11(b). It's clear that WCNS-C7-Z-Y gives correct mean velocity profile, which is very close to the experimental results. However, the results of WCNS-E5-Z deviate from the experimental results more obviously because it has larger error in predicting the skin friction velocity. Fig. 12 shows the Q iso-surface. It can be seen that the WCNS-C7-Z-Y captures more small scale structures in the simulations.

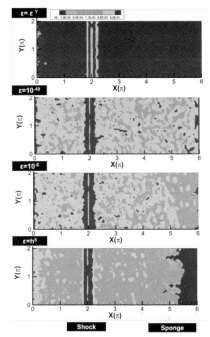

Fig. 10. Distributions of NI of schemes with different ϵ.

(colored picture attached at the end of the book)

and near the inflow boundary it is activated because there is a fierce transition from the initially unphysical flow field to a physical one, which can be reflected in Fig. 9. On the contrary, for other ϵ the nonlinear weights deviate from the optimal weights in most regions. In another word, these nonlinear weights wrongly treat turbulence as shocks and wrongly add large nonlinear dissipations to turbulence. In summary, the WCNS-C7-Z-Y has much better property in distinguishing turbulence from shocks and in simulating small scale structures.

5. Conclusions

In this paper, new 7th-order compact nonlinear interpolation method is developed on the same stencil as 5th-order CRWENO scheme. Proper nonlinear method based on Y type nonlinear weights are developed for the compact interpolation. The Y type nonlinear weights are also further analyzed in this paper. Through equivalent transformation, it is found that they are equivalent to the original nonlinear weights with an adaptive ϵ^Y. The ϵ^Y is very small near discontinuities and is much larger than other ϵ

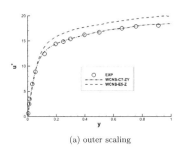

(a) outer scaling

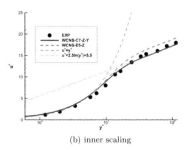

(b) inner scaling

Fig. 11. Mean streamwise velocity profile.

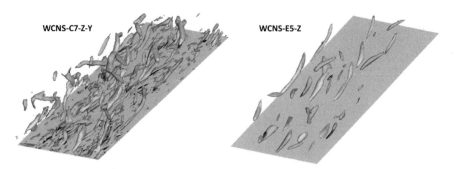

Fig. 12. Q iso-surface of results of different schemes. (colored picture attached at the end of the book)

tested in smooth regions. As a result, the scheme with ϵ^Y not only captures discontinuities without obvious oscillation but also has improved resolution and optimal accuracy. To improve the efficiency of the characteristic-wise compact interpolation, a new characteristic projection method is proposed, which reduces the computation cost by about 4 times.

Canonical test cases are used to verify the theoretical analysis. In the convergence test, WCNS-C7-Z-Y achieves optimal 7th order even near critical points. The good shock capturing ability of WCNS-C7-Z-Y is clearly shown in Lax problem and double Mach reflection problem. From the results of high frequency waves in Osher-Shu problem and vortices captured in double Mach problem, WCNS-C7-Z-Y exhibits higher resolution than other schemes tested. The decay curve and energy spectrum of homogeneous compressible turbulence problem show that WCNS-C7-Z-Y gets similar or better results on coarser grid and the efficiency is largely improved. Finally, shock turbulence interaction problem and turbulent channel flow problem are used to test the performance of WCNS-C7-Z-Y in compressible turbulence simulations with shocks and the ability to simulate turbulent boundary-layer. The results indicate that WCNS-C7-Z-Y has obvious advantages in distinguishing turbulence from shocks and has advantages in capturing small scale structures and boundary-layer.

Acknowledgments

This study was supported by National Natural Science Foundation of China (Grant Nos. 11572342, 11301525 and 11372342) and the foundation of State Key Laboratory of Aerodynamics.

References

[1] Ducros F, Ferrand V, Nicoud F, Weber C, Darracq D, Gacherieu C, et al. Large-eddy simulation of the shock/turbulence interaction. J Comput Phys 1999;152:517–49.
[2] Wang ZJ, Fidkowski K, Abgrall R, Bassi F, Caraeni D, Cary A, et al. High-order cfd methods: current status and perspective. Int J Numer Methods Fluids 2013;72:811–45.
[3] Liang X, Li X. Direct numerical simulation on mach number and wall temperature effects in the turbulent flows of flat-plate boundary layer. Commun Comput Phys 2015;17(1):189–212.
[4] Jiang G-S, Shu C-W. Efficient implementation of weighted eno schemes. J Comput Phys 1996;126(1):202–28.
[5] Deng XG, Zhang HX. Developing high-order weighted compact nonlinear schemes. J Comput Phys 2000;165(1):22–44.
[6] Jiang L, Shan H, Liu C. Weighted compact scheme for shock capturing. Int J Comut Fluid Dyn 2001;15:147–55.
[7] Harten A, Engquist B, Osher S, Chakravarthy SR. Uniformly high order essentially non-oscillatory schemes, iii. J Comput Phys 1987;71(2):231–303.
[8] Liu XD, Osher S, Chan T. Weighted essentially non-oscillatory schemes. J Comput Phys 1994;115(1):200–12.
[9] Henrick AK, Aslam TD, Powers JM. Mapped weighted essentially non-oscillatory schemes: achieving optimal order near critical points. J Comput Phys 2005;207:542–67.
[10] Borges R, Carmona M, Costa B, Don WS. An improved weighted essentially non-oscillatory scheme for hyperbolic conservation laws. J Comput Phys 2008;227:3191–211.
[11] Yamaleev NK, Carpenter MH. A systematic methodology for constructing high-order energy stable weno schemes. J Comput Phys 2009;228:4248–72.
[12] Castro M, Costa B, Don WS. High order weighted essentially non-oscillatory weno-z schemes for hyperbolic conservation laws. J Comput Phys 2011;230:1766–92.
[13] Don W-S, Borges R. Accuracy of the weighted essentially non-oscillatory conservative finite difference schemes. J Comput Phys 2013;250:347–72.
[14] Arandiga F, Marti MC, Mulet P. Weights design for maximal order weno schemes. J Sci Comput 2014;60:641–59.
[15] Yan Z, Liu H, Mao M, Zhu H, Deng X. New nonlinear weights for improving accuracy and resolution of weighted compact nonlinear scheme. Comput Fluids 2016;127:226–40.
[16] Lele SK. Compact finite-difference schemes with spectral-like resolution. J Comput Phys 1992;103(1):16–42.
[17] Deng XG, Jiang Y, Mao ML, Liu HY, Tu GH. Developing hybrid cell-edge and cell-node dissipative compact scheme for complex geometry flows. Sci Chin Technol Sci 2013a;56(10):2361–9.
[18] Deng XG, Jiang Y, Mao ML, Liu HY, Li S, Tu GH. A family of hybrid cell-edge and cell-node dissipative compact schemes satisfying geometric conservation law. Comput Fluids 2015;116:29–45.
[19] Cockburn B, Shu CW. Nonlinearly stable compact schemes for shock calculations. SIAM J Numer Anal 1994;31(3):607–27.
[20] Deng X, Maekawa H. Compact high-order accurate nonlinear schemes. J Comput Phys 1997;130(1):77–91.
[21] Fu H, Wang Z, Yan Y, Liu C. Modified weighted compact scheme with global weights for shock capturing. Comput Fluids 2014;96:165–76.
[22] Ghosh D, Baeder JD. Compact reconstruction schemes with weighted eno limiting for hyperbolic conservation laws. SIAM J Sci Comput 2012;34(3):A1678–706.
[23] Deng XG, Mao ML, Tu GH, Liu HY, Zhang HX. Geometric conservation law and applications to high-order finite difference schemes with stationary grids. J Comput Phys 2011a;230(4):1100–15.
[24] Deng XG, Min YB, Mao ML, Liu HY, Tu GH, Zhang HX. Further studies on geometric conservation law and applications to high-order finite difference schemes with stationary grids. J Comput Phys 2013b;239:90–111.
[25] Nonomura T, Iizuka N, Fujii K. Freestream and vortex preservation properties of high-order weno and wcns on curvilinear grids. Comput Fluids 2010;39(2):197–214.
[26] Wang SY, Deng XG, Wang GX, Xu D, Wang DF. Efficiency benchmarking of seventh-order tridiagonal weighted compact nonlinear scheme on curvilinear mesh. Int J Comut Fluid Dyn 2016;30(7–10):469–88.
[27] Deng XG, Mao ML, Jiang Y, Liu HY. New high-order hybrid cell-edge and cell-node weighted compact nonlinear schemes. In: 20th AIAA Computational Fluid Dynamics Conference, AIAA 2011-3857; 2011b. p. 1–10.
[28] Deng XG, Jiang Y, Mao ML, Liu HY, Tu GH. Developing hybrid cell-edge and cell-node dissipative compact scheme for complex geometry flows. In: The Ninth Asian Computational Fluid Dynamics Conference (Invited); 2012. p. 1–11.
[29] Tu GH, Deng XG, Min YB, Mao ML, Liu HY. Method for evaluating spatial accuracy order of cfd and applications to wcns on four typically distorted meshes. ACTA Aerodyn Sinica 2014;32(4):425–32.
[30] Shen Y, Zha G. Improved seventh-order weno scheme. 48th AIAA Aerospace Sciences Meeting Including the New Horizons Forum and Aerospace Exposition, AIAA 2010-1451; 2010.
[31] Hu XY, Adams NA. Scale separation for implicit large eddy simulation. J Comput Phys 2011;230:7240–9.

[32] Peer AAI, Dauhoo MZ, M B. A method for improving the performance of the weno5 scheme near discontinuities. Appl Math Lett 2009;22:1730-3.
[33] Deng XG. High-order accurate dissipative weighted compact nonlinear schemes. Sci China Ser A-Mathe Phys Astron 2002;45(3):356-70.
[34] Samtaney R, Pullin DI, Kosovic B. Direct numerical simulation of decaying compressible turbulence and shocklet statistics. Phys Fluids (1994-present) 2001;13(5):1415-30.
[35] Pirozzoli S. Conservative hybrid compact-weno schemes for shock-turbulence interaction. J Comput Phys 2002;178(1):81-117.
[36] Wu M, Martin MP. Direct numerical simulation of supersonic turbulent boundary layer over a compression ramp. AIAA J 2007;45(4):879-89.
[37] Lenormand E, Sagaut P, Phuoc LT. Large eddy simulation of subsonic and supersonic channel flow at moderate reynolds number. Int J Numer Methods Fluids 2000;32:369-406.
[38] Fang J, Yao Y, Li Z, Lu L. Investigation of low-dissipation monotonicity-preserving scheme for direct numerical simulation of compressible turbulent flows. Comput Fluids 2014;104:55-72.
[39] Ghosh D, Baeder JD. Weighted non-linear compact schemes for the direct numerical simulation of compressible, turbulent flows. J Sci Comput 2014;61(1):61-89.

New nonlinear weights for improving accuracy and resolution of weighted compact nonlinear scheme

Zhenguo Yan [a,*], Huayong Liu [a], Meiliang Mao [a,b], Huajun Zhu [a], Xiaogang Deng [c]

[a] State Key Laboratory of Aerodynamics, China Aerodynamics Research and Development Center, P.O. Box 211, Mianyang, Sichuan 621000, People's Republic of China
[b] Computational Aerodynamics Institute, China Aerodynamics Research and Development Center, P.O. Box 211, Mianyang, Sichuan 621000, People's Republic of China
[c] National University of Defense Technology, Changsha, Hunan 410073, People's Republic of China

ARTICLE INFO

Article history:
Received 7 August 2015
Revised 15 November 2015
Accepted 7 January 2016
Available online 14 January 2016

Keywords:
Hyperbolic conservation laws
High-order schemes
Weighted compact nonlinear scheme (WCNS)
Nonlinear weights

ABSTRACT

This paper proposes a new kind of nonlinear weights to improve accuracy and resolution of high-order weighted compact nonlinear scheme. The new nonlinear weights are constructed based on not only the ratios between different smoothness indicators but also their values. The values of smoothness indicators are explicitly considered in the basic formulas of the new nonlinear weights after a careful analysis of convergence accuracy and shock capturing property. The new nonlinear weights approach to the optimal weights as the smoothness indicators approaching zero. Thus, the new nonlinear weights are close to the optimal weights in smooth regions where the smoothness indicators are small. Therefore, optimal order accuracy is maintained in smooth regions. In addition, near discontinuities the new nonlinear weights degenerate to the original nonlinear weights used for designing. Thus, discontinuity capturing ability is ensured. Numerical results show that the weighted compact nonlinear scheme with the new nonlinear weights achieves optimal order accuracy even near high-order critical points, captures discontinuities sharply without obvious oscillation, has higher resolution and higher efficiency than other nonlinear schemes and has obvious advantage in capturing small scale structures.

© 2016 Elsevier Ltd. All rights reserved.

1. Introduction

High-order accurate and high-resolution schemes with discontinuity capturing ability are desired in simulating multi-scale flows which contain shock waves, such as direct numerical simulation (DNS) and large eddy simulation (LES) of high speed turbulence [1–4]. Many high-order discontinuity capturing schemes have been constructed. In 1980s, third-order essentially non-oscillatory (ENO) scheme was constructed by Harten and Engquist [5]. Later, Jiang and Shu [6] put forward weighted ENO (WENO) scheme by combining the weighting technique with the ENO scheme. Compared with the ENO scheme, the WENO scheme has similar discontinuity capturing property but higher-order accuracy and higher resolution. However, the WENO scheme proposed by Jiang and Shu [6] (WENO-JS) can not achieve optimal order accuracy near critical points of smooth solutions where some leading derivatives of the solution vanish [7,8].

To solve this problem, Henrick et al. [7] put forward a Mapped WENO scheme (WENO-M), which ensures optimal order accuracy at first-order critical points. The WENO-M scheme exhibits better resolution than the WENO-JS scheme, but its computation cost is about 25% higher than the latter. Borges and co-workers [8] solved this problem by proposing Z nonlinear weights. And the WENO scheme with the Z nonlinear weights (WENO-Z) can also achieve optimal order accuracy at first-order critical points [8,9]. Compared with the WENO-JS scheme, the WENO-Z scheme has similar computation cost, but can achieve higher resolution and capture discontinuities more sharply [8,9]. However, the WENO-M and WENO-Z schemes still suffer a loss in accuracy near second-order and higher-order critical points. To achieve optimal order accuracy near high-order critical points, Yamaleev and Carpenter [10,11] proposed some limitations of ϵ, a parameter originally introduced to avoid the denominator becoming zero. Castro, Don and coworkers [12,13] made some further studies on the limitations of ϵ for the WENO-Z scheme and proved that the WENO-Z scheme can achieve optimal order accuracy near high-order critical points if ϵ satisfies some carefully designed limitations.

Many investigations on the ϵ have shown that it is far more than a parameter to avoid the denominator becoming zero, it also affects the accuracy and resolution of the nonlinear schemes. Arandiga et al. [14,15] and Kolb [16] have made some valuable work on ϵ based on the work of Yamaleev and Carpenter [10,11]. Osher and Fedkiw [17] have pointed out that ϵ is a dimensional quantity and have proposed a new formula for ϵ which scales consistently with the local flow variables. Henrick et al. have found that ϵ has a dramatic effect on the convergence order of the WENO-JS scheme near critical points [7]. Their results indicate that the nonlinear weights have completely different performances for the case that β_k is much smaller than or comparable to ϵ and for the case that β_k is much larger than ϵ. In another word, the values of β_k have been implicitly introduced into the nonlinear weights if ϵ is much larger than the machine zero. Thus, the nonlinear weights in [6,10–13] are actually dependent on both the ratios between and the values of β_k. There is also some other work that makes the nonlinear effects related to the values of β_k or some other undivided flow variable derivatives to improve some properties of nonlinear schemes, such as the studies in [18–20]. The numerical results in these papers have revealed the fact that considering the values of β_k properly can improve the properties of nonlinear schemes.

Traditionally, the nonlinear weights are designed to use the ratios between β_k to detect discontinuities. However, the ratios between β_k may also be very large in smooth regions, for example, near critical points. Thus these points may be treated like discontinuities, which may lead to a loss in accuracy. Note that the values of β_k in smooth regions are much smaller than those near discontinuities. Thus, it is possible for nonlinear weights to distinguish critical points from discontinuities by considering the values of β_k.

However, the properties (resolution, for example) of the nonlinear weights in [6,10–13] can be improved only in a limited part of the smooth regions because the values of β_k are effective only if β_k is much smaller than or comparable to ϵ and ϵ has to be small enough to restrict the oscillations which may appear near discontinuities. Based on these observations, making full use of β_k to improve the properties of nonlinear schemes is the main motivation of the current work.

In this paper, a completely new method is put forward to explicitly consider the values of β_k in the basic formulas of the nonlinear weights, which is different from previous work that introduces the values of β_k implicitly by ϵ [6,10–13]. Besides achieving optimal order accuracy, the resolution of the corresponding nonlinear scheme is improved in the major smooth regions rather than only in a limited part of the smooth regions. The present work is based on weighted compact nonlinear scheme (WCNS) [21] which also uses nonlinear weights similar to those of the WENO schemes. The WCNS can easily satisfy geometric conservation law (GCL) [22,23] and has superiority in simulations on complex grids [24]. The new nonlinear weights can also be extended to the WENO schemes. Some canonical cases, such as Osher–Shu problem double Mach problem, shock turbulence interaction problem are used to test the shock capturing ability, high frequency wave simulating ability and turbulence simulating ability of the WCNS with the new nonlinear weights. It is shown that the new scheme not only achieves optimal order accuracy and captures discontinuities without obvious oscillation, but also has higher resolution and obvious advantage in simulating turbulence.

The organization of the paper is as follows. The WCNS is introduced in Section 2 which also contains a convergence analysis and three kinds of nonlinear weights. In Section 3, the new nonlinear weights are put forward, and their properties are analyzed. Section 4 presents numerical experiments to verify the theoretical analysis. At last, conclusions are drawn in Section 5.

2. WCNS and it convergence analysis

In Section 2.1, the framework of the WCNS is introduced briefly. Then convergence order of the WCNS is analyzed and sufficient conditions for optimal order accuracy are derived in Section 2.2. Finally, three typical kinds of nonlinear weights are presented and analyzed in Section 2.3.

2.1. WCNS

Hyperbolic conservation law has the following form

$$\frac{\partial u}{\partial t} + \frac{\partial f(u)}{\partial x} = 0, \quad (1)$$

where u is a conserved quantity, f describes its flux. Here, we restrict our discussion to the one dimensional scalar case.

Consider a uniform grid defined by $x_j = j\triangle x = jh$, $j = 0, \ldots, N$, where $\triangle x = h$ is the uniform grid spacing. The semi-discrete form of Eq. (1) yields an ordinary differential equation:

$$\frac{du_j(t)}{dt} = -F_j', \quad (2)$$

where $u_j(t)$ is a numerical approximation of $u(x_j, t)$, and F_j' is a spacial discretization of $\frac{\partial f}{\partial x}|_{x=x_j} = f_j'$. In this paper, the WCNS is used for spatial discretization.

The WCNS [21] was proposed by combining the weighted nonlinear interpolation with central compact schemes of Lele [25]. The WCNS can satisfy GCL easily by using symmetrical conservative metric method (SCMM) [23] to calculate grid metrics, and numerical tests have shown that WCNS is robust and can give accurate results on complex grids [24,26]. The WCNS consists of three parts [21,27]: high-order flux difference scheme for flux derivatives, numerical flux construction for cell-edge fluxes and high-order interpolation for cell-edge variable values.

In the first part, hybrid cell-edge and cell-node compact scheme (HCS) in [27,28] is adopted to calculate flux derivatives. HCS is an extension of the cell-node mesh compact scheme and cell-centered mesh compact scheme of Lele [25]. It uses both cell-edge and cell-node values on the right hand side of the scheme, which results in better spectral properties [27,29,30]. The general form of HCS reads

$$\gamma F_{j-2}' + \chi F_{j-1}' + F_j' + \chi F_{j+1}' + \gamma F_{j+2}'$$
$$= \frac{\varphi}{h}(\tilde{F}_{j+\frac{1}{2}} - \tilde{F}_{j-\frac{1}{2}}) + \frac{1}{h}\sum_{m=1}^{3} a_m(f_{j+m} - f_{j-m}). \quad (3)$$

where f_{j+m} is the flux at cell-node $j+m$,

$$\tilde{F}_{j+\frac{1}{2}} = \tilde{F}(\tilde{u}_{j+\frac{1}{2}}^R, \tilde{u}_{j+\frac{1}{2}}^L). \quad (4)$$

is the numerical flux at cell-edge $j+\frac{1}{2}$, $\tilde{u}_{j+\frac{1}{2}}^R$ and $\tilde{u}_{j+\frac{1}{2}}^L$ are cell-edge variable values calculated by upwind-like nonlinear interpolations. The parameters in Eq. (3) determine the accuracy and spectral property of the HCS. In this paper, we consider a special case (HCS-E6)

$$F_j' = \frac{\varphi}{h}\left(\tilde{F}_{j+\frac{1}{2}} - \tilde{F}_{j-\frac{1}{2}}\right) + \frac{192 - 175\varphi}{256h}\left(f_{j+1} - f_{j-1}\right)$$
$$+ \frac{-48 + 35\varphi}{320h}\left(f_{j+2} - f_{j-2}\right) + \frac{64 - 45\varphi}{3840h}\left(f_{j+3} - f_{j-3}\right). \quad (5)$$

Replacing $\tilde{F}_{j\pm\frac{1}{2}}$ by the exact fluxes, we can get the following estimation using Taylor series expansion

$$F_j' = f_j' + \left(\frac{1}{8960} - \frac{5}{65536}\varphi\right)f_j^{(6)}h^6 + O(h^8). \quad (6)$$

In general, Eq. (5) with exact cell-edge fluxes has sixth-order accuracy. In this paper, $\varphi = 256/175$ is taken, with which the HCS in Eq. (5) has eighth-order accuracy.

In the second part, the choice of numerical flux can be flexible. In this paper, Steger and Warming's numerical flux is used for its simplicity and good properties [31].

In the third part, nonlinear methods are used for cell-edge variable interpolations to capture discontinuities without obvious oscillation. When constructing the WCNS, Deng and Zhang [21] developed a flow variable interpolation method inspired by the weighted reconstruction of the WENO-JS scheme. Under the assumption $df/du > 0$, Eq. (4) now has the form $\tilde{F}_{j+\frac{1}{2}} = \tilde{F}(\tilde{u}^L_{j+\frac{1}{2}})$. And only formulas for $\tilde{u}^L_{j+\frac{1}{2}}$ is given. $\tilde{u}^R_{j+\frac{1}{2}}$ can be easily calculated since it is symmetry with $\tilde{u}^L_{j+\frac{1}{2}}$. In the following, L will be dropped for simplicity. Fifth-order weighted nonlinear interpolation has the following form

$$\tilde{u}_{j+\frac{1}{2}} = \sum_{k=0}^{2} \omega_k \tilde{u}^k_{j+\frac{1}{2}}. \tag{7}$$

where $\omega_k (k = 0, 1, 2)$ are nonlinear weights of the three sub-stencils, and $\tilde{u}^k_{j+\frac{1}{2}} (k = 0, 1, 2)$ are third-order linear interpolations of $u_{j+\frac{1}{2}}$ on the three sub-stencils with the following form

$$\begin{aligned}
\tilde{u}^0_{j+\frac{1}{2}} &= \frac{1}{8}(15u_j - 10u_{j-1} + 3u_{j-2}), \\
\tilde{u}^1_{j+\frac{1}{2}} &= \frac{1}{8}(3u_{j+1} + 6u_j - u_{j-1}), \\
\tilde{u}^2_{j+\frac{1}{2}} &= \frac{1}{8}(-u_{j+2} + 6u_{j+1} + 3u_j).
\end{aligned} \tag{8}$$

Expansions of Eqs. (8) in Taylor series give

$$\tilde{u}^k_{j+\frac{1}{2}} = u_{j+\frac{1}{2}} + A^k_j h^3 + O(h^4), \quad (k = 0, 1, 2), \tag{9}$$

where $A^0_j = -5u^{(3)}_j/16$, $A^1_j = u^{(3)}_j/16$ and $A^2_j = -u^{(3)}_j/16$. The nonlinear weights are designed such that they approach to the optimal weights ($d_0 = 1/16, d_1 = 10/16$ and $d_2 = 5/16$) as close as possible in smooth regions, meanwhile, they approach to appropriate values to prevent interpolations from crossing discontinuities. Various kinds of nonlinear weights can be used in Eq. (7), and some kinds of them are presented in Section 2.3. Setting $\omega_k = d_k$ in Eq. (7), we can achieve the optimal order accuracy of Eq. (7)

$$\tilde{u}_{j+\frac{1}{2}} = \sum_{k=0}^{2} d_k \tilde{u}^k_{j+\frac{1}{2}} = u_{j+\frac{1}{2}} + B_j h^5 + O(h^6), \tag{10}$$

where $B_j = -3u^{(5)}_j/256$.

2.2. Convergence analysis

In this section, convergence order of the WCNS is analyzed and sufficient conditions on the nonlinear weights for optimal order accuracy are derived, which will be basic constraints for designing the new nonlinear weights.

Using Eqs. (7), (9) and (10), it can be derived that

$$\tilde{u}_{j\pm\frac{1}{2}} = u_{j\pm\frac{1}{2}} + B_j h^5 + O(h^6) + u_{j\pm\frac{1}{2}} \sum_{k=0}^{2}(\omega^{\pm}_k - d_k)$$

$$+ h^3 \sum_{k=0}^{2} A^k_j (\omega^{\pm}_k - d_k) + \sum_{k=0}^{2}(\omega^{\pm}_k - d_k) O(h^4), \tag{11}$$

for $j+\frac{1}{2}$ and $j-\frac{1}{2}$. The superscript $\pm$ is added to ω_k to distinguish the values at different cell-edges while no $\pm$ is added to d_k,

B_j and A^k_j since they have the same value at the two cell-edges. Expansions of $\tilde{F}_{j\pm\frac{1}{2}}$ at $u_{j\pm\frac{1}{2}}$ give

$$\tilde{F}_{j\pm\frac{1}{2}} = f_{j\pm\frac{1}{2}} + \left(\frac{\partial f}{\partial u}\right)_{j\pm\frac{1}{2}} (\tilde{u}_{j\pm\frac{1}{2}} - u_{j\pm\frac{1}{2}})$$

$$+ \frac{1}{2}\left(\frac{\partial^2 f}{\partial u^2}\right)_{j\pm\frac{1}{2}} (\tilde{u}_{j\pm\frac{1}{2}} - u_{j\pm\frac{1}{2}})^2 + \cdots. \tag{12}$$

Then, substituting Eqs. (11) and (12) into Eq. (5) and ignoring high-order terms, we obtain

$$\begin{aligned}
F'_j =& f'_j + O(h^8) \\
&+ \varphi \frac{\left(\frac{\partial f}{\partial u}\right)_{j+\frac{1}{2}} u_{j+\frac{1}{2}} \sum_{k=0}^{2}(\omega^+_k - d_k) - \left(\frac{\partial f}{\partial u}\right)_{j-\frac{1}{2}} u_{j-\frac{1}{2}} \sum_{k=0}^{2}(\omega^-_k - d_k)}{h} \\
&+ \varphi\left[\left(\frac{\partial f}{\partial u}\right)_{j+\frac{1}{2}} \sum_{k=0}^{2}(\omega^+_k - d_k) O(h^3)\right. \\
&\left. -\left(\frac{\partial f}{\partial u}\right)_{j-\frac{1}{2}} \sum_{k=0}^{2}(\omega^-_k - d_k) O(h^3)\right] + h^5 B_j \left(\frac{\partial^2 f}{\partial u^2}\right)_j u'_j \\
&+ \left[h^2 \varphi\left(\frac{\partial f}{\partial u}\right)_j + O(h^4)\right] \sum_{k=0}^{2} A^k_j(\omega^+_k - \omega^-_k) \\
&+ \left[h^3 \varphi\left(\frac{\partial^2 f}{\partial u^2}\right)_j u'_j + O(h^5)\right] \\
&\times \left[\sum_{k=0}^{2} A^k_j(\omega^+_k - d_k) + \sum_{k=0}^{2} A^k_j(\omega^-_k - d_k)\right].
\end{aligned} \tag{13}$$

Thus, necessary and sufficient conditions for fifth-order convergence are

$$\sum_{k=0}^{2}(\omega^{\pm}_k - d_k) = O(h^6), \tag{14}$$

$$\sum_{k=0}^{2} A^k_j(\omega^+_k - \omega^-_k) = O(h^3), \tag{15}$$

$$\sum_{k=0}^{2} A^k_j(\omega^{\pm}_k - d_k) = O(h^2), \tag{16}$$

$$\omega^{\pm}_k - d_k = O(h^2). \tag{17}$$

Since condition (16) can be derived from condition (17), the conditions (14)–(17) are actually the same with the ones derived in [7,8,10] for the WENO schemes. The condition (14) is always satisfied, since $\sum \omega^{\pm}_k = \sum d_k = 1$ by construction. Overall, a simple sufficient condition for optimal order accuracy is

$$\omega^{\pm}_k - d_k = O(h^3). \tag{18}$$

In this paper, we define the nonlinear errors of the nonlinear weights as

$$e_{\omega_k} = \omega_k - d_k. \tag{19}$$

And we say that the nonlinear weights are r_{th}-order accurate, if

$$e_{\omega_k} = O(h^r). \tag{20}$$

From Eq. (13), we can see that the orders of the nonlinear weights not only affect the convergence order of the WCNS but also reflect the deviation of the overall scheme from the optimal linear one. Higher-order accuracy of the nonlinear weights usually means that the corresponding nonlinear scheme has smaller nonlinear errors in smooth region. This may explain why Yamaleev, Castro and Hu et al. tried to increase the orders of nonlinear weights in [10,12,18]. In this paper, we will also attempt to increase the orders of nonlinear weights as high as possible to decrease nonlinear errors.

2.3. Nonlinear weights

In this section, three different kinds of nonlinear weights are presented, which give direct inspirations of the current work. We also give an examination of whether these nonlinear weights can meet the sufficient conditions for optimal order accuracy.

The first kind of nonlinear weights is the JS nonlinear weights [6], which are

$$\omega_k = \frac{\alpha_k}{\sum_{l=0}^{2}\alpha_l}, \tag{21}$$

$$\alpha_k = \frac{d_k}{(\beta_k + \epsilon)^p}. \tag{22}$$

Here, ω_k is designed such that it approaches to d_k in smooth regions and interpolations across discontinuities are avoided. ϵ is originally introduced to avoid the denominator becoming zero, however it plays more role than that actually [7]. Parameter p can control the dissipation and discontinuity capturing ability, and $\epsilon = 10^{-6}$ and $p = 2$ are recommended in [6]. β_k in Eq. (22) is the smoothness indicator of the k_{th} sub-stencil. In this paper, β_k defined in [21] is used, namely,

$$\beta_k = (hg_j^k)^2 + (h^2 s_j^k)^2. \tag{23}$$

where g_j^k and s_j^k are numerical approximations of first and second derivatives on the k_{th} sub-stencil, respectively. Substituting the expressions of g_j^k and s_j^k into β_k, we get

$$\begin{aligned}\beta_0 &= \frac{1}{4}(3u_j - 4u_{j-1} + u_{j-2})^2 + (u_j - 2u_{j-1} + u_{j-2})^2,\\ \beta_1 &= \frac{1}{4}(u_{j+1} - u_{j-1})^2 + (u_{j+1} - 2u_j + u_{j-1})^2,\\ \beta_2 &= \frac{1}{4}(-u_{j+2} + 4u_{j+1} - 3u_j)^2 + (u_{j+2} - 2u_{j+1} + u_j)^2.\end{aligned} \tag{24}$$

Numerical results have shown that the WCNS with the above JS nonlinear weights (WCNS-JS) can capture discontinuities without obvious oscillation [21,27].

To analyze whether the WCNS-JS can achieve optimal order accuracy, we expand Eqs. (24) as

$$\begin{aligned}\beta_0 &= u_j'^2 h^2 + \left(u_j''^2 - \frac{2}{3}u_j' u_j'''\right)h^4 + \left(\frac{1}{2}u_j' u_j''' - 2u_j'' u_j'''\right)h^5 + O(h^6),\\ \beta_1 &= u_j'^2 h^2 + \left(u_j''^2 + \frac{1}{3}u_j' u_j'''\right)h^4 + O(h^6),\\ \beta_2 &= u_j'^2 h^2 + \left(u_j''^2 - \frac{2}{3}u_j' u_j'''\right)h^4 + \left(-\frac{1}{2}u_j' u_j''' + 2u_j'' u_j'''\right)h^5 + O(h^6).\end{aligned} \tag{25}$$

It can be easily deduced from Eqs. (25) that $\beta_k = u_j'^2 h^2 (1 + O(h^2))$ is satisfied if $u_j' \neq 0$. In addition, for the JS nonlinear weights, it is proved in [8] that

$$\beta_k = D(1 + O(h^r)) \quad implies \quad e_{\omega_k} = O(h^r), \tag{26}$$

where D is independent of k. As a result, the JS nonlinear weights are second-order accurate. Although the sufficient condition (18) is not satisfied, Borges et al. [8] pointed out that all of conditions (14)–(17) can be satisfied. So the WCNS-JS can achieve optimal order accuracy at non-critical points. However, if $u_j' = 0$ is encountered, β_k can only satisfy $\beta_k = D(1 + O(h))$, which results in a loss in accuracy.

For real calculations, the choice of $\epsilon = 10^{-6}$ may help the JS nonlinear weights obtain optimal order accuracy in some cases. For example, when $\beta_k \ll \epsilon$, the differences between different $\beta_k + \epsilon$ become negligible and optimal order accuracy can be obtained, even if $u_j' = 0$. Thus, the nonlinear weights have completely different performance for $\beta_k \ll \epsilon$ and for $\beta_k \gg \epsilon$, which means that the values of β_k have been implicitly introduced into the nonlinear weights. The numerical tests in [7] also show that the performance of the JS nonlinear weights is very sensitive to the ratios of β_k to ϵ. And ϵ can help the JS nonlinear weights improve convergence property and decrease errors only if β_k is comparable to or much smaller than ϵ. For a given problem, the JS nonlinear weights can not guarantee optimal order accuracy since the values of β_k are influenced by both the local flow field and the local grid resolution.

The second kind of nonlinear weights is the Z nonlinear weights [8], which is constructed by calculating Eq. (21) with the following α_k

$$\alpha_k = \frac{d_k}{\beta_k^Z} = d_k\left(1 + \left(\frac{\tau_5}{\beta_k + \epsilon}\right)^q\right), \tag{27}$$

$$\tau_5 = |\beta_2 - \beta_0|. \tag{28}$$

where β_k is the smoothness indicator of Eq. (23) and $\epsilon = 10^{-40}$ is used.

For the Z nonlinear weights, it can be easily derived [8] that

$$\alpha_k = d_k\left(1 + \left(\frac{\tau_5}{\beta_k + \epsilon}\right)^q\right) \quad implies \quad e_{\omega_k} = O\left(\left(\frac{\tau_5}{\beta_k + \epsilon}\right)^q\right). \tag{29}$$

Expanding Eqs. (24) and (28) in Taylor series, we have

$$\begin{aligned}\beta_k &= O(h^{2(n_{cp}+1)}),\\ \tau_5 &= \begin{cases}O(h^5), & n_{cp} = 0\\ O(h^{2(n_{cp}+1)+1}), & n_{cp} \geq 1\end{cases}\\ &= O(h^{\max(5,2(n_{cp}+1)+1)}),\end{aligned} \tag{30}$$

for n_{cp}th-order critical points ($u'(x_c) = \cdots = u^{(n_{cp})}(x_c) = 0, u^{(n_{cp}+1)}(x_c) \neq 0$), where n_{cp} is a non-negative integer. Substituting Eqs. (30) into Eq. (27) we can obtain

$$\alpha_k = d_k\left(1 + O\left(h^{q(\max(5,2(n_{cp}+1)+1)-2(n_{cp}+1))}\right)\right). \tag{31}$$

Both the $q = 1$ and $q = 2$ versions of the Z nonlinear weights are considered in [8]. From Eqs. (29) and (31), it can be verified that if $n_{cp} = 0$ (non-critical points), optimal order accuracy can be obtained for the nonlinear scheme with either $q = 1$ or $q = 2$. However, if n_{cp} becomes larger, the Z nonlinear weights will have lower order accuracy, which eventually leads to a loss in accuracy of the overall scheme. The WCNS-Z with $q = 1$ degenerates to fourth-order if $n_{cp} = 1$, for instance. Borges et al. [8] have pointed out that the nonlinear scheme achieves optimal order accuracy at first-order critical points with $q = 2$, while it exhibits smaller dissipation with $q = 1$. In this paper, the Z nonlinear weights with $q = 1$ is used, and the property of the Z nonlinear weights with $q = 2$ can be reflected from the following third kind of nonlinear weights.

The third kind of nonlinear weights are a modified version of the Z nonlinear weights to achieve optimal order accuracy near high-order critical points. In [10,11], Yamaleev and Carpenter found that with some limitations of ϵ, the nonlinear scheme always has optimal order accuracy even near high-order critical points. Don and Borges [13] used the similar idea of [10,11] in the Z nonlinear weights and proved that the Z nonlinear weights with $q = 2$ and $\epsilon = h^4$ can maintain optimal order accuracy at high-order critical points, which is called ZO nonlinear weights, a abbreviation for Z nonlinear weights with optimal order accuracy. For the ZO nonlinear weights, α_k has the form

$$\alpha_k = d_k\left(1 + \left(\frac{\tau_5}{\beta_k + h^4}\right)^2\right). \quad (32)$$

However, increasing q causes more numerical dissipation which may have more influence than the convergence order at critical points in some problems [8]. In the ZO nonlinear weights, $\epsilon = h^4$ works similarly to the $\epsilon = 10^{-6}$ of the JS nonlinear weights in helping the nonlinear scheme achieve optimal order accuracy. In addition, the $\epsilon = h^4$ also makes the nonlinear weights related to the values of β_k, and can help improve the properties of the nonlinear weights only if β_k is small, since ϵ is small to restrict the oscillations near discontinuities. The difference is that the ϵ in the ZO nonlinear weights is determined by the convergence criteria instead of a pre-set value in the JS nonlinear weights.

It is a highly desired property that the numerical scheme is invariant when the space and time are uniformly scaled by the same factor, or the flow variables are uniformly scaled while keeping the similarity parameters constant (Mach number and Reynolds number, for instance). The former property is known as self-similarity (or scale-invariant) property [10,11,13], and the latter property is called scale-invariant of flow variables property in this paper. The scaling of space and time or the scaling of the flow variables may be frequently encountered since it may be caused by different choices of units for computations without nondimensionalization or by different choices of nondimensionalization procedure for computations with nondimensionalization. It can be easily checked that the nonlinear weights even with $\epsilon = 10^{-6}$ are self-similar. However, they are not self-similar if $\epsilon = h^4$ is used, and possible ways to solve this problem were given in [10,11,13]. The scale-invariant of flow variables property is not satisfied if either $\epsilon = 10^{-6}$ or $\epsilon = h^4$ is used since they have implicitly introduced into the nonlinear weights the values of smoothness indicators which also changes as the flow variables are scaled. This problem is shared by all nonlinear weights that directly use the values of smoothness indicators, and some extra care is required to solve this problem. Some possible solutions have been given in [10,11,17].

In summary, the original JS nonlinear weights suffer a loss in accuracy near critical points. The Z nonlinear weights solve the problem of losing accuracy near first-order critical points. With the idea of limiting ϵ, Don and Borges finally managed to obtain optimal order accuracy near critical points of arbitrary order by developing the ZO nonlinear weights, which, however, have their own drawbacks. Firstly, increasing q leads to more numerical dissipation. Secondly, ϵ can help improve the properties of the nonlinear weights only if β_k is very small. As a result, the ZO nonlinear weights do not guarantee a better result than the Z nonlinear weights, which will be illustrated in Section 4.

3. Design and analysis of the new nonlinear weights

Usually, nonlinear weights are designed based on the ratios between the smoothness indicators (β_k). However, the values of smoothness indicators themselves can also reflect the smoothness of the flow field. Previous work indicates that some properties of the nonlinear weights can be improved with the help of the values of β_k. In this paper, we will try to introduce the values of β_k explicitly into the nonlinear weights and try to get new nonlinear weights with better properties in achieving optimal order accuracy and high resolution. To achieve this aim, a new kind of smoothness indicators is defined in the following form,

$$\beta_k^Y = (1-\phi)\bar{\beta} + \phi\beta_k, \; \phi = \phi(\bar{\beta}). \quad (33)$$

where $\bar{\beta}$ is the algebraic average of β_k. Replacing β_k by β_k^Y in the original nonlinear weights, we can get the new nonlinear weights.

In this way, the values of smoothness indicators have been introduced into the new nonlinear weights. The ϕ in Eq. (33) is vital in the property of the new nonlinear weights, the final formula of which is given in Eq. (53) at the end of Section 3. In the following, we will give detailed analysis of the new nonlinear weights and the step by step development of ϕ.

3.1. Analysis of the new nonlinear weights

Replacing β_k by β_k^Y in the Z nonlinear weights (Eqs. (27) and (28)), we get the formulas of the new Z-Y nonlinear weights

$$\alpha_k = d_k\left(1 + \frac{\tau_5^Y}{\beta_k^Y + \epsilon}\right) = d_k\left(1 + \frac{\phi|\beta_2 - \beta_0|}{\bar{\beta} + \phi(\beta_k - \bar{\beta}) + \epsilon}\right), \quad (34)$$

$$\tau_5^Y = |\beta_2^Y - \beta_0^Y| = \phi\tau_5, \quad (35)$$

where $\epsilon = 10^{-40}$ is used as in [8].

From Eq. (33), it can be easily seen that

$$\begin{aligned}\phi \to 0 &\Rightarrow \beta_k^Y \to \bar{\beta}, \\ \phi \to 1 &\Rightarrow \beta_k^Y \to \beta_k.\end{aligned} \quad (36)$$

Therefore, if ϕ satisfies the following property

$$\phi \to \begin{cases} 0, & \text{in smooth regions}, \\ 1, & \text{near discontinuities}, \end{cases} \quad (37)$$

β_k^Y will approach to the same value ($\bar{\beta}$) in smooth regions and the resulting nonlinear weights will approach to the optimal weights. Meanwhile, β_k^Y will degenerate to β_k near discontinuities and in this way discontinuity capturing ability is ensured.

Besides the property of Eq. (37), ϕ should also satisfy some other basic restrictions. Firstly, $0 \leq \phi \leq 1$ should be satisfied, so that β_k^Y is a convex combination of β_k and $\bar{\beta}$. Secondly, ϕ should be a monotone function, so that the ratios between β_k^Y are closer to one if $\bar{\beta}$ is smaller. Thirdly, $\phi(0) = 0$ and $\phi(+\infty) = 1$ should be satisfied, which can be deduced from the above two restrictions and the fact that $0 \leq \bar{\beta} < +\infty$.

3.2. The design of ϕ

In this paper, we take

$$\phi = \tanh(\kappa\bar{\beta}). \quad (38)$$

which satisfies all the restrictions above. Here, κ is added to make ϕ get reasonable value. In the following, the convergence order and the shock capturing ability of the Z-Y nonlinear weights are analyzed to give some guidance for the design of κ. Just like [7], ϵ is assumed to be zero in the theoretical analysis since it only plays the role of keeping the nonlinear weights bounded.

Note that $\bar{\beta}$ is of $O(h^2)$ in smooth regions, we can expand ϕ at zero

$$\phi(\bar{\beta}) = \kappa\bar{\beta} + O(\bar{\beta}^2). \quad (39)$$

Substituting Eq. (39) into Eq. (34) and using the estimations $\bar{\beta} = O(h^2)$, $\beta_k = O(h^2)$, $(\beta_k - \bar{\beta}) = O(h^4)$ and Eq. (30), we have

$$\alpha_k = d_k\left(1 + \kappa O\left(h^{\max(5, 2n_{cp}+3)}\right)\right). \quad (40)$$

Thus, according to Eqs. (18) and (29), a sufficient condition for the overall scheme to achieve optimal order accuracy is

$$\kappa \leq O\left(h^{-\max(2, 2n_{cp})}\right). \quad (41)$$

We can also conclude from Eqs. (29) and (40) that the nonlinear errors are monotonically related to κ, and larger κ results in larger nonlinear errors in smooth regions.

Near discontinuities, $\kappa\bar{\beta}$ should be large enough so that ϕ is very close to one, and in this way the original Z nonlinear weights

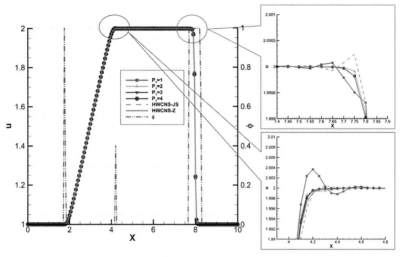

Fig. 1. Scalar case to determine p_1.
(colored picture attached at the end of the book)

are recovered. However, this may not be satisfied if κ is small near discontinuities. For example, in the Sod problem the $\bar{\beta}$ near shock decreases from $O(1)$ to $O(10^{-2})$ as the shock gradually gets smeared. In this case, ϕ can not be close to one unless κ is large. Therefore, κ should be large enough near discontinuities.

Based on the above analysis of smooth and discontinuous regions, one possible choice of κ is

$$\kappa = \left(\frac{\beta_{max} + h^5}{\beta_{min} + h^5}\right)^{p_1} \quad (42)$$

where $\beta_{max} = \max(\beta_0, \beta_1, \beta_2)$, $\beta_{min} = \min(\beta_0, \beta_1, \beta_2)$ and $p_1 > 0$.

Firstly, we consider the properties of κ in smooth regions. For non-critical points, it can be verified that $\kappa = O(1)$ by using the estimations in Eqs. (25). And the h^5 in Eq. (42) plays no more role than keeping κ bounded. For critical points, κ may become very large without the h^5. For example, when $n_{cp} = 2$, we have

$$\beta_0 = 10(u''')^2 h^6/9 + O(h^7),$$
$$\beta_1 = (u'')^2 h^6/36 + O(h^8), \quad (43)$$
$$\beta_2 = 10(u''')^2 h^6/9 + O(h^7),$$

and $\beta_{max}/\beta_{min} \approx 40$. And β_{max}/β_{min} grows even larger as n_{cp} increases. Then, the nonlinear errors will be greatly enlarged, although optimal order accuracy may still be achieved since the condition (41) for optimal order accuracy are relaxed with the increase of n_{cp}. Notice that $\beta_{max}/\beta_{min} \gg 1$ holds only if the leading term of β_k is of $O(h^6)$ or even smaller. Following the idea of [10], the h^5 term is added so that $\kappa = O(1)$ is maintained in all smooth regions. In summary, it can be verified that the Z-Y nonlinear weights satisfy the following error estimation

$$e_{\omega_k} = O\left(h^{\max(5, 2n_{cp}+3)}\right). \quad (44)$$

The Z-Y nonlinear weights have higher order accuracy than the Z nonlinear weights in the whole smooth regions, which not only maintains optimal order accuracy of the overall scheme but also leads to smaller nonlinear errors. It is worth noticing that the Z-Y nonlinear weights have even higher order accuracy near critical points rather than the loss in accuracy of traditional nonlinear weights.

Secondly, the properties of κ near discontinuities are analyzed and the parameter p_1 is determined. In smooth regions, p_1 does not affect the value of κ much, since $(\beta_{max} + h^5)/(\beta_{min} + h^5) = O(1)$. However, p_1 affects the value of κ near discontinuities greatly, since $(\beta_{max} + h^5)/(\beta_{min} + h^5)$ is very large. So p_1 should be determined by the performance of the nonlinear weights near discontinuities. A very simple 1D scalar test case based on Burgers' equation with $f(u) = u^2/2$, is used here to determine p_1. The initial conditions go as follows

$$u(x, 0) = \begin{cases} 2, & x \in [0, 5], \\ 1, & x \in [-10, 0] \cup [5, 10]. \end{cases} \quad (45)$$

$p_1 = 1, 2, 3, 4$ are adopted in the calculations. Fig. 1 shows the results at $t = 2.0$, which mainly consist of a shock wave and a rarefaction wave. The results of the WCNS-Z with $\epsilon = 10^{-40}$ and the WCNS-JS with $\epsilon = 10^{-6}$ are also presented for comparison. Near the shock, the result of the WCNS-Z-Y with $p_1 = 1$ and the result of the WCNS-JS exhibit some small oscillations, while no obvious oscillation is observed in the other results. Meanwhile, all schemes give accurate results near the rarefaction wave except for the WCNS-Z-Y with $p_1 = 1$. In this paper, we recommend the WCNS-Z-Y with $p_1 = 2$, which has better performance than the WCNS-JS in capturing shocks in this case. If better shock capturing ability is required, $p_1 = 3$ is recommended, which, however, introduces more nonlinear dissipation. From Fig. 1 we can see that the distribution of ϕ with $p_1 = 2$ satisfies the property of Eq. (37).

3.3. Scale-invariant of flow variables and self-similarity

In the design of the Z-Y nonlinear weights, the values of β_k are introduced to distinguish discontinuities from smooth regions (including critical points). However, the scale-invariant of flow variables property is not satisfied if the values of β_k are used directly. Fortunately, this problem has already been widely studied such as the studies in [10,11,17,20]. To prevent the nonlinear weights from violating the scale-invariant of flow variables caused by ϵ, Osher and Fedkiw [17] and Yamaleev and Carpenter [10,11] multiplied ϵ

by a factor C, which scales consistently with β_k. Take the JS nonlinear weights for example, after multiplying ϵ by C, the α_k becomes

$$\alpha_k = \frac{d_k}{(\beta_k + \epsilon C)^p}, \tag{46}$$

which is equivalent to

$$\alpha_k = \frac{d_k}{(\beta_k/C + \epsilon)^p}. \tag{47}$$

In another word, their work actually makes β_k be divided by C, and make the fraction β_k/C a constant when the flow variables get scaled. We adopt this idea in the Z-Y nonlinear weights. Dividing β_k by C in Eqs. (34) and (36), and using some equivalent transformations, we obtain

$$\alpha_k = d_k \left(1 + \frac{\phi|\beta_2 - \beta_0|}{\bar{\beta} + \phi(\beta_k - \bar{\beta}) + \epsilon C} \right), \tag{48}$$

$$\phi = \tanh\left(\frac{\bar{\beta}}{C} \left(\frac{\beta_{max} + Ch^5}{\beta_{min} + Ch^5} \right)^2 \right). \tag{49}$$

The C used in [10,11] is

$$C = \max_{\xi \neq \xi_d} \left(u_0(\xi)^2, u_0'(\xi)^2, \cdots, u_0^{(r)}(\xi)^2 \right), \tag{50}$$

where $r+1$ is the number of sub-stencils, ξ_d is a set of points at which the solution is discontinuous, u_0 represents initial variables, and u_0' and $u_0^{(r)}$ are initial variable derivatives. Our choice of C is based on Eq. (50) with some minor modifications. The C in Eq. (50) using the velocities may become zero and some extra techniques are needed to solve this problem [10,11]. In this paper, the speed of sound is used instead of the velocities to prevent C from becoming zero. In our tests, it is shown that searching only the maximum of $u_0(\xi)^2$ in Eq. (50) is sufficient to give good results for the Z-Y nonlinear weights, and it is unnecessary to distinguish continuous and discontinuous regions anymore since no derivative is used. In summary, the C in this paper is

$$C(\rho) = \max\left(\rho_0^2\right),$$
$$C(p) = \max\left(p_0^2\right), \tag{51}$$
$$C(v) = \max\left(c_0^2\right),$$

where ρ, p, c are the density, pressure and speed of sound, respectively and the subscript 0 indicates the initial conditions. For nonlinear interpolations based on characteristic variables [21], $C(\rho)$ is used for the first characteristic variable while $C(v)$ is used for the rest characteristic variables, because the characteristic projection matrices are chosen such that the first characteristic variable varies only if the density gets scaled, and the rest characteristic variables vary only if the velocities get scaled [32]. We only need to calculate Eqs. (51) once at the beginning of the computation, thus the computation cost does not increase much.

The self-similarity property is not satisfied with the h^5 term in Eq. (49) as stated in [10,11]. In our numerical tests, it is shown that the WCNS-Z-Y is not sensitive to the h^5 term in the test cases of Section 4, and for simplicity no extra technique is used to make the nonlinear weights self-similar. To make the Z-Y nonlinear weights self-similar, the method developed in [10,11] can be used, which uses $\Delta\xi = 1/N$ instead of h, where N is the total number of grid cells.

In summary, the final formulas for the new nonlinear are

$$\beta_k^Y = (1 - \phi)\bar{\beta} + \phi\beta_k, \tag{52}$$

$$\phi = \tanh\left(\frac{\bar{\beta}}{C} \left(\frac{\beta_{max} + Ch^5}{\beta_{min} + Ch^5} \right)^2 \right). \tag{53}$$

Replacing β_k by β_k^Y in the original nonlinear weights, we can get the new nonlinear weights. In Eqs. (52) and (53), $\bar{\beta} = (\beta_0 + \beta_1 + \beta_2)/3$, $\beta_{max} = \max(\beta_0, \beta_1, \beta_2)$, $\beta_{min} = \min(\beta_0, \beta_1, \beta_2)$, h is the grid spacing and C is introduced to make the scheme scale-invariant of flow variables.

From Eq. (52), it can be seen that a new kind of smoothness indicator is developed by a convex combination of the smoothness indicators and their average value. The ϕ is vital in achieving the design aim, which is to achieve optimal order and high resolution by considering the values of smoothness indicators. In the development of ϕ, the product of the average value of the smoothness indicators ($\bar{\beta}/C$) and the largest ratio between the smoothness indicators ($(((\beta_{max} + Ch^5)/(\beta_{min} + Ch^5))^2)$) is used to detect discontinuities. Then the Eq. (53) is obtained by mapping the product to [0, 1] using the tanh function. The h^5 term, which usually is very small, is added to avoid the denominator becoming zero and to improve the performance near high-order critical points. C is added to make the scheme scale-invariant of flow variables. The C is calculated using Eq. (51), which is obtained by searching the initial flow field, and no repeated cost is needed.

Eqs. (52) and (53) can also be used to develop some other nonlinear weights, the new nonlinear weights based on the JS nonlinear weights (JS-Y nonlinear weights) for example. And we present the equations and numerical results of the JS-Y nonlinear weights based on the WENO scheme in the Appendix A.

4. Numerical experiments

In this section, firstly the convergence tests of the WCNS-Z-Y are given. Then, some canonical cases, such as shock tube problems and double Mach problem are adopted to examine the performance of the WCNS-Z-Y in capturing shocks and high frequency waves. Finally, homogeneous compressible turbulence and shock turbulence interaction problems are adopted to demonstrate the potential of the WCNS-Z-Y in turbulence simulations. The new nonlinear weights are applied to all the test cases with the same formulas and parameters as given in Eqs. (52) and (53). In all the test cases, third-order Runge-Kutta scheme [21] is used to discrete the time derivatives, and CFL=0.3 is used. Nonlinear interpolations using characteristic variables are adopted for cases based on Euler and Navier–Stokes equations. And sixth-order central explicit scheme is adopted to discrete the viscous terms in the Navier–Stokes equations [33].

4.1. Convergence tests

Two test cases are used to verify the convergence analysis of Section 3.2. Consider the linear advection equation (taking $f(u) = u$ in Eq. (1)) with two different initial conditions

$$\text{Case 1}: u(x) = x^3 + \cos(x), \tag{54}$$

and

$$\text{Case 2}: u(x) = x^4 + x^3 \sin(x). \tag{55}$$

The point $x = 0$ is a critical point of first-order ($n_{cp} = 1$) for Eq. (54) and of third-order ($n_{cp} = 3$) for Eq. (55), respectively. For this problem, the C is simply set to be one, since we care about the performance at only one point and no obvious reference scale of u exists.

The errors of numerical flux derivative F' are given in Figs. 2 and 3. The WCNS-Z can maintain fourth-order for Case 1 but only third-order for Case 2. Result of the WCNS-JS exhibits some super-convergence phenomenon in Case 1, however its errors are much larger than these of the optimal linear scheme. In Case 2, the WCNS-JS achieves optimal order accuracy. The performances of the JS nonlinear weights is in agreement with the tests in [7], and they

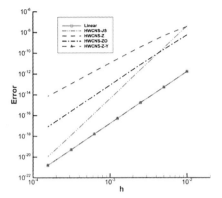

Fig. 2. First-order critical point convergence tests.
(colored picture attached at the end of the book)

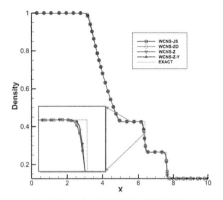

Fig. 4. Exact and computed solutions of Sod problem.
(colored picture attached at the end of the book)

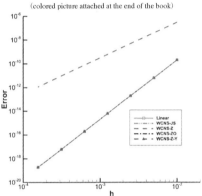

Fig. 3. Third-order critical point convergence tests.
(colored picture attached at the end of the book)

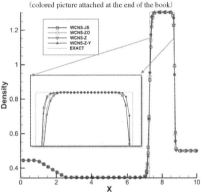

Fig. 5. Exact and computed solutions of Lax problem.
(colored picture attached at the end of the book)

are very sensitive to the relative magnitude of β_k to ϵ. The WCNS-ZO achieves optimal order accuracy in both cases, however the errors in Case 1 are much larger than the errors of the optimal linear scheme. In both cases, the errors of the WCNS-Z-Y coincide with the errors of the optimal linear scheme, which indicates that optimal order accuracy is achieved. These results illustrate that the WCNS-Z-Y can achieve optimal order accuracy at high-order critical points. In addition, the WCNS-Z-Y has smaller nonlinear errors than the WCNS-ZO in Case 1.

4.2. Shock tube problems

Two shock tube cases are used to test the discontinuity capturing ability of the WCNS-Z-Y. The first case is the Sod problem with initial conditions [8]

$$(\rho, u, p) = \begin{cases} (1, 0, 1) & x \in [0, 5], \\ (0.125, 0, 0.1) & x \in (5, 10]. \end{cases} \quad (56)$$

The second case is the Lax problem with initial conditions

$$(\rho, u, p) = \begin{cases} (0.445, 0.698, 3.528) & x \in [0, 5], \\ (0.5, 0, 0.571) & x \in (5, 10]. \end{cases} \quad (57)$$

These two cases are calculated using 200 grid points and results at $t = 1.5$ are given in Figs. 4 and 5. Essentially non-oscillatory property is achieved by all schemes, while the WCNS-JS and the WCNS-ZO are obviously more dissipative than the WCNS-Z and the WCNS-Z-Y near discontinuities.

4.3. Osher–Shu problem

There are shock/high frequency wave interactions in the Osher–Shu problem, which can be used to test the performance of schemes in capturing high frequency waves. The initial conditions are

$$(\rho, u, p) = \begin{cases} (3.87143, 2.629369, \frac{31}{3}) & x \in [-5, -4], \\ (1 + 0.2 \sin(kx), 0, 1) & x \in (-4, 5], \end{cases} \quad (58)$$

where $k = 5$. The problem is solved on 200 grid points and results at $t = 1.8$ are given in Fig. 6, in which the reference curve is the result of the WCNS-JS on 2000 grid points. It can be seen from Fig. 6a that result of the WCNS-Z-Y is the closest to the reference result, while the result of the WCNS-JS is the worst one. The WCNS-ZO exhibits more dissipation than the WCNS-Z and does not guarantee a better result as stated in Section 2.3. In Fig. 6b, the ϕ distribution of the WCNS-Z-Y is given. We can see that ϕ

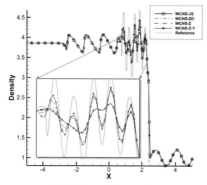

(a) Solutions of different schemes

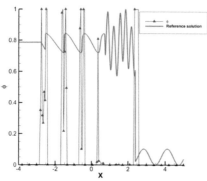

(b) ϕ distribution of WCNS-Z-Y

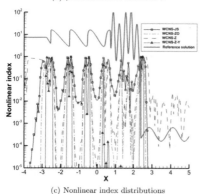

(c) Nonlinear index distributions

Fig. 6. Results of Osher–Shu problem.
(colored picture attached at the end of the book)

has values close to one near discontinuities and values close to zero in smooth regions even near the high frequency waves. From Eq. (36), it is implied that in smooth regions each β_k^Y is very close to the same $\bar{\beta}$ and the resulting nonlinear weights are very close

to the optimal weights. Nonlinear index in [19] is used to estimate the nonlinear errors of different schemes, which has the following form

$$NI = \frac{1}{\sqrt{r(r+1)}}\left[\sum_{k=0}^{r}\left(\frac{1/(r+1) - [(\omega_k/d_k)/\sum_{l=0}^{r}(\omega_l/d_l)]}{1/(r+1)}\right)^2\right]^{1/2}, \quad (59)$$

where $r+1$ is the number of sub-stencils, $NI \in [0, 1]$. The smaller NI is, the less nonlinear errors the scheme has. Fig. 6c shows the distributions of NI for different schemes. Comparing the NI distributions of the WCNS-Z with these of the WCNS-Z-Y, we can conclude that the new way of introducing the values of smoothness indicators can help the nonlinear weights decrease their nonlinear errors and improve their resolution even in the high frequency wave region, where the values of smoothness indicators are much larger than the $\epsilon = h^4$ and the $\epsilon = 10^{-6}$. The nonlinear errors of the WCNS-Z-Y are several orders of magnitude smaller than the errors of the other schemes in smooth regions, which means that the WCNS-Z-Y is very close to the optimal linear scheme in smooth regions. Thus, further improvement in resolution requires to improve the property of the corresponding optimal linear scheme, the study of which is beyond the scope of the current work.

4.4. Double Mach problem

Double Mach problem is widely used to test the performance of nonlinear schemes in capturing strong two-dimensional discontinuities. The initial conditions are

$$(\rho, u, v, p) = \begin{cases} (8.0, 7.14471, -4.125, 116.5), & x < \frac{1}{6} + \frac{y}{\sqrt{3}} \\ (1.4, 0, 0, 1.0), & x \geq \frac{1}{6} + \frac{y}{\sqrt{3}} \end{cases}, \quad (60)$$

with a Mach 10 shock reflected from the wall with an incidence angle of 60° [6]. The calculations are based on the two-dimensional Euler equations in $[0, 4] \times [0, 1]$ computation region with 960×240 grid points.

Density distribution of the WCNS-Z-Y is shown in Fig. 7a, it can be seen that the discontinuities are captured without obvious oscillation. Details of the region around the double Mach stems are shown in Fig. 7b. From the small vortices captured along the slip line, we can see that the WCNS-Z-Y has the highest resolution among the schemes tested.

4.5. Homogeneous compressible turbulence

The WCNS-Z-Y is now tested in freely decaying homogeneous compressible turbulence without shock in a periodic $[0, 2\pi] \times [0, 2\pi] \times [0, 2\pi]$ square box. The efficiency of the WCNS-Z-Y is also investigated using side-by-side comparisons of different schemes. The initial conditions are random, isotropic velocity fluctuations satisfying a prescribed energy spectrum [34]

$$E(k) = Ak^4 \exp\left(-2k^2/k_0^2\right), \quad (61)$$

where k is the wave number, k_0 is the wave number at which the spectrum peaks, and A is a constant chosen to get a specified initial kinetic energy. In this paper, we take $k_0 = 8.0$, $M_t = 0.3$ and $Re_\lambda = 72.0$, where M_t is the initial turbulence Mach number and Re_λ is the initial Taylor micro-scale Reynolds number. Firstly, the problem is simulated using the WCNS-Z-Y on a series of grids with different grid resolutions, and the results are given in Fig. 8, where τ is the large-eddy-turnover time [34] and the kinetic energy is normalized by the initial kinetic energy. As the grid gets refined, more kinetic energy is preserved in the simulations, and the decay curve with $256 \times 256 \times 256$ grid points is in good agreement with the DNS results in [34]. The preservation of kinetic energy is very important

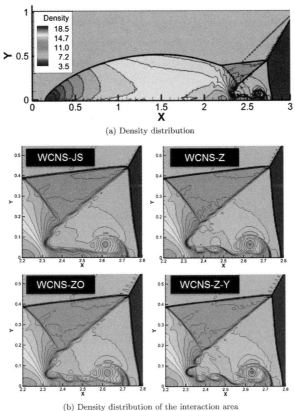

(a) Density distribution

(b) Density distribution of the interaction area

Fig. 7. Results of double Mach problem.
(colored picture attached at the end of the book)

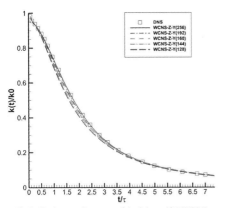

Fig. 8. Kinetic energy decay curve of simulations with WCNS-Z-Y.
(colored picture attached at the end of the book)

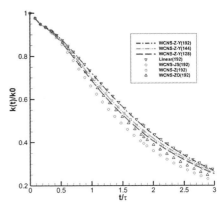

Fig. 9. Comparison of kinetic energy decay curves with different schemes.
(colored picture attached at the end of the book)

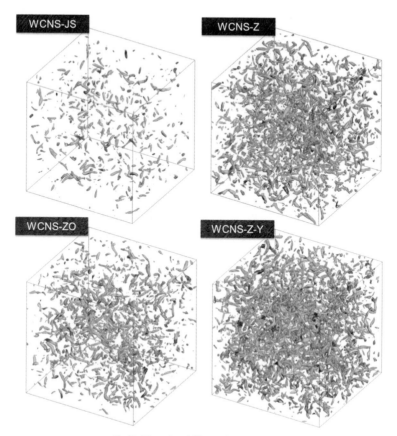

Fig. 10. Q iso-surface of different schemes at $t = 1.0$.
(colored picture attached at the end of the book)

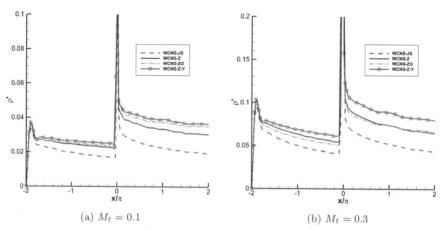

(a) $M_t = 0.1$ (b) $M_t = 0.3$

Fig. 11. Streamwise variation of density fluctuation.
(colored picture attached at the end of the book)

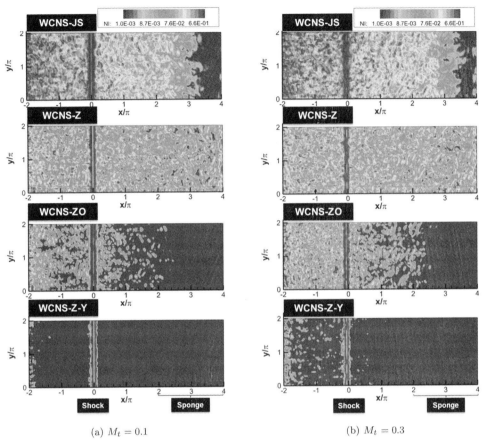

(a) $M_t = 0.1$ (b) $M_t = 0.3$

Fig. 12. Distributions of NI of different schemes. (colored picture attached at the end of the book)

in turbulence simulations and can reflect the turbulence simulation ability of numerical schemes.

In addition, side-by-side comparisons are conducted to investigate the efficiency of different schemes. As shown in Fig. 9, the WCNS-Z-Y preserves more kinetic energy on 128 × 128 × 128 grid than the WCNS-ZO does on 192 × 192 × 192 grid, and the WCNS-Z-Y preserves more kinetic energy on 144 × 144 × 144 grid than the WCNS-Z does on 192 × 192 × 192 grid, which indicate that the Z-Y nonlinear weights can largely improve the turbulence simulation ability of the WCNS. The computation cost (normalized by the computation cost of the WCNS-JS) is listed in Tab. 1. Although the WCNS-Z-Y is computationally more expensive on the same grid, it successfully gets better results and thus the overall efficiency is much higher. On the contrary, the WCNS-ZO dissipates even more kinetic energy than the WCNS-Z doses, despite its advantage in simulating critical points in smooth regions. The decay curve of the optimal linear scheme is also presented in Fig. 9, and it almost coincides with the curve of the WCNS-Z-Y, which implies that the WCNS-Z-Y has very small nonlinear errors in this problem. Fig. 10 shows the Q iso-surface (Q=40) of different schemes at $t = 1.0$ on 128 × 128 × 128 grid. It is clearly shown that the WCNS-Z-Y captures more flow details than the other schemes.

4.6. Shock turbulence interaction

Nonlinear schemes such as the WCNS and the WENO are widely used in turbulence simulations with strong shocks for their good shock capturing ability and relatively high resolution [4,19,35,36]. The interaction of an isotropic turbulent flow with a normal shock wave has been extensively studied [1,2], which can test the turbulence simulation ability at the present of shock. The computational domain is defined by $[-2\pi, 2\pi] \times [0, 2\pi] \times [0, 2\pi]$. And initially there is a stationary, Mach 2 shock at $x = 0$ with uniform flow upstream and downstream of the shock. Periodic boundary conditions are used in the y and z directions. At the inflow boundary the same procedure of Section 4.5 is adopted to produce the incident turbulence, which is added to the uniform supersonic inflow. An extra sponge zone is added at the outflow boundary ($x = 2\pi$),

Table 1
Computation cost of different schemes.

Grids	192^3				128^3	144^3
Schemes	WCNS-JS	WCNS-Z	WCNS-ZO	WCNS-Z-Y	WCNS-Z-Y	WCNS-Z-Y
Costs	1.00	0.99	1.00	1.16	0.23	0.38

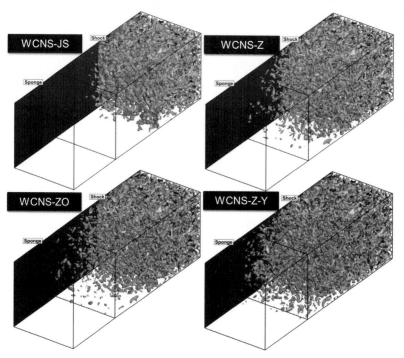

Fig. 13. Q iso-surface of results of different schemes.
(colored picture attached at the end of the book)

which is extended to $x = 4\pi$, where a sink term is added to the governing equations. The sink term has the form of $\sigma(u - u_{ps})$ where σ varies linearly from 0 at $x = 2\pi$ to 1 at $x = 4\pi$ and u_{ps} is the uniform post-shock flow [2]. Characteristic-based outflow boundary conditions are applied at the downstream end of the sponge zone [37]. The simulations are run with two different M_t ($M_t = 0.1$ and $M_t = 0.3$) while $k_0 = 8.0$ and $Re_\lambda = 72.0$ are used in all simulations. $192 \times 64 \times 64$ uniform grid points are used in the simulations (with $64 \times 64 \times 64$ for the sponge zone). The simulations are run long enough to obtain statistically relevant quantities and the procedure is detailed in [1]. No explicit sub-grid stress model is used.

As can be seen from Fig. 11, in both the $M_t = 0.1$ and $M_t = 0.3$ cases the WCNS-Z-Y preserves the most turbulence fluctuations and the WCNS-JS preserves the least. The WCNS-ZO gives better result than the WCNS-Z in the $M_t = 0.1$ case but worse result in the $M_t = 0.3$ case. The distributions of NI are shown in Fig. 12. In the $M_t = 0.1$ case, nonlinear weights of the WCNS-Z-Y are very close to the optimal weights through out the flow field except for the regions near the shock and near the inflow boundary. Near the shock, nonlinear effect is activated to maintain essentially non-oscillatory property, and near the inflow boundary it is activated because there is a fierce transition from the initially unphysical flow field to a physical one, which can be reflected in Fig. 11. In the $M_t = 0.3$ case, the nonlinear weights deviate from the optimal weights in a bit more regions, but they are still very close to the optimal weights in the major flow field. In both cases, the WCNS-Z-Y has much better property in distinguishing turbulence (with various kinds of critical points) from discontinuities. The Q iso-surface (Q = 0.08) of the $M_t = 0.3$ cases is shown in Fig. 13, and it can be seen that the WCNS-Z-Y captures more small scale structures in the simulation.

5. Conclusions

In this paper, Z-Y nonlinear weights are proposed to help weighted compact nonlinear scheme achieve optimal order accuracy and improve resolution. The Z-Y nonlinear weights are designed with the idea of introducing the values of β_k into the basic formulas of the nonlinear weights so that they are determined by both the ratios between β_k and the values of β_k. The Z-Y nonlinear weights tend to retrieve the optimal weights as β_k become smaller, and they degenerate to the Z nonlinear weights when the values of β_k and the ratios between β_k are large. As a result, the Z-Y nonlinear weights achieve optimal order accuracy near critical points of arbitrary order where the values of β_k are very small and have similar discontinuity capturing ability to the original nonlinear weights. In addition, methods are developed to make the Z-Y nonlinear weights scale-invariant of flow variables and self-similar.

Theoretical analysis and convergence tests show that the WCNS-Z-Y can achieve optimal order accuracy even near high-order critical points and its errors are smaller than the other nonlinear schemes tested in this paper. Results of shock tube problems, Osher-Shu problem and double Mach problem show that the WCNS-Z-Y can simulate discontinuities sharply without obvious oscillation. Meanwhile, the results of Osher-Shu problem and double Mach problem indicate that the resolution of the WCNS is improved by the Z-Y nonlinear weights even in high frequency wave regions. Finally, the results of homogeneous compressible turbulence and shock turbulence interaction problem illustrate that the WCNS-Z-Y has obvious advantage in distinguishing turbulent flow from discontinuities, has superiority in capturing small scale structures and is more efficient in turbulence simulations. The same idea is also applied to the original WENO scheme to develop WENO-JS-Y scheme, and results indicate that the WENO-JS-Y has similar advantages over the WENO-JS to those on the WCNS schemes.

Acknowledgments

This study was supported by the project of National Basic Research Program of China (Grant No. 2009CB723800), National Natural Science Foundation of China (Grant Nos. 11301525, 11372342 and 11572342) and the foundation of State Key Laboratory of Aerodynamics (Grant No. JBKY11030902). We would like to thank Dr. Guohua Tu for his valuable suggestions.

Appendix A. Implementing JS-Y nonlinear weights on WENO

Beside the WCNS, the JS-Y nonlinear weights can be easily applied to other weighted nonlinear schemes like the WENO scheme. In the WENO scheme, a numerical flux function, $H(x)$, is implicitly defined by

$$f(u(x)) = \frac{1}{h}\int_{x+h/2}^{x+h/2} H(\xi)d\xi. \tag{62}$$

so that

$$\frac{\partial f}{\partial x}|_{x=x_j} = \frac{H_{j+1/2} - H_{j-1/2}}{h}. \tag{63}$$

holds exactly. The spacial discretization

$$F'_j = \frac{\hat{f}_{j+1/2} - \hat{f}_{j-1/2}}{h}. \tag{64}$$

will be of high order if $\hat{f}_{j+1/2}$ can approximate $H_{j+1/2} = H(x+h/2)$ to a high order. When calculating $\hat{f}_{j+1/2}$, f is first split into two parts, f^+ and f^-, where $df^+/du \geq 0$ and $df^-/du \leq 0$. Then the WENO reconstruction is used to f^+ and f^- separately to get $\hat{f}^+_{j+1/2}$ and $\hat{f}^-_{j+1/2}$. Finally, $\hat{f}_{j+1/2}$ is calculated using

$$\hat{f}_{j+1/2} = \hat{f}^+_{j+1/2} + \hat{f}^-_{j+1/2}. \tag{65}$$

In this paper, the global Lax-Friedrichs flux splitting is used to split the flux. Only WENO reconstruction formulas for f^+ are given, and the superscript will be dropped for simplicity. 5th-order WENO reconstruction has the following form

$$\hat{f}_{j+\frac{1}{2}} = \sum_{k=0}^{2} \omega_k \hat{f}^k_{j+\frac{1}{2}}. \tag{66}$$

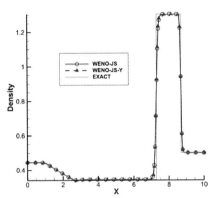

Fig. 14. Results of Lax problem using WENO schemes.
(colored picture attached at the end of the book)

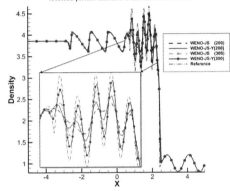

Fig. 15. Results of Osher-Shu problem using WENO schemes.
(colored picture attached at the end of the book)

where ω_k ($k = 0, 1, 2$) in Eq. (66) are nonlinear weights of the three sub-stencils, and $\hat{f}^k_{j+\frac{1}{2}}$ ($k = 0, 1, 2$) are 3rd-order linear reconstruction on the three sub-stencils with the following form

$$\hat{f}^0_{j+\frac{1}{2}} = \frac{1}{6}(11f_j - 7f_{j-1} + 2f_{j-2}),$$
$$\hat{f}^1_{j+\frac{1}{2}} = \frac{1}{6}(2f_{j+1} + 5f_j - f_{j-1}), \tag{67}$$
$$\hat{f}^2_{j+\frac{1}{2}} = \frac{1}{6}(-f_{j+2} + 5f_{j+1} + 2f_j).$$

It should be noted that the coefficients in Eq. (67) are different from those of Eq. (8), since Eq. (67) is used to reconstruct the implicitly defined $H(x)$ while the Eq. (8) is used to interpolate the flow variables. The same nonlinear weights of Eqs. (21) and (22) are used, while the optimal weights become $d_0 = 1/10, d_1 = 6/10$ and $d_2 = 3/10$ and the smoothness indicators become

$$\beta_0 = \frac{1}{4}(3f_j - 4f_{j-1} + f_{j-2})^2 + \frac{13}{12}(f_j - 2f_{j-1} + f_{j-2})^2,$$
$$\beta_1 = \frac{1}{4}(f_{j+1} - f_{j-1})^2 + \frac{13}{12}(f_{j+1} - 2f_j + f_{j-1})^2, \tag{68}$$
$$\beta_2 = \frac{1}{4}(-f_{j+2} + 4f_{j+1} - 3f_j)^2 + \frac{13}{12}(f_{j+2} - 2f_{j+1} + f_j)^2.$$

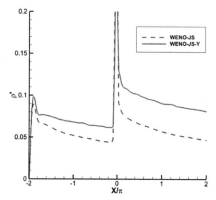

Fig. 16. Streamwise density fluctuation of shock turbulence interaction problem using WENO schemes.

For the JS-Y nonlinear weights, the β_k^Y are obtained by substituting Eqs. (68) into Eqs. (49) and (33). And the final JS-Y nonlinear weights are obtained by replacing β_k by β_k^Y in Eq. (22).

The Lax problem, Osher-Shu problem (using both 200 and 300 grid points) and shock turbulence interaction problem of Section 4 are adopted to illustrate the advantage of the new WENO-JS-Y scheme over the original WENO-JS scheme. From the results in Figs. 14–16, we can draw similar conclusions as in Section 4 that the newly developed Y type nonlinear scheme can simulate discontinuities sharply without obvious oscillation, has higher resolution, and has obvious advantage in capturing small scale structures.

References

[1] Ducros F, Ferrand V, Nicoud F, Weber C, Darracq D, Gacherieu C, et al. Large-eddy simulation of the shock/turbulence interaction. J Comput Phys 1999;152:517–49.
[2] Pirozzoli S. Conservative hybrid compact-WENO schemes for shock-turbulence interaction. J Comput Phys 2002;178(1):81–117.
[3] Wang ZJ, Fidkowski K, Abgrall R, Bassi F, Caraeni D, Cary A, et al. High-order CFD methods: current status and perspective. Int J Numer Methods Fluids 2013;72:811–45.
[4] Liang X, Li X. Direct numerical simulation on mach number and wall temperature effects in the turbulent flows of flat-plate boundary layer. Commun Comput Phys 2015;17(1):189–212.
[5] Harten A, Engquist B, Osher S, Chakravarthy SR. Uniformly high order essentially non-oscillatory schemes, iii. J Comput Phys 1987;71(2):231–303.
[6] Jiang G-S, Shu C-W. Efficient implementation of weighted ENO schemes. J Comput Phys 1996;126(1):202–28.
[7] Henrick AK, Aslam TD, Powers JM. Mapped weighted essentially non-oscillatory schemes: achieving optimal order near critical points. J Comput Phys 2005;207:542–67.
[8] Borges R, Carmona M, Costa B, Don WS. An improved weighted essentially non-oscillatory scheme for hyperbolic conservation laws. J Comput Phys 2008;227:3191–211.
[9] Arshed GM, Hoffmann KA. Minimizing errors from linear and nonlinear weights of WENO scheme for broadband applications with shock waves. J Comput Phys 2013;246:58–77.
[10] Yamaleev NK, Carpenter MH. A systematic methodology for constructing high-order energy stable WENO schemes. J Comput Phys 2009;228:4248–72.
[11] Yamaleev NK, Carpenter MH. Third-order energy stable WENO scheme. J Comput Phys 2009;228:3025–47.
[12] Castro M, Costa B, Don WS. High order weighted essentially non-oscillatory WENO-Z schemes for hyperbolic conservation laws. J Comput Phys 2011;230:1766–92.
[13] Don W-S, Borges R. Accuracy of the weighted essentially non-oscillatory conservative finite difference schemes. J Comput Phys 2013;250:347–72.
[14] Arandiga F, Baeza A, Belda AM, Mulet P. Analysis of WENO schemes for full and global accuracy. SIAM J Numer Anal 2011;49(2):893–915.
[15] Arandiga F, Marti MC, Mulet P. Weights design for maximal order WENO schemes. J Sci Comput 2014;60:641–59.
[16] Kolb O. On the full and global accuracy of a compact third order WENO scheme. SIAM J Numer Anal 2014;52(5):2335–55.
[17] Osher S, Fedkiw R. Level set methods and dynamic implicit surfaces. Applied Mathematical Sciences, vol. 153. New York: Springer; 2003.
[18] Hu XY, Adams NA. Scale separation for implicit large eddy simulation. J Comput Phys 2011;230:7240–9.
[19] Wu M, Martin MP. Direct numerical simulation of supersonic turbulent boundary layer over a compression ramp. AIAA J 2007;45(4):879–89.
[20] Li G, Qiu JX. Hybrid weighted essentially non-oscillatory schemes with different indicators. J Comput Phys 2010;229:8105–29.
[21] Deng XG, Zhang HX. Developing high-order weighted compact nonlinear schemes. J Comput Phys 2000;165(1):22–44.
[22] Deng XG, Mao ML, Tu GH, Liu HY, Zhang HX. Geometric conservation law and applications to high-order finite difference schemes with stationary grids. J Comput Phys 2011;230(4):1100–15.
[23] Deng XG, Min YB, Mao ML, Liu HY, Tu GH, Zhang HX. Further studies on geometric conservation law and applications to high-order finite difference schemes with stationary grids. J Comput Phys 2013;239:90–111.
[24] Nonomura T, Iizuka N, Fujii K. Freestream and vortex preservation properties of high-order WENO and WCNS on curvilinear grids. Comput Fluids 2010;39(2):197–214.
[25] Lele SK. Compact finite-difference schemes with spectral-like resolution. J Comput Phys 1992;103(1):16–42.
[26] Tu GH, Deng XG, Min YB, Mao ML, Liu HY. Method for evaluating spatial accuracy order of CFD and applications to WCNS scheme on four typically distorted meshes. ACTA Aerodyn Sinica 2014;32(4):425–32.
[27] Deng XG, Mao ML, Jiang Y, Liu HY. New high-order hybrid cell-edge and cell-node weighted compact nonlinear schemes. In: Proceedings of the 20th AIAA computational fluid dynamics conference. AIAA 2011-3857; 2011. p. 1–10.
[28] Deng XG, Jiang Y, Mao ML, Liu HY, Tu GH. Developing hybrid cell-edge and cell-node dissipative compact scheme for complex geometry flows. Sci China Technol Sci 2013;56(10):2361–9.
[29] Deng XG, Jiang Y, Mao ML, Liu HY, Tu GH. Developing hybrid cell-edge and cell-node dissipative compact scheme for complex geometry flows. In: Proceedings of the ninth asian computational fluid dynamics conference (invited); 2012. p. 1–11.
[30] Deng XG, Jiang Y, Mao ML, Liu HY, Li S, Tu GH. A family of hybrid cell-edge and cell-node dissipative compact schemes satisfying geometric conservation law. Comput. Fluids 2015;116:29–45.
[31] Tu GH, Zhao XH, Mao ML, Chen JQ, Deng XG, Liu HY. Evaluation of euler fluxes by a high-order CFD scheme: shock instability. Int J Comput Fluid Dyn 2014:1–16.
[32] Hirsch C. Numerical computation of internal and external flows (computational methods for inviscid and viscous Flows). Vol. 2. John Wiley & Sons; 1990.
[33] Deng XG. High-order accurate dissipative weighted compact nonlinear schemes. Sci China Ser A Math Phys Astron. 2002;45(3):356–70.
[34] Samtaney R, Pullin DI, Kosovic B. Direct numerical simulation of decaying compressible turbulence and shocklet statistics. Phys Fluids (1994-present) 2001;13(5):1415–30.
[35] Fujii K. CFD contributions to high-speed shock-related problems. Shock Waves 2008;18:145–54.
[36] Ishiko K, Ohnishi N, Ueno K, Sawada K. Implicit large eddy simulation of two-dimensional homogeneous turbulence using weighted compact nonlinear scheme. J Fluids Eng Trans Asme 2009;131(061401):1–14.
[37] Hoffmann KA, Chiang ST. Computational fluid dynamics; 4th ed.. Engineering Education System, Vol. 2. Wichita, Kansas; 2000.

Parallelizing a High-Order CFD Software for 3D, Multi-block, Structural Grids on the TianHe-1A Supercomputer

Chuanfu Xu[1], Xiaogang Deng[1], Lilun Zhang[1], Yi Jiang[2], Wei Cao[1], Jianbin Fang[3], Yonggang Che[1], Yongxian Wang[1], and Wei Liu[1]

[1] School of Computer, National University of Defense Technology, Changsha 410073, China
xuchuanfu@nudt.edu.cn
[2] State Key Laboratory of Aerodynamics, China Aerodynamics Research and Development Center, Mianyang 621000, China
[3] Parallel and Distributed Systems Group, Delft University of Technology, Delft 2628CD, The Netherlands

Abstract. In this paper, with MPI+CUDA, we present a dual-level parallelization of a high-order CFD software for 3D, multi-block structural girds on the TianHe-1A supercomputer. A self-developed compact high-order finite difference scheme HDCS is used in the CFD software. Our GPU parallelization can efficiently exploit both fine-grained data-level parallelism within a grid block and coarse-grained task-level parallelism among multiple grid blocks. Further, we perform multiple systematic optimizations for the high-order CFD scheme at the CUDA-device level and the cluster level. We present the performance results using up to 256 GPUs (with 114K+ processing cores) on TianHe-1A. We can achieve a speedup of over 10 when comparing our GPU code on a Tesla M2050 with the serial code on an Xeon X5670, and our implementation scales well on TianHe-1A. With our method, we successfully simulate a flow over a high-lift airfoil configuration using 400 GPUs. To the authors' best knowledge, our work involves the largest-scale simulation on GPU-accelerated systems that solves a realistic CFD problem with complex configurations and high-order schemes.

Keywords: GPU parallelization, high-order CFD, multi-block structural grid, heterogeneous system.

1 Introduction

Although low-order (e.g., second-order) schemes for computational fluid dynamics (CFD) have been widely used in engineering applications, they are insufficient in capturing small disturbances in an environment containing sharp gradients. To ensure the high-resolution and fidelity of a numerical simulation, it is imperative that CFD researchers develop and apply robust and high-order CFD methods that can deal with complex flows in complex domains. Among others, compact high-order finite difference schemes based on a compact stencil are very attractive for flows with multiscales (e.g., aeroacoustics and turbulence), due to their high formal order, good spectral resolution and flexibility[1].

Despite the rapid development of high-order algorithms in CFD, the applications of high-order finite difference schemes on complex configurations are still few. One main factor which hinders the widely application of these methods is the complexity of grids. Particularly, the Geometric Conservation Law (GCL)[2] and block-interface conditions, which can be neglected for low-order schemes must be treated carefully for high-order ones when the configurations are complex. In order to apply high-order finite difference schemes on complex multi-block grids, we have developed a Conservative Metric Method (CMM)[3] to calculate the grid derivatives, and employed a Characteristic-Based Interface Condition (CBIC)[4] to fulfill high-order multi-block computing. Based on CMM and CBIC, we have developed a Hybrid cell-edge and cell-node Dissipative Compact Scheme (HDCS)[5] recently and implemented it in our high-order CFD software HOSTA (High-Order SimulaTor for Aerodynamics) for multi-block structural grids (see Sect.2).

HOSTA has been successfully applied to a wide range of flow simulations so far, showing its flexibility and robustness[6]. However, to parallelize HOSTA on supercomputer like TianHe-1A[7] (a GPU-accelerated massive parallel processing system) is challenging. Developers often need to manage different levels of parallelisms using different parallel programming models (e.g., NVIDIA's Compute Unified Device Architecture (CUDA)[8] for GPUs and MPI or OpenMP for CPUs) for heterogeneous compute devices. Further, when performing complex high-order, multi-block grid CFD simulations, we need extensive implementation and optimization efforts to achieve high performance and efficiency.

The work that we present here demonstrates a comprehensive effort to efficiently accelerate **large-scale simulations** of **realistic CFD problems** using both **complex multi-block grids** and **high-order CFD schemes** on the **TianHe-1A supercomputer** (see Sect.3). In the past 8 months, we have successfully parallelized HOSTA (using the HDCS scheme) with MPI+CUDA. When parallelizing HOSTA on a single GPU, we exploit dual-level parallelisms: fine-grained parallelism by using a CUDA thread to compute a cell within a grid block, and coarse-grained parallelism by using multiple CUDA streams to compute multiple blocks. At the CUDA-device level, we also use a kernel-decomposition optimization to further enhance the performance. For efficient simulations on large-scale GPUs, we use non-blocking MPI, CUDA multi-stream and CUDA events to maximize the overlapping of kernel computation, intra-node data transfer and inter-node communication. The GPU-enabled HOSTA shows promising strong and weak scalability for large-scale parallel tests on TianHe-1A. Finally, we simulate a flow over a high-lift airfoil configuration to evaluate the high-order behavior of HDCS. To our best knowledge, this is the largest-scale simulation on GPU-accelerated systems that solves a realistic CFD problem with both complex configurations and high-order schemes so far.

The remainder of the paper is organized as follows. Section 2 briefly describes the numerical methods and HOSTA implementation. In Section 3, we detail our MPI-CUDA implementation and optimizations, and present the performance results and the validation results. In Section 4, we introduce some related work. Finally we conclude this paper in Section 5.

2 Numerical Methods and HOSTA Implementation

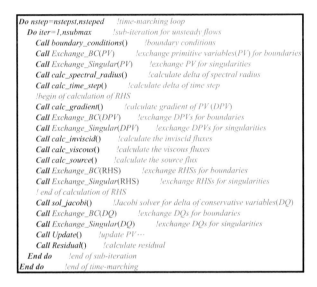

Fig. 1. The main pseudocode for the time-marching loop of HOSTA

In curvilinear coordinates the governing equations (Euler or Navier-Stokes) in strong conservative form are:

$$\frac{\partial \tilde{Q}}{\partial \tau} + \frac{\partial \tilde{F}}{\partial \xi} + \frac{\partial \tilde{G}}{\partial \eta} + \frac{\partial \tilde{H}}{\partial \zeta} = 0 \tag{1}$$

where $\tilde{F}$, $\tilde{G}$ and $\tilde{H}$ are the fluxes along the ξ, η and ζ direction respectively; $\tilde{Q}$ is the conservative variable.

Let us first consider the discretization of the inviscid flux derivative along the ξ direction. The discretization for the other inviscid fluxes can be computed in a similar way. The seventh order HDCS scheme called HDCS-E8T7 can be expressed as follows:

$$\frac{\partial \tilde{F}_i}{\partial \xi} = \frac{256}{175h}(\tilde{F}_{i+1/2} - \tilde{F}_{i-1/2}) - \frac{1}{4h}(\tilde{F}_{i+1} - \tilde{F}_{i-1}) \\ + \frac{1}{100h}(\tilde{F}_{i+2} - \tilde{F}_{i-2}) - \frac{1}{2100h}(\tilde{F}_{i+3} - \tilde{F}_{i-3}) \tag{2}$$

where h is the grid size, $\tilde{F}_{i\pm 1/2} = \tilde{F}(U_{i\pm 1/2})$ are the cell-edge fluxes, and $\tilde{F}_{i+m} = \tilde{F}(U_{i+m})$ are the cell-node fluxes. The cell-edge variables ($U_{i\pm 1/2}$) are interpolated, and the numerical flux $\tilde{F}_{i\pm 1/2}$ can be evaluated by cell-edge variables, which is similar to that of the WCNS (Weighted Compact Nonlinear Schemes)[9]:

$$\tilde{F}_{i\pm 1/2} = \tilde{F}(U^L_{i\pm 1/2}, U^R_{i\pm 1/2}) \qquad (3)$$

where $U^L_{i\pm 1/2}$ and $U^R_{i\pm 1/2}$ are the left-hand and right-hand cell-edge variables. For HDCS-E8T7, the seventh-order dissipative compact interpolation needs to solve the following system of tri-diagonal equations:

$$\begin{aligned}&\tfrac{5}{14}(1-\alpha)U^L_{i-1/2} + U^L_{i+1/2} + \tfrac{5}{14}(1+\alpha)U^L_{i+3/2} = \tfrac{25}{32}(U_{i+1}+U_i) \\ &+ \tfrac{5}{64}(U_{i+2}+U_{i-1}) - \tfrac{1}{448}(U_{i+3}+U_{i-2}) + \alpha[\tfrac{25}{64}(U_{i+1}-U_i) \\ &+ \tfrac{15}{128}(U_{i+2}-U_{i-1}) - \tfrac{5}{896}(U_{i+3}-U_{i-2})]\end{aligned} \qquad (4)$$

where α is the dissipative parameter to control numerical dissipation. For the viscous fluxes, we also use a sixth order central difference scheme (see [9] for more details).

HOSTA is a production-level in-house CFD software containing more than 25,000 lines of FORTRAN90 codes. We present the main pseudo-code for the timing-marching loop of HOSTA in Fig.1. The HDCS scheme is implemented when calculating the viscous (*calc_viscous*) fluxes and the inviscid fluxes (*calc_inviscid*). We have implemented explicit Runge-Kutta method and several implicit time marching methods in HOSTA, but in the following GPU parallelization, we focus on the Jacobi iterative method (*Sol_jacobi*). Note that in each time step HOSTA performs four exchanges of boundary and singularity data to ensure the robustness of high-order schemes (see Fig.1).

3 MPI-CUDA Implementation and Optimizations for HOSTA

When parallelizing HOSTA via MPI, we partition a complex grid with single or multiple computational domains into multiple grid blocks for better load balance and then distribute them to MPI processes (as shown in Fig.2(a)). For a MPI-CUDA implementation, a collection of grid blocks owned by a MPI process will be computed by a GPU. In this section, we begin by parallelizing and optimizing HOSTA on a single GPU, and then move forward to large-scale GPU-accelerated systems. Finally we evaluate and validate our solutions on TianHe-1A.

3.1 Parallelization and Optimizations on a Single GPU

On a single GPU, we present a dual-level parallelization: fine-grained data parallelism within a block and coarse-grained task parallelism among multiple blocks. For the fine-grained parallelism, we use two approaches (3D kernel configuration or 2D kernel configuration) according to data dependency among cells in a block.

If there is no data dependency between cells (e.g., the procedure *Sol_jacobi*), each GPU thread calculates a cell independently. A 3D grid block is mapped to a GPU grid in CUDA as illustrated in Fig.2(b): the processing of grid cells on *I,J* and *K* direction (coordinate) are mapped to GPU threads on *X,Y* and *Z* dimension respectively; a grid block of size (*NI,NJ,NK*) is logically decomposed to some *sub_blocks* of size (*x_blk,y_blk,z_blk*) , each *sub_block* is computed by a GPU thread block of the same size, and those thread blocks compose a GPU grid of size ($\lceil NI/x\_blk \rceil$,$\lceil NJ/y\_blk \rceil$, $\lceil NK/z\_blk \rceil$), where $\lceil x \rceil$ is the minimum integer that is larger than x. When implementing HOSTA, complex stencil computations for the viscous (*calc_viscous*) and inviscid (*calc_inviscid*) fluxes are decomposed along the *I, J* and *K* directions. Thus, data dependencies only exist in the corresponding direction. Since CUDA has no global synchronization, we choose to use a 2D kernel configuration or 2D decomposition for this case, i.e., we use one GPU thread to compute all the cells on a cell line. For example, in the *I* direction, the 2D thread block is configured as (1,*y_blk,z_blk*) and the size of GPU grid is (1,$\lceil NJ/y\_blk \rceil$, $\lceil NK/z\_blk \rceil$).

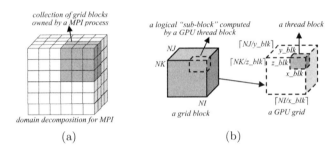

Fig. 2. Domain decomposition for MPI-CUDA parallelization

When there is a data dependency, we propose and use a kernel decomposition strategy to substitute a large 2D kernel with several small 3D kernels. A 3D configuration can ensure more GPU threads on the same dimension to access the global memory in a coalesced manner and thus improve performance. The kernel decomposition is mainly implemented in the computation of viscous and inviscid fluxes. For example, by carefully analyzing the data dependency, the initial 2D *I_viscous_kernel* to compute *I* direction's viscous fluxes is decomposed to three 3D kernels: *I_viscous_kernel_1* to compute the cell-edge metrics and primitives, and the cell-node viscous fluxes, *I_viscous_kernel_2* to compute the cell-edge viscous fluxes and *I_viscous_kernel_3* to compute the cell-node derivative of viscous fluxes. However, due to the interpolation in HDCS-E8T7, the initial 2D *I_inviscid_kernel* to compute *I* direction's inviscid fluxes is decomposed to four small kernels (three 3D kernels and one *semi*-3D kernel): *I_inviscid_kernel_1* to compute the cell-edge metrics, *I_inviscid_kernel_2* to reconstruct the left-hand and right-hand cell-edge primitives, *I_inviscid_kernel_3* to compute the cell-edge and cell-node inviscid fluxes and *I_inviscid_kernel_4* to compute the cell-node derivative of inviscid fluxes. We implement the *semi*-3D *I_inviscid_kernel_2* by changing data structure and inter-exchanging loops. As Fig.3(a) shows, initially separated primitive variables $U_m(i,j,k)$ ($1 \leq m \leq 5$) for the cell-node (i,j,k) form a system of tri-diagonal equations and there is a data dependency among neighboring statements. Fig.3(b) shows the new code snippet. A merged primitive variable $PV(m,i,j,k)$ replaces the five separate primitive variables to form another loop for m. Then we exchange loop m and loop i, and an independent "M" direction is available to be parallelized on GPU. Thus, a *semi*-3D kernel configuration of a 3D thread block $(5, y\_blk, z\_blk)$ and a 2D GPU grid $(1, \lceil NJ/y\_blk \rceil, \lceil NK/z\_blk \rceil)$ can be used for the decomposed kernel.

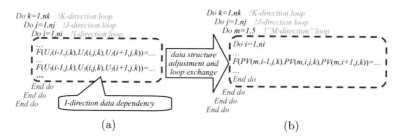

Fig. 3. Data structure adjustment and loop exchange for HDCS-E8T7

We implement the coarse-grained parallelism based on CUDA streams. Because there is no data dependency among grid blocks between exchanges of boundary/singularity data, we bind each block to a CUDA stream and issue all the streams simultaneously to the GPU. As Fig.4 shows, all the operations for block i such as computation, host to device (H2D) data copy and device to host (D2H) data copy are associated with stream i. The multi-stream implementation can fully exploit the potential power of modern GPU architecture. For example, when a stream is accessing global memory, the kernel engine can schedule and execute warps from other streams to hide memory access latency. More importantly, the kernel execution and PCI-E data transfer of different streams can be substantially overlapped, especially for GPUs with separate copy engines for H2D and D2H such as Tesla M2050 in TianHe-1A. Our multi-stream design is independent with the fine-grained GPU parallelization within a block. Furthermore, it can be used to overlap the GPU computation, data transfer and MPI communication as Sect.3.2 describes.

3.2 Moving Forward to Large-Scale GPU-Accelerated Systems

Since a compute node (of TianHe-1A) contains only one GPU, and thus, we choose to use one MPI process on each node for large-scale simulations on multiple GPUs. Considering a 3D block with six boundaries (or ghost zones), the data for a single boundary is continuously stored in the device memory, while the

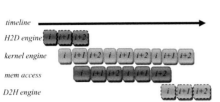
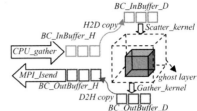

Fig. 4. Multi-stream execution for multiple grid blocks

Fig. 5. Gather-scatter optimization for boundary data transfer of a 3D block

data for different boundaries is not. Further, CUDA API for data transfer can only copy continuous data elements for a time. Thus, we use a Scatter-Gather optimization to minimize the times of data transfer for boundary/singularity data of a 3D grid block via PCI-E, as Fig.5 illustrated. Before performing D2H copy, we use a gather kernel (i.e., *Gather_kernel*) to collect all non-continuous data to a continuous device buffer *BC_OutBuffer_D*, and then the entire buffer is copied to host buffer *BC_OutBuffer_H*. Correspondingly, before performing H2D copy, we use a gather procedure (i.e., *CPU_gather*) on the CPU to pack all the updated boundary data to a host buffer *BC_InBuffer_H*, and then the entire buffer is transferred to device buffer *BC_InBuffer_D*. Finally, we use a scatter kernel (i.e., *Scatter_kernel*) to distribute the data elements to each boundary.

Furthermore, we overlap the kernel execution, data transfer and MPI communication using CUDA multi-stream, non-blocking MPI and CUDA events. Fig.6 shows a schematic for the computation of the gradient of primitives, the inviscid and viscous fluxes. When the GPU finishes all the stream/block's operations for the gradient of primitives, a CUDA event with the same stream ID is recorded to represent data dependency between MPI communication and the boundary data computed by the stream. Before the host calls *MPI_Isend* to send the boundary data for a block, it must query the device to make sure that the event associated with the block/stream has been executed. As we can see from Fig.6, the data copy for block *i*, the non-blocking MPI send for block (*i-1*) and the computation of the gradient of primitives for block (*i+1*) are largely overlapped. When the GPU is executing the kernel or performing data copy, the CPU can also call *MPI_Irecv* to receive boundary data from blocks on other nodes (e.g., block (*i+1*), as illustrated by Fig.6). Note that *MPI_Waitall* must be called to ensure that *MPI_Irecv* has finished receiving data.

3.3 Performance Evaluation and Validation

We ported HOSTA to the GPU using a CUDA C and Fortran90 mixed implementation. Thus, our performance results on the CPU are all obtained from the Fortran90 implementation. We ported all the procedures (more than 16000 lines of CUDA C codes) in the time-marching loop except the procedures for boundary conditions and MPI communication (red codes in Fig.1). In our implementation,

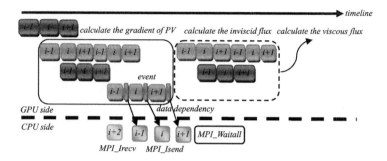

Fig. 6. The overlapping of kernel execution, data transfer and MPI communication when computing the gradient of primitives, the inviscid and viscous fluxes

we often maximize the kernel's performance when setting (x_blk, y_blk, z_blk) to (16,4,4) (i.e., the number of threads in a thread block is 256) and thus we choose it as the default setting in the following tests.

We use the TianHe-1A supercomputer as our test platform. Each of TianHe-1A compute nodes contains a Tesla M2050 and two six-core 2.93GHz Xeon X5670s. A customized high-speed interconnection network is used for the internode communication with a bi-directional bandwidth of 160Gbps and a latency of 1.57μs. For more information about TianHe-1A, please refer to [7]. We use MPICH2-GLEX for MPI communication. The CUDA version is 4.2, and the compilers for FORTRAN and C are icc11.1 and ifort 11.1. All the code is compiled with -O3 optimization flag.

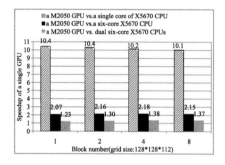
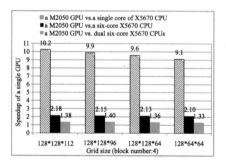

Fig. 7. Single GPU speedup for fixed grid size

Fig. 8. Single GPU speedup for different grid sizes

In Fig.7, we present the speedup of a M2050 GPU over a single core of X5670 CPU, and the speedups of a M2050 GPU over a six-core X5670 CPU using 6 OpenMP threads and dual six-core X5670 CPUs using 12 OpenMP threads. The grid size is 128*128*112, and the block number is varying from 1 to 8. We can see that our GPU code can achieve a speedup of more than 10 over the serial

CPU code. Note that the price of a M2050 is similar to the price of an X5670, and a speedup of about 2.1 when comparing a M2050 to a six-core X5670 is also comparable to the results of paper [10] as far as similar-priced comparison of CPU and GPU is concerned. A speedup of about 1.3 when comparing a M2050 to dual X5670s further validates GPU's cost-effectiveness. We also observed slight performance degradation for both GPU and CPU when the block number is increased for a fixed grid size. This can be explained by the fact that multiple blocks will incur extra OpenMP and kernel overheads.

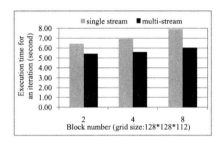
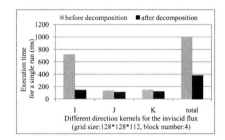

Fig. 9. Performance comparison for single stream and multi-stream implementation

Fig. 10. Performance comparison for before and after kernel decomposition

In Fig.8, we present performance results for different grid sizes and the block number is fixed to 4. We get better performance for larger problem sizes. This is because higher workloads can better overlap the computation and global memory access for GPUs. Fig.9 shows the results of the multi-stream optimization. We see a 20% to 30% performance enhancement for a whole iteration. But our multi-stream implementation requires multiple blocks on a GPU to be associated with multiple streams and the extra GPU memory space to store the intermediate results for simultaneously executed blocks. Fig.10 shows the results when the kernel decomposition is adopted for the computation of the inviscid fluxes. We can reduce about 80% execution time for the I direction kernel, but for the J and K directions, the performance improvements are only about 15%. This is because the decomposition in the I direction can ensure more GPU threads to access the global memory in a coalesced manner.

In Fig.11, we shows the strong scaling results with and without overlapping. We obtain the results without the overlapping described in Sect.3.2 by directly copying the whole grid block between GPU and CPU. We use the performance achieved on 32 GPUs as a baseline and each GPU simulates a 128*128*112 grid with 4 blocks for the baseline test. Thus our total grid size for strong scalability test is fixed to 58720256 (namely 32*128*128*112). The grid is evenly partitioned to grid blocks and distributed to GPUs. The GPU number is scaled from 32 to 256 and the block number per GPU is fixed to 4. We see that the overlapping of computation, data transfer and MPI communication plays an important role for good scalability. We observe a speedup of about 5.5 when scaling from 32 GPUs

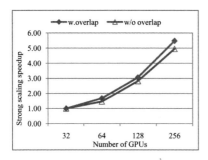 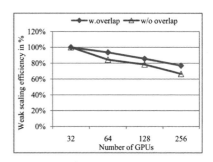

Fig. 11. Strong scaling speedup on TianHe-1A **Fig. 12.** Weak scaling efficiency on TianHe-1A

to 256 GPUs with overlapping, and a speedup of about 5.0 without overlapping, demonstrating a promising result for strong scalability test on large-scale GPU-accelerated systems. Fig.12 presents the weak scaling efficiency results with and without overlapping. The problem size for each GPU is fixed to 128*128*112 with 4 blocks. Again we use the performance achieved on 32 GPUs as a baseline and the GPU number is increased from 32 to 256. We lose about 23% (from the perfect weak scaling efficiency of 100%) using our overlapping strategy and 256 GPUs, while the efficiency loss is up to 33.5% for the non-overlapping one.

We use the same EET high-lift configuration and the same conditions as tested in the LTPT facility at NASA LaRC[11] to validate HDCS. Fig.13 shows the grid structure. The incoming Mach number $M=0.17$, and the chord length $c=0.457$m, the corresponding chord Reyonlds number is 1.71×10^6, the angle of attack is $4°$. The effects of explicit LES models are imitated by the truncation error of HDCS to simulate the turbulent flow. The computational grid contains approximately 800, 000, 000 grid points. The dual time stepping scheme with 60 subiterations based on Jacobi iterative method is used for the time integration. The computation has been successfully performed on TianHe-1A using 400 GPUs and advanced with a time step of 2.5×10^{-6}s for total 30,000 time steps. Flow statistics are collected for 10, 000 time steps. Fig.14 demonstrates the iso-surfaces of the second invariant colored by density contours for the instantaneous flow. The second invariant Q is:

$$Q = (\Omega_{ij}\Omega_{ij} + S_{ij}S_{ij})/2 \qquad (5)$$

where $\Omega_{ij} = (u_{i,j}-u_{j,i})/2$ and $S_{ij} = (u_{i,j}+u_{j,i})/2$. Due to the fine mesh, we can see the details about the vortex structure. The computational mean spanwise vorticity comparing with the experimental measurement from [11] is shown in Fig.15. We clearly see that the shear layer separating from the leading edge is demonstrated by the computation.

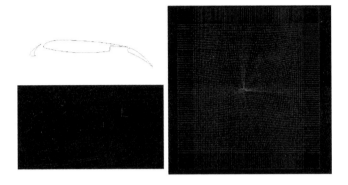

Fig. 13. Grid structure of the high-lift airfoil configuration
(colored picture attached at the end of the book)

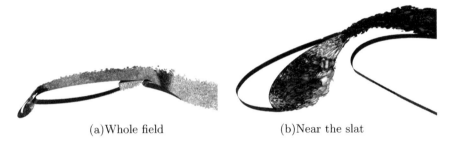

(a)Whole field   (b)Near the slat

Fig. 14. Iso-surfaces of the second invariant of velocity gradient tensor for instantaneous vortex structure (colored picture attached at the end of the book)

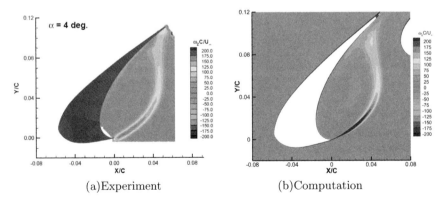

(a)Experiment   (b)Computation

Fig. 15. Time-averaged spanwise-vorticity distribution
(colored picture attached at the end of the book)

4 Related Work

In the past twenty years, there have been many studies in developing and applying high-order compact finite difference schemes for CFD. Lele[1] has developed several central compact schemes with spectral-like resolution. Visbal and Gaitonde[12] use filters to prevent numerical oscillations of central compact schemes. Recently, Lele's central compact schemes have been successfully applied by Rizzetta et al.[13] for the simulation of low speed flows. In order to deal with shock wave problems, Adams et al.[14] have developed a compact-ENO (Essentially Non-Oscillatory) scheme. Pirozzoli[15] has developed a compact-WENO (Weighted Essentially Non-Oscillatory) scheme which was further improved by Ren et al.[16]. Deng et al. have developed the WCNS[9]. But as mentioned by Deng et al.[6], meeting GCL for complex configurations is difficult, which limits the applications of high-order finite difference schemes. For this, HDCS with inherent dissipation was derived and implemented in HOSTA for the high-resolution flow simulation on complex geometry. HDCS employs a new central compact scheme to fulfil the GCL and adds dissipation on the central scheme by high-order dissipative interpolation of cell-edge variables (see Sect.2).

Prior work has also shown the experiences of porting CFD codes to GPUs, with impressive speed-ups. Phillips et al.[17] have developed a 2D compressible Euler solver on a cluster of GPUs for rapid aerodynamic performance prediction. Appa et al.[18] implemented and optimized an unstructured 3D explicit finite volume CFD code on multi-core CPUs and GPUs for efficient aerodynamic design. Corrigan et al.[19] have ported an adaptive, edge-based finite element code FEFLO to GPU clusters in a semi-automated fashion: the existing Fortran-MPI code is preserved while the translator inserts data transfer calls as required. Brandvik and Pullan[22] implemented a multi-GPU enabled Navier-Stokes solver for flows in turbomachines. They reported almost linear weak scaling for typical turbomachinery cases using up to 16 GPUs. All the above studies were tested on small-scale GPU platforms using low-order CFD methods. For large-scale GPU clusters, Jacobsen et al.[20] parallelized a CFD solver for incompressible fluid flows. They used up to 128 GPUs for parallel scalability test. They further demonstrated the large eddy simulation of a turbulent channel flow on 256 GPUs[21], but they only simulated a lid-driven cavity problem using low-order schemes.

Due to the complexity, high-order CFD schemes generally require extra implementation and optimization efforts on GPUs. Tutkun et al.[23] have implemented a six-order compact finite difference scheme on a single GPU. Antoniou et al.[24] implemented a high-order solver that can run on multi-GPUs for the compressible turbulence using WENO. But the solver can only run on a single node platform containing 4 GPUs for very simple domains like a 2D or 3D box. Results show that their single-precision implementation can achieve a speedup of 53 when comparing 4 Tesla C1070 GPUs with a single core of an Xeon X5450 CPU. Castonguay et al.[25] published the parallelization of the first high-order, compressible viscous flow solver for mixed unstructured grids by using MPI and CUDA, where the Vincent-Castonguay-Jameson-Huynh method is used. A flow over SD7003 airfoil and a flow over sphere is simulated using 32 GPUs. Appleyard et al.[26] reported a MPI-CUDA implementation to accelerate the solution of the level set equations for interface tracking using a HOUC (High-Order Upstream Central) scheme. But they only demonstrated performance results using 4 GPUs. Zaspel et al.[10] described the implementation of an incompressible double-precision two-phase solver on GPU clusters using a fifth-order WENO scheme. The test problem is a rising bubble of air inside a tank of water with surface tension effects and parallel performance results are reported using up to 48 GPUs. To summarize, we see that the above GPU-enabled CFD simulations using high-order schemes are still preliminary as far as the grid complexity, problem size and parallel scale are comprehensively concerned.

5   Conclusion and Future Work

The complexity of high-order CFD simulation for complex multi-block grids makes it very difficult to parallelize on large-scale heterogeneous HPC systems. In this work, we parallelize and optimize HOSTA, a high-order CFD software for multi-block structural grids. We successfully simulate a flow over a high-lift airfoil configuration on TianHe-1A using 400 GPUs. We conclude that TianHe-1A is a suitable platform to run the HOSTA-like CFD software in terms of performance and scalability, although programming it takes many efforts. Thus, we want to develop a heterogenous programming and auto-tuning framework for CFD-like scientific applications based on the current work. Furthermore, we plan to collaborate CPUs and GPUs for large-scale calculation on multi-block grids to further improve HOSTA's performance.

Acknowledgement. This paper was supported by the National Science Foundation of China under Grant No.11272352, the National Basic Research Program of China under Grant No. 2009CB723803 and the Open Research Program of China State Key Laboratory of Aerodynamics.

References

1. Lele, S.K.: Compact finite difference schemes with spectral-like resolution. J. Comput. Phys. 103, 16–42 (1992)
2. Trulio, J.G., Trigger, K.R.: Numerical solution of the one-dimensional hydrodynamic equations in an arbitrary time-dependent coordinate system. Technical Report UCLR-6522. University of California Lawrence Radiation laboratory (1961)
3. Deng, X.G., Mao, M.L., Tu, G.H., Liu, H.Y., Zhang, H.X.: Geometric conservation law and applications to high-order finite difference schemes with stationary grids. J. Comput. Phys. 230, 1100–1115 (2011)
4. Deng, X.G., Mao, M.L., Tu, G.H., et al.: Extending the fifth-order weighted compact nonlinear scheme to complex grids with characteristic-based interface conditions. AIAA Journal 48(12), 2840–2851 (2010)

5. Deng, X.G., Jiang, Y., Mao, M.L., Liu, H.Y., Tu, G.H.: Developing hybrid cell-edge and cell-node dissipative compact scheme for complex geometry flows. In: The Ninth Asian Computational Fluid Dynamics Conference (2012)
6. Deng, X.G., Mao, M.L., Tu, G.H., et al.: High-order and high accurate CFD methods and their applications for complex grid problems. Commun. Comput. Phys. 11, 1081–1102 (2012)
7. Yang, X.J., Liao, X.K., Lu, K., et al.: The TianHe-1A supercomputer: its hardware and software. Journal of Computer Science and Technology 26, 344–351 (2011)
8. NVIDIA CUDA: CUDA C programming guide v4.2 (2012)
9. Deng, X.G., Zhang, H.X.: Developing high-order weighted compact nonlinear schemes. J. Comput. Phys. 165, 22–44 (2000)
10. Zaspel, P., Griebel, M.: Solving incompressible two-phase flows on multi-GPU clusters. Comput. & Fluids (2012)
11. Jenkins, L.N., Khorrami, M.R., Choudhari, M.: Characterization of unsteady flow structures near leading-edge slat: part I. PIV measurements. AIAA paper 2004-2801 (2004)
12. Visbal, M.R., Gaitonde, D.V.: High-order accurate methods for complex unsteady subsonic Flows. AIAA Journal 37, 1231–1239 (1999)
13. Rizzetta, D., Visbal, M., Morgan, P.: A high-order compact finite difference scheme for large-eddy simulation of active flow control. Progress in Aerospace Sciences 44, 397–426 (2008)
14. Adams, N.A., Shariff, K.: A High-resolution hybrid compact-ENO scheme for shock-turbulence interaction problems. J. Comput. Phys. 127 (1996)
15. Pirozzoli, S.: Conservative hybrid compact-WENO schemes for shock-turbulence interaction. J. Comput. Phys. 179, 81–117 (2002)
16. Ren, Y., Liu, M., Zhang, H.: A characteristic-wise hybrid compact-WENO schemes for solving hyperbolic conservations. J. Comput. Phys. 192, 365–386 (2005)
17. Phillips, E.H., Zhang, Y., Davis, R.L., Owens, J.D.: Rapid aerodynamic performance prediction on a cluster of graphics processing units. AIAA paper 2009-565 (2009)
18. Appa, J., Sharpe, J., Moinier, P.: An unstructured 3D CFD code optimised for multicore and graphics processing units. In: MRSC 2010 (2010)
19. Corrigan, A., Lohner, R.: Porting of FEFLO to multi-GPU clusters. AIAA paper 2011-0948 (2011)
20. Jacobsen, D.A., Thibault, J.C., Senocak, I.: An MPI-CUDA implementation for massively parallel incompressible flow computations on multi-GPU clusters. AIAA paper 2010-0522 (2010)
21. DeLeon, R., Jacobsen, D., Senocak, I.: Large-eddy simulations of turbulent incompressible flows on GPU clusters. Computing in Science & Engine 15, 26–33 (2013)
22. Brandvik, T., Pullan, G.: An accelerated 3D Navier-Stokes solver for flows in turbomachines. In: ASME Turbo Expo 2009: Power for Land, Sea and Air (2009)
23. Tutkun, B., Edis, F.O.: A GPU application for high-order compact finite difference scheme. Computers & Fluids 55, 29–35 (2012)
24. Antoniou, A.S., Karantasis, K.I., Polychronopoulos, E.D.: Acceleration of a finite-difference WENO scheme for large-scale simulations on many-core architectures. AIAA paper 2010-0525 (2010)
25. Castonguay, P., Williams, D.M., Vincent, P.E., Lopez, M., Jameson, A.: On the development of a high-order, multi-GPU enabled, compressible viscous flow solver for mixed unstructured grids. AIAA paper 2011-3229 (2011)
26. Appleyard, J., Drikakis, D.: Higher-order CFD and interface tracking methods on highly-parallel MPI and GPU systems. Computers & Fluids 46, 101–105 (2011)

Collaborating CPU and GPU for large-scale high-order CFD simulations with complex grids on the TianHe-1A supercomputer

Chuanfu Xu [b,*], Xiaogang Deng [b], Lilun Zhang [b], Jianbin Fang [c],
Guangxue Wang [a], Yi Jiang [a], Wei Cao [b], Yonggang Che [b], Yongxian Wang [b],
Zhenghua Wang [b], Wei Liu [b], Xinghua Cheng [b]

[a] *State Key Laboratory of Aerodynamics, P.O. Box 211, Mianyang 621000, PR China*
[b] *College of Computer Science, National University of Defense Technology, Changsha 410073, PR China*
[c] *Parallel and Distributed Systems Group, Delft University of Technology, Delft 2628CD, The Netherlands*

ARTICLE INFO

Article history:
Received 23 September 2013
Received in revised form 13 August 2014
Accepted 17 August 2014
Available online 22 August 2014

Keywords:
GPU parallelization
CFD
CPU–GPU collaboration
High-order finite difference scheme
Multi-block structured grid

ABSTRACT

Programming and optimizing complex, real-world CFD codes on current many-core accelerated HPC systems is very challenging, especially when collaborating CPUs and accelerators to fully tap the potential of heterogeneous systems. In this paper, with a tri-level hybrid and heterogeneous programming model using MPI + OpenMP + CUDA, we port and optimize our high-order multi-block structured CFD software HOSTA on the GPU-accelerated TianHe-1A supercomputer. HOSTA adopts two self-developed high-order compact definite difference schemes WCNS and HDCS that can simulate flows with complex geometries. We present a dual-level parallelization scheme for efficient multi-block computation on GPUs and perform particular kernel optimizations for high-order CFD schemes. The GPU-only approach achieves a speedup of about 1.3 when comparing one Tesla M2050 GPU with two Xeon X5670 CPUs. To achieve a greater speedup, we *collaborate* CPU and GPU for HOSTA instead of using a naive *GPU-only* approach. We present a novel scheme to balance the loads between the *store-poor* GPU and the *store-rich* CPU. Taking CPU and GPU load balance into account, we improve the maximum simulation problem size per TianHe-1A node for HOSTA by 2.3×, meanwhile the collaborative approach can improve the performance by around 45% compared to the GPU-only approach. Further, to scale HOSTA on TianHe-1A, we propose a gather/scatter optimization to minimize PCI-e data transfer times for ghost and singularity data of 3D grid blocks, and overlap the collaborative computation and communication as far as possible using some advanced CUDA and MPI features. Scalability tests show that HOSTA can achieve a parallel efficiency of above 60% on 1024 TianHe-1A nodes. With our method, we have successfully simulated an EET high-lift airfoil configuration containing 800M cells and China's large civil airplane configuration containing 150M cells. To our best knowledge, those are the largest-scale CPU–GPU collaborative simulations that solve realistic CFD problems with both complex configurations and high-order schemes.

© 2014 Elsevier Inc. All rights reserved.

1. Introduction

The solution of the Navier–Stokes equations for CFD (Computational Fluid Dynamics) involves reasonably complex numerical methods and algorithms which are used in computational science today. In the past several decades, low-order (order of accuracy ≤2) CFD schemes have been widely used in engineering applications, but they are insufficient in simulation resolution and fidelity for complex flows (e.g., turbulence and many viscosity dominant flows such as boundary layer flows, vortical flows, shock–boundary layer interactions, heat flux transfers, etc.) containing sharp gradients and small disturbances. An effective approach to overcome the obstacle of accurate numerical simulations is to employ high-order methods [47]. Over the past 20 to 30 years, there have been many studies in developing and applying various kinds of high-order and high accurate schemes in CFD. Among them, *compact high-order finite difference schemes* are one of the most promising schemes [36]. Based on a given compact stencil, compact schemes can achieve relative higher order of accuracy with good spectral resolution and flexibility over traditional explicit finite difference schemes. Recently, to capture discontinuities in flows with shock wave, nonlinear formulation is added and various non-linear compact schemes such as Weighted Essentially Non-Oscillatory (WENO) [33] and Weighted Compact Non-linear Scheme (WCNS) [23] were developed. They have been extensively studied and widely used for incompressible, compressible and hypersonic flows, and are also very attractive for multi-scale flows (e.g., computational aeroacoustics, computational electromagnetics and turbulence).

Although many researchers claim that they have successfully applied high-order schemes in complex geometries, the grids in their tests are fairly simple compared to those in low-order scheme applications. Problems such as robustness and grid-quality sensitivity [47,24] still hinder the wide application of these schemes on complex meshes. Recently, we have shown that these problems can be largely mitigated by treating the Geometric Conservation Law (GCL) [43] and block-interface conditions carefully when employing high-order schemes on complex configurations. To this end, we have developed a Symmetrical Conservative Metric Method (SCMM) [22] to fulfill GCL by computing grid derivatives in a conservative form, with the same scheme used for fluxes. Further, we have employed a Characteristic-Based Interface Condition (CBIC) [19] to fulfill high-order multi-block computing by directly exchanging the spatial derivatives (i.e., RHS computed on each block) on each side of an interface. Based on SCMM and CBIC, we have successfully simulated a wide range of flow problems with complex grids using WCNS [23] and recently developed a new Hybrid cell-edge and cell-node Dissipative Compact Scheme (HDCS) [18]. The WCNS and HDCS high-order schemes, along with SCMM and CBIC, have been implemented in our in-house high-order CFD software HOSTA (High-Order SimulaTor for Aerodynamics) for 3D multi-block structured grids.

Some recent applications of HOSTA can be found in [25,19,21,20]. Meanwhile, running HOSTA with a fairly large grid on a local machine often takes several weeks or even months. Thus, it is essential and practical to port it onto modern supercomputers, often featuring many-core accelerators/co-processors (GPUs [26], MIC [29], or specialized ones [4]). These heterogeneous processors can dramatically enhance the overall performance of HPC systems with remarkably low total cost of ownership and power consumption, but the development and optimization of large-scale applications are also becoming exceptionally difficult. Researchers often need to use various programming models/tools (e.g., CUDA [1] for NVIDIA's GPUs, OpenMP and MPI for intra-/inter-node parallelization), and map a hierarchy of parallelism in problem domains to heterogeneous devices with different processing capabilities, memory availability, and communication latencies. Hence, for simplicity developers often choose a naive GPU-only approach in a GPU-accelerated systems: letting GPUs compute efficiently and CPUs only manage GPU computation and communication. This is obviously a vast waste of CPU capacity, especially for top supercomputers like TianHe-1A [49] equipped with powerful multi-core CPUs. Our previous GPU-only implementation [48] shows that on one TianHe-1A node (containing two Intel Xeon X5670 CPUs and one NVIDIA Tesla M2050 GPU) the achievable performance of HOSTA on X5670s is around 80% of that on a M2050. In other words, we can potentially obtain an 80% performance improvement by using both the CPUs and the GPU. Therefore, to tap the full potential of heterogeneous systems and maximize application performance, it is immensely important to collaborate CPU and GPU for a *CPU–GPU collaborative simulation* instead of a GPU-only simulation.

Nevertheless, making an efficient and large-scale collaboration for real-world complex CFD applications on heterogeneous supercomputers has been described as a problem as hard as any that computer science has faced. Besides the aforementioned programming challenge, we need to balance the use of unbalanced hardware resources: GPUs are relatively rich in compute capacity but poor in memory capacity and the opposite holds for CPUs; this is generally true for most current heterogeneous supercomputers. On each TianHe-1A node the M2050 contributes about 515 GFlops with less than 3 GB device memory, while the X5670s, with 32 GB shared memory, contribute only 140 GFlops. Our previous results [48] show that HOSTA achieves a speedup of about 1.3 when comparing one M2050 with two hexa-core X5670s, but it can hold a maximum of 2M grid cells on the M2050. Thus, taking CPU–GPU load balance into account, the maximum amount of cells on one TianHe-1A node is around 3.5M, with 2M cells on the GPU and 1.5M ($\approx \frac{2M}{1.3}$) cells on the CPUs. Therefore, HOSTA's simulation capacity on a single TianHe-1A node is bounded by the relatively small device memory on GPUs. For large-scale grids, we can either use more compute nodes to keep the amount of cells per node below 3.5M, or perform an inefficient and unbalanced collaboration. Both cases severely limit the efficiency of heterogeneous systems. Hence, achieving load balance is one of the most serious challenges we face when collaborating *store-poor* GPU with *store-rich* CPU for HOSTA on TianHe-1A. Another practical challenge is that HOSTA performs more extensive data exchanges (specifically four exchanges in each iteration, see 2.2) among neighboring grid blocks than traditional low order scheme applications to ensure the robustness of high-order simulations. Thus, we also need to design an effective mechanism for data transfers among different parallel levels of TianHe-1A when scaling HOSTA.

In this work, with MPI + OpenMP + CUDA, we collaborate CPU and GPU on TianHe-1A for massively parallel high-order CFD simulations. The aforementioned challenges are tackled by implementing a novel balancing scheme for intra-node CPU–GPU collaboration and overlapping CPU–GPU computation with communication. Specifically, our key contributions include:

- We present a dual-level parallelization scheme for efficient multi-block CFD computation on GPUs. We use multiple CUDA streams (*multi-stream*) to exploit the coarse-grained parallelism among grid blocks and CUDA thread blocks to exploit the fine-grained parallelism among cells within a data block. Further, with a kernel decomposition strategy, we dramatically optimize the time-consuming CUDA kernels calculating high-order flow fluxes. The GPU-only approach achieves a speedup of about 1.3 when comparing one Tesla M2050 with two hexa-core Xeon X5670s.
- On a single TianHe-1A node we collaborate the CPUs and the GPU using nested OpenMP and CUDA. The key idea is to surmount the GPU memory limit by a dynamic device memory use policy for selected flow variables, coupled with grouping and pipelining GPU-computed grid blocks using CUDA streams. The balanced collaborative approach can top up HOSTA's maximum simulation problem size on a single TianHe-1A node from 3.5M cells to 8M cells (2.3×), and meanwhile has up to 45% performance advantage over the GPU-only approach.
- To scale HOSTA on large-scale compute nodes, we propose a gather/scatter optimization to minimize PCI-e data transfer times for ghost and singularity data of 3D grid blocks which are discontinuously stored in the device memory. Further, we use non-blocking MPI, CUDA events and CUDA streams to perform inter-block data exchanges as early as possible and overlap multiple levels of computation and communication. Despite of the extensive data exchanges required in HOSTA, we achieve a parallel efficiency of about 60% on 1024 nodes.
- As a case study, we have successfully collaborated hundreds of TianHe-1A nodes to simulate the high-lift airfoil configuration *30p30n* containing about 800M cells and China's large civil airplane configuration *C919* containing about 150M cells. To our best knowledge, those are the **largest-scale CPU–GPU collaborative simulations** that solve **realistic CFD problems** with both **complex configurations** and **high-order schemes**.

The remainder of the paper is organized as follows. Section 2 briefly describes numerical methods and implementation in HOSTA. Section 3 presents the overall consideration for MPI + OpenMP + CUDA parallelization. Section 4 introduces the GPU implementation and optimization. In Section 5, we detail the intra-node CPU–GPU collaboration and the balancing scheme. Section 6 scales HOSTA to a large number of TianHe-1A nodes. Section 7 presents performance results and two test cases are shown in Section 8. Finally in Section 9, we introduce some related work and conclude in Section 10.

2. Numerical methods and HOSTA implementation

In this section, we briefly introduce some numerical methods used in this work and then give the implementation of HOSTA.

2.1. Numerical methods

In Cartesian coordinates the governing equations (Euler or Navier–Stokes) in strong conservative form are

$$\frac{\partial Q}{\partial t} + \frac{\partial F}{\partial x} + \frac{\partial G}{\partial y} + \frac{\partial H}{\partial z} = 0 \tag{1}$$

the equations are transformed into curvilinear coordinates by introducing the transformation $(x, y, z, t) \rightarrow (\xi, \eta, \zeta, \tau)$

$$\frac{\partial \tilde{Q}}{\partial \tau} + \frac{\partial \tilde{F}}{\partial \xi} + \frac{\partial \tilde{G}}{\partial \eta} + \frac{\partial \tilde{H}}{\partial \zeta} = 0 \tag{2}$$

where

$$\tilde{F} = \tilde{\xi}_t Q + \tilde{\xi}_x F + \tilde{\xi}_y G + \tilde{\xi}_z H, \quad \tilde{\xi}_x = J^{-1}\xi_x \tag{3}$$

and with similar relations for the other terms.

The fifth-order WCNS scheme WCNS-E-5 and the seventh-order HDCS scheme HDCS-E8T7 were used in this work. Here we only consider the discretization of the inviscid flux derivative along the ξ direction. The discretization along the other directions can be dealt in a similar manner. The interior scheme of WCNS-E-5 can be expressed as

$$\frac{\partial \tilde{F}_i}{\partial \xi} = \frac{75}{64h}(\tilde{F}_{i+1/2} - \tilde{F}_{i-1/2}) - \frac{25}{384h}(\tilde{F}_{i+3/2} - \tilde{F}_{i-3/2}) + \frac{3}{640h}(\tilde{F}_{i+5/2} - \tilde{F}_{i-5/2}) \tag{4}$$

and the interior scheme of HDCS-E8T7 can be expressed as

$$\frac{\partial \tilde{F}_i}{\partial \xi} = \frac{256}{175h}(\tilde{F}_{i+1/2} - \tilde{F}_{i-1/2}) - \frac{1}{4h}(\tilde{F}_{i+1} - \tilde{F}_{i-1}) + \frac{1}{100h}(\tilde{F}_{i+2} - \tilde{F}_{i-2}) - \frac{1}{2100h}(\tilde{F}_{i+3} - \tilde{F}_{i-3}) \tag{5}$$

where h is the grid size, and $\tilde{F}_{i+1/2} = \tilde{F}(U^L_{i+1/2}, U^R_{i+1/2}, \tilde{\xi}_{x,i+1/2}, \tilde{\xi}_{y,i+1/2}, \tilde{\xi}_{z,i+1/2})$ is the cell-edge flux computed by some flux-splitting methods that can be found in [23]. $U_i = U(\xi_i, t)$ is the flow variables. $U^L_{i+1/2}$ and $U^R_{i+1/2}$ are the left-hand and right-hand cell-edge variables. For WCNS-E-5, they can be obtained by a high-order nonlinear weighted interpolation of cell-node variables. For HDCS-E8T7, we use the following seventh-order dissipative compact interpolation

$$\frac{5}{14}(1-\alpha)U^L_{i-1/2} + U^L_{i+1/2} + \frac{5}{14}(1+\alpha)U^L_{i+3/2}$$
$$= \frac{25}{32}(U_{i+1} + U_i) + \frac{5}{64}(U_{i+2} + U_{i-1}) - \frac{1}{448}(U_{i+3} + U_{i-2})$$
$$+ \alpha\left[\frac{25}{64}(U_{i+1} - U_i) + \frac{15}{128}(U_{i+2} - U_{i-1}) - \frac{5}{896}(U_{i+3} - U_{i-2})\right] \quad (6)$$

where $\alpha < 0$ is the dissipative parameter to control numerical dissipation. $U^R_{i+1/2}$ can be obtained by setting $\alpha > 0$. The boundary and near-boundary schemes of WCNS-E-5 and HDCS-E8T7 can be found in [23] and [18] respectively. For the viscous fluxes, we use the same sixth order central difference scheme as [23].

The GCL includes the volume conservation law (VCL) and the surface conservation law (SCL). For stationary grids, VCL naturally holds

$$I_t = (1/J)_\tau + (\tilde{\xi}_t)_\xi + (\tilde{\eta}_t)_\eta + (\tilde{\zeta}_t)_\zeta = 0 \quad (7)$$

if SCL is satisfied, then

$$I_x = (\tilde{\xi}_x)_\xi + (\tilde{\eta}_x)_\eta + (\tilde{\zeta}_x)_\zeta = 0$$
$$I_y = (\tilde{\xi}_y)_\xi + (\tilde{\eta}_y)_\eta + (\tilde{\zeta}_y)_\zeta = 0$$
$$I_z = (\tilde{\xi}_z)_\xi + (\tilde{\eta}_z)_\eta + (\tilde{\zeta}_z)_\zeta = 0 \quad (8)$$

High-order schemes with their low dissipative property usually bear more risk from the SCL-related errors than that of low-order schemes. The SCMM ensures the SCL in high-order difference schemes at the following two aspects:

- First, metrics are acquired through the "conservative forms"

$$\tilde{\xi}_x = \frac{1}{2}[(zy_\eta)_\zeta + (yz_\zeta)_\eta - (zy_\zeta)_\eta - (yz_\eta)_\zeta]$$
$$\tilde{\eta}_x = \frac{1}{2}[(zy_\zeta)_\xi + (yz_\xi)_\zeta - (zy_\xi)_\zeta - (yz_\zeta)_\xi]$$
$$\tilde{\zeta}_x = \frac{1}{2}[(zy_\xi)_\eta + (yz_\eta)_\xi - (zy_\eta)_\xi - (yz_\xi)_\eta] \quad (9)$$

and with similar relations for the remain items.

- Second, on each grid direction, the algorithm of the derivatives in Eq. (9) shall be identical to that of flow fluxes where the metrics are re-discretized in combination with the flow fluxes. For example, the re-discretization scheme of WCNS-E-5 is

$$\frac{\partial a_i}{\partial \xi} = \frac{75}{64h}(a_{i+1/2} - a_{i-1/2}) - \frac{25}{384h}(a_{i+3/2} - a_{i-3/2}) + \frac{3}{640h}(a_{i+5/2} - a_{i-5/2}) \quad (10)$$

where the cell-edge values can be computed by high-order linear interpolation.

Finally, we briefly introduce the CBIC. Suppose RHS be the right-hand-side of the n-th time step

$$RHS = -J\left[\left(\frac{\partial \tilde{F}}{\partial \xi} + \frac{\partial \tilde{G}}{\partial \eta} + \frac{\partial \tilde{H}}{\partial \zeta}\right)\right]^n \quad (11)$$

define the transformation matrix P_{QV_C} in terms of conservative variables Q and characteristic variables V_C

$$P_{QV_C} = \frac{\partial Q}{\partial V_C} \quad (12)$$

the CBIC are

$$\left.\frac{\partial Q}{\partial t}\right|_L = \left(A_s^+\right)\big|_L (RHS)\big|_L + \left(A_s^-\right)\big|_L (RHS)\big|_R$$
$$\left.\frac{\partial Q}{\partial t}\right|_R = \left(A_s^+\right)\big|_R (RHS)\big|_R + \left(A_s^-\right)\big|_R (RHS)\big|_L \quad (13)$$

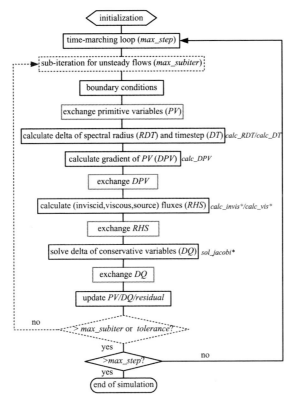

Fig. 1. Flowchart of HOSTA. Some subroutines implementing the steps are also shown. Time consuming subroutines are indicated by asterisk signs. HDCS and WCNS are implemented in `calc_vis` and `calc_invis`, indicated by red asterisk signs. Note that each iteration has four exchanges (denoted by red rectangular) for ghost/singularity data to ensure the robustness of high-order schemes.

where $(A_s^+) = P_{QV_C} \mathrm{diag}[(1+\mathrm{sign}(\lambda_i))/2]P_{QV_C}^{-1}$, $(A_s^-) = P_{QV_C} \mathrm{diag}[(1-\mathrm{sign}(\lambda_i))/2]P_{QV_C}^{-1}$. $(A_s^+)|_{L,R}$ are calculated by variables on the interface, and $(RHS)|_{L,R}$ can be calculated by one-side differencing or other methods. Here, the subscripts "L" and "R" denote the variables and terms on an interface but belong to the left-hand-side sub-block and the right-hand-side sub-block, respectively. One can verify that $\frac{\partial Q}{\partial t}|_L = \frac{\partial Q}{\partial t}|_R$. Please refer to [19] for more details about the CBIC.

2.2. HOSTA implementation

HOSTA is implemented in Fortran 90 and we present its main flowchart in Fig. 1. At the beginning of each iteration (a time step or a sub-iteration of unsteady simulations), boundary conditions are applied. Then data exchange for primitive variables of ghost/singularity cells is performed. Singularity cells are those cells that belong to several grid blocks. They are often unavoidable in complex multi-block structured grids. HOSTA calculates the delta of spectral radius (`calc_RDT`) and time-step before it calculates and exchanges the gradient of primitive variables (`calc_DPV`). HDCS and WCNS are implemented when calculating viscous (`calc_vis`) and inviscid (`calc_invis`) fluxes for RHS. The RHS is also exchanged. Implicit methods including LU-SGS, PR-SGS, Jacobi and explicit Runge–Kutta method are implemented for time marching. In this paper, we use the Jacobi iterative method (`sol_jacobi`) on GPUs. After HOSTA exchanges the delta of conservative variables, it updates the primitive variables and residuals and then completes the iteration. Flow variables are declared as global variables and thus can be used in several subroutines conveniently. Those variables are allocated and initialized prior to the main time-marching loop, and are deallocated at the end of the simulation. In the MPI implementation, generally a large multi-block complex 3D grid is repartitioned, along three grid directions, into many grid blocks and those blocks are then grouped and distributed among MPI processes according to a load balance model [50]. Non-blocking MPI is used to overlap message passing and computation.

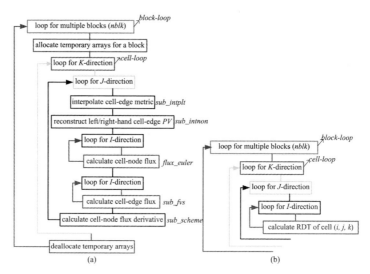

Fig. 2. A closer look of typical loop skeletons in HOSTA. (a) The *I*-direction loop skeleton in `calc_invis`. It clearly shows how high-order schemes such as HDCS and WCNS are implemented for calculating inviscid fluxes. Some steps are encapsulated in Fortran subroutines. (b) The loop skeleton in `calc_RDT`.

Fig. 2 presents a closer look of two typical loop skeletons to further illustrate multi-block calculating procedures in HOSTA. Generally, HOSTA organizes its calculation in a four-level loop: an outer *block-loop* specifying block index and a three-level inner *cell-loop* specifying cell index in the block along the three directions. For complex kernels like `calc_vis` and `calc_invis`, we implement three different four-level loops for the three directions. Fig. 2(a) shows the *I*-direction loop skeleton in `calc_invis`. Firstly, the cell-edge metrics are interpolated and the left-hand and right-hand cell-edge primitive variables are reconstructed. Then, the cell-node fluxes are calculated and the cell-edge fluxes are also calculated using flux-splitting methods such as Roe, Steger-warming and etc. Finally, the cell-node flux derivatives are calculated. The whole calculation procedure is very complex, involving the algorithms and methods described in Section 2.1. HOSTA uses temporary arrays to hold intermediate results in the procedure, and those arrays are allocated and deallocated dynamically for each grid block in each iteration. Data dependence only exists in the inner *I*-loop, i.e., the calculation of cell (i, j, k) depends on its neighboring cells $(i \pm m, j, k)$, where m is the width of the stencil. The complexity of the loop skeletons in `calc_vis` and `calc_invis` requires our extra implementation and optimization efforts when porting HOSTA to GPUs. Fig. 2(b) shows the loop skeleton in `calc_RDT`. It is relatively simple: each cell uses flow variables on itself and there is no data dependence. To summarize, we can see that in HOSTA there exists a two-level parallelism, i.e., a coarse-grained parallelism (corresponding to the block-loop) among multiple grid blocks and a fine-grained parallelism (corresponding to the cell-loop) among cells in a grid block.

3. Overall MPI + OpenMP + CUDA parallelization

In this section, we briefly describe the TianHe-1A supercomputer and the overall MPI + OpenMP + CUDA parallelization of HOSTA on TianHe-1A.

3.1. The TianHe-1A supercomputer

TianHe-1A [49] is a GPU-accelerated supercomputer developed by National University of Defense Technology (NUDT), China, and was ranked No. 1 in the 36th Top500 list [5] for HPC systems in Nov., 2010. Each TianHe-1A node (see Fig. 3) contains two Intel hexa-core Xeon X5670s sharing a 32 GB main memory and one NVIDIA Tesla M2050 with a 3 GB device memory. Each GPU contains 14 Streaming Multiprocessors (SMs) and each SM has 32 CUDA cores. The theoretical double-precision performance of the two X5670s is about 140 GFlops, which is about 27% of the M2050's 515 GFlops. However, the memory bandwidth of the two X5670s is about 64 GB/s, which is about 43% of the M2050's 148 GB/s. Since CFD is memory-bounded, the difference of memory bandwidth rather than that of the computation is expected to be the dominant factor of the achievable performance difference between CPU and GPU. Nodes are connected via a self-developed fat-tree interconnection network, with a latency of about 1.57 μs and a bidirectional bandwidth of about 20 GB/s.

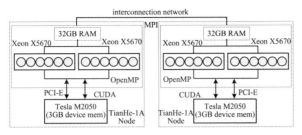

Fig. 3. Compute nodes and heterogeneous programming in TianHe-1A. Each TianHe-1A node has two hexa-core Xeon X5670 CPUs and one Tesla M2050 GPU, contributing totally about 656 GFlops theoretical performance in double-precision.

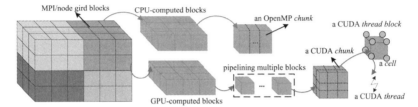

Fig. 4. Overall domain decomposition for MPI + OpenMP + CUDA parallelization. Grid blocks are first distributed among MPI processes/nodes and then on each node are further allocated between CPU and GPU.

With regard to programming models, we use CUDA on GPUs, OpenMP on CPUs. We also use MPI to enable inter-node communication. CUDA functions that run on GPUs are called *kernels*. A kernel is implemented using the SPMD (Single Program Multiple Data) model and executed in parallel by many GPU threads on CUDA cores using the SIMD (Single Instruction Multiple Data) model. GPU threads are grouped into *thread blocks* which are scheduled to a multiprocessor. All thread blocks compose a *thread grid*. Before the kernel is launched, developers configure the number of threads in a thread block and the number of thread blocks in a thread grid. Each CUDA thread can use the built-in dim3 type variables *threadIdx* and *blockIdx* to specify its index in the local thread block and the global thread grid. Applications could manage concurrency through CUDA *streams*. A CUDA stream is simply a sequence of operations that are performed in order on the device. Operations from different streams can be interleaved on devices that are capable of concurrent copy and execute, and it is also possible to overlap host computation with asynchronous data transfers and device computations.

3.2. Overall MPI + OpenMP + CUDA parallelization

Fig. 4 shows the overall domain decomposition for MPI + OpenMP + CUDA parallelization. The domain decomposition for MPI parallelization is same as described in Section 2.2. We create one MPI process per node (of TianHe-1A) to manage the two CPUs and the GPU. The *nblk* grid blocks on a MPI process/node is further allocated between the intra-node CPUs and GPU (see Section 5). The OpenMP directives are added over the outer-most cell-loop, i.e., each CPU-computed grid block is logically partitioned to many *OpenMP chunks* and each chunk is calculated by an OpenMP thread. GPU-computed grid blocks are issued simultaneously by multiple CUDA streams, and each grid block is logically partitioned into many *CUDA chunks*; each CUDA chunk is updated by a CUDA thread block and finally, each cell in the chunk is calculated by a CUDA thread.

4. GPU-only parallelization and optimization

Before we detail the collaborative approach, in this section we parallelize and optimize HOSTA on a single GPU.

4.1. Dual-level GPU parallelization

As described in Section 2.2, in HOSTA a multi-block loop can be viewed as two-level: block-loop and cell-loop. This two-level parallelism is mapped to two CUDA constructs, i.e., CUDA stream and kernel, in our dual-level GPU parallelization (see Fig. 5). Each cell-loop is implemented as a CUDA kernel dedicated to the calculation in a grid block. The block-loop is replaced with a CUDA *stream-loop* in which a stream ID is used in a CUDA kernel to index a specific grid block.

For different CUDA kernels, we employ different CUDA implementations and configurations according to data dependence in cell-loops. If there is no data dependence between cells (e.g., `sol_jacobi` and `calc_RDT`), we adopt a 3D CUDA kernel configuration. For example, consider a 3D grid block of size (NI, NJ, NK) (i.e., there are NI, NJ and NK cells on I, J and

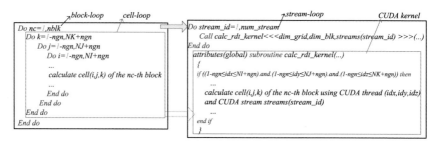

Fig. 5. Dual-level GPU parallelization for multi-block calculation. The block-loop and cell-loop are parallelized with CUDA streams and CUDA kernels respectively. Since HOSTA is implemented in Fortran, for simplicity and clarification in this paper we use Fortran and CUDA Fortran [2] features to describe code skeletons or pseudocodes.

K directions respectively) with *ngn* ghost cells on each direction, we use (*x_blk, y_blk, z_blk*) to specify a 3D CUDA thread block and define two CUDA built-in type dim3 variables `dim_blk` and `dim_grid` for CUDA thread block and thread grid configuration respectively

```
dim_blk=dim3(x_blk,y_blk,z_blk)
dim_grid=dim3(⌈NI + 2 × ngn/x_blk⌉, ⌈NJ + 2 × ngn/y_blk⌉, ⌈NK + 2 × ngn/z_blk⌉)
```

where $\lceil x \rceil$ is the minimum integer no smaller than x. For a 3D kernel configuration, 3D grid cells on I, J and K directions are mapped to GPU threads on X, Y and Z dimensions correspondingly. The 3D CUDA kernel is implemented as an elemental function. Each GPU thread executes its own copy of the kernel code, uses its thread ID (*idx, idy, idz*) to index a cell and calculates the cell independently, where

$$idx = threadIdx.x + blockDim.x \times (blockIdx.x) + 1 - ngn$$

$$idy = threadIdx.y + blockDim.y \times (blockIdx.y) + 1 - ngn$$

$$idz = threadIdx.z + blockDim.z \times (blockIdx.z) + 1 - ngn$$

However, as illustrated in Fig. 2, in HOSTA there also exist more complex cell-loops (e.g., `calc_invis` and `calc_vis`) with data dependence and subroutine calling. As explained in Section 2.2, HOSTA decomposes those complex high-order flux calculation kernels along individual directions, thus data dependence and subroutine calling only exists in one direction, i.e., the inner-most loop. Since CUDA has no global synchronization, our initial GPU implementation uses a 2D kernel configuration to satisfy the data dependence and takes subroutines in the cell-loop as CUDA device kernels. Take the I direction cell-loop of `calc_invis` as an example, the 2D CUDA kernel configuration is

```
dim_blk=dim3(1,y_blk,z_blk)
dim_grid=dim3(1, ⌈NJ + 2 × ngn/y_blk⌉, ⌈NK + 2 × ngn/z_blk⌉)
```

Thus, each CUDA thread calculates all cells on the I direction.

Since there is no data dependence in a block-loop (i.e., among grid blocks), it can be parallelized with CUDA streams. Fig. 5 shows a simple multi-stream implementation. We create the same number of CUDA streams as that of grid blocks before the time loop and destroy them at the end of the simulation. Thus, the block-loop is converted into the stream-loop. In the stream-loop, we bind each grid block with a CUDA stream and issue all the streams to the GPU. All the GPU operations of the grid block including kernel execution, host to device (H2D) and device to host (D2H) data copy are performed with the associated CUDA stream. Fig. 7(a) shows the timeline for multi-stream when streaming 4 grid blocks. The multi-stream scheme can eliminate the serial execution semantics between grid blocks in the block-loop, which is implicitly added by programming languages such as Fortran and C, and better exploit the parallel potential of modern GPU architecture. For example, when a stream is accessing the global memory, the kernel engine can schedule and execute GPU warps from other streams to hide memory access latency. More importantly, the kernel execution and PCI-e data transfers of different streams/blocks can be substantially overlapped, especially for GPUs with separate H2D and D2H copy engines such as the Tesla M2050 in TianHe-1A. Furthermore, as we will describe in the following sections, the multi-stream scheme can be extended for better load balance in a collaborative simulation and overlapping CPU/GPU computation and communication.

4.2. Kernel decomposition

Generally, a 3D kernel configuration can ensure more GPU threads to run simultaneously and may allow threads to access the global memory in a coalesced manner, and thus improves performance. In HOSTA, high-order flux calculation kernels

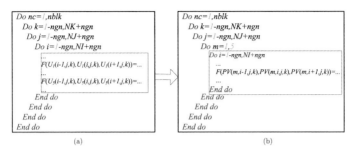

Fig. 6. Code transformation of data structures and inner-loops in HDCS-E8T7.

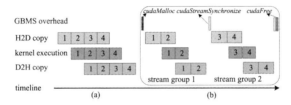

Fig. 7. Different schemes when pipelining 4 grid blocks: (a) multi-stream uses 4 streams; (b) GBMS uses 2 stream groups and each group contains 2 streams. It calls `cudaStreamSynchronize` to synchronize streams in each group and also calls `cudaMalloc/cudaFree` (and may also perform H2D/D2H copy) for selected flow variables.

(i.e., kernels in `calc_invis` and `calc_vis`) are the most time-consuming ones. But they are initially implemented with a 2D configuration, due to data dependence. In this paper, we optimize those kernels with a functional decomposition strategy. By carefully analyzing the data dependence and inlining device kernels, a large complex 2D kernel can be broken into several small 3D kernels. For example, the initial 2D kernel `calc_I_invis` calculating the *I* direction inviscid fluxes based on WCNS-E-5 is decomposed to four 3D kernels: `calc_I_invis_1` calculating the cell-edge metrics, `calc_I_invis_2` reconstructing the left-hand and right-hand cell-edge primitives, `calc_I_invis_3` calculating the cell-edge and cell-node inviscid fluxes and `calc_I_invis_4` calculating the cell-node derivative of inviscid fluxes.

For HDCS-E8T7, we use a *semi-3D* kernel configuration for `calc_I_invis_2` because its interpolation scheme requires solving a system of tri-diagonal equations (see Eq. (6)). The semi-3D kernel is implemented based on a code transformation of data structures and inner-loops. Fig. 6(a) shows the initial code snippet which has five separated primitive variables U_m ($1 \leq m \leq 5$). Fig. 6(b) shows the new code snippet with a merged primitive variable *PV* replacing the five and forming a new inner loop for *m*. Then we exchange loop *m* and loop *i*, and an independent "*M*" direction is available to be parallelized on the GPU. Thus, a semi-3D kernel configuration of a 3D thread block (5,*y_blk*,*z_blk*) and a 2D GPU grid (1, $\lceil NJ + 2 \times ngn/y\_blk \rceil$, $\lceil NK + 2 \times ngn/z\_blk \rceil$) can be used in `calc_I_invis_2`.

5. Collaborating intra-node CPU and GPU

5.1. Intra-node collaborative programming

Since most OpenMP compilers support nested parallelism, we use this feature to coordinate the OpenMP parallelized CPU code and the CUDA parallelized GPU code. Fig. 8 illustrates the collaborative programming as well as the balancing scheme (see Section 5.2) for calculating the inviscid fluxes (`calc_invis`). The other calculating procedures can be implemented in a similar manner. To implement a collaborative calculation, we develop two different versions of `calc_invis`: the GPU version `calc_invis_gpu` wrapping CUDA kernels and the CPU version `calc_invis_cpu` parallelized with OpenMP. We create two OpenMP threads at the first-level parallel region. One thread calls `calc_invis_gpu` to launch CUDA kernels, dealing with the GPU-computed grid blocks. The other thread calls `calc_invis_cpu`, dealing with the CPU-computed grid blocks. In `calc_invis_cpu`, for each block we fork a second-level nested OpenMP threads (11 threads) on the multi-core CPUs. We call `cudaDeviceSynchronize` to perform a CPU–GPU synchronization before exchanging ghost/singularity data. During collaboration, heterogeneous devices are concurrently working on their own grid blocks.

We employ a static task/load partition for the intra-node CPUs and GPU: on each side all kernels always calculate the same data objects (specifically grid blocks in HOSTA) until the end of a simulation. This is based on an observation: in HOSTA (and also many other heterogeneous applications) the GPU code employs a static but efficient memory use policy similar to the CPU code. All flow variables of GPU-computed grid blocks are allocated and initialized before the time-marching loop,

```
streams(:) : CUDA stream arrays
gpu_blk=ceiling(gpu_ratio*nblk)
...
!$omp parallel num_threads(2)
If (omp_get_thread_num().eq.0) then
    If GBMS is defined then    !GBMS implementation
        allocate device mem of num_grpstream blocks for mb_var
        Do stream_grp_id=1,num_grp
            Do stream_id=1,num_grpstream
                nc=(stream_grp_id-1)*num_grpstream+stream_id
                If (nc<gpu_blk) then
                    H2D copy the nc-th mb_var using streams(stream_id)
                    Call calc_invis_gpu(...) using streams(stream_id)
                    D2H copy the nc-th mb_var using streams(stream_id)
                End If
            End Do
            synchronize streams
        End Do
        deallocate mb_var
    Else !multi-stream implementation
        Do stream_id=1,gpu_blk
            Call calc_invis_gpu(...) using streams(stream_id)
        End Do
    End If
End If
If (omp_get_thread_num().eq.1) then
    Call calc_invis_cpu(...)
End If
!$omp end parallel
...
```

Fig. 8. Pseudocode illustrating collaborative programming and GBMS in `calc_invis.mb_var` is a selected flow variable. The host and device kernels are encapsulated in `calc_invis_cpu` and `calc_invis_gpu` respectively.

and deallocated until the end of the loop (see Section 2.2). This can avoid unnecessary PCI-e data transfers during iterations, and we only need to transfer four flow variables (i.e., RHS, DQ, PV and DPV) when their ghost/singularity data exchanges are needed. The partition is implemented at the granularity of grid blocks. We use a parameter *gpu_ratio* to adjust the number of grid blocks simulated by each side

$$gpu\_blk = ceiling(gpu\_ratio \times nblk) \qquad (14)$$

The GPU calculates [1, *gpu_blk*] blocks and the others are calculated on the CPUs. *gpu_ratio* is configured by profiling HOSTA's sustainable performance of one iteration on different devices. We set *gpu_ratio* = 0 to perform a CPU-only simulation and then set *gpu_ratio* = 1.0 to perform a GPU-only simulation. Taking the CPU performance as the baseline, *gpu_ratio* is evaluated as $\frac{SP_{GPU}}{(1.0+SP_{GPU})}$ (suppose the achieved speedup of GPU to CPU is SP_{GPU}).

Note that it is possible to implement a more fine-grained or dynamic partition like the approaches presented in [35,7]. For example, in HOSTA we can precisely partition specific number of cells to different devices, or even assign different number of cells for different kernels according to their execution characteristics on different devices. Those approaches generally require developers to reconstruct data structures, and meanwhile carefully manage data dependence among kernels and data transfers among devices. However, HOSTA has so many data objects/kernels implemented by different programming models on different devices that the changes will definitely make programming more difficult and may result in uncertain overall performance benefits. Therefore, the approaches in [35,7] are often tested and applied in heterogeneous code implemented using a portable and unified language such as OpenCL [3] on different devices. Our partition approach in Eq. (14) is fairly simple but generic for multi-block CFD applications. Furthermore, we can adjust the partition granularity (e.g., mesh block size) by independent mesh repartition tools [46].

5.2. Balancing intra-node CPU and GPU

Our results show that HOSTA runs about 0.3 times faster on the M2050 than on the two X5670s (*gpu_ratio* = $\frac{1.3}{2.3} \approx 0.57$), and thus we tend to distribute more grid blocks on the GPU. Suppose we have a large grid and each node is expected to simulate 6M cells, ideally we let the GPU simulate 6M × 0.57 ≈ 3.4M cells and the CPUs simulate 2.6M (= 6M − 3.4M) cells according to their achievable performance. However, the M2050 can only hold 2M cells and the other 4M cells have to be distributed to the X5670s. Consequently, without a load-balancing scheme, for this case there is a significant imbalance of resource utilization: the GPU stays idle for most of the time while the CPUs are busy. Therefore, HOSTA's performance on a single TianHe-1A node is bounded by the relatively small device memory on GPUs.

To mitigate the limit, we need to reduce the maximum device memory requirement of GPU-computed grid blocks. For this end, we extend the multi-stream scheme [48] in two steps. As a first step, we group multiple CUDA streams in multi-stream to form a Group-Based Multiple Streams (GBMS) scheme. In multi-stream, we use *gpu_blk* streams to issue the same number of grid blocks (see Fig. 5 and Fig. 7(a)). In GBMS, we divide `gpu_blk` streams into `num_grp` stream groups and each group contains `num_grpstream` streams (see Fig. 5 and Fig. 7(b)), i.e., $gpu\_blk = num\_grp \times num\_grpstream$. As a second step, we add a dynamic device memory use policy in GBMS for particularly *selected flow variables* (e.g., `mb_var` in Fig. 5). For the GPU-computed grid blocks, `mb_var` is also stored in the host main memory. The device memory of `mb_var` is allocated/deallocated dynamically before/after a CUDA kernel referencing it. Before executing the kernel, we copy it from the host memory to the device memory. After executing the kernel, we copy it back to the host memory if it is changed in the kernel. Fig. 5 shows the pseudocode illustrating how GBMS is implemented in a collaborative simulation.

Compared with the multi-stream scheme, using GBMS can significantly reduce the requirement of device memory. Suppose for each block, `mb_var` requires D device memory, thus we need $gpu\_blk \times D$ device memory in multi-stream since those blocks are simultaneously executed on the GPU. With GBMS, the device memory requirement of `mb_var` is reduced to $num\_grpstream \times D$. Furthermore, in multi-stream the device memory of `mb_var` is persistent until the end of a simulation, while in GBMS the device memory of `mb_var` is only needed during the execution of a kernel and thus is temporary. In other words, the device memory of `mb_var` is deallocated immediately after the kernel finishes its execution; if the to-be-executed kernel do not use `mb_var`, we do not need to keep its device memory. We can use different (num_grp, $num\_grpstream$) configurations to further adjust the device memory requirement. For example, when pipelining 8 grid blocks on the GPU, (num_grp, $num\_grpstream$) can be set to $(2, 4)$ or $(4, 2)$. Obviously, the $(4, 2)$ configuration needs less device memory. For a large grid, users can adjust the configuration to check if the device memory is enough.

We specify selected flow variables based on the following principles:

- Variables used by a small number of CUDA kernels.
- Variables that have to be transferred for ghost/singularity data exchanges.
- Variables that occupy relatively large memory spaces.
- Variables used in the CUDA kernels that are far faster than their counterpart Fortran subroutines.

Those principles allow us to specify fewer selected flow variables, avoid significantly changing HOSTA code, and possibly minimize GBMS overhead (see next paragraph). The more selected flow variables we specify, the less device memory we require. In our implementation, we carefully specify 9 flow variables as selected flow variables, which accounts for about 36% of the total 25 flow variables and about 50% of the total device memory requirement. As a result, with GBMS we increase the maximum simulation problem size of HOSTA on the M2050 from 2M cells to about 4M cells.

At the same time, GBMS incurs an overhead on the GPU, due to dynamic device memory allocation/deallocation, H2D/D2H copy and stream synchronization between stream groups to ensure the completeness of asynchronous H2D/D2H copy. As Section 7.3 shows, the relative speedup of one M2050 to two X5670s drops from 1.3 to around 1.0 in GBMS. GBMS also needs additional host memory to hold selected flow variables for GPU-computed grid blocks. The overhead is reasonable, mainly attributed to the employment of CUDA streams and carefully specifying selected flow variables, both of which can hide or minimize the GBMS overhead properly. GBMS provides a solid base for more flexible task partition and better load balance in a collaborative simulation. In the previous example where each node is expected to simulate 6M cells, we can perform a well balanced collaborative simulation using GBMS: both the CPUs and the GPU simulates roughly 3M cells (i.e., $gpu\_ratio = 0.5$).

6. Scaling HOSTA on TianHe-1A

6.1. Minimizing PCI-e data transfer times

Grid blocks need to exchange ghost and singularity data with their neighboring blocks. On heterogeneous systems, when neighboring blocks reside on different compute nodes or devices, data exchanges may involve both intra-node PCI-e data transfer and inter-node MPI communication (see Fig. 9). In CUDA, `cudaMemcpy`/`cudaMemcpyAsync` can only copy continuous data elements at a time. Considering a 3D grid block with six ghost zones, the data for a single ghost zone is continuously stored in the device memory, while the data for different ghost zones is not. Thus, we can either call `cudaMemcpy` many times for the six ghost zones or call `cudaMemcpy` one time to transfer the whole 3D grid block rather than just ghost zones. Both solutions will result in extra costs on PCI-e data transfers. We present a gather/scatter optimization to minimize transfer times for discontinuous ghost and singularity data (see Fig. 10). Before performing D2H copy, a CUDA kernel `GPU_gather` collects all non-continuous data to a continuous device buffer `OutBuffer_D`, and then the entire buffer is copied to a host buffer `OutBuffer_H`. Correspondingly, before performing H2D copy, a Fortran subroutine `CPU_gather` packs all the updated data to a host buffer `InBuffer_H`, and then the entire buffer is transferred to a device buffer `InBuffer_D`. Finally, we use a scatter kernel `Scatter_kernel` to distribute the data elements in `InBuffer_D` to ghost and singularity cells.

6.2. Overlapping computation and communication

HOSTA has already implemented the overlapping of CPU computation and MPI communication by non-blocking MPI. For example, after calculating `DPV`, HOSTA calls `MPI_Isend` and `MPI_Irecv` to exchange `DPV` and the non-blocking

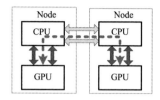

Fig. 9. Inter-node and intra-node data exchanges among heterogeneous devices.

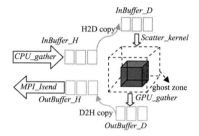

Fig. 10. Gather/scatter optimization.

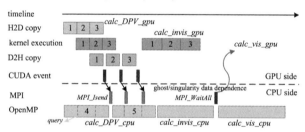

Fig. 11. Overlapping the collaborative CPU–GPU computation and communication for `calc_DPV`, `calc_invis` and `calc_vis`. Grid block *1*, *2* and *3* are calculated by the GPU. Grid block *4* and *5* are calculated by the CPUs.

MPI communication will be overlapped with the following `calc_invis`. Before using `DPV` in `calc_vis`, HOSTA calls `MPI_Waitall` to ensure that `MPI_Irecv` has finished receiving `DPV`.

In this paper, using non-blocking MPI, CUDA multiple streams and CUDA events, we implemented an *as early as possible* non-blocking communication mechanism to further overlap the collaborative computation with PCI-e data transfers and MPI communication. Fig. 11 illustrates the overlapping in `calc_DPV`, `calc_invis` and `calc_vis`. On the GPU side, when a stream finishes its operations of the associated grid block for `calc_DPV`, a CUDA event with the same stream ID is recorded to represent the data dependence between MPI communication and ghost/singularity data calculated by the stream. On the CPU side, we enable a counter in `calc_DPV`. During the execution of `calc_DPV`, for each CPU-computed gird block we query CUDA events several times (e.g., 3 times) to check if there are any GPU-computed grid blocks that have finished their calculations and intra-node data transfers, and also check the state of CPU-computed grid blocks. Then we call `MPI_Isend` or `MPI_Irecv` for ready grid blocks on both sides. Hence, we perform non-blocking communication for each grid block as early as possible, and this helps us better overlap the collaborative computation and the extensive high-order inter-block communication.

7. Performance results

7.1. HOSTA implementation on TianHe-1A

The GPU code is implemented using CUDA C. All CPU subroutines in the time-marching loop except those implementing MPI communications are rewritten as CUDA C kernels. Those kernels are wrapped in C functions to be called from Fortran code. We extend the main program loop to glue the GPU code with the CPU code for collaborative calculations. Since HOSTA is often used for high-resolution flow simulations, we always use a double precision implementation with ECC on. In total,

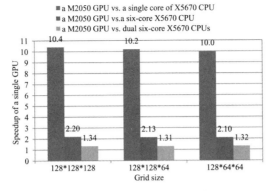

Fig. 12. Single GPU speedup for different grid sizes.

we (4 full-time researchers and several part-time researchers) spent roughly 8 months for the MPI–CUDA version and another 3 months for the collaborative version. As we have mentioned, it is because we need to use diverse programming models/tools. Another important reason is that HOSTA is rather than a *toy* project, but a large-scale software containing more than 25 000 lines of code. Totally we have added more than 16 000 lines of CUDA C/C code and Fortran wrapper code for the final collaborative version. We believe it is also due to the time spent on finding and reducing bugs. By direct comparison of corresponding data fields, we make sure that the GPU-only version and the collaborative version always produces the same results as the CPU version up to machine accuracy. Debugging tends to be harder when various components/routines are tightly coupled, as changes in one may cause bugs to emerge in another. Therefore, making parallelization and optimization on heterogeneous supercomputers is still an exceptionally time-consuming and challenging job.

7.2. Experimental setup and metrics

We use CUDA of version 5.0, Intel icc11.1 for C code and Intel ifort11.1 for Fortran code. All code is compiled with the -O3 option. We use MPICH2-GLEX for MPI communication. We use automatically generated 3D airfoil configurations to facilitate performance evaluation. For each airfoil grid, users can specify the numbers of cells and partition parameters on three grid directions and obtain grid blocks of equal size. This can save a lot of grid generation time for complex geometries. We also present detailed timing results of *C919*. The total double precision floating point operation count on X5670s is measured by Intel Vtune amplifier XE 2011. Since we have no tools to count the GPU operations, we use the CPU operation count as a reference to estimate the GPU and the CPU + GPU machine efficiencies. To collect performance data, we simulate 10 time steps ($max\_step = 10$), with each having 50 sub-iterations ($max\_subiter = 50$), and average the execution time. We define the collaborative efficiency (*CE*) to evaluate the efficiency loss in CPU–GPU collaboration

$$CE = \frac{SP_{CPU+GPU}}{SP_{CPU} + SP_{GPU}} \times 100\% \qquad (15)$$

$SP_{CPU+GPU}$ is the achieved collaborative speedup, taking the performance of the two X5670s as the baseline. For example, if $SP_{CPU+GPU} = 1.8$ and $SP_{GPU} = 1.3$, then *CE* is $\frac{1.8}{1.0+1.3} \times 100\% \approx 78.3\%$, which represents an efficiency loss of around 22% in the collaboration.

Since we achieve a speedup of about 1.3/1.0 when comparing the M2050 to the X5670s without/with GBMS, *gpu_ratio* in Eq. (14) is set to 0.57 and 0.5 respectively. For simplicity, in default we fix the number of grid blocks (*#block*) on each node to be 8.[1] Correspondingly, we set (*num_grp*, *num_grpstream*) = (2, 2) in GMBS.[2]

7.3. Node-level performance

In this subsection, we present performance results on one TianHe-1A node. Fig. 12 presents the speedup of a M2050 GPU over a single core of X5670 CPU using 1 OpenMP thread, a six-core X5670 CPU using 6 OpenMP threads and dual six-core X5670 CPUs using 12 OpenMP threads respectively. We evaluated grids of three different problem sizes. We can see that our GPU code achieves a speedup of about 10 over the serial CPU code and the speedup is about 2.1 when comparing a M2050 to an X5670. Note that the price of a M2050 is similar to that of an X5670, and the results are comparable to the speedup

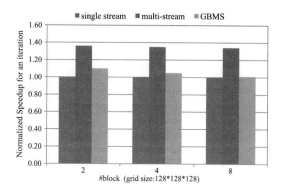

Fig. 13. Performance comparison of different CUDA stream implementation.

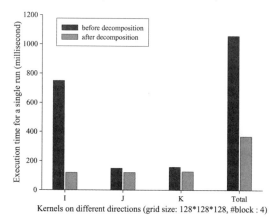

Fig. 14. Performance for `calc_invis` before/after kernel decomposition.

in paper [51] as far as similar-priced comparison of CPU and GPU is concerned. Also note that the power consumption of a M2050 is roughly equal to that of two X5670s and the speedup is about 1.3 when comparing a M2050 to dual X5670s, which further validates GPU's cost-effectiveness. This also motivates us for CPU–GPU collaboration to maximize HOSTA's performance at the node-level, since the sustainable performance of HOSTA on the two CPUs is not negligible (about 80% of the GPU). We get better performance on larger problem sizes, because higher workloads can better overlap the GPU computation and device memory accesses.

Fig. 13 shows the performance comparison of different streaming implementation for one iteration. The grid size is fixed to $128 \times 128 \times 128$ with the number of grid blocks varied from 2 to 8. We take the single CUDA stream performance as the baseline. We see a 25% to 30% performance enhancement when using CUDA streams to simultaneously execute grid blocks on GPU in multi-stream. The overhead of GBMS is indeed a little significant and we lose about 28% performance from multi-stream, due to CPU–GPU synchronization and PCI-e data transfers for selected flow variables. However, the compensation of this overhead is also substantial: we improve HOSTA's maximum simulation capacity on a M2050 from 2M cells to 4M cells. With GBMS, we successfully trade rich compute capacity for poor memory footprint in a GPU.

Fig. 14 shows the performance comparison between before and after the kernel decomposition in `calc_invis`. We reduce about 80% execution time for the *I* direction inviscid fluxes calculation when using decomposed kernels, but for the *J* and *K* directions, the performance improvements are both around 15%. This is because the decomposition in the *I* direction can ensure more GPU threads to access the global memory in a coalesced manner.

Fig. 16 presents the intra-node speedup and *CE* with/without GBMS. The GPU-only results are only available for the four small grids, due to the device memory limit. For $256 \times 128 \times 128$ (roughly 4M), we have to use GBMS and SP_{GPU} drops from 1.3 to 1.05. Although losing about 25% performance due to the GBMS overhead, one M2050 is still comparable to two X5670s. For those four small problems, we need not use GBMS in collaboration, and the average $SP_{CPU+GPU}$ and *CE* are 1.81

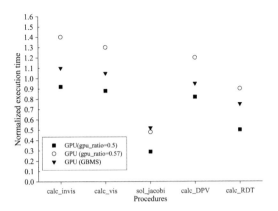

Fig. 15. Normalized execution times of different CPU/GPU procedures.

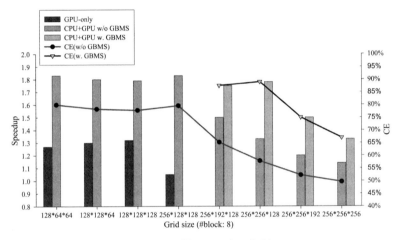

Fig. 16. Intra-node collaborative speedup and efficiency.

and 79% respectively. The collaborative approach has up to 45% performance advantage over the GPU-only approach. For larger problem sizes (≥4M), GBMS plays a very important role for improving both the collaborative speedup and efficiency. Taking 256 × 256 × 128 (roughly 8M) as an example, without GBMS the GPU can only simulates 2M cells and the other 6M cells have to be simulated on the CPUs. Due to the severe load imbalance, $SP_{CPU+GPU}$ is only about 1.3 and CE is only about 57%. With GBMS each side simulates 4M cells, $SP_{CPU+GPU}$ and CE are increased to about 1.79 and 89% respectively. 8M cells is HOSTA's maximum simulation capacity on one TianHe-1A node for a balanced collaboration, and this is an improvement of 2.3× compared to 3.5M cells without GBMS. As Fig. 16 shows, further increase in problem size leads to a significant decrease in the collaborative speedup and efficiency. This is because the GPU simulation workload is still limited (≤4M) even using GBMS. Despite of this, the GBMS balanced collaboration always has a notable advantage over the collaborative approach without GBMS.

We note that we lose around 20% speedup from ideal speedup (we get 1.8 of 2.3) in the intra-node collaboration. This is mainly due to the difference in sustainable performance between different kernels on CPU and GPU. As we mentioned, HOSTA has various complex numerical procedures consisting of many kernels, with each implemented in two versions. Those kernels may exhibit different characteristics on different devices. Some kernels run faster on the M2050, and others perform better on the X5670s. Fig. 15 shows the normalized execution time of several time-consuming procedures in HOSTA, taking CPU procedure as the baseline. For an equal problem size on both devices (i.e., $gpu\_ratio = 0.5$), the difference is very significant. The GPU version `sol_jacobi` and `calc_RDT` far outperform their CPU counterparts (around 3.5× and 2× respectively); while for `calc_invis` and `calc_vis` the two versions almost have the same performance. Even the

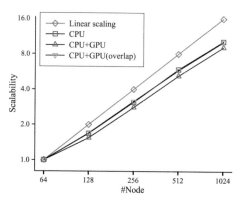

Fig. 17. Strong scalability results.

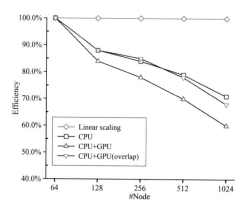

Fig. 18. Weak scalability results.

problem size is adjusted according to Eq. (14) (e.g., $gpu\_ratio = 0.57$ in Fig. 15) to maximize HOSTA's performance of a whole iteration, the difference is still obvious. GBMS to some extent levels off the difference, because the GBMS overhead is carefully added to CUDA kernels (i.e., `sol_jacobi`) that run far faster than their counterpart Fortran subroutines.

The kernel characteristic difference is a practical bottleneck for efficient heterogeneous collaboration in multi-kernel applications [35], because kernels on different devices must wait for each other to finish. A possible optimization is to employ a more fine-grained (e.g., at the granularity of grid cells) partition or tailoring partition for different kernels, as described in Section 5. To level off the performance difference, we may also focus on optimizing kernels with larger performance difference on different devices, or use a larger OpenMP parallel region and put the unbalanced and independent kernels together.

7.4. Large-scale performance

In the following scalability tests, the 64 node results are used as the baseline. Fig. 17 shows strong scalability results. The problem size of the baseline test is $256 \times 128 \times 128$ per node, and the total problem size is $64 \times 256 \times 128 \times 128$ (about 268M). Grids are evenly distributed among compute nodes. We achieve a speedup of about 10 when using 1024 nodes for CPU-only results, with an efficiency of around 64%. We use "*CPU + GPU*" to denote the collaborative results without the overlapping optimization. Since the largest problem size per node (i.e., the 64 node test) is about 4M, we need not use GBMS. The collaborative speedup and efficiency decrease to about 9 and 56% respectively on 1024 nodes without the overlapping optimization. We turn on the overlapping optimization to evaluate its effectiveness. The optimized scalability results are similar to the CPU-only results. Fig. 18 presents weak scalability results. The problem size per node is fixed to $256 \times 128 \times 128$. We lose about 27% efficiency from the perfect efficiency of 100% for CPU-only simulations when scaling

Table 1
The execution times (in seconds), GFlops and machine efficiencies (E) for HOSTA when simulating C919 on TianHe-1A.

#node	64	128	256	512
CPU	124	72	44	30
GPU	104	61	37	26
CPU + GPU (CE)	70 (77%)	43 (73%)	27 (71%)	19 (68%)
CPU GFlops (E)	780 (8.7%)	1344 (7.5%)	2186 (6.1%)	3226 (4.5%)
GPU GFlops (E)	928 (2.8%)	1586 (2.4%)	2580 (2.0%)	3710 (1.4%)
CPU + GPU GFlops (E)	1382 (3.3%)	2250 (2.7%)	3562 (2.1%)	5094 (1.5%)

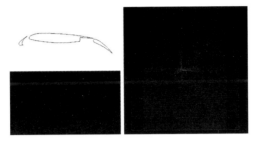

Fig. 19. The grid structure of *30p30n*. (colored picture attached at the end of the book)

to 1024 nodes. The efficiency loss is up to 32% (with the overlapping) and 40% (without the overlapping) for 1024 node collaboration, again demonstrating the effectiveness of the overlapping optimization.

To summarize, heterogeneous applications involve more complex and expensive data movements between devices at different levels than traditional CPU applications, which indicates that achieving a good parallel scalability and a high efficiency in large-scale systems is a tough challenge. Although the overlapping of communication and computation in HOSTA is fairly effective, we note that four data exchanges in one iteration is indeed a bottleneck in large-scale simulations. We are working on an improved implementation that can avoid some ghost/singularity communication.

Table 1 shows the detailed timing results of *C919*. *C919* is China's large civil airplane which currently is still under development. Since the grid contains about 150M cells, the 64 node GPU-only results are obtained with GBMS. For other node scales or collaborations, we need not use GBMS. The collaborative efficiency and the parallel efficiency are lower than that of the airfoil configuration, especially for large-number of nodes. This is because *C919* has a more complex geometry which is hard to be partitioned into grid blocks with a similar size. When distributing those blocks to large amount of TianHe-1A nodes, and further allocating them between intra-node CPU and GPU, we cannot achieve as good a load balance (of grid cells) as the less complex configurations such as airfoil. Nevertheless, the benefit of collaboration when simulating large grids with fairly complex geometries like *C919* is still reasonable: generally our collaborative approach can obtain the same performance, but uses only 50% less compute nodes than the CPU-/GPU-only approach on TianHe-1A. This will significantly save the simulation cost of HOSTA.

Note that the machine efficiencies are not high: less than 10% for CPU and less than 3.0% for GPU. For HOSTA-like implicit CFD code, the flop per byte is quite low, but modern cache-based CPU architecture needs to execute large amount of operations per data item for high efficiency. This mismatch can explain the low efficiency (often less than 20% and even less than 10%) of real-world CFD applications [27]. For GPU, it is more difficult to exploit its full potential Flops and requires more tuning work to achieve a higher efficiency.

8. Case study

8.1. HDCS-E8T7

To validate HDCS, we perform the Large-Eddy Simulation (LES) of a compressible viscous flow over the EET high-lift airfoil configuration *30p30n* as tested in the LTPT facility at NASA LaRC [32]. The conditions are same as [32]. The incoming Mach number $Ma = 0.17$, and the chord length $c = 0.457$ m, the corresponding chord Reynolds number is 1.71×10^6, the angle of attack is 4°. Fig. 19 shows the grid structure. The *30p30n* configuration used in this work contains 13 computational domains (with approximately 800M cells and 2400 grid blocks) and highly stretched meshes are used. The spacing on the surface is 0.0002 (normalized by the reference length $L\_ref = 0.0254$ m) in the normal direction. At the outer boundary, far-field boundary conditions based on the LODI approximation [40] and sponge technique [11] are prescribed. The no-slip condition is invoked on the airfoil surface, together with fifth-order accurate approximations for an adiabatic wall and zero normal pressure gradient. At the spanwise boundaries, periodicity is applied.

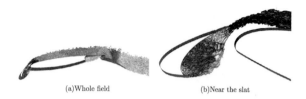

Fig. 20. Iso-surfaces of the second invariant of velocity gradient tensor for instantaneous vortex structure. (colored picture attached at the end of the book)

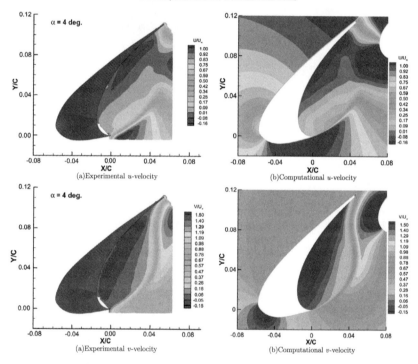

Fig. 21. Contours of the velocity. (colored picture attached at the end of the book)

HDCS-E8T7 uses the same idea as Monotone Integrated Large-Eddy Simulation (MILES) [12] to simulate the turbulent flow, i.e., the effects of explicit LES models are imitated by the truncation error. The computation has been advanced with a time step of 2.5×10^{-6} s for total 30 000 time steps. Fig. 20 demonstrates the iso-surfaces of the second invariant colored by density contours for the instantaneous flow. The second invariant Q is:

$$Q = (\Omega_{ij}\Omega_{ij} + S_{ij}S_{ij})/2 \tag{16}$$

where $\Omega_{ij} = (u_{i,j} - u_{j,i})/2$ and $S_{ij} = (u_{i,j} + u_{j,i})/2$. Due to the fine mesh, we can see the details about the vortex structure. Fig. 21 shows the mean velocity contours near the slat comparing with the experimental PIV measurements. We can see a qualitative agreement between the computation and the experiment with all salient features in the experiment being reproduced in the computation. Fig. 22 compares the mean velocity-vector near the slat between computation and experiment. The computational mean spanwise-vorticity comparing with the experimental measurement is shown in Fig. 23. We clearly see that the shear layer separating from the leading edge is demonstrated by the computation.

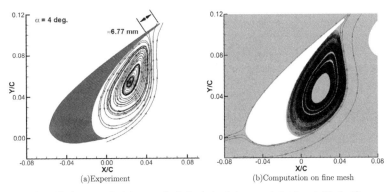

Fig. 22. The time-averaged velocity-vector distribution. (colored picture attached at the end of the book)

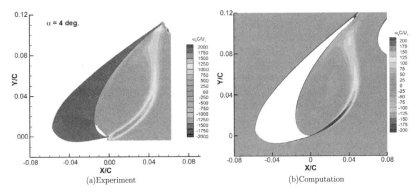

Fig. 23. The time-averaged spanwise-vorticity distribution. (colored picture attached at the end of the book)

Fig. 24. The grid structure of C919. (colored picture attached at the end of the book)

8.2. WCNS-E-5

C919 airliner is China's large civil airplane which currently is still under development. We perform the high-order aerodynamic simulation using WCNS-E-5 to evaluate its takeoff state and configuration. The body surface grid of the whole airplane is shown in Fig. 24. The geometry is very complex and contains the wing, the body, the horizontal tail, the vertical fin, the pylon, the nacelle, the winglet, the leading edge slat and the trailing edge flap. The grid points near the leading and trailing wing edges, the post-wing, the horizontal and vertical tails are refined, with a 0.00001 m first-level mesh spacing. The reference area is 120 m^2 and the distance between the reference point of torque and airplane nose is 18.35 m. The incoming Mach number $Ma = 0.2$, and the chord length $c = 4.8$ m, the corresponding chord Reynolds number is 2.0×10^7, the angle of attack is 8°. We use the Roe flux-splitting scheme and the two-equation turbulence model in the simulation.

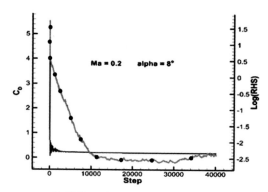

Fig. 25. The aerodynamic coefficient and the residual.

Fig. 26. The equivalent total pressure for different cross sections.
(colored picture attached at the end of the book)

Fig. 27. Streamline of the wing tip, the side edge of wing flap and the nacelle.
(colored picture attached at the end of the book)

Fig. 25 shows convergence curves of the aerodynamic coefficient and the residual. The residual decreases about five order of magnitude and the drag coefficient keeps stable after 40 thousands step calculations, indicating the convergence of pressure and flow field in the simulation. In Fig. 26, the nephogram for the equivalent total pressure of different cross sections is shown. Fig. 27 shows the streamline of the wing tip, the side edge of wing flap and the nacelle. We clearly see the vortex at the wing tip, the slide and the trailing edge. The results indicate a very subtle numerical simulation based on the WCNS-E-5 high-order scheme.

9. Related work

In the past twenty years, there have been many studies in developing and applying high-order compact finite difference schemes in CFD. In 1981, Harten et al. [28] developed a fourth order accurate implicit finite difference scheme for shock capturing. Lele [36] developed several central compact schemes with spectral-like resolution. Cockburn et al. [15] developed a fourth order compact TVD scheme for shock calculations. Visbal and Gaitonde [45] use filters to prevent numerical oscillations of central compact schemes. Recently, Lele's central compact schemes have been successfully applied by Rizzetta et al. [42] in the simulation of low speed flows. Adams et al. [6] developed a compact-ENO (Essentially Non-Oscillatory) scheme. Pirozzoli [39] developed a compact-WENO (Weighted Essentially Non-Oscillatory) scheme which was further improved by Ren et al. [41]. The WCNS scheme developed by us in 2000 has been successfully applied in many complex flows. In [37], Nonomura shows the excellent freestream and vortex preservation properties of WCNS on curvilinear grids, compared with those of WENO. Recently we have shown the importance of fulfilling GCL when applying high-order finite difference schemes on complex grids. Based on the idea of SCMM, we have proposed the HDCS scheme with inherent dissipation that can conveniently fulfill the GCL.

Many prior work has shown the experiences of porting CFD codes to GPUs, with impressive speed-ups. Phillips et al. [38] developed a 2D compressible Euler solver on a GPU cluster. Appa et al. [9] implemented and optimized an unstructured 3D explicit finite volume CFD code on multi-core CPUs and GPUs. Corrigan et al. [16] ported an adaptive, edge-based finite element code FEFLO to GPU clusters in a semi-automated fashion: the existing Fortran–MPI code is preserved while the translator inserts data transfer calls as required. Brandvik and Pullan [13] implemented a multi-GPU enabled Navier–Stokes solver for flows in turbomachines. They reported almost linear weak scaling using up to 16 GPUs. All the above studies were tested on small-scale GPU platforms using low-order CFD methods. For large-scale GPU clusters, Ali et al. [34] parallelized an incompressible solver for turbulence with a second-order scheme and reported scalability results on 64 GPUs. But no validation is presented. Jacobsen et al. [31] parallelized a CFD solver for incompressible fluid flows using up to 128 GPUs. They further demonstrated the large eddy simulation of a turbulent channel flow on 256 GPUs [17], but they only simulated a lid-driven cavity problem using 1D decomposition and low-order schemes. They also compare the scalability between the tri-level MPI + OpenMP + CUDA and the dual-level MPI + CUDA implementation [30], but they only use OpenMP to substitute intra-node MPI communication rather than collaborating CPU and GPU for computation as we have done in our work.

All the above studies use low-order CFD methods. Due to the complexity, high-order CFD schemes generally require extra implementation and optimization efforts on GPUs. Tutkun et al. [44] implemented a six-order compact finite difference scheme on a single GPU. Antoniou et al. [8] implemented a high-order solver on multi-GPUs for the compressible turbulence using WENO. But the solver can only run on a single node platform containing 4 GPUs for very simple domains like a 2D or 3D box. Their single-precision implementation achieve a speedup of 53 when comparing 4 Tesla C1070s with a single core of an Xeon X5450. Castonguay et al. [14] parallelized the first high-order, compressible viscous flow solver for mixed unstructured grids with MPI and CUDA, where the Vincent-Castonguay-Jameson-Huynh method is used. A flow over SD7003 airfoil and a flow over sphere is simulated using 32 GPUs. With MPI + CUDA, Appeyard et al. [10] accelerated the solution of the level set equations for interface tracking using an HOUC (High-Order Upstream Central) scheme. But they only demonstrated performance results on 4 GPUs. Zaspel et al. [51] implemented an incompressible double-precision two-phase solver on GPU clusters using a fifth-order WENO scheme. The test problem is a rising bubble of air inside a tank of water with surface tension effects, and parallel performance results are reported using up to 48 GPUs. In [48], with MPI and CUDA, we parallelize HOSTA on TianHe-1A using HDCS, but no CPU–GPU collaboration as well as heterogeneous load balance were discussed or implemented. This work was a major extension based on [48]. To summarize, the above GPU-enabled CFD simulations using high-order schemes are still preliminary as far as grid complexity, problem size and parallel scale are comprehensively concerned.

10. Conclusion and future work

The employment of high-order CFD schemes is an effective approach to obtain high-resolution and high fidelity simulation results for complex flow problems. In the past twenty years, we have developed several high-order compact finite difference schemes including WCNS and HDCS. To tackle the problems challenging the application of high-order schemes in complex grids, we have also proposed SCMM to fulfill GCL and CBIC for high-order block-interface conditions. Those high-order CFD schemes and methods have been successfully used in many complex flow problems with reasonably complex geometries, showing promising perspective for engineering applications.

Large-scale high-order CFD simulations often need to exploit powerful supercomputers to accelerate their running. In this work, with MPI + OpenMP + CUDA, we ported and optimized HOSTA, our in-house high-order CFD software for 3D multi-block structured grids, onto the GPU-accelerated supercomputer TianHe-1A. We presented some novel techniques to achieve balanced and scalable high-order CPU–GPU collaborative simulations on TianHe-1A. The approach and implementation of GPU parallelization, CPU–GPU collaboration as well as balancing scheme provide a fairly general experience of similar porting and optimizing efforts for multi-block CFD applications. As a case study, we have successfully performed CPU–GPU collaborative high-order accurate simulations with two large-scale complex grids using hundreds of TianHe-1A nodes.

For future work, besides fine tuning of the aforementioned bottlenecks in the collaborative code, we are planning to port HOSTA to China's new supercomputer TianHe-2, which was ranked No. 1 in Jun., 2013 for large-scale simulations such as direct numerical simulations of turbulence. Due to the usage of traditional programming models, we expect that the work would take us relatively less time. Since porting and optimizing CFD code often involves some common programming efforts such as restructuring kernel skeletons and data structures, in the long run we will encapsulate our experience learned in this paper and further work into an application programming and auto-tuning infrastructure for multi-block CFD simulations on heterogeneous systems.

Acknowledgements

This paper was supported by the National Science Foundation of China under Grant No. 11272352 and No. 61379056, and the Open Research Program of China State Key Laboratory of Aerodynamics under Grant No. SKLA20130105.

References

[1] NVIDIA Corp., CUDA C Programming Guide v4.2, 2012.
[2] Cuda Fortran, available online, http://www.pgroup.com/cudafortran, 2013.
[3] Khronos OpenCL working group, http://www.khronos.org/opencl, 2013.
[4] Mont-Blanc project home page, available online, http://www.montblanc-project.eu/, 2013.
[5] Top500 cite, available online, http://top500.org/, 2013.
[6] N. Adams, K. Shariff, A high-resolution hybrid compact-ENO scheme for shock-turbulence interaction problems, J. Comput. Phys. 127 (1996).
[7] E. Albayrak, Improving application behavior on heterogeneous manycore systems through kernel mapping, Parallel Comput. (2013).
[8] A.S. Antoniou, K.I. Karantasis, E.D. Polychronopoulos, Acceleration of a finite-difference WENO scheme for large-scale simulations on many-core architectures, AIAA Paper 2010-0525, 2010.
[9] J. Appa, J. Sharpe, P. Moinier, An unstructured 3D CFD code optimised for multicore and graphics processing units, in: MRSC2010 Proceedings.
[10] J. Appleyard, D. Drikakis, Higher-order CFD and interface tracking methods on highly-parallel MPI and GPU systems, Comput. Fluids 46 (2011) 101-105.
[11] D.J. Bodony, Analysis of sponge zones for computational fluid mechanics, J. Comput. Phys. 212 (2006) 681-702.
[12] J. Boris, F. Grinstein, E. Oran, R. Kolbe, New insights into large eddy simulation, Fluid Dyn. Res. 10 (1992).
[13] T. Brandvik, G. Pullan, An accelerated 3D Navier-Stokes solver for flows in turbomachines, in: ASME Turbo Expo 2009: Power for Land, Sea and Air.
[14] P. Castonguay, D.M. Williams, P.E. Vincent, A. Jameson, On the development of a high-order, multi-GPU enabled, compressible viscous flow solver for mixed unstructured grids, AIAA Paper 2011-3229, 2011.
[15] B. Cockburn, C.W. Shu, Nonlinearly stable compact schemes for shock calculations, SIAM J. Numer. Anal. 31 (1994) 607-630.
[16] A. Corrigan, R. Lohner, Porting of FEFLO to multi-GPU clusters, AIAA Paper 2011-0948, 2011.
[17] R. DeLeon, D. Jacobsen, I. Senocak, Large-eddy simulations of turbulent incompressible flows on GPU clusters, Comput. Sci. Eng. 15 (2013) 26-33.
[18] X. Deng, Y. Jiang, M. Mao, H. Liu, G. Tu, Developing hybrid cell-edge and cell-node dissipative compact scheme for complex geometry flows, in: The Ninth Asian Computational Fluid Dynamics Conference Proceedings.
[19] X. Deng, M. Mao, G. Tu, et al., Extending the fifth-order weighted compact nonlinear scheme to complex grids with characteristic-based interface conditions, AIAA J. 48 (2010) 2840-2851.
[20] X. Deng, M. Mao, G. Tu, et al., High-order and high accurate CFD methods and their applications for complex grid problems, Commun. Comput. Phys. 11 (2012) 1081-1102.
[21] X. Deng, M. Mao, G. Tu, H. Liu, H. Zhang, Geometric conservation law and applications to high-order finite difference schemes with stationary grids, J. Comput. Phys. 230 (2011) 1100-1115.
[22] X. Deng, Y. Min, M. Mao, H. Liu, G. Tu, H. Zhang, Further studies on Geometric conservation law and applications to high-order finite difference schemes with stationary grids, J. Comput. Phys. 239 (2013) 90-111.
[23] X. Deng, H. Zhang, Developing high-order weighted compact nonlinear schemes, J. Comput. Phys. 165 (2000) 22-44.
[24] J. Ekaterinaris, High-order accurate, low numerical diffusion methods for aerodynamics, Prog. Aerosp. Sci. 41 (2005) 192-300.
[25] K. Fujii, T. Nonomura, S. Tsutsumi, Toward accurate simulation and analysis of strong acoustic wave phenomena—a review from the experience of our study on rocket problems, Int. J. Numer. Methods Fluids 64 (2010) 1412-1432.
[26] GPGPU, General-purpose computation on graphics hardware, available online, http://gpgpu.org/, 2013.
[27] W.D. Gropp, D.K. Kaushik, D.E. Keyes, B.F. Smith, Towards realistic performance bounds for implicit CFD codes, in: ParCFD.
[28] A. Harten, H. Tal-Ezer, On a fourth order accurate implicit finite difference scheme for hyperbolic conservation laws, II. Five-point schemes, J. Comput. Phys. 41 (1981) 329-356.
[29] Intel, Many integrated core (MIC) architecture, available online, http://www.intel.com/content/www/us/en/architecture-and-technology/many-integrated-core/intel-many-integrated-core-architecture.html, 2013.
[30] D.A. Jacobsen, I. Senocak, Scalability of incompressible flow computations on multi-GPU clusters using dual-level and tri-level parallelism, AIAA Paper 2011-947, 2011.
[31] D.A. Jacobsen, J.C. Thibault, I. Senocak, An MPI-CUDA implementation for massively parallel incompressible flow computations on multi-GPU clusters, AIAA Paper 2010-0522, 2010.
[32] L. Jenkins, M. Khorrami, M. Choudhari, Characterization of unsteady flow structures near leading-edge slat: Part I. PIV measurements, AIAA Paper 2004-2801, 2004.
[33] G. Jiang, C. Shu, Efficient implementation of weighted ENO schemes, J. Comput. Phys. 126 (1996) 202-228.
[34] A. Khajeh-Saeed, J.B. Perot, Computational fluid dynamics simulations using many graphics processors, Comput. Sci. Eng. (2012).
[35] K. Kofler, I. Grasso, B. Cosenza, T. Fahringer, An automatic input-sensitive approach for heterogeneous task partitioning, in: ICS.
[36] S. Lele, Compact finite difference schemes with spectral-like resolution, J. Comput. Phys. 103 (1992) 16-42.
[37] T. Nonomura, Free-stream and vortex preservation properties of high-order WENO and WCNS on curvilinear grids, Comput. Fluids 39 (2010) 197-214.
[38] E.H. Phillips, Y. Zhang, R.L. Davis, J.D. Owens, Rapid aerodynamic performance prediction on a cluster of graphics processing units, AIAA Paper 2009-565, 2009.
[39] S. Pirozzoli, Conservative hybrid compact-WENO schemes for shock-turbulence interaction, J. Comput. Phys. 179 (2002) 81-117.
[40] T. Poinsot, S.K. Lele, Boundary conditions for direct simulations of compressible viscous flows, J. Comput. Phys. 101 (1992) 104-129.
[41] Y. Ren, M. Liu, H. Zhang, A characteristic-wise hybrid compact-WENO schemes for solving hyperbolic conservations, J. Comput. Phys. 192 (2005) 365-386.
[42] D. Rizzetta, M. Visbal, P. Morgan, A high-order compact finite difference scheme for large-eddy simulation of active flow control, Prog. Aerosp. Sci. 44 (2008) 397-426.
[43] J. Trulio, K. Trigger, Numerical solution of the one-dimensional hydrodynamic equations in an arbitrary time-dependent coordinate system, Technical Report UCLR-6522, University of California, Lawrence Radiation Laboratory, 1961.
[44] B. Tutkum, F.O. Edis, A GPU application for high-order compact finite difference scheme, Comput. Fluids 55 (2012) 29-35.
[45] M. Visbal, D. Gaitonde, High-order accurate methods for complex unsteady subsonic flows, AIAA J. 37 (1999) 1231-1239.

[46] Y. Wang, L. Zhang, Y. Che, W. Liu, C. Xu, H. Liu, TH-meshsplit: a multi-block grid repartitioning tool for parallel CFD applications on heterogeneous CPU/GPU supercomputer, in: ParCFD.
[47] Z. Wang, High-order methods for the Euler and Navier–Stokes equations on unstructured grids, Prog. Aerosp. Sci. 43 (2007) 1–41.
[48] C. Xu, L. Zhang, X. Deng, Y. Jiang, W.C.J. Fang, Y. Che, Y. Wang, W. Liu, Parallelizing a high-order CFD software for 3D, multi-block, structural grids on the TianHe-1A supercomputer, in: International Supercomputing Conference.
[49] X. Yang, X. Liao, K. Lu, et al., The TianHe-1A supercomputer: its hardware and software, J. Comput. Sci. Technol. 26 (2011) 344–351.
[50] A. Ytterstrom, A tool for partitioning structured multiblock meshes for parallel computational mechanics, Int. J. High Perform. Comput. Appl. 11 (1997) 336–343.
[51] P. Zaspel, M. Griebel, Solving incompressible two-phase flows on multi-GPU clusters, Comput. Fluids (2012).

Parallelizing and optimizing large-scale 3D multi-phase flow simulations on the Tianhe-2 supercomputer

Dali Li[1], Chuanfu Xu[1,2,*,†], Yongxian Wang[1,2], Zhifang Song[1], Min Xiong[1], Xiang Gao[1] and Xiaogang Deng[3]

[1]*College of Computer, National University of Defense Technology, ChangSha 410073, China*
[2]*National Laboratory for Parallel and Distributed Processing, National University of Defense Technology, ChangSha 410073, China*
[3]*National University of Defense Technology, ChangSha 410073, China*

SUMMARY

The lattice Boltzmann method (LBM) is a widely used computational fluid dynamics method for flow problems with complex geometries and various boundary conditions. Large-scale LBM simulations with increasing resolution and extending temporal range require massive high-performance computing (HPC) resources, thus motivating us to port it onto modern many-core heterogeneous supercomputers like Tianhe-2. Although many-core accelerators such as graphics processing unit and Intel MIC have a dramatic advantage of floating-point performance and power efficiency over CPUs, they also pose a tough challenge to parallelize and optimize computational fluid dynamics codes on large-scale heterogeneous system.

In this paper, we parallelize and optimize the open source 3D multi-phase LBM code *openlbmflow* on the Intel Xeon Phi (MIC) accelerated Tianhe-2 supercomputer using a hybrid and heterogeneous MPI +OpenMP+Offload+single instruction, mulitple data (SIMD) programming model. With cache blocking and SIMD-friendly data structure transformation, we dramatically improve the SIMD and cache efficiency for the single-thread performance on both CPU and Phi, achieving a speedup of 7.9X and 8.8X, respectively, compared with the baseline code. To collaborate CPUs and Phi processors efficiently, we propose a load-balance scheme to distribute workloads among intra-node two CPUs and three Phi processors and use an asynchronous model to overlap the collaborative computation and communication as far as possible. The collaborative approach with two CPUs and three Phi processors improves the performance by around 3.2X compared with the CPU-only approach. Scalability tests show that *openlbmflow* can achieve a parallel efficiency of about 60% on 2048 nodes, with about 400K cores in total. To the best of our knowledge, this is the largest scale CPU-MIC collaborative LBM simulation for 3D multi-phase flow problems. Copyright © 2015 John Wiley & Sons, Ltd.

Received 11 August 2015; Revised 12 October 2015; Accepted 12 October 2015

KEY WORDS: heterogeneous system; intel xeon phi; Tianhe-2; multi-phase flow; LBM

1. INTRODUCTION

The lattice Boltzmann method (LBM) is an alternative to classical Navier–Stokes solvers for computational fluid dynamics (CFD) simulations [1]. From the microscopic perspective of view, LBM regards fluids as Newtonian fluids, divides flow field into small lattices (mass points), and simulates fluid evolution dynamics through collision models (lattices' collision and streaming). LBM can simulate a variety of sophisticated fluid problems, including multi-phase,

---

Li Dali, Xu Chuanfu, Wang Yongxian, Song Zhifang, Xiong Min, Gao Xiang and Deng Xiaogang. Parallelizing and optimizing large-scale 3D multi-phase flow simulations on the Tianhe-2 supercomputer. Concurrency and Computation: Practice and Experience, 2016, 28(5): 1678-1692.

multi-component and thermal fluids [2]. Meanwhile, LBM can be easily applied to complex geometries with various boundary conditions as well. LBM codes have also been used in problems such as geofluidic flows, or magma flow through porous media, macro-scale solute transport, the dispersion of airborne contaminants in an urban environment, impact effects of tsunamis on near-shore infrastructure, melting of solids and resultant fluid flow in ambient air, and to determine permeability of materials [3].

Large-scale LBM simulations with increasing resolution and extending temporal range require massive high-performance computing resources. Thus, it is essential and practical to port LBM codes onto modern supercomputers, often featuring many-core accelerators/coprocessors (GPU[3], Intel MIC [4, 5], or specialized ones). These heterogeneous processors can dramatically enhance the overall performance of HPC systems with remarkably low total cost of ownership and power consumption. Although key kernels and procedures in LBM codes are naturally discrete, completely calculation independent, and ready for parallel processing, researchers often need to use various programming models/tools (e.g., Offload [4] for Intel MIC, OpenMP and MPI for intra-node/inter-node parallelization) on heterogeneous supercomputers, thus making the development and optimization of large-scale LBM applications exceptionally difficult. This is especially true when collaborating host CPUs and accelerators/coprocessors to maximize application performance. Taking Tianhe-2 (Milky Way-2), the present No. 1 supercomputer, as an example, each compute node contains 2 Xeon E5-2692 v2 CPUs and three Xeon Phi 31S1P coprocessors. The 2 CPUs and 3 MICs contribute 422 GFLOPS and 3 TFLOPS double-precision floating point performance, respectively. Our experience shows that the realistic performance of LBM codes on two Xeon E5-2692 CPUs is comparable with that of single Xeon Phi 31S1P, because the host CPU has advanced instruction issuing and scheduling mechanism with higher average memory bandwidth per core [6]. Thus, instead of using a CPU-only or MIC-only approach, it is necessary to perform CPU/MIC collaborative computing on Tianhe-2. However, to achieve efficient CPU/MIC collaboration, besides the aforementioned programming challenge, developers need to map a hierarchy of parallelism in problem domains to heterogeneous devices with different processing capabilities, memory availability, and communication latency. As for Tianhe-2, we must carefully balance the use of two CPUs and three Phi coprocessors, design mechanisms to effectively overlap and minimize the interaction costs between them, and finally scale LBM codes to large scale compute nodes.

Based on our previous experience of CFD codes on graphical processing unit (GPU)-accelerated supercomputers[6], in this paper, we parallelize and optimize a popular open source LBM code *openlbmflow* for large-scale 3D multi-phase flow simulations on the Tianhe-2 supercomputer. We tackle the challenge of heterogeneous CPU/MIC programming and collaboration for LBM codes. In particular, we make the following main contributions:

- With an in-depth analysis of both characteristics of LBM simulations and multi-core/many-core processors on Tianhe-2, we substantially enhance the cache efficiency and single instruction, mulitple data (SIMD) utilization of key LBM kernels in *openlbmflow* using cache-blocking and SIMD-friendly data layout transformation. As a result, we achieve a speedup of 7.9X and 8.8X on Xeon and Xeon Phi, respectively.
- On a single Tianhe-2 node, we collaborate CPU and Phi using offload and OpenMP. To do so, we propose a load-balance scheme to distribute workloads among CPUs and Phi processors and use an asynchronous model to overlap collaborative computation and communication on both devices. The intra-node CPU-MIC collaborative performance for *openlbmflow* is improved by 3.2X compared with the CPU-only approach.
- We further scale *openlbmflow* up on Tianhe-2 for massively parallel LBM simulations. Scalability tests show that the final MPI+OpenMP+Offload+SIMD parallelized code achieves a parallel efficiency of above 60% on 2048 Tianhe-2 nodes, with around 400K cores in total. To the best of our knowledge, this is the largest scale CPU-MIC collaborative LBM simulation for 3D multi-phase flow problems.

The remainder of the paper is organized as follows: Section 2 briefly describes LBM method and *openlbmflow* implementation. Section 3 presents the overall parallelization of *openlbmflow* on Tianhe-2. Section 4 presents hybrid MPI+OpenMP implementation and optimization for *openlbmflow*.

Section 5 presents the collaboration of CPU and Phi and scaling to large scale on Tianhe-2. Section 6 presents performance results. Finally, in Section 7, we introduce some related work and conclude in Section 8.

2. LBM METHOD AND *OPENLBMFLOW* IMPLEMENTATION

2.1. LBM method in openlbmflow

The *openlbmflow* program uses the popular D2Q9/D3Q19 (as shown in Figure 1) Lattice Boltzmann discretization model and Shan-Chen BGK single relaxation time collision model for multi-phase flows. It can simulate both 2D/3D single-phase or multi-phase flow problems with periodic and/or bounce-back boundary conditions. In this paper, we use 3D multi-phase flow problem as our simulation case. External-body-force feature enables *openlbmflow* to simulate gravity-like effects, and top/bottom-wall-speed feature is used for lid-driven flow. Further information about *openlbmflow* is available from its official website [7].

The particle distribution function of LBM obeys the Lattice Boltzmann equation (LBE). LBE performs in a micro-kinetics point of view and considers the macroscopic flow phenomenon as statistically averaged outcome of massive microscopic particles' movements. The particle collision in *openlbmflow* multi-phase flow is a simplified Shan-Chen BGK model. The particle distribution function $f(r, u, t)$ represents the density of particles at the moment t and spatial point r, whose speed is between u and $u + du$ according to statistical kinetics. The evolution equation of f is

$$\frac{\partial f}{\partial t} + u \frac{\partial f}{\partial r} = \Omega(f) + F, \tag{1}$$

where F and $\Omega(f)$ stand for external force and collision term, respectively. In linear BGK collision model, $\Omega(f)$ satisfies the following formula:

$$\Omega(f) = -\frac{1}{\tau}(f - f^{eq}), \tag{2}$$

where f^{eq} is the equilibrium distribution function, and τ is the dimensionless relaxation time. And f is discretized as follows:

$$f_i(x + e_i, t + \Delta t) - f_i(x, t) = -\frac{1}{\tau}[f_i(x, t) - f_i^{eq}(x, t)], \tag{3}$$

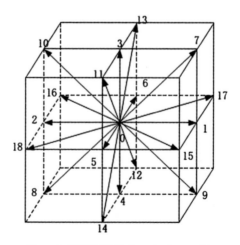

Figure 1. D3Q19 LBM discretization model.

where i represents the moving direction of a given particle, Δt is the time step, e_i is the unit length vector of each velocity direction of particle movements. Fluid density ρ and momentum ρu are described in the following formula, and the pressure **p** can be obtained through density.

$$\rho = \sum_i f_i = \sum_i f_i^{eq} \tag{4}$$

$$\rho u = \sum_i e_i f_i = \sum_i e_i f_i^{eq}, \tag{5}$$

the equilibrium distribution function f_i^{eq} and e_i are calculated as follows:

$$\begin{aligned}
f_0^{eq} &= \frac{1}{3}\rho\left(1 - \frac{3}{2c^2}u^2\right) \\
f_i^{eq} &= \frac{1}{18}\rho\left[1 + \frac{3}{c^2}(e_i u) + \frac{9}{2c^4}(e_i u)^2 - \frac{3}{2c^2}u^2\right] \quad (i = 1, 2 \cdots 6) \\
f_i^{eq} &= \frac{1}{36}\rho\left[1 + \frac{3}{c^2}(e_i u) + \frac{9}{2c^4}(e_i u)^2 - \frac{3}{2c^2}u^2\right] \quad (i = 7, 8 \cdots 18),
\end{aligned} \tag{6}$$

$$e_i = \begin{cases} (0,0,0) & i = 0 \\ (\pm 1, 0, 0)c, (0, \pm 1, 0)c, (0, 0, \pm 1)c & i = 1, 2 \cdots 6 \\ (\pm 1, \pm 1, 0)c, (\pm 1, 0, \pm 1)c, (0, \pm 1, \pm 1)c & i = 7, 8 \cdots 18 \end{cases}, \tag{7}$$

where $c = \frac{\Delta x}{\Delta t}$, Δx is the spatial grid stride, Δt is the temporal stride.

As for boundary conditions, periodic boundary lattices are treated as normal inner lattices, while bounce-back boundary lattices are updated after inner lattices and periodic boundary lattices during collision and streaming procedure. Figure 2 shows the bounce-back boundary mechanism in a 2D case. The gray boxes and white boxes are bounce-back boundary lattices and near boundary inner lattices, respectively; the arrows represent the distribution function components of each lattice. First, regular inner lattices' components are all updated when inner lattice collision and streaming are finished, as shown in Figure 2(a); meanwhile, the outward components (bold-black arrow) of bounce-back boundary lattices are also updated, but the back boundary components of near boundary lattices remain unchanged. Next, during the solid boundary lattice collision step in Figure 2(b), the outward components change their direction in a reflex manner, imitating the bounce-back effect of solid surface. At last the reversed components of bounce-back boundary lattices migrate to the corresponding components of near boundary lattices, see Figure 2(c).

2.2. *openlbmflow* implementation

The serial baseline LBM code *openlbmflow* is one of the most popular open-source LBM packages. We use the latest version (v1.0.1) in this work as of July 2015. Figure 3 shows the flowchart of *openlbmflow*. It mainly consists of three phases: initialization, time iteration, and post-processing. During initialization phase, the geometry of the flow field, flow density, and the distribution function are initialized. *Time* iteration phase includes three important procedures: inter-particular force calculation (as well as velocity and density), collision, and streaming. Because of the bounce-back boundary mechanism, procedures for boundary and inner lattices are implemented in separate kernels.

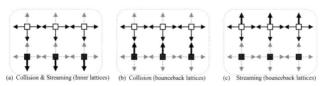

(a) Collision & Streaming (Inner lattices) (b) Collision (bounceback lattices) (c) Streaming (bounceback lattices)

Figure 2. Example of 2D bounceback boundary.

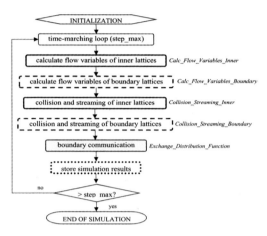

Figure 3. Flowchart of *openlbmflow*.

Procedures for collision and streaming are implemented together (*Collision_Streaming_Inner* for inner lattices and *Collision_Streaming_Boundary* for boundary lattices), and the equilibrium distribution calculation is also implemented in the collision code. Similarly, flow variables are updated for inter-particular force calculation in *Calc_Flow_Variables_Inner* and *Calc_Flow_Variables_Boundary*, respectively. In post-processing phase, simulation results are stored. *Collision_Streaming_Inner* and *Calc_Flow_Variables_Inner* for inner lattices are the most time-consuming kernels, and the time cost for boundary lattice kernels are negligible compared with inner-lattice kernels.

3. OVERALL HETEROGENEOUS AND COLLABORATIVE PARALLELIZATION

3.1. The Tianhe-2 supercomputer

Tianhe-2 (Milky Way-2) [8] is a heterogeneous supercomputer developed by National University of Defense Technology (NUDT), China. Tianhe-2 has been ranked No.1 since June, 2013. It consists of 16,000 compute nodes in total. As shown in Figure 4, each compute node has 2 Xeon E5-2692 v2 (ivy bridge) CPUs with 64 GB shared memory and 3 Xeon Phi 31S1P MIC coprocessors with 8 GB memory for each. All nodes are connected through self-developed THExpress-2 interconnection with a peak network bandwidth of 160 Gbps. Xeon Phi is connected to the host CPU using PCI-e 2.0 with a data transfer rate of 10 Gbps. Each Xeon CPU has 12 cores and the two CPUs deliver 422 GFLOPS peak performance in double precision. Each Xeon Phi coprocessor has up to 57 cores, with 4 hardware threads on each core, delivering 1 TFLOPS peak performance in double precision. Note that both processors have extended math unit (EMU) and vector processing unit (VPU) with long

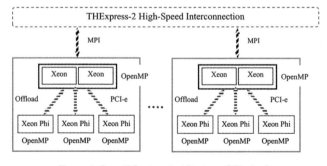

Figure 4. Overall System Architecture of Tianhe-2.

vector width. The vector length for Xeon CPU and Xeon Phi are 256 and 512 bit, respectively. Thus, it is fundamental to efficiently utilize the wide vector unit to tap the full potential of both processors.

Xeon has 3 cache levels: L1 cache (32KB for data and instruction, respectively) and L2 cache (256KB) are both core-private; L3 cache (30MB) is shared by all the 12 cores within a socket. Xeon Phi has 2 cache levels: L1 cache (32KB for data and 32KB for instruction) is core-private; the total capacity for L2 cache is 28.5MB and shared by all 57 cores using Core-Ring Interface (i.e., 512KB per core) with a bidirectional ring bus. The peak memory bandwidth of the 2 socket Xeon CPUs and a Xeon Phi within a node are about 204.8GB/s and 352GB/s, respectively.

We use the *roofline* model and operation intensity [9] to briefly analyze the performance characteristics of *openlbmflow* on Tianhe-2. According to the *roofline* model, if the operation intensity of a kernel is larger than the operation intensity of a processor, then the kernel is expected to be compute-bounded on the processor, otherwise it is most likely to be memory-bounded. With in-depth analysis, we find that the operational intensity of *Calc_Flow_Variables_Inner* and *Collision_Streaming_Inner* are 0.14 FLOP/Byte and 0.12 FLOP/Byte, respectively, while the operational intensity of Xeon is nearly 2.07 FLOP/Byte, and that of Xeon Phi is up to 2.84 FLOP/Byte. Therefore, *openlbmflow* is expected to be severely memory-bounded on both Xeon CPU and Xeon Phi coprocessor of Tianhe-2.

3.2. Overall parallel decomposition on Tianhe-2

Figure 5 shows the overall domain decomposition for MPI+OpenMP+Offload heterogeneous parallelization of *openlbmflow* on Tianhe-2. We decompose the original computational domain into many sub-blocks evenly and distribute them among MPI processes. For simplicity, we create one MPI process per node and assign one sub-block to it. We extend each sub-block with two layers of ghost lattices for data exchanging. On each node, we decompose the sub-block into four block chunks along X dimension by a load balancing factor (See Section 5 for more detail), and one block chunk is calculated by the 2 Xeon CPUs, and the other three chunks are offloaded to Xeon Phi coprocessors, with each coprocessor calculating one chunk. The load-balance scheme can successfully avoid frequent data movements between CPU and Phi, and only the boundary regions need to be updated after each time-step iteration.

The calculation of each lattice is completely independent; therefore, we use OpenMP to exploit the multi-core/many-core performance of both processors. On CPU the workload is parallelized in a piece-wise manner, the computation of block chunk is processed piece by piece along x dimension. While on Xeon Phi coprocessor, the workload is parallelized in a fine-grained line-wise manner to fulfill its massive amount of concurrent threads. To exploit fine-grained parallelism, we use the *collapse(2)* OpenMP clause to unfold the outer two layers of the loop region.

4. HYBRID MPI+OPENMP IMPLEMENTATION

4.1. Optimizing cache and SIMD efficiency

As described in Section 3, *openlbmflow* is severely memory-bounded on both Xeon and Xeon Phi coprocessor. Before performing heterogeneous parallelization and collaboration, we focus on improving the single core/thread performance of *openlbmflow* by optimizing Cache and SIMD efficiency.

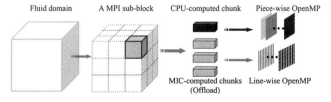

Figure 5. Overall domain decomposition for MPI+OpenMP+Offload parallelization of *openlbmflow* on Tianhe-2.

Data layout, that is, data structure, has significant impacts on SIMD performance and cache efficiency. In *openlbmflow*, it originally uses multi-dimensional pointers for flow variables. Thus, flow variables might be non-contiguous in memory resulting in poor data locality and further degrading cache efficiency. It will also cause extra vector gather/scatter operations when the SIMD feature is enabled and thus degrade the vectorization performance of Xeon and Xeon Phi. We reconstruct the initial multi-dimensional pointer using a plain one-dimensional pointer for flow variables. The SIMD-friendly data structure can achieve better data locality and higher cache efficiency and also dramatically reduces gather and scatter operations.

Furthermore, we employ 3D cache-blocking in memory-intensive kernels such as *Collision_Streaming_Inner* to enhance Cache efficiency. As shown in Figure 6, the original tri-loop is divided into many small chunks on each dimension (BX, BY, and BZ for X, Y, and Z dimension, respectively). By modifying BX, BY, and BZ, we can fit the small chunk into L3/L2 cache perfectly. As a result, the expensive memory access operations can be reduced significantly. In our tests, the optimal cache-blocking dimension size is 48 (BX=BY=BZ=48) for *openlbmflow* program on both Xeon and Xeon Phi.

To take the full advantage of VPU, especially for the Xeon Phi coprocessor with 512-bit vectors, we enable the SIMD capability with mandatory SIMD directive (*!$DIR SIMD*) and eliminate the fake vector dependence between pointers of distribution function among two iterations using *!$DIR ivdep* directive.

4.2. Hybrid MPI/OpenMP parallelization

Before collaborating CPU and MIC, we implement a hybrid MPI/OpenMP parallelization on CPUs. In the MPI implementation, the original computational domain is decomposed for large amounts of nodes. For the test case in this work, we divide the cubic flow field domain with periodic or bounce-back boundaries in a progressive bisection manner, that is, dividing the domain along each coordinates orderly and repeatedly. Each sub-block is distributed to one Tianhe-2 node/ MPI process. As described in Section 2, a lattice requires communicating with its direct neighbors. Domain decomposition for parallelization will change the processing of original boundary conditions and introduce new ghost boundaries among sub-blocks. We expand sub-blocks with two layers of ghost lattices for data exchanges among neighboring sub-blocks. The reason of using two-layer ghost lattices rather than one layer is that the periodic boundary ghost lattices streaming procedure needs an extra layer of ghost lattices.

After each iteration, ghost lattices need to be updated. Figure 7 expresses the data exchanges of boundary and corner ghost lattices in 2D case. Figure 7(a) shows the data exchange of boundary ghost lattices. Source sub-block boundary lattices (denoted by solid squares) will exchange and update the ghost lattices of its destination sub-block (denoted by shadowed squires). Figure 7(b) shows the data exchange of domain corner lattices. Domain corner lattices (solid color squares) will exchange and update the ghost corner lattices (shadowed squares) that denoted by the same number. For instance, the left-bottom shadowed square (3) is updated by the right-top solid square (3). The corner lattices and corner ghost lattices data exchange relations among sub-blocks are shown in Figure 7(c).

```
#pragma omp parallel for
for(i = xs;i<xe;i++) //outer loop for thread parallel
for(j = ys;j<ye;j++) //middle loop for thread collapse
for(k = zs;k<ze;k++) // inner loop for SIMD
{
    fn = F(fp, phi,...); //collision and streaming
}
```
(a) Original Tri-loop

```
for(b_xs=xs;b_xs<xe;b_xs+=BX)
{ b_xe = min(b_xs+BX,xe);
    for(b_ys=ys;b_ys<ye;b_ys+=BY)
    { b_ye = min(b_ys+BY,ye);
        for(b_zs=zs;b_zs<ze;b_zs+=BZ)
        { b_ze = min(b_zs+BZ,ze);
        #pragma omp parallel for
        for (x=b_xs; x<b_xe; x++)
            for (y=b_ys; y<b_ye; y++)
                for (z=b_zs; z<b_ze; z++)
                {
                    fn = F(fp, phi,...);
                }
} } }
```
(b) Tri-loop with cache-blocking

Figure 6. Cache-blocking for tri-loop kernels.

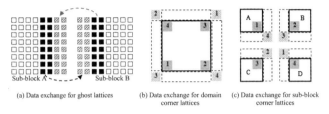

Figure 7. Data exchange after domain decomposition.

Within a node, shared-memory thread-level parallelization using OpenMP directives for time-consuming kernels is implemented on both Xeon CPU and Xeon Phi. As shown in Figure 8, OpenMP directives are added in the outer *for* loop. On Xeon CPU, the 24 cores/threads can be easily satisfied by the outer loop range (xs to xe). However, on Xeon Phi, the amount of concurrency threads (4 per core, up to 228 in total) is close to the outer loop range, so we have to make further efforts to exploit the parallelism in the middle loop, whereas the inner loop should be reserved for vectorization, to fully saturate Xeon Phi and avoid potential load imbalance problem. This can be achieved by using OpenMP *collapse* clause on Xeon Phi, as demonstrated in Figure 8(b).

5. COLLABORATING CPU AND XEON PHI

5.1. Intra-node collaborative programming

As described in 3.2, on each sub-block, one chunk is calculated by OpenMP threads on CPUs, and another three chunks are offloaded to the three Phi coprocessors. Figure 9 illustrates the intra-node

```
#pragma omp parallel for
for(i = xs;i<xe;i++)    //outer loop for thread parallel
for(j = ys;j<ye;j++)
for(k = zs;k<ze;k++)    // inner loop for SIMD
{
    fn = F(fp, phi,...)   //collision and streaming
}
        (a)  OpenMP code on Xeon
```

```
#pragma omp parallel for collapse(2)
for(i = xs;i<xe;i++)    // outer loop for thread parallel
for(j = ys;j<ye;j++)    // middle loop for thread collapse
for(k = zs;k<ze;k++)    // inner loop for SIMD

    fn = F(fp, phi,...)   //collision and streaming

  (b) OpenMP code with loop collapse on Xeon Phi
```

Figure 8. OpenMP implementation on Xeon and Xeon Phi.

```
// MIC calculation
for(d_id=0;d_id<MICNUM;d_id++){
    #pragma offload target (mic:d_id) in() signal(&sig[d_id])
//Offload asynchronously
    {......
        #pragma omp parallel for collapse(2) //MIC threads
        Tri-loop-kernel;
    ......}
}; // end for & return immediately
// CPU calculation
#pragma omp parallel for    // CPU threads
Tri-loop-kernel;
......
// MIC-CPU synchronization
for(d_id=0;d_id<MICNUM;d_id++){
    #pragma offload_wait target(d_id) wait(&sig[d_id])
    #pragma offload_transfer target(d_id) out()
}; //end for
```

Figure 9. Pseudo-code illustrating collaborative programming.

collaborative programming model. We use an asynchronous model to overlap CPU and MIC computation. Before starting calculations on CPU, we launch the Offload code on Phi, and it will execute asynchronously and return to the CPU code immediately. We perform synchronization to make sure that both devices have finished their computations. In this way, we overlap the computation on both sides.

The key issue for intra-node collaboration is to balance workloads among multiple CPUs and Xeon Phi coprocessors. We use a load balancing factor α to adjust workloads according to realistic performance of *openlbmflow* for different problem sets on both processors. Here, α is defined as follows:

$$\alpha = \frac{P_{mic}}{P_{cpu} + N_{mic}*P_{mic}}$$
$$W_{cpu} = W_{total}*(1 - N_{mic}*\alpha),$$
$$W_{mic} = W_{total}*\alpha$$
(8)

where W_{total}, W_{cpu}, and W_{mic} denote the total workload, CPU workload, and MIC workload, respectively. The number of Xeon Phi coprocessors is denoted as N_{mic}. The practical performance of CPU(P_{cpu}) and MIC(P_{mic}) can be obtained by sample running. The offload data transfer mechanism requires the transferred data to be in a continuous buffer, we divide the workload along the **x** dimension, that is, the outer layer of global array, eliminating additional memory copy operations for offload implementation.

The workload on each processor will keep constant during simulations to avoid frequent expensive data movements, and only boundary lattices need to be exchanged in each iteration. We collect the outer two-layer lattices of each block chunk into a temporal memory buffer for data transfers between CPU and Phi. This can dramatically reduce the quantity and frequency of PCI-e data transfers.

Figure 10 shows the collaborative CPU and multi-MIC computation and communication of synchronous and asynchronous Offload. Compared with the synchronous Offload, the asynchronous Offload can successfully overlap the computation of CPU and MICs and dramatically improve the intra-node performance. We also overlap the PCI-e data transfers between CPU and Phi with CPU end computation to further alleviate the intra-node communication costs.

We use a double buffering mechanism to update distribution function. Two duplicate distribution function arrays, *fp* for the present time step and *fn* for the next time step, need to exchange pointer addresses after each time-step iteration. The two pointers are associated to memory addresses on CPU and Xeon Phi during data transferring. Therefore, after exchanging pointer addresses, the pointers might be associated with wrong addresses; this is especially tricky when involving multiple Xeon Phi coprocessors. We use a delicate combined virtual pointer and solid pointer method by associating the pointers on Xeon Phi memory space with a solid pointer on CPU memory space for data exchanges and then only exchanging the addresses of virtual pointers for time iterations.

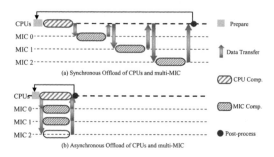

Figure 10. Overlapping the collaborative CPU-MIC computation and communication.

5.2. Scaling openlbmflow on large-scale Tianhe-2 nodes

To scale *openlbmflow* on large-scale Tianhe-2 nodes, we need to tackle the communication challenge among CPUs and Phi coprocessors on different nodes. The communication is especially complex for Tianhe-2 when utilizing multiple CPUs and Phi coprocessors on each node. Generally, running large-scale *openlbmflow* simulations on Tianhe-2 will involve the following communication patterns: (1) intra-processor communication within CPU/MIC, (2) intra-node communication of CPU-MIC and MIC-MIC, and (3) inter-node communication of CPU-CPU, CPU-MIC, and MIC-MIC. To simplify the MPI+OpenMP+Offload parallel implementation, we design a two-level hierarchical communication procedure and it contains only two explicit communications: the intra-node CPU-MIC communication (PCI-e data transfer) using offload directives and the inter-node CPU-CPU communication using MPI. Intra-processor communications are accomplished using shared OpenMP. Other types of communication can be implemented using the aforementioned two basic types of communications. Firstly, Xeon Phi processors transfer their updated boundaries to host CPU through the PCI-e channel. Then, all nodes update their ghost layers from neighbors by MPI communication. Finally, CPUs transfer the updated ghost layers to Xeon Phi processors.

6. PERFORMANCE RESULTS

6.1. Experimental setup and metrics

We use Intel *icc v11.1* for *openlbmflow* with the -O3 option. We use MPICH2-GLEX for MPI communication, and the Xeon Phi coprocessor shares the same compiler tool-chain with the Xeon CPU. In this work, we use the test case from *openlbmflow* website for performance tests. The case simulates the 3D drop on drop impact multi-phase (liquid and gases) problem with gravity effect. As shown in Figure 11, the fluid domain is a cubic space, the invisible background is filled with air, and there are two drops inside: one is on the floor, and the other one is in the air (Figure 11(a)). Because of the effect of gravity, the upper drop will fall down (Figure 11(b)) and fuses with the drop on the floor into one bigger drop (Figure 11(c)).

The problem size for single-core and node-level performance test is $512 \times 256 \times 256$. For large-scale tests, we extend the three dimensions of the fluid domain orderly. All the performance results are obtained using the average value of 10 effective sample simulations with 10-time iterations.

6.2. Single-core performance

In this section, we evaluate the performance optimization for *openlbmflow* on a single core of Xeon CPU and Xeon Phi. As shown in Figure 12, the 'OPT_Layout' denotes the optimization of LBM data layout using cache blocking and data structure transformation, and this will benefit both Cache and SIMD efficiency. Compared with the baseline code, we achieve a speedup of 3.0X and 3.4X on Xeon CPU and Xeon Phi, respectively. The performance is further improved by 1.7X on Xeon CPU and more significantly 2.6X on Xeon Phi when the SIMD feature is enabled (denoted as 'OPT_SIMD' in Figure 12). This is because the vector unit of Xeon Phi 31S1P is two times wider than that of Xeon E5-2692. As a result, the single-core performance for optimized LBM code on Xeon Phi is comparable with that of a Xeon CPU.

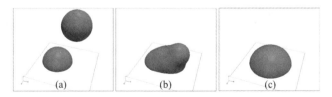

Figure 11. Test case - A 3D drop on drop impact multi-phase problem with gravity effect.

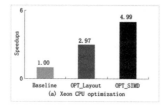
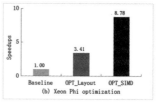

Figure 12. Performance optimization on a single core.

6.3. Node-level performance

In this section, we evaluate the shared-memory OpenMP performance for *openlbmflow* on both Xeon and Xeon Phi, as well as the collaborative performance on a single Tianhe-2 node. Figure 13 shows the percentage of execution time for the four kernels in each iteration on the Xeon CPU, with the number of OpenMP thread scaled to 24. *Collision_Streaming_Inner* is the most time-consuming kernel in *openlbmflow*, and it takes more than 60% execution time of the whole iteration. The percentage rises slightly with more than 8 threads, because of the increasing memory bandwidth requirement compared with other LBM kernels.

Figure 14 shows the speedup for the LBM code and the four kernels, respectively, on Xeon CPU with up to 24 OpenMP threads. Both *Calc_Flow_Variables_Inner* and *Collision_Streaming_Boundary* can achieve a maximum speedup of 11 using 16 OpenMP threads, demonstrating a good shared memory parallel scalability. And the *Calc_Flow_Variables_Boundary* kernel has a maximum speedup of 7.7 at 24 threads. However, limited by the memory bandwidth and its irregular memory footprint, the *Collision_Streaming_Inner* kernel achieves the maximum speedup of merely 6.5 at 8 threads and

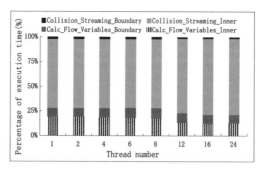

Figure 13. Percentages of different kernels on Xeon CPU.

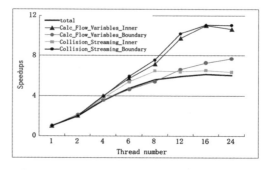

Figure 14. Speedup of different kernels on Xeon CPU.

decreases slightly afterwards. As a result, the speedup for the whole LBM code is about 6 with 24 threads. Besides these four primary kernels, there are some sequential sections such as initialize stage, and boundary data exchange stage. Limited by these sequential codes, the total speedup here is unsurprisingly lower than these four primary kernels.

We also evaluate the performance of Offload calculation on multiple Xeon Phi processors within a Tianhe-2 node. In the test, we use the performance on a single Xeon Phi as our baseline, and all the LBM calculations are offloaded to Xeon Phi. As shown in Figure 15, we achieve a speedup of 1.7 and 2.1 using two and three Xeon Phi processors, respectively. We further evaluate the collaborative speedup using both Xeon CPUs and Xeon Phis on a Tianhe-2 node. As shown in Figure 16, compared with the CPU-only approach, which merely utilizes both two CPUs, the collaborative approach can achieve a speedup of 1.9, 2.7, and 3.2 when we use one, two, and up to all three Xeon Phi processors, respectively. The results notably validate the effectiveness of our collaborative LBM simulation. The efficiency decreases of multi-MIC in previous tests are mainly caused by sequential sections of the program including initialize state, boundary data preparation stage before offloading, and boundary data harvest stage after asynchronous executing.

6.4. Large-scale performance

In the following large-scale parallel scalability tests, we use the performance on two nodes as our baseline, and the problem set is the same as before. Figure 17 shows the weak scalability and parallel efficiency of the MPI+OpenMP hybrid *openlbmflow* implementation using up to 4096 Tianhe-2 compute nodes. On each node, we create one MPI process and 24 OpenMP threads. As shown in Figure 17(a), although the MPI communication time changes a lot with the varying of node number, the computational time is always the dominating part of *openlbmflow* simulations. The communication time also keeps stable when we use more than 64 nodes and increases obviously after 512 nodes. The parallel efficiency of MPI+OpenMP approach is presented in Figure 17(b), the hybrid parallel version obtains a parallel efficiency of nearly 70% when scaling up to 4096 nodes, which demonstrates fairly good hybrid MPI+OpenMP parallel scalability.

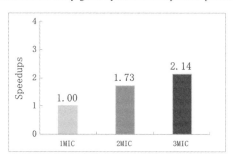

Figure 15. Intra-node multi-MIC speedups.

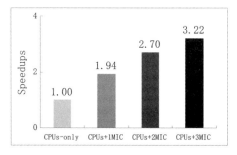

Figure 16. Intra-node CPU+MIC speedups.

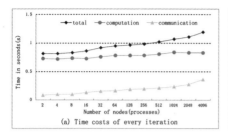
(a) Time costs of every iteration

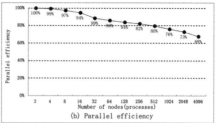
(b) Parallel efficiency

Figure 17. Weak scalability for MPI+OpenMP hybrid parallel implementation of *openlbmflow*.

Further, we present the weak scalability of MPI+OpenMP+Offload+SIMD *openlbmflow* on up to 2048 Tianhe-2 compute nodes, and each node utilizes all 2 CPUs and 3 MIC coprocessors. From Figure 18(a), we can see that the *openlbmflow* program has a good scalability at large scale. The computation performance remains steady with increasing node number, while the total simulation time increases remarkably with communication performance when compute nodes scale up. The communication time starts to overtake computation time and play a dominant role when scales up to 1024 and beyond. However, Figure 18(b) shows that the parallel efficiency of heterogeneous approach remains 60% with 2048 nodes, which demonstrates the effectiveness of overlapping collaborate computation and communication. Based on the obtained test results, we can infer that when the number of compute nodes scales beyond 2048 the parallel efficiency will surely decrease gradually but will still be acceptable for practical engineering applications.

7. RELATED WORK

Lattice Boltzmann method is considered to be very promising for large-scale computing. There have been many studies in parallelizing and optimizing LBM codes on HPC systems. Williams S. *et al.* [10] presented an auto-tuning approach to optimize LBM on multi-core architectures including Intel, AMD as well as Sun CPU processors. They developed a code generator to identify the specific CPU platform and generate highly optimized LBM program, with a speedup of over 15 times. Liu Z. *et al.* [11] evaluated the performance of a MPI+OpenMP hybrid parallel LBM program performance on the 'Ziqiang 4000' cluster in Shanghai University. Mountrakis L. *et al.* [12] evaluated the MPI performance for the open-source LBM framework Palabos. Their results show excellent weak and strong scalability with 8192 CPU cores. Pananilath I. *et al.* [13] developed an automated code generator for LBM simulations, featuring optimization techniques of tiling, load balancing, SIMD, etc. to boost LBM codes' performance. Generally their optimization can outperform Palabos by 3 times on Intel Xeon Sandy-bridge CPU. To summarize, previous studies on multi-core CPUs show that LBM has excellent parallel potential and achieves fairly well performance scalability.

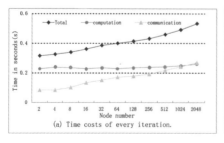
(a) Time costs of every iteration.

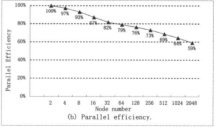
(b) Parallel efficiency.

Figure 18. Weak scalability of MPI+OpenMP+Offload+SIMD heterogeneous parallel *openlbmflow*.

Much prior work has shown the experience of porting various LBM codes to many-core processors such as GPUs, with impressive speed-ups. Wei C. et al. [14] implemented an improved LBM program to utilize both CPUs and GPUs of Tianhe-1A, and they proposed a new load-balance scheme on the heterogeneous system. Kanoria A. A. et al. [15] implemented a parallel LBM program using MATLAB, and they evaluated the implementation three different GPU products (i.e., C2070, GTX 680, and K20). Feichtinger C. et al. [16] implemented a LBM code on GPU-accelerated Tsubame 2.0 supercomputer. They proposed a performance model for the communication overhead and evaluated the LBM program performance on more than 1000 GPUs. Koda Y. et al. [17] developed a LBM code to simulate turbulent flow problems using large eddy simulations (LES) on GPUs. They achieved the speedup of over 150 using 2 GPUs. LBM codes achieve remarkable speedups using GPU accelerators, and researches about large scale CPU-GPU cluster are also presented and gain good parallel efficiency.

Intel MIC products were available until the year of 2013, there are very few related studies involving porting and optimizing scientific codes including LBM codes on MIC architecture, especially for large-scale applications. Crimi G. et al. [18] ported a GPU-accelerated 2D LBM code to Xeon Phi coprocessor and compared with previous GPU implementation and state-of-the-art CPUs by sustained performance and a defined ξ metric that provides a fair comparison of performances across architectures with widely different number of cores and vector sizes. MIC shows higher sustained performance but lower ξ parameter value. Rosales C. et al. [19] demonstrated the challenges such as tradeoff between performance and portability when porting and optimizing codes on Xeon Phi with several micro-benchmarks including a LBM code. Their work also shows that data transfers using Offload and MPI exchanges are a fundamental bottleneck for large-scale codes on HPC systems accelerated by Xeon Phi. McIntosh-Smith S. et al. [20] implemented a performance-portable OpenCL version of LBM program and evaluated its performance on modern processors including CPUs, NVIDIA GPUs as well as Intel Xeon Phi. Results show that the OpenCL LBM code achieves both good performance portability and competitive performance compared with the hand-tuned version on different architectures. Bortolotti G. et al. [21] evaluated the performance of a LBM code on both Xeon Phi (Knight) and NVIDIA K20X (Kepler) GPU. Specific optimizations on both architectures were stated in detail, and the optimized program on accelerators achieved 2 to 3 times speedups compared with Intel Sandy Bridges CPUs. Calore E. et al. [22] also demonstrated the tradeoff of portability and performance on different high-performance processors, including NVIDIA GPUs, Intel Xeon Phi, and Intel Ivy Bridge and Opteron CPUs, using a OpenCL LBM code. To the best of our knowledge, optimization study of LBM program on heterogeneous CPU and multi-MIC system at large scale has not been noted.

8. CONCLUSION AND FUTURE WORK

In this paper, with MPI+OpenMP+Offload+SIMD, we have parallelized and optimized an open-source LBM code *openlbmflow* on the Intel Xeon Phi-accelerated Tianhe-2 supercomputer. We have performed particular optimizations to boost achievable performance for D3Q19 LBM kernels on leading-edge multi-core and many-core architectures with wide vector units and complex Cache hierarchy. To achieve a greater speedup and fully tap the potential of Tianhe-2 compute node, we have collaborated CPU and Xeon Phi for *openlbmflow* instead of using a naive MIC-only approach. We have presented a flexible load-balance scheme to distribute the loads among intra-node CPUs and Phi processors and an asynchronous model to overlap the collaborative computation and communication as far as possible. With our method, we successfully simulate a 3D multi-phase flow problem on the Tianhe-2 supercomputer using 2048 compute nodes with more than 400K cores and obtain a parallel efficiency of 60%.

With the increasing popularity of Xeon Phi in HPC systems, it is practical and essential to port and optimize real-world complex LBM codes on the many-core architecture. The approach and implementation of CPU-MIC collaboration as well as the balancing scheme presented in this work provide a fairly general experience of similar efforts for LBM codes on heterogeneous supercomputers using Xeon Phi.

ACKNOWLEDGEMENTS

This paper was supported by the Basic Research Program of National University of Defense Technology under Grant No. ZDYYJCYJ20140101, the Open Research Program of China State Key Laboratory of Aerodynamics under Grant No. SKLA20140104, the IAPCM Application Research Program for High Performance Computing under Grant No. R2015-0402-01, and the National Science Foundation of China under Grant No. 11502296. We would like to thank Guangzhou Supercomputer Center for providing us with Tianhe-2 system for performance evaluation and giving us technical supports.

REFERENCES

1. Godenschwager C, Schornbaum F, Bauer M, Köstler H, Rüde U. A framework for hybrid parallel flow simulations with a trillion cells in complex geometries. In Proceedings of the *International Conference on High Performance Computing, Networking, Storage and Analysis* (p. 35). ACM, 2013.
2. Biferale L, Mantovani F, Pivanti M, Pozzati F, Sbragaglia M, Scagliarini A, … Tripiccione R. Optimization of multi-phase compressible lattice Boltzmann codes on massively parallel multi-core systems. *Procedia Computer Science* 2011; **4**: 994–1003.
3. Bailey P, Myre J, Walsh SD, Lilja DJ, Saar MO. Accelerating lattice Boltzmann fluid flow simulations using graphics processors. In *Parallel Processing*, 2009. ICPP'09. International Conference on (pp. 550–557). IEEE, 2009.
4. Fang J, Sips H, Zhang L, Xu C, Che Y, Varbanescu AL. Test-driving intel xeon phi. In Proceedings of the *5th ACM/SPEC International Conference on Performance Engineering* (pp. 137–148). ACM, 2014.
5. Fang J. Towards a Systematic Exploration of the Optimization Space for Many-Core Processors (*Doctoral dissertation*, TU Delft, Delft University of Technology), 2014.
6. Xu C, Deng X, Zhang L, Fang J, Wang G, Jiang Y, Cheng X. Collaborating CPU and GPU for large-scale high-order CFD simulations with complex grids on the TianHe-1A supercomputer. *Journal of Computational Physics* 2014; **278**:275–297.
7. Official website of *openlbmflow*. http://www.lbmflow.com. [10 August 2015]
8. Introduction of Tianhe-2 on Top500 website. http://www.top500.org/system/177999. [10 August 2015]
9. Williams S, Waterman A, Patterson D. Roofline: an insightful visual performance model for multicore architectures. *Communications of the ACM* 2009; **52**(4):65–76.
10. Williams S, Carter J, Oliker L, Shalf J, Yelick K. Optimization of a lattice boltzmann computation on state-of-the-art multicore platforms. *Journal of Parallel and Distributed Computing* 2009; **69**(9):762–777.
11. Liu Z, Song A, Xu L, Feng W, Zhou L, Zhang W. A High Scalable Hybrid MPI/OpenMP Parallel Model of Multiple-relaxation-time Lattice Boltzmann Method*. *Journal of Computational Information Systems* 2014; **10**(20):10147–10157.
12. Mountrakis L, Lorenz E, Malaspinas O, Alowayyed S, Chopard B, Hoekstra AG. Parallel performance of an IB-LBM suspension simulation framework. *Journal of Computational Science* 2015; **9**:45–50.
13. Pananilath I, Acharya A, Vasista V, Bondhugula U. An Optimizing Code Generator for a Class of Lattice-Boltzmann Computations, 2014.
14. Wei C, Zhenghua W, Zongzhe L, Lu Y, Yongxian W. An improved LBM approach for heterogeneous GPU-CPU clusters. In *Bio]medical Engineering and Informatics (BMEI)*, 2011 4th International Conference on (Vol. 4, pp. 2095–2098). IEEE, 2011.
15. Kanoria AA, Damodaran M. Parallel Matlab implementation of the lattice boltzmann method on GPUs, 2014.
16. Feichtinger C, Habich J, Köstler H, Rüde U, Aoki T. Performance Modeling and Analysis of Heterogeneous Lattice Boltzmann Simulations on CPU-GPU Clusters. *Parallel Computing*, 2014.
17. Koda Y, Lien FS. The lattice Boltzmann method implemented on the GPU to simulate the turbulent flow over a square cylinder confined in a channel. *Flow, Turbulence and Combustion*, 2014, 1–18.
18. Crimi G, Mantovani F, Pivanti M, Schifano SF, Tripiccione R. Early experience on porting and running a lattice Boltzmann code on the Xeon-Phi co-processor. *Procedia Computer Science* 2013; **18**:551–560.
19. Rosales C. Porting to the intel Xeon phi: opportunities and challenges. In *Extreme Scaling Workshop (XSW)*, 2013 (pp. 1–7). IEEE, 2013.
20. McIntosh-Smith S, Curran D. Evaluation of a performance portable lattice Boltzmann code using OpenCL. In Proceedings of the *International Workshop on OpenCL 2013 & 2014* (p. 2). ACM, 2014.
21. Bortolotti G, Caberletti M, Crimi G, Ferraro A, Giacomini F, Manzali M, … Zanella M. Computing on Knights and Kepler Architectures. In *Journal of Physics: Conference Series* 2014; **513**(5):52032–52038.
22. Calore E, Schifano SF, Tripiccione R. A portable OpenCL lattice Boltzmann code for multi-and many-core processor architectures. *Procedia Computer Science* 2014; **29**:40–49.

Performance modeling and optimization of parallel LU-SGS on many-core processors for 3D high-order CFD simulations

Dali Li[1] · Chuanfu Xu[1,2] · Bin Cheng[1] · Min Xiong[1] · Xiang Gao[1] · Xiaogang Deng[3]

Published online: 16 December 2016
© Springer Science+Business Media New York 2016

Abstract As a typical Gauss–Seidel method, the inherent strong data dependency of lower-upper symmetric Gauss–Seidel (LU-SGS) poses tough challenges for shared-memory parallelization. On early multi-core processors, the pipelined parallel LU-SGS approach achieves promising scalability. However, on emerging many-core processors such as Xeon Phi, experience from our in-house high-order CFD program show that the parallel efficiency drops dramatically to less than 25%. In this paper, we model and analyze the performance of the pipelined parallel LU-SGS algorithm, present a two-level pipeline (TL-Pipeline) approach using nested OpenMP to further exploit fine-grained parallelisms and mitigate the parallel performance bottlenecks. Our TL-Pipeline approach achieves 20% performance gains for a regular problem ($256 \times 256 \times 256$) on Xeon Phi. We also discuss some practical problems including domain decomposition and algorithm parameters tuning for realistic CFD simulations. Generally, our work is applicable to the shared-memory parallelization of all Gauss–Seidel like methods with intrinsic strong data dependency.

Keywords LU-SGS · Multi-/many-core processor · Xeon Phi · Pipeline · Shared-memory parallelization · CFD

Li Dali, Xu Chuanfu, Cheng Bin, Xiong Min, Gao Xiang and Deng Xiaogang. Performance modeling and optimization of parallel LU-SGS on many-core processors for 3D high-order CFD simulations . Journal of Supercomputing, 2017, 73(6): 2506-2524.

1 Introduction

The lower-upper symmetric Gauss–Seidel (LU-SGS) [4,24] algorithm is a popular implicit method for solving large sparse linear equations in computational fluid dynamics (CFD) and many other PDE (partial differential equation)-based computational science areas. LU-SGS combines the LU factorization and the Gauss–Seidel relaxation, demonstrating high algorithmic efficiency with excellent convergence rate in many realistic CFD simulations. On the other hand, as other Gauss–Seidel like methods, it has strong data dependency. As Fig. 1 shows for a 3D structured grid, generally the forward (lower) sweep calculation of grid point (i, j, k) needs the data from grid points $(i-1, j, k)$, $(i, j-1, k)$ and $(i, j, k-1)$, and the data dependence of backward (upper) sweep is vice versa. This characteristic poses tough challenge for shared-memory parallelization of LU-SGS algorithm using OpenMP as what we usually do for typical loops [26].

In CFD applications, researchers have proposed the hyper-plane/hyper-line approach (for 3D/2D grids) and the pipeline approach, for shared-memory parallelization of LU-SGS. The key idea is to exploit parallelism among independent grid points from different grid lines/planes. The hyper-plane/hyper-line strategy is based on the fact that grid points with the same index sum can be updated in parallel, and the pipeline strategy resolves the data dependency through thread synchronization. Experience shows that generally the pipeline approach largely outperforms the hyper-plane/hyper-line approach [25]. The same performance gap is also discovered in our in-house high-order CFD program. By carefully constructing a pipeline with each parallel thread/core acting as a pipeline stage, the LU-SGS computation on a grid is decomposed into many sub-tasks/sub-grids, and they will be scheduled to execute on the pipeline at different time according to the data dependency among them. The pipeline approach achieves promising scalability results on early multi-core processors and SMP systems. However, on latest multi-core processors, especially many-core processors with tens to hundreds of concurrent cores/threads, the pipeline LU-SGS algorithm performance is far from satisfactory [3]. Our experiences from the pipeline parallel LU-SGS kernel in our in-house high-order 3D structured CFD software HOSTA [22] show that the parallel efficiency is still over 70% on two shared-memory Xeon E5-2692 v2 (24 cores in total) using 24 OpenMP threads. However, on emerging many-core processors such as Xeon Phi 31S1P with 57 cores, the efficiency drops dramatically to less than 25% when using all 228 OpenMP threads.

Fig. 1 Data dependence of LU-SGS forward sweep

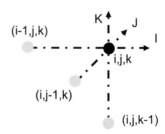

The scalability degradation can be explained from two aspects: (1) with more pipeline stages or threads/cores (i.e., increasing the pipeline depth) on many-core processors, the overall startup and finishing overhead for the pipeline increases, for mesh block with relatively small outer k dimension, probably even there is not enough sub-tasks in the task queue to fully load the pipeline; and (2) the load balance of pipeline LU-SGS computation among tens or even hundreds of threads on many-core processors will generally be worse.

In this paper, we model and optimize the performance of the pipeline parallel LU-SGS algorithm for 3D structured grids on modern multi-/many-core processors. The main contributions of this paper are as follows:

- We present a performance model for the pipeline parallel LU-SGS on 3D grids. We use two performance metrics to evaluate the pipeline efficiency: the ratio of all pipeline stages under full-load operations (RPFO) to estimate the overall startup and finishing overheads, and the ratio of upper-/lower-bound load (RULL) between pipeline stages to estimate load imbalance. With the model, we analyze the performance behavior of the pipeline approach on Xeon CPU and Xeon Phi many-core processor, achieving similar scalability trends in line with the realistic test results.
- Guided by the model, we present a two-level pipeline (TL-Pipeline) approach for 3D problems and extend the performance model accordingly. A sub-pipeline (second-level pipeline) is implemented in each original (first-level) pipeline-stage to further exploit fine-grained parallelisms among grid dimensions (sub-planes) in a grid-plane. Particularly, the original pipeline parallel LU-SGS algorithm can be regarded as a special case of TL-Pipeline LU-SGS algorithm. The TL-Pipeline approach is implemented using nested OpenMP in the LU-SGS kernel of HOSTA. We model and analyze the performance of TL-Pipeline using a fixed problem size ($256 \times 256 \times 256$) with various pipeline depths for the second-level pipeline and the first-level pipeline. Compared to the original pipeline approach, realistic tests show that on Xeon Phi many-core processor TL-Pipeline can achieve a performance gain of 20%.
- Furthermore, we discuss how grid dimension size and the pipeline depth could impact the parallel performance, and much more impressive performance gains are obtained. We also study the feedback effects of improved two-level pipeline approach on domain decomposition. At last we give some useful suggestions in algorithm parameter configuration, making our method more practical in realistic CFD simulations. Generally, our work provides a common practice and is applicable to the shared-memory parallelization of all Gauss–Seidel like methods with intrinsic strong data dependency.

The remainder of this paper is organized as follows. We first present some related work in Sect. 2, and briefly introduces the LU-SGS algorithm and its parallel computing in Sect. 3. We model and analyze the performance of the pipelined parallel LU-SGS in Sect. 4. In Sect. 5, we detail the two-level pipeline approach. Several algorithmic problems are further discussed in Sect. 6. Finally, we conclude in Sect. 7.

2 Related work

Due to its algorithmic efficiency and convergence rate, LU-SGS has been very popular in CFD since it was first proposed by Seokkwan Yoon et al. [24]. Seokkwan Yoon presented a vectorizable and unconditionally stable LU-SGS method, achieving a 30% speedup with respect to the LU implicit scheme for transonic flow simulations. For specific CFD application problems, researchers also developed some improved LU-SGS variants. For example, Chen et al. [4] developed a block LU-SGS (BLU-SGS) for unstructured meshes, and at the cost of 20–30% more memory usage, BLU-SGS increases converges many times faster than conventional LU-SGS approaches in the simulations of transonic flows over the NACA 0012 airfoil and ONERA M6 wing, as well as supersonic flows over a 3D forebody. Yuzhi Sun et al. [20] developed an implicit nonlinear LU-SGS solver for high-order spectral difference Navier–Stokes problems, and achieved a speedups of 1 to 2 orders of magnitude over a Runge–Kutta scheme for inviscid flow and steady viscous flow simulations. Gang Wang et al. [12] proposed an improved LU-SGS scheme to meet the needs of high Reynolds number problems, and gained a significantly efficiency increase in the simulations of turbulent flows around the NACA 0012 airfoil, RAE 2822 airfoil and LANN wing. Besides, LU-SGS is also used as an efficient preconditioner for Krylov sub-space iterative algorithms [9,15,16,19] and multigrid iterative methods [14,17,18]. Generally, our parallel approach is applicable to shared-memory parallelization of LU-SGS and all its variants in various CFD applications.

To enable MPI/OpenMP hybrid parallelization of implicit CFD codes based on LU-SGS on large-scale multi-core HPC systems, researchers have proposed the hyper-plane/hyper-line approach and the pipeline approach for multithreading LU-SGS. Djomehri et al. [8] implemented both the hyper-plane and the pipeline strategies for hybrid MPI+OpenMP CFD simulations on Cray SX6, IBM Power3 and Power4, and SGI origin3000. In [25], Seokkwan Yoon investigated the performance of the hyper-plane and the pipeline strategies for real gas flow simulations, and the results show that the pipeline approach achieves better scalabilities on SMP platforms. Rupak Biswas et al. [1,2] implemented the LU-SGS linear solver of OVERFLOW-D using the pipeline approach on the Columbia supercomputer. Satoru Yamamoto et al. [23] developed a parallel "Numerical Turbine" to simulate 3D multistage stator-rotor cascade flows, using a pipelined parallel LU-SGS for implicit time-integration. All of the above works are implemented and evaluated on early multi-core processors and SMP systems.

With the development and complexity of both CFD applications and computer architectures, especially the shift from multi-core technology to many-core technology in HPC systems, it is essential to understand the behavior of the pipelined LU-SGS on modern multi-core HPC processors or even many-core processors such as GPU and MIC. Recently, Yonggang Che et al. [3] implemented a pipelined parallel LU-SGS method for supersonic combustion simulations on an compute node of Intel Xeon CPUs, observing a severe drop of parallel efficiency of 33.8% with 24 threads, and the poor scalability may be attributed to their relatively small test problem (only 812835 cells) according to our performance model. Due to the strong data dependency of LU-SGS, researchers tend to adopt more parallel-friendly time advancing methods on many-core platforms, e.g., Runge–Kutta method [21] and Jacobi iterative method

[22], and we seldom see researchers choose LU-SGS when porting their codes onto GPUs. To our knowledge, this is the first paper that presents a performance model and an improved parallel implementation for LU-SGS on Xeon Phi.

3 LU-SGS and its parallel computing

3.1 The LU-SGS algorithm

In CFD, after discretization and linearization, we need to solve a large equation system,

$$Ax = b, \tag{1}$$

where A is the left-hand-side (LHS) matrix, b is the right-hand-side (RHS) vector, and x is the solution vector. In LU-SGS, a lower-upper splitting or decomposition is applied on A, and in each iteration the solution is separated into a forward sweep and a backward sweep. For example, in the lower-upper decomposition method, the LHS matrix A is decomposed and approximated as follows:

$$\begin{aligned} A &= D + L + U = D(I + D^{-1}L + D^{-1}U) \\ &= D\left(I + D^{-1}L\right)\left(I + D^{-1}U\right) + LD^{-1}U \\ &\approx D\left(I + D^{-1}L\right)\left(I + D^{-1}U\right) \\ &= (D + L) D^{-1}(D + U), \end{aligned} \tag{2}$$

where D, L and U are the diagonal part, the lower triangular part and the upper triangular part of A, respectively. Thus the Eq. (1) can be rewritten as

$$(D+L)D^{-1}(D+U)x = b. \tag{3}$$

An intermediate variable y is introduced into Eq. (3) to form two symmetric Gauss–Seidel sweep stages, i.e., a forward sweep and a backward sweep.

$$\begin{aligned} (D+L)\,y &= b \\ (D+U)\,x &= Dy, \end{aligned} \tag{4}$$

where y and x can be solved by back substitution.

In CFD, the entries of LHS matrix A (also called the Jacobian matrix) are usually approximated using grid points and their neighboring grid points. In structured CFD applications, those grid points form a regular computational stencil. Typically, the 7-point stencil is used in three-dimensional cases and the matrix A can be illustrated as follows:

$$\begin{bmatrix} \ddots & & \ddots & & \ddots & \ddots & \ddots & & \ddots & & \ddots & \\ 0 & l_{k-1} & 0 & l_{j-1} & 0 & l_{i-1} & d_{ijk} & u_{i+1} & 0 & u_{j+1} & 0 & u_{k+1} & 0 \\ & \ddots & & \ddots & & \ddots & \ddots & \ddots & & \ddots & & \ddots & \end{bmatrix}, \qquad (5)$$

where i, j, and k represent the indexes of three spatial dimensions, respectively. Elements of l and u in the same line are neighbors of d in physical and logical space, but in memory space with linear storage feature, they are of different distances to d. All entries of A are zero, except the entries at the diagonal and the six symmetric off-diagonals.

3.2 The pipeline approach

In CFD, other researches [25] and our experience show that the pipeline approach generally outperforms the hyper-plane/hyper-line approach for shared-memory parallelization of LU-SGS. Thus, we focus on the parallel performance modeling and optimization of the pipeline approach. Algorithm 1 presents the pseudo-code skeleton of forward LU-SGS sweep with a pipelined OpenMP parallel implementation. Due to the data dependency, we cannot simply add some OpenMP directives on the $I/J/K$ loop for multithreading. The key idea of the pipelined parallel LU-SGS is to construct a pipeline on the K dimension, with each parallel thread acting as a pipeline-stage, and on the J dimension, the LU-SGS computation is statically decomposed into many sub-tasks (i.e., $I-J$ grid sub-planes) and scheduled to the threads, the variable $flag$ is used for thread synchronization.

Algorithm 1 The pseudo-code skeleton of forward sweep of pipelined LU-SGS

```
01  !$OMP PARALLEL NUM_THREADS(d_p)
02  idt=omp_get_thread_num();
03  flag(idt) = 0;
04  !$OMP BARRIER
05  DO k = 1, nk
06    IF(idt > 0) THEN
07      Wait until flag(idt − 1) = 1;
08      flag(idt − 1) = 0;
09    END IF
10    !$OMP DO SCHEDULE(STATIC)
11    DO j = 1, nj
12      DO i = 1, ni
13        v_{i,j,k} = F(v_{i,j,k}, v_{i−1,j,k}, v_{i,j−1,k}, v_{i,j,k−1});
14      END DO
15    END DO
16    !$OMP END DO NOWAIT
17    IF(idt < d_p − 1) flag(idt) = 1;
18  END DO
19  !$OMP END PARALLEL
```

Figure 2 shows a schematic illustration for both task scheduling and the execution timeline of the pipeline with 4 threads (pipeline stages). For the mth layer (i.e., $k = m$),

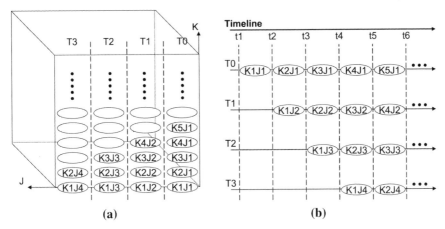

Fig. 2 Schematic illustration of the forward sweep stage of pipelined LU-SGS. **a** Workload partitioning and task scheduling. **b** Execution timeline of sub-tasks

the $I - J$ plane is divided into n ($n = 4$ for this example) sub-planes/sub-tasks (denoted as $K_m J_n$). Obviously, the sub-task $K_m J_n$ depends on $K_{m-1} J_n$ and $K_m J_{n-1}$ in the task queue, due to the data dependency. Those sub-tasks are carefully scheduled to threads in the pipeline, and in each sub-task/thread the grid points are calculated along the I dimension serially. Figure 2b shows the execution timeline, in this way, dependent sub-tasks from different K and J are pipelined with multithreading and parallelisms along $K - J$ dimensions are exploited. As we will analyze in the following sections, the pipeline efficiency will vary dramatically for different problems with various K/J and pipeline depth (i.e., the number of threads).

3.3 Implementation and experimental setup

The traditional pipeline approach and our two-level pipeline approach are implemented in the LU-SGS kernel from our in-house high-order 3D structured software HOSTA [22]. HOSTA solves the Navier–Stokes equations of fluid flow in compressible and incompressible forms with a self-developed weighted compact nonlinear scheme (WCNS) [5–7], and it has been used extensively for aerodynamic research and design optimization of realistic aircrafts such as China's civil large airplane C919. In HOSTA, LU-SGS is a very popular solving method with both favorable algorithmic efficiency and convergence rate.

The test platform we use contains on Intel many integrated cores (MIC) coprocessor, specifically Intel Xeon Phi Knight's Corner (KNC) 31S1P [10,11,13], and two shared memory Xeon E5-2692 v2 (Ivy Bridge) CPUs. The Xeon Phi 31S1P has 57 cores, with 4 hardware threads per core (i.e., supporting 228 concurrent threads in total), delivering 1 TFLOPS peak performance in double precision. The coprocessor has an on-chip memory of 8 GB, with a peak memory bandwidth of 352GB/s. Each CPU has 12 cores and the two CPUs share a 64 GB memory with a peak memory bandwidth of 204.8 GB/s, delivering 422 GFLOPS peak performance in double precision. Both the

CPUs and the Xeon Phi have extended math unit (EMU) and vector processing unit (VPU) with 256 and 512 bit vector width, respectively. We use the native programming model on Xeon Phi and compile our code using Intel *ifort* with *-mmic* option. The baseline 3D problem size is $(n_i, n_j, n_k) = (256, 256, 256)$, and other problem sizes are evaluated by altering $I/J/K$ dimensions arbitrarily.

4 Performance modeling of pipelined LU-SGS

4.1 Performance metrics

It is clearly stated in [25] that the pipeline efficiency is limited by the startup and finishing procedures. We define the ratio of all pipeline stages under full-load operations (RPFO) to evaluate the overhead of the startup and finishing. For a 3D problem with (n_i, n_j, n_k) grid points on a pipeline with d_p pipeline stages, suppose all sub-tasks are well balanced on every pipeline stages, and RPFO is defined as

$$\text{RPFO} = \begin{cases} \dfrac{(n_k - d_p + 1)}{(n_k + d_p - 1)} \times 100\% , & n_k \geq d_p \\ 0\% , & n_k < d_p, \end{cases} \quad (6)$$

where $n_k + d_p - 1$ is the number of pipeline cycles including the startup and finishing, and $n_k - d_p + 1$ is the cycles when the whole pipeline is fully occupied. If there are not enough sub-tasks in the task queue (i.e., $n_k < d_p$), the pipeline cannot be under full-load (i.e., RPFO = 0).

As we can see from Eq. (6), on early multi-core processors with only a few cores, d_p is generally much less than n_k, and correspondingly RPFO will be close to 100%, indicating a high pipeline efficiency. However, on modern multi-/many-core processors, the number of threads is likely comparable to n_k, and RPFO will decrease dramatically.

With the increase of the number of threads, the load balance among threads (pipeline stages) could be worse. Since we assume a static workload allocation among the J dimension of threads, we define the ratio of upper-/lower-bound load (RULL) between pipeline stages to estimate load imbalance as follows:

$$\text{RULL} = \left\lfloor \dfrac{n_j}{d_p} \right\rfloor \Big/ \left\lceil \dfrac{n_j}{d_p} \right\rceil \times 100\%, \quad (7)$$

where n_j/d_p is the ideal workload size for each thread, and "$\lfloor \ \rfloor$" and "$\lceil \ \rceil$" represent rounding to the floor and ceiling, respectively. Assume $n_j = d_p \times t + r (0 \leq r < dp)$, i.e., for r threads their workload size is $t + 1$, and for the other $d_p - r$ threads, their workload size is t. When the pipeline depth d_p is much less than n_j or n_j can be divided by d_p with no remainder, RULL is nearly or equals to 100%. However, when d_p is comparable to n_j, and n_j cannot be exactly divided by d_p, RULL becomes worse. An extreme case is that r equals to 1 and t is very small, the total performance would be limited by the single thread with a workload size $t + 1$, indicating a poor load balance.

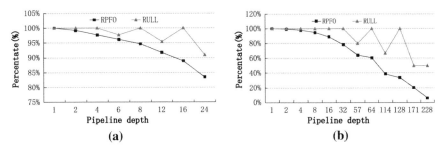

Fig. 3 Performance metrics of the pipelined LU-SGS algorithm. **a** Xeon CPUs. **b** Xeon Phi

Figure 3a, b shows the variation of RPFO and RULL on the Xeon CPUs and Xeon Phi coprocessor, respectively, for a (256, 256, 256) problem. Both metrics drop with the increase of the number of threads (pipeline depth). On the two Xeon CPUs, RPFO drops from 100% to less than 85%, and RULL drops from 100% to about 90%. On Xeon Phi, the two metrics drops significantly after the pipeline depth scales beyond 32. When all 228 threads are used (i.e. $d_p = 228$), RPFO drops sharply to less than 10% and RULL drops to 50%. Consequently, we estimate a severe parallel scalability loss for the pipelined parallel LU-SGS algorithm on many-core processors, such as Xeon Phi, with tens or even hundreds of threads according to our performance metrics.

4.2 Performance issues

Suppose the computational cost of LU-SGS for each grid point is an equal unit, then for a given (n_i, n_j, n_k) problem, the wall-time cost for a serial LU-SGS calculation WT_s is $n_i \times n_j \times n_k$, and for a pipelined implementation, the wall-time cost WT_{pp} is $(n_k + d_p - 1) \times \lceil n_j/d_p \rceil \times n_i$. We derive the speedup of the pipelined parallel LU-SGS S_{pp} as follows:

$$S_{pp} = \frac{WT_s}{WT_{pp}} = \frac{n_k \times n_j \times n_i}{(n_k + d_p - 1) \times \left\lceil \dfrac{n_j}{d_p} \right\rceil \times n_i} = \frac{n_k \times n_j}{(n_k + d_p - 1) \times \left\lceil \dfrac{n_j}{d_p} \right\rceil}. \quad (8)$$

This performance model combines the aforementioned two performance metrics and ignores the impact of memory bandwidth and cache efficiency for simplicity.

We compare the predicted speedup based on Eq. (8) with the test results and the ideal/linear speedup to validate our performance model. Figure 4 shows the comparison on Xeon CPUs and Xeon Phi coprocessor for a (256, 256, 256) problem.

As Fig. 4 indicates, the predicted speedup (S_{pp}) is smaller than the ideal speedup (S_{pp}^*), this is due to the overhead of pipeline mechanism. The performance gap between S_{pp} and test speedup ($\hat{S}_{pp}$) may be attributed to the memory bandwidth limitation and other potential hardware limitations. On Xeon CPUs (Fig. 4a), the three speedups are fairly close when the number of threads (pipeline depth) is relatively small (6 or less). A detaching point between S_{pp} and $\hat{S}_{pp}$ (DP$_2$) occurs at 6 threads, and another detaching

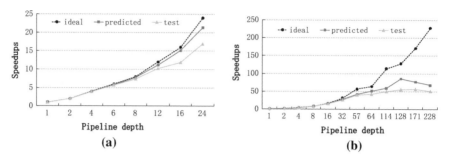

Fig. 4 The contrast of ideal, predicted and test speedups of the pipelined LU-SGS algorithm. **a** On Xeon CPUs node. **b** On Xeon Phi

point between S_{pp}^* and S_{pp}(DP$_1$) occurs at 8 threads. In Fig. 4a, the gap between S_{pp}^* and S_{pp} is much smaller than the gap between S_{pp} and $\hat{S}_{pp}$, indicating that the pipeline approach is almost optimal on Xeon CPUs, and the parallel performance is limited by the memory bandwidth and other hardware limitations.

On Xeon Phi coprocessor (Fig. 4b) DP$_1$ and DP$_2$ appear at 16 and 57 threads, respectively. Unlike the results in Fig. 4a, the gap between S_{pp}^* and S_{pp} is much larger than the gap between S_{pp} and $\hat{S}_{pp}$. This means that the pipeline approach has a severe adaptability problem on the Xeon Phi coprocessor. Because of the relatively larger memory bandwidth and lower core/thread average floating performance, the DP$_1$ and DP$_2$ on Xeon Phi appears later than on Xeon CPU node.

5 The two-level pipeline LU-SGS approach

5.1 Fundamental ideas and implementation

As both the predicted and test results show in Sect. 3, although we could utilize more available cores for pipelined parallelization on many-core processors, the speedup drops significantly due to a long pipeline depth. Based on an in-depth analysis, we propose a two-level pipeline (TL-pipeline) approach for multithreading LU-SGS. The fundamental idea is to further exploit parallelism in each 2D sub-task/sub-plane for 3D LU-SGS problems. This is accomplished by transforming a long deep original pipeline (with depth d_p) into a relatively short first-level pipeline (with depth d_{p1}), with each pipeline-stage containing a second-level sub-pipeline (with depth d_{p2}). Obviously, $d_p = d_{p1} \times d_{p2}$ holds. The sub-pipeline is constructed in the J dimension, and each sub-task is statically decomposed along the I grid line. Figure 5 shows the task schedule for TL-Pipeline with both $d_{p1} = 4$ and $d_{p2} = 4$. In this case, $K_2 J_4$ is decomposed into some more fine-grained sub-tasks/grid-lines $J_m I_n$ to be scheduled on the sub-pipeline. Similarly, Fig. 6 presents the execution timeline for the TL-Pipeline model. The TL-Pipeline approach is implemented using nested OpenMP. For each level of OpenMP threads, we need a separate *flag* array for synchronization. We customize the number of threads of each pipeline level (d_{p1} and d_{p2}) using *num_threads()* directive clause.

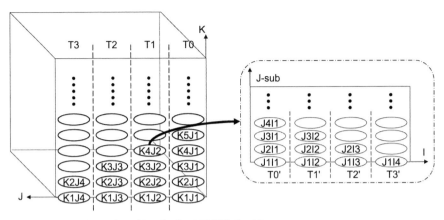

Fig. 5 Workload dispatching of TL-Pipeline LU-SGS algorithm

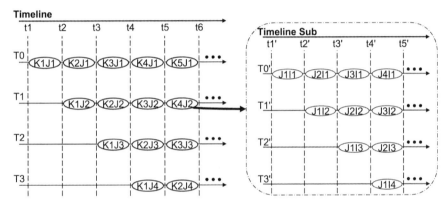

Fig. 6 Execute procedure of TL-Pipeline LU-SGS algorithm

5.2 Performance evaluation

We extend the performance model in Sect. 3 for our TL-pipeline approach. The two performance metrics, RPFO and RULL, are evaluated according to Table 1. For the first-level pipeline, $RPFO_1$ and $RULL_1$ are calculated in the same way as Eq. (8) with a pipeline depth d_{p1}. On a sub-pipeline with depth d_{p2}, sub-tasks from a $I - J$ sub-plane of $(n_i, n_j/d_{p1})$ are scheduled, thus the $RPFO_2$ and $RULL_2$ are slightly different.

Figure 7 shows the variation of the two metrics for our TL-Pipeline on Xeon CPUs and Xeon Phi for a (256, 256, 256) problem. On Xeon CPUs, appropriate d_{p1} and d_{p2} configuration ($d_{p1} \times d_{p2}$) can enhance both RPFO and RULL of the two pipelines to a nearly ideal level (i.e., 100%). For example, for a $d_{p1} \times d_{p2} = 4 \times 6$ case on Xeon CPUs with 24 cores, all metrics except $RPFO_2$ are all between 95 and 100%. The superiority of TL-Pipeline is particularly ' outstanding on Xeon Phi coprocessor with hundreds of threads. $RPFO_1$ can be increased from under 10% to nearly 100%. Although $RPFO_2$ decreases rapidly, the overall performance is a combined effect of both pipelines and its impact is limited to a certain extent.

Table 1 Performance metrics of the TL-pipeline LU-SGS algorithm

Level	RPFO	RULL
1	$\begin{cases} \dfrac{n_k - d_{p1} + 1}{n_k + d_{p1} - 1} \times 100\%, & n_k \geq d_{p1} \\ 0\%, & n_j < d_{p1} \end{cases}$	$\left\lfloor \dfrac{n_j}{d_{p1}} \right\rfloor / \left\lceil \dfrac{n_j}{d_{p1}} \right\rceil \times 100\%$
2	$\begin{cases} \dfrac{\left\lceil \dfrac{n_j}{d_{p1}} \right\rceil - d_{p2} + 1}{\left\lceil \dfrac{n_j}{d_{p1}} \right\rceil + d_{p2} - 1} \times 100\%, & \left\lceil \dfrac{n_j}{d_{p1}} \right\rceil \geq d_p \\ 0\%, & \left\lceil \dfrac{n_j}{d_{p1}} \right\rceil < d_p \end{cases}$	$\left\lfloor \dfrac{n_i}{d_{p2}} \right\rfloor / \left\lceil \dfrac{n_i}{d_{p2}} \right\rceil \times 100\%$

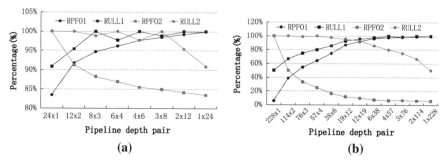

Fig. 7 Performance metrics of the TL-Pipeline LU-SGS algorithm. **a** Xeon CPUs. **b** Xeon Phi

Similar to Eq. (8), the speedup of our TL-Pipeline approach compared to a serial LU-SGS, S_{tlp}, can be evaluated as follows:

$$S_{tlp} = \frac{WT_s}{WT_{tlp}} = \frac{n_k \times n_j \times n_i}{(n_k + d_{p1} - 1) \times \left(\left\lceil \dfrac{n_j}{d_{p1}} \right\rceil + d_{p2} - 1 \right) \times \left\lceil \dfrac{n_i}{d_{p2}} \right\rceil}. \quad (9)$$

Equation (9) is the same as Eq. (8) if $d_{p2} = 1$, which indicates that the original pipeline approach is a special case of our TL-Pipeline approach.

Figure 8 shows the predicted speedups (S_{tlp}) and test speedups ($\hat{S}_{tlp}$) of the TL-Pipeline approach with different $d_{p1} \times d_{p2}$ configurations for a (256, 256, 256) problem. In Fig. 8a, both the S_{tlp} and $\hat{S}_{tlp}$ vary slightly with $d_{p1} \times d_{p2}$ on Xeon CPUs. Note that the test speedup for the 1×24 case drops a lot, this is mainly due to a worse data locality and cache efficiency when decomposing the I dimension with large numbers of threads ($d_{p2} = 24$). Although for this regular problem with the same size of $n_i/n_j/n_k$, TL-Pipeline has little performance advantage over the original pipeline approach (i.e., 24×1), for other problem with various $n_i/n_j/n_k$, the benefit of TL-Pipeline is certain, as we will discuss in Sect. 5. In Fig. 8b, the S_{tlp} has a maximum improvement of nearly 70% ($d_{p1} \times d_{p2} = 6 \times 38$) compared to the original approach on Xeon Phi coprocessor, while the $\hat{S}_{tlp}$ has a much more moderate improvement of

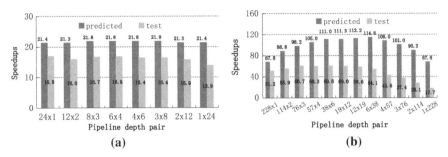

Fig. 8 The contrast of predicted and test speedups of TL-Pipeline LU-SGS. **a** Xeon CPUs. **b** Xeon Phi

nearly 20% (76 × 3, 57 × 4 and 38 × 6), because of memory bandwidth and other hardware limitations. Both the S_{tlp} and the $\hat{S}_{tlp}$ start to drop when $d_{p2} > 38$, and this is mainly caused by the increase of the overhead of the sub-pipeline (RULL$_2$) as well as the performance problems such as cache efficiency and nested OpenMP costs.

6 Further analyses and discussions

As we can see in the previous sections, the problem size (n_i, n_j, n_k) and the depth of the two pipelines (d_{p1} and d_{p2}) have an direct impact on the performance of TL-Pipeline. This section presents further discussions and suggestions helpful to achieve optimal performance in practical CFD applications.

6.1 The impact of each dimension size

Previous results are obtained using a regular problem size (i.e., $n_i = n_j = n_k = 256$); however in practice, both the problem size ($n_i \times n_j \times n_k$) and the ($n_i, n_j, n_k$) dimension size vary significantly for complex multi-block structured grids. For example, a 3D delta wing grid used in our daily CFD simulations has 30 grid blocks with 16 million grid points in total. Figure 9 presents the dimension sizes and the geometric average dimension size ($\sqrt[3]{n_i \times n_j \times n_k}$) for each grid block. We can find that the dimension sizes of each block vary in a large range. In realistic applications with various (n_i, n_j, n_k) sizes on different multi-/many-core processors, if it is not convenient to decide an optimal $d_{p1} \times d_{p2}$ configuration for each grid block, users can achieve a roughly optimal configuration according to our performance model.

We change on dimension in the range of $0.1 \times n_c$ to $10 \times n_c$ (n_c is the number of cores/threads of a given processor), with the other two dimensions fixed to 256, to analyze its independent performance effect on TL-Pipeline. Figures 10 and 11 present the predicted and test performance increases for different scale ratios (from 0.1 to 10) compared to the traditional pipeline approach on Xeon CPUs and Xeon Phi, with each dimension in a sub-figure. For the K dimension on Xeon CPUs (Fig. 10a), we can see that the predicted increase is remarkable (up to 1000%) for small-scale ratios, and drops to less than 10% with large-scale ration. The test results show a similar increase, indicating that our TL-Pipeline is much more superior for the case with

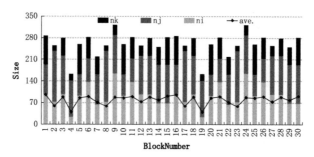

Fig. 9 Dimension sizes of a Delta Wing mesh with 30 blocks

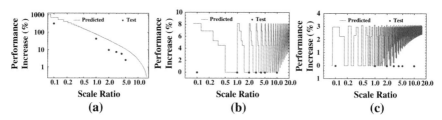

Fig. 10 The performance impact of each dimension size on Xeon CPUs node. **a** n_k, **b** n_j, **c** n_i

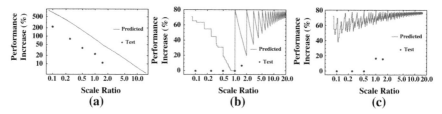

Fig. 11 The performance impact of each dimension size on Xeon Phi. **a** n_k, **b** n_j, **c** n_i

smaller K dimension. On the other hand, for the J and I dimension (Fig. 10b, c), the variation only has slight impact on performance: less than 8 and 3%, respectively, for the predicted performance and no improvement observed for the test performance. On Xeon Phi, we observe a similar variation as that of the result on Xeon CPUs: for the K dimension (Fig. 11a), both the predicted and test results have a significant increase up to 200 and 500%; for the other two dimensions (Fig. 11b, c), the predicted increases indicate a up to 80% increase, and the test result show a up to 20 and 40% increase. This is reasonable since our TL-Pipeline approach is much more superior with more thread numbers on Xeon Phi. Limited by its relatively small memory space, further tests with larger scale ratios on Xeon Phi are unavailable.

6.2 The impact of workload shape

Solving realistic CFD problems on HPC systems often involves partitioning the original grid blocks into many sub-blocks and mapping them to large-scale paral-

Table 2 The relative speedups of TL-pipeline for different workload shapes on Xeon CPUs node

	1/16	1/8	1/4	1/2	1	2	4	8	16 (α)
(a) Predicted values									
1/16	–	–	–	–	0.728	0.724	0.716	0.701	**0.672**
1/8	–	–	–	0.73	0.728	0.726	0.721	0.712	0.694
1/4	–	–	0.971	0.966	0.955	0.934	0.896	0.833	0.806
1/2	–	0.974	0.971	0.966	0.956	0.949	0.926	0.924	0.929
1	1.059	1.053	1.042	1.019	**1**	0.996	1	1	1.006
2	1.059	1.054	1.049	1.045	1.037	1.029	1.043	1.043	–
4	1.072	1.07	1.066	1.058	1.042	1.042	1.066	–	–
8	1.082	1.078	1.07	1.054	1.053	1.053	–	–	–
16 (β)	**1.084**	1.075	1.059	1.028	1.059	–	–	–	–
(b) Test results									
1/16	–	–	–	–	0.839	0.75	0.809	0.72	0.739
1/8	–	–	–	0.849	0.951	0.83	0.813	0.808	0.653
1/4	–	–	1.181	0.868	1.19	1.046	0.97	0.613	0.74
1/2	–	1.034	0.872	1.043	1.039	1.015	0.954	**0.607**	0.76
1	1.002	1.194	**1.219**	1.09	**1**	0.998	0.857	0.782	0.832
2	0.95	1.037	1.059	1.03	0.983	0.845	0.848	0.783	–
4	0.938	0.975	0.991	0.921	0.831	0.826	0.79	–	–
8	0.807	0.942	0.652	0.624	0.763	0.766	–	–	–
16 (β)	0.81	0.756	0.805	0.718	0.761	–	–	–	–

Bold values highlight relative speedups of the baseline workload shape, the best performance workload shape, and the worst performance workload shape

lel processes. The (n_i, n_j, n_k) size of sub-blocks is determined according to the average workload of each process. As we can see in Sect. 5.1, the performance of TL-Pipeline varies a lot with different (n_i, n_j, n_k) sizes. In this subsection, given a fixed overall workload $n_i \times n_j \times n_k$, we discuss some guidelines to partition optimal workload/sub-blocks for TL-Pipeline. We use the cubic grid block (256, 256, 256) as our baseline workload and compare the performance of TL-Pipeline for various grid blocks with the same workload and with scale factors α, β and γ for n_i, n_j and n_k, i.e., $(\alpha \times n_i) \times (\beta \times n_j) \times (\gamma \times n_k) = 256 \times 256 \times 256$ and $\alpha \times \beta \times \gamma = 1$. Tables 2 and 3 present both the predicted result and test result on Xeon CPUs and Xeon Phi, respectively, assuming α, β and γ ranging from 1/16 to 16 and γ determined by α and β according to the identical equation.

On Xeon CPUs (Table 2a, b), compared to the baseline with ($\alpha = 1, \beta = 1, \gamma = 1$), we observe a speedup of up to $1.084 \times$ ($\alpha = 1/16, \beta = 16, \gamma = 1$) for predicted results, and a speedup of up to $1.219 \times$ ($\alpha = 1/4, \beta = 1, \gamma = 4$) for test results. On the other hand, both results show cases with significant performance degradation: $0.672 \times$ for the predicted performance and $0.607 \times$ for the test performance, respectively. The results have a strong implication for grid partition strategies in CFD simulations for achieving an optimal parallel performance using TL-Pipeline. On Xeon Phi, the variation of speedup is even more significant. In Table 3a, the max-

Table 3 The relative speedups of TL-pipeline for different workload shapes on Xeon Phi

	1/16	1/8	1/4	1/2	1	2	4	8	16 (α)
(a) Predicted values									
1/16	–	–	–	–	0.138	0.137	0.136	0.134	**0.131**
1/8	–	–	–	0.275	0.272	0.267	0.26	0.245	0.245
1/4	–	–	0.543	0.534	0.516	0.485	0.436	0.436	0.436
1/2	–	1.057	1.004	0.913	0.773	0.713	0.711	0.715	0.715
1	1.10	1.085	1.057	1.008	**1**	1.033	1.037	1.051	1.051
2	1.447	1.41	1.34	1.282	1.276	1.317	1.334	1.373	–
4	1.611	1.532	1.417	1.444	1.455	1.487	1.528	–	–
8	**1.650**	1.501	1.483	1.473	1.489	1.53	–	–	–
16 (β)	1.562	1.526	1.537	1.492	1.425	–	–	–	–
(b) Test results									
1/16	–	–	–	–	0.187	0.181	0.239	0.238	**0.22**
1/8	–	–	–	0.351	0.356	0.369	0.402	0.373	0.327
1/4	–	–	0.656	0.564	0.66	0.631	0.598	0.43	0.509
1/2	–	0.208	0.932	1.086	1.02	0.868	0.761	0.65	0.743
1	0.184	0.808	1.102	1.036	**1**	0.97	0.984	0.97	0.973
2	0.408	**1.383**	1.272	1.163	1.129	1.074	1.155	1.24	–
4	1.032	1.294	1.246	1.206	1.108	1.151	1.297	–	–
8	1.323	1.174	1.097	1.058	1.119	1.178	–	–	–
16 (β)	1.105	1.104	1.025	1.062	1.088	–	–	–	–

Bold values highlight relative speedups of the baseline workload shape, the best performance workload shape, and the worst performance workload shape

imum speedup is 1.650 × ($\alpha = 1/16, \beta = 8, \gamma = 2$) and the minimum speedup is 0.131 × ($\alpha = 16, \beta = 1/16, \gamma = 1$) for the predicted performance; in Table 3b, the maximum speedup is 1.383 × ($\alpha = 1/16, \beta = 8, \gamma = 2$) and the minimum speedup is 0.181 × ($\alpha = 2, \beta = 1/16, \gamma = 8$) for the test performance.

According to the above results, generally we prefer bigger n_k and smaller n_i for optimal performance using TL-Pipeline with a fixed workload. The conclusion has another implication beneficial to structured CFD: the stripe of the LHS matrix A equals to the product of the middle and inner dimension sizes (i.e., $n_i \times n_j$ in our case) and primitive flow variable number for 3D problems. Since TL-Pipeline tends to perform better with large n_k and small n_i, this will result in a small stripe and consequently, a relatively higher memory access efficiency and a better convergence rate for iterative methods.

6.3 The configuration of d_{p1} and d_{p2}

As we can see from the performance model and realistic test results, the configuration of d_{p1} and d_{p2} in TL-Pipeline also has a complicated impact on the performance. In practical CFD simulations with varying number of threads and grid sizes, choosing appropriate d_{p1} and d_{p2} is definitely a non-trivial problem. It is not possible to provide a universal optimal configuration of d_{p1} and d_{p2} for all cases. According to previous test results and our performance model, we suggest the following guidelines to configure d_{p1} and d_{p2}:

- $d_{p1} \times d_{p2}$ should be equal to or at least as close as possible to n_c (the number of cores/threads), utilizing as more available cores/threads as possible;
- d_{p1} should be much smaller than the K dimension size n_k;
- d_{p1} should be a factor of the J dimension size n_j, and d_{p2} should be much smaller than n_j/d_{p1};
- d_{p2} should be a factor of the I dimension size n_i, and n_i/d_{p2} should be rationally large enough for better cache efficiency;
- Generally d_{p1} should be larger than d_{p2}, unless the K dimension size n_k is extremely small.

Besides, since CFD simulation often involves numerous time steps of running, we can also profile simulations in the first several time steps with all possible $d_{p1} \times d_{p2}$ combinations, and use an optimal one in the following time steps.

7 Conclusion and future work

In this paper, we first discuss the strong data-dependent feature of LU-SGS algorithm and its tough challenges for parallel computing. After that, we introduce two existing parallel LU-SGS algorithms (hyper-plane and pipeline) on shared-memory platforms, and compare the merits and drawbacks of them. Then, we analyze the performance factors of pipeline LU-SGS algorithm, which usually has better performance than hyper-plane, extract two performance metrics RPFO and RULL to reflect the performance problems, and build a performance model of naïve parallel LU-SGS algorithm. Through these analyses, we discover that on latest multi-/many-core processors the pipeline depth (number of cores/threads) is commensurate with the sizes of realistic workload dimensions. This would cause worse performance metrics of the original pipeline parallel algorithm and become the main performance bottleneck.

In order to alleviate the performance problems of pipeline LU-SGS algorithm on latest multi-core especially many-core processors, we propose a novel Two-Level Pipeline LU-SGS (TL-Pipeline LU-SGS) algorithm. The TL-Pipeline LU-SGS algorithm further exploits the fine-grained parallelism of 3-dimensional workloads and organizes the cores/threads hierarchically in nested two pipeline layers. We further evaluate the performance metrics of TL-Pipeline LU-SGS algorithm on Xeon CPU node and Xeon Phi with the given workload ($256 \times 256 \times 256$), and build the performance model of TL-Pipeline LU-SGS algorithm as well. Emphatically, the basic idea of TL-Pipeline LU-SGS algorithm is not limited to the specific LU-SGS algorithm, and it can be easily extended to other strong data-dependent algorithms in various 3-dimensional applications, including the whole Gauss–Seidel algorithm family.

We implement the TL-Pipeline LU-SGS algorithm in a domestic in-house high-order accuracy CFD program, and we evaluate and contrast the performances of TL-Pipeline LU-SGS versus naïve pipeline LU-SGS algorithm on both Xeon CPU node and Xeon Phi. Theoretically, for the given workload (256 × 256 × 256), the TL-Pipeline LU-SGS algorithm has 2 and 70% performance increases on Xeon CPU node and Xeon Phi, respectively. Our program test results draw a similar conclusion as model predicted, our TL-Pipeline LU-SGS algorithm has a performance gain of moderately 20% on Xeon Phi. Afterwards, we analyze the effects of workload sizes on algorithm performance, and discover that the k dimension size n_k has the most significant effect on algorithm performance, and the performance gain of TL-Pipeline LU-SGS algorithm on Xeon Phi is more promising than that on Xeon CPU node. We also discuss the algorithm feedback effects on domain decomposition to generate performance-friendly workloads. And finally we offer two useful strategies for the optimal or nearly optimal (d_{p1}, d_{p2}) pair determination.

Both Xeon CPU node and Xeon Phi have powerful wide vector processing ability, but the strong data-dependent feature causes complicated vector dependency and makes it difficult to exploit the performance potential of vectorization (SIMD). If this problem can be settled, the performance benefits would be impressive on these processors. Memory access and cache efficiency are two constant key points to performance optimization, especially for high memory bandwidth bounded stencil computing applications. Our TL-Pipeline LU-SGS algorithm can greatly enlarge the potential performance increases space of memory and cache optimizations.

Acknowledgements This paper was supported by the Basic Research Program of National University of Defense Technology under Grant No. ZDYYJCYJ20140101, the Open Research Program of China State Key Laboratory of Aerodynamics under Grant No. SKLA20160104, the Defense Industrial Technology Development Program under Grant No. C1520110002, and the National Science Foundation of China under Grant Nos. 11502296 and 61561146395.

References

1. Aftosmis M, Berger M, Biswas R, Djomehri MJ, Hood R, Jin H, Kiris C (2006) A detailed performance characterization of columbia using aeronautics benchmarks and applications. In: Proc. 44th AIAA Aerospace Sciences Meeting & Exhibit
2. Biswas R, Djomehri MJ, Hood R, Jin H, Kiris C, Saini S (2005) An application-based performance characterization of the columbia supercluster. In: Proceedings of the 2005 ACM/IEEE conference on Supercomputing, p 26. IEEE Computer Society
3. Che Y, Cheng X, Xu C, Zhu X, Wang Z (2015) Performance engineering of a supersonic combustion simulator on heterogeneous platforms. In: Proceedings of 27th International Conference on Parallel Computational Fluid Dynamics
4. Chen R, Wang Z (2000) Fast, block lower-upper symmetric gauss-seidel scheme for arbitrary grids. AIAA j 38(12):2238–2245
5. Deng X, Mao M (1997) Weighted compact high-order nonlinear schemes for the euler equations. AIAA paper, pp 97–1941
6. Deng X, Mao M, Jiang Y, Liu H (2011) New high-order hybrid cell-edge and cell-node weighted compact nonlinear schemes. AIAA Pap 3857:2011
7. Deng X, Zhang H (2000) Developing high-order weighted compact nonlinear schemes. J Comput Phys 165(1):22–44
8. Djomehri MJ, Jin HH, Biegel B (2002) Hybrid mpi+ openmp programming of an overset cfd solver and performance investigations. Tech. rep., NASA Ames Research Center, NAS Technical Report, NAS-02-002
9. Economon TD, Palacios F, Alonso JJ, Bansal G, Mudigere D, Deshpande A, Heinecke A, Smelyanskiy M (2015) Towards high-performance optimizations of the unstructured open-source su2 suite. AIAA SciTech AIAA Pap 1949:2015

10. Fang J (2014) Towards a Systematic Exploration of the Optimization Space for Many-Core Processors. Delft University of Technology, Delft
11. Fang J, Sips H, Zhang L, Xu C, Che Y, Varbanescu AL (2014) Test-driving intel xeon phi. In: Proceedings of the 5th ACM/SPEC international conference on Performance engineering. ACM, pp 137–148
12. Gang W, Jiang Y, Zhengyin Y (2012) An improved lu-sgs implicit scheme for high reynolds number flow computations on hybrid unstructured mesh. Chin J Aeronaut 25(1):33–41
13. Li D, Xu C, Wang Y, Song Z, Xiong M, Gao X, Deng X (2015) Parallelizing and optimizing large-scale 3d multi-phase flow simulations on the tianhe-2 supercomputer. Practice and Experience, Concurrency and Computation
14. Li R, Wang X, Zhao W (2008) A multigrid block lu-sgs algorithm for euler equations on unstructured grids. Numer Math Theory Methods Appl 1:92–112
15. Liu W, Zhang L, Zhong Y, Wang Y, Che Y, Xu C, Cheng X (2015) Cfd high-order accurate scheme jacobian-free newton krylov method. Comput Fluids 110:43–47
16. Luo H, Sharov D, Baum JD, Löhner R (2003) Parallel unstructured grid gmres+ lu-sgs method for turbulent flows. AIAA Pap 273:2003
17. Otero E, Eliasson P (2011) Convergence acceleration of the cfd code edge by lu-sgs. In: 3rd CEAS European Air & Space Conference. CEAS/AIDAA, pp 606–611
18. Parsani M, Van den Abeele K, Lacor C (2007) Implicit lu-sgs time integration algorithm for high-order spectral volume method with p-multigrid strategy. In: West-East High-Speed Flow Field Conference, Moscow, Russia
19. Sharov D, Luo H, Baum JD, Löhner R (2000) Implementation of unstructured grid gmres+ lu-sgs method on shared-memory, cache-based parallel computers. AIAA Pap 927:2000
20. Sun Y, Wang Z, Liu Y (2009) Efficient implicit non-linear lu-sgs approach for compressible flow computation using high-order spectral difference method. commun. Comput Phys 5(2–4):760–778
21. Wang YX, Zhang LL, Che YG, Xu CF, Liu W, Cheng XH (2015) Efficient parallel computing and performance tuning for multi-block structured grid cfd applications on tianhe supercomputer. Tien Tzu Hsueh Pao/acta Electronica Sinica 43(1):36–44
22. Xu C, Deng X, Zhang L, Fang J, Wang G, Jiang Y, Cao W, Che Y, Wang Y, Wang Z et al (2014) Collaborating cpu and gpu for large-scale high-order cfd simulations with complex grids on the tianhe-1a supercomputer. J Comput Phys 278:275–297
23. Yamamoto S, Sasao Y, Sato S, Sano K (2007) Parallel-implicit computation of three-dimensional multistage stator-rotor cascade flows with condensation. In: Proc. 18th AIAA Computational Fluid Dynamics Conference, AIAA Paper, vol 4460, p 2007
24. Yoon S, Jameson A (1988) Lower-upper symmetric-gauss-seidel method for the euler and navier-stokes equations. AIAA J 26(9):1025–1026
25. Yoon S, Jost G, Chang S (2005) Parallelization of gauss-seidel relaxation for real gas flow. Tech. rep., NAS Technical Report, NAS-05-011
26. Zhang L, Wang Z (2004) A block lu-sgs implicit dual time-stepping algorithm for hybrid dynamic meshes. Comput Fluids 33(7):891–916

高精度格式的具体应用

High-Order and High Accurate CFD Methods and Their Applications for Complex Grid Problems

Xiaogang Deng*, Meiliang Mao, Guohua Tu, Hanxin Zhang and Yifeng Zhang

State Key Laboratory of Aerodynamics, Aerodynamics Research & Development Center, Mianyang, 621000, P.R. China.

Received 15 May 2010; Accepted (in revised version) 15 May 2011

Available online 29 November 2011

Abstract. The purpose of this article is to summarize our recent progress in high-order and high accurate CFD methods for flow problems with complex grids as well as to discuss the engineering prospects in using these methods. Despite the rapid development of high-order algorithms in CFD, the applications of high-order and high accurate methods on complex configurations are still limited. One of the main reasons which hinder the widely applications of these methods is the complexity of grids. Many aspects which can be neglected for low-order schemes must be treated carefully for high-order ones when the configurations are complex. In order to implement high-order finite difference schemes on complex multi-block grids, the geometric conservation law and block-interface conditions are discussed. A conservative metric method is applied to calculate the grid derivatives, and a characteristic-based interface condition is employed to fulfil high-order multi-block computing. The fifth-order WCNS-E-5 proposed by Deng [9,10] is applied to simulate flows with complex grids, including a double-delta wing, a transonic airplane configuration, and a hypersonic X-38 configuration. The results in this paper and the references show pleasant prospects in engineering-oriented applications of high-order schemes.

AMS subject classifications: 76M20

Key words: WCNS, complex configurations, geometric conservation law, conservative metric methods, characteristic-based interface conditions.

Contents

1 Introduction   1082

2	High-order and high accurate CFD methods for complex grid problems	1083
3	Governing equations and the fifth-order weighted compact nonlinear scheme	1085
4	Geometric conservative law and the conservative metric method	1086
5	Characteristic-based interface conditions	1087
6	Applications and discussions	1088
7	Concluding remarks	1096

1 Introduction

Over the past 20 to 30 years, there have been a lot of studies in developing and applying high-order and high accurate numerical methods for computational fluid dynamics (CFD). Although low-order schemes (second-order schemes) are widely used for engineering applications, they are insufficient for turbulence, aeroacoustics, and many viscosity dominant flows, such as boundary layer flows, vortical flows, shock-boundary layer interactions, heat flux transfers, etc. An effective approach to overcome the obstacle of accurate numerical simulations is to employ high-order methods [1]. In [2,3], Shu and Cheng gave profound reviews on high-order weighted essentially non-oscillatory (WENO) schemes and discontinuous Galerkin (DG) schemes. A comprehensive review was also given by Ekaterinaris [4] for high-order difference schemes, ENO and WENO schemes, DG schemes, and spectral volume (SV) schemes.

It is generally believed that the accurate simulation of fluid flow with multiple and wide range of spatial scales and structures is a difficult task except through spectral approximations. However, the use of spectral approximations is limited to simple geometries with generally periodic boundary conditions. Compact schemes make it possible to devise, on a given stencil, finite difference schemes that have much better resolution properties than conventional explicit finite difference schemes of comparable order of accurate. Compact schemes with spectral-like resolution properties are more convenient to use than spectral and pseudo-spectral schemes, and are easier to handle, especially when nontrivial geometries are involved [5]. Deng et al. [6] have proposed a type of one-parameter linear dissipative compact schemes (DCS). DCSs are derived for high-order accurate simulation of shock-free problems while damping out the dispersive and parasite errors in the high-wave-number regions. Visbal and Gaitonde [7] use filters to prevent numerical oscillations of central compact schemes. In order to dealing with shock wave problems, Adams and Shariff [74] have developed a compact-ENO scheme. Pirozzoli [75] have developed a compact-WENO scheme which was further improved by Ren et al. [73]. Deng et al. have developed compact nonlinear schemes [8] and weighted compact nonlinear schemes (WCNS) [9, 10]. The WCNSs have been successfully applied to a wide range of flow simulations so far to show its flexibility and robustness [11, 16, 32, 99, 100]. Some results on using the fifth-order WCNS (WCNS-E-5) [9, 10] for complex grid problems are presented in this paper to further show its engineering prospects.

2 High-order and high accurate CFD methods for complex grid problems

The state-of-the-art applications of high-order and high-accurate methods in complex grids are still limited, while there are many kinds of high-order schemes, such as compact schemes [5–7,33,35], WCNSs [9,10], ENO [37,38] and WENO schemes [2,3,39,52,53], DG schemes [40–44,48,49,51,54], spectral methods (including spectral element (SE) [55], SV [56,57,110], spectral difference (SD) [58], staggered spectral methods [60]), gas-kinetic schemes (GKS) [61,63–65], dispersion-relation-preserving (DRP) schemes [67], low dissipative high-order schemes [68], monotonicity preserving (MP) schemes [69,70], group velocity control (GVC) schemes [71,72], SPH methods [45,62], ALE methods [64–66,81], methods for multiphase flows [42,46,47], and many hybrid ones [73–76]. In most published articles and reviews, the flows are complex with shock waves, vortices, and turbulent structures, while the grids are relatively simple. Despite many authors say in their articles that they have successfully applied high-order schemes in complex geometries, the grids in their test cases are still very simple when compared to those frequently appear in low-order scheme applications. Taking WENO schemes for example, as one of the most widely popularized high-order schemes, there are hundreds of articles on WENO schemes or employing WENO schemes for complex flows. However, It is not easy to find a grid or configuration which can compare with the complexity of any grids or configurations applied in the four series of AIAA CFD Drag Prediction Workshops (DPW, DPW I $\sim$ DPW IV) [77].

Although the finite volume methods (FVM), the finite element methods (FEM) and the DG schemes with unstructured grid system have advantages in treating complex geometries, the finite difference methods (FDM) with the structured grid system are superior in boundary layer simulations, computing costs, and convenience. Thus high-order multi-block FDM techniques are still attractive at present and in the future. Perhaps, many undesirable effects will not occur or can be neglected for lower-order methods, but they should be considered carefully for high-order methods. The interface conditions for multi-block grids, geometric conservation law (GCL), viscous treatments and convergence accelerating are some of such factors that must be reevaluated for high-order schemes.

Generally speaking, it is difficult to generate a high-quality structured single-block grid system for a complex configuration. Visbal and Gaitonde [7,13], Delfs [14], Sherer et al. [15] and many other researchers use overset grid strategy for high-order FDM. However, the inherent interpolations in overset grids may cause numerical instability and loss of global accuracy, especially when there are shock waves. Furthermore, the grid topologies and numerical procedures for overset grid approach are generally complicated and not easy to implement for 3D complex configurations. Point-matched multi-block structured grids, or patched grids, which do not require overlapped zones, are widely adopted choices for complex configurations. Multi-block structured grid technique makes it possible to run high-order finite difference schemes on each individual block, and the information transmission between neighbouring blocks and the propagation throughout the flow field can be realized by some kinds of interface conditions. Rai [17] used a flux interpolation to construct coupled conditions on mid-node with Beam-Warming and Osher scheme. Lerat and Wu [18] adopted local flux construction to establish conservative and unconditionally stable interface conditions. Huan, Hicken and Zingg [19, 20] proposed a kind of high-order interface boundary schemes which combine a conventional scheme and a summation-by-parts (SBP) [21–23] scheme with simultaneous approximation terms (SATs) [24–26]. The SBP operators are derived from the energy method, first by Kreiss and Scherer in 1974 [21] for low orders, and extended to high-order by Strand [27] and Jurgens [28]. The SBP operators allow an energy estimate of the discrete form, potentially guaranteeing strict stability for hyperbolic problems. However, boundary operators are needed, which can destroy the SBP property [22]. With the aim of preserving the SBP property of the overall scheme, the projection method [23] and the SAT approach [24] have been developed. However, the projection method becomes unstable for the linear convection-diffusion equation despite that it is strictly stable for hyperbolic PDEs [22]. Then, this kind of interface method is mainly adopted in linear hyperbolic systems and further improvement and examinations are necessary to upgrade this kind of interface boundary schemes suitable for fluids dynamics (Navier-Stokes systems). In content of CFD, Kim and Lee [29] and Sumi et al. [12, 30, 31] employed characteristic interface conditions (CIC) or generalized characteristic interface conditions (GCIC) which show some attractive performance in practice. We have derived some new interface approaches (CBIC) which directly exchange the spatial derivatives (i.e. RHS, computed on each block) on each side of an interface by means of a characteristic-based projection [32]. The CBIC have shown excellent performance in complex grids.

Traditionally, high-order FDM require smooth grids, and the smoothness of the grids shall be comparable with the order of the accuracy of the schemes [3]. This premise is usually far more rigorous for complex configurations even when multi-block decomposition techniques are applied. One may feel despair on using high-order methods for complex configurations after reading the comments above. Fortunately, Visbal and Gaitonde [78] have shown that high-order schemes can be applied in low quality grids such as extremity deforming grids and nondifferentiable grids if some special techniques are applied to get rids of the GCL-related errors and numerical oscillations. The studies of Visbal et al. [78] and Nonomura et al. [79] indicate that the GCL is very important in ensuring freestream preservation. Our experiences show that low quality grids bear large GCL-related errors than that of high-quality grids. A general condition to satisfy the GCL in FDM has been derived by Deng et al. [100].

High-order accuracy requires high-order evaluations for both the inviscid and viscous terms. However, most efforts, including many in the references mentioned above, mainly focus on the inviscid terms to resolve discontinuities and small scale structures. In order to supplement 5th-order WENO scheme with high-order viscous formula, Shen et al. [101] developed a 4th-order conservative scheme with the stencil less than that of the WENO scheme. Deng [10] listed some viscous derivative and interpolation formulas for WCNSs. Tu et al. [94] proposed a staggered compact difference scheme to avoid the odd-even oscillations which may emerge when high-order schemes are used. High-order viscous flux discretizations are also very important for DG, SE, SV and SD. Sun et al. [105] employed the local DG (LDG) approach in their SV method to discretize viscous terms. A penalty-like method, which is more symmetrical than LDG and better suitable for unstructured and non-uniform grids, was developed by Kannan and Wang [102] by applying the penalty method of Bassi and Rebay [87, 88]. Kannan and Wang [102] also conducted some Fourier analysis and accuracy studies for a variety of viscous flux formulations and showed that the penalizing schemes may enhance fidelity. Very recently, Kannan and Wang [103, 104] implemented the SV method for the Navier-Stokes equations using the LDG2 (which is an improvised variant of the LDG approach) [103] and the direct DG approaches [104]. However, the researches on viscous terms are still limited, and many aspects, such as stability and conservation, are still need to be further investigated for high-order and high accurate methods.

It is commonly encountered that the convergence property of high-order methods is general inferior to that of low-order methods especially when the grid quality is low. Kitamura et al. [59] employed the preconditioned LU-SGS for low speeds flows. Zhang et al. [106] investigated several implicit time matching methods, including lower-upper symmetric Gauss-Seidel (LU-SGS), generalized minimum residual (GMRES), Gauss-Seidel method with point relaxation and line relaxation. The results indicate that GMRES can considerably improve the convergence rate of WCNSs for the flows they simulated. Implicit LU-SGS is also successfully used for SD and SV, such as the methods in [107, 108]. P-multigrid approach, where p is the order of polynomial degree, is widely employed in DG [111,112], SV [108], SD [113], SE [114], and some similar schemes to accelerate convergence rate. It is worth to note that Kannan et al. [107, 108, 113] have successfully blended p-multigrid approach with pre-conditions or implicit LU-SGS, and the convergence rate is drastically improved for bad and skewed unstructured grids. For more details about p-multigrid methods, please refer to [108, 111–114] and the references therein.

3 Governing equations and the fifth-order weighted compact nonlinear scheme

In Cartesian coordinates the governing equations (Euler or Navier-Stokes) in strong conservative form are

$$\frac{\partial Q}{\partial t}+\frac{\partial F}{\partial x}+\frac{\partial G}{\partial y}+\frac{\partial H}{\partial z}=0. \tag{3.1}$$

The equations are transformed into curvilinear coordinates by introducing the transformation $(x,y,z,t) \to (\xi,\eta,\zeta,\tau)$

$$\frac{\partial \tilde{Q}}{\partial \tau}+\frac{\partial \tilde{F}}{\partial \xi}+\frac{\partial \tilde{G}}{\partial \eta}+\frac{\partial \tilde{H}}{\partial \zeta}=0, \tag{3.2}$$

where

$$\tilde{F} = \tilde{\xi}_t Q + \tilde{\xi}_x F + \tilde{\xi}_y G + \tilde{\xi}_z H, \quad \tilde{\xi}_x = J^{-1}\xi_x,$$

and with similar relations for the other terms.

Let us first consider the discretization of the inviscid flux derivative along the ξ direction. The discretization for other inviscid fluxes can be computed by similar procedures. Suppose $U_i = U(\xi_i, t)$ be the flow variables, the fifth-order weighted compact nonlinear scheme (WCNS-E-5) can be expressed as

$$\frac{\partial \tilde{F}_i}{\partial \xi} = \frac{75}{64h}(\tilde{F}_{i+1/2} - \tilde{F}_{i-1/2}) - \frac{25}{384h}(\tilde{F}_{i+3/2} - \tilde{F}_{i-3/2}) + \frac{3}{640h}(\tilde{F}_{i+5/2} - \tilde{F}_{i-5/2}), \quad (3.3)$$

where h is the grid size, and

$$\tilde{F}_{i+1/2} = \tilde{F}\left(U^L_{i+1/2}, U^R_{i+1/2}, \tilde{\xi}_{x,i+1/2}, \tilde{\xi}_{y,i+1/2}, \tilde{\xi}_{z,i+1/2}\right) \quad (3.4)$$

is computed by some kind of flux-splitting method which can be found in [9]. $U^L_{i+1/2}$ and $U^R_{i+1/2}$ are the left-hand and right-hand cell-edge flow variables, which are obtained by a high-order nonlinear weighted interpolation. The idea is that the stencil to interpolate the cell-edge values contains several substencils, each of the substencils is assigned a weight factor, which determines its contribution to the final approximation of the cell-edge values. The weights are designed in such a way that in the smooth region they approach the optimal weights to achieve fifth-order accuracy, whereas in the regions near the discontinuities, the weight of the substencil, which contains the discontinuities, is assigned nearly zero. Therefore the weighted interpolations can prevent numerical oscillations around discontinuities. For more details, see, [9, 10].

4 Geometric conservative law and the conservative metric method

The concept of the geometric conservative law (GCL) was first introduced in 1961 by Trulio and Trigger [80]. In 1978, Pulliam & Steger [84] observed that the metric discretizations will lead to the nonconservation of flow fields. Thomas & Lombard [85] extended the conception of the GCL to general applications in CFD. The effect of the GCL can be evidently noted in finite difference system when curvilinear coordinate transformation is applied. Eqs. (3.2) is equivalent to (3.1) only if the following GCL items are all zero.

$$I_t = (1/J)_\tau + (\tilde{\xi}_t)_\xi + (\tilde{\eta}_t)_\eta + (\tilde{\zeta}_t)_\zeta, \quad (4.1)$$

$$I_x = (\tilde{\xi}_x)_\xi + (\tilde{\eta}_x)_\eta + (\tilde{\zeta}_x)_\zeta, \quad I_y = (\tilde{\xi}_y)_\xi + (\tilde{\eta}_y)_\eta + (\tilde{\zeta}_y)_\zeta, \quad I_z = (\tilde{\xi}_z)_\xi + (\tilde{\eta}_z)_\eta + (\tilde{\zeta}_z)_\zeta. \quad (4.2)$$

The first item (I_t) constitutes a differential statement of volume conservation, and I_x, I_y and I_z express surface conservation. The GCL usually is divided into the volume conservation law (VCL) and the surface conservation law (SCL) as discussed by Zhang et al. [93]. In fact, $I_t = I_x = I_y = I_z = 0$ analytically provided that the grids are differentiable. However, it may not be true in discretized forms because of numerical errors. Even when low-order schemes are used, Étienne et al. [86] showed that a numerical method satisfying the VCL will generally allow a much larger computational time step than its counterpart violating the VCL. To ensure the VCL, the volume conservation item is usually applied to calculate the time derivative of Jacobian following

$$(1/J)_\tau = -[(\tilde{\xi}_t)_\xi + (\tilde{\eta}_t)_\eta + (\tilde{\zeta}_t)_\zeta]$$

by many authors [78, 86]. The errors in I_x, I_y and I_z are usually related to grid quality, such as smoothness, uniformity, orthogonality, and stretch rate. High-order schemes with their low dissipative property usually bear more risk from the SCL-related errors than that of low-order schemes. This difficult is one of the key obstacles which hinder the widely applications of high-order schemes for complex grid problems. In order to use high-order schemes for low quality grids, a conservative metric method (CMM) which can ensure the SCL has been derived by Deng et al. [100]. The CMM contains the following two aspects:

(i) First, the metrics are acquired through the 'conservative forms'

$$\tilde{\xi}_x = (y_\eta z)_\zeta - (y_\zeta z)_\eta, \quad \tilde{\eta}_x = (y_\zeta z)_\xi - (y_\xi z)_\zeta, \quad \tilde{\zeta}_x = (y_\xi z)_\eta - (y_\eta z)_\xi \quad (4.3)$$

and with similar relations for the remain items.

(ii) Second, for each grid (coordinate) direction, the algorithm for the derivatives in Eq. (4.3) shall be identical to that of flow fluxes where the metrics are re-discretized in combination with the flow fluxes.

When the flux derivatives are discretized, the metrics are re-discretized because that the fluxes contain metrics. Then we can get equivalent schemes for the re-discretizations of the metrics. The equivalent schemes are called the re-discretization schemes in this paper. For WCNS-E-5, the re-discretization scheme is

$$\frac{\partial a_i}{\partial \xi} = \frac{75}{64h}[a_{i+1/2} - a_{i-1/2}] - \frac{25}{384h}[a_{i+3/2} - a_{i-3/2}] + \frac{3}{640h}[a_{i+5/2} - a_{i-5/2}], \quad (4.4)$$

where the cell-edge values can be acquired by high-order linear interpolations. In this paper, we use

$$a_{i+1/2} = \frac{1}{16}(-a_{i-1} + 9a_i + 9a_{i+1} - a_{i+2}). \quad (4.5)$$

5 Characteristic-based interface conditions

Characteristic boundary conditions have been widely studied by many researchers, such as Thompson [89] and Poinsot and Lele [90]. Characteristic interface conditions for multi-block grids have also been studied by Kim and Lee [29], Sumi et al. [30, 31] and

Deng et al. [32]. The characteristic-based interface condition (CBIC), which was developed by Deng et al. [32], will be briefly introduced below.

Defining the transformation matrix P_{QV_C} in terms of conservative variables Q, and characteristic variables V_C,

$$P_{QV_C} = \frac{\partial Q}{\partial V_C}. \tag{5.1}$$

Let

$$RHS = -J\left(\frac{\partial \tilde{F}}{\partial \xi} + \frac{\partial \tilde{G}}{\partial \eta} + \frac{\partial \tilde{H}}{\partial \zeta}\right). \tag{5.2}$$

The characteristic-based interface conditions (CBIC) are

$$\begin{cases} \left.\dfrac{\partial Q}{\partial t}\right|_L = (A_s^+)|_L (RHS)|_L + (A_s^-)|_L (RHS)|_R, \\ \left.\dfrac{\partial Q}{\partial t}\right|_R = (A_s^+)|_R (RHS)|_R + (A_s^-)|_R (RHS)|_L, \end{cases} \tag{5.3}$$

where

$$A_s^+ = P_{QV_C} diag\left[(1+sign(\lambda_i))/2\right] P_{QV_C}^{-1}, \quad A_s^- = P_{QV_C} diag\left[(1-sign(\lambda_i))/2\right] P_{QV_C}^{-1},$$

$(A_s^+)|_{L,R}$ are calculated by the variables on the interface, and $(RHS)|_{L,R}$ can be calculated by one-side differencing or other methods. Here, the subscripts, "L" and "R" denote the variables and terms on an interface but belong to the left-hand-side sub-block and the right-hand-side sub-block, respectively. One can verify that $(\partial Q/\partial t)|_L = (\partial Q/\partial t)|_R$. Please refer to [32] for more details about the CBIC, such as the implementations.

6 Applications and discussions

6.1 A hypersonic cylinder problem with interfaces on the shock wave

The effectiveness of the CBIC has been shown for a wide range of flows except shock waves by Deng et al. [32]. This case is chosen to show that the CBIC works well across shock waves provided that the RHS_L and RHS_R are oscillation-free. The flow is supposed to be laminar, and the incoming flow conditions for computing are: $Ma = 8.03$, $Re = 1.835 \times 10^5$, $T_\infty = 124.94K$, $T_w = 294.44K$, and the minimal grid size $h_{min} = 20/Re$.

The whole computed domain (121×81 notes) is decomposed into 7 sub-domains which are shown in Fig. 1. Three interfaces with a triple connecting point and a quadruple connecting point are purposively set on the static shock wave to check performances of the CBIC. Although the flow field is relatively simple, it is a tough case for an interface management technique because the strong shock wave will not only cross some of the interfaces, but also settle on some other interfaces ultimately. The computation begins at the uniform inflow conditions, then the shock wave first forms at the solid wall and then

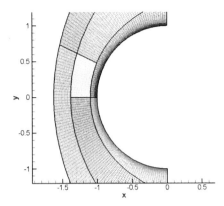

Figure 1: The grids over the cylinder. (colored picture attached at the end of the book)

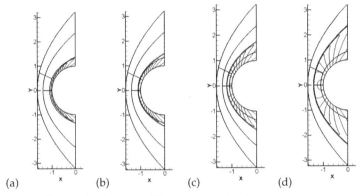

Figure 2: The bow shock wave at different computational moment, pressure contours.
(colored picture attached at the end of the book)

marches its way to the final position. Fig. 2(a) shows pressure contours at the moment when the shock wave is near the wall. Fig. 2(b) shows result at the moment when the shock wave is traversing the first three interfaces which are 'parallel' to the wall. Fig. 2(c) shows the result at the moment after the shock wave successfully traversed the first three wall-parallel interfaces. Fig. 2(d) shows the finial computed result with the shock wave settles on the other three wall-parallel interfaces. Although the decomposed domains are asymmetric to the y-axis, the result shown in Fig. 2 is still symmetric. Fig. 3 shows the enlarged views at two of the triple connecting points and one of the quadruple connecting points. It can be seen that the results are smooth as if there are no interfaces. The wall pressure distribution and wall heat transfer rate distribution are shown in Fig. 4, which shows that the computed results are consistent with the experimental results of Wieting [95].

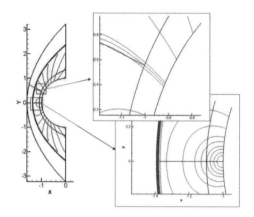

Figure 3: The enlarged parts at multi-connected points, pressure contours. (colored picture attached at the end of the book)

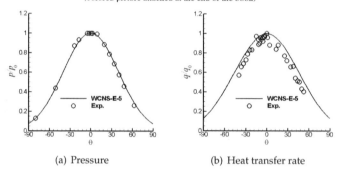

(a) Pressure (b) Heat transfer rate

Figure 4: The wall pressure and heat transfer rate. (colored picture attached at the end of the book)

6.2 The GCL for a blunt configuration

A two-dimensional blunt configuration with dual backward-facing steps is shown in Fig. 5(a). The flows conditions is set to be the same as that of [100], namely, $Ma = 0.3$, $Re = 2.7 \times 10^6$ (base on the height of one step), and $T_\infty = 300K$. The solid wall is assumed to be adiabatic. The flow is supposed to be full turbulence and the Spalart-Allmaras turbulence model [91] is applied here.

Although it is a two-dimensional problem, the mesh is deemed as a three-dimensional one by equally extending five grid levels in the transverse (z) direction. Before starting the simulation, a numerical analysis of the three SCL items of the GCL, namely I_x, I_y and I_z, is given to show the effectiveness of the CMM. Table 1 lists the SCL-related errors (the violation of the SCL), where N is the total grid number. Five schemes are tested here to check the characteristics of the CMM. The five schemes are the second-order explicit

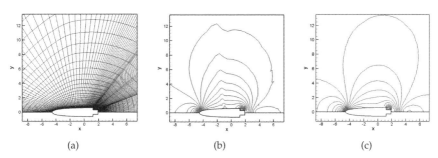

Figure 5: Grids and pressure contours. The flow fluxes are solved by WCNS-E-5. Middle: The metrics are solved by the traditional standard metrics of (6.1). Right: The metrics are solved by the CMM.

central scheme (EC2), the forth-order explicit central scheme (EC4), the forth-order Padé-type central compact scheme (CC4), the sixth-order Padé-type central compact scheme (CC6), and the scheme defined by Eq. (4.4) and Eq. (4.5) (WCNS-E-5). Table 1 indicates that the violation of the SCL is negligible if the CMM is applied. However, it can be seen from Table 2 that, if the schemes for the metrics are different from the re-discretization schemes for the grid-related items, the metric cancellation errors are much greater than those with the CMM, despite that the 'conservative forms' of Eqs. (4.3) are applied here. Another case is that the schemes for the metrics are same to the re-discretization schemes for the grid-related items, while the metrics are calculated by the traditional standard metric method. The errors shown in Table 3 are larger than that in Table 1. We note here that the traditional standard metrics are referred as followings

$$\tilde{\xi}_x = J^{-1}\xi_x = y_\eta z_\zeta - z_\eta y_\zeta, \quad \tilde{\eta}_x = J^{-1}\eta_x = y_\zeta z_\xi - z_\zeta y_\xi, \quad \tilde{\zeta}_x = J^{-1}\zeta_x = y_\xi z_\eta - z_\xi y_\eta, \cdots. \quad (6.1)$$

Table 1: I_x, I_y and I_z calculated by the CMM with different schemes.

	EC2	EC4	CC4	CC6	WCNS-E-5
$\frac{1}{N}\sum|I_x|$	1.14E-18	9.63E-17	1.56E-16	2.16E-16	2.04E-16
$\frac{1}{N}\sum|I_y|$	1.96E-17	2.39E-16	1.88E-16	5.93E-16	9.47E-16
$\frac{1}{N}\sum|I_z|$	8.51E-17	6.28E-16	6.58E-16	1.26E-15	1.28E-15
$\max(|I_x|,|I_y|,|I_z|)$	5.20E-14	2.88E-13	2.43E-13	3.74E-13	1.71E-13

Table 2: The maximal errors of I_x, I_y and I_z computed by different schemes with the conservative form (4.3).

		δ_1 schemes				
		EC2	EC4	CC4	CC6	WCNS-E-5
	ECS2	5.20E-14	2.42E-02	2.99E-02	3.84E-02	1.90E-02
	ECS4	2.42E-02	2.88E-13	2.33E-02	2.70E-02	1.89E-02
δ_2	CCS4	2.99E-02	2.33E-02	2.43E-13	1.44E-02	1.18E-02
schemes	CCS6	3.84E-02	2.70E-02	1.44E-02	3.74E-13	2.51E-02
	WCNS-E-5	1.90E-02	1.89E-02	1.18E-02	2.51E-02	1.71E-13

Table 3: Same schemes applied both to the metrics and their re-discretization with the traditional standard metrics (6.1).

	EC2	EC4	CC4	CC6	WCNS-E-5
$\frac{1}{N}\sum |I_x|$	1.82E-16	1.24E-14	1.06E-14	1.89E-14	1.03E-15
$\frac{1}{N}\sum |I_y|$	1.01E-16	6.60E-16	6.04E-16	1.28E-15	1.15E-15
$\frac{1}{N}\sum |I_z|$	3.63E-12	1.50E-12	1.31E-12	2.06E-12	8.23E-13
$\max(|I_x|,|I_y|,|I_z|)$	8.03E-10	7.52E-10	4.33E-10	8.02E-10	2.11E-10

Our previous study shows that high-order traditional standard metric methods (CC6, [100]) may cause numerical oscillations when we conduct high-order flow computing. In this paper, we can find that the oscillations are still obvious (Fig. 5(b)) even when the metrics are calculated by the 2nd-order scheme (EC2) through the traditional standard form of Eq. (6.1). We have proved in [100] that this kind of oscillations can be successfully avoided by applying the CMM notwithstanding that the high-order schemes are applied. The result in [100] is also quoted as Fig. 5(c).

6.3 The application of the GCL for vortex preservation

A vortex convecting problem is chosen to show the performance of the GCL on vortex preservation property of high-order schemes. Similar researches can be found in Nonomura's work [79]. The initial conditions are defined by adding a vortex in a mean flow [90].The mesh contains random perturbations from a normal Cartesian grid as shown in Fig. 6(a). A nominal 50×50 uniform mesh is generated in the domain $x\in[-6,6]$, $y\in[-6,6]$. Interior points are then perturbed by 20% of the nominal spacing in a randomly chosen direction. In order to simplify the boundary treatment, four points near the boundaries of each grid line are left unperturbed. There are no analytical metric derivatives because of the discontinuities of the grid. The quality of the mesh is far lower to meet the requirements of many spatial schemes if the GCL is not satisfied during computing. For example, Visbal & Gaitonde [78] showed that the CC4 scheme and the EC4 scheme were unstable even at extremely small time steps, and strong filters are necessary to stabilize the computing. Nonomura [79] employs the same method as the CMM to calculate the metrics except that Nonomura adopts a six-order interpolation other than the forth-order interpolation (Eq. (4.5)). Nonomura's results and ours show that a vital aspect for the accurate and stable simulations of flows on this extremely lower quality grid is the SCL (namely the GCL of stationary grids). This conclusion can be easily obtained by comparing the results in Figs. 6(b), 6(c) and 6(d) in which the results at $t=12$ are shown. The CMM, which satisfies the SCL, is much superior to the traditional standard metrics in the condition of lower quality grids. If the SCL is satisfied on every discrete point, the WCNS-E-5 with its nonlinear mechanism can successfully damp out the numerical oscillations on the low quality mesh without any other extra treatments such as filters.

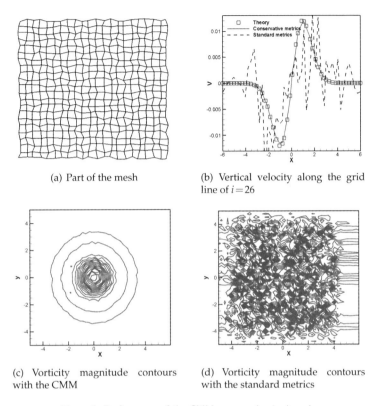

(a) Part of the mesh

(b) Vertical velocity along the grid line of $i=26$

(c) Vorticity magnitude contours with the CMM

(d) Vorticity magnitude contours with the standard metrics

Figure 6: Performance of the CMM on a randomized mesh.

6.4 Applications of the high-order WCNS-E-5 to complex grid problems

The latest CFD progresses, which include some complex applications of high-order schemes (referred as high resolution schemes by Fujii), such as linear compact schemes, WENO, and WCNS, has been summarized by in [96, 97]. Fujii et al. [16] also found that the WCNS resolves flow structure with 8-10 points per wave, and has 2-4 times higher spatial resolution than the conventional second-order TVD scheme for the simulation of acoustics from supersonic jets. The WCNS reduce the total number of grid points 10-50 times in three dimensions, which saves the same order of computer time. The applications of WCNS for implicit large eddy simulations of turbulence are tested by Ishiko et al. [98]. One aim of our studies is to apply the WCNS-E-5 to flows over complex configurations. Deng et al. [32] have successfully applied the WCNS-E-5 in simulating the complex flows over a two-element NLR7301 airfoil, a three-element 30P-30N airfoil, and a three-dimensional wing-body configuration. The followings continue to show some engineering-oriented applications of the WCNS-E-5.

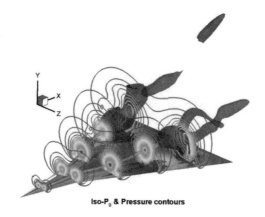

Figure 7: Iso-p_0 and pressure contours of a $Ma=0.3$ double-delta wing, 4.26 million grids.
(colored picture attached at the end of the book)

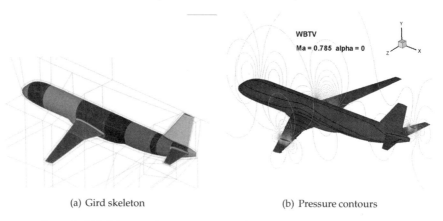

(a) Gird skeleton (b) Pressure contours

Figure 8: Grid skeleton and pressure contours of a transonic airplane configuration.
(colored picture attached at the end of the book)

Usually, the implementations of high-order schemes for complex three-dimensional grids are much more difficult than that of low-order schemes as numerical oscillations and instabilities will cause the failure of high-order applications. This shortage of high-order finite difference schemes can be largely mitigated for subsonic problems by applying the CBIC and the CMM mentioned above. From the isotonic total-pressure surface and the pressure contours of the double-delta wing shown in Fig. 7, it can be found that the result is oscillation free and the main vortex core, the secondary vortex core, as well as the breakdown of the main vortex are clearly resolved. The CMM can also enhance the capability of the WCNS for complex shock wave problems. Fig. 8 shows the grid skeleton and the pressure contours of a transonic airplane configuration. Fig. 9 shows the grid

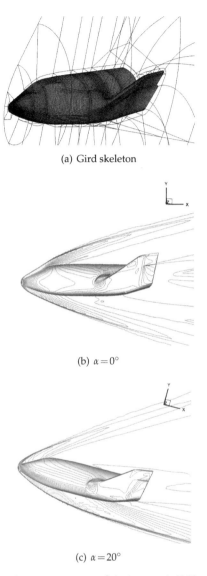

(a) Gird skeleton

(b) $\alpha=0°$

(c) $\alpha=20°$

Figure 9: Grid skeleton and pressure contours of the hypersonic X-38 configuration ($\alpha=0°$).
(colored picture attached at the end of the book)

skeleton and the pressure contours of the hypersonic X-38 configuration. It can be seen from the two figures that the flows are smooth despite that the grids are complex with many block interfaces.

7 Concluding remarks

In contrast to the rapid development of high-order algorithms in CFD, the applications of high-order and high accurate methods on complex configurations are still limited. One of the reasons, which baffle the wide applications of high-order schemes in engineering-oriented problems, is mainly derived from the complexity of grids.

In order to apply high-order schemes for multi-block grids, the characteristic-based interface conditions (CBIC) are employed to conduct the transmission of flow information. Our results indicate that the CBIC can handle complex multi-block grids conveniently and effectively. For the purpose of satisfying the three surface conservation items of the GCL (or SCL) in discretized forms, a conservative metric method (CMM) was employed for curvilinear coordinates. The CMM, which is independent to the accuracy of a scheme, can eliminate the GCL-related errors derived from the discretization of flow fluxes, while the traditional standard metric method can cause unacceptable GCL-related errors because of discretization errors when the grid quality is low. Numerical tests demonstrate that the GCL is very important to ensure the freestream conservation, as well as to help to prevent numerical oscillations.

Finally, the successful applications of the high-order WCNS-E-5 to complex problems are briefly shown. The complex problems include a double-delta wing, a transonic airplane configuration, and a hypersonic X-38 configuration. Some other complex applications of WCNS can be found in [16, 32, 79, 97–99]. These researches show pleasant prospects in engineering-oriented applications of high-order WCNS schemes especially for detailed simulations.

Acknowledgments

This study was supported by the project of National Natural Science Foundation of China (Grant 11072259 and 10621062) and National Basic Research Program of China (Grant No. 2009CB723800). The authors would like to thank Dr. Huayong Liu, and Assistant Researcher Guangxue Wang of State Key Laboratory of Aerodynamics for their contributions.

References

[1] Wang, Z.J., High-Order Methods for the Euler and Navier-Stokes Equations on Unstructured Grids, Progress in Aerospace Sciences, 43 (2007), pp. 1-41.

[2] Shu, C.-W., High-Order Finite Difference and Finite Volume WENO Schemes and Discontinuous Gakerkin Methods for CFD, Int. J. Comput. Fluid Dyn., 17 (2003), pp. 107-118.

[3] Cheng, J., and Shu, C.-W., High Order Schemes for CFD: a Review, Chinese J. Comput. Phys., 26(5) (2009), pp. 633-655.

[4] Ekaterinaris, J.A., High-Order Accurate, Low Numerical Diffusion Methods for Aerodynamics, Progress in Aerospace Sciences, 41 (2005), pp. 192-300.

[5] Lele, S.K., Compact Finite Difference Schemes with Spectral-Like Resolution, J. Comput. Phys., 103 (1992), pp. 16-42.
[6] Deng, X., Maekawa, H., and Shen, Q., A Class of High Order Dissipative Compact Schemes, AIAA Paper, 96-1972, (1996).
[7] Visbal, M.R., and Gaitonde, D.V., High-Order Accurate Methods for Complex Unsteady Subsonic Flows, AIAA Journal, 37(10) (1999), pp. 1231-1239.
[8] Deng, X., Maekawa, H., Compact High-Order Accurate Nonlinear Schemes, J. Comput. Phys., 130 (1997), pp. 77-91.
[9] Deng, X., and Zhang, H., Developing High-Order Accurate Nonlinear Schemes, J. Comput. Phys., 165 (2000), pp. 22-44.
[10] Deng, X., High-Order Accurate Dissipative Weighted Compact Nonlinear Schemes, Science in China (Serial A), 45(3) (2002), pp. 356-370.
[11] Liu, X., Deng, X., and Mao, M., High-Order Behaviors of Weighted Compact Fifth-Order Nonlinear Schemes, AIAA Journal, 45(8) (2007), pp. 2093-2097.
[12] Sumi, T., Kurotaki, T., and Hiyama, J., Generalized Characteristic Interface Conditions with High-Order Interpolation Method, AIAA Paper, 2008-752, (2008).
[13] Gaitonde, D.V., and Visbal, M.R., Padé-Type High-Order Boundary Filters for the Navier-Stokes Equations, AIAA Journal, 18(11) (2000), pp. 2103-2112.
[14] Delf, J.W., Sound Generation From Gust-Airfoil Interaction Using CAA-Chimera Method, AIAA Paper, 2001-2136, (2001).
[15] Sherer, S.E., Gordnier, R.E., and Visbal, M.R., Computational Study of a UCAV Configuration Using a High-Order Overset-Grid Algorithm, AIAA Paper, 2008-626, (2008).
[16] Fujii, K., Nonomura, T., and Tsutsumi, S. Toward Accurate Simulation and Analysis of Strong Acoustic Wave Phenomena-A Rview Fom the Eperience of Our Study on Rocket Problems, Int. J. Numer. Meth. Fluids, 64 (2010), pp. 1412-1432.
[17] Rai, M.M., A Relaxation Approach to Patched Grid Calculations with the Euler Equations, J. Comput. Phys., 66 (1986), pp. 99-131.
[18] Lerat, A., and Wu, Z.N., Stable Conservative Multidomain Treatments for Implicit Euler Equations, J. Comput. Phys., 123 (1996), pp. 45-64.
[19] Huan, X., Hicken, J.E., and Zingg, D.W., Interface and Boundary Schemes for High-Order Methods, AIAA Paper 2009-3658, (2009).
[20] Hicken, J.E., and Zingg, D.W., Parallel Newton-Krylov Solver for the Euler Equations Discretized Using Simultaneous-Approximation Terms, AIAA Journal, 46(11) (2008), pp. 2773-2786.
[21] Kreiss, H.-O., and Scherer, G., Finite Element and Finite Difference Methods for Hyperbolic Partial Differential Equations, Mathematical Aspects of Finite Elements in Partial Differential Equations, Academic Press, New York, (1974).
[22] Mattsson, K., Boundary Procedures for Summation-by-Parts Operators, SIAM J. Sci. Comput., 18(1) (2003), pp. 133-153.
[23] Olsson, P., Summation by Parts, Projections, and Stability. I, Math. Comput., 64(211) (1995), pp. 1035-1065.
[24] Carpenter, M.H., Gottlieb, D., and Abarbanel, S., Time-Stable Boundary Conditions for Finite-Difference Schemes Solving Hyperbolic Systems, J. Comput. Phys., 111 (1994), pp. 220-236.
[25] Nordström, J. and Carpenter, M.H., Boundary and Interface Conditions for High-Order Finite-Difference Methods Applied to the Euler and Navier-Stokes Equations, J. Comput. Phys., 148 (1999), pp. 621-645.

[26] Carpenter, M.H., Nordstrom, J., and Gottlieb, D., A Stable and Conservative Interface Treatment of Arbitrary Spatial Accuracy, SIAM J. Sci. Comput., 148 (1999), pp. 341-365.
[27] Strand, B., Summation by Parts for Finite Difference Approximations for d/dx, J. Comput. Phys., 110 (1994), pp. 47-67.
[28] Jurgens, H.M. and Zingg, D.W., Numerical Solution of the Time-Domain Maxwell Equations Using High-Accuracy Finite-Difference Methods, SIAM J. Sci. Comput., 22(5) (2001), pp. 1675-1696.
[29] Kim, J.W., and Lee, D.J., Characteristic Interface Conditions for Multi-Block High-Order Computation on Singular Structured Grid, AIAA Paper, 2003-3122, (2003).
[30] Sumi, T., Kurotaki, T., and Hiyama, J., Generalized Characteristic Interface Conditions for Accurate Multi-Block Computation, AIAA Paper, 2006-1272, (2006).
[31] Sumi, T., Kurotaki. T., and Hiyama, J., Practical Multi-Block Computation with Generalized Characteristic Interface Conditions Around Complex Geometry, AIAA Paper, 2007-4471, (2007).
[32] Deng, X., Mao, M., and Tu, G., et al., Extending the fifth-order weighted compact nonlinear scheme to complex grids with characteristic-based interface conditions, AIAA Journal, 48(12) (2010), pp. 2840-2851.
[33] Fu, D., Ma, Y., Analysis of Super Compact Finite Difference Method and Application to Simulation of Vortex-Shock Interaction, Int. J. Numer. Meth. Fluids, 36(7) (2001), pp. 773-805.
[34] Uzgoren, E., Sim, J., Shyy, W., Marker-Based, 3-D Adaptive Cartesian Grid Method for Multiphase Flow Around Irregular Geometries, Commun. Comput. Phys., 5 (2009), pp. 1-41.
[35] Yee, H.C., Explicit and Implicit Multidimensional Compact High-Resolution Shock-Capturing Methods: Formulation, J. Comput. Phys., 131 (1997), pp. 216-232.
[36] Sjogreen, B., Yee, H.C., Variable High Order Multiblock Overlapping Grid Methods for Mixed Steady and Unsteady Multiscale Viscous Flows, Commun. Comput. Phys., 5 (2009), pp. 730-744.
[37] Osher, S., Efficient Implementation of High Order Accurate Essentially Non-Oscillatory Shock Capturing Algorithms Applied to Compressible Flow, Numerical Methods for Compressible Flows-Finite Difference, Element and Volume Techniques. Anaheim, Ca, UAS: ASME, New York, NY, UAS, (1986), pp. 127-128.
[38] Chakravarthy, S.R., Harten, A., and Osher, S., Essentially Non-Oscillatory Shock-Capturing Schemes of Arbitrarily-High Accuracy, AIAA-86-0339 (1986).
[39] Shu, C.-W., High Oder Weighted Essentially Non-Oscillatory Schemes for Convection Dominated Problems, SIAM Rev., 51 (2009), pp. 82-126.
[40] Cockburn, B., and Shu, C.-W., The Runge-Kutta Discontinuous Galerkin Finite Element Method for Conservation Laws V: Multidimensional Systems, J. Comput. Phys., 141 (1998), pp. 199-224.
[41] Bassi, F., and Rebay, S., Numerical Evaluation of Two Discontinuous Galerkin Methods for the Compressible Navier-Stokes Equations, Int. J. Numer. Meth. Fluids, 40(1) (2002), pp. 197-207.
[42] Dolejsi, V., Semi-Implicit Interior Penalty Discontinuous Galerkin Methods for Viscous Compressible Flows, Commun. Comput. Phys., 4 (2008), pp. 231-274.
[43] Luo, H., Baum, J.D., and Lohner, R., A Discontinous Galerkin Method Based on a Taylor Basis for the Compressible Flows on Arbitrary Grids, J. Comput. Phys., 227 (2008), pp. 8875-8893.

[44] Dumbser, M., Balsara, D.S., Toro, E.F., and Munz, C.D., A Unified Framework for the Construction of One-Sep Finite Volume and Discontinuous Galerkin Schemes on Unstructured Meshes, J. Comput. Phys., 227 (2008), pp. 8209-8253.
[45] Ferrari, A., Dumbser, M., Toro, E.F., Armanini, A., A New Stable Version of the SPH Method in Lagrangian Coordinates, Commun. Comput. Phys., 4 (2008), pp. 378-404.
[46] Grosso, G., Antuono, M., Toro, E.F., The Riemann Problem for the Dispersive Nonlinear Shallow Water Equations, Commun. Comput. Phys., 7 (2010), pp. 64-102.
[47] Dumbser, M., Toro, E.F., On Universal Osher-Type Schemes for General Nonlinear Hyperbolic Conservation Laws, Commun. Comput. Phys., 10 (2011), pp. 635-671.
[48] Wang, L., and Mavriplis, D.J., Implicit Solution of the Unsteady Euler Equations for High-Order Accurate Discontinuous Galerkin Discretiztions, J. Comput. Phys., 225 (2007), pp. 1994-2015.
[49] Nastase, C.R., and Mavriplis, D.J., High-Order Discontinuous Galerkin Methods Using a Hp-Multigrid Approach, J. Comput. Phys., 213 (2006), pp. 330-357.
[50] Abgrall, R., Shu, C.-W., Development of Residual Distribution Schemes for the Discontinuous Galerkin Method: The Scalar Case with Linear Elements, Commun. Comput. Phys., 5 (2009), pp. 376-390.
[51] Xia, Y., Xu, Y., Shu, C.-W., Application of the Local Discontinuous Galerkin Method for the Allen-Cahn/Cahn-Hilliard System, Commun. Comput. Phys., 5 (2009), 821-835.
[52] Zhang, Z.-T., Shu, C.-W., Third Order WENO Scheme on Three Dimensional Tetrahedral Meshes, Commun. Comput. Phys., 5 (2009), pp. 836-848.
[53] Zhang, R., Zhang, M., Shu, C.-W., On the Order of Accuracy and Numerical Performance of Two Classes of Finite Volume WENO Schemes, Commun. Comput. Phys., 9 (2011), pp. 807-827.
[54] Kroll, N., ADIGMA - A European Project on the Development of Adaptive Higher Order Variational Methods for Aerospace Applications, European Conference on Computational Fluid Dynamics, Eccomas CDF, (2006).
[55] Patera, A.T., A Spectral Element Method for Fluid Dynamics: Laminar Flow in a Channel Expansion, J. Comput. Phys., 54 (1984), pp. 468.
[56] Liu, Y., Vinokur, M. and Wang, Z.J., Spectral (Finite) Volume Method for Conservation Laws on Unstructured Grids V: Extension to Three-Dimensional Systems, J. Comput. Phys., 215 (2006), pp. 454-472.
[57] Sun, Y., Wang, Z.J., and Liu, Y., Spectral (finite) Volume Method for Conservation Laws on Unstructured Grids VI: Extension to Viscous Flow, J. Comput. Phys., 215 (2006), pp. 41-58.
[58] Sun, Y., Wang, Z.J., and Liu, Y., High-Order Multi-Domain Spectral Difference Method for the Navier-Stokes Equations on Unstructured Hexahedral Grids, Commun. Comput. Phys., 2(2) (2007), pp. 310-333, and also AIAA Paper 2006-301, (2006).
[59] Kitamura, K., Shima, E., Fujimoto, K., Wang, Z.J., Performance of Low-Dissipation Euler Fluxes and Preconditioned LU-SGS at Low Speeds, Commun. Comput. Phys., 10 (2011), pp. 90-119.
[60] Kopriva, D.A., A Staggered-Grid Multidomian Spectral Method for the Compressible Navier-Stokes Equations, J. Comput. Phys., 143 (1998), pp. 125-158.
[61] Xu, K., Jin, C., A Unified Moving Grid Gas-Kinetic Method in Eulerian Space for Viscous Flow Computation, J. Comput. Phys., 222 (2007), pp. 155-175.
[62] Jin, C., Xu, K., Numerical Study of the Unsteady Aerodynamics of Freely Falling Plates, Commun. Comput. Phys., 3 (2008), pp. 834-851.
[63] Xu, K., A Gas-Kinetic BGK Scheme for the Navier-Stokes Equations and its Connection

with Artificial Dissipation and Godunov Method, J. Comput. Phys., 171 (2001) 289-335.
[64] Xu, K., Liu, H., A Multiple Temperature Kinetic Model and its Application to Near Continuum Flows, Commun. Comput. Phys., 4 (2008), pp. 1069-1085.
[65] Ni, G., Jiang, S., and Xu, K. Remapping-Free ALE-Type Kinetic Method for Flow Computations, J. Comput. Phys., 228 (2009), pp. 3154-3171.
[66] Chen, Y., Jiang, S., An Optimization-Based Rezoning for ALE Methods, Commun. Comput. Phys., 4 (2008), pp. 1216-1244.
[67] Tam, C.K.W., and Webb, J.C., Dispersion-Relation-Preserving Finite Difference Schemes for Computational Acoustics, J. Comput. Phys., 107 (1993), pp. 262-281.
[68] Yee, H.C., Sandham, N. D., Hadjadj, A., Progress in the Development of a Class of Efficient Low Dissipative High Order Shock-Capturing Methods, Proceeding of the Symposium in Computational Fluid Dynamics for the 21st Century, July, (2000), Kyoto, Japan.
[69] Suresh, A., and Huynk, H.T., Accurate Monotonicity Preserving Scheme with Runge-Kutta Time Stepping, J. Comput. Phys., 136 (1997), pp. 83-99.
[70] Daru, V., and Tenaud, C., High Order One Step Monotonicity-Preserving Schemes for Unsteady Compressible Flow Calculations, J. Comput. Phys., 193, (2004), pp. 563-594.
[71] Trefethen, L.N., Group Velocity in Finite Difference Schemes, SIAM Rev., 24(2) (1982), pp. 113-136.
[72] Ma, Y., and Fu, D., Forth Order Accurate Compact Scheme with Group Velocity Control (GVC), Science in China (Serial A), 44(9) (2001), pp. 1197-1204.
[73] Ren, Y., Liu, M., and Zhang, H., A Characteristic-Wise Hybrid Compact-WENO Schemes for Solving Hyperbolic Conservations, J. Comput. Phys., 192 (2005), pp. 365-386.
[74] Adams, N.A., and Shariff, K., A High-Resolution Hybrid Compact-ENO Scheme for Shock-Turbulence Interaction Problems, J. Comput. Phys., 127 (1996), pp. 27.
[75] Pirozzoli, S., Conservative Hybrid Compact-WENO Schemes for Shock-Turbulence Interaction, J. Comput. Phys., 179 (2002), pp. 81-117.
[76] Wang, Z., and Huang, G., An Essentially Nonoscillatory High-Order Padé-Type (ENO-Padé) Scheme, J. Comput. Phys., 177 (2002), pp. 37-58.
[77] http://aaac.larc.nasa.gov/tsab/cfdlarc/aiaa-dpw
[78] Visbal, R.M., Gaitonde, D.V., On the Use of Higher-Order Finite-Difference Schemes on Curvilinear and Deforming Meshes, J. Comput. Phys., 181 (2002), pp. 155-185.
[79] Nonomura, N., Iizuka, N., and Fujiji, K., Freestream and Vortex Preservation Properties of High-Order WENO and WCNS on Curvilinear Grids, Comput. Fluids, 39 (2010), pp. 197-214.
[80] Trulio, J.G., and Trigger, K.R., Numerical Solution of the One-Dimensional Hydrodynamic Equations in an Arbitrary Time-Dependent Coordinate System, Technical Report UCLR-6522, University of California Lawrence Radiation laboratory, (1961).
[81] Cheng, J., Shu, C.-W., A Third Order Conservative Lagrangian Type Scheme on Curvilinear Meshes for the Compressible Euler Equations, Commun. Comput. Phys., 4 (2008), pp. 1008-1024.
[82] Qiu, J.-M., Shu, C.-W., Conservative Semi-Lagrangian Finite Difference WENO Formulations with Applications to the Vlasov Equation, Commun. Comput. Phys., 10 (2011), pp. 979-1000.
[83] Liu, W., Yuan, L., Shu, C.-W., A Conservative Modification to the Ghost Fluid Method for Compressible Multiphase Flows, Commun. Comput. Phys., 10 (2011), pp. 785-806.
[84] Pulliam, T.H., and Steger, J.L., On Implicit Finite-Difference Simulations of Three-Dimensional Flow, AIAA Paper 78-10, (1978).

[85] Thomas, P.D., and Lombard, C.K., Geometric Conservation Law and Its Application to Flow Computations on Moving Grids, AIAA Journal, 17(10) (1979), pp. 1030-1037.
[86] Étienne, S., Garon, A., and Pelletier, D., Perspective on the Geometric Conservation Law and Finite Element Methods for ALE Simulations of Incompressible Flow, J. Comput. Phys., 228 (2009), pp. 2313-2333.
[87] Bassi, F., and Rebay, S., A High-Order Accurate Discontinuous Finite Element Method for the Numerical Solution of the Compressible Navier-Stokes Equations, J. Sci. Comput., 131 (1997), pp. 267-279.
[88] Bassi, F., and Rebay, S., GMRES Discontinuous Galerkin Solution of the Compressible Navier-Stokes Equations, In Karniadakis Cockburn and Shu, eds., Discontinuous Galerkin Methods: Theory, Computation and Applications, pp. 197-208. Springer, Berlin, 2000.
[89] Thompson, K.W., Time Dependent Boundary Conditions for Hypersonic System, II, J. Comput. Phys., 89, (1990), pp. 439-461.
[90] Poinsot, T.J., and Lele, S.K., Boundary Conditions for Direction Simulations of Compressible Viscous Flows, J. Comput. Phys., 101 (1992), pp. 104-129
[91] Spalart, P.R., and Allmaras, S.R., A One-Equation Turbulence Model for Aerodynamic Flows, AIAA Paper 92-0439, (1992).
[92] Zhang, H., and Zhuang, F., NND Schemes and Their Applications to Numerical Simulation of Two- and Three-Dimensional Flows, Adv. Appl. Mech., 29 (1991), pp. 193-256
[93] Zhang, H., Reggio, M., Trépanier, J. Y., Camarero, R.,Discrete Form of the GCL for Moving Meshes and its Implementation in CFD Shemes, Comput. Fluids, 22 (1993), pp. 9-23.
[94] Tu, G., Deng, X., Mao, M., A Staggered Non-Oscillatory Finite Difference Method for High-Order Discretization of Viscous Terms, Acta Aerodynamica Sinica, 29(1) (2011), pp. 10-15.
[95] Wieting, A. R., Experimental Study of Shock Wave Interference Heating on a Cylindrical Leading Edge, NASA TM-100484, (1987).
[96] Fujii, K. Progress and Future Prospects of CFD in Aerospace-Wind Tunnel and Beyond, Progress in Aerospace Sciences, 41 (2005), pp. 455-470.
[97] Fujii, K. CFD Contributions to High-Speed Shock-Related Problems: Examples Today and New Features Tomorrow, Shock Waves, 18 (2008), pp. 145-154.
[98] Ishiko, K., Ohnishi, N., Ueno, K., et al., Implicit Large Eddy Simulation of Two-Dimensional Homogeneous Turbulence Using Weighted Compact Nonlinear Scheme, J. Fluid Eng., 131(6) (2009), pp. 1-14.
[99] Deng, X., Liu, X., Mao, M. et al., Advances in High-Order Accurate Weighted Compact Nonlinear Schemes. Adv. Mech., 37(3) (2007), pp.417-427.
[100] Deng, X., Mao, M., Tu, G., et al., Geometric Conservation Law and Applications to High-Order Finite Difference Schemes with Stationary Grids, J. Comput. Phys., 230(4) (2011), pp. 1100-1115.
[101] Shen, Y., Zha, G., and Chen, X., High Order Conservative Differencing for Viscous Terms and the Application to Vortex-Induced Vibration Flows, J. Comput. Phys., 228 (2009), pp. 8283-8300.
[102] Kannan, R., and Wang, Z.J., A Study of Viscous Flux Formulations for a p-Multigrid Spectral Volume Navier Stokes Solver, J. Sci. Comput., 41(2) (2009), pp. 165-199.
[103] Kannan, R., and Wang, Z.J., LDG2: A Variant of the LDG Flux Formulation for the Spectral Volume Method, J. Sci. Comput., 46(2) (2010), pp. 314-328.
[104] Kannan, R., and Wang, Z.J., The Direct Discontinuous Galerkin (DDG) Viscous Flux Scheme for the High Order Spectral Volume Method, Comput. Fluids, 39(10) (2010), pp. 2007-2021

[105] Sun, Y., Wang, Z.J., Liu, Y., Spectral (Finite) Volume Method for Conservation Laws on Unstructured Grids VI: Extension to Viscous Flow, J. Comput. Phys., 215 (2006), pp. 41-58.

[106] Zhang, Y., Deng, X., Mao, M. et al., Investigation of Convergence Acceleration for High-Order Scheme (WCNS) in 2D Supersonic Flows, Acta Aerodynamica Sinica, 26(1) (2008), pp. 14-18.

[107] Parsani, M., Ghorbaniasl, G., Lacor, C., et al., An Implicit High-Order Spectral Difference Approach for Large Eddy Simulation, J. Comput. Phys., 229(14) (2010), pp. 5373-5393.

[108] Kannan, R., An implicit LU-SGS Spectral Volume Method for the Moment Models in Device Simulations: Formulation in 1D and Application to A P-Multigrid Algorithm, Int. J. Numer. Methods Biomed. Eng., (Article online in advance of print) n/a. doi: 10.1002/cnm.1359.

[109] Sun, Y., Wang, Z.J., Liu, Y., Efficient Implicit Non-linear LU-SGS Approach for Compressible Flow Computation Using High-Order Spectral Difference Method, Commun. Comput. Phys., 5 (2009), pp. 760-778.

[110] Haga, T., Sawada, K., Wang, Z.J., An Implicit LU-SGS Scheme for the Spectral Volume Method on Unstructured Tetrahedral Grids, Commun. Comput. Phys., 6 (2009), pp. 978-996.

[111] Luo, H., Baum, J. D., Löhner, R., A P-Multigrid Discontinuous Galerkin Method for the Euler Equations on Unstructured Grids, J. Comput. Phys., 211 (2006), pp. 767-783.

[112] Fidkowski, K.J., Oliver, T.A., Lu, J., et al., P-Multigrid Solution of High-Order Discontinuous Galerkin Discretizations of the Compressible Navier-Stokes Equations, J. Comput. Phys., 207 (2005), pp. 92-113.

[113] Liang, C., Kannan, R., Wang, Z.J., A P-Multigrid Spectral Difference Method with Explicit and Implicit Smoothers on Unstructured Triangular Grids, Comput. Fluids, 38(2) (2009), pp. 254-265.

[114] Ronquist, E.M., Patera, A.T., Spectral Element Multigrid, I. Formulation and Numerical Results, J. Sci. Comput., 2(4) (1987), pp. 389-406.

Numerical investigation on body-wake flow interaction over rod–airfoil configuration

Yi Jiang[1,2,†], Mei-Liang Mao[1,2], Xiao-Gang Deng[3] and Hua-Yong Liu[1]

[1]State Key Laboratory of Aerodynamics, China Aerodynamics Research and Development Center,
PO Box 211, Mianyang 621000, PR China

[2]Computational Aerodynamics Institute, China Aerodynamics Research and Development Center,
PO Box 211, Mianyang 621000, PR China

[3]National University of Defense Technology, Changsha, Hunan 410073, PR China

(Received 29 December 2014; revised 11 July 2015; accepted 18 July 2015;
first published online 14 August 2015)

Numerical investigations of body-wake interactions were carried out by simulating the flow over a rod–airfoil configuration using high-order implicit large eddy simulation (HILES) for the incoming velocity $U_\infty = 72$ m s^{-1} and a Reynolds number based on the airfoil chord 4.8×10^5. The flow over five different rod–airfoil configurations with different distances of $L/d = 2, 4, 6, 8$ and 10, respectively, were calculated for the analysis of body-wake interaction phenomena. Various fundamental mechanisms dictating the intricate flow phenomena including force varying regulation, flow structures and flow patterns in the interaction region, turbulent fluctuations and their suppression, noise radiation and fluid resonant oscillation, have been studied systematically. Due to the airfoil downstream, a relatively higher base pressure is exerted on the surface of the cylinder upstream, and the pressure fluctuation on the surface of the rod–airfoil configuration with $L/d = 2$ is significantly suppressed, resulting in a reduction of the fluctuating lift. Following the distance between the cylinder and airfoil strongly decreases, Kármán-street shedding is suppressed due to the blocking effect. The flow in this interaction region has two opposite tendencies: the influence of the airfoil on the steady flow is to accelerate it and the counter-rotating vortices connecting with the leading edge of the airfoil tend to slow the flow down. There may be two flow patterns associated with the interference region, i.e. the Kármán-street suppressing mode and the Kármán-street shedding mode. The primary vortex shedding behind the cylinder upstream, and the shedding wake impingement onto the airfoil downstream, play a dominant role in the production of turbulent fluctuations. When primary vortex shedding is suppressed, the intensity of impingement is weakened, resulting in a significant suppression of the turbulent fluctuations. Due to these factors, a special broadband noise without a manifestly distinguishable peak is radiated by the rod–airfoil configuration with $L/d = 2$. The fluid resonant oscillation within the flow interaction between the turbulent wake and the bodies was further investigated by adopting a feedback model, which confirmed that the effect of fluid resonant oscillation becomes stronger when $L/d = 6$ and 10.

Jiang Yi, Mao Meiliang, Deng Xiaogang, Liu Huayong. Numerical investigation on body-wake flow interaction over rod-airfoil configuration. Journal of Fluid Mechanics, 2015, 779: 1-35.

The results obtained in this study provide physical insight into the understanding of the mechanisms relevant to the body-wake interaction.

Key words: aeroacoustics, flow-structure interactions, turbulence simulation

1. Introduction

The interaction between the turbulent wake and other bodies has attracted much attention in recent decades because of its obvious importance in a wide range of applications (e.g. Ljungkrona, Norberg & Sunden 1991; Mahir & Rockwell 1996; Zdravkovich 2003; Munekata et al. 2008). These researchers used a cylinder to create a turbulent wake which interacted with another cylinder or an airfoil downstream, and mainly focused on the flow features in the interaction region such as the vortex shedding suppressing and flow patterns. Moreover, the body-wake interaction involves some important and complicated phenomena related to the Kármán-street and wake impingement, such as noise radiation and fluid resonant oscillation (e.g. Boudet, Grosjean & Jacob 2005; Munekata et al. 2006; Jiang, Li & Zhou 2011; Hutcheson & Brooks 2012). However, the physical mechanisms dictating these phenomena are still unclear and are of great interest for further detailed studies.

The wake of a circular cylinder has been extensively studied by Zdravkovich (1997), who summarized the different states of flow for smooth cylinders in a uniform flow according to: laminar, transition-in-wake, subcritical (transition to turbulence in the free shear layer), critical (transition to turbulence in the boundary layer) and fully turbulent. The subcritical, critical and turbulent states of flow are of most interest for body-wake interaction studies due to the Reynolds number range it encompasses. The range of Reynolds numbers based on the cylinder diameter and the incoming velocity of the uniform flow are $(350\text{--}400) < Re < (10^5\text{--}2 \times 10^5)$ and $(10^5\text{--}2 \times 10^5) < Re$ for the subcritical and critical states, respectively. The required Reynolds number is still unknown at present for the activation of a fully turbulent state of flow. The Reynolds number is expected to be the dominant parameter to determine the flow state for the smooth cylinders in a uniform flow, however, the structure of the flow may be significantly changed and transitions can occur at relatively lower values of Re if surface roughness or free stream turbulence are introduced. The noise radiation from the turbulent wake of a cylinder has been well recognized and carefully examined by Schlinker, Fink & Amiet (1976). They observed that the radiation noise may decrease sharply in amplitude following an increase in Reynolds number.

Two cylinders in tandem is a common configuration for the study of body-wake interactions having important implications for flow-induced vibration and noise generation. Representative applications include heat exchanger tubes, adjacent tall buildings, bundled transmission lines and the piles of offshore platforms (Mahir & Rockwell 1996). The characteristics of the flow around this configuration has been comprehensively reviewed by Zdravkovich (1977), who has shown that vortex formation in the inter-cylindrical region is not present until the non-dimensional distance is longer than approximately four diameters. According to the inter-cylindrical distance, Zdravkovich (2003) has categorized the different possible flow regimes associated with the wake interference as follows: (i) a single vortex street formed by the cylinder upstream; (ii) a shear layer separated from the cylinder upstream reattaches intermittently onto the cylinder downstream, and vortex shedding occurs only from the downstream cylinder; (iii) vortex shedding from the cylinder downstream and intermittent vortex shedding from the cylinder upstream; (iv) vortex streets from the upstream and downstream cylinders synchronized in phase and frequency; and (v) uncoupled vortex streets take place behind both cylinders. Based on the experimental measurements on the flow interactions for tandem cylinders, Ljungkrona et al. (1991) concluded that there was no significant vortex action between the cylinders when the non-dimensional distance is less than 3.4–3.8 diameters. The flow feature in the inter-cylindrical region was also investigated by Mahir & Rockwell (1996), who found an absence of vortex shedding for small inter-cylinder distance. Moreover, some other researchers (e.g. Fitzpatrick 2003; King & Pfizenmaier 2009) studied the influence of inter-cylindrical distance and rod diameter on the Strouhal numbers and sound pressure levels (SPLs). Recently, Hutcheson & Brooks (2012) have performed extensive acoustic measurements on multiple rod configurations to study the effect of Reynolds number, free stream turbulence and wake interference on the radiated noise.

The rod–airfoil configuration is another benchmark model for the study of body-wake interactions. A rod in this configuration is embedded upstream of the airfoil, so the turbulent wake is formed and convects downstream, which then impinges onto the airfoil and partly splits at the leading edge. Numerous studies have been performed to examine the flow characteristics for the rod–airfoil configuration including flow patterns altering, noise radiation suppressing and the fluid resonant oscillations due to the interval between the cylinder and the airfoil varying or the attack angle of the airfoil changing. Few studies however appear to have investigated the forces and turbulent fluctuations, moreover, these studies (e.g. Boudet et al. 2005; Caraeni, Dai & Caraeni 2007) examined the two features for the rod–airfoil configuration at a fixed interval. Munekata et al. (2006) performed experimental measurements on the rod–airfoil configuration to investigate the effects of the interval between the cylinder and the airfoil on the characteristics of the aerodynamic sound. They found that vortex shedding from the cylinder upstream is suppressed for a short enough interval, and that simultaneously, the interaction between the turbulent wake from the cylinder upstream and the airfoil downstream is weakened and the level of noise radiation due to the interaction is correspondingly decreased. They observed the fluids resonant oscillations when the interval is varied. Munekata et al. (2008) further investigated the effects of the attack angle of the airfoil located downstream on the characteristics of the aerodynamic sound and wake structure at a given interval between the cylinder and the airfoil. It was found that the SPL decreases with increasing attack angle of the airfoil because of the diffusive wake structure caused by the blocking effect of the airfoil. Jacob et al. (2005) conducted a measurement of the interaction flow field of the rod/NACA0012 airfoil configuration, and also obtained the far field noise spectra which showed that the body-wake interaction is mainly responsible for the radiation of noise. Their experimental results provide a detailed database for the validation of the numerical simulation. More recently, Li et al. (2014) investigated experimentally the body-wake interaction noise radiated from the rod–airfoil configuration. They mainly focused on the noise control concept and found that the noise radiated is suppressed by two control methods including 'air blowing' on the upstream rod and a soft-vane leading edge on the airfoil.

The rod–airfoil case has also been investigated widely by numerical approaches, such as the Reynolds-averaged Navier–Stokes method (RANS) (e.g. Casalino, Jacob & Roger 2003; Jacob et al. 2005), the large eddy simulation method (LES) (e.g. Casalino et al. 2003; Magagnato, Sorgüven & Gabi 2003; Boudet et al. 2005;

Jacob et al. 2005; Greschner et al. 2008; Agrawal & Sharma 2014; Giret et al. 2015) and the detached eddy simulation method (DES) (e.g. Creschner et al. 2004; Caraeni et al. 2007; Gerolymos & Vallet 2007; Greschner et al. 2008). However, reliable results are not obtained using RANS in these papers. Jiang et al. (2011) investigated interaction phenomena for a rod–airfoil configuration with different cylinder positions and attack angles of the airfoil using an experimental method and RANS. In their study, significant differences became apparent when the numerical noise spectra were compared to the experimental data. This behaviour may be explained by the fact that it is difficult for RANS to resolve the strong unsteady phenomena resulting from the impingement of the turbulent wake and the airfoil, which is mainly responsible for the noise radiation. Daude et al. (2012) performed a LES on the prediction of noise radiation from a rod–airfoil configuration, and observed a good agreement between the numerical and experimental results.

In this paper, a LES technique is utilized to simulate the flow over a rod–airfoil configuration and to investigate the body-wake interactions within the flow field. The purpose is to achieve an improved understanding of some of the fundamental phenomena associated with this flow, including force varying regulation, flow structures and flow patterns in the interaction region, turbulent fluctuations and their suppression and noise radiation. Special attention is given to the fluid resonant oscillation caused by the varying interval between the cylinder upstream and airfoil downstream.

This paper is organized as follows. The mathematical formulation and numerical methods are presented in §2. The computational overview and validation are described in §3. Detailed results got body-wake flow interaction over a rod–airfoil configuration are then given in §4 and the concluding remarks are given in §5.

2. Mathematical formulation and numerical methods

2.1. *Governing equations*

The governing equations are the three-dimensional compressible Navier–Stokes equations in computational coordinates, these equations may be written

$$\frac{\partial \widetilde{U}}{\partial t} + \frac{\partial \widetilde{E}}{\partial \xi} + \frac{\partial \widetilde{F}}{\partial \eta} + \frac{\partial \widetilde{G}}{\partial \zeta} = \frac{1}{Re}\left(\frac{\partial \widetilde{E}_v}{\partial \xi} + \frac{\partial \widetilde{F}_v}{\partial \eta} + \frac{\partial \widetilde{G}_v}{\partial \zeta}\right), \tag{2.1}$$

where,

$$\left.\begin{array}{ll} \widetilde{U} = U/J, & \\ \widetilde{E} = (\xi_t U + \xi_x E + \xi_y F + \xi_z G)/J, & \widetilde{E}_v = (\xi_x E_v + \xi_y F_v + \xi_z G_v)/J, \\ \widetilde{F} = (\eta_t U + \eta_x E + \eta_y F + \eta_z G)/J, & \widetilde{F}_v = (\eta_x E_v + \eta_y F_v + \eta_z G_v)/J, \\ \widetilde{G} = (\zeta_t U + \zeta_x E + \zeta_y F + \zeta_z G)/J, & \widetilde{G}_v = (\zeta_x E_v + \zeta_y F_v + \zeta_z G_v)/J. \end{array}\right\} \tag{2.2}$$

The details of the governing equations (2.1) are given in Jiang et al. (2014a), J is the Jacobian of the grid transformation, $\xi_t, \xi_x, \xi_y, \xi_z, \eta_t, \eta_x, \eta_y, \eta_z, \zeta_t, \zeta_x, \zeta_y$ and ζ_z are grid derivatives. The grid metric derivatives have a conservative form of

$$\begin{aligned}
\tilde{\xi}_x &= \xi_x/J = (y_\eta z)_\zeta - (y_\zeta z)_\eta, & \tilde{\xi}_y &= \xi_y/J = (z_\eta x)_\zeta - (z_\zeta x)_\eta, \\
\tilde{\xi}_z &= \xi_z/J = (x_\eta y)_\zeta - (x_\zeta y)_\eta, & & \\
\tilde{\eta}_x &= \eta_x/J = (y_\zeta z)_\xi - (y_\xi z)_\zeta, & \tilde{\eta}_y &= \eta_y/J = (z_\zeta x)_\xi - (z_\xi x)_\zeta, \\
\tilde{\eta}_z &= \eta_z/J = (x_\zeta y)_\xi - (x_\xi y)_\zeta, & & \\
\tilde{\zeta}_x &= \zeta_x/J = (y_\xi z)_\eta - (y_\eta z)_\xi, & \tilde{\zeta}_y &= \zeta_y/J = (z_\xi x)_\eta - (z_\eta x)_\xi, \\
\tilde{\zeta}_z &= \zeta_z/J = (x_\xi y)_\eta - (x_\eta y)_\xi. & &
\end{aligned} \quad (2.3)$$

Large eddy simulation is employed in the present study for turbulence closure. In a standard compressible LES, the governing equations are filtered using a grid-filtering function and Favre-averaged variables are introduced, the small-scale structures are left unresolved and are accounted for by a subgrid scale (SGS) turbulence model (Xu, Chen & Lu 2010). In this work, however, an alternative approach is employed where the truncation error of the high-order discretization itself is used to model the effects of the unresolved scales and is referred to as a HILES (Jiang *et al.* 2014*a*). The concept of this approach is the same as that of the monotone integrated LES (MILES) proposed by Boris *et al.* (1992), and a new seventh-order hybrid cell-edge and cell-node dissipative compact scheme (HDCS-E8T7) is used (Deng *et al.* 2013*b*) for spatial discretization in the HILES.

Numerous high-order schemes have been adopted for implicit LES, e.g. the high-order monotone upstream-centered schemes for conservation laws (MUSCL) (Van Leer 1977) employed by Thornber & Drikakis (2008) and Hahn *et al.* (2011), the high-order weighted essentially non-oscillatory schemes (WENO) (Jiang & Shu 1996) employed by Drikakis *et al.* (2009) and the cell-node type central compact scheme (CCSN) (Lele 1992) employed by Visbal & Rizzetta (2002). Compared with the MUSCL and WENO schemes, the present HDCS-E8T7 is a linear scheme without nonlinear mechanics capturing the discontinuities, and is expected to have a higher resolution for the same order of accuracy. Moreover, the HDCS-E8T7 is a dissipative scheme with inherent dissipation and the filter operations for the CCSN are not needed.

2.2. *Numerical procedure*

The temporal integration for solving the governing equations (2.1) is performed using a dual time stepping approach (John & Jameson 2002) with Newton-like subiterations (Gordnier & Visbal 1993). The convective and viscous terms are discretized by the HDCS-E8T7, which employs the concept of the dissipative compact scheme (DCS) (Deng, Maekawa & Shen 1996) for simulating subsonic flow on a complex geometry. Although some aeroacoustic benchmark problems have been simulated successfully by the DCS, applications of this scheme on complex grids may pose serious problems (Deng *et al.* 2011). The HDCS-E8T7 has demonstrated a promising ability in solving complex flow problems because the surface conservation law (SCL) is satisfied, which will be further discussed in § 2.3. For the convenience of understanding our high-order strategy in the present study, details of the HDCS-E8T7 will be given in the following.

Considering discretization of the convective terms,

$$\frac{\partial \widetilde{U}}{\partial t} + \frac{\partial \widetilde{E}}{\partial \xi} + \frac{\partial \widetilde{F}}{\partial \eta} + \frac{\partial \widetilde{G}}{\partial \zeta} = 0, \quad (2.4)$$

and their semi-discrete approximation,

$$\frac{\partial \widetilde{U}}{\partial t} = -\delta_1^\xi \widetilde{E} - \delta_1^\eta \widetilde{F} - \delta_1^\zeta \widetilde{G}. \tag{2.5}$$

The discretization δ_1^ξ, δ_1^η and δ_1^ζ are the same, thus we only give the discretization in the ξ direction. The δ_1^ξ of the HDCS-E8T7 is

$$\delta_1^\xi \widetilde{E}_j = \frac{256}{175h}(\hat{E}_{j+1/2} - \hat{E}_{j-1/2}) - \frac{1}{4h}(\widetilde{E}_{j+1} - \widetilde{E}_{j-1}) \\ + \frac{1}{100h}(\widetilde{E}_{j+2} - \widetilde{E}_{j-2}) - \frac{1}{2100h}(\widetilde{E}_{j+3} - \widetilde{E}_{j-3}), \tag{2.6}$$

where, $\hat{E}_{j\pm 1/2} = \widetilde{E}(\hat{U}_{j\pm 1/2}, \hat{\xi}_{x,j\pm 1/2}, \hat{\xi}_{y,j\pm 1/2}, \hat{\xi}_{z,j\pm 1/2})$ and $\widetilde{E}_{j+m} = \widetilde{E}(\hat{U}_{j+m}, \hat{\xi}_{x,j+m}, \hat{\xi}_{y,j+m}, \hat{\xi}_{z,j+m})$ are the fluxes at the cell edges and the cell nodes, respectively. The numerical flux $\hat{E}_{j\pm 1/2}$ is evaluated by the variables at the cell edges,

$$\hat{E}_{j\pm 1/2} = \widetilde{E}\left(\hat{U}_{j\pm 1/2}^L, \hat{U}_{j\pm 1/2}^R, \hat{\xi}_{x,j\pm 1/2}, \hat{\xi}_{y,j\pm 1/2}, \hat{\xi}_{z,j\pm 1/2}\right), \tag{2.7}$$

where, $\hat{U}_{j\pm 1/2}^L$, $\hat{U}_{j\pm 1/2}^R$ are variables at the cell edge.

$$\frac{5}{14}(1-\alpha)\hat{U}_{j-1/2}^L + \hat{U}_{j+1/2}^L + \frac{5}{14}(1+\alpha)\hat{U}_{j+3/2}^L \\ = \frac{25}{32}(U_{j+1} + U_j) + \frac{5}{64}(U_{j+2} + U_{j-1}) - \frac{1}{448}(U_{j+3} + U_{j-2}) \\ + \alpha\left[\frac{25}{64}(U_{j+1} - U_j) + \frac{15}{128}(U_{j+2} - U_{j-1}) - \frac{5}{896}(U_{j+3} - U_{j-2})\right], \tag{2.8}$$

where $\alpha < 0$ is the dissipative parameter to control dissipation in the HDCS-E8T7. The corresponding $\hat{U}_{j+1/2}^R$ can be obtained easily by setting $\alpha > 0$. In figure 1, the modified wavenumber ω^* of the HDCS-E8T7 with different dissipative parameters is compared with that of the eighth-order central compact scheme (CCSN-8) proposed by Lele (1992). According to the discussion given by Lele (1992), the real part of ω^*, namely, ω_r^* denotes the resolution power of a scheme, and the imaginary part of ω^*, namely, ω_i^* denotes the dissipation intensity of a scheme. It can be seen that the dissipative parameter α has effect on the resolution power of the HDCS-E8T7, but it has no influence on the order of the truncation error of the HDCS-E8T7 according to Taylor series expansion (Deng et al. 2015). If a proper dissipative parameter is chosen, the resolution power of the HDCS-E8T7 can be higher than that of the CCSN-8. In order to obtain fine resolution power, the HDCS-E8T7 has been optimized by following the concept of dispersion-relation-preserving (DRP) (Tam & Webb 1993) and adjusting the value of α. The optimized dissipative parameter is equal to 0.3 (Deng et al. 2015), which will be adopted in this paper.

Seven-point stencil is used by the HDCS-E8T7, thus three levels of boundary and near-boundary schemes are required. The details of the boundary and near boundary schemes are given by Jiang et al. (2014a). For computation of the viscous terms, the primitive variables, u, v, w, T, are first differentiated to form the components of the stress tensor and the heat flux vector. The viscous flux derivatives are then computed by a second application of the HDCS-E8T7.

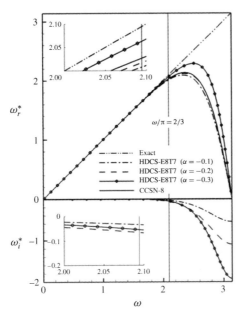

FIGURE 1. Modified wavenumber of the HDCS-E8T7.

2.3. *Calculation of the grid metric derivatives*

According to the study of Deng *et al.* (2011), the SCL is always satisfied on uniform meshes, however this may not be so on curvilinear meshes. If the SCL has not been satisfied, numerical instabilities and even computing collapse may occur on complex curvilinear grids during numerical simulation. To aid in ensuring that the SCL is satisfied, the CCSNs have been successfully applied to various flow simulations by Visbal & Gaitonde (2002) and Rizzetta, Visbal & Morgan (2008) on complex grids. In this section, we will discuss the calculation of the grid metric derivatives according to the principle of satisfying the SCL on curvilinear meshes. The grid metric derivatives of governing equations (2.1) have a conservative form of

$$\left.\begin{array}{l}\tilde{\xi}_x = \delta_{\text{II}}^{\zeta}((\delta_{\text{III}}^{\eta}y)z) - \delta_{\text{II}}^{\eta}((\delta_{\text{III}}^{\zeta}y)z), \quad \tilde{\xi}_y = \delta_{\text{II}}^{\eta}((\delta_{\text{III}}^{\zeta}z)x) - \delta_{\text{II}}^{\zeta}((\delta_{\text{III}}^{\eta}z)x), \\ \tilde{\xi}_z = \delta_{\text{II}}^{\zeta}((\delta_{\text{III}}^{\eta}x)y) - \delta_{\text{II}}^{\eta}((\delta_{\text{III}}^{\zeta}x)y), \\ \tilde{\eta}_x = \delta_{\text{II}}^{\xi}((\delta_{\text{III}}^{\zeta}y)z) - \delta_{\text{II}}^{\zeta}((\delta_{\text{III}}^{\xi}y)z), \quad \tilde{\eta}_y = \delta_{\text{II}}^{\zeta}((\delta_{\text{III}}^{\xi}z)x) - \delta_{\text{II}}^{\xi}((\delta_{\text{III}}^{\zeta}z)x), \\ \tilde{\eta}_z = \delta_{\text{II}}^{\xi}((\delta_{\text{III}}^{\zeta}x)y) - \delta_{\text{II}}^{\zeta}((\delta_{\text{III}}^{\xi}x)y), \\ \tilde{\zeta}_x = \delta_{\text{II}}^{\eta}((\delta_{\text{III}}^{\xi}y)z) - \delta_{\text{II}}^{\xi}((\delta_{\text{III}}^{\eta}y)z), \quad \tilde{\zeta}_y = \delta_{\text{II}}^{\xi}((\delta_{\text{III}}^{\eta}z)x) - \delta_{\text{II}}^{\eta}((\delta_{\text{III}}^{\xi}z)x), \\ \tilde{\zeta}_z = \delta_{\text{II}}^{\eta}((\delta_{\text{III}}^{\xi}x)y) - \delta_{\text{II}}^{\xi}((\delta_{\text{III}}^{\eta}x)y),\end{array}\right\} \quad (2.9)$$

where $\delta_{\text{II}}^{\xi}, \delta_{\text{II}}^{\eta}, \delta_{\text{II}}^{\zeta}$ and $\delta_{\text{III}}^{\xi}, \delta_{\text{III}}^{\eta}, \delta_{\text{III}}^{\zeta}$ are numerical derivative operators used for the metric calculations in the ξ, η and ζ coordinate directions, respectively. For applications on curvilinear meshes, a finite difference scheme should satisfy the SCL which means

$I_x = I_y = I_z = 0$ (Deng *et al.* 2011),

$$\left.\begin{aligned}I_x &= \delta_I^\xi(\tilde{\xi}_x) + \delta_I^\eta(\tilde{\eta}_x) + \delta_I^\zeta(\tilde{\zeta}_x), \\ I_y &= \delta_I^\xi(\tilde{\xi}_y) + \delta_I^\eta(\tilde{\eta}_y) + \delta_I^\zeta(\tilde{\zeta}_y), \\ I_z &= \delta_I^\xi(\tilde{\xi}_z) + \delta_I^\eta(\tilde{\eta}_z) + \delta_I^\zeta(\tilde{\zeta}_z).\end{aligned}\right\} \qquad (2.10)$$

In order to fulfill the SCL, the grid metrics should be calculated with a conservative form by the same schemes used for flux derivative calculations, i.e. $\delta_I = \delta_{II}$, to implement the conservative metric method (CMM) (Deng *et al.* 2011). It has been proved that the CMM can be easily applied in high-order schemes if the difference operator δ_I of flux derivatives is not split, while it is difficult to apply in schemes where the δ_I is split into two upwind operators as δ_I^+ and δ_I^-. Although the inner-level difference operators δ_{III} in the conservative metrics have no effect on the SCL, the constraint $\delta_{III} = \delta_{II}$ is recommended by Deng *et al.* (2011). More recently, the constraint $\delta_{III} = \delta_{II}$ has been explained from a geometry viewpoint using a symmetrical conservative metric method (SCMM) (Deng *et al.* 2013a), which can evidently increase the numerical accuracy on irregular grids. The SCMM needs symmetrical conservative metrics, which can be written as follows,

$$\left.\begin{aligned}\tilde{\xi}_x &= \tfrac{1}{2}\left(\delta_{II}^\zeta((\delta_{III}^\eta y)z) - \delta_{II}^\eta((\delta_{III}^\zeta y)z) + \delta_{II}^\eta((\delta_{III}^\zeta z)y) - \delta_{II}^\zeta((\delta_{III}^\eta z)y)\right), \\ \tilde{\xi}_y &= \tfrac{1}{2}\left(\delta_{II}^\zeta((\delta_{III}^\eta z)x) - \delta_{II}^\eta((\delta_{III}^\zeta z)x) + \delta_{II}^\eta((\delta_{III}^\zeta x)z) - \delta_{II}^\zeta((\delta_{III}^\eta x)z)\right), \\ \tilde{\xi}_z &= \tfrac{1}{2}\left(\delta_{II}^\zeta((\delta_{III}^\eta x)y) - \delta_{II}^\eta((\delta_{III}^\zeta x)y) + \delta_{II}^\eta((\delta_{III}^\zeta y)x) - \delta_{II}^\zeta((\delta_{III}^\eta y)x)\right), \\ \tilde{\eta}_x &= \tfrac{1}{2}\left(\delta_{II}^\xi((\delta_{III}^\zeta y)z) - \delta_{II}^\zeta((\delta_{III}^\xi y)z) + \delta_{II}^\zeta((\delta_{III}^\xi z)y) - \delta_{II}^\xi((\delta_{III}^\zeta z)y)\right), \\ \tilde{\eta}_y &= \tfrac{1}{2}\left(\delta_{II}^\xi((\delta_{III}^\zeta z)x) - \delta_{II}^\zeta((\delta_{III}^\xi z)x) + \delta_{II}^\zeta((\delta_{III}^\xi x)z) - \delta_{II}^\xi((\delta_{III}^\zeta x)z)\right), \\ \tilde{\eta}_z &= \tfrac{1}{2}\left(\delta_{II}^\xi((\delta_{III}^\zeta x)y) - \delta_{II}^\zeta((\delta_{III}^\xi x)y) + \delta_{II}^\zeta((\delta_{III}^\xi y)x) - \delta_{II}^\xi((\delta_{III}^\zeta y)x)\right), \\ \tilde{\zeta}_x &= \tfrac{1}{2}\left(\delta_{II}^\eta((\delta_{III}^\xi y)z) - \delta_{II}^\xi((\delta_{III}^\eta y)z) + \delta_{II}^\xi((\delta_{III}^\eta z)y) - \delta_{II}^\eta((\delta_{III}^\xi z)y)\right), \\ \tilde{\zeta}_y &= \tfrac{1}{2}\left(\delta_{II}^\eta((\delta_{III}^\xi z)x) - \delta_{II}^\xi((\delta_{III}^\eta z)x) + \delta_{II}^\xi((\delta_{III}^\eta x)z) - \delta_{II}^\eta((\delta_{III}^\xi x)z)\right), \\ \tilde{\zeta}_z &= \tfrac{1}{2}\left(\delta_{II}^\eta((\delta_{III}^\xi x)y) - \delta_{II}^\xi((\delta_{III}^\eta x)y) + \delta_{II}^\xi((\delta_{III}^\eta y)x) - \delta_{II}^\eta((\delta_{III}^\xi y)x)\right).\end{aligned}\right\} \quad (2.11)$$

The randomized grids and a grid around wingtip are used by Deng *et al.* (2013a) to show the improvement of the SCMM over the CMM. These grids are representative of the configuration considered in this study, thus the SCMM is a better choice for the present investigation. To eliminate the SCL errors on a curvilinear mesh, here we calculate the grid metrics with the symmetrical conservative form (2.11) by the same schemes as used for flux derivative calculations, i.e. $\delta_I = \delta_{II} = \delta_{III}$, to implement the SCMM (Deng *et al.* 2013a).

3. Computational overview and validation

3.1. *Computational overview*

We consider the rod–airfoil configuration to numerically investigate the interaction between bodies and the turbulent wake. An experiment for this configuration was

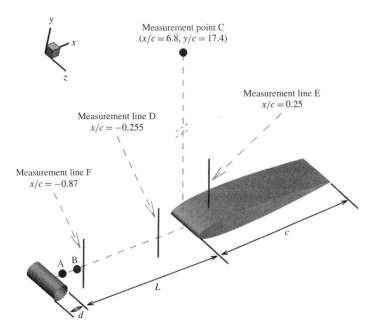

FIGURE 2. (Colour online) Schematic diagram of rod–airfoil configuration for the numerical simulation and experimental measurement. Here, the solid dots A and B denote the pressure probe locations in the present simulation, the point C is the far-field acoustic monitoring location in the experiment (Jacob et al. 2005), lines D, E and F represent the locations of the particle image velocimetry (PIV) measurement (Jacob et al. 2005).

carried out in the large anechoic wind tunnel at Ècole Centrale de Lyon (ECL) (Jacob et al. 2005). A sketch of the configuration is shown in figure 2. A NACA0012 airfoil of chord $c = 0.1$ m is located one chord downstream of a cylinder (diameter: $d = 0.01$ m). This configuration is placed in a uniform airflow with the incoming flow conditions $U_\infty = 72$ m s^{-1}, $T_\infty = 293$ K, and $\rho_\infty = 1.2$ kg m^{-3}. The Reynolds number based on the rod diameter and the airfoil chord are $Re_d = 4.8 \times 10^4$ and $Re_c = 4.8 \times 10^5$, respectively. For comparison, a sketch of locations in the measurement (Jacob et al. 2005) is also shown in figure 2.

In order to investigate the interaction phenomena in the rod–airfoil configuration, the simulations are performed with the changing interval L between the cylinder and the airfoil. The incoming flow conditions for the experiment of Jacob et al. (2005) are adopted for the present simulations. Five cases with $L/d = 2, 4, 6, 8$ and 10 are performed to reveal the body-wake interaction phenomena within the rod–airfoil configuration. The simulation for a single cylinder is also carried out under the same incoming flow conditions as the rod–airfoil configuration to present the interaction phenomena.

In this rod–airfoil configuration, the rod wake that impinges the airfoil contains both periodic and broadband turbulent disturbances. For the simulation of turbulent flow, which is largely responsibile for the noise radiated by this rod–airfoil configuration, the HILES (Jiang et al. 2014a) based on the HDCS-E8T7 scheme is employed. The far-field noise is calculated by applying the Ffowcs-Williams and Hawking (FW-H)

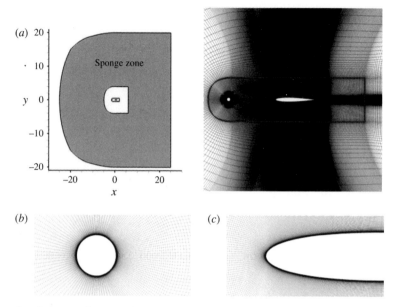

FIGURE 3. (Colour online) Sectional view of the computational mesh in the central plane. (a) Mesh global view and location of the FW-H surface (red solid line). (b) Zoom-in on the rod. (c) Zoom-in on the airfoil leading edge.

aeroacoustic analogy (Lyrintzis 2003) to the FW-H integration surface, and no volume integration is performed.

The grid topology is carefully designed to ensure computation accuracy. For the rod–airfoil configuration with $L/d = 10$, figure 3 outlines the computational domain, which contains approximately 16 000 000 grid points. The meshes used for the simulation of the five cases have a similar topology, though the grid points are slightly different from each other. Considering the case of $L/d = 10$, grid stretching is employed to increase the grid resolution near the body surface and in the interference region, ensuring that there are 481 nodes located in the streamwise direction between the rod upstream and the airfoil downstream. The minimum size of the grid in the wall-normal direction is 1.0×10^{-5} (normalized by the reference length $L_{ref} = 0.1$ m). The rod and the airfoil surfaces are meshed with 241 and 1081 circumferential points respectively, at each spanwise location, and with 45 points along the span. The spanwise grid is divided uniformly, and the spanwise length is chosen as $3d$. On the airfoil, the grid density is characterized by: $\Delta x^+ < 90$ (<20 for $x/c < 0.1$, i.e. in the leading edge region), $\Delta y^+ = 0.24$ and $\Delta z^+ = 161.8$, where Δx^+, Δy^+ and Δz^+ are the near wall mesh spacing tangent to wall, normal to the wall and spanwise, in wall units, respectively. Wall units are defined in terms of the time-averaged friction velocity at the middle of airfoil upper surface. The expansion ratio is set to 1.05 for the first 121 mesh layers from the rod and airfoil surfaces. The mesh resolution on the FW-H surface can be described as $\Delta x/L_{ref} < 0.03$ and $\Delta y/L_{ref} < 0.02$. Outside the FW-H surface, the mesh is then slightly coarsened using a stretching ratio of 1.1 until the sponge zone is met, where the stretching ratio is 1.02.

To justify the choice of the spanwise length, the two-point correlations are calculated in terms of the formulation (Pirozzoli, Grasso & Gatski 2004). Figure 4

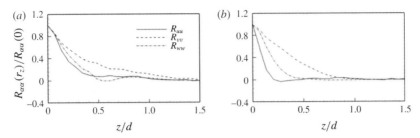

FIGURE 4. Distributions of the two-point correlations of the velocity components in the spanwise direction for the case with $L/d = 10$ at $y/c = 0$ and (a) $x/c = -0.96$ and (b) $x/c = -0.87$.

shows the two-point correlations $R_{\alpha\alpha}(r_z)$ for the case of $L/d = 10$ in the spanwise direction, i.e. z direction, where α represents the fluctuations of any one of the velocity components u_i (or u, v and w). The correlations decay towards zero which means that the two-point correlations are sufficiently decorrelated over a distance $1.5d$. This ensures that the spanwise computational domain is sufficiently wide not to inhibit the turbulence dynamics.

At the outer boundary, far-field boundary conditions based on the LODI approximation (Poinsot & Lele 1992) and sponge technique (Daniel 2006) are adopted. The no-slip condition is invoked on the rod and airfoil surfaces, together with fifth-order accurate approximations for an adiabatic wall and zero normal pressure gradient. At the spanwise boundaries, periodicity condition is applied. The non-dimensional time step for the present simulations is $\Delta t = 0.001c/U_\infty$ with U_∞ being the incoming velocity, corresponding to a physical time of $\Delta t = 1.389 \times 10^{-6}$ s. The time to eliminate initial transients is chosen as $100d/U_\infty$ and is very similar to that commonly used in simulations of the flow past a straight circular cylinder (Gallardo, Andersson & Pettersen 2014). The flow statistics are collected for 20 000 time steps, corresponding either to a physical time of $T = 0.02778$ s or to 20 flow-over times from the leading edge to the trailing edge of the airfoil. The sampling time is chosen to obtain statistically meaningful turbulence properties in a temporal averaging operation. We took samples every time step, and the sampling frequencies in the present simulations are 720 kHz. For periodic boundaries in the spanwise directions, flow statistics are averaged in the z direction.

3.2. Validation

To validate the present simulation, we compare the numerical results with the experimental measurements (Jacob et al. 2005). Figure 5 shows the mean velocity normalized by the incoming velocity at two locations $x/c = -0.255$ and $x/c = 0.25$. At the location $x/c = -0.255$, the mean streamwise velocity $x/c = -0.87$ near the center-line is underpredicted (the maximal relative error is 7.5%), which is consistent with the overprediction (the maximal relative error is 9.7%) of the mean streamwise velocity near the wall at location $x/c = 0.25$, since the flow predicted moves more slowly between the rod and airfoil. A similar phenomenon is observed from the LES prediction of Boudet et al. (2005). Moreover, as shown in figure 6, the predicted profiles of the root-mean-square (r.m.s.) value of streamwise fluctuating velocity at the two locations $x/c = -0.255$ and $x/c = 0.25$ have good agreement with those from

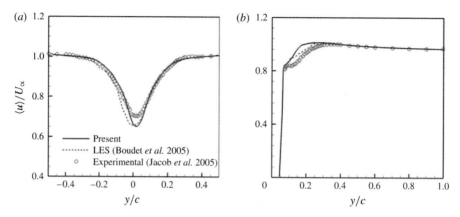

FIGURE 5. (Colour online) Spanwise averaged mean streamwise velocity distributions at locations $x/c = -0.255$ (a) and $x/c = 0.25$ (b).

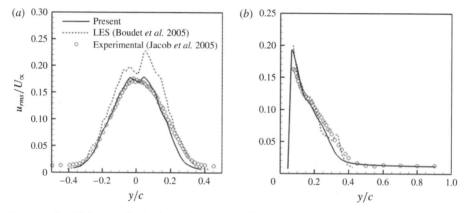

FIGURE 6. (Colour online) Spanwise averaged r.m.s. value of the fluctuating velocity distributions at locations $x/c = -0.255$ (a) and $x/c = 0.25$ (b).

the experimental data. This seems to indicate that the turbulence intensities are well predicted.

For the location closer to the cylinder, i.e. $x/c = -0.87$, figure 7 the mean streamwise velocity and r.m.s. value of streamwise fluctuating velocity compared with the corresponding data from the experimental measurements (Jacob et al. 2005) and the LES prediction of Agrawal & Sharma (2014). The present simulation provides similar mean streamwise velocities in comparison with the LES results of Agrawal & Sharma (2014). Although both the experimental and numerical results exhibit the wake velocity deficit, i.e. the mean streamwise velocity is at a minimum on the center-line, a very large discrepancy in the mean velocity is observed. Agrawal & Sharma (2014) are suspicious of the experimental mean velocity data at $x/c = -0.87$ because the velocity deficit in the wake is expected to reduce with distance away from the rod (Agrawal & Sharma 2014; Giret et al. 2015). This behaviour can be identified from the present simulation by comparing the mean streamwise velocity

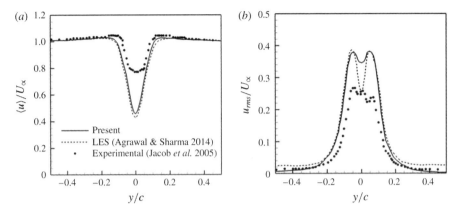

FIGURE 7. Spanwise averaged distributions at locations $x/c = -0.87$ for mean streamwise velocity (a) and r.m.s. value of the fluctuating velocity (b).

at $x/c = -0.255$ (see figure 5) with at $x/c = -0.87$ (see figure 7), however, the experimental data show the velocity deficit increasing with downstream distance. Compared with the experimental data, the r.m.s. value of streamwise fluctuating velocity is overpredicted. Yet, the two numerical solutions are close to each other, with relatively obvious differences near the center-line, indicating that a relatively higher level of streamwise fluctuating velocity is predicted by the present simulation near the center-line at $x/c = -0.87$. Since different methods and meshes are adopted, a dispersion of the numerical fluctuating velocity can be observed near the center-line (Giret et al. 2015).

Other evidence demonstrating the reliability of the present simulation is the resolved energy spectrum behind the cylinder, which is given in figure 8. The resolved scales appear to reach an inertial subrange reasonably close to a $St^{-5/3}$ scaling (Xu et al. 2010). St is the characteristic Strouhal number defined as $St = fd/U_\infty$ with f being the frequency. The illustrated slope indicates that the turbulence spectrum is captured reliably. The capability of the present numerical method for resolving energy spectra has been demonstrated in the previous study (Jiang et al. 2014a).

Figure 9 shows the SPL spectra at the location ($x = 0.68$ m, $y = 1.74$ m) calculated from the pressure fluctuations on the FW-H integration surface. The peak frequency is predicted quite well. Furthermore, the broadband spectrum is also fairly well described around the peak frequency. However, the spectrum at high frequencies is overpredicted, which is similar to LES results of Boudet et al. (2005). A possible explanation given by Boudet et al. (2005) for this phenomenon is that the dipole cancellation of some quadrupole terms may be momentarily ineffective, which could lead to more efficient sources. Nevertheless, a good broadband sound is predicted. It can be seen that the peak frequency in figure 9 is almost identical to the primary shedding frequency of the cylinder, which will be calculated from the power spectral densities (PSD) of the time-dependent lift coefficient in figure 13. This behaviour shows that the shedding vortex from the cylinder plays a dominant role in the flow over the rod–airfoil configuration.

The present high-order numerical strategy has been applied successfully to simulate a wide range of turbulent cases such as transition and turbulence decay in the Taylor–Green vortex (Jiang et al. 2014a), channel flow (Jiang et al. 2015), subsonic flow over a circular cylinder (Jiang et al. 2013) and a three-dimensional delta wing

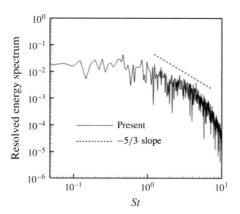

FIGURE 8. Resolved energy spectrum behind the cylinder at the location $x = -6.5d$, $y = 0$, $z = 1.5d$.

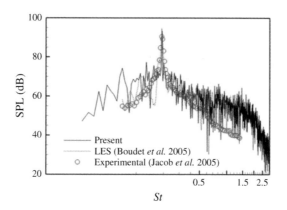

FIGURE 9. (Colour online) The SPL spectra at the location ($x = 0.68$ m, $y = 1.74$ m) calculated from the pressure perturbations on the FW-H integration surface.

(Deng *et al.* 2013*b*), stall characteristic of flow past a thin airfoil (Jiang *et al.* 2014*a*) and noise radiation from a jet nozzle (Jiang *et al.* 2014*b*; Mao *et al.* 2016). We have carefully examined the numerical strategy used in this study and have verified that the numerical solutions are reliable.

4. Results and discussion

4.1. *Force behaviours*

4.1.1. *Surface pressure and friction*

The behaviour of the forces exerted on the bodies, which are contributed by the pressure and viscous shear stress, is an important issue associated with the interaction between the body and turbulent wake. An illustration of the force behaviour of the rod–airfoil configuration is given by the distribution of the mean pressure on the surface of the cylinder and airfoil, shown in figure 10(*a*,*b*), respectively. As shown in figure 10(*a*), the airfoil behind the cylinder leads to the pressure increase in the base

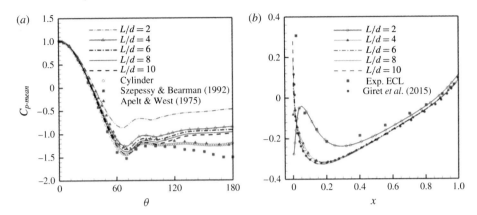

FIGURE 10. (Colour online) The calculated mean pressure distributions compared with the experimental data and LES results: (a) on the surface of the cylinder, (b) on the surface of the airfoil. Here, the experimental data from Apelt & West (1975) and Szepessy & Bearman (1992) are measured for the circular cylinder with $Re_d = 4 \times 10^4$; the experimental data from ECL (Jacob et al. 2005) and the LES results of Giret et al. (2015) are obtained for the rod–airfoil configuration with $L/d = 10$.

range of the cylinder. When the distance between the airfoil and the cylinder continues to increase, the influence of the airfoil is weakened correspondingly on the surface pressure distributions of the cylinder, and gradually converges to the distribution on the single cylinder. Moreover, in figure 10(a), the present mean pressure distributions on the surface of the cylinder are compared with the experimental results of Apelt & West (1975), and those of Szepessy & Bearman (1992) to complement the verification of the present simulation. The present distributions from the simulation of the single cylinder demonstrate that the angular position (70°) for the mean pressure minimum is very close to that of the experimental measurement of Szepessy & Bearman (1992). Although the present mean pressure distributions in the base region provide an unsatisfactory comparison with the experimental data, a significant dispersion of the experimental results is observed.

As shown in figure 10, the pressure distributions with $L/d = 2$ are obviously different from the others. This behaviour is related to the flow phenomena of the separated shear layers from the upstream cylinder impinging onto the leading edge of the airfoil. The main separated vortices impact the leading edge of the airfoil directly, and von Kármán vortex shedding is suppressed, which will be discussed in detail in §4.2. The increasing L, followed by the weakening interaction between the turbulent wake and airfoil, is also consistent with the pressure distributions on the airfoil surface tending to be identical. In figure 10(b), the present mean pressure distributions on the airfoil surface are also compared with the experimental data from ECL (Jacob et al. 2005) and the LES results of Giret et al. (2015). It can be seen that the calculated solutions for the rod–airfoil configuration with $L/d = 10$ are very close to the LES results of Giret et al. (2015), however, a large gap between the numerical results displayed and the experimental data can be seen. According to the discussion of Giret et al. (2015), the experimental dip is most likely caused by a curvature discontinuity in the mockup leading-edge region.

Continuing the investigation of the skin friction, figure 11(a) shows the present skin friction distributions on the cylinder, together with the experimental data of

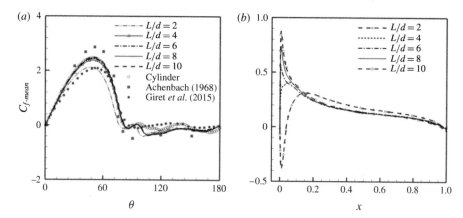

FIGURE 11. (Colour online) Mean skin friction distributions on the surface of bodies: (a) cylinder and (b) airfoil. Here, the calculated distributions on the surface of the cylinder are compared with the experimental data of Achenbach (1968) for the circular cylinder with $Re_d = 10^5$ and the LES results of Giret et al. (2015) for the rod–airfoil configuration with $L/d = 10$; moreover, the mean skin friction in this figure is non-dimensional according to the formulation given by Achenbach (1968).

Achenbach (1968) and the LES results of Giret et al. (2015). It shows that the skin friction increases continuously from the stagnation point until the maximum value is obtained at the location near $\theta = 45°$, this trend then reverses, and it finally achieves a zero or negligible negative value near $\theta = 80°$ where the flow is separated. The positive skin friction occurs obviously in the attached flow region and drops quickly at the mean separation location. Moreover, the absolute value of the skin friction observed in the attached flow region is usually much higher than that in the separated region. Although the calculated skin friction maximum of the single cylinder is underpredicted, with a relative error of 11 %. It can be seen that the present simulations for the single cylinder and the rod–airfoil configuration with $L/d \geqslant 4$ provide similar shapes and maximum positions, as well as mean separation positions, in comparison with the experimental data of Achenbach (1968) and the LES results of Giret et al. (2015). For the case of $L/d = 2$, however, the peak skin friction is lower in the present simulations, furthermore, the mean separation in this case is earlier than in the other cases.

To clearly demonstrate the effect of the distance L on the skin friction of the airfoil, figure 11(b) shows the mean friction distributions on the airfoil surface. It is not surprising to see that the distributions are highly distinguishable at the leading edge with $L/d = 2$. A similar phenomenon is also identified in the surface pressure distributions shown in figure 10(b). The distinguishable distributions, i.e. the negative mean friction values observed in figure 11(b) are caused by the large separated region in the rod–airfoil configuration with $L/d = 2$, which is in figure 14. The main vortex separated from the cylinder with negative z-vorticity entrains the boundary layer and causes early separation at the leading edge with $L/d = 2$. With increment of the distance L, the skin friction tends to be indistinguishable. When $L/d \geqslant 4$, the skin friction reaches its maximum value at the leading edge, and then drops quickly, furthermore, the maximum skin friction increases as the distance L is enlarged.

Cases	St	$\langle C_D \rangle_t$	$\langle C_L \rangle_t$	C_{Lrms}
(a)		Experiments		
Gerrard (1961) $Re_d = 4.8 \times 10^4$	—	—	—	0.4–0.8
Achenbach (1968) $Re_d = 6 \times 10^4$	—	1.0–1.3	—	—
Apelt & West (1975) $Re_d = 4 \times 10^4$	0.19	1.2	—	—
Szepessy & Bearman (1992) $Re_d = 4.3 \times 10^4$	0.19	1.2–1.5	—	0.4–0.7
Unsteady RANS on structured grid				
Casalino et al. (2003)	0.24	0.8	—	—
Boudet et al. (2005)	0.24	1.03	—	0.76
LES/DES on unstructured grid				
Schell (2013)	0.19	—	—	—
Giret et al. (2015)	0.19	1.19	—	0.60
LES/DES on structured grid				
Magagnato et al. (2003)	0.19–0.203	—	—	—
Boudet et al. (2005)	0.19	1.17	—	0.57
Greschner et al. (2008)	0.185	0.81	—	0.42
Present study	0.2	1.23	7.4×10^{-3}	0.57
(b)				
Distance	$\langle C_{D\text{-}rod} \rangle_t$	$\langle C_{D\text{-}tol} \rangle_t$	$\langle C_{L\text{-}tol} \rangle_t$	$C_{Lrms\text{-}tol}$
$L/d = 2$	0.72	0.64	1.0×10^{-2}	0.47
$L/d = 4$	0.97	0.89	-1.0×10^{-2}	1.63
$L/d = 6$	1.01	0.95	7.3×10^{-2}	1.75
$L/d = 8$	1.04	1.01	7.7×10^{-2}	1.32
$L/d = 10$	1.06	1.04	5.0×10^{-2}	1.38

TABLE 1. The mean and fluctuating integral force for the single cylinder and rod–airfoil system. (a) The calculated mean and fluctuating integral force on the single cylinder compared with experimental data and other numerical results. Here, $\langle C_D \rangle_t$ represents the time-averaged drag coefficient, $\langle C_L \rangle_t$ denotes the time-averaged lift coefficient and C_{Lrms} is the r.m.s. of the lift coefficient, St of the present study is calculated from the figure 13. (b) The force behaviour of the rod–airfoil configuration. Here, $\langle C_{D\text{-}rod} \rangle_t$ represents the time-averaged drag coefficient on the cylinder, as well as $\langle C_{D\text{-}tol} \rangle_t$, $\langle C_{L\text{-}tol} \rangle_t$ and $C_{Lrms\text{-}tol}$ are the time-averaged drag coefficient, the time-averaged lift coefficient and the r.m.s. of the lift coefficient for the whole configuration, respectively.

This behaviour may correspond to the phenomenon of the influence of the turbulent wake on the airfoil fading gradually as the distance L increases.

4.1.2. *Mean and fluctuating integral force*

To assess quantitatively the integral force on the rod–airfoil system, table 1(*a*) lists the calculated mean and fluctuating integral force on the single cylinder compared with the experimental data and other numerical results, while table 1(*b*) lists the force behaviour of the rod–airfoil system. There is, in general, a considerable scattering in

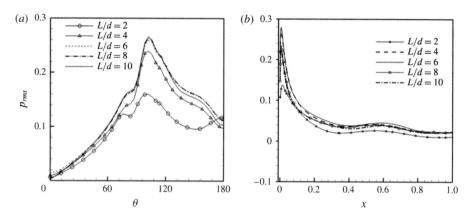

FIGURE 12. Root-mean-square value of pressure fluctuation on the surface of bodies: (*a*) cylinder and (*b*) airfoil.

the results of the force on the cylinder surface, the present solutions however are in good agreement with the experimental data of Szepessy & Bearman (1992). As the Reynolds number considered here is $O(10^5)$, the pressure force plays a dominant role in the total forces on the cylinder, additionally, the pressure increases in the base region as the distance L is shortened (see figure 10*a*), therefore we can see in table 1(*b*) that the mean drag of the cylinder falls as the airfoil approaches the cylinder. The drag varys on the whole rod–airfoil system, as shown in table 1(*b*). The regulation of the drag varying of the is same as that of the cylinder in the configuration when the distance L increases.

Under the circumstances, it is obvious that the drag of cylinder plays a dominant role in the total drag of the rod–airfoil configuration.

As shown in table 1(*b*), the mean lift of the rod–airfoil system almost vanishes, however, the instability feature of this configuration can be investigated by the r.m.s. value of the lift fluctuation $C_{Lrms\text{-}tol}$. When $L/d = 2$, the value of $C_{Lrms\text{-}tol}$ is the lowest of all the five rod–airfoil configurations, and even less than that of a single cylinder, see table 1(*a*). This demonstrates that the lift fluctuation can be suppressed if the distance L is very short. According to the analysis above, the total forces are dominated by the surface pressure and it is natural to investigate the lift fluctuation by examining the pressure oscillation on the body surface. Figure 12 shows the distributions of r.m.s. value of the pressure fluctuation on the cylinder and airfoil. The pressure fluctuation on the rod–airfoil configuration with $L/d = 2$ is suppressed, which is consistent with the least lift fluctuation shown in table 1(*b*). Moreover, it is interesting to find that a hump is shown in the distributions of the pressure fluctuations on the airfoil between $x/c = 0.4$ and 0.8 (see figure 12*b*). A similar shape can be observed from the typical vortex-blade interactions in counter-rotating open rotors (e.g. Roger & Carazo 2010; Carazo, Roger & Omais 2011).

The PSD of the time-dependent lift coefficient of the rod–airfoil and cylinder are shown in figure 13. As exhibited in this figure, the smallest peak value corresponding to the least lift fluctuation given in table 1(*b*) is obtained by the configuration with $L/d = 2$. Usually, the force fluctuation on a cylinder is associated with vortex shedding in the wake (e.g. Oertel & Affiliation 1990; Owen & Bearman 2001), and the power spectral density of the cylinder in figure 13 can be used to identify the frequency

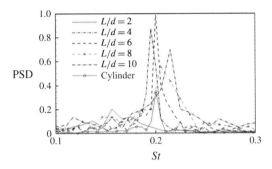

FIGURE 13. Profiles of power spectral density of time-dependent lift coefficient on the rod–airfoil and cylinder. Here, the time-dependent lift coefficients are collected every 10 time steps in the last 20 000 time steps for all the cases.

of vortex shedding (Xu *et al.* 2010). The primary frequency corresponding to the highest peak for the single cylinder is approximately 0.2. For the configurations with $L/d \geqslant 4$, it is interesting to find that the peak value in figure 13 periodically increases and decreases with the distance L increasing. This behaviour is most likely caused by the fluid resonant oscillations, which will be discussed in §4.3.2. Furthermore, the St corresponding to the peak value increases with increments in the distance L. This feature can be reasonably related to the higher-base-pressure distribution on the cylinder with the shorter distance L in figure 10(a), and the vortex shedding suppression in the interaction region of $L/d = 2$, which will be discussed in §4.2.

Based on the above analysis, the force on the surface of the rod–airfoil configuration is affected by the distance L. In particular, the force is obviously different when $L/d = 2$. The differences involving surface pressure and friction lead to the reduction of mean and fluctuating forces for the flow past the rod–airfoil configuration. The distance L may also have an effect on the flow structures and we thus pay more attention to the relevant flow characteristics in the next section.

4.2. *Flow structures and turbulent fluctuations*

Despite the complexity associated with a turbulent wake and two bodies, the dynamics of the flow in the present cases are largely determined by the interaction in the region between the cylinder and airfoil. In this section, we investigate the flow features in the interference region by first looking into the flow structures based on the time-averaged quantities. Then, the interference flow patterns are discussed using the instantaneous quantities. Finally, based on the statistical quantities, we will proceed to study the turbulent fluctuations and their suppression.

4.2.1. *Mean flow structures in the interference region*

We investigate the mean flow structures by first examining the vorticity consisting of the streamwise ω_x, vertical ω_y and spanwise ω_z vorticity. The time-averaged ω_y becomes zero because of the periodic boundary condition applied in the spanwise direction. The increasing vertical gradients of the streamwise u and spanwise w velocity, together with the gradients of vertical velocity v in the horizontal (x, z)-plane, give rise to a streamwise ω_x and spanwise ω_z vorticity. It is important to note that

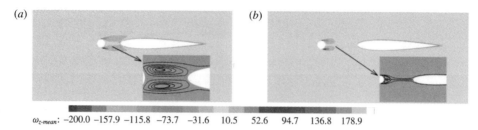

FIGURE 14. (Colour online) Time-averaged spanwise vorticity and streamlines in the interaction region taken from the flow field with $U_\infty = 72$ m s^{-1}, $Re_d = 4.8 \times 10^4$, $L/d = 2$ (a) and $L/d = 4$ (b). (colored picture attached at the end of the book)

these two quantities will appear in the instantaneous turbulent flow field, but in the present cases, the time-averaged quantity ω_x is negligible compared with that of ω_z.

Time-averaged spanwise vorticity $\omega_{z\text{-}mean}$ together with the corresponding streamlines projected onto the (x, y)-plane in the interference region for $L/d = 2$ and $L/d = 4$ are shown in figure 14. Clearly, two counter-rotating vortices evolve along the axis of the cylinder. The mean streamlines are symmetric about the horizontal center-line, whereas the $\omega_{z\text{-}mean}$ contours exhibit an antisymmetry. Visualizations of the flow field in the interference region have revealed that at short enough distances, the Kármán-street type of shedding is replaced by a non-shedding mode (see e.g. Munekata et al. 2006), with two counter-rotating vortices with opposite circulation connecting to each side of the leading edge of the airfoil. In the non-shedding mode, the negative mean friction values are observed at the leading edge (see figure 11b) due to the main vortex separated from the cylinder with a negative z-vorticity connecting to the airfoil. Kármán-street shedding is also observed in the other three cases. Two patterns which will be further discussed in § 4.2.2 are then outlined: Kármán-street shedding for the long distances and Kármán-street suppressing for the short distances.

Due to the fact that counter-rotating vortices are connected to the airfoil, the distributions of $\omega_{z\text{-}mean}$ for $L/d = 2$ are evidently different from those of other cases at the leading edge of the airfoil. This also explains why the distributions of pressure and friction, shown in figures 10(b) and 11(b) respectively, are obviously distinguishable on the surface of the airfoil for $L/d = 2$. When the cylinder is set away from the airfoil, the counter-rotating vortices with opposite circulation emanate from each side of the cylinder and do not touch the airfoil. The interaction is relatively weak between the turbulent wake and airfoil, reasonably related to the fact that the distributions of mean friction tend to be identical on the surface of the cylinder (see figure 11a) and the leading edge of the airfoil (see figure 11b).

Figure 15(a) shows the contours of time-averaged streamwise velocity u_m in the interference region, where the counter-rotating vortices connecting with the leading edge of airfoil lead to negligible or negative values of u_m for the case of $L/d = 2$; the flow behind the counter-rotating vortices is accelerated until it impinges on the leading edge of airfoil for the cases with $L/d \geqslant 4$ (the other three cases are not shown in figure 15a). The phenomena exhibited in this figure can be explained by the fact that in this interaction region, the flow is the result of two tendencies. The influence of the airfoil on the steady flow tends to accelerate it, whilst the counter-rotating vortices are connected with the leading edge of airfoil due to the approaching cylinder, which tends to slow the flow down.

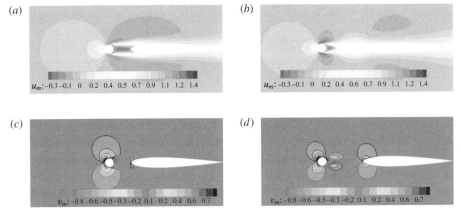

FIGURE 15. (Colour online) Contours of time-averaged velocity taken from the flow field with $U_\infty = 72$ m s^{-1}, $Re_d = 4.8 \times 10^4$, $L/d = 2$ (a,c) and $L/d = 4$ (b,d). (a,b) Streamwise component u_m; (c,d) vertical component v_m, here solid lines denote positive values and dashed lines negative ones. (colored picture attached at the end of the book)

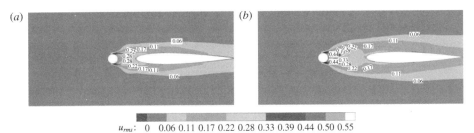

FIGURE 16. (Colour online) The r.m.s. value of streamwise velocity taken from the flow field with $U_\infty = 72$ m s^{-1}, $Re_d = 4.8 \times 10^4$, $L/d = 2$ (a) and $L/d = 4$ (b).
(colored picture attached at the end of the book)

In figure 15(b), we plot the contours of time-averaged vertical velocity v_m for $L/d = 2$ and $L/d = 4$. It can be seen that the positive v_m presents below the leading edge of the airfoil with $L/d = 2$, and the negative one is observed above. However, the contrary distribution is shown in this region with $L/d = 4$. When the incoming flow approaches the upper part of the leading edge of the airfoil with $L/d = 2$, it experiences the downdraft depicted by the contours of v_m (see also the streamlines in figure 14), which produces negative horizontal gradients in v_m in this region, thereby giving rise to the negative $\omega_{z\text{-}mean}$. The positive one however is generated symmetrically about the center-line, as illustrated by the contours of $\omega_{z\text{-}mean}$ in figure 14.

Finally, the contours of r.m.s. value of streamwise velocity u_{rms} are plotted in figure 16 for $L/d = 2$ and $L/d = 4$. It can be seen that the u_{rms} values of $L/d = 2$ behind the cylinder and near the central line $y = 0$ are much lower than those of $L/d = 4$. This phenomenon is consistent with the fact that the non-shedding mode is observed in the rod–airfoil configuration with $L/d = 2$, and the negligible or negative u_m generated in the interaction region (see figure 15a) is caused by the main separated

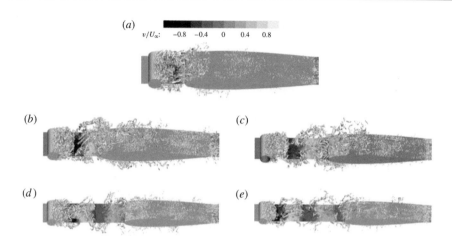

FIGURE 17. (Colour online) Vortical structures by iso-surface of the Q-criterion ($Q = 1000$). The instantaneous vertical velocity v at the symmetry x, z-plane ($y = 0$) is included to highlight the alternating pattern of the main vortices. (a) Taken from the flow field with $U_\infty = 72$ m s^{-1}, $Re_d = 4.8 \times 10^4$, $L/d = 2$; (b) $L/d = 4$; (c) $L/d = 6$; (d) $L/d = 8$; (e) $L/d = 10$. (colored picture attached at the end of the book)

vortices connecting to the airfoil (see figure 14) in this case. It is expected that the turbulent fluctuations in the interaction region of $L/d = 4$ are much more intensive than those of $L/d = 2$, and the underlying reason will be further discussed in §4.2.3 for the generation of lower-level turbulent fluctuations in the rod–airfoil configuration with $L/d = 2$.

4.2.2. *Instantaneous flow structures and interference flow patterns*

Figure 17 shows instantaneous snapshots of the flow field depicted by iso-surface of the Q criterion (Jeong & Hussain 1995)

$$Q = (\Omega_{ij}\Omega_{ij} + S_{ij}S_{ij})/2, \quad (4.1)$$

where $\Omega_{ij} = (u_{i,j} - u_{j,i})/2$ and $S_{ij} = (u_{i,j} + u_{j,i})/2$ are the antisymmetric and the symmetric components of the curl of the velocity, respectively. A positive value of Q represents the regions in which the rotation exceeds the strain. It should be recalled that the criterion (4.1) is only applied to the resolved scales obtained by the HILES and that the vortical structures could be different if the whole flow field is considered. It can be seen that the vortices in the rod wake impinge onto the airfoil, and then partly split at the leading edge. The flow topology exhibits two different patterns in the interaction region depicted by figure 17. The first one observed in this region with $L/d = 2$ includes only the primary vortical structures, roughly oriented along the axis of the cylinder. These primary structures originate from the rolling-up of the detached shear layers at each side of the cylinder. The other one presented by the rod–airfoil configuration with $L/d = 4$, 6, 8 and 10 comprises both the primary vortical structures and also secondary vortical structures, exhibiting a wide range of scales and orientations. These secondary structures come from the Kármán-street vortex shedding. For rod–airfoil configurations, Munekata *et al.* (2006) also observed

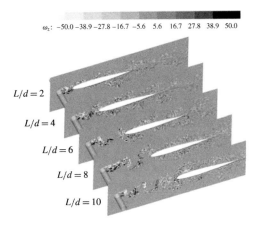

FIGURE 18. (Colour online) Juxtapositional view of the unsteady spanwise vorticity ω_z of vertical (x, y)-planes taken from the flow field of five rod–airfoil configurations at the central slice $z = 0.015$ m of the computational domain.

(colored picture attached at the end of the book)

two flow patterns. In the first, the Kármán-street is not generated for a short enough distance L, whereas in the other, the Kármán vortex is steadily generated for a long distance L. Based on this observation, the authors suggested that changing the distance L led to the flow pattern altering. In figure 17, the instantaneous vertical velocity v, highlighting the alternating pattern of the main vortices, indicates that two flow patterns exist in the present study, supporting the observation made by Munekata *et al.* (2006) who observed the pattern dependence of the interference flow on the distance L.

Based on the discussion above, a possible categorization for the flow patterns associated with the interference region is given as follow: (i) for $L/d \leq 2$, there is the Kármán-street suppressing mode. The eddies propagate downstream without being shed in this mode. In the present study, the Kármán-street is not generated by the free shear layers separated from the upstream cylinder in the interference region with $L/d = 2$; (ii) for $L/d \geq 4$, Kármán-street occurs. This is classified as the Kármán-street shedding mode and occurs in the present study at $L/d = 4, 6, 8$ and 10. As the Reynolds number considered here is large enough, the eddies arise from the rolling-up of the shear layers at an almost fixed position and the vortex shedding then occurs when one eddy becomes strong enough to cut the opposite eddy from the supply of the circulation from the shear layer (Zdravkovich 1997), furthermore, the distance L is too long to suppress the vortex shedding in the interaction region; (iii) for $2 < L/d < 4$, the altering mode may exist. This flow pattern is not observed in the present study, however, it has been reported by Munekata *et al.* (2006) that the two flow patterns, i.e. Kármán-street shedding mode and Kármán-street suppressing mode alternate at aperiodic time intervals when $L/d = 3.1$.

A different illustration of the instantaneous flow structures in the interaction region is given by the snapshots of the spanwise vorticity ω_z in the (x, y)-plane, shown in figure 18. This component of the vorticity complements the Q-iso-surfaces by enabling the visualization of the shear layers that detach from both sides of the cylinder and their further roll-up to form primary vortices, as well as the turbulent wake impinging onto the airfoil and then partly splitting at the leading edge. In the latter process, the

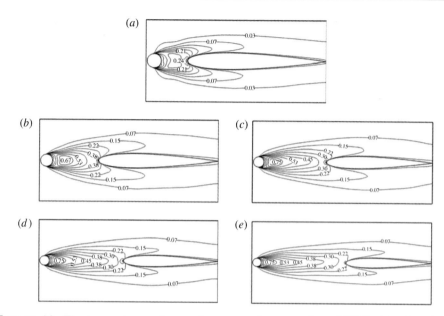

FIGURE 19. The iso-contours of spanwise averaged mean value of TKE. (*a*) taken from the flow field with $L/d = 2$; (*b*) $L/d = 4$; (*c*) $L/d = 6$; (*d*) $L/d = 8$; (*e*) $L/d = 10$.

turbulent wake containing Kármán-street shedding is clearly seen in all the vertical planes in figure 18, with the exception for the case of $L/d = 2$, suggesting that with reducing distance L, the shear layer is less susceptible to rolling up.

In figure 18, the four vorticity planes with $L/d \geqslant 4$ exhibit in the interaction region a clear Kármán-street shedding. Due to entrainment of free stream fluid, this turbulent wake expands as it evolves downstream until it hits the leading edge of the airfoil. For $L/d = 2$, however, the Kármán-street shedding in the interaction region seems to be suppressed due to the short distance L. These observations suggest that vortex shedding in the present cases is influenced by the distance L between the two bodies. It is obvious that the wake from the cylinder upstream is blocked by the airfoil downstream. The blocking effect is minor for long distances between the two bodies, but becomes more evident at short distances, which may lead to the base pressure rising abnormally on the cylinder upstream (see figure 10*a*). It is natural to assume that the blocking effect takes main responsibility for the Kármán-street suppression in the rod–airfoil configuration.

4.2.3. *Turbulent fluctuations and their suppression*

The turbulent fluctuations around a body are closely associated with the fluctuating forces exerted on it (Wu, Lu & Zhuang 2007). To describe the turbulent fluctuations around the rod–airfoil configurations, the iso-contours of spanwise averaged mean value of turbulent-kinetic-energy (TKE), i.e. TKE_m, are shown in figure 19. The TKE_m is relatively smaller in the region around the airfoil with $L/d = 2$. This feature leads to correspondingly lower fluctuating surface pressure (see figure 12*b*). Moreover, the suppression of the turbulent fluctuations is also observed in the interaction region.

Additionally, the profiles of TKE_m in the interaction region along the line of symmetry of the cylinder are shown in figure 20. From the distribution behind the

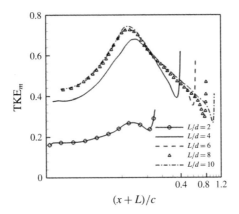

FIGURE 20. Profiles of TKE_m in the interaction region along the symmetry line of the cylinder.

cylinder with $L/d = 2$, the TKE_m is significantly lower in the range of approximately $x/c < 0.1$, and increases gradually because of the development of primary vortical structures, reaching a peak value in the range of approximately $x/c = 0.15$. It then falls as the wake propagates downstream, finally undergoing a severe jump and gaining a maximal value since the vortices in turbulent wake impinge onto the airfoil and partly split at the leading edge. It is expected in figure 20 that the reduction and jump are also present in the distributions of TKE_m from the flow fields with $L/d = 4, 6, 8, 10$. However, it may be noted that in these distributions, the maximal value of TKE_m is not observed after the significant jump, but is exhibited at the first peak. This difference indicates that the suppression is associated with the primary vortical structures.

To understand the characteristics relating to the suppression of turbulent fluctuations of the rod–airfoil configuration with $L/d = 2$, we analyse the underlying reason for the generation of higher-level turbulent fluctuations in the other four cases. As shown in figure 17, for the configuration with $L/d = 4, 6, 8, 10$, the vortex shedding occurs when one eddy becomes strong enough to cut the opposite eddy off from the supply of the circulation from the shear layer, and the formed vortex is released into the wake. The flow structures appear to be stronger in the vortex formation region which corresponds approximately to the location of the primary vortex. This observation is consistent with numerical simulation of a turbulent wake behind a curved circular cylinder (Gallardo et al. 2014), and correspondingly, the maximal value of turbulent fluctuations is present at this location (see figure 20). Further downstream, but before the airfoil, the size of the structures tends to increase as their concentration decreases, this behaviour is related to the varying of TKE_m in figure 20 which goes down until the blocking effect is met, also consistent with the report of Mansy, Yang & Williams (1994) that the amplification of the larger scales and attenuation of the smaller scales occur as the wake propagates downstream. The shedding wake containing massive turbulent fluctuations requires impingement onto the airfoil to create the second peak of turbulent fluctuations. The larger before the impact the TKE_m is, the higher the second peak is, as shown in figure 20. It is clear that the primary vortex shedding and the shedding wake impingement onto the airfoil play a dominant role in the production of turbulent fluctuations. However, for the case of $L/d = 2$, primary vortex shedding

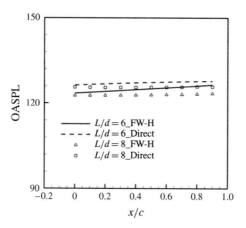

FIGURE 21. Comparison between the OASPLs calculated directly from the HILES solutions and provided by the FW-H analogy on the line $y = 0.4$ m for the configuration with $L/d = 6$ and $L/d = 8$.

is suppressed, correspondingly, the intensity of impingement is weakened, resulting in a significant suppression of the turbulent fluctuations.

4.3. Acoustic analysis and fluid resonant oscillation

4.3.1. Acoustic analysis

From the preceding analysis of the flow structures and turbulent fluctuations, it is reasonably well understood that the distance L plays an important role in the flow features relevant to the flow pattern altering and the turbulent fluctuations being suppressed. The flow features are closely associated with the noise radiation. We further analyse the influence of distance L on the noise radiation.

For rod–airfoil configurations with $L/d = 6$ and $L/d = 8$, figure 21 shows the comparison between the overall SPLs (OASPLs) calculated directly from the HILES solutions and provided by the FW-H analogy to complement the verification with respect to the acoustics outcome. The OASPLs on the line $y = 0.4$ m are chosen to ensure the mesh resolution is fine enough for the HILES with the HDCS-E8T7 capturing the acoustic noise. Although small differences are observed between the two results, the same varying trend of the OASPLs is predicted by the two methods. In the following, we will apply the FW-H analogy to discuss the acoustic behaviour in the far-field.

In order to show the dominant noise radiation directions of different configurations, figure 22 plots the directivities of OASPL at 1.85 m for all five rod–airfoil configurations. It is clear that the dominant noise radiation directions are between the observation angles of $60°$ and $90°$. For a further investigation of the noise radiation in the dominant radiation direction, figure 23 shows the SPL spectra at the location ($x = 0.68$ m, $y = 1.74$ m), i.e. the observation angle $70°$ for all the five rod–airfoil configurations. Although a relatively narrow range around the peak frequency of the noise spectra is presented in this figure, it is clearly observed that the noise spectra is separated into two different types. The first one is a broadband noise without a manifestly distinguishable peak, which is radiated by the rod–airfoil configuration

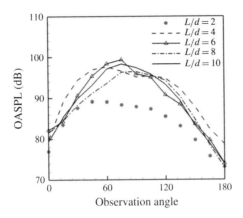

FIGURE 22. (Colour online) Directivities for the OASPL of the rod–airfoil configurations at 1.85 m. Here, the location for the observation angle 0 is $x = 1.85$ m, $y = 0$.

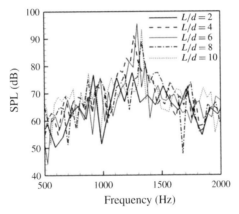

FIGURE 23. The SPL spectra at the location ($x = 0.68$ m, $y = 1.74$ m) for all the five rod–airfoil configurations.

with $L/d = 2$. The second type is present when $L/d \geqslant 4$. With a dominant peak, the broadband noise and the tonal noise are exhibited simultaneously in this type. Moreover, the levels of the peaks in the second type are much higher than that in the first, which is not surprising since the noise radiation in the far-field is closely associated with the flow features in the near-field. As reported by Boudet et al. (2005), the interaction between the wake from the cylinder upstream and the airfoil downstream is mainly responsibile for the noise radiation of the rod–airfoil configuration. In particular, the spectrum of a single cylinder may be approximately 10 dB lower than that of the rod–airfoil system (Jacob et al. 2005), which is also supported by the observation of Munekata et al. (2006). Based on the discussion of flow features above, vortex shedding is suppressed for a very short distance $L/d = 2$, leading to significant suppression of the pressure fluctuations and the turbulent fluctuations, as well as the weakening of the impingement intensity. Thus, the lower levels of noise are observed for the rod–airfoil configuration with $L/d = 2$.

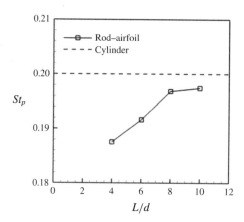

FIGURE 24. Effect of L/d on St based on peak frequency.

To exhibit the influence of the distance L on the peak frequency of the radiative noise, a peak Strouhal number, St_p, is analysed as performed by Munekata *et al.* (2006) and is defined as $St_p = f_p d/U_\infty$, where f_p represent the peak frequency given in figure 23. Quantitatively, compared with the primary frequency of the vortex shedding behind a single cylinder, represented by the St which is calculated from the PSD of the time-dependent lift coefficient (see figure 13), the St_ps of all the rod–airfoil configurations are linked to form a line are and plotted in figure 24 with a exception of the case of $L/d = 2$, in which there is no manifestly distinguishable peak frequency. The St_p increases with the distance L increasing, finally approaching the primary frequency of the single cylinder, supporting the observation made by Munekata *et al.* (2006) who observed the peak frequency dependence of the noise radiation on L/d. This feature may be reasonably related to the higher-base-pressure distribution on the cylinder with shorter distance L in figure 10(*a*), and vortex shedding suppressing in the interaction region in figure 17.

4.3.2. *Fluid resonant oscillation*

The fluid resonant oscillation within the flow interaction between the turbulent wake from a cylinder and the bodies have been observed in previous studies (e.g. Mochizuki *et al.* 1994). This feature is usually associated with the impingement of the turbulent wake onto the surface of the body and a feedback loop system (e.g. Mochizuki *et al.* 1994; Fitzpatrick 2003; Munekata *et al.* 2006). The fluid resonant oscillation means that the vortex generation in the rod–airfoil configuration is promoted by the acoustic field induced from the wake interference between the shedding vortex behind the cylinder upstream and the airfoil downstream (Munekata *et al.* 2006). In the quantitative observation made by Munekata *et al.* (2006), it was found that the peak SPL depended on L/d, and increases and decreases periodically, relating to the fluid resonant oscillation. Mochizuki *et al.* (1994) also observed the fluid resonant oscillation in the configuration of two circular cylinders in tandem, and these authors proposed a model to describe the feedback loop system as illustrated in figure 25. According to this feedback model (Mochizuki *et al.* 1994), the feedback frequency f_f, defined by the inverse of time taken to feed back for one loop, and is

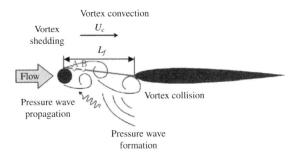

FIGURE 25. (Colour online) Schematic diagram of feedback system for fluid resonant oscillation.

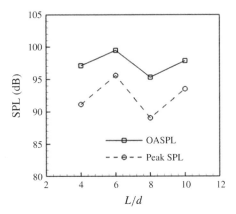

FIGURE 26. The peak SPL and OASPL for the second type of noise spectra.

described by the following equation:

$$f_f = \frac{1}{L_f/U_c + L_f/c}, \qquad (4.2)$$

where L_f represents the characteristic length of feedback loop, U_c is the convective velocity of shedding vortex and c is the speed of sound. If the constraint $f_p = nf_f$ is satisfied, the effect of fluid resonant oscillation becomes stronger. Here, n is a positive integer.

As shown in figure 23, there are two types of noise spectra due to the varying of the distance L. The first one is caused by the flow features related to vortex shedding suppression in the interaction region. Regarding the second one, although a similar dominant peak is present in this type of noise spectra, the value of the dominant peak varies periodically. To show this varying regularity clearly, figure 26 plots the peak SPL and OASPL for the second type of noise spectra exhibited in figure 23. The periodic increasing and decreasing of the peak SPL and OASPL are clearly exhibited in this figure and are similar to the phenomenon of fluid resonant oscillation observed by Munekata *et al.* (2006). It should be noted that the regulation of the peak SPL from all the five rod–airfoil configurations in the dominant noise radiation directions,

Distance	L_f	L_{AB}	U_c	f_f (Hz)
$L/d = 4$	$4.445d$	$0.5779d$	$0.77U_\infty$	1063.07
$L/d = 6$	$6.44d$	$0.5779d$	$0.72U_\infty$	692.74
$L/d = 8$	$8.439d$	$0.9632d$	$0.70U_\infty$	515.96
$L/d = 10$	$10.438d$	$0.9632d$	$0.70U_\infty$	417.15

TABLE 2. The corresponding parameters of the feedback model for fluid resonant oscillation.

i.e. between an observation angle of 60° and 90° is similar to that at the observation angle of 70°, however, in the other directions, this regulation is not observed.

Based on the above analysis, we suggest that fluid resonant oscillation is observed in the present study. To further describe the fluid resonant oscillation, the feedback model (4.2) is adopted to analyse this phenomenon. The characteristic length L_f in (4.2) is reasonably well given by the horizontal distance between the leading edge and the mean separation location on the cylinder surface, corresponding to the resolved skin friction stress vanishing (Munekata et al. 2006), and can be determined from the figure 11(a). The values of L_f are listed in table 2 for $L/d = 4$, 6, 8 and 10, respectively.

We here deal with the convective velocity of the shedding vortex, which is used in the feedback model of the fluid resonant oscillation. Two-point cross-correlation of the unsteady pressure can be used to quantitatively determine the propagation speed of the pressure disturbances along a given path (Xu et al. 2010). A covariance coefficient C_{ij} for two pressure signals $p_i(t)$ and $p_j(t)$ with time delay τ can be defined as

$$C_{ij}(\tau) = \frac{\langle (p_i(t) - \langle p_i \rangle_t)(p_j(t-\tau) - \langle p_j \rangle_t) \rangle_t}{\langle (p_i(t) - \langle p_i \rangle_t)^2 (p_j(t) - \langle p_j \rangle_t)^2 \rangle_t}, \quad (4.3)$$

where $\langle \ \rangle_t$ denotes time average. Within the interaction region, the cross-correlation analysis is conducted for probes A and B shown in figure 25 and the results are given in figure 27. Positive time delays are obtained, indicating that the pressure disturbances within the interaction region propagate downstream toward the airfoil leading edge. The convective velocity of the shedding vortex U_c can then be calculated by dividing the horizontal distances between the neighbouring probes by the time delays between the peaks of the corresponding cross-correlations. Due to the differences of the grid point distributions, the horizontal distances between the probes A and B, i.e. L_{AB} may be different. For $L/d = 4$, $L/d = 6$, $L/d = 8$ and $L/d = 10$, the values of L_{AB} are listed in table 2, where the corresponding speed U_c and the calculated feedback frequency f_f are also presented.

Figure 28 shows the relation between f_p and f_f for the four different interaction distances. When $L/d = 6$ and 10, the exhibited data are periodically close to the solid lines, which means that the effect of fluid resonant oscillation becomes stronger. This behaviour corresponds to the periodic change of the peak SPL and OASPL shown in figure 26. We recognize the limitations of the simplified analysis of the fluid resonant oscillation based on the feedback model; nevertheless, the results obtained from the model are of help in understanding the physical mechanism of the fluid resonant oscillation involved in this flow.

高精度格式的具体应用

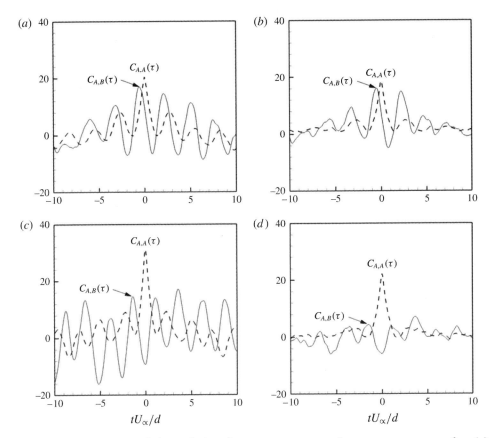

FIGURE 27. Cross-correlation of the downstream propagating pressure waves for (a) $L/d = 4$; (b) $L/d = 6$; (c) $L/d = 8$ and (d) $L/d = 10$.

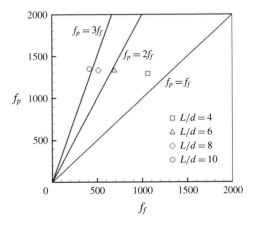

FIGURE 28. The relation between f_p and f_f for the four different interaction distances. Here, the solid lines mean the equation of resonant condition.

5. Concluding remarks

Numerical investigation on body-wake interaction was carried out by means of the HILES for flow over five rod–airfoil configurations with the incoming velocity $U_\infty = 72$ m s^{-1}, a Reynolds number based on the airfoil chord 4.8×10^5 and interaction distances $L/d = 2, 4, 6, 8$ and 10. Due to the varying of the distance L, two flow patterns have been observed associated with the interference region, i.e. the Kármán-street suppressing mode for $L/d = 2$ and the Kármán-street shedding mode for $L/d = 4, 6, 8$ and 10 in the present study.

In the Kármán-street suppressing mode, the flow in the interaction region has two opposite tendencies: the influence of the airfoil on the steady flow is to accelerate it; the counter-rotating vortices connecting with the leading edge of airfoil tend to slow the flow down. Moreover, the forces exerted on the bodies, the flow structures and turbulent fluctuations as well as the noise radiation are closely associated with each other. Due to the blocking effect, the base pressure rises abnormally on the cylinder upstream, furthermore, the high base pressure leads to that von Kármán vortex shedding suppression in the region between the cylinder upstream and the airfoil downstream, correspondingly, the intensity of impingement is weakened, resulting in the significant suppression of the turbulent fluctuations and the pressure fluctuations. The obviously lower levels of noise are observed for this rod–airfoil configuration because of the significant suppression of the pressure fluctuations and the turbulent fluctuations, as well as the weakening of the impingement intensity.

In the Kármán-street shedding mode, vortex shedding occurs, and the distance L is too long to suppress the vortex shedding in the interaction region. Large velocity fluctuations and turbulent fluctuations are generated by the primary vortex shedding and the shedding wake impinging onto the airfoil. The broadband noise and tonal noise with a dominant peak are exhibited simultaneously in the SPL spectra. The value of the dominant peak varies periodically in the dominant noise radiation directions, which is recognized as a phenomenon caused by the fluid resonant oscillation within the flow interaction between the turbulent wake and the bodies. Based on the feedback model proposed by Mochizuki *et al.* (1994) and the pressure signal in the flow field, the fluid resonant oscillation has been further investigated, and it is confirmed that the effect of fluid resonant oscillation becomes stronger when $L/d = 6$ and 10, corresponding to the periodic change of the peak SPL and OASPL observed in the present study.

Due to the lack of data for further analysis, it cannot be determined that there is a transitional flow pattern before the Kármán-street shedding mode and after the Kármán-street suppressing mode. It would have required further simulations to reveal the flow pattern for the interaction distance $2 < L/d < 4$. Even if the hump of the pressure fluctuation distributions on the airfoil is recognized as a typical phenomenon caused by body-wake interaction, more efforts are required to clarify the effect of the body-wake interaction on the airfoil. In addition, further investigations are expected to show the sensitivity of the solution to the grid resolution and the FW-H surface type, location and shape.

Acknowledgements

The authors are very grateful to the anonymous referees for their significant and detailed comments as well as very helpful suggestions made by Professor L. P. Zhang in China Aerodynamics Research and Development Center. This study was supported by the National Natural Science Foundation of China (grant nos 11372342 and 11202226).

REFERENCES

ACHENBACH, E. 1968 Distribution of local pressure and skin friction around a circular cylinder in cross-flow up to $Re = 5 \times 10^6$. *J. Fluid Mech.* **34**, 625–639.

AGRAWAL, B. R. & SHARMA, A. 2014 Aerodynamic noise prediction for a rod–airfoil configuration using large eddy simulations. *AIAA Paper* 2014-3295.

APELT, C. J. & WEST, G. S. 1975 The effects of wake splitter plates on bluff-body flow in the range $10^4 < Re < 5 \times 10^4$. Part 2. *J. Fluid Mech.* **71**, 145–160.

BORIS, J. P., GRINSTEIN, F. F., ORAN, E. S. & KOLBE, R. L. 1992 New insights into large eddy simulation. *Fluid Dyn. Res.* **10**, 199–228.

BOUDET, J., GROSJEAN, N. & JACOB, M. C. 2005 Wake-airfoil interaction as broadband noise source: a large-eddy simulation study. *Intl J. Aeroacoust.* **4** (1), 93–116.

CARAENI, M., DAI, Y. & CARAENI, D. 2007 Acoustic investigation of rod airfoil configuration with DES and FWH. *AIAA Paper* 2007-4016.

CARAZO, A., ROGER, M. & OMAIS, M. 2011 Analytical prediction of wake-interaction noise in counter-rotating open rotors. *AIAA Paper* 2011-2758.

CASALINO, D., JACOB, M. C. & ROGER, M. 2003 Prediction of rod airfoil interaction noise using the FWH analogy. *AIAA J.* **41** (2), 182–191.

CRESCHNER, B., THIELE, F., CASALINO, D. & JACOB, M. C. 2004 Influence of turbulence modelling on the broadband noise simulation for complex flows. *AIAA Paper* 2004-2926.

DANIEL, J. B. 2006 Analysis of sponge zones for computational fluid mechanics. *J. Comput. Phys.* **212**, 681–702.

DAUDE, F., BERLAND, J., EMMERT, T., LAFON, P., CROUZET, F. & BAILLY, C. 2012 A high-order finite-difference algorithm for direct computation of aerodynamic sound. *Comput. Fluids* **61**, 46–63.

DENG, X. G., JIANG, Y., MAO, M. L., LIU, H. Y., LI, S. & TU, G. H. 2015 A family of hybrid cell-edge and cell-node dissipative compact schemes satisfying geometric conservation law. *Comput. Fluids* **116**, 29–45.

DENG, X. G., JIANG, Y., MAO, M. L., LIU, H. Y. & TU, G. H. 2013b Developing hybrid cell-edge and cell-node dissipative compact scheme for complex geometry flows. *Sci. China Technol. Sci.* **56**, 2361–2369.

DENG, X. G., MAEKAWA, H. & SHEN, Q. 1996 A class of high-order dissipative compact schemes. *AIAA Paper* 1996-1972.

DENG, X. G., MAO, M. L., TU, G. H., LIU, H. Y. & ZHANG, H. X. 2011 Geometric conservation law and applications to high-order finite difference schemes with stationary grids. *J. Comput. Phys.* **230**, 1100–1115.

DENG, X. G., MIN, Y. B., MAO, M. L., LIU, H. Y., TU, G. H. & ZHANG, H. X. 2013a Further studies on geometric conservation law and applications to high-order finite difference schemes with stationary grids. *J. Comput. Phys.* **239**, 90–111.

DRIKAKIS, D., HAHN, M., MOSEDALE, A. & THORNBER, B. 2009 Large eddy simulation using high-resolution and high-order methods. *Phil. Trans. R. Soc. Lond.* A **367**, 2985–2997.

FITZPATRICK, J. A. 2003 Flow/acoustic interactions of two cylinders in cross-flow. *J. Fluids Struct.* **17**, 97–113.

GALLARDO, J. P., ANDERSSON, H. I. & PETTERSEN, B. 2014 Turbulent wake behind a curved circular cylinder. *J. Fluid Mech.* **742**, 192–229.

GEROLYMOS, G. A. & VALLET, I. 2007 Influence of temporal integration and spatial discretization on hybrid RSM-VLES computations. *AIAA Paper* 2007-4094.

GERRARD, J. H. 1961 An experimental investigation of the oscillating lift and drag of a circular cylinder shedding turbulent vortices. *J. Fluid Mech.* **11**, 244–256.

GIRET, J. C., SENGISSEN, A., MOREAU, S., SANJOSÉ, M. & JOUHAUD, J. C. 2015 Noise source analysis of a rod–airfoil configuration using unstructured large eddy simulation. *AIAA J.* **53** (4), 1062–1077.

GORDNIER, R. E. & VISBAL, M. R. 1993 Numerical simulation of delta-wing roll. *AIAA Paper* 1993-0554.

GRESCHNER, B., THIELE, F., JACOB, M. C. & CASALINO, D. 2008 Prediction of sound generated by a rod–airfoil configuration using EASM DES and the generalised Lighthill/FW-H analogy. *Comput. Fluids* **37**, 402–413.

HAHN, M., DRIKAKIS, D., YOUNGS, D. L. & WILLIAMS, R. J. R. 2011 Richtmyer–Meshkov turbulent mixing arising from an inclined material interface with realistic surface perturbations and reshocked flow. *Phys. Fluids* **23**, 046101.

HUTCHESON, F. V. & BROOKS, T. F. 2012 Noise radiation from single and multiple rod configurations. *Intl J. Aeroacoust.* **11**, 291–334.

JACOB, M. C., BOUDET, J., CASALINO, D. & MICHARD, M. 2005 A rod–airfoil experiment as benchmark for broadband noise modeling. *J. Theor. Comput. Fluid Dyn.* **19** (3), 171–196.

JEONG, J. & HUSSAIN, F. 1995 On the identification of a vortex. *J. Fluid Mech.* **285**, 69–94.

JIANG, M., LI, X. D. & ZHOU, J. J. 2011 Experimental and numerical investigation on sound generation from airfoil-flow interaction. *Appl. Math. Mech.* **32** (6), 765–776.

JIANG, Y., MAO, M. L., DENG, X. G. & LIU, H. Y. 2013 Effect of surface conservation law on large eddy simulation based on seventh-order dissipative compact scheme. *Appl. Mech. Mater.* **419**, 30–37.

JIANG, Y., MAO, M. L., DENG, X. G. & LIU, H. Y. 2014a Large eddy simulation on curvilinear meshes using seventh-order dissipative compact scheme. *Comput. Fluids* **104**, 73–84.

JIANG, Y., MAO, M. L., DENG, X. G. & LIU, H. Y. 2015 Extending seventh-order dissipative compact scheme satisfying geometric conservation law to large eddy simulation on curvilinear grids. *Adv. Appl. Maths Mech.* **7** (4), 407–429.

JIANG, Y., MAO, M. L., DENG, X. G., LIU, H. Y. & YAN, ZH. G. 2014b Numerical prediction of jet noise from nozzle using seventh-order dissipative compact scheme satisfying geometric conservation law. *Appl. Mech. Mater.* **574**, 259–270.

JIANG, G. & SHU, C. 1996 Efficient implementation of weighted ENO. *J. Comput. Phys.* **181**, 202–228.

JOHN, M. H. & JAMESON, A. 2002 An implicit–explicit hybrid scheme for calculating complex unsteady flows. *AIAA Paper* 2002-0714.

KING, W. F. N. & PFIZENMAIER, E. 2009 An experimental study of sound generated by flows around cylinders of different cross-section. *J. Sound Vib.* **328**, 318–337.

LELE, S. K. 1992 Compact finite difference schemes with spectral-like resolution. *J. Comput. Phys.* **103**, 16–42.

LI, Y., WANG, X. N., CHEN, ZH. W. & LI, ZH. CH. 2014 Experimental study of vortex-structure interaction noise radiated from rod–airfoil configurations. *J. Fluids Struct.* **51**, 313–325.

LJUNGKRONA, L., NORBERG, CH. & SUNDEN, B. 1991 Free-stream turbulence and tube spacing effects on surface pressure fluctuations for two tubes in an in-line arrangement. *J. Fluids Struct.* **5**, 701–727.

LYRINTZIS, A. S. 2003 Surface integral methods in computational aeroacoustics – from the (CFD) near-field to the (acoustic) far-field. *Intl J. Aeroacoust.* **2** (2), 95–128.

MAGAGNATO, F., SORGÜVEN, E. & GABI, M. 2003 Far field noise prediction by large eddy simulation and Ffowcs-Williams Hawkings analogy. *AIAA Paper* 2003-3206.

MAHIR, N. & ROCKWELL, D. 1996 Vortex shedding from a forced system of two cylinders. Part I: tandem arrangement. *J. Fluids Struct.* **9**, 473–489.

MANSY, H., YANG, P. M. & WILLIAMS, D. R. 1994 Quantitative measurements of three-dimensional structures in the wake of a circular cylinder. *J. Fluid Mech.* **270**, 277–296.

MAO, M. L., JIANG, Y., DENG, X. G. & LIU, H. Y. 2016 Noise prediction in subsonic flow using seventh-order dissipative compact scheme on curvilinear mesh. *Adv. Appl. Maths Mech.* doi:10.4208/aamm.2014.m459.

MOCHIZUKI, M., KIYA, M., SUZUKI, T. & ARAI, T. 1994 Vortex-shedding sound generated by two circular cylinders arranged in tandem. *Trans. JSME B* **60** (578), 3223–3229.

MUNEKATA, M., KAWAHARA, K., UDO, T., YOSHIKAWA, H. & OHBA, H. 2006 An experimental study on aerodynamic sound generated from wake interference of circular cylinder and airfoil vane in tandem. *J. Therm. Sci.* **15** (4), 342–348.

MUNEKATA, M., KOSHIISHI, R., YOSHIKAWA, H. & OHBA, H. 2008 An experimental study on aerodynamic sound generated from wake interaction of circular cylinder and airfoil with attack angle in tandem. *J. Therm. Sci.* **17** (3), 212–217.

OERTEL, H. & AFFILIATION, J. 1990 Wakes behind blunt bodies. *Annu. Rev. Fluid Mech.* **22**, 539–564.

OWEN, J. C. & BEARMAN, P. W. 2001 Passive control of VIV with drag reduction. *J. Fluids Struct.* **15**, 597–605.

PIROZZOLI, S., GRASSO, F. & GATSKI, T. B. 2004 Direct numerical simulation and analysis of a spatially evolving supersonic turbulent boundary layer at $M = 2.25$. *Phys. Fluids* **16**, 530–545.

POINSOT, T. & LELE, S. K. 1992 Boundary conditions for direct simulations of compressible viscous flows. *J. Comput. Phys.* **101**, 104–129.

RIZZETTA, D. P., VISBAL, M. R. & MORGAN, P. E. 2008 A high-order compact finite-difference scheme for large-eddy simulation of active flow control. *Prog. Aerosp. Sci.* **44**, 397–426.

ROGER, M. & CARAZO, A. 2010 Blade-geometry considerations in analytical gust-airfoil interaction noise models. *AIAA Paper* 2010-3799.

SCHELL, A. 2013 Validation of a direct noise calculation and a hybrid computational aeroacoustics approach in the acoustic far field of a rod–airfoil configuration. *AIAA Paper* 2013-2122.

SCHLINKER, R. H., FINK, M. R. & AMIET, R. K. 1976 Vortex noise from non-rotating cylinders and airfoils. *AIAA Paper* 1976-81.

SZEPESSY, S. & BEARMAN, P. W. 1992 Aspect ratio and end plate effects on vortex shedding from a circular cylinder. *J. Fluid Mech.* **234**, 191–217.

TAM, C. K. W. & WEBB, J. C. 1993 Dispersion-relation-preserving finite difference schemes for computational acoustics. *J. Comput. Phys.* **107**, 262–281.

THORNBER, B. & DRIKAKIS, D. 2008 Implicit large eddy simulation of a deep cavity using high-resolution methods. *AIAA J.* **46**, 2634–2645.

VAN LEER, B. 1977 Towards the ultimate conservative difference scheme. IV. A new approach to numerical convection. *J. Comput. Phys.* **23**, 276–299.

VISBAL, M. R. & GAITONDE, D. V. 2002 On the use of higher-order finite-difference schemes on curvilinear and deforming meshes. *J. Comput. Phys.* **181**, 155–185.

VISBAL, M. R. & RIZZETTA, D. P. 2002 Large-eddy simulation on curvilinear grids using compact differencing and filtering schemes. *Trans. ASME J. Fluids Engng* **124**, 836–847.

WU, J. Z., LU, X. Y. & ZHUANG, L. X. 2007 Integral force acting on a body due to local flow structures. *J. Fluid Mech.* **576**, 265–286.

XU, C. Y., CHEN, L. W. & LU, X. Y. 2010 Large-eddy simulation of the compressible flow past a wavy cylinder. *J. Fluid Mech.* **665**, 238–273.

ZDRAVKOVICH, M. M. 1977 Review of flow interference between two circular cylinders in various arrangements. *Trans. ASME J. Fluids Engng* **99**, 618–633.

ZDRAVKOVICH, M. M. 1997 *Flow Around Circular Cylinder, Vol. 1 Fundamentals*. Oxford University Press.

ZDRAVKOVICH, M. M. 2003 *Flow Around Circular Cylinder, Vol. 2 Fundamentals*. Oxford University Press.

Validation of a RANS transition model using a high-order weighted compact nonlinear scheme

TU GuoHua[1,2*], DENG XiaoGang[1,3] & MAO MeiLiang[1,4]

[1] *State Key Laboratory of Aerodynamics, China Aerodynamics Research & Development Center, Mianyang 621000, China;*
[2] *Science and Technology on Scramjet Laboratory, Hypervelocity Aerodynamics Institute of CARDC, Mianyang 621000, China;*
[3] *National University of Defense Technology, Changsha 410073, China;*
[4] *Computational Aerodynamics Institute, China Aerodynamics Research & Development Center, Mianyang 621000, China*

Received October 24, 2012; Accepted December 24, 2012; published online March 4, 2013

A modified transition model is given based on the shear stress transport (SST) turbulence model and an intermittency transport equation. The energy gradient term in the original model is replaced by flow strain rate to saving computational costs. The model employs local variables only, and then it can be conveniently implemented in modern computational fluid dynamics codes. The fifth-order weighted compact nonlinear scheme and the fourth-order staggered scheme are applied to discrete the governing equations for the purpose of minimizing discretization errors, so as to mitigate the confusion between numerical errors and transition model errors. The high-order package is compared with a second-order TVD method on simulating the transitional flow of a flat plate. Numerical results indicate that the high-order package give better grid convergence property than that of the second-order method. Validation of the transition model is performed for transitional flows ranging from low speed to hypersonic speed.

hypersonic boundary layer transition, transition model, intermittency factor, high-order schemes

PACS number(s): 47.11.Bc, 47.27.Cn, 47.27.em, 47.27.nb, 47.40.Ki

Citation: Tu G H, Deng X G, Mao M L. Validation of a RANS transition model using a high-order weighted compact nonlinear scheme. Sci China-Phys Mech Astron, 2013, 56: 805–811, doi: 10.1007/s11433-013-5037-1

1 Introduction

The transition from laminar to turbulence has a major impact on many fluid engineering applications. The accurate prediction of transition is particularly important for skin friction prediction, heat transfer prediction as well as separation flow prediction. A classical and efficient attempt to coach transition in computational fluid dynamics (CFD) is to correlate the transition location with a turbulence model for the Reynolds-averaged Navier-Stokes (RANS) equations. Fu and Wang [1] categorize the RANS transition models in three types: the low-Reynolds number model, the intermittency model, and the local-variables-based model. A transition model should be carefully and thoroughly calibrated and validated before applications in that the flow transition can be affected by many factors, such as Mach number, Reynolds number, transverse and stream-wise curvature, pressure gradient and temperature. A transition process should include the receptivity mechanisms, the growth of instable disturbances, and/or bypass mechanisms. However, many mechanisms leading to transition are still poorly understood, which makes transition prediction difficult. Furthermore, the knowledge about hypersonic transition is far less than that of low speed transition not only because wind tunnel experiments of hypersonic transition are deficient and expensive together with flaw of the high background noise level, but also because theoretical analysis and

---

Tu Guohua, Deng Xiaogang, Mao Meiliang. Validation of a RANS transition model using a high-order weighted compact nonlinear scheme. Science China(Physics, Mechanics & Astronomy), 2013, 56(4): 805-811.

numerical simulations are also difficult because of the possible existence of shock waves, shock/boundary layer interactions, and contact layers. Many transition prediction methods, which work well for low speed flows, may be unsuitable for high speed flows.

Using the RANS approach, Wang and Fu [2,3] have developed a transition model for hypersonic flows. This model can take into account the effects of different instability modes associated with the flow property of different boundary layers. The model is based on k-ω-γ three-equation eddy-viscosity concept with k representing the fluctuating (including non-turbulent fluctuating) kinetic energy, ω denoting the specific dissipation rate, and γ denoting the intermittency factor. The model employs the local variables only, and can avoid the use of the integral parameters, such as the boundary layer thickness δ, which are often cost-ineffective with the modern CFD methods [3]. Recently, Wang et al. [4] have extended the transition model to capture the bypass and separation-induced transition in turbomachinery flows.

The fundamental principle of the transition model is similar to that of Warren and Hassan [5] and Papp et al. [6], where the turbulence viscosity μ_t is replaced by the "effective" eddy viscosity μ_{eff} as:

$$\mu_{eff} = (1-\gamma)\mu_{nt} + \gamma\mu_t, \quad (1)$$

where γ is the intermittency factor, and the subscript 'nt' designates non-turbulent fluctuations. Definition of the eddy viscosity μ_{nt} is the same as that of refs. [5,6], namely:

$$\mu_{nt} = \rho C_\mu k_{nt} \tau_{nt}, \quad (2)$$

where ρ is the density, k_{nt} is the non-turbulent fluctuating kinetic energy per unit mass, τ_{nt} is the characteristic time scale of the fluctuations, and the parameter C_μ is 0.09.

It is worth noting that there are three noticeable differences among the methods in refs. [2–6]. The first difference is that different transition-onset triggers are applied in different models. The second and third differences are that the intermittency equations and the turbulence models are different, respectively. We choose the transition model of Wang and Fu to begin the present work primarily because of the following three reasons: first, the intermittency transport equation is relatively simple; second, the transition onset is naturally implied in the transport equation; third, the shear stress transport (SST) turbulence model of Menter [7] is adopted by the researchers. Some widely spread turbulence models had been extensively reviewed by Roy and Blottner [8] and Celic and Hirschel [9], and their results indicate that the SST model generally produce better results for a wide range of flow conditions.

Some modifications are made to the transition model to saving computational costs. The Renewed model is validated by comparing the numerical results with several subsonic and hypersonic experimental data. The fifth-order weighted compact nonlinear scheme (WCNS-E-5) of Deng et al. [10,11] is applied to minimize the confusion between numerical errors and model errors. It was reported that the WCNS-E-5 is suitable for a wide range of flows without much requirement on grid quality [12,13]. Nonomura et al. [14] showed that the WCNS-E-5 is superior to the fifth-order weighted essentially non-oscillatory (WENO) scheme in free-stream and vortex preservation on curvilinear grids. The non-oscillatory and non-free-parameter dissipative (NND) scheme of Zhang and Zhuang [15], which is a second-order TVD scheme, is also applied in this paper. A comparison between the fifth-order scheme and second-order scheme are also made to show the advantages of the high-order scheme on solving transitional problems.

2 k-ω-γ transition prediction method

The basic for the k-ω-γ transition prediction method mainly contains the following four aspects: (1) k includes the non-turbulent as well as turbulent fluctuations; (2) Different transition mechanisms (or instability modes) are represented by different τ_{nt}, the characteristic time scale of fluctuations; (3) The intermittency transport equation is proposed here with a source term set to trigger the transition onset; (4) In the fully turbulent region, the model retreats to the original SST turbulence model.

2.1 k and ω equations and the SST turbulence model

The SST model of Menter [7] can be written for transitional problems as:

$$\frac{D}{Dt}(\rho k) = \tau_{ij}\frac{\partial u_i}{\partial x_j} + \frac{\partial}{\partial x_j}\left[(\mu + \mu_{eff}\sigma_k)\frac{\partial k}{\partial x_j}\right] - \beta^* \rho \omega k, \quad (3)$$

$$\frac{D}{Dt}(\rho \omega) = \frac{\gamma \rho \tau_{ij}}{\mu_t}\frac{\partial u_i}{\partial x_j} + \frac{\partial}{\partial x_j}\left[(\mu + \sigma_\omega \mu_{eff})\frac{\partial \omega}{\partial x_j}\right] - \beta \rho \omega^2 + 2(1-F_1)\frac{\rho \sigma_{\omega 2}}{\omega}\frac{\partial k}{\partial x_j}\frac{\partial \omega}{\partial x_j}, \quad (4)$$

where μ_{eff} is defined by eq. (1). Eqs. (3) and (4) are the same as the original SST model [7] except that μ_t is replaced by μ_{eff}. If γ is equal to unity, eqs. (3) and (4) are identical to the original SST model. If γ is equal to zero, then eq. (3) becomes the transport equation governing the non-turbulent fluctuating kinetic energy. Then, k is identical to k_{nt} which denotes the energy of non-turbulent fluctuations.

2.2 γ equation and transition onset

The γ equation developed by Wang and Fu [2,3] is adopted here

$$\frac{\partial(\rho\gamma)}{\partial t} + \frac{\partial(\rho u_j \gamma)}{\partial x_j} = \frac{\partial}{\partial x_j}\left\{(\mu + \mu_{eff})\frac{\partial \gamma}{\partial x_j}\right\} + P_\gamma - \varepsilon_\gamma. \quad (5)$$

Refer to further details elsewhere [2,3].

Let us consider:

$$E_u = \frac{1}{2}(u - u_w)_i (u - u_w)_i, \quad (6)$$

where u_w is the wall velocity.

For the γ equation, $|\nabla E_u|$ is applied in the P_γ terms and the F_{onset} terms. In the present paper, the $|\nabla E_u|$ is replaced by strain rate S, and the P_γ and F_{onset} are redefined as the following:

$$P_\gamma = C_4 \rho F_{onset} S \sqrt{2E_u} \sqrt{-\ln(1-\gamma)}\left(1 + C_s \sqrt{\frac{k}{2E_u}}\right)\frac{d}{v}, \quad (7)$$

$$F_{onset} = 1 - \exp\left(-C_6 \frac{\rho \zeta_{eff} \sqrt{k} |\nabla k|}{\mu S \sqrt{2E_u}}\right). \quad (8)$$

It should be noted that the strain rate S is also appears in the SST turbulence model. Then the new version of the k-ω-γ transition model possesses the advantage of saving computational costs when comparing with the original model, for that the CPU time and the memory for computing and saving the $|\nabla E_u|$ are avoided.

3 High-order numerical method based on WCNS-E-5

The RANS equations are chosen as the governing equations. The explicit fifth-order weighted compact nonlinear scheme (WCNS-E-5) developed by Deng et al. [10,11] is adopted for inviscid terms, for example, $\partial \tilde{E}/\partial \xi$ can be acquired by the following procedures:

$$\frac{\partial \tilde{E}_{i,j,k}}{\partial \xi} = \frac{75}{64\Delta\xi}(\tilde{E}_{i+1/2,j,k} - \tilde{E}_{i-1/2,j,k})$$
$$- \frac{25}{384\Delta\xi}(\tilde{E}_{i+3/2,j,k} - \tilde{E}_{i-3/2,j,k})$$
$$+ \frac{3}{640\Delta\xi}(\tilde{E}_{i+5/2,j,k} - \tilde{E}_{i-5/2,j,k}), \quad (9)$$

where the flux at a cell-edge is given as:

$$\tilde{E}_{i+1/2,j,k} = \tilde{E}^*(Q^*_{Li+1/2,j,k}, Q^*_{Ri+1/2,j,k}, \tilde{\xi}_{xi+1/2,j,k}, \tilde{\xi}_{yi+1/2,j,k}, \tilde{\xi}_{zi+1/2,j,k}), \quad (10)$$

and the cell-edge variables $Q^*_{Li+1/2,j,k}$ and $Q^*_{Ri+1/2,j,k}$ may be computed by the fifth-order weighted interpolations that have been derived by Deng and Zhang [10]. Here $\tilde{E}^*$ are the flux splittings which can also be found in ref. [10]. The cell-edge metrics can be obtained by the 4th-order interpolation, for example:

$$\tilde{\xi}_{xi+1/2,j,k} = \frac{1}{16}(-\tilde{\xi}_{xi-1,j,k} + 9\tilde{\xi}_{xi,j,k} + 9\tilde{\xi}_{xi+1,j,k} - \tilde{\xi}_{xi+2,j,k}). \quad (10)$$

The cell-node metrics, such as $\tilde{\xi}_{xi,j,k}$ and $\tilde{\xi}_{yi,j,k}$, are calculated by the conservative metric method (CMM) devised by Deng et al. [16].

The numerical method of viscous terms is the fourth-order staggered central scheme which has been developed by Tu et al. [17] for preventing numerical oscillations which are common for high-order methods. The LU-SGS method is applied for the temporal discretization.

4 Numerical results

4.1 Low speed flat plate

The unite Reynolds number is given as 1.6×10^6/m, the turbulence intensity (Tu) before the plate is 0.03%, and the Mach number is 0.147. Figure 1 shows the 281×141 (Stream-wise × Wall normal) mesh, among which 201 points are allocated on the plate, 40 points are set before the plate, and the other 40 points are set behind the plate. A coarse mesh of 141 × 71 points and a fine mesh of 561 × 281 points are also applied here to study grid convergence. Figure 2 compares the computed skin friction coefficients, C_f, with the experimental data of Schubauer and Klebanoff [18]. It can be found that the two results of WCNS on the fine mesh and the medium mesh are almost identical, and the predicted transition locations agree well with the

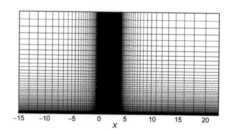

Figure 1 Grid of the flat plate (unit: m).

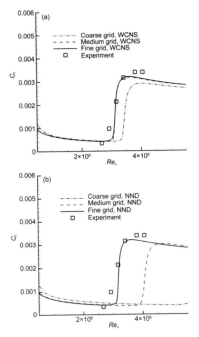

Figure 2 (Color online) Results of the flat plate. (a) WCNS; (b) NND (second-order TVD).

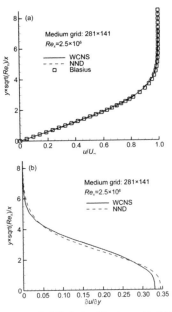

Figure 3 (Color online) Distribution of the stream-wise velocity and its first partial derivative at $Re_x=2.5\times10^6$. (a) velocity u; (b) $\partial u/\partial y$.

experimental data. The transition location of WCNS on the coarse grid is slightly downstream delayed, which indicate the grid density of the coarse grid is insufficient. The results in Figure 2(b) indicate that the second-order TVD (NND) scheme is more sensitive to grid resolution than that of the high-order scheme. Only the fine grid can ensure the NND scheme to give a reasonable result. As the grid become sparse, the transition location would be delayed.

A possible reason for the delay of transition location on crude grid is that transition and turbulence intensity is sensitive to the first derivatives of flow variables. For example, the sheer strain rate is contained in the source terms of both the SST model and the γ equation. However, the sheer strain is a combination of the flow velocity derivatives which are very arduous to be accurately simulated by common CFD methods. Figure 3 gives the stream-wise velocity and partial derivative at the location $Re_x=2.5\times10^6$ where the flow is laminar. It can be found that the velocity profile of NND is almost identical to that of WCNS, and both are consistent to the Blasius profile. However, the difference in $\partial u/\partial y$ between the two numerical results is apparent. This difference may explain why on the same grid different order scheme may produce different transiton location and low-order schemes are more sensitive to grid resolution than high-order schemes. In the following texts, the WCNS is applied without further announcement.

4.2 The S809 airfoil

The experimental test for the S809 airfoil was done in the National Renewable Energy Laboratory. We conduct our computing in according to the experimental results of Somers [19]: the chord Reynolds number is 2×10^6, Mach number is 0.1, and the turbulence intensity is 0.2%. The 433 × 141 mesh shown in Figure 4 is used. Figure 4 also shows the computed intermittency contours for three different angles of attack ($\alpha=0°$, $8°$ and $-8°$). The lift coefficient, drag coefficient as well as pitching moment are shown in Figure 5. The results obtained with a fully turbulent computation (denoted by 'SST') are also plotted in Figure 5. It can be found that significant improvements in the drag coefficient and pitch moment predictions are acquired by the transition prediction method. The transition onset locations are given in Figure 6, which indicates that qualitative agreements between the computation and the experiment are acquired.

4.3 Hypersonic sharp nose cone

Hypersonic flows present unique challenges in both model-

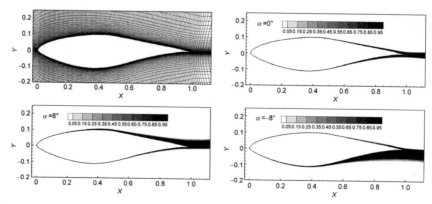

Figure 4 Grid and the intermittency contours of the S809 airfoil.

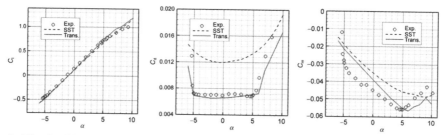

Figure 5 (Color online) Lift coefficient, drag coefficient and pitching moment of the S809 airfoil.

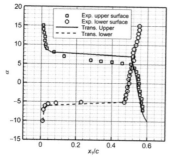

Figure 6 Transition-onset locations on the upper and lower surface of the S809 airfoil.

ing and experimentation [1,20]. This test is chosen to study the behavior of the present transition model in hypersonic transitional flows. The experimental data are collected by Kimmel [21] in the AEDC tunnel B on sharp nose cones at Mach 7.93. The simulation is performed on a 7° half angle and 40 inches (1016.3 mm) length cone. The wall temperature is set to be $0.42T_0$, where T_0 is the free-stream stagnation temperature. The Reynolds number and inlet Tu are set to be the same as that of ref. [20], 6.8×10^6/m and 1.25%, respectively.

Figure 7 shows the intermittency contours both in 3D and 2D vision. Kimmel [21] defined the Stanton number (St) as:

$$St = \frac{\dot{q}_w}{\rho_\infty U_\infty \left(h(T_0) - h(T_w) \right)}, \qquad (20)$$

here h is the enthalpy. Figure 8 compares the Stanton numbers, acquired by the present methods, with that of the "Production Term Modifier (PTM)" by Denissen et al. [20] and the experimental data [21]. There is a free parameter, C_{PTM}, suggesting $C_{PTM} = 1–2$. It can be found from Figure 8 that "$C_{PTM} = 2$" can capture the transition onset, and "$C_{PTM} = 1$" can capture the transition end. However, the transition region predicted by the PTM is overly abrupt. The transition region predicted by the present k-ω-γ method is more consistent to the experimental results.

4.4 Hypersonic blunt nose cone

The blunt nose cone as computed by Yang et al. [22] is

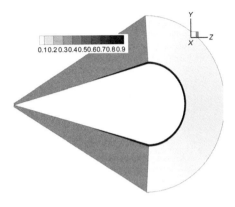

The wall temperature is set to be $3.72T_\infty$. The grids that can avoid the odd axis are shown in Figure 9. The computed intermittency contours are show in Figure 10, and the heat transfer rate is show in Figure 11. It can be seen that the agreement of heat transfer rate between the computation and experiment in laminar region is good, so as to the transition location. In the turbulence region, the computed heat transfer rate is about 10% greater than that of experiment.

5 Concluding remarks

The $k\text{-}\omega\text{-}\gamma$ transition model of Wang and Fu [2,3] has been slightly modified to saving computational costs. In order to minimize numerical errors, the fifth-order WCNS-E-5 scheme [10,11] is applied to discretizing the inviscid fluxes

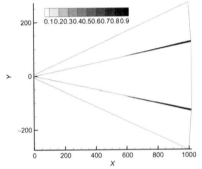

Figure 7 Intermittency contours of the sharp nose cone.

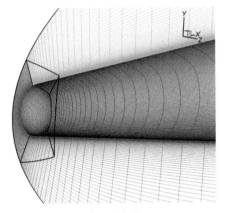

Figure 9 Topology and mesh near the the blunt nose.

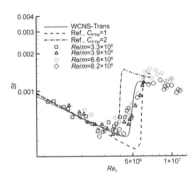

Figure 8 (Color online) Stanton number distribution along the generatrix of the sharp nose cone. Symbols: experiment; Ref.: Denissen et al. [20].

selected to further validate the preset transition model. The blunt nose cone with 5° half angle is 800 mm long, and its nose bluntness is 1mm. The Reynolds number is given as 1.3×10^7/m. The free-stream static temperature, T_∞, is 79 K.

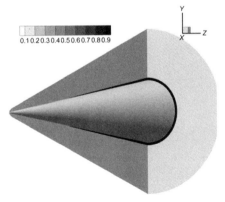

Figure 10 Intermittency contours of the blunt nose cone.

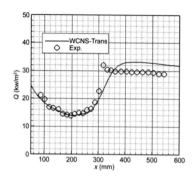

Figure 11 Heat transfer rate of the blunt nose cone

and the fourth-order stager central scheme [17] is applied to discretizing the viscous terms. This high-order package can effectively reduce the grid sensitivity when compared to a second-order TVD scheme. The renewed model is validated against experimental data for cases ranging from low speed to hypersonic flows over a range of inlet turbulence intensities. As the configurations in this paper are two-dimensional or axisymmetric, first mode and second mode instabilities of boundary layer are only considered. Then, the transition model in its present form can be applied to predict the natural transition of two-dimensional and axisymmetric boundary layer flows. Additionally, other transition mechanisms, such as crossflow instability, bypass mechanism, and other more complex ones, will be explored in future studies.

It should be noted that the convergence speed of the γ equation is much slower than the RANS equations in our numerical tests. This shortcoming may derive from the source terms in the right-hand-side of eq. (5). Although quite important, the study on accelerating the convergence speed of the γ equation is beyond the scope of the present report. Another important aspect is the compressibility corrections for hypersonic complex flows. As reported by Tu et al. [23], the density corrections of Catris and Aupoix [24] may suitable for the present model.

The authors are grateful to Prof. FU Song for his useful suggestions and discussions. This study was supported by the National Basic Research Program of China (Grant No. 2009CB723800) and the National Natural Science Foundation of China (Grand No. 11072259).

1. Fu S, Wang L. RANS modeling of high-speed aerodynamic flow transition with consideration of stability theory. Prog Aerospace Sci, 2012, http://dx.doi.org/10.1016/j.paerosci.2012.08.004
2. Wang L, Fu S. Modelling flow transition in a hypersonic boundary layer with Reynolds-averaged Navier-Stokes approach. Sci China Ser G-Phys Mech Astron, 2009, 52(5): 768–774
3. Wang L, Fu S. Development of an intermittency equation for the modeling of the supersonic/hypersonic boundary layer flow transition. Flow Turbulence Combust, 2011, 87: 165–187
4. Wang L, Fu S, Carnarius A, et al. A modular RANS approach for modelling laminar-turbulent transition in turbomachinery flows. Int J Heat Fluid Flow, 2012, 34: 62–69
5. Warren E S, Hassan H A. A transition closure model for predicting transition onset. J Aircraft, 1998, 35(5): 769–775
6. Papp J L, Kenzakowski D C, Dash S M. Extensions of a rapid engineering approach to modeling hypersonic laminar to turbulent transitional flows. AIAA Paper, 2005, AIAA-2005-892
7. Menter F R. Zonal two equation k-ω turbulence models for aerodynamic flows. AIAA Paper, 1993, AIAA-93-2906
8. Roy C J, Blottner F G. Review and assessment of turbulence models for hypersonic flows. Prog Aerospace Sci, 2006, 42: 469–530
9. Celic A, Hirschel E H. Comparison of eddy-viscosity turbulence models in flows with adverse pressure gradient. AIAA J, 2006, 44(10): 2156–2169
10. Deng X, Zhang H. Developing high-order accurate nonlinear schemes. J Comput Phys, 2000, 165: 22–44
11. Deng X. High-order accurate dissipative weighted compact nonlinear schemes. Sci China Ser A-Math Phys Astron, 2002, 45(3): 356–370
12. Deng X, Mao M, Tu G, et al. Extending weighted compact nonlinear schemes to complex grids with characteristic-based interface conditions. AIAA J, 2010, 48(12): 2840–2851
13. Deng X, Mao M, Tu G, et al. High-order and high accurate CFD methods and their applications for complex grid problem. Commun Comput Phys, 2012, 11(4): 1081–1102
14. Nonomura T, Iizuka N, Fujii, K. Freestream and vortex preservation properties of high-order WENO and WCNS on curvilinear grids. Comput Fluids, 2010, 39: 197–214
15. Zhang H, Zhuang F. NND schemes and their applications to numerical simulation of two- and three-dimensional flows. Adv Appl Mech, 1991, 29: 193–256
16. Deng X, Mao M, Tu G, et al. Geometric conservation law and applications to high-order finite difference schemes with stationary grids. J Comput Phys, 2011, 230(4): 1100–1115
17. Tu G, Deng X, Mao M. A staggered non-oscillatory finite difference method for high-order discretization of viscous terms (in Chinese). Acta Aerodyn Sin, 2011, 29(1): 6–11
18. Schubauer G B, Klebanoff P S. Contribution on the Mechanics of Boundary Layer Transition. NACA TN 3489, 1955
19. Somers D M. Design and Experimental Results for the s809 Airfoil. SR-440-6918, National Renewable Energy Laboratory, 1997
20. Denissen N A, Yoder D A, Georgiadis N J. Implementation and Validation of a Laminar-to-Turbulent Transition Model in the Wind-US code. NASA/TM-2008-215451, 2008
21. Kimmel R L. The effect of pressure gradients on transition zone length in hypersonic boundary layers. J Fluids Eng, 1997, 119: 36–41
22. Yang Y, Shen Q, Zhan H, et al. Investigation on asymmetric transition about hypersonic boundary layer over a slight blunt cone (in Chinese). J Astronaut, 2008, 29(1): 34–39
23. Tu G, Deng X, Mao M. Assessment of two turbulence models and some compressibility corrections for hypersonic compression corners by high-order difference schemes. Chin J Aeronaut, 2012, 25(1): 25–32
24. Catris S, Aupoix B. Density corrections for turbulence models. Aerospace Sci Tech, 2000, 4(1): 1–11

Osher Flux with Entropy Fix for Two-Dimensional Euler Equations

Huajun Zhu[1,*], Xiaogang Deng[2], Meiliang Mao[1,3], Huayong Liu[1] and Guohua Tu[1]

[1] *State Key Laboratory of Aerodynamics, China Aerodynamics Research and Development Center, Mianyang, Sichuan 621000, China*
[2] *National University of Defense Technology, Changsha, Hunan 410073, China*
[3] *Computational Aerodynamics Institute, China Aerodynamics Research and Development Center, Mianyang, Sichuan 621000, China*

Received 21 January 2014; Accepted (in revised version) 3 July 2015

Abstract. We compare in this paper the properties of Osher flux with O-variant and P-variant (Osher-O flux and Osher-P flux) in finite volume methods for the two-dimensional Euler equations and propose an entropy fix technique to improve their robustness. We consider both first-order and second-order reconstructions. For inviscid hypersonic flow past a circular cylinder, we observe different problems for different schemes: A first-order Osher-O scheme on quadrangular grids yields a carbuncle shock, while a first-order Osher-P scheme results in a dislocation shock for high Mach number cases. In addition, a second-order Osher scheme can also yield a carbuncle shock or be unstable. To improve the robustness of these schemes we propose an entropy fix technique, and then present numerical results to show the effectiveness of the proposed method. In addition, the influence of grid aspects ratio, relative shock position to the grid and Mach number on shock stability are tested. Viscous heating problem and double Mach reflection problem are simulated to test the influence of the entropy fix on contact resolution and boundary layer resolution.

AMS subject classifications: 65M06, 65M70

Key words: Osher flux, entropy fix, Euler equation, finite volume method, "carbuncle" shock.

1 Introduction

Upwind schemes based on the characteristic theory have been widely used in solving the Euler equations for their capability of capturing discontinuities. Upwind schemes are usually classified into flux-vector and flux-difference splitting schemes. Flux-difference splitting schemes such as Osher scheme [1,2] and Roe scheme [3] have gained popularity as they are robust and can capture steady discontinuities. Since the Osher scheme can capture steady discrete shock sharply [4–7], it has been widely used in fluid dynamics simulations [8–11]. There have been many numerical evidences showing the superiority of the Osher flux. For instance, Qiu et al. has done lots of work in [12–14] to compare the Osher fluxes with others recently, including Lax-Friedrich (LF), Godunov, Harten-Lax-van Leer (HLL), HLLC which is a modification of the HLL. They found that among these considered fluxes the Osher flux is the most robust one and has the best resolution near discontinuities. In addition, the Osher flux is also compared with the LF flux and the Vijayasundaram flux by Felcman and Havle in [15], which also shows that the numerical results obtained by the Osher scheme is the most accurate one.

For a supersonic blunt body problem, it was shown that the Roe's splitting produces a "carbuncle" schock (see [16, 17]). To solve this problem many efforts have been made. For instance, Lin proposed a cure by adding dissipation in linear waves to Roe's Riemann solver [18]. Sanders developed H-corrections based on the difference of the characteristic speeds on five interfaces [19]. Pandolfi and Ambrosio added appropriate dissipation according to the maximum difference between the characteristic speeds across the four interfaces [20]. Ren proposed a rotated Roe scheme based on the rotated Roe's approximate Riemann solver [21]. For Euler equations, the Osher flux takes a path integral of flux derivative in phase space. The integral subpaths are constructed based on the eigenvalues in decreasing order by Osher [4] and in increasing order by Spekreijse [22], which are referred to as the Osher-O flux and the Osher-P flux, respectively. There are already some investigations about the properties of these two different fluxes. For instance, Jacobs found that it is difficult to apply the Osher scheme for strong shock computations and proposed a 3-stage approximate Riemann solver at the cost of computations [23]. Amaladas and Kamath compared six kinds of upwind schemes based on steger-Warming, van Leer, Roe, Osher-P, AUSM and HUS, respectively [24]. It is indicated that the Osher schemes with O-variant and P-variant have different performance in inviscid hypersonic flow past a circular cylinder under structured mesh and the Osher scheme with O-variant can not result in a converged solution. However, detail comparisons of the two fluxes and the treatment of instability have not been investigated yet, which is the motivation of this paper.

The purpose of this paper is to compare the performance of the Osher-O and Osher-P fluxes for the two-dimensional (2D) Euler equation and to improve their shock robustness. Firstly, the Osher flux is combined with first-order and second-order reconstructions, respectively, to construct finite volume methods. Secondly, for inviscid hypersonic flow past a circular cylinder, the Osher schemes are investigated from the aspects of different grids and different Mach numbers. We find that the first-order Osher scheme will produce a "carbuncle" shock or "dislocation" shock and the second-order Osher scheme becomes unstable at the head of the cylinder. In order to improve the robustness of the above numerical methods, we propose an entropy fix technique, which adds dissipation based on the maximum difference of the characteristic speeds across the four interfaces.

Numerical results show its effectiveness. We also tested the influence of grid aspect ratio, relative shock position to grid and Mach number on shock instability. Thirdly, to test the influence of the entropy fix on contact resolution and boundary layer resolution, Osher schemes with entropy fix are applied to simulate viscous heating problem and double Mach reflection problem.

The rest of the paper is organized as follows. In Section 2, we state the finite volume discretization with a second-order upwind linear reconstruction and the Osher flux. In Section 3, we present numerical investigations to show the performance of the Osher flux. Moreover, in order to improve the numerical results an entropy fix technique is proposed, which is the most important contribution of this paper. Finally, conclusions drawn from the results are given in Section 4.

2 Finite volume discretization

The 2D Euler equation can expressed as the following conservation form:

$$U_t + f(U)_x + g(U)_y = 0, \tag{2.1}$$

where U is the vector of conserved variable, f and g are convective fluxes

$$U = \begin{pmatrix} \rho \\ \rho u \\ \rho v \\ e \end{pmatrix}, \quad f(U) = \begin{pmatrix} \rho u \\ \rho u^2 + p \\ \rho u v \\ u(e+p) \end{pmatrix}, \quad g(U) = \begin{pmatrix} \rho v \\ \rho u v \\ \rho v^2 + p \\ v(e+p) \end{pmatrix}.$$

Here ρ, u and v denotes the density, velocity components in x and y directions, respectively, p is the pressure and e is the total energy. For the finite volume method, let us integrate the 2D Euler equation in a finite volume cell Ω_i. Then by using the Green formula, we obtain

$$\frac{\partial \overline{U_i}}{\partial t} + \frac{1}{|\Omega_i|} \int_{\partial \Omega_i} \vec{F} \cdot \vec{n} \, dl = 0, \tag{2.2}$$

where $\vec{F} = (f,g)$, $\overline{U_i} = \frac{1}{|\Omega_i|} \int_{\Omega_i} U dx dy$ denotes cell average of U over the cell Ω_i, $|\Omega_i|$ is the area of the cell Ω_i, $\partial \Omega_i$ the boundary of Ω_i, and $\vec{n}$ the outward unit normal. Here "·" represents the inner product algorithm. Suppose the cell Ω_i has m edges. Then the integral of the flux in the second term of Eq. (2.2) can be evaluated as

$$\int_{\partial \Omega_i} \vec{F} \cdot \vec{n} \, dl = \sum_{j=1}^{m} h_{ij}(U_{ij,L}, U_{ij,R}) l_{ij}, \tag{2.3}$$

where h_{ij} is a numerical approximation of $\vec{F} \cdot \vec{n}$, $U_{ij,L}$ and $U_{ij,R}$ are the left and right Riemann states of the j_{th} edge of $\partial \Omega_i$, respectively, and l_{ij} is the length of the j_{th} edge. Here $U_{ij,L}$ and $U_{ij,R}$ can be evaluated by reconstruction functions. In the following sections, we will present an upwind limited linear reconstruction procedure and introduce the Osher flux. Then, we finally construct a second order finite volume method by combining the Osher flux and the reconstruction.

2.1 Reconstruction

In this paper, we will consider a constant reconstruction and a linear reconstruction. In a constant reconstruction, the values at any points in the cell are equal to the cell average values. The linear reconstruction is given in the following.

Let us set a finite volume cell Ω to be a triangle $\triangle_{A_1A_2A_3}$ with three vertices A_1, A_2 and A_3. And take u as an example to illustrate the reconstruction procedure. Now consider a linear reconstruction function in $\triangle_{A_1A_2A_3}$ with the form

$$u(X) = \bar{u}_{A_1A_2A_3} + \nabla u_{A_1A_2A_3} \cdot (X - X_{A_1A_2A_3}), \tag{2.4}$$

where $\bar{u}_{A_1A_2A_3}$ is the cell average of u on $\triangle_{A_1A_2A_3}$, and $\nabla u_{A_1A_2A_3}$ the gradient of u at the cell center $X_{A_1A_2A_3}$, evaluated as

$$\nabla u_{A_1A_2A_3} \approx \frac{1}{|\Omega|} \oint_{\partial \Omega} u \vec{n} \, dl. \tag{2.5}$$

Firstly, the values at each vertex is reconstructed by using the nodal averaging procedure [25], that is

$$u_{A_i} = \frac{\sum_{j=1}^{N} u_{i,j}/r_{i,j}}{\sum_{j=1}^{N} 1/r_{i,j}}, \tag{2.6}$$

where N denotes the number of cells which share the vertex A_i, $u_{i,j}$ is the cell average of u on the j_{th} cell, and $r_{i,j}$ the distance between the cell center of the j_{th} cell and the vertex A_i. Then, the gradient $\nabla u_{A_1A_2A_3}$ can be given by the midpoint-trapezoidal rule, i.e.,

$$\nabla u_{A_1A_2A_3} \approx \frac{1}{|\Omega|} \left(\frac{u_{A_1}+u_{A_2}}{2} l_{12} \vec{n}_{12} + \frac{u_{A_2}+u_{A_3}}{2} l_{23} \vec{n}_{23} + \frac{u_{A_3}+u_{A_1}}{2} l_{31} \vec{n}_{31} \right). \tag{2.7}$$

From [25] we know that the above reconstruction has upwind properties.

Secondly, the reconstruction values $\tilde{u}_{A_i}$ at the vertex A_i is obtained by substituting Eq. (2.7) into Eq. (2.4) and setting $X = X_{A_i}$. To avoid numerical oscillations, a slope limiter is necessary in the reconstruction procedure. Here to make the reconstruction function to be monotonous, we use the following slope limiter

$$\phi_{A_i} = \begin{cases} \min\left\{1, \frac{u_{\max} - \bar{u}_{A_1A_2A_3}}{\tilde{u}_{A_i} - \bar{u}_{A_1A_2A_3}}\right\}, & \tilde{u}_{A_i} - \bar{u}_{A_1A_2A_3} > 0, \\ \min\left\{1, \frac{u_{\min} - \bar{u}_{A_1A_2A_3}}{\tilde{u}_{A_i} - \bar{u}_{A_1A_2A_3}}\right\}, & \tilde{u}_{A_i} - \bar{u}_{A_1A_2A_3} < 0, \\ 1, & \tilde{u}_{A_i} - \bar{u}_{A_1A_2A_3} = 0, \end{cases} \tag{2.8}$$

which is a generalization from the van Leer's 1D limiter [26,27] to 2D unstructured mesh. Here $u_{\min}$ and $u_{\max}$ are the minimum and the maximum of the cell averages surrounding the vertex A_i, respectively.

Finally, the linear reconstruction function with limiter reads

$$u(X) = \bar{u}_{A_1A_2A_3} + \phi_{A_1A_2A_3} \nabla u_{A_1A_2A_3} \cdot (X - X_{A_1A_2A_3}), \tag{2.9}$$

where $\phi_{A_1A_2A_3} = \min_{i=1,2,3} \{\phi_{A_i}\}$.

2.2 Osher flux

For an arbitrary edge noted as AB of the finite volume cell Ω, let us denote the outward normal vector of the edge as $\vec{n} = (n_x, n_y)$, and the complex flux as $F(U) = (f,g) \cdot \vec{n} = f(U)n_x + g(U)n_y$. To simplify the question, we take a coordinate transformation as follws,

$$\begin{pmatrix} \xi \\ \eta \end{pmatrix} = \begin{pmatrix} xn_x + yn_y \\ yn_x - xn_y \end{pmatrix}. \tag{2.10}$$

After transformation, the outward nornal vector of the edge becomes $\vec{n_2} = (1,0)$ in the (ξ, η) coordinate system. In addition, the Euler equation (2.1) can be rewritten as

$$U_t + \tilde{f}_\xi + \tilde{g}_\eta = 0, \tag{2.11}$$

where $\tilde{f} = \xi_x f(U) + \xi_y g(U)$ and $\tilde{g} = \eta_x f(U) + \eta_y g(U)$. According to the transformation (2.10), we have $\xi_x = n_x$ and $\xi_y = n_y$, thus $\tilde{f} = F(U)$. Now the main task is to compute $\tilde{f}$ along the edge. If we walk along the boundary $\partial \Omega$ anticlockwise, the inside of the cell is on the left and the outside of the cell is on the right. Assume the left state to be U_L and the right state be U_R. Then it can be treated as a Riemann problem in the ξ direction with flux function $\tilde{f}$. Hence the one-dimensional Osher flux can be used here. In the following we just present the essential steps of the implementation of Osher flux. We refer the interested readers to [4,22,28] for more details.

The Jacobian matrix of the complex flux is

$$\frac{\partial F(U)}{\partial U} = n_x \frac{\partial f(U)}{\partial U} + n_y \frac{\partial g(U)}{\partial U}, \tag{2.12}$$

which has eigenvalues $\lambda_1 = \tilde{u} \mp c$, $\lambda_2 = \tilde{u}$ and $\lambda_3 = \tilde{u} \pm c$. Here the upper and lower signs are used for the O- and P-variants, respectively. It is referred to as the O-variant when the eigenvalues λ_3, λ_2 and λ_1 are in decreasing order, and as the P-variant when they are in increasing order. The two independent contravariant velocities for the Jacobian matrix are $\tilde{u} = un_x + vn_y$ and $\tilde{v} = vn_x - un_y$, respectively. Set Γ_1, Γ_2, Γ_3 be the integral curves corresponding to λ_3, λ_2, λ_1. The Riemann invariants corresponding to $\tilde{u} - c$ are $\tilde{u} + \frac{2}{\gamma-1}c$, p/ρ^γ and $\tilde{v}$. As for $\tilde{u} + c$, the Riemann invariants are $\tilde{u} - \frac{2}{\gamma-1}c$, p/ρ^γ and $\tilde{v}$. As for $\tilde{u}$, the Riemann invariants are $\tilde{u}$ and p. Let the two intermediate states be $U_{1/3} =$

$(\rho_{1/3}, u_{1/3}, v_{1/3}, p_{1/3})$ and $U_{2/3} = (\rho_{2/3}, u_{2/3}, v_{2/3}, p_{2/3})$, which can be determined by solving the following equations,

$$\rho_{2/3} = \left\{ \frac{(\gamma-1)(\tilde{u}_R - \tilde{u}_L)/2 \pm (c_R + c_L)}{c_R [1 + (p_L/p_R)^{1/(2\gamma)} (\rho_L/\rho_R)^{-1/2}]} \right\}^{\frac{2}{\gamma-1}} \rho_R, \tag{2.13a}$$

$$\rho_{1/3} = \left\{ \frac{(\gamma-1)(\tilde{u}_R - \tilde{u}_L)/2 \pm (c_R + c_L)}{c_L [1 + (p_R/p_L)^{1/(2\gamma)} (\rho_R/\rho_L)^{-1/2}]} \right\}^{\frac{2}{\gamma-1}} \rho_L, \tag{2.13b}$$

$$p_{1/3} = p_L \left(\frac{\rho_{1/3}}{\rho_L}\right)^\gamma, \quad c_{1/3} = \sqrt{\frac{\gamma p_{1/3}}{\rho_{1/3}}}, \quad c_{2/3} = \sqrt{\frac{\gamma p_{2/3}}{\rho_{2/3}}}, \quad \tilde{u}_{1/3} = \tilde{u}_L \mp \frac{2}{\gamma-1}(c_L - c_{1/3}), \tag{2.13c}$$

$$p_{2/3} = p_{1/3}, \quad \tilde{u}_{2/3} = \tilde{u}_{1/3}, \quad \tilde{v}_{2/3} = \tilde{v}_R, \quad \tilde{v}_{1/3} = \tilde{v}_L. \tag{2.13d}$$

Then $u_{1/3}$ and $v_{1/3}$ can be obtained from $\tilde{u}_{1/3}$ and $\tilde{v}_{1/3}$. Similarly, $u_{2/3}, v_{2/3}$ can be obtained from $\tilde{u}_{2/3}, \tilde{v}_{2/3}$. Thus, $U_{1/3}$ and $U_{2/3}$ are determined.

For the integral curves Γ_1 and Γ_3, we calculate the sonic points $U_{s_1} = (\rho_{s_1}, u_{s_1}, v_{s_1}, p_{s_1})$ and $U_{s_2} = (\rho_{s_2}, u_{s_2}, v_{s_2}, p_{s_2})$, which are determined by the following equations

$$\tilde{u}_{s_1} = \frac{\gamma-1}{\gamma+1} \tilde{u}_L \mp \frac{2}{\gamma+1} c_L, \qquad c_{s_1} = \mp \tilde{u}_{s_1}, \tag{2.14a}$$

$$\rho_{s_1} = \left(\frac{c_{s_1}}{c_L}\right)^{2/(\gamma-1)} \rho_L, \qquad p_{s_1} = p_L \left(\frac{\rho_{s_1}}{\rho_L}\right)^\gamma, \tag{2.14b}$$

$$\tilde{v}_{s_1} = \tilde{v}_L, \qquad \tilde{u}_{s_2} = \frac{\gamma-1}{\gamma+1} \tilde{u}_R \pm \frac{2}{\gamma+1} c_R, \tag{2.14c}$$

$$c_{s_2} = \pm \tilde{u}_{s_2}, \qquad \rho_{s_2} = \left(\frac{c_{s_2}}{c_R}\right)^{2/(\gamma-1)} \rho_R, \tag{2.14d}$$

$$p_{s_2} = p_R \left(\frac{\rho_{s_2}}{\rho_R}\right)^\gamma, \qquad \tilde{v}_{s_2} = \tilde{v}_R. \tag{2.14e}$$

Finally, the Osher flux approximation to the complex flux F is

$$\frac{1}{2} \left[F(U_L) + F(U_R) - \int_{U_L}^{U_R} |\partial_U F(U)| dU \right]$$
$$= \frac{1}{2} [1 + \text{sign}(\lambda_3(U_0))] F_0 + \frac{1}{2} [\text{sign}(\lambda_3(U_{1/3})) - \text{sign}(\lambda_3(U_0))] F_{s_1}$$
$$+ \frac{1}{2} [\text{sign}(\tilde{u}_{1/3}) - \text{sign}(\lambda_3(U_{1/3}))] F_{1/3} + \frac{1}{2} [\text{sign}(\lambda_1(U_{2/3})) - \text{sign}(\tilde{u}_{1/3})] F_{2/3}$$
$$+ \frac{1}{2} [\text{sign}(\lambda_1(U_1)) - \text{sign}(\lambda_1(U_{2/3}))] F_{s_2} + \frac{1}{2} [1 - \text{sign}(\lambda_1(U_1))] F_1, \tag{2.15}$$

where $F_{s_1} = F(U_{s_1})$, $F_0 = F(U_L)$. Since the intermediate states $U_{1/3}$ and $U_{2/3}$ are different for O-variant and P-variant, the resulted flux are different, which will be verified by numerical investigation in Section 3.

2.3 Osher flux with entropy fix

In this subsection, we give small perturbation analysis for Osher-O flux and Osher-P flux to explain instabilities of such fluxes. Then we propose an entropy fix technique to improve the robustness of Osher flux.

2.3.1 Small perturbation analysis

We assume initial values at two adjacent cells

$$\rho_L = 1-\hat{\rho}, \quad u_L = u_0 - \hat{u}, \quad v_L = 0, \quad p_L = 1-\hat{p},$$
$$\rho_R = 1+\hat{\rho}, \quad u_R = u_0 + \hat{u}, \quad v_R = 0, \quad p_R = 1+\hat{p},$$

as shown in Fig. 1. Since perturbations $\hat{\rho}$, $\hat{u}$, $\hat{p}$ are small, we can obtain

$$c_L = \sqrt{\gamma}\left(1 - \frac{\hat{p}}{2} + \frac{\hat{\rho}}{2}\right), \quad c_R = \sqrt{\gamma}\left(1 + \frac{\hat{p}}{2} - \frac{\hat{\rho}}{2}\right),$$

and the two intermediate states

$$\rho_{1/3} = \left(1 - \hat{\rho} + \frac{\hat{p}}{\gamma}\right), \quad u_{1/3} = u_0 - \hat{u}, \quad v_{1/3} = v_{2/3} = \frac{\hat{p}}{\sqrt{\gamma}}, \quad c_{1/3} = \sqrt{\gamma}\left(1 + \frac{\hat{\rho}}{2} - \frac{\hat{p}}{2\gamma}\right),$$
$$\rho_{2/3} = \left(1 + \hat{\rho} - \frac{\hat{p}}{\gamma}\right), \quad u_{2/3} = u_0 + \hat{u}, \quad p_{2/3} = p_{1/3} = 1, \quad c_{2/3} = \sqrt{\gamma}\left(1 - \frac{\hat{\rho}}{2} + \frac{\hat{p}}{2\gamma}\right).$$

Thus the fluxes can be evaluated as

$$F(U_{1/3}) = F(U_{2/3}) = \begin{pmatrix} -\dfrac{\hat{p}}{\sqrt{\gamma}} \\ -u_0 \dfrac{\hat{p}}{\sqrt{\gamma}} \\ -1 \\ -\left(\dfrac{\gamma}{\gamma-1} + \dfrac{u_0^2}{2}\right)\dfrac{\hat{p}}{\sqrt{\gamma}} \end{pmatrix}, \quad F(U_L) = F(U_R) = \begin{pmatrix} 0 \\ 0 \\ -1 \\ 0 \end{pmatrix}.$$

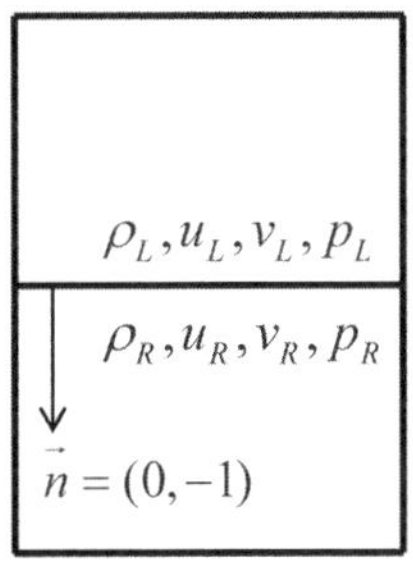

Figure 1: Initial values with small perturbation.

Finally, the Osher-P flux is

$$\frac{1}{2}\{F(U_L)+F(U_R)-[(F(U_L)-F(U_{1/3}))+(F(U_R)-F(U_{2/3}))]\}$$

$$=\frac{1}{2}(F(U_{1/3})+F(U_{2/3}))=\begin{pmatrix}-\dfrac{\hat{p}}{\sqrt{\gamma}}\\ -u_0\dfrac{\hat{p}}{\sqrt{\gamma}}\\ -1\\ -\left(\dfrac{\gamma}{\gamma-1}+\dfrac{u_0^2}{2}\right)\dfrac{\hat{p}}{\sqrt{\gamma}}\end{pmatrix}$$

and Osher-O flux is

$$\frac{1}{2}\{F(U_L)+F(U_R)-[(F(U_{1/3})-F(U_L))+(F(U_{2/3})-F(U_R))]\}$$

$$=F(U_L)+F(U_R)-\frac{1}{2}(F(U_{1/3})+F(U_{2/3}))=\begin{pmatrix}\dfrac{\hat{p}}{\sqrt{\gamma}}\\ u_0\dfrac{\hat{p}}{\sqrt{\gamma}}\\ -1\\ \left(\dfrac{\gamma}{\gamma-1}+\dfrac{u_0^2}{2}\right)\dfrac{\hat{p}}{\sqrt{\gamma}}\end{pmatrix}.$$

We can see that both the Osher-P flux and the Osher-O flux are only affected by the pressure perturbation ($\hat{p}\neq 0$) but not affected by density or shear velocity perturbations ($\hat{\rho}\neq 0$, $\hat{u}\neq 0$). According to the work of Pandolfi in [20], such kind of fluxes are strongly affected by numerical instabilities and adding appropriate dissipation is a good technique to deal with this problem.

2.3.2 Entropy fix

We propose entropy fix technique for the interface face0, as shown in Fig. 2. It is based on a detection routine as in [20]. Here the maximum difference of the characteristic speeds across the four interfaces connected to the interface face0 is evaluated by U_L and U_R directly but not by the intermediate states,

$$\eta=\max\{\epsilon_i\}_{i=1,2,3,4},\quad \epsilon_i=|\tilde{u}_i^L-\tilde{u}_i^R|+|c_i^L-c_i^R|, \tag{2.16}$$

where $\tilde{u}_i=un_x+vn_y$ is the velocity components normal to the interface face$i$. Then the global flux F_{ef} across the interface face0 is evaluated as

$$F_{ef}=F_{Osher}-\eta(U_0^R-U_0^L),$$

where U_0^R and U_0^L are the reconstructed variables at the interface face0. Here F_{Osher} is the Osher flux computed by (2.15). The entropy fix is added for original Osher flux F_{Osher} but not the intermediate flux $F_{1/3}$ or $F_{2/3}$.

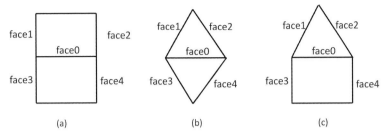

Figure 2: Four interfaces connected to the interface face0.

2.4 Boundary conditions

For the side at the solid boundary of the computational area, we take the physical solid wall boundary condition. Denote the cell average value of the cell containing this side as $U_{1/2}$. Then we can find the boundary value U_0 by using the integral curve of $\tilde{u}-c$ which connects $U_{1/2}$ to U_0. And U_0 is calculated according to the Riemann invariants

$$\frac{2}{\gamma-1}c_0 = \tilde{u}_{1/2} + \frac{2}{\gamma-1}c_{1/2}, \tag{2.17a}$$

$$\tilde{u}_0 = 0, \tag{2.17b}$$

$$\frac{p_0}{\rho_0^\gamma} = \frac{p_{1/2}}{\rho_{1/2}^\gamma}, \tag{2.17c}$$

$$\tilde{v}_0 = \tilde{v}_{1/2}. \tag{2.17d}$$

Hence c_0 is simply determined by Eq. (2.17a). By noting that $c_0 = \sqrt{\gamma p_0/\rho_0}$, it is straightforward to solve from Eq. (2.17c) that

$$\rho_0 = \left(\frac{c_0}{c_{1/2}}\right)^{\frac{2}{\gamma-1}} \rho_{1/2}, \quad p_0 = \left(\frac{\rho_0}{\rho_{1/2}}\right)^\gamma p_{1/2}. \tag{2.18}$$

Therefore, the values of U_0 are determined. If a sonic point exists in the integral curve of $\tilde{u}-c$, which denoted by $U_{1/6}$, then the corresponding state can be obtained from Riemann invariants, i.e.,

$$\tilde{u}_{1/6} = c_{1/6} = \frac{\gamma-1}{\gamma+1}\left(\tilde{u}_{1/2} + \frac{2}{\gamma-1}c_{1/2}\right), \quad \rho_{1/6} = \left(\frac{c_{1/6}}{c_{1/2}}\right)^{\frac{2}{\gamma-1}} \rho_{1/2}, \tag{2.19a}$$

$$p_{1/6} = \left(\frac{\rho_{1/6}}{\rho_{1/2}}\right)^\gamma p_{1/2}, \quad \tilde{v}_{1/6} = \tilde{v}_{1/2}. \tag{2.19b}$$

So we have

$$\int_{U_{1/2}}^{U_0} (\partial F)^- dU = \begin{cases} F(U_0) - F(U_{1/6}), & \tilde{u}_{1/2} - c_{1/2} > 0, \\ F(U_0) - F(U_{1/2}), & \tilde{u}_{1/2} - c_{1/2} \leq 0, \end{cases} \tag{2.20}$$

and
$$\int_{U_{1/2}}^{U_0} (\partial F)^+ dU = \begin{cases} F(U_{1/6}) - F(U_{1/2}), & \tilde{u}_{1/2} - c_{1/2} > 0, \\ 0, & \tilde{u}_{1/2} - c_{1/2} \leq 0. \end{cases} \quad (2.21)$$

Therefore, the Osher flux for the solid boundary is

$$\frac{1}{2}\left[F(U_{1/2}) + F(U_0) - \int_{U_{1/2}}^{U_0} |\partial F| dU\right] = \begin{cases} F_0 + F_{1/2} - F_{1/6}, & \tilde{u}_{1/2} - c_{1/2} > 0, \\ F_0, & \tilde{u}_{1/2} - c_{1/2} \leq 0. \end{cases} \quad (2.22)$$

2.5 Time discretization

We take TVD Runge-Kutta scheme for time discretization, i.e.,

$$U^{(1)} = U^n + \Delta t L(U^n), \quad (2.23a)$$

$$U^{n+1} = \frac{U^n}{2} + \frac{1}{2}\left[U^{(1)} + \Delta t L(U^{(1)})\right], \quad (2.23b)$$

where Δt is the time-step. In this paper we choose $\Delta t = \frac{CFl \cdot \Delta x}{\max\{|u|+c,|v|+c\}}$ for Euler equations, where Δx is the space-step and $CFL = 0.1$ is fixed for convenience.

3 Numerical investigation

3.1 Inviscid hypersonic flow past a circular cylinder

We present in this section numerical experiments for inviscid hypersonic flow past a circular cylinder. We consider different grid aspect ratios, different relative shock positions to the grid and different Mach numbers.

We mainly consider three grids with different grid density, which are the grids with 40×80 (Grid1) intervals, 30×240 intervals (Grid2) and 87×100 intervals (Grid3), as shown in Fig. 3. Here the grid with 40×80 intervals, for instance, means that it is composed of 40 radials and 80 tangential intervals. In addition, grid points on the boundary are equally located for Grid1.

The numerical results are categorized according to three properties which are symmetry, convergence and shock location. We use "1" and "0" to denote the good performance and bad performance of the property as shown in Table 1. Solution is convergent if the density residual $Res^n = \max(|\rho^n - \rho^{n-1}|)$ reduces by at least fifth orders of magnitude. To evaluate the quality of the numerical solution, we divide the 8 results into three classes, which contains good results with symmetric convergent solution marked by black (class A), bad results with shock breakdown marked by red (class B) and the other results marked by blue (class C), as shown in Table 2.

Firstly, we apply first order and second order schemes based on Osher-O and Osher-P to solve the problem on the three grids for the case $M_\infty = 20$. The numerical results are given in Table 3 and the results on Grid1 are also shown in Fig. 4. We can see that first-order Osher-O scheme, second-order Osher-O and second-order Osher-P scheme yield

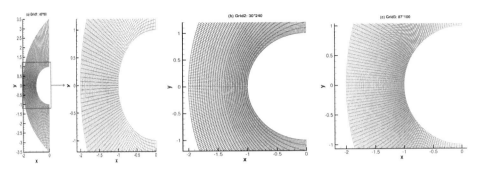

Figure 3: Grid1, Grid2 and Grid3.

"carbuncle" shocks at the head of cylinder and the first-order Osher-P scheme yields a "dislocation" shock at the angle $\theta \approx 45^0$ near the bow shock. After entropy fix technique in Section 2.3.2 is taken, all of the four schemes result in convergent solutions and are free from "carbuncle" shock or "dislocation" shock. These results show the effectiveness of the entropy fix technique. As for Roe flux with Harten's entropy fix, from Table 3 we can see that convergent solutions on these three grids can only be obtained by first order schemes.

Table 1: Three properties of the numerical solution.

Shock location	Symmetry	Convergence
1: regular	1: symmetric	1: convergent
0: breakdown	0: asymmetric	0: not convergent

Table 2: Quality evaluation of the numerical solution.

Classes	A	B			C			
Case	7	6	5	4	3	2	1	0
Solution	1,1,1	1,1,0	1,0,1	1,0,0	0,1,1	0,1,0	0,0,1	0,0,0

Table 3: Test results for different numerical fluxes on grids with different grid ratio ($M_\infty = 20$).

Schemes	Grid1	Grid2	Grid3
1st Osher-O	0	0	0
1st Osher-P	3	3	3
1st Osher-O(E-fix)	7	7	7
1st Osher-P(E-fix)	7	7	7
1st Roe(E-fix)	7	7	7
2nd Osher-O	0	0	0
2nd Osher-P	2	3	4
2nd Osher-O(E-fix)	7	7	7
2nd Osher-P(E-fix)	7	7	7
2nd Roe(E-fix)	0	0	0

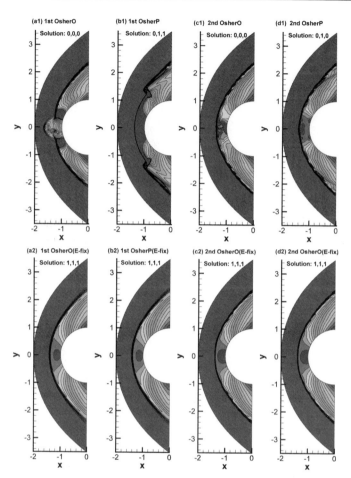

Figure 4: Density contours for schemes without entropy fix (Up) and with entropy fix (Down) on Grid1.
(colored picture attached at the end of the book)

Secondly, we investigate the influence of grid aspect ratio, relative shock position to the grid (shock-grid-position) and Mach number on shock stability. Using the Grid1, we can obtain grids with different aspect ratio by refining tangential intervals and can obtain a series of grids with different shock-grid-position by moving the middle 11 grid points in the symmetric line towards the head of the cylinder with distances $\alpha \cdot \Delta x$ ($\alpha = 0.0, 0.1, 0.2, \cdots, 1.0$), as shown in Fig. 5. Tables 4 and 5 give numerical results for $M_\infty = 20$ while Table 6 gives the numerical results for the case $M_\infty = 8$. For the flux with entropy fix, shock stability are mainly influenced by grid aspect ratio for first order schemes but mainly by shock-grid-position and Mach number for second order schemes. The different performance between first order scheme and second order scheme reveal that reconstruc-

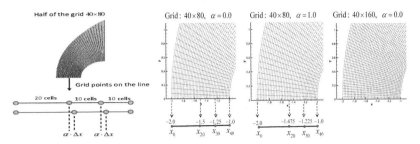

Figure 5: Grid generation based on Grid1.

tion function together with the slope limiter also affect dissipation properties of the final schemes and thus influence shock stability.

Compare the results in Table 5 and Table 6, we can see that the first-order Osher-P scheme converges to smooth solutions on 7 grids for $M_\infty=8$ but results in a "dislocation" shock as shown in Fig. 4(b1) on 6 grids for $M_\infty=20$. Mach number has influence on shock stability of first-order Osher-P scheme. However, first-order Osher-O schemes can always yields a "carbuncle" shock for both $M_\infty=8$ and $M_\infty=20$. For second order schemes, both Osher-P and Osher-O can not obtain correct solutions. In fact, Osher-P and Osher-O has different performances. After using entropy fix technique, the numerical results are improved. For Osher-O (E-fix) and Osher-P (E-fix), first order schemes can obtain good results in most grids and second order schemes can obtain correct convergent results on three grids ($\alpha=0.0$, 0.1 and 0.2 for the case $M_\infty=20$; $\alpha=0.0$, 0.8 and 0.9 for the case $M_\infty=8$). Second order Osher-O (E-fix) and Osher-P (E-fix) has similar numerical performances. Thus, the entropy fix largely improve the robustness of Osher-O flux and Osher-P flux. For Roe flux, numerical results of first order schemes has similar performance like Osher flux, but the second order Roe scheme yields carbuncle on many grids. Thus, Osher-O (E-fix) flux and Osher-P (E-fix) flux may be more robust than Roe (E-fix).

We also apply our entropy fix technique on the original Roe flux and call this resultant

Table 4: Test results for different numerical fluxes on grids with different grid aspect ratio ($M_\infty=20$).

Schemes Grid	40×40	40×80	40×120	40×160	40×200	40×240
1st Osher-O	0	0	0	0	0	0
1st Osher-P	5	3	3	3	3	3
2nd Osher-O	0	0	0	0	0	0
2nd Osher-P	3	2	0	0	0	0
1st Osher-O(E-fix)	5	7	7	7	7	7
1st Osher-P(E-fix)	7	7	7	4	4	6
2nd Osher-O(E-fix)	7	7	7	7	7	7
2nd Osher-P(E-fix)	7	7	7	7	7	7
1st Roe(E-fix)	7	7	4	7	7	7
2nd Roe(E-fix)	2	0	0	0	0	0

Table 5: Test results for different numerical fluxes on grids with different shock-grid-position ($M_\infty = 20$).

Schemes Grid (α)	$\alpha=0.0$	0.1	0.2	0.3	0.4	0.5	0.6	0.7	0.8	0.9	1.0
1st Osher-O	0	0	0	0	0	0	0	0	0	0	0
1st Osher-P	3	3	3	3	3	3	5	4	7	7	7
2nd Osher-O	0	0	0	0	0	0	0	0	0	0	0
2nd Osher-P	2	2	2	1	1	2	0	0	2	2	2
1st Osher-O(E-fix)	7	7	7	7	7	7	7	7	7	7	7
1st Osher-P(E-fix)	7	7	7	7	7	7	7	7	7	7	7
2nd Osher-O(E-fix)	7	7	7	5	5	5	5	5	4	4	5
2nd Osher-P(E-fix)	7	7	7	5	5	5	4	5	4	4	4
1st Roe(E-fix)	7	7	7	7	7	7	7	7	7	5	7
1st Roe(Zhu E-fix)	5	5	5	7	7	7	7	7	5	5	5
1st Roe(HZ E-fix)	7	7	7	7	7	7	7	7	7	7	7
2nd Roe(E-fix)	0	0	7	0	0	5	5	0	0	0	0
2nd Roe(Zhu E-fix)	3	0	4	5	5	5	1	0	2	2	0
2nd Roe(HZ E-fix)	3	2	1	3	5	5	5	4	5	5	5

Table 6: Test results for different numerical fluxes on grids with different shock-grid-position ($M_\infty = 8$).

Schemes	$\alpha=0.0$	0.1	0.2	0.3	0.4	0.5	0.6	0.7	0.8	0.9	1.0
1st Osher-O	0	0	0	0	0	0	0	0	0	0	0
1st Osher-P	5	7	7	7	7	7	7	5	5	5	7
2nd Osher-O	0	0	0	0	0	0	0	0	0	0	0
2nd Osher-P	5	4	5	4	3	3	2	3	0	5	5
1st Osher-O(E-fix)	7	7	7	7	7	7	7	7	7	7	7
1st Osher-P(E-fix)	7	7	7	7	7	7	7	7	7	7	7
2nd Osher-O(E-fix)	7	4	5	5	5	4	4	4	7	7	5
2nd Osher-P(E-fix)	7	5	5	5	5	4	4	4	7	7	5
1st Roe(E-fix)	7	7	7	7	7	7	4	5	5	7	7
1st Roe(Zhu E-fix)	7	7	7	7	5	5	4	4	5	7	7
1st Roe(HZ E-fix)	7	7	7	7	7	7	7	7	7	7	7
2nd Roe(E-fix)	0	7	5	0	0	2	2	2	0	5	7
2nd Roe(Zhu E-fix)	5	5	7	4	0	0	0	0	1	4	5
2nd Roe(HZ E-fix)	4	5	7	4	0	0	5	6	2	4	5

flux as Roe (Zhu E-fix). From Table 5 and Table 6, we can see that Roe (Zhu E-fix) is not as good as Osher flux with entropy fix. The reason may lies in the difference between the original 1D Roe flux and 1D original Osher flux. Then the fix technique is used for Roe (E-fix) and the resultant Roe (HZ E-fix) performs better than Roe (E-fix) for first order schemes but also can not obtain robust results for second order schemes.

3.2 Viscous heating problems

We consider hypersonic heating on the wall of a two-dimensional blunt body with $r = $ 20mm radius mounted in Nagoya University Shock Tunnel [30], in which the freestream

Table 7: Quality evaluation of the numerical solution.

Grid	2nd Osher-O(E-fix)		2nd Osher-P(E-fix)	
	pressure error	heating error	pressure error	heating error
Grid 4: 150×80	1.11%	4.09%	0.79%	3.29%
Grid 5: 150×240	2.37%	3.83%	2.48%	6.12%
Grid 6: 147×100	0.37%	1.34%	0.37%	1.38%
Grid 7: 16242	0.19%	0.46%	0.17%	0.81%

conditions are $M_\infty = 8.1$, $P_\infty = 370.6\text{Pa}$, $Re = 1.3 \times 10^5$, $T_\infty = 63.73\text{K}$ and $T_{wall} = 300\text{K}$. This problem is also considered by Kitamura in [31,32]. From Table 3 in Section 3.1, we have tested that second order Osher-O (E-fix) and second order Osher-P (E-fix) schemes can obtain correct solutions on Grid1, Grid2 and Grid3. To compute heating, we generate another three grids based on the Grid1, Grid2 and Grid3 by taking grid refinement only near the wall so that the cell Reynolds number is $Re_{cell} = 1.3$. The new grids has 150×80 cells (Grid4), 150×240 cells (Grid5) and 147×100 cells (Grid6) correspondingly with local plots shown in Fig. 6.

The derivatives in viscous term are evaluated as

$$\nabla u_L = \nabla u_{GHIJ} = \frac{1}{S_{GHIJ}} \left[u_B l_{GH} \vec{n}_{GH} + u_N l_{HI} \vec{n}_{HI} + u_E l_{IJ} \vec{n}_{IJ} + u_M l_{JG} \vec{n}_{JG} \right],$$

where u_M and u_N are cell averages, u_B and u_E are values at vertices, as shown in Fig. 13.

Second order schemes based on Osher-O (E-fix), Osher-P (E-fix) and Roe (E-fix) are tested. Pressure distributions are displayed in Fig. 6. We can see that Osher-O (E-fix) and Osher-P (E-fix) obtain similar numerical results and can obtain smooth flowfields. The surface pressure and heating profiles are shown in Fig. 7. Here we use theoretical stagnation pressure $P_{10}/P_\infty = 84.9$ and accurate stagnation heating $q_{FR} = 17.5\text{W}/\text{cm}^2$, which are the same as that in [31,32]. The surface pressure and heating profile are also smooth. It is concluded that both of shock stability and boundary layer resolution are crucial for heating computations in [32]. The errors in resultant pressure and heating for Osher-O (E-fix) and Osher-P (E-fix) on three grids are small as shown in Table 7, which may indicate that Osher flux with entropy fix has good shock stability and good boundary layer resolution. The results on the Grid6 are more accurate than that on the Grid4 and Grid5, which reveal that grid density near shock and grid distribution near the wall will influence pressure distribution and heating profiles. Roe (E-fix) yields carbuncle on Grid4 and Grid5, and shows weak asymmetry in flowfields on Grid6. For all three grids, Roe (E-fix) has obvious asymmetric patterns in surface heating.

3.3 Double Mach reflection

Double Mach reflection problem described in [29] is solved by Osher schemes. Results on a grid with 240×960 intervals at time $t = 0.2$ are displayed in Fig. 8. We can see that both first order Osher-O and first order Osher-P yield carbuncle for the principal Mach shock

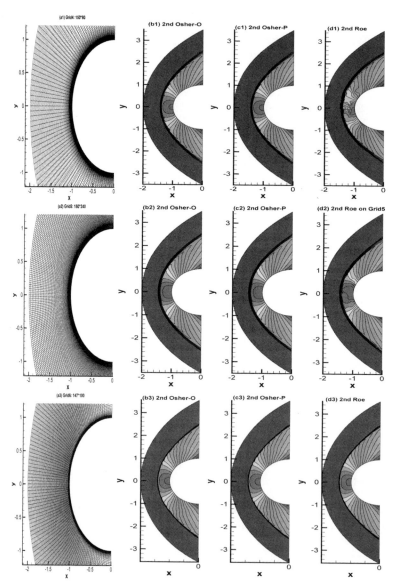

Figure 6: Pressure distributions of schemes with entropy fix on Grid4 (Up), Grid5 (Middle) and Grid6 (Down).

near $x=2.8$, as shown in Figs. 8(a1) and (b1). After taking entropy fix, the carbuncle phenomena disappear, as shown in Figs. 8(a2) and (b2). For second order Osher-O scheme and second order Osher-P scheme, the shape of the principal Mach stem appears to be in-

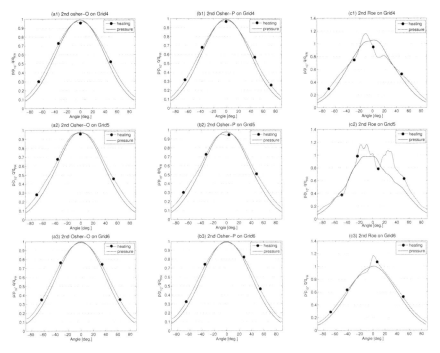

Figure 7: Surface pressure and heating profiles of schemes with entropy fix.

correct and a piece of jet has struck the principle Mach shock, as shown in Figs. 6(c1) and (d1). Second order Osher schemes with entropy fix are free from carbuncle phenomenon.

In order to verify that the entropy fix won't influence the robustness of shock, we give the contour plots of the ratio of the maximum difference η in (2.16) to $\lambda_{\max} = \max\{|\tilde{u}_L| + c_L, |\tilde{u}_R| + c_R\}$ for interfaces with $n_x = 0$ and $n_y = 0$ respectively, as shown in Fig. 9. We can see that near the Mach shock the ratio $\frac{\eta}{\lambda_{\max}}$ is much larger for the interfaces with $n_x = 0$ than those with $n_y = 0$, which means the entropy fix is mainly given for the interfaces perpendicular to the Mach shock and thus the entropy fix may not influence shock robustness. We also make a test of only taking entropy fix for the interfaces with $n_x = 0$. We find that the numerical result is also free from carbuncle phenomenon and is almost the same with the result in Fig. 8(a2). We also show the density distribution along lines $x = 2.4$ and $y = 0.2$ in Fig. 10. These results indicate that adding appropriate dissipation for the interfaces perpendicular to the shock is the reason for the entropy fix to prevent carbuncle phenomenon.

From Fig. 9, we can see that the ratio is less than 0.1 for contact discontinuity, which means the added dissipation is very small and may not influence contact resolution. Fig. 11 gives comparisons for second order schemes. Entropy fix also did not influence the robustness of the Mach shock. In addition, for contact discontinuity second order

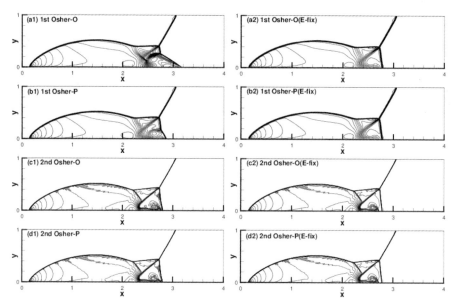

Figure 8: Density distributions at time 0.2 for the double Mach reflection problem.

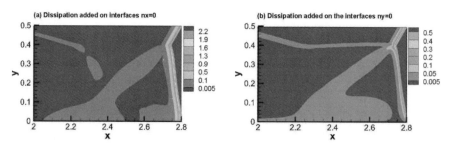

Figure 9: Contour plots of the ratio $\frac{\eta}{\lambda_{\max}}$ for interfaces with $n_x=0$ (left plot) and $n_y=0$ (right plot).
(colored picture attached at the end of the book)

Osher schemes with entropy fix locates between second order Osher schemes without entropy fix and second order schemes with Roe flux.

3.4 Tests on unstructured grids

In this section, we make tests for unstructured grids. Firstly, we generate triangular grids based on a series of quadrilateral grids with different shock-grid-location in Section 3.1, see one such grid in Fig. 12. From Table 8, we can see that the entropy fix also improve the robustness of Osher scheme on triangular grids. Secondly, we recompute the vicous heating problem in Section 3.2 on an unstructured grid. The derivatives at the interfaces

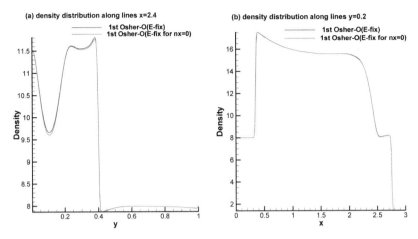

Figure 10: Comparison between Osher-O with entropy fix for all interfaces and only for interfaces with $n_x=0$.

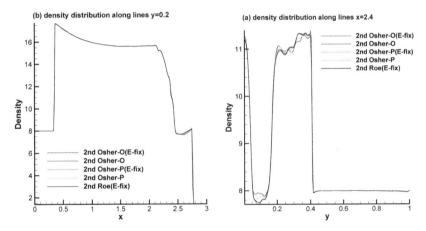

Figure 11: Comparison between second order Osher-O and Osher-P schemes.

in Figs. 13(b) and (c) are evaluated as

$$\nabla u_L = \frac{1}{|S_{MBNE}|} \{ u_{MB} l_{MB} \vec{n}_{MB} + u_{BN} l_{BN} \vec{n}_{BN} + u_{NE} l_{NE} \vec{n}_{NE} + u_{EM} l_{EM} \vec{n}_{EM} \}, \quad (3.1)$$

where $u_{MB} = \frac{u_M + u_B}{2}$, u_M is cell averages, u_B is the value at vertices.

A hybrid grid is generated based on Grid6 by keeping quadrilateral grids in the area near the wall and using triangular grids in the left area, as shown in Fig. 14(a). From Figs. 14(b) and (c), we can see that second order Osher-O (E-fix) and second order Osher-

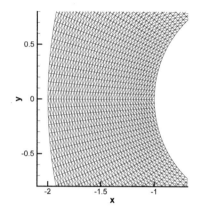

Figure 12: An triangular grid obtained based on the quadrilateral grid with $\alpha=0.0$.

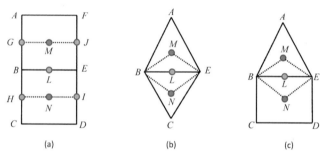

Figure 13: Integral path for computing derivatives in viscous term.

(colored picture attached at the end of the book)

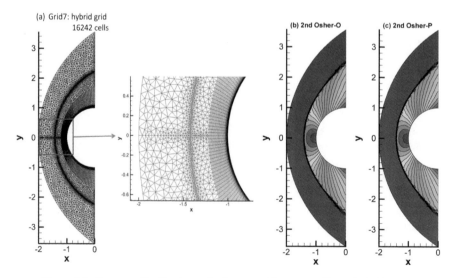

Figure 14: Pressure distributions of schemes with entropy fix on hybrid grid.

(colored picture attached at the end of the book)

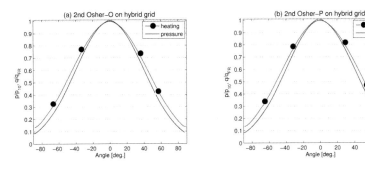

Figure 15: Surface pressure and heating profiles on hybrid grid.

P (E-fix) obtain similar numerical results and can obtain smooth flowfields. In addition, Fig. 15 shows that the surface pressure and heating profile are also smooth. Table 7 also shows that the errors in resultant pressure and heating are small.

Table 8: Test results for Osher fluxes on triangular grids based on quadrilateral grids with different shock-grid-position ($M_\infty = 20$).

Schemes Grid (α)	$\alpha=0.0$	0.1	0.2	0.3	0.4	0.5	0.6	0.7	0.8	0.9	1.0
1st Osher-O	5	5	5	5	0	0	0	0	0	0	0
1st Osher-P	3	3	3	3	0	3	1	1	1	3	3
1st Osher-O(E-fix)	7	7	7	5	5	5	5	7	7	7	7
1st Osher-P(E-fix)	7	7	7	7	7	7	7	5	7	7	7

4 Conclusions

In this paper we have compared the Osher-O and the Osher-P schemes for the 2D Euler equation and proposed an entropy fix technique to improve their robustness. For an inviscid hypersonic flow past a circular cylinder, it has been shown that the first-order Osher-O scheme based on quadrangular grids always yields a "carbuncle" shock in front of the cylinder and the first-order Osher-P scheme results in a "dislocation" shock for the cases with high Mach number. The second-order Osher schemes will always fail to obtain correct convergent solutions. To improve the robustness of the Osher schemes, we have proposed an entropy fix technique. Numerical tests on a series of grids verify that the first-order Osher schemes with entropy fix can obtain correct convergent solutions for most grids while second order Osher schemes can obtain correct convergent solutions for some grids. Thus, the entropy fix largely improve the robustness of Osher-O flux and Osher-P flux. Numerical results on hypersonic heating on the wall of blunt-body and double Mach reflection show that influence of the added dissipation on boundary-layer resolution and contact resolution is very small. The methods proposed in this paper can be generalized to moving grids by changing the outer normal of each edge at each time step. The future work will involve the shock-robust combination of Osher flux with appropriate slope limiters.

Acknowledgments

H. Zhu was supported by the Natural Science Foundation of China (Grant No. 11301525). We would like to thank Dr. Xinghua Chang for providing some unstructured grids.

References

[1] B. ENGQUIST AND S. OSHER, *Stable and entropy satisfying approximations for transonic flow calculations*, Math. Comput., 34 (1980), pp. 45–75.
[2] S. OSHER AND F. SOLOMON, *Upwind difference schemes for hyperbolic systems of conservation laws*, Math. Comput., 38 (1982), pp. 339–374.
[3] P. ROE, *Approximate Riemann solvers, parameter vectors, and difference schemes*, J. Comput. Phys., 43 (1981), pp. 357–372.
[4] S. OSHER AND S. CHAKRAVARTHY, *Upwind schemes and boundary conditions with applications to Euler equations in general geometries*, J. Comput. Phys., 50 (1983), pp. 447–481.
[5] S. CHAKRAVARTHY AND S. OSHER, *Numerical experiments with the Osher upwind scheme for the Euler equations*, AIAA J., 21 (1983), pp. 1241–1248.
[6] S. OSHER, *Riemann solvers, the entropy condition, and difference approximations*, SIAM J. Numer. Anal., 21 (1984), pp. 217–235.
[7] S. OSHER AND S. CHAKRAVARTHY, *High resolution schemes and the entropy condition*, SIAM J. Numer. Anal., 21 (1984), pp. 955–984.
[8] D. LANSER, J. BLOM AND J. VERWER, *Spatial discretization of the shallow water equations in spherical geometry using Osher's scheme*, J. Comput. Phys., 165 (2000), pp. 542–565.
[9] R. FEDKIW, G. SAPIRO AND C. SHU, *Shock capturing, level sets, and PDE based methods in computer vision and image processing: a review of Osher's contributions*, J. Comput. Phys., 185 (2003), pp. 309–341.
[10] L. QIANG AND W. ZHAO, *Computation of turbulence flow on hybrid grids using k-w turbulence model and Osher scheme*, Trans. Nanjing University Aeronaut. Astronaut., 21 (2004), pp. 94–97.
[11] G. COCLITE, S. MISHRA AND N. RISEBRO, *Convergence of an Engquist-Osher scheme for a multi-dimensional triangular system of conservation laws*, Math. Comput., 79 (2010), pp. 71–94.
[12] J. QIU, B. KHOO AND C. SHU, *A numerical study for the performance of the Runge Kutta discontinuous Galerkin method based on different numerical fluxes*, J. Comput. Phys., 212 (2006), pp. 540–565.
[13] J. QIU, *Development and comparison of numerical fluxes for LWDG methods*, J. Comput. Phys., 1 (2008), pp. 1–32.
[14] C. LU, J. QIU AND R. WANG, *A numerical study for the performance of the WENO schemes based on different numerical fluxes for the shallow water equations*, J. Comput. Math., 28 (2010), pp. 807–825.
[15] J. FELCMAN AND O. HAVLE, *On a numerical flux for the shallow water equations*, Appl. Math. Comput., 217 (2011), pp. 5160–5170.
[16] K. PEERY AND S. IMLAY, *Blunt-body flow simulations*, AIAA, (1988), 2904.

[17] M. LIOU AND C. STEFFEN, *A new flux splitting scheme*, J. Comput. Phys., 107 (1993), pp. 23–39.
[18] H. LIN, *Dissipation additions to flux-difference splitting*, J. Comput. Phys., 117 (1995), pp. 20–27.
[19] R. SANDERS, E. MORANO AND M. DRUGUETZ, *Multidimensional dissipation for upwind schemes: stability and applications to gas dynamics*, J. Comput. Phys., 145 (1998), pp. 511–537.
[20] M. PANDOLFI AND D. D'AMBROSIO, *Numerical instabilities in upwind methods: analysis and cures for the "carbuncle" phenomenon*, J. Comput. Phys., 166 (2001), pp. 271–301.
[21] Y. REN, *A robust shock-capturing scheme based on rotated Riemann solvers*, Comput. Fluids, 32 (2003), pp. 1379–1403.
[22] S. SPEKREIJSE, *Multiple grid and Osher's scheme for the efficient solution of the steady Euler equations*, Appl. Numer. Math., 2 (1986), pp. 475–493.
[23] P. JACOBS, *Approximation Riemann solver for hypervelocity flows*, AIAA J., 30 (1992), pp. 2558–2561.
[24] J. AMALADAS AND H. KAMATH, *Accuracy assessment of upwind algorithms for steady-state computations*, Comput. Fluids, 27 (1998), pp. 941–962.
[25] N. FRINK, *Upwind scheme for solving the Euler equations on unstructured tetrahedral meshes*, AIAA J., 30 (1991), pp. 70–77.
[26] B. VAN LEER, *Towards the ultimate conservative difference scheme IV. a new approach to numerical convection*, J. Comput. Phys., 23 (1977), pp. 276–299.
[27] B. VAN LEER, *Towards the ultimate conservative difference scheme V. a second-order sequel to Godunov's method*, J. Comput. Phys., 32 (1979), pp. 101–136.
[28] E. TORO, Riemann Solvers and Numerical Methods for Fluid Dynamics, Springer, 1997.
[29] P. WOODWARD AND P. COLELLA, *The numerical simulation of two-dimensional fluid flow with strong shocks*, J. Comput. Phys., 54 (1984), pp. 115–173.
[30] A. NISHINO, T. ISHIKAWA, K. KITAMURA AND Y. NAKAMURA, *Effect of the clearance between two bodies on TSTO aerodynamic heating*, J. Japan Soc. Aer. Space Sci., 53 (2005), pp. 503–509.
[31] K. KITAMURA, E. SHIMA, Y. NAKAMURA AND P. L. ROE, *Evaluation of Euler fluxes for hypersonic heating computations*, AIAA J., 48 (2010), pp. 763–776.
[32] K. KITAMURA, *A further survey of shock capturing methods on hypersonic heating issues*, AIAA paper, 2013–2698.

基于雷诺应力模型的高精度分离涡模拟方法*

王圣业[1])　王光学[2])　董义道[1])　邓小刚[1])†

1)(国防科学技术大学航天科学与工程学院,长沙　410073)

2)(中山大学物理学院,广州　510275)

(2017年3月20日收到; 2017年5月14日收到修改稿)

基于Speziale-Sarkar-Gatski/Launder-Reece-Rodi (SSG/LRR)-ω 雷诺应力模型发展了一类分离涡模拟方法,结合高精度加权紧致非线性格式在典型翼型及三角翼算例中进行了验证,并和传统基于线性涡粘模型的分离涡模拟方法进行了对比.结果表明:基于SSG/LRR-ω 模型的分离涡模拟方法,提高了原雷诺应力模型对非定常分离湍流的模拟能力;同时相比于传统基于线性涡粘模型的分离涡模拟方法,尤其是在翼型最大升力迎角和三角翼涡破裂迎角附近,该方法在平均气动力预测的准确度、分离湍流模拟的精细度等方面更加优秀.

关键词: 湍流流动, 雷诺应力模型, 分离涡模拟, 加权紧致非线性格式

PACS: 47.27.-i, 47.27.em, 47.27.ep, 47.11.Bc　　　　　　　　　　**DOI**: 10.7498/aps.66.184701

1　引　言

准确预测翼型和机翼在接近甚至超过失速迎角时的气动特性,已成为现代飞行器设计的重要方面.然而,对于高雷诺数大迎角条件下的分离流动,精确的计算流体力学(computational fluid dynamics, CFD)模拟仍十分困难[1].

任何流动模拟的结果都依赖于所代表的物理模型的准度以及求解相关方程的数值方法的精度[2].为了精确预测流动物理,主要是湍流,最好是直接求解所有尺度的湍流脉动甚至是最小尺度,即直接数值模拟;或求解大尺度湍流而模型化最小尺度,即大涡数值模拟(large eddy simulation, LES).然而,在真实飞行雷诺数下,即使附着边界层内的最大尺度脉动也会变得很小,这将使计算花费巨增而难以承担.因此,基于雷诺平均纳维-斯托克斯方程(Reynolds average Navier-Stokes, RANS)的湍流模型方法仍然是工业应用CFD的中坚力量[3].

混合雷诺平均/大涡模拟(hybrid RANS/LES)方法结合了大涡模拟方法在分离流动区域的高分辨率与雷诺平均方法在附着边界层中的高效率,在大迎角分离模拟中受到了广泛的关注[4−6].而在hybrid RANS/LES方法中, Spalart 等[7]于1997年提出的分离涡模拟(detached eddy simulation, DES)方法由于其简单、高效等特点得到了广泛的发展和应用[8].然而,最初版本的DES方法基于一方程Spalart-Allmaras (SA)模型,在大规模分离流动支配的算例中表现良好,但对于由逆压梯度造成的初始尾缘分离等问题却很难准确估计.因此,为了更准确地模拟翼型和机翼的失速分离问题,尤其在失速迎角附近, DES的RANS部分需要采用更先进的湍流模型[9]. 2001年, Strelets[10]提出了基于二方程剪切压力传输(shear stress transport, SST)模型的DES方法,对翼型最大升力迎角附近时的分离问题有一定的提高.然而,由于线性涡粘模型自身的限制,其效果仍不理想. 2006年, Greschner 等[11,12]发展了基于EASM LL k-ε 模型(一类非线性涡粘模型)的DES方法,并在串联圆柱-翼型算例中与传统k-ε-DES方法进行了对比.结果表明,

---

王圣业,王光学,董义道,邓小刚.基于雷诺应力模型的高精度分离涡模拟方法.物理学报,
2017, 66(18): 113–129.

EASM-DES 方法能够更精细地分辨脱落涡结构, 并且预测的气动力更接近试验值. 但由于 EASM 模型仍基于 Boussinesq 假设框架, 其对雷诺应力的求解仍然不直接, 因此模拟效果对算例的敏感性较大.

雷诺应力模型 (Reynolds stress model, RSM) 是更高阶矩封闭的湍流模型, 采用六个方程求解雷诺应力项和一个方程求解尺度约束变量 (例如湍动能耗散率 ε 或比耗散率 ω), 能够直接估计雷诺应力各向异性、流线弯曲效应等[13]. 2011 年, Probst 等[9] 基于 ε 型 RSM 模型, 发展了 ε^h-DES 类方法, 并在 HGR-01 翼型上取得了很好的结果. 但由于 ε 型 RSM 模型鲁棒性较差, 其适应性将不如 ω 型 RSM 模型[14].

近年来, 德国宇航局的 Eisfeld 等[13,15,16] 发展了一种鲁棒的 RSM 模型, Speziale-Sarkar-Gatski/Launder-Reece-Rodi (SSG/LRR)-ω 模型, 并在一系列复杂的航空分离流动中证明了其优势. 美国航空航天局的 Rumsey 等[2] 在湍流模型资源 (turbulence modeling resource, TMR) 网站上将 SSG/LRR-ω 模型设为最推荐的 RSM, 并开展了广泛的验证与确认工作. 国内, 董义道等[17] 基于 TMR 网站, 成功开展了 SSG/LRR-ω 模型的初步应用研究, 但在应用过程中发现基于 RANS 方法的 SSG/LRR-ω 模型对翼型大迎角状态的模拟仍不理想. 本文基于上述工作, 在 SSG/LRR-ω 模型上发展了一类 DES 方法, 并在典型航空算例 (17°, 45° 及 60° 迎角的 NACA0012 翼型; 12° 迎角的 NACA4412 翼型; 24.6° 迎角的钝前缘三角翼) 中进行了验证. 为了进行对比研究, 本文还同时采用了传统基于线性涡粘模型的分离涡模拟方法, SST-DES 方法.

对于 DES 方法, 在 RANS 部分采用更先进的湍流模型, 而在分离区 LES 部分则对数值精度的要求更高. 相比传统二阶方法, 高阶方法在采用较少计算花费获得更低误差方面具有更大的潜力[18−20]. 由于在计算效率和计算精度上的优势, 高阶方法受到了广泛关注, 并被推荐用于 LES 或 hybrid RANS/LES 方法中[21]. 本文中对 RANS 方程和模型方程的离散均采用了高阶精度的加权紧致非线性格式 (weighted compact nonlinear scheme, WCNS)[22,23], 并且在计算网格导数时采用了满足自由流保持的对称网格导数算法 (symmetrical conservative metric method, SCMM)[24].

2 湍流模型

2.1 SSG/LRR-ω 模型

对 Navier-Stokes 方程进行 Favre 平均后出现雷诺应力项. 传统线性涡粘模型, 如 SA 模型和 SST 模型, 对雷诺应力项的求解均基于 Boussinesq 假设, 而 RSM 则直接对其求解.

$$\frac{\partial(\bar{\rho}\tilde{R}_{ij})}{\partial t} + \frac{\partial(\bar{\rho}\tilde{u}_k\tilde{R}_{ij})}{\partial x_k} = \bar{\rho}P_{ij} + \bar{\rho}\Pi_{ij} - \bar{\rho}\varepsilon_{ij} + \bar{\rho}D_{ij}, \quad (1)$$

其中 t, x_k ($k=1,2,3$) 分别为时间和空间坐标分量; $\bar{\rho}$, $\tilde{u}_k$ 和 $\tilde{R}_{ij}$ 分别表示 Favre 平均的密度、速度分量和雷诺应力张量, 而全雷诺应力 $\tau_{ij} = -\bar{\rho}\tilde{R}_{ij}$. 不加特殊说明, 下面出现的变量均为 Favre 平均后的量. (1) 式右端第一项为生成项, 可准确求解:

$$\bar{\rho}P_{ij} = -\bar{\rho}\tilde{R}_{ik}\frac{\partial \tilde{u}_j}{\partial x_k} - \bar{\rho}\tilde{R}_{jk}\frac{\partial \tilde{u}_i}{\partial x_k}; \quad (2)$$

右端第三项为耗散项,

$$\bar{\rho}\varepsilon_{ij} = \frac{2}{3}\bar{\rho}\varepsilon\delta_{ij}, \quad (3)$$

其中 ε 表示各向同性耗散率, 需要通过额外的湍流尺度方程求解得到, δ_{ij} 为 Kronecker 符号. 右端第四项为输运项,

$$\bar{\rho}D_{ij} = \frac{\partial}{\partial x_k}\left[\left(\bar{\mu}\delta_{kl} + D\frac{\bar{\rho}\tilde{k}\tilde{R}_{kl}}{\varepsilon}\right)\frac{\partial \tilde{R}_{ij}}{\partial x_l}\right], \quad (4)$$

其中 $\tilde{k} = \tilde{R}_{ii}/2$ 表示湍动能, $\bar{\mu}$ 为平均动力黏性系数, D 表示湍流输运系数. 下面给出雷诺应力模型的压力应变联项的建模, 即 (1) 式右端第二项.

$$\begin{aligned}\bar{\rho}\Pi_{ij} = & -\left(C_1\bar{\rho}\varepsilon + \frac{1}{2}C_1^*\bar{\rho}P_{kk}\right)\tilde{a}_{ij} + C_2\bar{\rho}\varepsilon(\tilde{a}_{ik}\tilde{a}_{kj} \\ & - \frac{1}{3}\tilde{a}_{kl}\tilde{a}_{kl}\delta_{ij}) \\ & + (C_3 - C_3^*\sqrt{\tilde{a}_{kl}\tilde{a}_{kl}})\bar{\rho}\tilde{k}\tilde{S}_{ij}^* \\ & + C_4\bar{\rho}\tilde{k}(\tilde{a}_{ik}\tilde{S}_{jk} + \tilde{a}_{jk}\tilde{S}_{ik} - \frac{2}{3}\tilde{a}_{kl}\tilde{S}_{kl}\delta_{ij}) \\ & + C_5\bar{\rho}\tilde{k}(\tilde{a}_{ik}\tilde{W}_{jk} + \tilde{a}_{jk}\tilde{W}_{ik}).\end{aligned} \quad (5)$$

其中各向异性张量

$$\tilde{a}_{ij} = \frac{\tilde{R}_{ij}}{\tilde{k}} - \frac{2}{3}\delta_{ij}. \quad (6)$$

表 1 SSG/LRR-ω 模型封闭系数中 SSG 和 LRR 部分的贡献
Table 1. Values of closure coefficients for the SSG and the LRR contributions to the SSG/LRR-ω redistribution term.

	C_1	C_1^*	C_2	C_3	C_3^*	C_4	C_5	D
$\phi^{(\text{SSG})}$	1.7	0.9	1.05	0.8	0.65	0.625	0.2	0.22
$\phi^{(\text{LRR})}$	1.8	0	0	0.8	0	$(9c_2^{(\text{LRR})}+6)/11$	$(-7c_2^{(\text{LRR})}+10)/11$	$0.75C_\mu$

注：$c_2^{(\text{LRR})} = 0.52$, $C_\mu = 0.09$.

其他各项定义如下：

$$\tilde{S}_{ij} = \frac{1}{2}\left(\frac{\partial \tilde{u}_i}{\partial x_j} + \frac{\partial \tilde{u}_j}{\partial x_i}\right),$$

$$\widetilde{S_{ij}^*} = \tilde{S}_{ij} - \frac{1}{3}\tilde{S}_{kk}\delta_{ij},$$

$$\tilde{W}_{ij} = \frac{1}{2}\left(\frac{\partial \tilde{u}_i}{\partial x_j} - \frac{\partial \tilde{u}_j}{\partial x_i}\right). \quad (7)$$

方程中的系数 $\phi = C_1, C_1^*, C_2, C_3, C_3^*, C_4, C_5, D$ 通过混合函数 F_1 计算得到，具体系数值见表 1。

$$\phi = F_1\phi^{(\text{LRR})} + (1-F_1)\phi^{(\text{SSG})}. \quad (8)$$

为了封闭上述模型，需要额外引入尺度方程，SSG/LRR-ω 模型借鉴了 Menter 的思路，对 ε 方程和 ω 方程进行了混合[15]：

$$\frac{\partial(\bar{\rho}\omega)}{\partial t} + \frac{\partial}{\partial x_k}(\bar{\rho}\omega \tilde{u}_k)$$
$$= -\alpha_\omega \frac{\omega}{\tilde{k}}\frac{\bar{\rho}P_{kk}}{2} - \beta_\omega \bar{\rho}\omega^2 + \frac{\partial}{\partial x_k}\left[\left(\bar{\mu} + \sigma_\omega \frac{\bar{\rho}\tilde{k}}{\omega}\right)\frac{\partial \omega}{\partial x_k}\right]$$
$$+ \sigma_d \frac{\bar{\rho}}{\omega}\max\left(\frac{\partial \tilde{k}}{\partial x_k}\frac{\partial \omega}{\partial x_k}, 0\right). \quad (9)$$

方程 (9) 的变量为比耗散率 ω。各向同性耗散率通过下式计算：

$$\varepsilon = C_\mu \tilde{k}\omega. \quad (10)$$

方程中的系数 $\phi = \alpha_\omega, \beta_\omega, \sigma_\omega, \sigma_d$ 通过混合函数计算得到：

$$\phi = F_1\phi^{(\omega)} + (1-F_1)\phi^{(\varepsilon)}. \quad (11)$$

混合函数定义如下：

$$F_1 = \tanh(\zeta^4),$$
$$\zeta = \min\left[\max\left(\frac{\sqrt{\tilde{k}}}{C_\mu \omega d_w}, \frac{500\bar{\mu}}{\bar{\rho}\omega d_w^2}\right), \frac{4\sigma_\omega^{(\varepsilon)}\bar{\rho}\tilde{k}}{(CD)d_w^2}\right],$$
$$CD = \sigma_d^{(\varepsilon)}\frac{\bar{\rho}}{\omega}\max\left(\frac{\partial \tilde{k}}{\partial x_k}\frac{\partial \omega}{\partial x_k}, 0\right). \quad (12)$$

其中，d_w 为距壁面法向距离，其他系数具体见表 2。

表 2 ω 方程系数中 ε 和 ω 部分的贡献
Table 2. Values of coefficients of ω-equation corresponding to the ε and ω parts.

	α_ω	β_ω	σ_ω	σ_d
$\phi^{(\varepsilon)}$	0.44	0.0828	0.856	1.712
$\phi^{(\omega)}$	0.5556	0.075	0.5	0

2.2 SSG/LRR-DES 类方法

DES 方法由 Spalart 等[7]提出，将 LES 方法和 SA 模型结合，并通过比较当地网格尺度与壁面距离实现两种方法的自动切换。其后，Strelet[10]借鉴 Spalart 的思想，通过比较当地网格尺度与湍流长度尺度，将 DES 方法引入 SST 模型。本文采用的 SSG/LRR-ω 模型与 SST 模型均为基于比耗散率 ω 的湍流模型，因此在构造 SSG/LRR-DES 方法时将主要借鉴 SST-DES 的构造思路。

首先定义网格长度尺度

$$L_g = \max(\Delta x, \Delta y, \Delta z) \quad (13)$$

和湍流长度尺度

$$L_t = \sqrt{\tilde{k}}/(C_\mu \omega), \quad (14)$$

这里湍动能 $\tilde{k}$ 不像 SST-DES 中可直接引用，而是由雷诺正应力得到：$\tilde{k} = \tilde{R}_{ii}/2$。

然后得到 SSG/LRR-DES 的限制器

$$L_{\text{DES}} = \min(L_t, C_{\text{DES}}L_g). \quad (15)$$

对雷诺应力方程耗散项中的各向同性耗散率 (10) 式进行修正：

$$\varepsilon = \tilde{k}^{3/2}/L_{\text{DES}}, \quad (16)$$

这里 C_{DES} 为 DES 常数，通过 Menter 的混合函数 (12) 式得到

$$C_{\text{DES}} = (1-F_1)C_{\text{DES}}^{(\varepsilon)} + F_1 C_{\text{DES}}^{(\omega)}, \quad (17)$$

其中 $C_{\text{DES}}^{(\varepsilon)}$ 和 $C_{\text{DES}}^{(\omega)}$ 对于不同软件平台取值略有差异, 需通过衰减各向同性湍流[25]等算例进行标定. 本文计算采用的内部代码中两个系数的取值分别为 0.60 和 0.78.

原始 DES 方法的限制器在复杂网格上处理 RANS 和 LES 的转换过程中过于生硬, 会造成模化应力损耗 (modeled stress depletion, MSD) 等问题[8]. 2006 年, Spalart 等[26]借鉴 Menter 的 SST 模型构造思想, 采用"延迟 LES 函数"改善了原始版本的 MSD 问题, 发展出了延迟分离涡模拟 (delayed DES, DDES) 方法. 2008 年, Shur 等[27]将 DDES 方法和壁面模型大涡模拟 (wall-modeled LES, WMLES) 方法结合, 发展了强化延迟分离涡模拟 (improved DDES, IDDES) 方法, 并在非定常分离流动中得到了广泛的应用. 本文的计算中均采用 IDDES 形式, 其构造过程如下.

首先采用 WMLES 中网格尺度的加权思想, 定义新的 IDDES 限制器:

$$L_{\text{DES}} = \tilde{f}_d \left(1 + f_e\right) L_t + \left(1 - \tilde{f}_d\right) C_{\text{DES}} L_g, \quad (18)$$

式中, $\tilde{f}_d$ 为转换函数,

$$\tilde{f}_d = \max\left((1 - f_d), f_B\right), \quad (19)$$

其中, f_d 为原 DDES 中的转换函数, $f_d = 1 - \tanh(8r_d)^3$, f_B 为原 WMLES 中的转换函数, $f_B = \min(2\exp(-9\alpha^2), 1)$, 这里 $\alpha = 0.25 - d_w/h_{\max}$; f_e 为壁面模拟的控制函数, 其作用是保证在壁面附近网格分辨率满足 LES 要求时, 忽略过渡区雷诺应力的影响.

$$f_e = \max\left((f_{e1} - 1), 0\right) f_{e2}, \quad (20)$$

其中 f_{e1} 定义为

$$f_{e1} = \begin{cases} 2\exp(-11.09\alpha^2), & \alpha \geq 0, \\ 2\exp(-9.0\alpha^2), & \alpha < 0; \end{cases} \quad (21)$$

f_{e2} 定义为

$$f_{e2} = 1.0 - \max\left(\tanh((c_t^2 r_{\text{dt}})^3), \tanh((c_l^2 r_{\text{dl}})^{10})\right). \quad (22)$$

(22) 式中 r_d, r_{dt} 和 r_{dl} 均与文献[27]中相同.

另外, 在计算当地网格尺度时, Shur 等[27]还引进了壁面距离的影响, 新的 L_g 为

$$L_g = \min\left(\max\left(C_w d_w, C_w h_{\max}, h_{\text{wn}}\right), h_{\max}\right), \quad (23)$$

其中 $h_{\max} = \max(\Delta x, \Delta y, \Delta z)$, h_{wn} 为壁面垂直方向的网格步长. IDDES 中的系数为 $C_w = 0.15$, $c_t = 1.87$ 和 $c_l = 5.0$.

3 数值方法

3.1 高精度数值方法

WCNS 格式由 Deng 等[22]在 2000 年提出. 其后, 不同学者[20,28,29]发展了多种形式的 WCNS 格式, 并在广泛的流动问题中证明了该格式的优势. 本文采用的是 WCNS 系列格式中一种典型的五阶显式离散格式 WCNS-E-5[23]. WCNS-E-5 格式由其低耗散、高鲁棒和优秀的自由流与涡保持特性, 被广泛应用于各种实际流动问题的高精度数值模拟中. 其中冈敦殿等[30]将 WCNS-E-5 格式应用于平板圆台突起物绕流的 LES 中, 并和采用平面激光散射技术的试验结果进行了对比, 证明了格式精细模拟湍流流动的可行性.

另外本文在计算网格导数及雅克比时, 采用了满足几何守恒律的对称守恒网格导数算法[24], 有利于提高高精度有限差分方法的鲁棒性并减小数值误差. 本文中时间推进均采用子迭代步基于 LU-SGS (lower-upper symmetric-Gauss-Seidal) 方法的双时间步法.

3.2 对雷诺应力方程的高精度离散

较差的数值稳定性是限制 RSM 使用的障碍, 尤其是结合高精度数值方法时. 在迭代的过程中, 雷诺应力项很可能不满足 Schumann[31]提出的现实性限制. Chassaing 等[32]提出一种鲁棒的隐式格式, 依赖于使用当地双时间步技术和显式增强现实性限制. Yossef[33]发展了一种无条件正收敛隐式格式和一种现实性限制, 保证了任意时间步内雷诺正应力项为正值. 本文中, 在雷诺应力方程的迭代过程中添加如下限制条件:

$$\text{if}: \begin{cases} \tilde{R}_{ij} < 0 \text{ for } i = j \\ \text{or}: \tilde{R}_{ij}^2 - \tilde{R}_{ii}\tilde{R}_{jj} > 0 \text{ for } i \neq j \\ \text{or}: \det[\tilde{R}_{ij}] < 0 \end{cases};$$

$$\text{then}: \begin{cases} \tilde{R}_{ij} \leftarrow \frac{2}{3}\tilde{k} \text{ for } i = j \\ \tilde{R}_{ij} \leftarrow 0 \text{ for } i \neq j \end{cases}. \quad (24)$$

除了时间迭代方面,也应同时关注雷诺应力方程的空间离散.对于加权型有限差分格式,非线性权是影响数值稳定性的主要方面.需要指出的是,在对雷诺应力方程采用 WCNS 类格式离散时,加权公式中的小量 ε_{IS} 应取为 10^{-18},而非文献 [23] 中对 Navier-Stokes 方程离散时的 10^{-6}.

4 计算结果

4.1 NACA0012 翼型大迎角气动特性模拟

NACA0012 翼型大迎角分离算例是推动 DES 发展的经典算例. 2001 年, Strelets 等[7] 在提出 SST-DES 类方法时,就在该算例上与原始的 SA-DES 类方法进行了对比,但发现并无本质的差异. 2016 年, Yang 和 Zha[34] 结合高精度 WENO (weighted essentially nonoscillatory) 格式, 采用改进的 SA-IDDES 方法对该算例进行了模拟, 但发现在中等迎角下 SA-IDDES 方法还没有 SA-URANS 方法准确. 本文为了更好地对 SSG/LRR-IDDES 方法进行对比研究, 计算条件和网格均参考 Yang 和 Zha 的工作[34], 并同时采用了 SST-IDDES 方法.

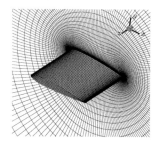

图 1 NACA0012 翼型三维粗糙网格

Fig. 1. Three-dimensional coarse grid of the NACA0012 airfoil.

NACA0012 翼型基于弦长 c 的雷诺数为 1.3×10^6, 基于自由流速度的马赫数为 0.5. 在 $0°\sim 90°$ 迎角范围内, 选取 $17°$ (大范围边界层分离), $45°$ 和 $60°$ (大规模脱体分离) 三个典型状态进行模拟. 由于该算例在高雷诺数 ($Re > 1.0 \times 10^5$) 下受雷诺数影响很小, 因此升力和阻力系数的参考值选取 $Re = 2.0 \times 10^6$ 的试验值[34,35]. 本文计算采用粗和密两套 O 型网格, 网格单元数分别为 $192 \times 102 \times 30$ 和 $288 \times 102 \times 30$. 第一层网格 Δy_1^+ 小于 1; 远场长度约为 $100c$; 展向长度为 $1.0c$. 图 1 为 NACA0012 翼型三维网格示意图. 时间推进采用双时间步方法, 周期为 $T = c/U_\infty$, 时间步长 $0.01T$. 时间平均统计从 $50T$ 时开始, 持续 $50T$.

图 2 展示了 3 个典型迎角下统计时间的平均升、阻力系数. 在 $17°$ 迎角时, 翼型处于失速状态, 所有方法得到的阻力系数相近且与试验吻合较好; 但对于升力系数, SSG/LRR-IDDES 在粗网格上略低于试验值, 而在密网格上与试验值吻合很好; 而其他方法均明显高于试验值. Yang 和 Zha[34] 在 $17°$ 迎角的计算中, 得出了 WENO+SA-IDDES 结果明显差于 WENO+SA-URANS 的现象, 但并未给出解释. 而本文采用的线性涡粘模型、SST 模型并未出现该现象. 在 $45°$ 和 $60°$ 迎角时, 翼型处于过失速状态, 所有 IDDES 方法的结果均明显优于 URANS 方法. 同时对比两类 URANS 方法, SSG/LRR-ω 模型略优于 SST 模型, 而该现象与相关湍流模型在分离涡处对雷诺应力预测的能力有关.

图 3 展示了 $17°$ 迎角时升、阻力系数在 $100T$ 内的变化过程, 其中 AoA 表示迎角. 可以看到在 $25T$ 后, SST-URANS 和 SSG/LRR-URANS 得到的升、阻力系数出现类似简谐振荡的发展过程. 而 SST-IDDES 和 SSG/LRR-IDDES 方法得到的升、阻力系数均为无周期振荡, 表明其能够描述分离湍流的随机性. 对比试验值, SST-IDDES 得到的升力系数明显偏高, 并且随网格加密无明显改善, 反映了基于线性涡粘模型的 DES 方法在翼型最大升力迎角附近模拟的局限.

经典的涡拉伸原理[36] 表明, 三维性是湍流最本质的特性之一. 图 4 给出的粗网格上 $100T$ 时刻 Q 判据为 0 的三维等值面图中, SST-URANS 预测的涡脱落过程明显为二维过程; SSG/LRR-URANS 能够预测出上翼面附近较小尺度的涡, 但整个涡脱落过程仍未表现出明显的三维性; 而 SST-IDDES 和 SSG/LRR-IDDES 均得到了高度混乱的三维涡结构, 但 SSG/LRR-IDDES 对前缘涡拉伸-弯曲-破裂的过程则模拟的更为精细.

图 5 对比了不同方法在粗网格上 $100T$ 时刻 0.5 展长处的展向速度, 其中 SST-URANS 得到的展向流动十分微弱, 也印证了其为二维涡脱落过程; SSG/LRR-URANS 得到的展向流动稍强于 SST-URANS, 但也未表现出明显的三维性; 而 SST-IDDES 和 SSG/LRR-IDDES 均预测出了明显的展向流动.

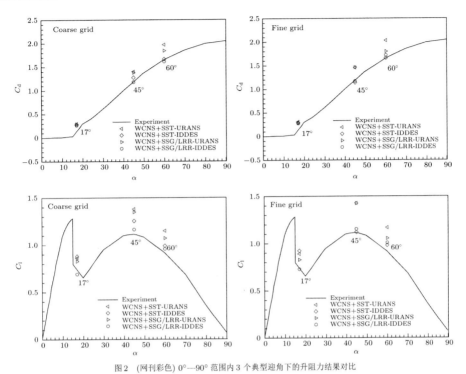

图 2 (网刊彩色) 0°—90° 范围内 3 个典型迎角下的升阻力结果对比

Fig. 2. (color online) Comparisons of lift and drag coefficients at 3 typical attack angles of among 0°–90°.

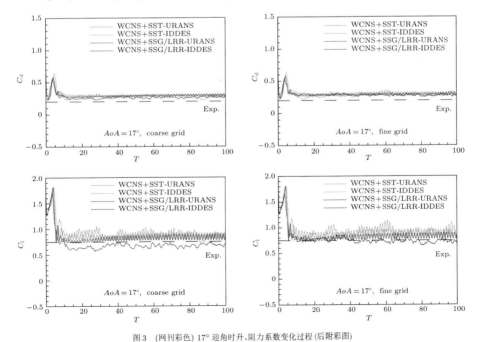

图 3 (网刊彩色) 17° 迎角时升、阻力系数变化过程 (后附彩图)

Fig. 3. (color online) Lift and drag coeffcient history at attack angle of 17°.

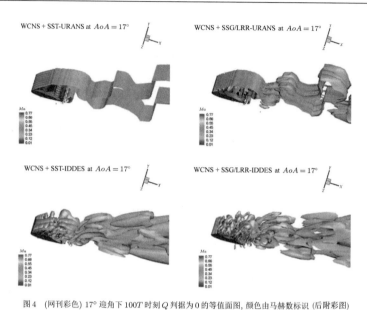

图 4 (网刊彩色) 17° 迎角下 100T 时刻 Q 判据为 0 的等值面图，颜色由马赫数标识 (后附彩图)

Fig. 4. (color online) Iso-surface of the Q-criterion = 0 at 100T and attack angle of 17°, colored by Mach number.

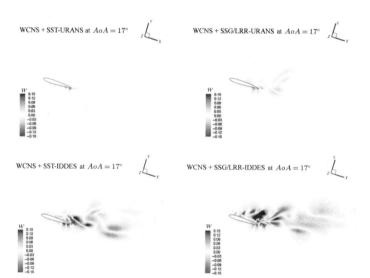

图 5 (网刊彩色) 17° 迎角下 100T 时刻, 0.5 展长处展向速度云图 (后附彩图)

Fig. 5. (color online) Spanwise velocity contours of 50% span at 100T and attack angle of 17°.

图 6 展示了 45° 迎角时升、阻力系数在 100T 内的变化过程. 可以看到在 25T 后, SST-URANS 得到的升、阻力系数出现类似简谐振荡的发展过程, 而 SSG/LRR-URANS 并未得到有规律的简谐振荡过程, 这与 RSM 能够估计雷诺应力的各向异性有关. 而 SST-IDDES 和 SSG/LRR-IDDES 方法得到的升、阻力系数均在试验值附近无规律振荡.

高精度格式的具体应用

来流流过机翼后形成脱体的分离湍流,并将产生明显的展向流动. 图7给出了粗网格上45°迎角, 100T时刻, Q判据为0的三维等值面图. SST-URANS仅预测出了大尺度的展向涡,其涡脱落过程仍为二维过程; SSG/LRR-URANS能够预测出少量的流向和法向涡,但仍未表现出明显的三维过

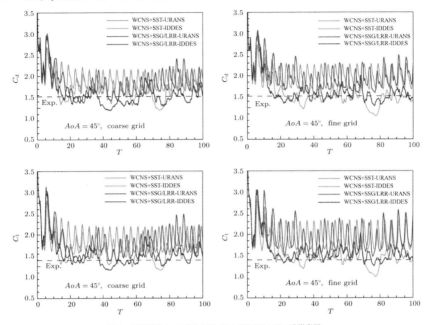

图 6 (网刊彩色) 45°迎角时升、阻力系数变化过程 (后附彩图)
Fig. 6. (color online) Lift and drag coeffcient history at attack angle of 45°.

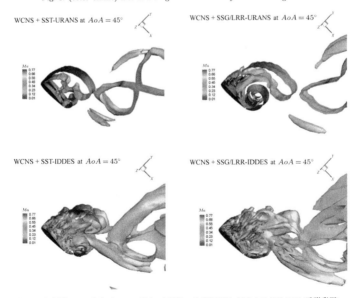

图 7 (网刊彩色) 45°迎角下100T时刻Q判据为0的等值面图,颜色由马赫数标识 (后附彩图)
Fig. 7. (color online) Iso-surface of the Q-criterion = 0 at 100T and attack angle of 45°, colored by Mach number.

程;而SST-IDDES和SSG/LRR-IDDES得到了高度混乱的大小涡结构,预测出了明显的三维效应. 图8对比了不同方法在粗网格上100T时刻0.5展长处的展向速度,更加清晰地表明各方法对于分离湍流的预测能力. 这也是图3中SST-IDDES和SSG/LRR-IDDES的结果优于SSG/LRR-URANS、更优于SST-URANS的原因.

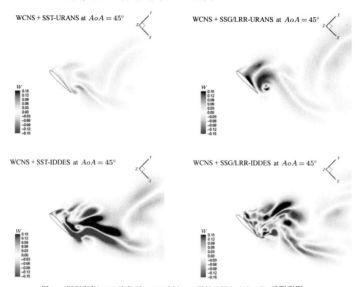

图 8 (网刊彩色) 45°迎角下100T时刻, 0.5展长处展向速度云图 (后附彩图)
Fig. 8. (color online) Spanwise velocity contours of 50% span at 100T and attack angle of 45°.

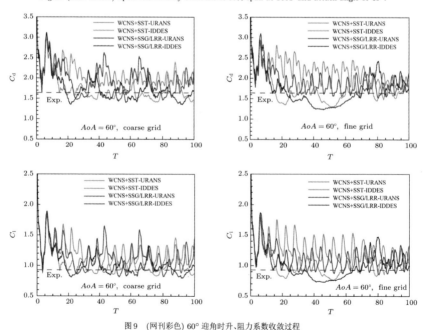

图 9 (网刊彩色) 60°迎角时升、阻力系数收敛过程
Fig. 9. (color online) Lift and drag coeffcient history at attack angle of 60°.

图9展示了60°迎角时升、阻力系数的变化过程. 与45°迎角时的结果类似, 在25T后, SST-URANS得到的升、阻力系数出现明显的周期性; 而SSG/LRR-URANS, SST-IDDES和SSG/LRR-IDDES呈现出无规律振荡, 但SSG/LRR-URANS得到的升、阻力系数平均值明显高于试验值. 图10

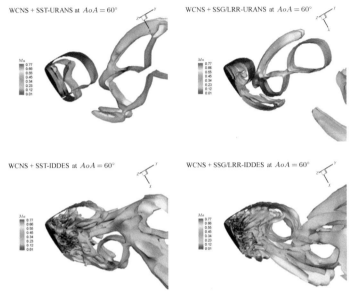

图 10 (网刊彩色) 60°迎角下100T时刻Q判据为0的等值面图, 颜色由马赫数标识 (后附彩图)
Fig. 10. (color online) Iso-surface of the Q-criterion $= 0$ at $100T$ and attack angle of $60°$, colored by Mach number.

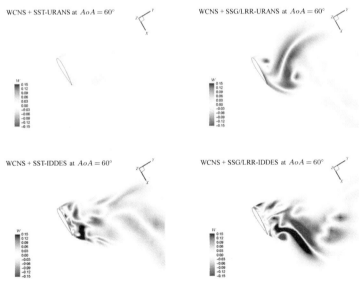

图 11 (网刊彩色) 60°迎角下100T时刻, 0.5展长处展向速度云图 (后附彩图)
Fig. 11. (color online) Spanwise velocity contours of 50% span at $100T$ and attack angle of $60°$.

给出了 60° 迎角 $100T$ 时刻下, Q 判据为 0 的三维等值面图. SST-URANS 预测的涡脱落过程仍为二维过程; SSG/LRR-URANS 得到的结果也仍未表现出明显的三维过程; 而 SST-IDDES 和 SSG/LRR-IDDES 得到了高度混乱的大小涡结构, 预测出了明显的三维效应. 图 11 给出了 60° 迎角 $100T$ 时刻 0.5 展长处的展向速度云图, 结果也与 45° 迎角时类似.

4.2 NACA4412 翼型尾缘分离模拟

NACA4412 翼型是经典的低速湍流验证算例[37], 有多种条件下的详细试验数据进行对比. 本文选择的是以 Wadcock[38] 实施的最大升力构型 (12° 迎角) 试验为参考的算例. 试验条件为: 翼型弦长 0.9 m, 雷诺数 1.64×10^6, 自由流速度 29.1 m/s 以及马赫数 0.085. 该条件下翼型尾缘开始出现分离, 对任何 hybrid RANS/LES 方法都是巨大的挑战[39].

本文计算的条件设置与试验相同. 采用粗和密两套 O 型网格, 展向拉伸 0.1 m, 网格单元数分别为 $169 \times 123 \times 30$ 和 $249 \times 123 \times 30$. 第一层网格 Δy_1^+ 均小于 1. 图 12 为 NACA4412 翼型三维网格示意图. 时间推进采用双时间步法, 周期为 $T = c/U_\infty$, 时间步长 $0.01T$. 时间平均统计从 $40T$ 时开始, 持续 $40T$.

表 3 列出了平均升力系数和分离位置的计算值与试验值. 可以看到, SST-URANS、SSG/LRR-URANS 和 SST-IDDES 得到的结果均与试验值偏差较大, 并且随着网格加密未有改善. 而 SSG/LRR-IDDES 的结果与试验吻合较好.

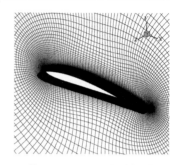

图 12 NACA4412 翼型三维粗网格

Fig. 12. Three-dimensional coarse grid of the NACA4412 airfoil.

表 3 NACA4412 翼型平均升力系数和分离位置的计算值与试验值对比

Table 3. Comparation of computional results of lift coefficient and location of separation for NACA4412 airfoil case.

	C_l		$x/c_{\text{separation}}$	
	粗网格	密网格	粗网格	密网格
SST-URANS 方法	1.7087	1.6782	0.9698	0.9827
SST-IDDES 方法	1.7125	1.6405	0.9674	0.9747
SSG/LRR-URANS 方法	1.7089	1.6845	0.9599	0.9713
SSG/LRR-IDDES 方法	1.5093	1.4864	0.8313	0.8366
试验值	1.4500		0.8150	

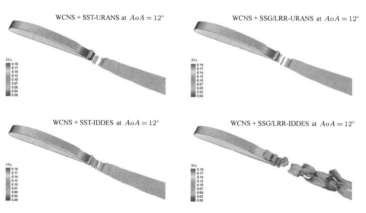

图 13 (网刊彩色) NACA4412 翼型 $80T$ 时刻, Q 判据为 0 等值面图, 颜色由马赫数标识 (后附彩图)

Fig. 13. (color online) Iso-surface of the Q-criterion = 0 at $80T$ for NACA4412 airfoil case, colored by Mach number.

图 13 给出了粗网格上 $80T$ 时刻 Q 判据为 0 的等值面图. 在翼型尾缘处, 流动开始分离并产生湍流尾迹涡. SSG/LRR-IDDES 成功模拟了该过程, 并且分离位置与试验吻合. 图 14 给出了该网格上得到的平均压力系数分布和试验值的对比. SSG/LRR-IDDES 在前缘得到的吸力值明显低于另三种方法, 并且与试验值更接近, 这也是其得到更准确升力系数的原因. 在翼型尾缘处, 由于 SSG/LRR-IDDES 成功模拟了尾迹涡结构, 其得到的压力系数也更接近试验值. 图 15 展示了翼型尾缘 ($x/c = 0.952$) 和尾迹 ($x/c = 1.282$) 两个典型站位处的流向速度分布对比. 在 $x/c = 0.952$ 处, SSG/LRR-IDDES 成功捕捉到了逆向速度, 计算结果与试验吻合. 表明 SSG/LRR-IDDES 能在边界层附近更好地感受逆压梯度的影响, 对于弱非定常流动较传统 SST-IDDES 方法有一定提高. 而仅采用 URANS 方式会引入过多湍流黏性, 使弱非定常运动的发展受到抑制. 在 $x/c = 1.282$ 处, SSG/LRR-IDDES 得到的流向速度略大于试验值, 且与其他三种方法无明显区别.

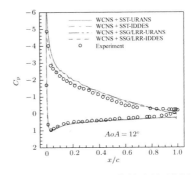

图 14 (网刊彩色) NACA4412 翼型表面压力系数分布

Fig. 14. (color online) Distribution of pressure coefficient on the surface of NACA4412 airfoil.

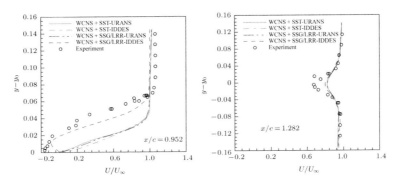

图 15 (网刊彩色) NACA4412 翼型流向速度分布比较

Fig. 15. (color online) Comparison of streamwise velocity profiles for NACA4412 airfoil.

4.3 钝前缘三角翼涡破裂模拟

现代战斗机和导弹多采用三角翼布局, 以获得良好的飞行品质和机动性能. 然而在大迎角下, 三角翼会产生涡破裂现象对气动特性造成影响. 本文采用的计算模型为 $65°$ 后掠三角翼, 是 NASA Langley 中心为研究该外形的雷诺数、马赫数影响而完成的试验模型[40]. 该模型分为四部分: 前缘、平板部分、后缘及整流罩. 前缘部分提供 4 种可替换外形, 本文选用中等半径钝前缘外形. 来流条件为: $Ma = 0.85$, $Re = 6 \times 10^6$. 迎角选择 $24.6°$, 为涡破裂现象发生的临界迎角.

本文采用的计算网格自主生成, 网格单元数约 600 万, 网格结构为多块对接网格. 图 16 给出了三角翼的表面网格. 网格拓扑采用 C-H 型, 以避免翼尖奇性轴的产生. 计算区域的远场边界取为 50 倍根弦长. 壁面的第一排网格达到了 10^{-6} 弦长, 网格在背风区和剪切层附近均进行了适当的加密, 以保证分离区、附面层内和剪切层的数值模拟精度. 时间推进采用双时间步法, 周期为 $T = c_{\text{ref}}/U_\infty$, 其中 c_{ref} 为参考气动弦长, 时间步长取 $0.01T$. 时间平均统计从 $10T$ 时开始, 持续 $20T$.

图 16 钝前缘三角翼计算网格

Fig. 16. Computational mesh for the blunt-edge deltawing.

图 17 展示了钝前缘三角翼不同站位处压力分布与试验[41]的对比, 其中 $\eta = 2z/b_l$, b_l 为当地展长, c_r 为翼根弦长. 回顾钝前缘三角翼分离的基本特性. 相比传统尖前缘, 钝前缘三角翼在前缘处有转捩过程[42], 但本文计算的马赫数较高, 迎角也较大, 因此转捩过程很短, 分离涡为全湍流状态. 主涡旋转产生展向流动, 在旋涡下部、三角翼的上翼面加速, 出现负压力峰值, 该点沿展向到翼边缘是逆压力梯度, 将诱导边界层分离, 产生二次分离和二次旋涡, 在主涡负压力峰值外侧又出现二次负压力峰值. 观察 $x/c_r = 0.4$ 站位的压力分布, SST-IDDES, SSG/LRR-URANS 和 SSG/LRR-IDDES 均成功捕捉到了二次涡结构, 而其中 SSG/LRR-IDDES 与试验值吻合最好.

当迎角大到一定程度时, 主涡开始破裂. 本文选择的 24.6° 迎角, 为涡破裂现象发生的临界迎角, 而 $x/c_r = 0.6$ 是该迎角下涡破裂发生前的临界站位. 因此该站位的模拟对 CFD 方法是个挑战. 观察本文计算值与试验值的对比, 四种方法均产生了不同程度的偏差. 其中 SSG/LRR-IDDES 在吸力峰处与试验吻合很好, 但在翼根处吻合较差. $x/c_r = 0.6$ 站位的翼根处临近整流罩头部(存在速度驻点), 压力系数由负值迅速变为接近 1 从而形成较大的压力梯度, 因此给数值模拟带来很大困难.

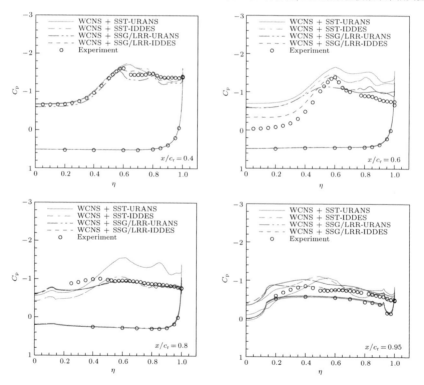

图 17 (网刊彩色) 钝前缘三角翼不同站位处压力分布对比

Fig. 17. (color online) Comparisons of surface pressure with experiments for blunt-edge deltawing at typical stations.

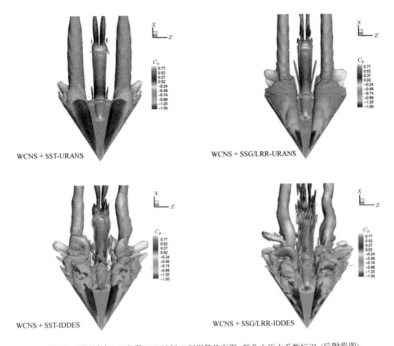

图 18 (网刊彩色) 三角翼 $20T$ 时刻 Q 判据等值面图, 颜色由压力系数标识 (后附彩图)

Fig. 18. (color online) Iso-surface of the Q-criterion at $20T$ for deltawing case, colored by pressure coefficient.

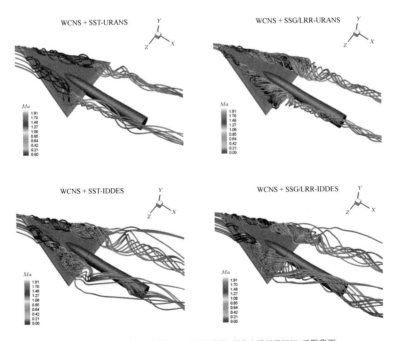

图 19 (网刊彩色) 三角翼 $20T$ 时刻流线图, 颜色由马赫数标识 (后附彩图)

Fig. 19. (color online) Streamlines at $20T$ for deltawing case, colored by Mach number.

在 $x/c_r = 0.8$ 处，涡破裂已发生，上翼面吸力峰消失. SST-IDDES, SSG/LRR-URANS 和 SSG/LRR-IDDES 均成功地预测了该现象，而 SST-URANS 仍得到了主涡及二次涡结构. 结合其他站位上 SST-URANS 得到的压力分布，可以看出在 24.6° 迎角下，其并未预测到涡破裂现象，即推迟了涡破裂的发生. 而该现象的产生明显由 SST-URANS 在分离区过大的湍流黏性导致. 由于来流为跨声速，前缘涡引起的上翼面流动加速，形成了局部超音速区域并产生了激波. 对于跨声速流动，Rumsey[14] 在 ONERO M6 机翼和 NASA CRM 翼身组合构型中均表明，SSG/LRR-ω 模型能更好地预测激波附近雷诺应力的剧烈变化，对激波诱导分离的模拟能力明显优于 SST 等涡粘模型. 本文发展的 SSG/LRR-IDDES 方法继承了 RSM 的优势，在非定常区 ($x/c_r = 0.8$ 和 $x/c_r = 0.95$) 得到了优于 SST-IDDES 方法的结果.

图 18 和图 19 分别展示了 $20T$ 时刻三角翼 Q 判据等值面图和流线图. SST-URANS 得到的仍为定常状态的前缘分离涡，在 24.6° 迎角时未发生涡破裂. SSG/LRR-URANS 虽然预测到了涡破裂，但涡破裂后的流动未表现出明显的非定常特性. 表明对涡破裂后形成的非定常运动，需采用诸如 DES、LES 等的尺度模拟方法. SST-IDDES 和 SSG/LRR-IDDES 均较好地模拟了涡破裂后的非定常流动状态，但就精细度而言，SSG/LRR-IDDES 要更为优秀.

5 结 论

本文借鉴 SST-IDDES 方法的构造方式，在 SSG/LRR-ω RSM 上发展了 SSG/LRR-IDDES 分离涡模拟方法. 通过结合高精度 WCNS 格式在 NACA 系列翼型及钝前缘三角翼算例上进行了验证，并和传统 SST-IDDES 方法以及 SSG/LRR-URANS 和 SST-URANS 方法进行了对比.

为说明 SSG/LRR-IDDES 方法在翼型大迎角气动力模拟方面的提升，选取 NACA0012 翼型 17°, 45° 和 60° 三个典型状态进行模拟. 在 17° 迎角时，翼型边界层分离扩大到前缘附近，升力系数接近最小. 此时，采用 SST-URANS 和 SSG/LRR-URANS 均会过高估计升力，而采用传统基于线性涡粘模型的 SST-IDDES 方法也并未明显改善. 这说明对于传统基于线性涡粘模型的 IDDES 方法，由于边界层附近 RANS 部分本身对逆压梯度引起的分离的模拟能力较差，使得整个方法较 URANS 并无明显提高. 而基于 RSM，可以更准确地估计逆压梯度区的雷诺应力变化，使得结合 IDDES 时得到了很好的结果. 到 45° 和 60° 迎角时，翼型产生大范围脱体涡，升力系数回升. 此状态是传统 IDDES 方法模拟的优势方面，而 SSG/LRR-IDDES 也继承了该特性，并较 SSG/LRR-URANS 有较大提高.

为说明 SSG/LRR-IDDES 方法在翼型最大升力迎角附近的模拟能力，选取 NACA4412 翼型 12° 状态进行模拟. 此时，翼型尾迹由于逆压梯度影响开始分离并产生湍流尾迹涡. 采用 SST-URANS 和 SSG/LRR-URANS 均会过高估计升力，并延迟了尾迹分离；而采用 SST-IDDES 也未明显改善. 但是采用 SSG/LRR-IDDES 方法给出了与试验吻合很好的气动力特性和准确的分离过程，表明该方法对模拟翼型失速迎角附近特性的能力提高.

为进一步说明 SSG/LRR-IDDES 方法在三维机翼失速迎角附近的模拟能力，选取钝前缘三角翼 24.6° 状态进行模拟. 该迎角下，三角翼主涡由于激波诱导的影响，在 $x/c_r = 0.6$ 站位后发生非定常涡破裂现象. 采用 SST-URANS 方法无法成功预测涡破裂现象；而采用 SSG/LRR-URANS 方法虽然预测到了涡破裂，但涡破裂后的流动未表现出明显的非定常特性. 采用 SST-IDDES 方法和 SSG/LRR-IDDES 方法均较好地模拟了涡破裂后的非定常流动状态，但在表面压力分布、流场精细度等方面，SSG/LRR-IDDES 方法更加优秀.

另外，本文采用的 WCNS-E-5 格式能够在粗糙网格上获得较高的保真度，体现了高精度数值方法在模拟分离流动中的效率优势. 但在加密网格计算时，也出现了网格收敛性差的现象，如 NACA0012 翼型 60° 迎角时，SSG/LRR-IDDES 在粗网格下得到的平均气动力系数略优于密网格. 该现象的产生与 IDDES 方法基于网格长度尺度来调整湍流黏性有关，也是 DES 类方法需要克服的问题之一[8]. 下一步，将继续结合高精度格式对 SSG/LRR-IDDES 方法在更广泛的流动中进行验证并提高网格收敛性.

感谢中山大学国家超级计算广州中心在计算资源方面提供的帮助.

参考文献

[1] Slotnick J, Khodadoust A, Alonso J, Darmofal D, Gropp W, Lurie E, Mavriplis D 2014 *CFD Vision 2030 Study: A Path to Revolutionary Computational Aerosciences* (Washington, DC: Langley Research Center, NASA) Tech. Rep. NASA/CR-2014-218178

[2] Eisfeld B, Rumsey C, Togiti V 2016 *AIAA J.* **54** 1524

[3] Rumsey C 2014 *52nd Aerospace Sciences Meeting* National Harbor, Maryland, January 13–17, 2014 AIAA 2014-0201

[4] Tucker P 2006 *Int. J. Numer. Meth. Fluids* **51** 261

[5] Richez F, Pape A, Costes M 2015 *AIAA J.* **53** 3157

[6] Xu G, Jiang X, Liu G 2016 *Acta Mech. Sin.* **32** 588

[7] Spalart P, Jou W H, Strelets M, Allmaras S 1997 *Comments on the Feasibility of LES for Wings, and on Hybrid RANS/LES Approach* (Columbus: Greyden Press)

[8] Spalart P 2009 *Annu. Rev. Fluid Mech.* **41** 181

[9] Probst A, Radespiel R, Knopp T 2011 *20st AIAA Computational Fluid Dynamics Conference* Honolulu, Hawaii, June 27–30, 2011 AIAA 2011-3206

[10] Strelets M 2001 *39th AIAA Aerospace Sciences Meeting and Exhibit* Reno, NV, 8-11 January 2001, AIAA 2001-0879

[11] Greschner B, Thiele F, Gurr A, Casalino D, Jacob M 2006 *12th AIAA/CEAS Aeroacoustics Conference* Cambridge, Massachusetts, May 8–10, 2006 AIAA 2006-2628

[12] Greschner B, Thiele F, Jacob M, Casalino D 2008 *Comput. Fluids* **37** 402

[13] Cécora R D, Radespiel R, Eisfeld B, Probst A 2015 *AIAA J.* **53** 739

[14] Rumsey C 2015 in Eisfeld B (ed.) *Differential Reynolds Stress Modeling for Separating Flows in Industrial Aerodynamics* (Springer Tracts Mechanical Engineering) p19

[15] Eisfeld B, Brodersen O 2005 *23rd AIAA Applied Aerodynamics Conference* Toronto, Ontario Canada, June 6–9, 2005 AIAA 2005-4727

[16] Togiti V, Eisfeld B, Brodersen O 2014 *J. Aircraft* **51** 1331

[17] Dong Y D, Wang D F, Wang G X, Deng X G 2016 *J. National Univ. Defense Technol.* **38** 46 (in Chinese) [董义道, 王东方, 王光学, 邓小刚 2016 国防科技大学学报 **38** 46]

[18] Shu C 2003 *Int. J. Comput. Fluid* **17** 107

[19] Wang Z, Fidkowski K, Abgrall R, Bassi F, Caraeni D, Cary A, Deconinck H, Hartmann R, Hillewaert K, Huynh H, Kroll N, May G, Persson P O, van Leer B, Visbal M 2013 *Int. J. Numer. Meth. Fluids* **72** 811

[20] Wang S, Deng X, Wang G, Xu D, Wang D 2016 *Int. J. Comput. Fluid D* **30** 469

[21] Georgiadis N, Rizzetta D, Fureby C 2010 *AIAA J.* **48** 1772

[22] Deng X, Zhang H 2000 *J. Comput. Phys.* **165** 22

[23] Deng X, Liu X, Mao M, Zhang H 2005 *17th AIAA Computational Fluid Dynamics Conference* Toronto, Ontario Canada, June 6–9, 2005 AIAA 2005-5246

[24] Deng X, Mao M, Tu G, Liu H, Zhang H 2011 *J. Comput. Phys.* **230** 1100

[25] Bellot G, Corrsin S 1971 *J. Fluid Mech.* **48** 273

[26] Spalart P, Deck S, Shur M, Squires K, Strelets M, Travin A 2006 *Theor. Comput. Fluid Dyn.* **20** 181

[27] Shur M, Spalart P, Strelets M, Travin A 2008 *Int. J. Heat Fluid Fl.* **29** 1638

[28] Nonomura T, Fujii K 2009 *J. Comput. Phys.* **228** 3533

[29] Liu H, Ma Y, Yan Z, Mao M, Deng X 2014 *8th International Conference on Computational Fluid Dynamics* Chengdu, China, July 14–18, 2014 ICCFD8-2014-0082

[30] Gang D D, Yi S H, Zhao Y F 2015 *Acta Phys. Sin.* **64** 054705 (in Chinese) [冈敦殿, 易仕和, 赵云飞 2015 物理学报 **64** 054705]

[31] Schumann U 1977 *Phys. Fluids* **20** 721

[32] Chassaing J, Gerolymos G, Vallet I 2003 *AIAA J.* **41** 763

[33] Yossef Y 2014 *J. Comput. Phys.* **276** 635

[34] Yang Y, Zha G 2016 *46th AIAA Fluid Dynamics Conference* Washington, D.C., USA, June 13–17, 2016 AIAA 2016-3185

[35] Shur M, Spalart P, Strelets M, Travin A 1999 *Proceedings of the 4th International Symposium on Engineering Turbulence Modelling and Measurements* Corsica, France, May 24–26, 1999 p669

[36] Chen M Z 2002 *Fundamentals of Viscous Fliud Dynamics* (Beijing: Higher Education Press) p239 (in Chinese) [陈懋章 2002 粘性流体动力学基础 (北京: 高等教育出版社) 第239页]

[37] Chen Y, Guo L D, Peng Q, Chen Z Q, Liu W H 2015 *Acta Phys. Sin.* **64** 134701 (in Chinese) [陈勇, 郭隆德, 彭强, 陈志强, 刘卫红 2015 物理学报 **64** 134701]

[38] Wadcock A 1987 *Investigation of Low Speed Turbulent Separated Flow Around Airfoils* (Washington, DC: Ames Research Center, NASA) Tech. Rep. NASA-CR-177450

[39] Roy R, Stoellinger M 2015 *53rd AIAA Aerospace Sciences Meeting* Kissimmee, Florida, January 5–9, 2015 AIAA 2015-1982

[40] Luckring M, Hummel D 2013 *Aerosp. Sci. Technol.* **24** 77

[41] Chu J, Luckring M 1996 *Experimental Surface Pressure Data Obtained on 65 deg Delta Wing Across Reynolds Number and Mach Number Ranges. Vol. 3: Medium-Radius Leading Edge* (Washington, DC: Ames Research Center, NASA) NASA-TM-4645-Vol-3

[42] Luckring M 2013 *Aerosp. Sci. Technol.* **24** 10

High-order detached-eddy simulation method based on a Reynolds-stress background model*

Wang Sheng-Ye[1])   Wang Guang-Xue[2])   Dong Yi-Dao[1])   Deng Xiao-Gang[1])†

1) (College of Aerospace Science and Engineering, National University of Defense Technology, Changsha 410073, China)

2) (School of Physics, Sun Yat-sen University, Guangzhou 510275, China)

(Received 20 March 2017; revised manuscript received 14 May 2017)

Abstract

Referring to the construction of shear stress transport-improved delayed detached-eddy simulation (SST-IDDES) method, a variant of IDDES method based on the Speziale-Sarkar-Gatski/Launder–Reece–Rodi (SSG/LRR)-ω Reynolds-stress model (RSM) as Reynolds-averaged Navier-Stokes (RANS) background model, is proposed. Through combining high-order weighted compact nonlinear scheme (WCNS), the SSG/LRR-IDDES method is applied to three aeronautic cases and compared with traditional methods: SST-unsteady Reynolds-averaged Navier-Stokes (URANS), SSG/LRR-URANS, and SST-IDDES. To verify the SSG/LRR-IDDES method in simulating airfoil stalled flow, NACA0012 airfoil is adopted separately at attack angles of 17°, 45° and 60°. At the attack angle of 17°, SST-URANS, SSG/LRR-URANS, and SST-IDDES methods each predict a higher lift coefficient than the experimental data, while the SSG/LRR-IDDES method obtains a better lift coefficient result and a higher fidelity vortical flow structure. It indicates that the RSM can improve the prediction of RANS-mode for pressure-induced separations on airfoil surfaces in detached-eddy simulation. At the attack angles of 45° and 60°, the SSG/LRR-IDDES method captures the massively separated flow with three-dimensional vortical structures and obtains a good result, which is the same as that from the traditional SST-IDDES method. To indicate the improvement of the SSG/LRR-IDDES method in simulating airfoil trailing edge separation, NACA4412 airfoil is adopted. At the attack angle of 12° (maximum lift), the trailing edge separation is mainly induced by pressure gradient. The SSG/LRR-IDDES method can predict the separation process reasonably and obtains a good lift coefficient and location of separation compared with experimental results. However, none of other methods can predict trailing edge separation. It confirms that when RSM is adopted as RANS background model in detached-eddy simulation, the ability to predict pressure-induced separation on airfoil surface is improved. For further verifying the SSG/LRR-IDDES method for simulating three-dimensional separated flow, blunt-edge deltawing at the attack angle of 24.6° is adopted. At this attack angle, the primary vortex will break, which is difficult to predict by using the SST-URANS method. For the SSG/LRR-URANS method, it predicts the vortex breakdown successfully, but the breakdown process does not show any significant unsteady characteristic. The SST-IDDES and the SSG/LRR-IDDES methods both predict a significant unsteady vortex breakdown. But in terms of the accuracy of surface pressure and the fidelity of unsteady flow, the result obtained by the SSG/LRR-IDDES method is better than by the SST-IDDES method.

Keywords: turbulence flows, Reynolds stress model, detached eddy simulation, weighted compact nonlinear scheme

PACS: 47.27.–i, 47.27.em, 47.27.ep, 47.11.Bc   **DOI:** 10.7498/aps.66.184701

邓小刚院士及合作者发表的论文

[1] 贾志刚, 吕志咏, 邓小刚. 均匀来流中旋转振荡圆柱绕流的数值研究. 航空学报, 1999, 20(5): 389-392

[2] Deng Xiaogang, Zhang Hanxin. Developing high-order weighted compact nonlinear schemes. Journal of Computational Physics, 2000, 165(1): 22-44

[3] 毛枚良, 邓小刚. 高阶精度线性耗散紧致格式的渐近稳定性. 空气动力学学报, 2000, 18(2): 165-171

[4] 邓小刚. 高阶精度耗散加权紧致非线性格式. 中国科学(A 辑), 2001, 31(12): 1104-1117

[5] 毛枚良, 董维中, 邓小刚等. 强激光与高超声速球锥流场干扰数值模拟研究. 空气动力学学报, 2001, 19(2): 172-176

[6] 毛枚良, 邓小刚, 向大平. 分区对接网格算法的应用研究. 空气动力学学报, 2002, 20(2): 179-183

[7] 向大平, 邓小刚, 毛枚良. 低马赫数流动数值模拟方法的研究. 空气动力学学报, 2002, 20(4): 373-378

[8] 向大平, 邓小刚, 毛枚良. 低马赫数流动分区并行计算研究. 空气动力学学报, 2002, 20: 77-81

[9] 邓小刚, 叶友达, 黎作武等. 研究方向: 复杂流动机理研究及其数值模拟. 科技和产业, 2002, 10: 44-55

[10] 邓小刚. High-order accurate dissipative weighted compact nonlinear schemes. 中国科学 A 辑(英文版), 2002, 45(3): 356-370

[11] 邓小刚, 庄逢甘. A novel slightly compressible model for low mach nimber perfect gas flow calculation. Acta Mechanica Sinica, 2002, 18(3): 193-207

[12] 毛枚良, 陈坚强, 邓小刚, 向大平. 高超声速流动分区对接网格算法研究. 空气动力学学报, 2003, 21(2): 173-181

[13] 宗文刚, 邓小刚, 张涵信. 双重加权实质无波动激波捕捉格式. 空气动力学学报, 2003, 21(2): 218-225

[14] 宗文刚, 邓小刚, 张涵信. 双重加权实质无波动激波捕捉格式的改进和应用. 空气动力学学报, 2003, 21(4): 399-407

[15] 屠恒章, 邓小刚. 空气动力学试验与计算. 科技和产业, 2004, 4(1): 36-38

[16] 刘昕, 邓小刚, 毛枚良等. 高阶精度非线性格式 WCNS-E-5 在二维流动中的应用研究. 空气动力学学报, 2004, 22(2): 206-210

[17] 向大平, 邓小刚, 毛枚良. 微可压缩模型(SCM)与可压缩 NS 方程数值计算对比研究. 空气动力学学报, 2005, 23(2): 195-199

[18] 赫新, 陈坚强, 邓小刚. NND 格式在多维理想磁流体方程组中的应用. 空气动力学学报, 2005, 23(2): 267-273

[19] 毛枚良, 徐昆, 邓小刚. 动能 BGK 算法在近连续流模拟中的应用. 空气动力学学报, 2005,

23(3): 317-321

[20] 刘昕, 邓小刚, 毛枚良, 宗文刚. 高精度格式 WCNS-E-5 计算物面热流. 计算物理, 2005, 22(5): 393-398

[21] 赫新, 陈坚强, 毛枚良, 邓小刚. 多块对接网格技术在电磁场散射问题中的应用. 计算物理, 2005, 22(5): 465-470

[22] 刘昕, 邓小刚, 毛枚良. 高阶精度格式 WCNS-E-5 在亚跨声速流动中的应用研究. 空气动力学学报, 2005, 23(4): 425-430

[23] 何开锋, 董维中, 陈坚强, 邓小刚. 超声速钝体空气和钾元素混合物喷流流场的数值研究. 空气动力学学报, 2006, 24(1): 90-94

[24] 刘昕, 邓小刚, 毛枚良. 高超声速飞行器外形热流密度分布计算的高精度方法研究. 宇航学报, 2006, 27(2): 157-161

[25] 陈坚强, 赫新, 张毅锋, 邓小刚. 跨大气层飞行器 RCS 干扰数值模拟研究. 空气动力学学报, 2006, 24(2): 182-186

[26] 刘昕, 邓小刚, 毛枚良. 高精度格式 WCNS-E-5 的 Fourier 分析与应用. 应用力学学报, 2006, 23(3): 329-333

[27] 毛枚良, 邓小刚, 向大平, 陈坚强. 辉光放电等离子体对边界层流动控制的机理研究. 空气动力学学报, 2006, 24(3): 269-274

[28] 袁先旭, 陈坚强, 邓小刚. 化学氧碘激光(COIL)三维混合反应流场数值模拟研究. 空气动力学学报, 2006, 24(4): 444-449

[29] 刘昕, 邓小刚, 毛枚良. 高精度非线性格式 WCNS 的分析研究与其应用. 计算力学学报, 2007, 24(3): 264-268

[30] 邓小刚, 宗文刚, 张来平等. 计算流体力学中的验证与确认. 力学进展, 2007, 37(2): 279-288

[31] 邓小刚, 刘昕, 毛枚良等. 高精度加权紧致非线性格式的研究进展. 力学进展, 2007, 37(3): 417-427

[32] 张毅锋, 邓小刚, 毛枚良等. 一种可压缩流动的高阶加权紧致非线性格式(WCNS)的加速收敛方法. 计算物理, 2007, 24(6): 698-704

[33] Liu Xin, Deng Xiaogang, Mao Meiliang. High-order behaviors of weighted compact fifth- order nonlinear schemes. AIAA Journal, 2007, 45(8): 2093-2097

[34] 刘伟, 杨小亮, 张涵信, 邓小刚. 大攻角运动时的机翼摇滚问题研究综述. 力学进展, 2008, 38(2): 214-228

[35] 毛枚良, 邓小刚, 陈坚强. 常气压辉光放电等离子体控制翼型失速的数值模拟研究. 空气动力学学报, 2008, 26(3): 334-338

[36] 张毅锋, 邓小刚, 毛枚良, 陈坚强. 高阶加权紧致非线性格式(WCNS)在二维流动计算中的加速收敛研究. 空气动力学学报, 2008, 26(3): 349-355

[37] 毛枚良, 邓小刚, 陈亮中, 陈坚强. 常气压辉光放电等离子体对边界层流动的影响. 计算物理, 2009, 26(1): 57-63

[38] 毛枚良, 邓小刚, 李松. 耗散紧致格式的频谱特性研究与应用. 计算物理, 2009, 26(3): 371-377

[39] 杨明智, 袁先旭, 谢昱飞, 张来平, 邓小刚. 前体涡非对称分离机理及前缘吹气控制研究. 空气动力学学报, 2009, 27(2): 186-192

[40] 毛枚良, 江定武, 邓小刚. 高超声速层流气动热预测混合算法研究. 空气动力学学报, 2009, 27(3): 275-280

[41] 毛枚良, 姜屹, 邓小刚. 基于 DCS5 格式的 LDDRK 算法. 计算物理, 2010, 27(2): 159-167

[42] 张来平, 邓小刚, 张涵信. 动网格生成技术及非定常计算方法进展综述. 力学进展, 2010, 40(4): 424-447

[43] 张来平, 刘伟, 贺立新, 邓小刚. 基于静动态混合重构的 DG/FV 混合格式. 力学学报, 2010, 42(6): 1013-1022

[44] Deng Xiaogang, Mao Meiliang, Tu Guohua, Zhang Yifeng, Zhang Hanxin. Extending weighted compact nonlinear schemes to complexgrids with characteristic-based interface conditions. AIAA Journal, 2010, 48(12): 2840-2851

[45] 涂国华, 邓小刚, 毛枚良. 消除粘性项高阶离散数值振荡的半结点-结点交错方法. 空气动力学学报, 2011, 29(1): 10-15

[46] 彭强, 邓小刚, 廖达雄, 符澄. 半柔壁喷管气动设计关键控制参数研究. 空气动力学学报, 2011, 29(1): 39-46

[47] 张来平, 刘伟, 贺立新, 赫新, 邓小刚. 基于静动态重构的高阶 DG/FV 混合格式在二维非结构网格中的推广. 计算物理, 2011, 28(2): 188-198

[48] 毛枚良, 江定武, 陈亮中, 邓小刚. 受 DBD 等离子体控制的低速流动数值模拟方法研究. 空气动力学学报, 2011, 29(2): 129-134

[49] 丛成华, 邓小刚, 毛枚良. 微可压缩模型预处理技术研究. 力学学报, 2011, 43(4): 775-779

[50] 张来平, 刘伟, 贺立新, 邓小刚, 张涵信. 一种新的间断侦测器及其在 DGM 中的应用. 空气动力学学报, 2011, 29(4): 401-406

[51] Li ZhiHui, Peng AoPing, Zhang HanXin, Deng Xiaogang. Numerical study on the gas-kinetic high-order schemes for solving Boltzmann model equation. Science China (Physics, Mechanics & Astronomy), 2011, 54(9): 1687-1701

[52] Deng Xiaogang, Mao Meiliang, Tu Guohua, Liu Huayong, Zhang Hanxin. Geometric conservation law and applications to high-order finitedifference schemes with stationary grids. Journal of Computational Physics, 2011, 230: 1100-1115

[53] 王新光, 毛枚良, 邓小刚, 涂国华. 基于 5 阶精度格式 WCNS-E-5 的 p-multigrid 方法研究. 空气动力学学报, 2012, 31(1): 1-6

[54] 王光学, 邓小刚, 刘化勇, 王运涛. 高阶精度格式 WCNS 在三角翼大攻角模拟中的应用研究. 空气动力学学报, 2012, 30(1): 28-33

[55] 王光学, 邓小刚, 王运涛, 刘化勇. 三角翼涡破裂的高精度数值模拟. 计算物理, 2012, 29(4): 489-494

[56] 王运涛, 王光学, 徐庆新, 邓小刚. 基于结构网格的大规模并行计算研究. 计算机工程与科学, 2012, 34(8): 63-68

[57] 王光学, 张玉伦, 李松, 邓小刚. WCNS 高精度并行软件的大规模计算研究. 计算机工程与科学, 2012, 34(8): 125-130

[58] 张理论, 邓小刚. 戈登奖——分析与思考. 计算机工程与科学, 2012, 34(8): 44-52

[59] 姜屹, 毛枚良, 邓小刚. DCS5 在计算气动声学中的应用研究. 空气动力学学报, 2012, 30(4): 431-436

[60] 赫新, 张来平, 赵钟, 邓小刚. 大型通用 CFD 软件体系结构与数据结构研究. 空气动力学学报, 2012, 30(5): 557-565

[61] 涂国华, 邓小刚, 毛枚良. 5 阶非线性 WCNS 和 WENO 差分格式频谱特性比较. 空气动力学学报, 2012, 30(6): 709-712

[62] Tu Guohua, Deng Xiaogang, Mao Meiliang. Assessment of two turbulence models and some

compressibility corrections for hypersonic compression corners by high-order difference schemes. Chinese Journal of Aeronautic, 2012, 25: 25-32

[63] Zhang Laiping, Liu Wei, He Lixin, Deng Xiaogang. A class of hybrid DG/FV methods for conservation laws III: two-dimensional euler equations. CCP, 2012, 12(1): 284-314

[64] Deng Xiaogang, Mao Meiliang, Tu Guohua, Zhang Hanxin, Zhang Yifeng. High-order and high accurate CFD methods and their applications for complex grid problems. CCP, 2012, 11(4): 1081-1102

[65] Zhang Laiping, Wei Liu, Lixin He, Xiaogang Deng, Hanxin Zhang. A class of hybrid DG/FV methods for conservation laws II: Two-dimensional cases. Journal of Computational Physics, 2012, 231(4): 1104-1120

[66] Zhang Laiping, Wei Liu, Lixin He, Xiaogang Deng, Hanxin Zhang. A class of hybrid DG/FV methods for conservation laws I: Basic formulation and one-dimensional systems. Journal of Computational Physics, 2012, 231(4): 1081-1103

[67] 王运涛, 张玉伦, 王光学, 邓小刚. 三角翼布局气动特性及流动机理研究. 空气动力学学报, 2013, 31(5): 554-558

[68] 孙岩, 邓小刚, 张征宇, 王超. 跨声速风洞模型变形测量实验中标记点影响研究. 空气动力学学报, 2013, 31(6): 769-775

[69] 王运涛, 孟德虹, 邓小刚. 多段翼型高精度数值模拟技术研究. 空气动力学学报, 2013, 31(1): 88-93

[70] 乔宇航, 马东立, 邓小刚. 基于升力线理论的机翼几何扭转设计方法. 北京航空航天大学学报, 2013, 39(3): 320-324

[71] 孙岩, 张征宇, 邓小刚, 杨党国, 周桂宇. 风洞模型静弹性变形对气动力影响研究. 空气动力学学报, 2013, 31(3): 294-300

[72] 李锦, 江定武, 毛枚良, 邓小刚. BGK-NS 格式在复杂流动中的应用研究. 空气动力学学报, 2013, 31(4): 449-461

[73] Deng Xiaogang, Jiang Yi, Mao MeiLiang, Liu HuaYong, Tu GuoHua. Developing hybrid cell- edge and cell-node dissipative compact scheme for complex geometry flows. Science China (Technological Sciences), 2013, 56(10): 2361-2369

[74] Tu GuoHua, Deng Xiaogang; Mao MeiLiang. Validation of a RANS transition model using a high-order weighted compact nonlinear scheme. Science China (Physics, Mechanics & Astronomy), 2013, 56(4): 805-811

[75] Deng Xiaogang, Min Yaobing, Mao Meiliang, Liu Huayong, Tu Guohua, Zhang Hanxin. Further studies on geometric conservation law and applications to high-order finite difference schemeswith stationary grids. Journal of Computational Physics, 2013, 239: 90-111

[76] Tu Guohua, Deng Xiaogang, Mao Meiliang. Implementing high-order weighted compact nonlinear scheme on patched grids with a nonlinear interpolation. Computers and Fluids, 2013, 77: 181-193

[77] Zhang Shuhai, Deng Xiaogang, Mao Meiliang, Wang Chi. Shu Improvement of convergence to steady state solutions of Euler equations with weighted compact nonlinear schemes. Acta Mathematicae Applicatae Sinica, 2013, 29(3): 449-464

[78] Xu Chuanfu, Deng Xiaogang, Zhang Lilun, Jiang Yi, Wei Cao, Jianbin Fang, Yonggang Che, Yongxian Wang, Wei Liu. Parallelizing a high-order CFD software for 3D, multi-block,

structural grids on the tianHe-1A supercomputer. Supercomputing, 2013, 7925: 26-39

[79] 孙岩, 邓小刚, 王光学, 王运涛, 毛枚良. 基于径向基函数改进的 Delaunay 图映射动网格方法. 航空学报, 2014, 35(3): 727-735

[80] 燕振国, 刘化勇, 毛枚良, 邓小刚. 朱华君基于高阶耗散紧致格式的 GMRES 方法收敛特性研究. 航空学报, 2014, 35(5): 1181-1192

[81] 涂国华, 邓小刚, 闵耀兵, 毛枚良, 刘化勇. CFD 空间精度分析方法及 4 种典型畸形网格中 WCNS 格式精度测试. 空气动力学学报, 2014, 32(4): 425-432

[82] 李松, 王光学, 王运涛, 张玉伦, 邓小刚. WCNS 格式在梯形翼高升力构型模拟中的应用. 研究空气动力学学报, 2014, 32(4): 439-445

[83] 孙岩, 邓小刚, 王运涛, 王光学. RBF_TFI 结构动网格技术在风洞静气动弹性修正中的应用. 工程力学, 2014, 31(10): 228-233

[84] 姜屹, 邓小刚, 毛枚良, 刘化勇. 基于 HDCS-E8T7 格式的气动噪声数值模拟方法研究. 进展空气动力学学报, 2014, 32(5): 559-574

[85] 赵云飞, 刘伟, 刘绪, 邓小刚. 非定常运动下的激波/边界层干扰分离特性研究. 空气动力学学报, 2014, 32(5): 610-617

[86] Jiang Yi, Mao Meiliang, Deng Xiaogang, Liu Huayong. Large eddy simulation on curvilinear meshes using seventh-orderdissipative compact scheme. Computers & Fluids, 2014, 104: 73-84

[87] Xu Chuanfu, Deng Xiaogang, Zhang Lilun, etc. Collaborating CPU and GPU for large-scale high-order CFD simulations with complex grids on the TianHe-1A supercomputer. Journal of Computational Physics, 2014, 278: 275-297

[88] Tu Guohua, Zhao Xiaohui, Mao Meiliang, Chen Jianqiang, Deng Xiaogang, Liu Huayong. Evaluation of Euler fluxes by a high-order CFD scheme: shock instability. International Journal of Computational Fluid Dynamics, 2014, 28(5): 171-186

[89] 江定武, 毛枚良, 李锦, 邓小刚. 气体动理学统一算法中相容性条件不满足引起的数值误差及其影响研究. 力学学报, 2015, 47(1): 163-168

[90] 毛枚良, 燕振国, 刘化勇, 朱华君, 邓小刚. 高阶加权非线性格式的拟线性频谱分析方法研究. 空气动力学学报, 2015, 33(1): 1-9

[91] 马燕凯, 刘化勇, 燕振国, 毛枚良, 邓小刚. 基于 HWCNS 格式的紧致插值方法研究. 计算力学学报, 2015, 32(3): 388-393

[92] 赵云飞, 刘伟, 冈敦殿, 易仕和, 邓小刚. 粗糙物面引起的超声速边界层转捩现象研究. 宇航学报, 2015, 36(6): 739-746

[93] 孙岩, 邓小刚, 王光学, 毛枚良, 张玉伦. 一种基于约束框架的棱柱网格生成方法. 空气动力学学报, 2015, 33(3): 319-324

[94] 毛枚良, 江定武, 李锦, 邓小刚. 气体动理学统一算法的隐式方法研究. 力学学报, 2015, 47(5): 822-829

[95] 张扬, 张来平, 赫新, 邓小刚. 基于非结构/混合网格的脱体涡模拟算法. 航空学报, 2015, 36(9): 2900-2910

[96] Deng Xiaogang, Jiang Yi, Mao Meiliang, Liu Huayong, Li Song, Tu Guohua. A family of hybrid cell-edge and cell-node dissipative compact schemessatisfying geometric conservation law. Computers & Fluids, 2015, 116: 29-45

[97] Jiang Yi, Mao Meiliang, Deng Xiaogang and Liu Huayong. Extending seventh-order dissipative compact scheme satisfying geometric conservation law to large eddy simulation on curvilinear

grids. advances in Applied Mathematics and Mechanics, 2015, 7(4): 407-429

[98] Mao Meiliang, Zhu Huajun, Deng Xiaogang, Min Yaobing and Liu Huayong. Effect of geometric conservation law on improving spatial accuracy for finite difference schemes on two-dimensional nonsmooth grids. Communications in Computational Physics, 2015, 18(3): 673-706

[99] Zhang Yang, Zhang Laiping, Hea Xin, Deng Xiaogang. Detached-eddy simulation of subsonic flow past a delta wing. Procedia Engineering, 2015, 126: 584-587

[100] Jiang Yi, Mao MeiLiang, Deng Xiaogang, Liu Hua Yong. Numerical investigation on body-wake flow interaction over rod-airfoil configuration Journal of Fluid Mechanics, 2015, 779: 1-35

[101] 张扬, 张来平, 赫新, 邓小刚. 基于自适应混合网格的脱体涡模拟. 航空学报, 2016, 37(12): 3605-3614

[102] 孙岩, 孟德虹, 王运涛, 邓小刚. 基于径向基函数与混合背景网格的动态网格变形方法. 航空学报, 2016, 37(5): 1462-1472

[103] 徐丹, 王东方, 陈亚铭, 邓小刚. 高精度有限体积格式在三维曲线坐标系下的应用. 国防科技大学学报, 2016, 38(2): 56-60

[104] 王圣业, 王光学, 董义道, 邓小刚. 基于高精度加权紧致非线性格式的γ-Re_θ转捩模型标定与应用. 国防科技大学学报, 2016, 38(4): 14-20

[105] 董义道, 王东方, 王光学, 邓小刚. 雷诺应力模型的初步应用. 国防科技大学学报, 2016, 38(4): 46-53

[106] 周云龙, 刘伟, 董义道, 王光学, 邓小刚. 五阶HWCNS在低速复杂流场中的应用. 国防科技大学学报, 2016, 38(4): 1-7

[107] 王光学, 王圣业, 王东方, 邓小刚. 应用加权紧致非线性格式的VFE-2钝前缘三角翼转捩模拟. 国防科技大学学报, 2016, 38(4): 8-13

[108] 王东方, 邓小刚, 王光学, 刘化勇. 高超声速尖双锥流动高精度数值模拟. 国防科技大学学报, 2016, 38(4): 54-63

[109] 张来平, 邓小刚, 何磊, 李明, 赫新. E级计算给CFD带来的机遇与挑战. 空气动力学学报, 2016, 34(4): 405-417

[110] Zhang Yang, Zhang Laiping, He Xin, Deng Xiaogang. An improved second-order finite-volume algorithmfor detached-eddy simulation based on hybrid grids. Communications in Computational Physics, 2016, 20(2): 459-485

[111] Yan Zhenguo, Liu Huayong, Mao Meiliang, Zhua Huajun, Deng Xiaogang. New nonlinear weights for improving accuracy and resolution of weighted compact nonlinear scheme. Computers and Fluids, 2016, 127: 226-240

[112] Zhu Huajun, Deng Xiaogang, Mao Meiliang, Liu Huayong and Guohua Tu. Osher flux with entropy fix for two-dimensional euler equations. Advances in Applied Mathematics and Mechanics, 2016, 8(4): 670-692

[113] Mao Meiliang, Jiang Yi, Deng Xiaogang and Liu Huayong. Noise Prediction in subsonic flow using seventh-order dissipative compact scheme on curvilinear mesh. Advances in Applied Mathematics and Mechanics, 2016, 8(2): 236-256

[114] Xu Dan, Deng Xiaogang, Chen Yaming, Dong Yidao, Wang Guangxue. On the freestream preservation of finite volume method in curvilinear coordinates. Computers & Fluids, 2016, 129: 20-32

[115] Wang Dongfang, Deng Xiaogang, Wang Guangxue, Dong Yidao. Developing a hybrid flux

function suitable for hypersonic flow simulation with high-order methods. International Journal for Numerical Methods in Fluids, 2016, 81(5): 309-327

[116] Wang Shengye, Deng Xiaogang, Wang Guangxue, Xu Dan, Wang Dongfang. Efficiency benchmarking of seventh-order tri-diagonal weighted compact nonlinear scheme on curvilinear mesh. International Journal of Computational Fluid Dynamics, 2016, 30: 7-10

[117] Li Dali, Xu Chuanfu, Wang Yongxian, Song Zhifang, Xiong Min, Gao Xiang, Deng Xiaogang. Parallelizing and optimizing large-scale 3D multi-phase flow simulations on the Tianhe-2 supercomputer. Concurrency and Computation: Practice and Experience, 2016, 28(5): 1678-1692

[118] 程彬，李大力，徐传福，刘巍，王光学，邓小刚. 面向高阶精度 CFD 的 JFNK 算法及其并行计算. 计算机科学与探索，2017, 11(1): 61-69

[119] Gao Xiang, Dong Yidao, Xu Chuanfu, Xiong Min, Wang Zhenghua, Deng Xiaogang. Developing a new mesh deformation technique based on support vector machine. International Journal of Computational Fluid Dynamics, 2017, 31: 246-257

[120] Yan Zhen Guo, Liu Huayong, Ma Yankai, Mao Meiliang, Deng Xiaogang. Further improvement of weighted compact nonlinear scheme using compact nonlinear interpolation. Computers and Fluids, 2017, 156: 135-145

[121] Xu Dan, Deng Xiaogang, Chen Yaming, Wang Guangxue, Dong Yidao. Effect of nonuniform grids on high-order finite difference method. Adv. Appl. Math. Mech, 2017, 9(4): 1012-1034

[122] Dong Yidao, Deng Xiaogang, Xu Dan, Wang Guangxue. Reevaluation of high-order finite difference and finite volume algorithms with freestream preservation satisfied. Computers and Fluids, 2017, 156: 343-352

[123] Li Dali, Xu Chuanfu, Cheng Bin, Xiong Min, Gao Xiang, Deng Xiaogang. Performance modeling and optimization of parallel LU-SGS on many-core processors for 3D high-order CFD simulations. The Journal of Supercomputing, 2017, 73(6): 2506-25

彩 图

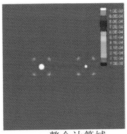
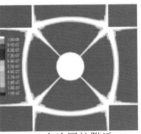

(a) 整个计算域　　　　(b) 左边圆柱附近

(第 20 页) 图 5 $\sigma=1$ 时 HDCS-E8T7A 格式的 $|I_x+I_y|$ 误差

Fig. 5　SCL error ($|I_x+I_y|$) of HDCS-E8T7A with $\sigma=1$

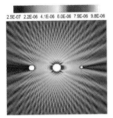
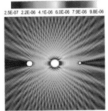
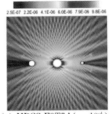
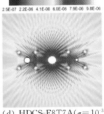

(a) HDCS-E8T7　　(b) HDCS-E8T7A($\sigma=10^{-5}$)　　(c) HDCS-E8T7A($\sigma=10^{-4}$)　　(d) HDCS-E8T7A($\sigma=10^{-3}$)

(第 21 页) 图 6　HDCS-E8T7 和 HDCS-E8T7A 计算得到的脉动压力均方根值云图

Fig. 6　Mean-squared fluctuating pressure contours of HDCS-E8T7 and HDCS-E8T7A

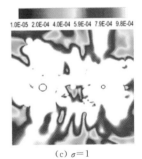

(a) $\sigma=10^{-2}$　　　　(b) $\sigma=10^{-1}$　　　　(c) $\sigma=1$

(第 21 页) 图 7　HDCS-E8T7A 格式计算得到的脉动压力均方根云图

Fig. 7　Mean-squared fluctuating pressure contours of HDCS-E8T7A

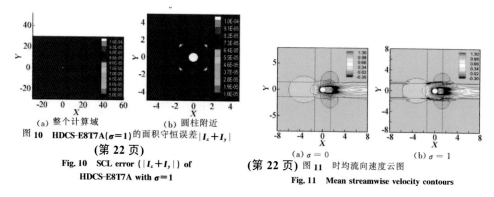

(a) 整个计算域　　(b) 圆柱附近

图 10　HDCS-E8T7A($\sigma=1$)的面积守恒误差$|I_x+I_y|$
（第 22 页）

Fig. 10　SCL error ($|I_x+I_y|$) of HDCS-E8T7A with $\sigma=1$

(a) $\sigma=0$　　(b) $\sigma=1$

（第 22 页）图 11　时均流向速度云图

Fig. 11　Mean streamwise velocity contours

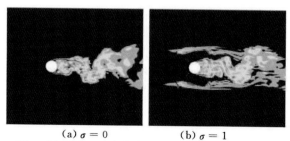

(a) $\sigma=0$　　(b) $\sigma=1$

（第 22 页）图 13　瞬时涡量云图

Fig. 13　Instantaneous vorticity contours

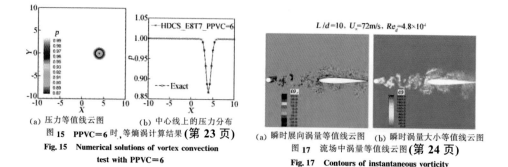

(a) 压力等值线云图　　(b) 中心线上的压力分布

图 15　PPVC=6 时，等熵涡计算结果（第 23 页）

Fig. 15　Numerical solutions of vortex convection test with PPVC=6

(a) 瞬时展向涡量等值线云图　　(b) 瞬时涡量大小等值线云图

图 17　流场中涡量等值线云图（第 24 页）

Fig. 17　Contours of instantaneous vorticity

（第 25 页）图 21　速度张量的第二不变量等值面

Fig. 21　Iso-surfaces of the second invariant of velocity gradient tensor for instantaneous vortex structure

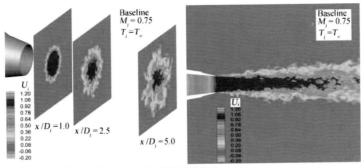

(第 26 页) 图 26 瞬时流向速度云图
Fig. 26 Contours of instantaneous axial velocity

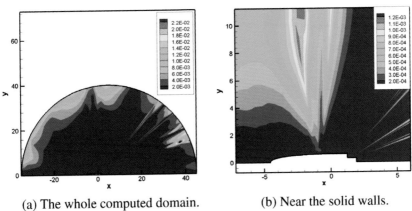

(a) The whole computed domain.   (b) Near the solid walls.

(第 91 页) Fig. 7. The SCL errors ($|I_x| + |I_y| + |I_z|$) of 6th-order traditional standard metrics.

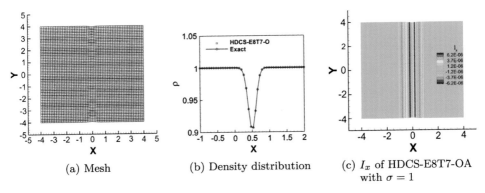

(a) Mesh   (b) Density distribution   (c) I_x of HDCS-E8T7-OA with $\sigma = 1$

(第 106 页) Fig. 10. Performance on a curvilinear mesh.

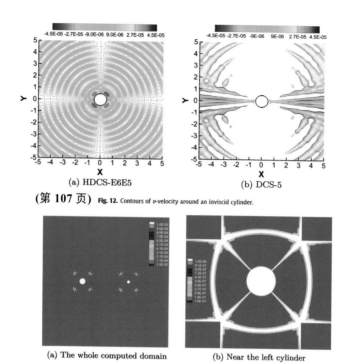

(第 107 页) **Fig. 12.** Contours of v-velocity around an inviscid cylinder.

(a) The whole computed domain (b) Near the left cylinder

(第 108 页) **Fig. 15.** The SCL error ($|I_x| + |I_y|$) of HDCS-E8T7-OA with $\sigma = 1$.

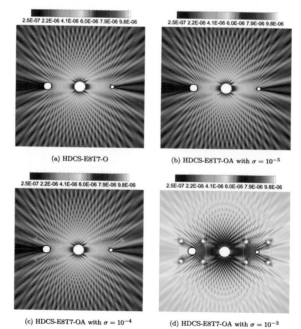

(第 109 页) **Fig. 16.** Mean-squared fluctuating pressure contours of HDCS-E8T7-O and HDCS-E8T7-OA.

(a) HDCS-E8T7-OA with $\sigma = 10^{-2}$
(b) HDCS-E8T7-OA with $\sigma = 10^{-1}$
(c) HDCS-E8T7-OA with $\sigma = 1$

(第 109 页) **Fig. 17.** Mean-squared fluctuating pressure contours of HDCS-E8T7-OA.

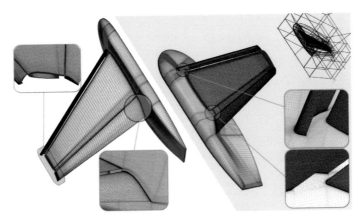

(第 110 页) **Fig. 19.** Gird system of trapezoidal wing case.

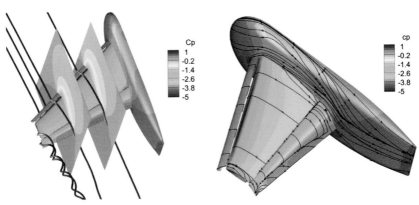

(第 111 页) **Fig. 21.** Pressure contours and streamlines.

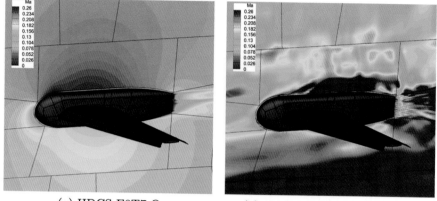

(a) HDCS-E8T7-O (b) HDCS-E8T7-OA with $\sigma = 1$

(第 111 页) **Fig. 22.** The Mach number on the symmetry surface.

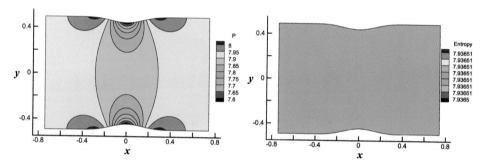

(a) The distribution of the pressure (b) The distribution of the entropy

(第 143 页) **Fig. 7.** Distributions of the pressure and entropy in the two-dimensional channel flow using Method-A.

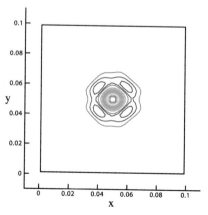

(第 144 页) **Fig. 9.** The vorticity magnitude distribution of the vortex transport problem on a Cartesian mesh.

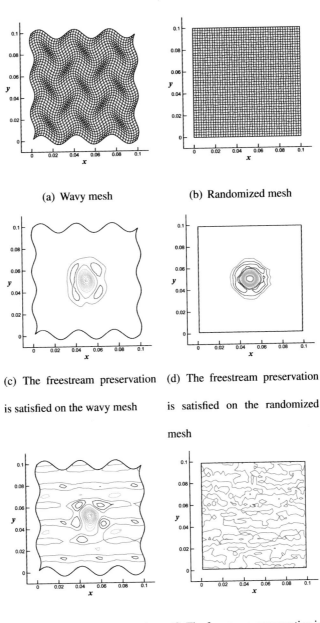

Fig. 10. Nonuniform meshes and vorticity magnitude distributions of the results of the vortex transport problem with 16 contours.

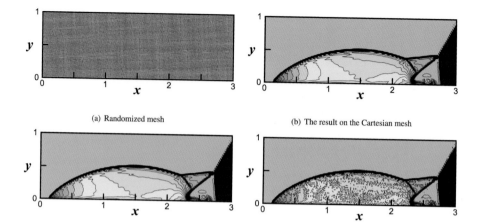

(a) Randomized mesh (b) The result on the Cartesian mesh

(c) The freestream preservation is satisfied on the randomized mesh (d) The freestream preservation is not satisfied on the randomized mesh

(第 145 页) Fig. 11. The mesh and distributions of the density in the double Mach reflection problem with 40 contours.

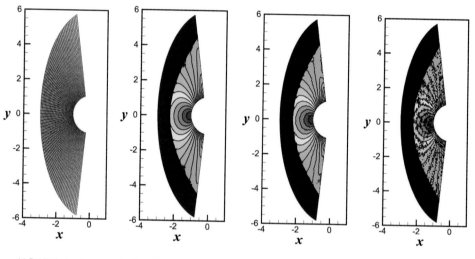

(a) Randomized mesh (b) The result on the original mesh (c) The freestream preservation is satisfied on the randomized mesh (d) The freestream preservation is not satisfied on the randomized mesh

(第 146 页) Fig. 12. The mesh and distributions of the pressure in the supersonic flow past a cylinder problem with 20 contours.

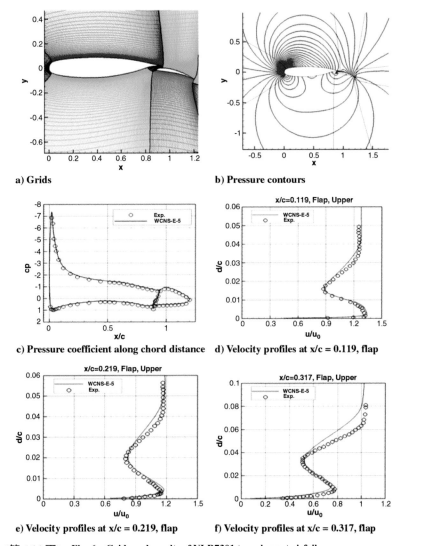

(第 166 页)　Fig. 6　Grids and results of NLR7301 two-element airfoil.

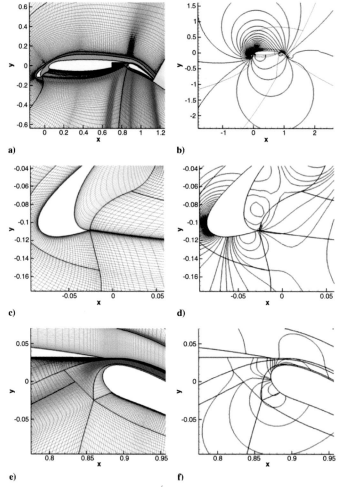

(第 167 页) Fig. 7 Grids and pressure contours of 30P-30N three-element airfoil.

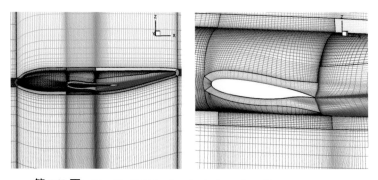

(第 169 页) Fig. 10 Grids for the wing-body configuration.

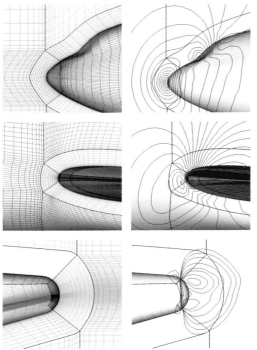

(第 170 页) Fig. 12 Grids and pressure contours at the nose (top), the wing-body junction (middle), and the tail (bottom). $\alpha = 0°$.

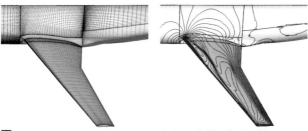

(第 170 页) Fig. 13 Grids and pressure contours on the leeward of the wing ($\alpha = 0°$).

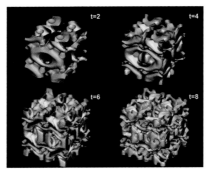

(第 181 页) Fig. 3. Iso-surfaces of the second invariant of the velocity gradient tensor colored by streamwise velocity.

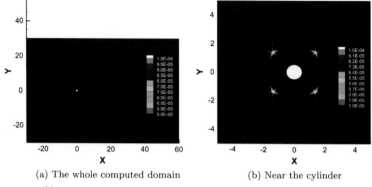

(a) The whole computed domain (b) Near the cylinder

(第 183 页) **Fig. 8.** The SCL error ($|I_x|+|I_y|$) of HDCS-E8T7A with $\sigma = 1$.

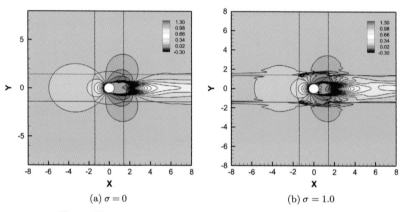

(a) $\sigma = 0$ (b) $\sigma = 1.0$

(第 184 页) **Fig. 10.** Mean streamwise velocity contours of HILES.

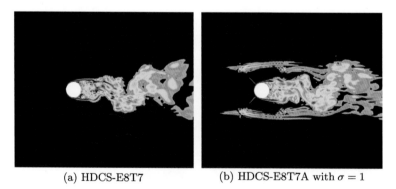

(a) HDCS-E8T7 (b) HDCS-E8T7A with $\sigma = 1$

(第 186 页) **Fig. 15.** The instantaneous vorticity contours.

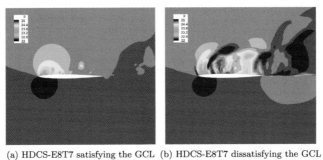

(a) HDCS-E8T7 satisfying the GCL (b) HDCS-E8T7 dissatisfying the GCL

(第 186 页) Fig. 17. Instantaneous pressure contours of HILES with HDCS-E8T7 and HDCS-E8T7A.

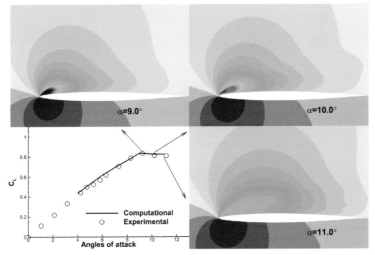

(第 187 页) Fig. 20. The time-averaged lift coefficient vs angles of attack and pressure contours at 9°, 10° and 11° angles of attack.

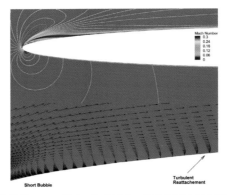

(第 188 页) Fig. 21. Time-averaged Mach number distribution and velocity vectors at 5.5° angle of attack

(第 188 页) **Fig. 22.** The instantaneous pressure distribution over airfoil surface and iso-surface of the second invariant of the velocity gradient tensor at 5.5° angle of attack.

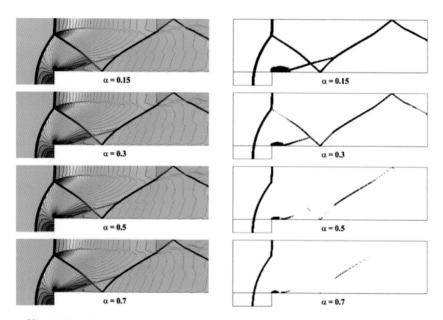

(第 198 页) Figure 10. Density contours and switch function at $t = 4$.

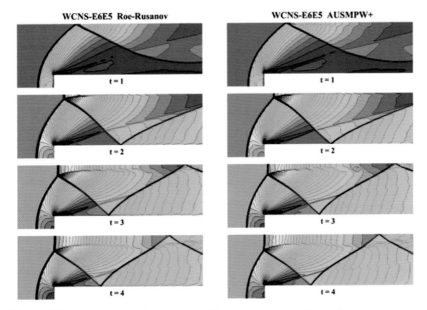

(第 198 页) Figure 11. Density contours of flow past a step at different time.

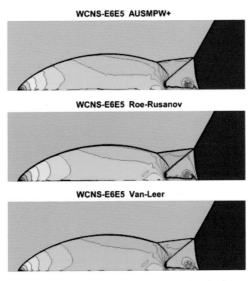

(第 199 页) Figure 13. Density contours of double Mach reflection.

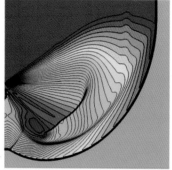

(第 200 页) Figure 16. Density contours of shock diffraction.

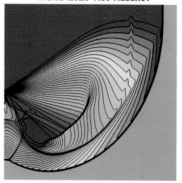

(第 201 页) Figure 17. Density contours of shock diffraction.

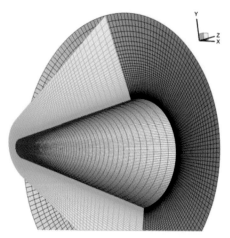

(第 203 页) Figure 21. Computational grids for blunt cone.

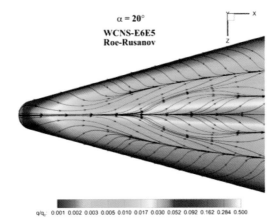

(第 203 页) Figure 22. Upper-surface streamlines and wall heat transfer rate contours.

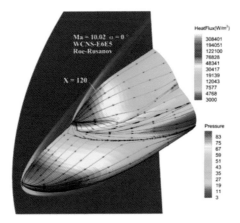

(第 205 页) Figure 26. Wall heat transfer rate contours, streamlines, and pressure contours in the symmetry plane.

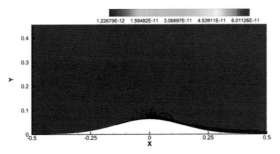

(第 215 页) **Figure 3.** Contour plot of the entropy error for WCNS-E8T7 on the extra-fine (512 × 256) grid.

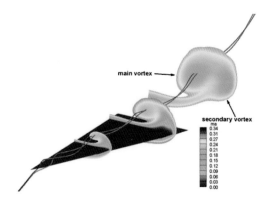

(第 219 页) **Figure 13.** Streamlines and slices of Mach number contours along and behind the delta wing.

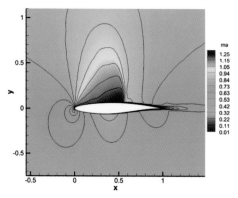

(第 221 页) **Figure 19.** Contours of the Mach number on the coarse grid obtained by WCNS-E8T7 for the RAE2822 airfoil.

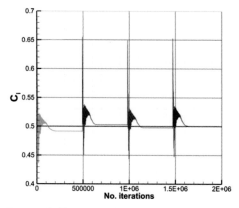

(第 224 页) **Figure 24.** The lift convergence history on coarse grid with WCNS-E8T7.

(第 224 页) **Figure 25.** Contours of the pressure coefficient on the medium grid obtained by WCNS-E8T7 for the wing-body configuration.

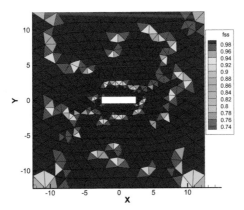

(第 232 页) **Figure 2.** The original mesh of rectangle rotation and translation.

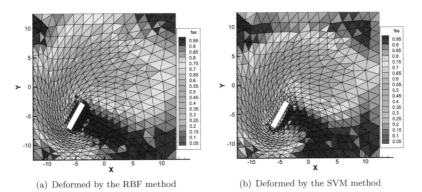

(a) Deformed by the RBF method (b) Deformed by the SVM method

(第 233 页) **Figure 4.** Final meshes using the RBF and SVM method after 64 steps.

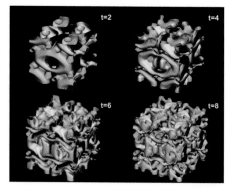

(第 243 页) **Fig. 3.** Iso-surfaces of the second invariant of the velocity gradient tensor colored by streamwise velocity.

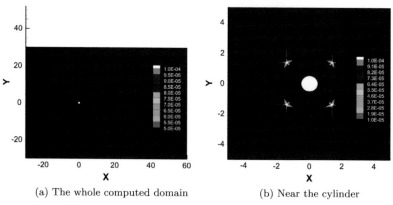

(a) The whole computed domain

(b) Near the cylinder

(第 245 页) **Fig. 8.** The SCL error ($|I_x| + |I_y|$) of HDCS-E8T7A with $\sigma = 1$.

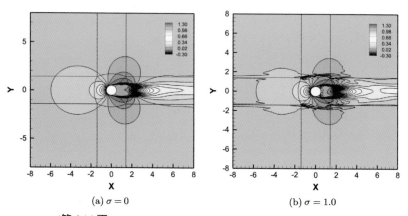

(a) $\sigma = 0$

(b) $\sigma = 1.0$

(第 246 页) **Fig. 10.** Mean streamwise velocity contours of HILES.

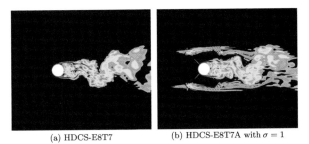

(a) HDCS-E8T7   (b) HDCS-E8T7A with $\sigma = 1$

(第 248 页) **Fig. 15.** The instantaneous vorticity contours.

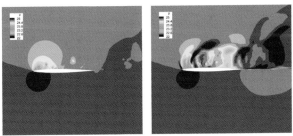

(a) HDCS-E8T7 satisfying the GCL   (b) HDCS-E8T7 dissatisfying the GCL

(第 248 页) **Fig. 17.** Instantaneous pressure contours of HILES with HDCS-E8T7 and HDCS-E8T7A.

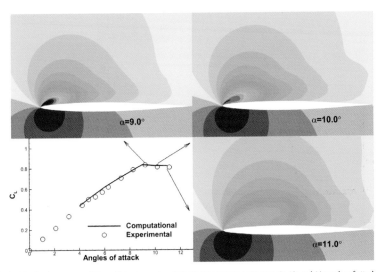

(第 249 页) **Fig. 20.** The time-averaged lift coefficient vs angles of attack and pressure contours at 9°, 10° and 11° angles of attack.

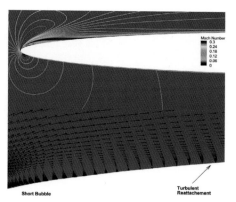

(第 250 页) **Fig. 21.** Time-averaged Mach number distribution and velocity vectors at 5.5° angle of attack.

(第 250 页) **Fig. 22.** The instantaneous pressure distribution over airfoil surface and iso-surface of the second invariant of the velocity gradient tensor at 5.5° angle of attack.

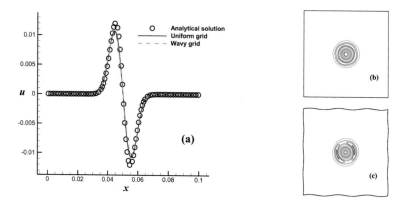

Figure 9: Numerical results on the uniform and wavy grids with grid number 120×120. (a) u alone the vertical centerline. (b) Vorticity magnitude contours on uniform grid. (c) Vorticity magnitude contours on wavy grid.

(第 268 页)

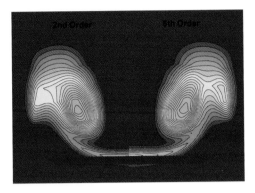

Figure 12: Comparison of the Mach number contours of the 2nd-order and 5th-order schemes at $x=2.6c$.
(第 270 页)

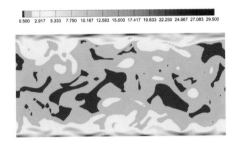

(第 285 页)　Figure 5: Instantaneous vorticity magnitude in the (y,z)-plane (fine grid).

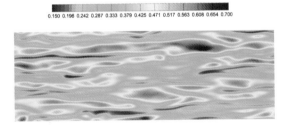

Figure 6: Instantaneous streamwise velocity in the (x,z)-plane at $y=0.03$ (fine grid).
(第 285 页)

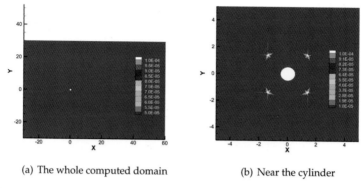

(a) The whole computed domain

(b) Near the cylinder

(第 287 页) Figure 8: The SCL error ($|I_x|+|I_y|$) of HDCS-E8T7A with $\sigma=1$.

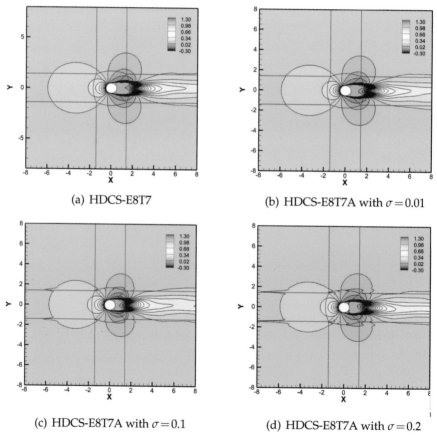

(a) HDCS-E8T7

(b) HDCS-E8T7A with $\sigma=0.01$

(c) HDCS-E8T7A with $\sigma=0.1$

(d) HDCS-E8T7A with $\sigma=0.2$

(第 288 页) Figure 12: Mean streamwise velocity contours of HILES.

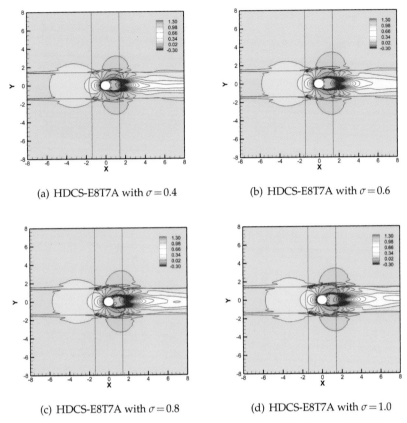

(a) HDCS-E8T7A with $\sigma=0.4$
(b) HDCS-E8T7A with $\sigma=0.6$
(c) HDCS-E8T7A with $\sigma=0.8$
(d) HDCS-E8T7A with $\sigma=1.0$

(第 289 页) Figure 13: Mean streamwise velocity contours of HILES with HDCS-E8T7A.

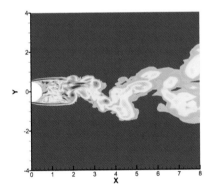

(第 290 页) Figure 18: The instantaneous vorticity contours.

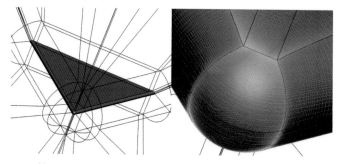

(第 291 页) Figure 19: Delta wing grid structure

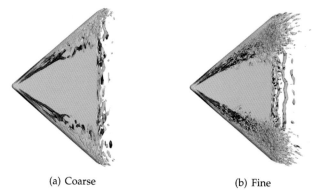

(a) Coarse (b) Fine

Figure 20: Iso-surfaces of the second invariant of velocity gradient tensor for instantaneous vortex structure.
(第 291 页)

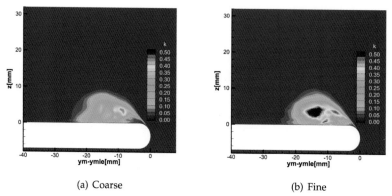

(a) Coarse (b) Fine

(第 292 页) Figure 21: Contours of the turbulent kinetic energy in the plane $x_m = 0.5$.

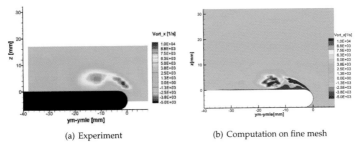

(a) Experiment (b) Computation on fine mesh

(第 292 页) Figure 22: Time-averaged axial-vorticity distribution, $x_m = 0.5$.

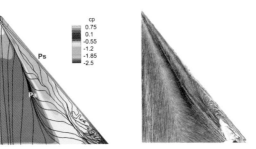

(a) Computational mean limiting streamline pattern and surface pressure coefficient

(b) Oil-flow pattern

(第 293 页) Figure 24: Mean flow patterns on the surface.

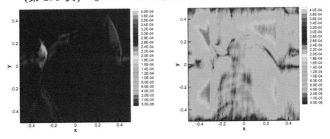

(a) Error distribution of density for GCL-FD4 (b) Error distribution of density for NGCL-FD4

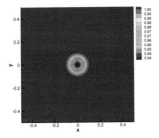

(c) Density distribution for NGCL-FD4

(第 324 页) Figure 4: Error distribution of density for GCL-FD4 and NGCL-FD4 at time $T=2$ in solving vortex problem on 500×500 grid with $k=1$.

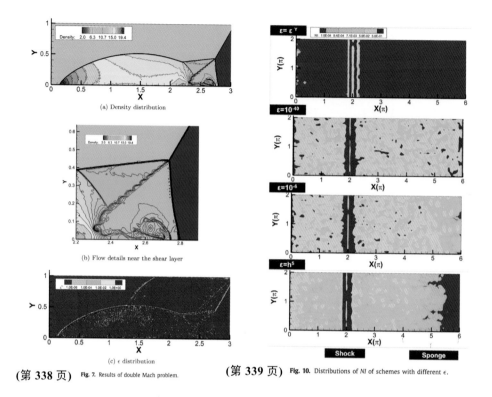

(第 338 页) **Fig. 7.** Results of double Mach problem.

(第 339 页) **Fig. 10.** Distributions of NI of schemes with different ϵ.

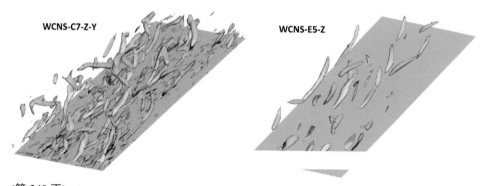

(第 340 页) **Fig. 12.** Q iso-surface of results of different schemes. (colored picture attached at the end of the book)

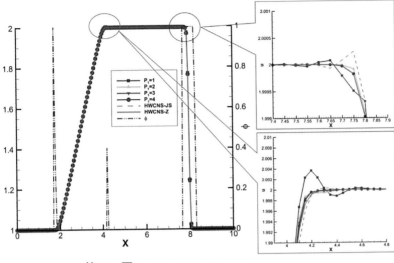

(第 347 页) Fig. 1. Scalar case to determine p_1.

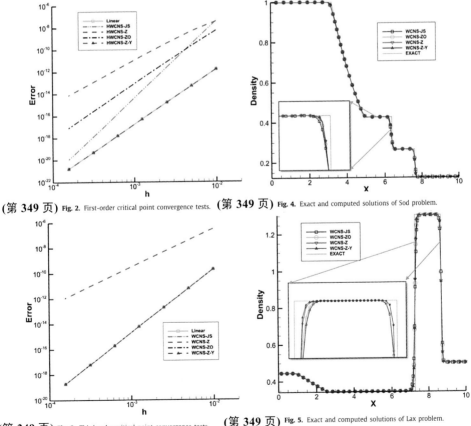

(第 349 页) Fig. 2. First-order critical point convergence tests.

(第 349 页) Fig. 4. Exact and computed solutions of Sod problem.

(第 349 页) Fig. 3. Third-order critical point convergence tests.

(第 349 页) Fig. 5. Exact and computed solutions of Lax problem.

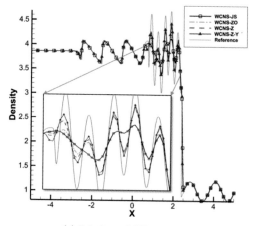

(a) Solutions of different schemes

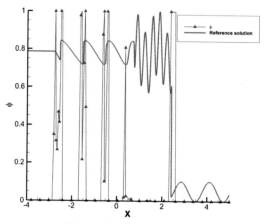

(b) ϕ distribution of WCNS-Z-Y

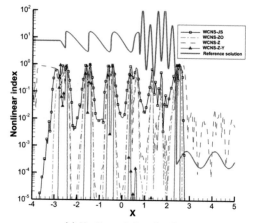

(c) Nonlinear index distributions

(第 350 页) **Fig. 6.** Results of Osher-Shu problem.

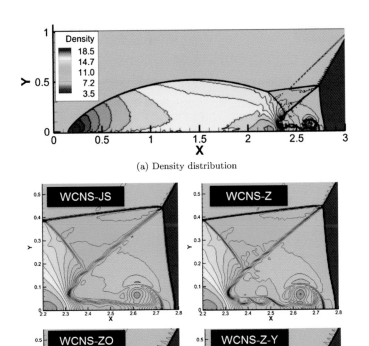

(a) Density distribution

(b) Density distribution of the interaction area

Fig. 7. Results of double Mach problem.

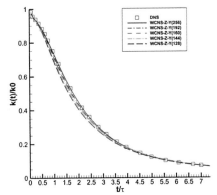

Fig. 8. Kinetic energy decay curve of simulations with WCNS-Z-Y.

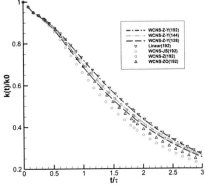

Fig. 9. Comparison of kinetic energy decay curves with different schemes.

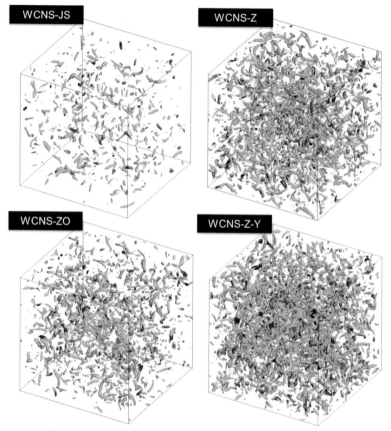

(第 352 页) **Fig. 10.** Q iso-surface of different schemes at $t = 1.0$.

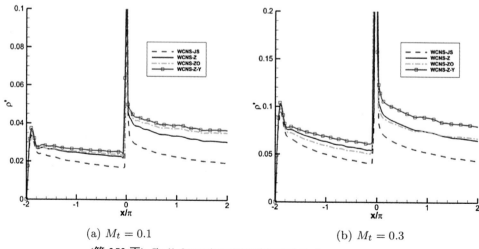

(a) $M_t = 0.1$ (b) $M_t = 0.3$

(第 352 页) **Fig. 11.** Streamwise variation of density fluctuation.

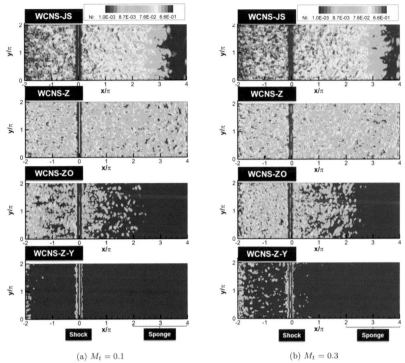

(a) $M_t = 0.1$　　　(b) $M_t = 0.3$

Fig. 12. Distributions of NI of different schemes.

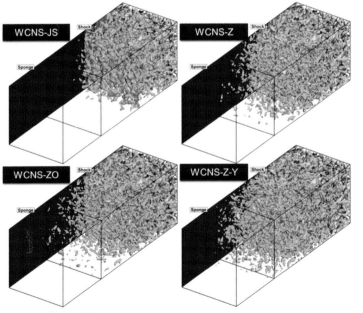

Fig. 13. Q iso-surface of results of different schemes.

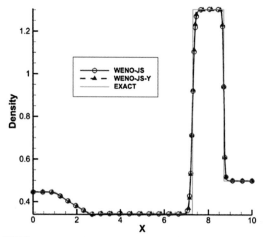

(第 355 页) Fig. 14. Results of Lax problem using WENO schemes.

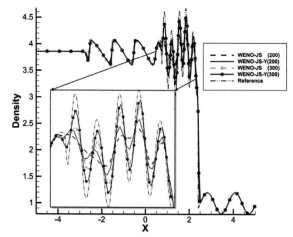

(第 355 页) Fig. 15. Results of Osher–Shu problem using WENO schemes.

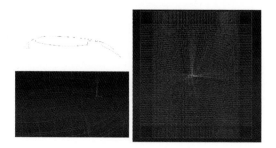

(第 367 页) Fig. 13. Grid structure of the high-lift airfoil configuration

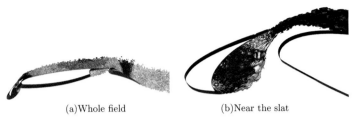

(a)Whole field (b)Near the slat

(第 367 页) **Fig. 14.** Iso-surfaces of the second invariant of velocity gradient tensor for instantaneous vortex structure

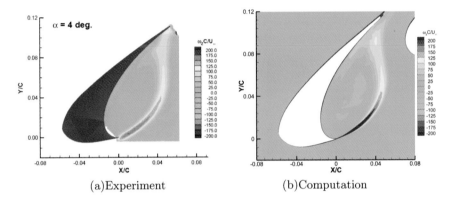

(a)Experiment (b)Computation

(第 367 页) **Fig. 15.** Time-averaged spanwise-vorticity distribution

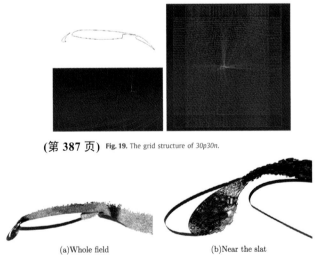

(第 387 页) **Fig. 19.** The grid structure of *30p30n*.

(a)Whole field (b)Near the slat

(第 388 页) **Fig. 20.** Iso-surfaces of the second invariant of velocity gradient tensor for instantaneous vortex structure.

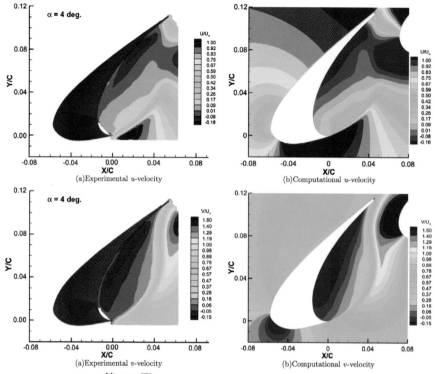

(第 388 页) **Fig. 21.** Contours of the velocity.

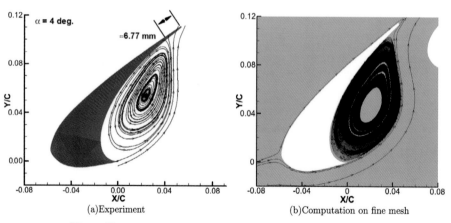

(第 389 页) **Fig. 22.** The time-averaged velocity-vector distribution.

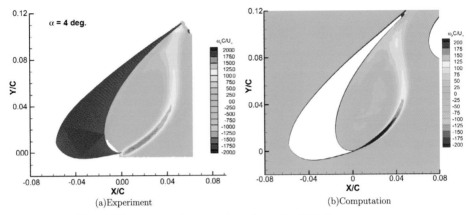

(第 389 页) **Fig. 23.** The time-averaged spanwise-vorticity distribution.

(第 389 页) **Fig. 24.** The grid structure of C919.

(第 390 页) **Fig. 26.** The equivalent total pressure for different cross sections.

(第 390 页) **Fig. 27.** Streamline of the wing tip, the side edge of wing flap and the nacelle.

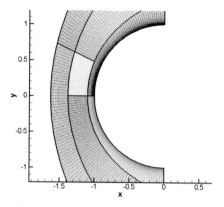

(第 439 页) Figure 1: The grids over the cylinder.

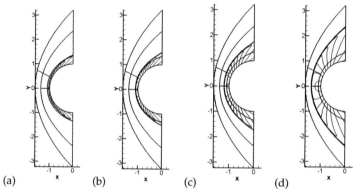

(第 439 页) Figure 2: The bow shock wave at different computational moment, pressure contours.

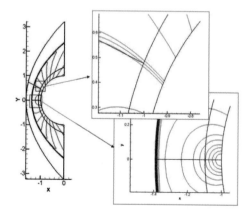

(第 440 页) Figure 3: The enlarged parts at multi-connected points, pressure contours.

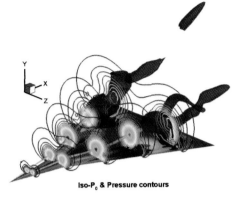

Figure 7: Iso-p_0 and pressure contours of a $Ma=0.3$ double-delta wing, 4.26 million grids.

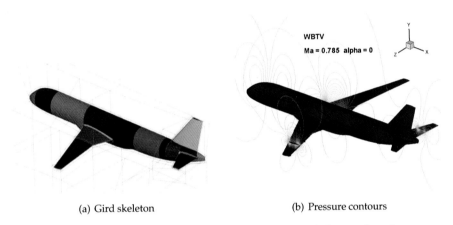

(a) Gird skeleton

(b) Pressure contours

Figure 8: Grid skeleton and pressure contours of a transonic airplane configuration.

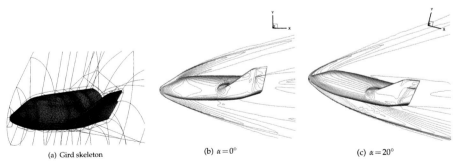

(a) Gird skeleton

(b) $\alpha=0°$

(c) $\alpha=20°$

Figure 9: Grid skeleton and pressure contours of the hypersonic X-38 configuration ($\alpha=0°$).

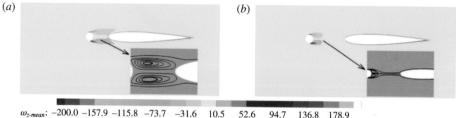

$\omega_{z\text{-}mean}$: −200.0 −157.9 −115.8 −73.7 −31.6 10.5 52.6 94.7 136.8 178.9

FIGURE 14. (Colour online) Time-averaged spanwise vorticity and streamlines in the interaction region taken from the flow field with $U_\infty = 72$ m s^{-1}, $Re_d = 4.8 \times 10^4$, $L/d = 2$ (a) and $L/d = 4$ (b). (第 472 页)

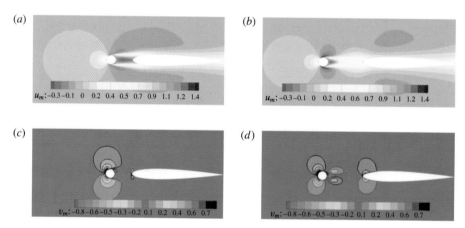

FIGURE 15. (Colour online) Contours of time-averaged velocity taken from the flow field with $U_\infty = 72$ m s^{-1}, $Re_d = 4.8 \times 10^4$, $L/d = 2$ (a,c) and $L/d = 4$ (b,d). (a,b) Streamwise component u_m; (c,d) vertical component v_m, here solid lines denote positive values and dashed lines negative ones. (第 473 页)

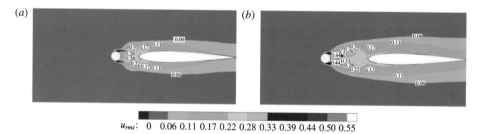

u_{rms}: 0 0.06 0.11 0.17 0.22 0.28 0.33 0.39 0.44 0.50 0.55

FIGURE 16. (Colour online) The r.m.s. value of streamwise velocity taken from the flow field with $U_\infty = 72$ m s^{-1}, $Re_d = 4.8 \times 10^4$, $L/d = 2$ (a) and $L/d = 4$ (b).
(第 473 页)

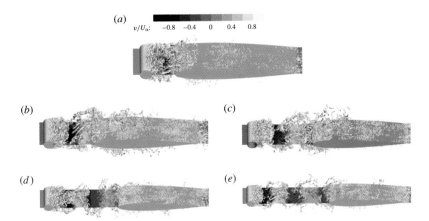

FIGURE 17. (Colour online) Vortical structures by iso-surface of the Q-criterion ($Q = 1000$). The instantaneous vertical velocity v at the symmetry x,z-plane ($y=0$) is included to highlight the alternating pattern of the main vortices. (a) Taken from the flow field with $U_\infty = 72$ m s^{-1}, $Re_d = 4.8 \times 10^4$, $L/d = 2$; (b) $L/d = 4$; (c) $L/d = 6$; (d) $L/d = 8$; (e) $L/d = 10$. (第 474 页)

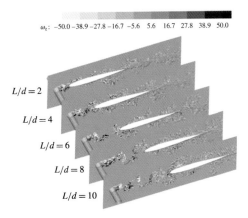

FIGURE 18. (Colour online) Juxtapositional view of the unsteady spanwise vorticity ω_z of vertical (x, y)-planes taken from the flow field of five rod–airfoil configurations at the central slice $z = 0.015$ m of the computational domain.

(第 475 页)

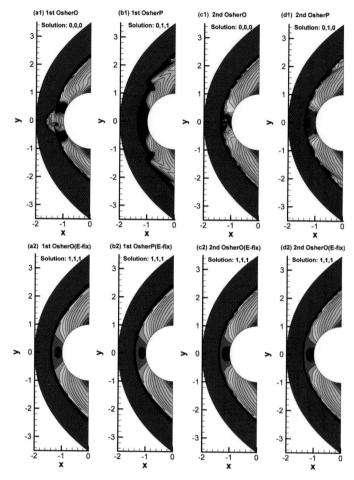

Figure 4: Density contours for schemes without entropy fix (Up) and with entropy fix (Down) on Grid1.
(第 506 页)

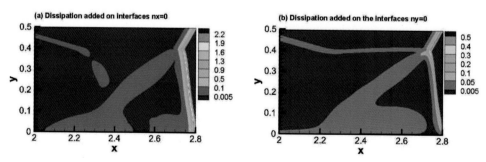

Figure 9: Contour plots of the ratio $\frac{\eta}{\lambda_{\max}}$ for interfaces with $n_x=0$ (left plot) and $n_y=0$ (right plot).
(第 512 页)

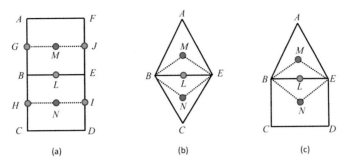

Figure 13: Integral path for computing derivatives in viscous term.

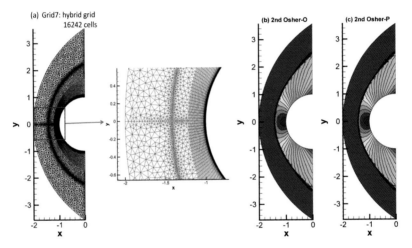

Figure 14: Pressure distributions of schemes with entropy fix on hybrid grid.

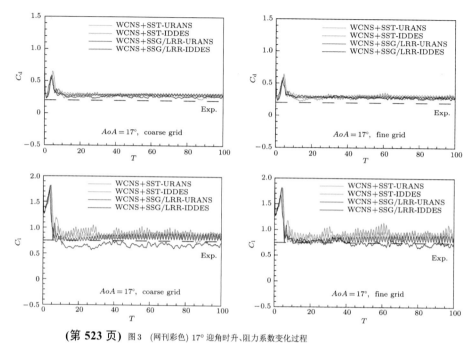

(**第 523 页**) 图 3 （网刊彩色）17° 迎角时升、阻力系数变化过程

Fig. 3. (color online) Lift and drag coeffcient history at attack angle of 17°.

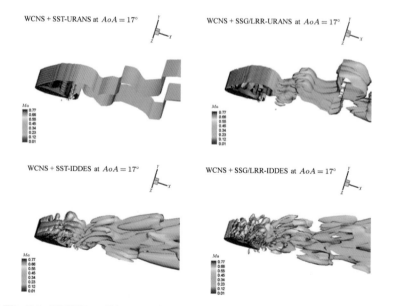

(**第 524 页**) 图 4 （网刊彩色）17° 迎角下 100T 时刻 Q 判据为 0 的等值面图，颜色由马赫数标识

Fig. 4. (color online) Iso-surface of the Q-criterion = 0 at 100T and attack angle of 17°, colored by Mach number.

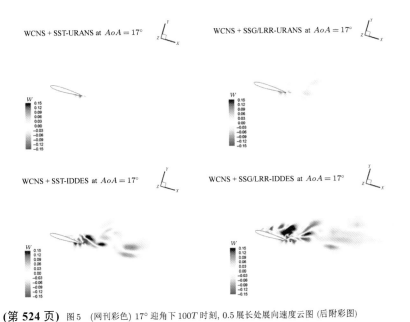

(第 524 页) 图 5 (网刊彩色) 17° 迎角下 100T 时刻, 0.5 展长处展向速度云图 (后附彩图)

Fig. 5. (color online) Spanwise velocity contours of 50% span at $100T$ and attack angle of $17°$.

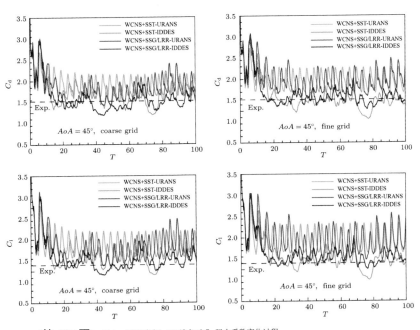

(第 525 页) 图 6 (网刊彩色) 45° 迎角升、阻力系数变化过程

Fig. 6. (color online) Lift and drag coeffcient history at attack angle of $45°$.

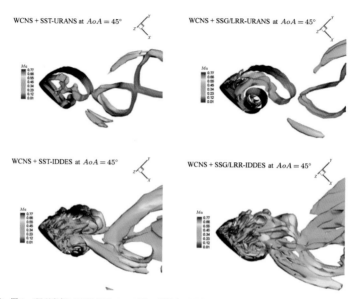

(第 525 页) 图 7 (网刊彩色) 45° 迎角下 100T 时刻 Q 判据为 0 的等值面图, 颜色由马赫数标识

Fig. 7. (color online) Iso-surface of the Q-criterion = 0 at 100T and attack angle of 45°, colored by Mach number.

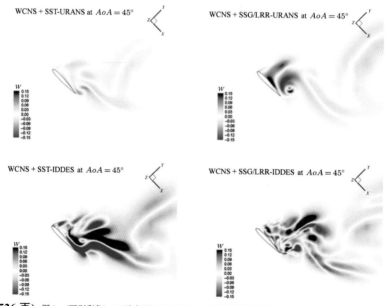

(第 526 页) 图 8 (网刊彩色) 45° 迎角下 100T 时刻, 0.5 展长处展向速度云图

Fig. 8. (color online) Spanwise velocity contours of 50% span at 100T and attack angle of 45°.

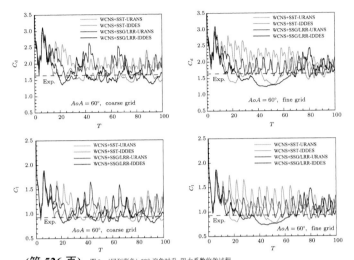

图 9 (网刊彩色) 60° 迎角时升、阻力系数收敛过程

Fig. 9. (color online) Lift and drag coeffcient history at attack angle of 60°.

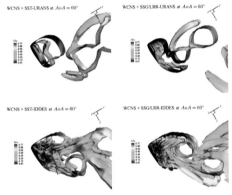

图 10 (网刊彩色) 60° 迎角下 100T 时刻 Q 判据为 0 的等值面图，颜色由马赫数标识

Fig. 10. (color online) Iso-surface of the Q-criterion $= 0$ at 100T and attack angle of 60°, colored by Mach number.

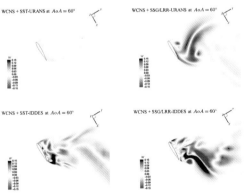

图 11 (网刊彩色) 60° 迎角下 100T 时刻，0.5 展长处展向速度云图

Fig. 11. (color online) Spanwise velocity contours of 50% span at 100T and attack angle of 60°.

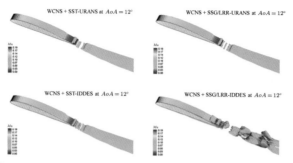

(第 528 页) 图 13 （网刊彩色）NACA4412 翼型 80T 时刻，Q 判据为 0 等值面图，颜色由马赫数标识

Fig. 13. (color online) Iso-surface of the Q-criterion = 0 at $80T$ for NACA4412 airfoil case, colored by Mach number.

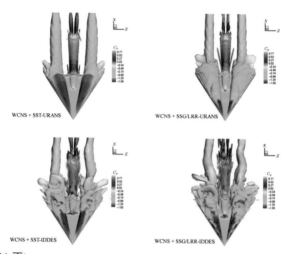

(第 531 页) 图 18 （网刊彩色）三角翼 20T 时刻 Q 判据等值面图，颜色由压力系数标识

Fig. 18. (color online) Iso-surface of the Q-criterion at $20T$ for deltawing case, colored by pressure coefficient.

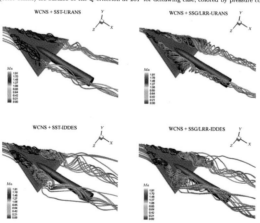

(第 531 页) 图 19 （网刊彩色）三角翼 20T 时刻流线图，颜色由马赫数标识

Fig. 19. (color online) Streamlines at $20T$ for deltawing case, colored by Mach number.